Other conversions

Engineering gas constant, R	1 ft·lb/(slug·°R) = 0.167 226 N·m/(kg·K)
Heat	
Metric	1 cal = 4.184* J (heat required to raise 1.0 g of water 1.0 K)
English	1 Btu = 251.996 cal (heat required to raise 1.0 lb of water 1.0°R)
Temperature	
Metric	K = 273.15° + °C
English	°R = 459.67° + °F
	ΔT of 1°C = ΔT of 1 K = ΔT of 1.8°F = ΔT of 1.8°R

* Exact conversion.

Relationships between temperatures
[celsius (°C), fahrenheit (°F), kelvin (K), and rankine (°R)]
°C = (5/9)(°F − 32); °F = 32 + (9/5)(°C); K = (5/9)(°R); °R = (9/5)(K)

°C	−30	−20	−10	0	10	20	30	40	50	60	70	80	90	100
°F	−22	−4	14	32	50	68	86	104	122	140	158	176	194	212
K	243	253	263	273	283	293	303	313	323	333	343	353	363	373
°R	438	456	474	492	510	528	546	564	582	600	618	636	654	672

Definition of metric quantities

Hectare (ha)	area	10^4 m^2 = (100 m square)
Joule (J)	energy (or work)	newton·meter (N·m)
Liter (L)	volume	10^{-3} m^3
Newton (N)	force (or weight)	1 kg × 1 m/s^2
Pascal (Pa)	pressure	$newton/meter^2$ (N/m^2)
Poise (P)	viscosity	10^{-1} $N{\cdot}s/m^2$
Stoke (St)	kinematic viscosity	10^{-4} m^2/s
Watt (W)	power	newton·meter/second (N·m/s)

Commonly used prefixes for SI units

Factor by which unit is multiplied	Prefix	Symbol
10^9	giga	G
10^6	mega	M
10^3	kilo	k
10^{-2}	centi	c**
10^{-3}	milli	m
10^{-6}	micro	μ
10^{-9}	nano	n

** Avoid if possible.

Important quantities

	BG (English) unit	SI (metric) unit
Acceleration due to gravity (sea level), g	32.1740 ft/sec^2	9.806 65 m/s^2
for most calculations, use	32.2 ft/sec^2	9.81 m/s^2
Density of water (39.2°F, 4°C), ρ	1.940 $slug/ft^3$ = 1.940 $lb{\cdot}sec^2/ft^4$	1 000 kg/m^3 = 1.0 g/cm^3 = 1.0 Mg/m^3
Specific weight of water (40°F, 5°C), γ	62.43 lb/ft^3	9807 N/m^3 or 9.807 kN/m^3
for most calculations, use	62.4 lb/ft^3	9.81 kN/m^3
Standard atmosphere at sea level (59°F, 15°C)	14.696 psia	101.325 kPa abs
	29.92 inHg	760 mmHg
	33.91 ft H_2O	10.34 m H_2O
	2116.2 psfa	1013.25 millibars abs

Fluid Mechanics with Engineering Applications

Fluid Mechanics with Engineering Applications

NINTH EDITION

Joseph B. Franzini, Ph.D.

Professor Emeritus of Civil Engineering
Stanford University

E. John Finnemore, Ph.D.

Professor of Civil Engineering
Santa Clara University

Boston, Massachusetts Burr Ridge, Illinois Dubuque, Iowa
Madison, Wisconsin New York, New York San Francisco, California St. Louis, Missouri

WCB/McGraw-Hill

A Division of The ***McGraw-Hill*** *Companies*

Fluid Mechanics with Engineering Applications

This book is printed on acid-free paper.

3 4 5 6 7 DOC/DOC 9 9

ISBN 0-07-021914 1.

This book was set in Times Roman by The Universities Press (Belfast) Ltd.
The editors were B. J. Clark and John M. Morriss;
the production supervisor was Leroy Young.
Project supervision was done by The Universities Press (Belfast) Ltd.

Cover photo credit: © Gary Braasch, Woodfin Camp & Associates, Inc.

Library of Congress Catalog Card Number preassigned is: 97-070011

INTERNATIONAL EDITION

When ordering this title, use ISBN 0-07-114214-2

http://www.mhcollege.com

2 3 4 5 6 7 DOC/DOC 9 9

ABOUT THE AUTHORS

Joseph B. Franzini is Professor Emeritus of Civil Engineering at Stanford University. Born in Las Vegas, New Mexico, he received B.S. and M.S. degrees from the California Institute of Technology in 1942 and 1943, and a Ph.D. from Stanford University in 1950. All his degrees are in civil engineering. Franzini served on the faculty at Stanford University from 1950 to 1986. There he taught courses in fluid mechanics, hydrology, sedimentation, and water resources, and also did research on a number of topics in those fields. Since retirement from Stanford, he has been active as an engineering consultant and an expert witness. He is coauthor of the authoritative and widely used textbook, *Water Resources Engineering,* and of its predecessor, *Elements of Hydraulic Engineering*. Through the years, Franzini has been active as a consultant to various private organizations and governmental agencies in both the United States and abroad; he was associated with Nolte and Associates, a consulting civil engineering firm in San Jose, California, for over 30 years. He is a Fellow of the American Society of Civil Engineers and a registered civil engineer in California. He lives with his wife Gloria in Palo Alto, California.

E. John Finnemore is Professor of Civil Engineering at Santa Clara University, California. Born in London, England, he received a B.Sc. (Eng.) degree from London University in 1960, and M.S. and Ph.D. degrees from Stanford University in 1966 and 1970, all in civil engineering. Finnemore worked with consulting civil engineers in England and Canada for five years before starting graduate studies, and for another seven years in California after completing his doctorate. He served one year on the faculty of Pahlavi University in Shiraz, Iran, and he has been a member of the faculty at Santa Clara University since 1979. He has taught courses in fluid mechanics, hydraulic engineering, hydrology, and water resources engineering, and has authored numerous technical articles and reports in several related fields. His research has often involved environmental protection, such as in stormwater management and onsite wastewater disposal. Professor Finnemore has served on governmental review boards and as a consultant to various private concerns. He is a Fellow of the American Society of Civil Engineers and a registered civil engineer in Britain and California. He lives with his wife Gulshan in Cupertino, California.

To that great love which encourages humanity
in all its noble endeavors

and

to Gloria and Gulshan
for their loving support

CONTENTS

PREFACE

Philosophy and History

This ninth edition of the classic textbook, *Fluid Mechanics with Engineering Applications,* continues and improves on its tradition of explaining the physical phenomena of fluid mechanics and applying its basic principles in the simplest and clearest possible manner without the use of complicated mathematics. It focuses on civil, environmental, and agricultural engineering problems, although mechanical and aerospace engineering topics are also strongly represented. The book has been written to serve as a text for a first course in fluid mechanics for engineering students, with sufficient breadth of coverage that it can be used in a number of ways for a second course if desired.

Thousands of engineering students and practitioners throughout the world have used this book for over 80 years; it has now moved into its third generation of authorship. Though this ninth edition is very different from the first edition, it retains the same basic philosophy and presentation of fluid mechanics as an engineering subject that was originally developed by Robert L. Daugherty over his many years of teaching at Cornell University, Rensselaer Polytechnic Institute, and the California Institute of Technology. The first edition, authored by Professor Daugherty, was published in 1916 with the title *Hydraulics.* He revised the book four times. On the fifth edition (fourth revision) he was assisted by Dr. Alfred C. Ingersoll, and the title of the book was changed to *Fluid Mechanics with Engineering Applications.* The sixth and seventh editions were entirely the work of Professor Franzini. Listed as the senior author of this ninth edition, Franzini had been a student of Daugherty's at Caltech and had received his first exposure to the subject of fluid mechanics from the fourth edition of the book. Professor Franzini enlisted the services of Professor Finnemore, a former student of Franzini's at Stanford, to assist him with the eighth edition. This ninth edition is the work of Dr. Finnemore, with the exception of Chapters 15 and 16, which were written by Dr. Franzini.

The Book, its Organization

We feel it is most important that the engineering student clearly visualize the physical situation under consideration. Throughout the book, therefore, there is considerable emphasis on physical phenomena of fluid mechanics. We stress the governing principles, the assumptions made in their development, and their limits of applicability, and show how the principles can be applied to the solution of practical engineering problems. The emphasis is on teachability for the instructor and on clarity for both the instructor and the student, so that basic principles and applications can be readily grasped. Numerous worked sample problems are presented to demonstrate the application of basic principles. These sample problems also help to clarify the text. Drill exercises with answers provided follow most sections to help students rapidly reinforce their understanding of the subjects and concepts. The end-of-chapter problems presented for assignment purposes have been carefully selected to provide the student with a thorough workout in the application of basic principles. Only by working numerous exercises and problems will the student experience the evolution so necessary to the learning process. Ways to study fluid mechanics and to approach problem solving are recommended in Chapter 1.

The book is essentially "self-contained". The treatment is such that an instructor generally need not resort to another reference to answer any question that a student might normally be expected to ask. This has required more detailed discussion than would have been necessary if the presentation of certain topics had been more superficial. A list of selected references is provided at the end of the book to serve as a guide for those students who wish to probe deeper into the various fields of fluid mechanics. The appendix contains information on physical properties of fluids and other useful tables, Chapter 1 contains information on dimensions and units, and, for convenient reference, the insides of the covers contain conversion factors and important quantites and definitions.

Even though British Gravitational (BG) units (feet, slugs, seconds, pounds) are used as the primary system of units, the corresponding SI units are given in the text. Sample problems, and exercises and problems are given in BG and in SI units in almost equal numbers. Every effort is made to ease the changeover from BG units to SI units; a discussion of unit systems and conversion of units is presented in Chapter 1. We encourage instructors to assign problems in each system so that students become conversant with both.

Improvements to This Edition

This revision is considerably more extensive than previous revisions. The authors have concentrated on improving, simplifying, clarifying, and modernizing the presentation of the material, and on improving the coverage of the subject. The text is more readable and more accurate as a result. In particular, we have added 50 new worked sample problems. About 75% of the 1318 problems are new or changed from the previous edition. With respect to applications, in the more advanced chapters we introduce the use of various programming techniques in problem solving. These techniques include mathematics software such as Mathcad, spreadsheets, and advanced programmable calculators, and use of these is included in worked examples and in problems. Also, the use of environmental examples and applications has been increased.

New features in the book include:

- A new introductory chapter, which emphasizes, and recommends approaches to, problem solving;
- Drill exercises at the end of most sections. These are simple, straightforward problems on the current section, for which answers are provided at the back of the book;
- The use of color in most figures; and
- Appendix B on programming and computer applications.

The text includes new treatments of: the relationship between a flow system and a control volume as a basis for the equations of continuity, energy, and momentum; the standard atmosphere; plumes and ocean outfalls; the Navier–Stokes equations; shear-stress (friction) velocity; wide and shallow channels; stilling basins; streamlined weirs; computer techniques for water hammer; and flow through porous media. Topics that have been considerably revised include: manometry; eddies; pipe flow equations; drag coefficients for vehicles; hydraulic jumps, standing waves, and transitions; surge tanks; and hydraulic machinery, emphasizing performance characteristics and machine selection.

Use of the Book, Course Planning

An excellent, brief first course in fluid mechanics can be achieved by covering Chapters 1 through 8; however, one might wish to include parts of Chapters 11 (Fluid Measurements) and 14 (Ideal Flow Mathematics) in a first course. Schools having stringent requirements in fluid mechanics might wish to cover the entire text in their course or courses required of all engineers. At other schools only partial coverage of the text might suffice for the course required of all engineers, and other portions of the text might be covered in a second course for students in a particular branch of engineering. Thus civil, environmental, and agricultural engineers would emphasize Chapter 10 and perhaps Chapter 12 in a second course, while mechanical engineers would probably include Chapters 9 and 13 in a second course. The book has been used at a number of schools for courses in hydraulic machinery.

For instructors only, a companion Solutions Manual is available from McGraw-Hill that contains typed and carefully explained solutions to all the exercises and end-of-chapter problems in the book; for convenience, the problem statements and problem figures (if any) are repeated with the solutions. It contains suggestions on how to use the manual most effectively to select problems for assignment, and a Problem Selection Guide for each chapter categorizes the problems by their difficulty, length, units used, and any special features.

Acknowledgements

The authors wish to acknowledge the many comments and suggestions that they have received from users of the book throughout the years, and from numerous anonymous-in-depth reviews arranged by the publishers. These have influenced the content and mode of presentation of the material. Further comments and suggestions for future editions of the book are always welcome.

The authors gratefully acknowledge the suggestions for revisions to Chapter 13 provided by Professor Michel A. Saad, the author of a number of books about compressible flow and thermodynamics.

We particularly appreciate the caring assistance, guidance, and encouragement given by our project editor, Marian Provenzano, and our executive editor, B. J. Clark, at McGraw-Hill. Also, we wish to thank the staff of The Universities Press (Belfast) Ltd. headed by Wallace McKee for their cooperation and diligence in shepherding the book through the production process.

Joseph B. Franzini
E. John Finnemore

LIST OF SYMBOLS

The following table lists the letter symbols generally used throughout the text. Because there are so many more concepts than there are English and suitable Greek letters, certain conflicts are unavoidable. However, where the same letter has been used for different concepts, the topics are so far removed from each other that no confusion should result. Occasionally a particular letter will be used in one special case only, but this local deviation from the table will be clearly indicated, and the usage will not be employed elsewhere. The customary units of measurement for each item are given in the British Gravitational (BG) system while the corresponding SI unit is given in parentheses or brackets.

For the most part, we have attempted to adhere to generally accepted symbols, but not always.

A = any area, ft^2 (m^2)

= cross-sectional area of a stream normal to the velocity, ft^2 (m^2)

= area in turbines or pumps normal to the direction of absolute velocity of the fluid, ft^2 (m^2)

A_c = circumferential flow area, ft^2 (m^2)

A_s = area of a liquid surface as for a tank or reservoir, ft^2 or acre (m^2 or hectare)

a = area in turbines or pumps normal to the relative velocity of the fluid, ft^2 (m^2)

= linear acceleration, ft/sec^2 (m/s^2)

B = any width, ft (m)

= width of open channel at water surface, ft (m)

b = bottom width of open channel, ft (m)

C = cavitation number $= (p - p_v)/(\frac{1}{2}\rho V^2)$ [dimensionless]

C = any coefficient [dimensionless]

= Chézy coefficient [ft$^{1/2}$sec^{-1} (m$^{1/2}$s^{-1})]

C_c = Coefficient of contraction

C_d = coefficient of discharge

C_v = coefficient of velocity

} for orifices, tubes, and nozzles [all dimensionless]

C_D = drag coefficient [dimensionless]

C_f = average friction-drag coefficient for total surface [dimensionless]

C_{HW} = Hazen–Williams pipe roughness coefficient, ft$^{0.37}$/sec (m$^{0.37}$/s)

C_L = lift coefficient [dimensionless]

C_p = pressure coefficient $= \Delta p/(\frac{1}{2}\rho V^2)$ [dimensionless]

c = specific heat of liquid, Btu/(slug·°R) [cal/(g·K) or N·m/(kg·K)]

= wave velocity (celerity), fps (m/s)

c = sonic (i.e., acoustic) velocity (celerity), fps (m/s)

c_f = local friction-drag coefficient [dimensionless]

c_P = velocity (celerity) of pressure wave in elastic pipe, ft/sec (m/s)
c_p = specific heat of gas at constant pressure, ft·lb/(slug·°R)[N·m/(kg·K)]
c_v = specific heat of gas at constant volume, ft·lb/(slug·°R) [N·m/(kg·K)]
D = diameter of pipe, turbine runner, or pump impeller, ft or in (m or mm)
$D''V$ = product of pipe diameter in inches and mean flow velocity in fps
$\mathbf{E}$ = Euler number = $V/\sqrt{2\,\Delta p/\rho}$ [dimensionless]
E = specific energy in open channels = $y + V^2/2g$, ft (m)
= linear modulus of elasticity, psi (N/m^2)
E_c = combined volume modulus for fluid in an elastic pipe, psi (N/m^2)
E_v = volume modulus of elasticity, psi (N/m^2)
e = height of surface roughness projections, ft (mm)
$\mathbf{F}$ = Froude number = $V/\sqrt{gL}$ [dimensionless]
F = any force, lb (N)
F_D = drag force, lb (N)
F_L = lift force, lb (N)
f = friction factor for pipe flow [dimensionless]
G = weight flow rate = γQ, lb/sec (N/s)
g = acceleration due to gravity = 32.1740 ft/sec^2 (9.806 65 m/s^2) (standard)
= 32.2 ft/sec^2 (9.81 m/s^2) for usual computation
H = total energy head = $p/\gamma + z + V^2/2g$, ft (m)
= head on weir or spillway, ft (m)
h = any head, ft (m)
= enthalpy (energy) per unit mass of gas = $i + p/\rho$, ft·lb/slug (N·m/kg)
h' = minor head loss, ft (m)
h_a = accelerative head = $(L/g)(dV/dt)$, ft (m)
h_c = depth to centroid of area, ft (m)
h_L = head lost in friction, ft (m)
h_M = energy added by a machine per unit weight of flowing fluid, ft·lb/lb (N·m/N)
h_O = stagnation (or total) enthalpy of a gas = $h + \frac{1}{2}V^2$, ft·lb/slug (N·m/kg)
h_p = depth to center of pressure, ft (m)
= head put into flow by pump, ft (m)
h_t = head taken from flow by turbine, ft (m)
I = moment of inertia of area, ft^4 or in^4 (m^4 or mm^4)
= internal thermal energy per unit weight, ft·lb/lb (N·m/N)
I_c = moment of inertia about centroidal axis, ft^4 or in^4 (m^4 or mm^4)
i = internal thermal energy per unit mass = gI, ft·lb/slug (N·m/kg)

K = any constant [dimensionless]
k = any loss coefficient [dimensionless]
= specific heat ratio = c_p/c_v [dimensionless]
L = length, ft (m)
L_r = $1/\lambda$ = scale ratio = L_p/L_m [dimensionless]
l = mixing length, ft or in (m or mm)
M = Mach number = V/c [dimensionless]
M = molecular weight
MR = manometer reading, ft or in (m or mm)
m = mass = W/g, slugs (kg)
$\dot{m}$ = mass flow rate = dm/dt, slugs/sec (kg/s)
N = a dimensionless number
N_s = specific speed = $n_e\sqrt{\text{gpm}}/h^{3/4}$ for pumps [dimensionless]
= specific speed = $n_e\sqrt{\text{bhp}}/h^{5/4}$ for turbines [dimensionless]
NPSH = net positive suction head, ft (m)
n = an exponent or any number in general
= Manning coefficient of roughness, sec/ft$^{1/3}$ (s/m$^{1/3}$)
= revolutions per minute, min^{-1}
n_e = rotative speed of hydraulic machine at maximum efficiency, rev/min
P = power, ft·lb/sec (N·m/s)
= height of weir or spillway crest above channel bottom, ft (m)
= wetted perimeter, ft (m)
p = fluid pressure, lb/ft^2 or psi (N/m^2 = Pa)
p_{atm} = atmospheric pressure, psia (N/m^2 abs)
p_b = back pressure in gas flow, psf or psi (Pa)
p_O = stagnation pressure, psf or psi (Pa)
p_v = vapor pressure, psia (N/m^2 abs)
Q = volume rate of flow, cfs (m^3/s)
Q_H = heat added per unit weight of fluid, ft·lb/lb (N·m/N)
q = volume rate of flow per unit width of rectangular channel, cfs/ft = ft^2/sec (m^2/s)
q_H = heat transferred per unit mass of fluid, ft·lb/slug (N·m/kg)
R = Reynolds number = $LV\rho/\mu$ = LV/ν [dimensionless]
R = gas constant, ft·lb/(slug·°R) or N·m/(kg·K)
R_h = hydraulic radius = A/P, ft (m)
R_0 = universal gas constant = 49,709 ft·lb/(slug·°R) [8312 N·m/(kg·K)]
r = any radius, ft or in (m or mm)
r_0 = radius of pipe, ft or in (m or mm)
S = slope of energy grade line = h_L/L [dimensionless]
S_c = critical slope of open channel flow [dimensionless]
S_0 = slope of channel bed [dimensionless]
S_w = slope of water surface [dimensionless]
s = specific gravity of a fluid = ratio of its density to that of a standard fluid (water, air, or hydrogen) [dimensionless]

T = temperature, °F or °R (°C or K)
= period of time for travel of a pressure wave, sec (s)
= torque, ft·lb (N·m)
T_O = stagnation temperature of a gas = $T + \frac{1}{2}V^2/c_p$, °F or °R (°C or K)
T_r = travel time (pulse interval) of a pressure wave, sec (s)
t = time, sec (s)
= thickness, ft or in (m or mm)
t_c = time for complete or partial closure of a valve, sec (s)
U, U_0 = uniform velocity of fluid, fps (m/s)
u = velocity of a solid body, fps (m/s)
= tangential velocity of a point on a rotating body = $r\omega$, fps (m/s)
= local velocity of fluid, fps (m/s)
u' = turbulent velocity fluctuation in the direction of flow, fps (m/s)
u_* = shear stress velocity or friction velocity = $\sqrt{\tau_0/\rho}$, ft/sec (m/s)
V = mean velocity of fluid, fps (m/s)
= absolute velocity of fluid in hydraulic machines, fps (m/s)
V_c = critical mean velocity of open channel flow, fps (m/s)
V_j = jet velocity, fps (m/s)
V_m = meridional velocity, fps (m/s)
V_r = radial component of velocity = $V \sin \alpha$ = $v \sin \beta$, fps (m/s)
V_u = tangential component of velocity = $V \cos \alpha$ = $u + v \cos \beta$, fps (m/s)
$\forall$ = any volume, ft^3 (m^3)
v = relative velocity of fluid in hydraulic machines, fps (m/s)
= specific volume = $1/\rho$, ft^3/slug (m^3/kg)
v_r = radial component of relative velocity = $v \sin \beta$, fps (m/s)
v_u = tangential component of relative velocity = $v \cos \beta$, fps (m/s)
v' = turbulent velocity fluctuation normal to the direction of flow, fps (m/s)
u, v, w = components of velocity in x, y, z, directions, fps (m/s)
W = Weber number = $V/\sqrt{\sigma/\rho L}$ [dimensionless]
W = total weight, lb (N)
= work, ft·lb (N·m)
x = a distance, usually parallel to flow, ft (m)
x_c = distance from leading edge to point where boundary layer becomes turbulent, ft (m)
y = a distance along a plane in hydrostatics, ft (m)
= total depth of open channel flow, ft (m)
y_c = critical depth of open channel flow, ft (m)
= distance to centroid, ft (m)
y_h = hydraulic (mean) depth = A/B, ft (m)
y_0 = depth for uniform flow in open channel (normal depth), ft (m)

y_p = distance to center of pressure, ft (m)
z = elevation above any arbitrary datum plane, ft (m)

α (alpha) = an angle; between V and u in rotating machinery, measured between their positive directions
= kinetic energy correction factor [dimensionless]
β (beta) = an angle; between v and u in rotating machinery, measured between their positive directions
= momentum correction factor [dimensionless]
Γ (gamma) = circulation, ft^2/sec (m^2/s)
γ (gamma) = specific weight, lb/ft^3 (N/m^3)
δ (delta) = thickness of boundary layer, in (mm)
δ_l = thickness of viscous sublayer in turbulent flow, in (mm)
δ_t = thickness of transition boundary layer in turbulent flow, in (mm)
ε (epsilon) = kinematic eddy viscosity, ft^2/sec (m^2/s)
η (eta) = eddy viscosity, $lb \cdot sec/ft^2$ ($N \cdot s/m^2$)
= efficiency of hydraulic machine
θ (theta) = any angle
λ (lambda) = model ratio or model scale = 1/(scale ratio) = L_m/L_p [dimensionless]
μ (mu) = absolute or dynamic viscosity, $lb \cdot sec/ft^2$ ($N \cdot s/m^2$)
ν (nu) = kinematic viscosity, = μ/ρ, ft^2/sec (m^2/s)
ξ (xi) = vorticity, sec^{-1} (s^{-1})
Π (pi) = dimensionless parameter
ρ (rho) = density, mass per unit volume = γ/g, $slug/ft^3$ (kg/m^3)
ρ_O = stagnation density of a gas, $slug/ft^3$ (kg/m^3)
Σ (sigma) = summation
σ (sigma) = surface tension, lb/ft (N/m)
= cavitation parameter in turbomachines [dimensionless]
σ = submergence of weir = h_d/h_u [dimensionless]
σ_c = critical cavitation parameter in turbomachines [dimensionless]
τ (tau) = shear stress, lb/ft^2 (N/m^2)
τ_0 = shear stress at a wall or boundary, lb/ft^2 ($N \cdot m^2$)
ϕ (phi) = any function
ϕ = velocity potential, ft^2/sec (m^2/s) for two-dimensional flow
ϕ = peripheral-velocity factor = $u_{periph}/\sqrt{2gh}$ [dimensionless]
ϕ_e = peripheral-velocity factor at point of maximum efficiency [dimensionless]
ψ (psi) = stream function, ft^2/sec (m^2/s) for two-dimensional flow
ω (omega) = angular velocity = u/r = $2\pi n/60$, rad/sec (rad/s)

Values at specific points will be indicated by suitable subscripts. In the use of subscripts 1 and 2, the fluid is always assumed to flow from 1 to 2.

LIST OF ABBREVIATIONS

abs = absolute
atm = atmospheric
avg = average
bhp = brake (or shaft) horsepower
Btu = British thermal unit
cal = calorie
cfm = cubic feet per minute
cfs = cubic feet per second
cm = centimeter
d = day or days (SI)
fpm = feet per minute
fps = feet per second
ft = foot or feet
g = gram or grams
gal = gallon
gpm = gallons per minute
h = hour or hours (SI)
ha = hectare
hp = horsepower
hr = hour or hours (BG)
Hz = hertz (cycles per second)
in = inch or inches
J = joules = N·m = W·s
kg = kilograms = 10^3 grams
kgf = kilogram force
kgm = kilogram mass
km = kilometer
L = liter
lb = pounds of force (*not* lbs)
lbf = pound force
lbm = pound mass
ln = $\log_e$
log = $\log_{10}$
m = meter or meters
mb = millibars = 10^{-3} bar
mb abs = millibars, absolute
mgd = million gallons per day
min = minute or minutes (BG and SI)
mL = milliliter
mm = millimeters = 10^{-3} meter
mph = miles per hour
N = newton or newtons = kg·m/s^2
N/m^2 abs = newtons per square meter, absolute
oz = ounce
P = poise = 0.10 N·s/m^2
Pa = pascal = N/m^2
pcf = pounds per cubic foot
psf = pounds per square foot
psfa = pounds per square foot, absolute
psfg = pounds per square foot, gage
psi = pounds per square inch
psia = pounds per square inch, absolute
psig = pounds per square inch, gage
rpm = revolutions per minute
rps = revolutions per second
s = second or seconds (SI)
sec = second or seconds (BG)
St = stoke = cm^2/s
W = watt or watts = J/s
y = year or years (SI)
yr = year or years (BG)

CHAPTER 1

Introduction

In preparing this ninth edition of *Fluid Mechanics with Engineering Applications* we have strived to present the material in such a way that you, the student, can readily learn the fundamentals of fluid mechanics and see how those fundamentals can be applied to practical engineering problems. Only by working many problems can one master the application of the fundamentals.

1.1 SCOPE OF FLUID MECHANICS

Undoubtedly you have observed the movement of clouds in the atmosphere, the flight of birds through the air, the flow of water in streams, and the breaking of waves at the seashore. Fluid mechanics phenomena are involved in all of these. Fluids include gases and liquids, with air and water as the most prevalent. Some of the many other aspects of our lives that involve fluid mechanics are flow in pipelines and channels, movements of air and blood in the body, air resistance or drag, wind loading on buildings, motion of projectiles, jets, shock waves, lubrication, combustion, irrigation, sedimentation, and meteorology and oceanography. The motions of moisture through soils and oil through geologic formations are other applications. A knowledge of fluid mechanics is required to properly design water supply systems, wastewater treatment facilities, dam spillways, valves, flow meters, hydraulic shock absorbers and brakes, automatic transmissions, aircraft, ships, submarines, breakwaters, marinas, rockets, computer disk drives, windmills, turbines, pumps, heating and air conditioning systems, bearings, artificial organs, and even sports items like golf balls, yachts, race cars, and hang gliders. It is clear that everybody's life is affected by fluid mechanics in a variety of ways. All engineers should have at least a basic knowledge of fluid phenomena.

Fluid mechanics is the science of the mechanics of liquids and gases, and is based on the same fundamental principles that are employed in the mechanics of solids. Fluid mechanics is a more complicated subject, however, because with solids one deals with separate and tangible elements, while with fluids there are no separate elements to be distinguished.

Fluid mechanics may be divided into three branches: ***fluid statics*** is the study of the mechanics of fluids at rest; ***kinematics*** deals with velocities and streamlines without considering forces or energy; and ***fluid dynamics*** is concerned with the relations between velocities and accelerations and the forces exerted by or upon fluids in motion.

Classical ***hydrodynamics*** is largely a subject in mathematics, since it deals with an imaginary ideal fluid that is completely frictionless. The results of such studies, without consideration of all the properties of real fluids, are of limited practical value. Consequently, in the past, engineers turned to experiments, and from these developed empirical formulas that supplied answers to practical problems. When dealing with liquids, this subject is called ***hydraulics***.

Empirical hydraulics was confined largely to water and was limited in scope. With developments in aeronautics, chemical engineering, and the petroleum industry, the need arose for a broader treatment. This has led to the combining of classical hydrodynamics with the study of real fluids, both liquids and gases, and this is called ***fluid mechanics***. In modern fluid mechanics the basic principles of hydrodynamics are combined with experimental data. The experimental data can be used to verify theory or to provide information supplementary to mathematical analysis. The end product is a unified body of basic principles of fluid mechanics that can be applied to the solution of fluid-flow problems of engineering significance. With the advent of the computer, during the past 20 years the entirely new field of ***computational fluid dynamics*** has developed. Various numerical methods such as finite differences, finite elements, boundary elements, and analytic elements are used to solve advanced problems in fluid mechanics.

1.2 HISTORICAL SKETCH OF THE DEVELOPMENT OF FLUID MECHANICS[1]

From time to time we discover more about the knowledge that ancient civilizations had about fluids, particularly in the areas of irrigation channels

[1] See also Rouse, H., and S. Ince, *History of Hydraulics*, Dover, New York, 1963.

and sailing ships. The Romans are well known for their aqueducts and baths, many of which were built in the fourth century B.C., with some still operating today. The Greeks are known to have made quantified measurements, the best known being those of Archimedes who discovered and formulated the principles of buoyancy in the third century B.C.

We know of no basic improvements to the understanding of flow until Leonardo da Vinci (1452–1519), who performed experiments, investigated, and speculated on waves and jets, eddies and streamlining, and even on flying. He contributed to the one-dimensional equation for conservation of mass.

Isaac Newton (1642–1727), by formulating his laws of motion and his law of viscosity, in addition to developing the calculus, paved the way for many great developments in fluid mechanics. Using Newton's laws of motion, numerous 18th century mathematicians solved many frictionless (zero-viscosity) flow problems. However, most flows are dominated by viscous effects, so engineers of the 17th and 18th centuries found the inviscid flow solutions unsuitable, and by experimentation they developed empirical equations, thus establishing the science of hydraulics.

Late in the 19th century the importance of dimensionless numbers and their relationship to turbulence was recognized, and dimensional analysis was born. In 1904 Ludwig Prandtl published a key paper, proposing that the flow fields of low-viscosity fluids be divided into two zones, namely a thin, viscosity-dominated ***boundary layer*** near solid surfaces, and an effectively inviscid outer zone away from the boundaries. This concept explained many former paradoxes, and enabled subsequent engineers to analyze far more complex flows. However, we still have no complete theory for the nature of turbulence, and so modern fluid mechanics continues to be a combination of experimental results and theory.

1.3 THE BOOK, ITS CONTENTS, AND HOW TO BEST STUDY FLUID MECHANICS

In this introductory chapter we attempt to give you some insight into what fluid mechanics is all about. In the previous sections we discussed the scope of fluid mechanics and the historical development of the subject. This and the next section explain how to best use this book to study fluid mechanics. The last section of this chapter discusses the importance of dimensions and units.

You can get a feel for the contents of the book and the variety of topics it covers by reviewing the Table of Contents at the front of the text. The powerful analytical techniques of similitude and dimensional analysis build on the knowledge of dimensions to be reviewed in Sec. 1.5. Most of the other subject titles are self-explanatory.

Because problem solving is such an important part of the study of fluid mechanics, before beginning you should make yourself very familiar with the supporting resources available. You will often be expected to know where to find such information, without any direct reference. For convenience, many unit conversions and related data have been collected on the inside covers of the book and the facing pages. Many data on material properties, often needed, are collected into the tables of Appendix A; but some are also in the chapters, such as Figs. 2.1 and 2.3–2.4 for example. The lists of symbols and abbreviations preceding this chapter are also a useful resource. As you progress you will increasingly realize how helpful programming procedures can be in solving many fluid mechanics problems, such as flow in pipes and pipe networks, water surface profiles in open channels and culverts, and unsteady flow problems. The most convenient of these procedures are in mathematics software packages such as Mathcad, and equation solvers on programmable scientific calculators; use of these is described particularly in Chaps. 8 and 10. The major different types of programming procedures available are described in Appendix B, and problems for which these might be helpful are indicated by a ▣ preceding the problem number. To help you broaden your horizons by reading books on subjects related to those in this text, a list of such references is provided in Appendix C.

Throughout the book we strive to develop basic concepts in a logical manner so that you can readily read the material and understand it. Material is divided into "building blocks" within separate sections of the chapters. Once the basic concepts are developed, we often provide sample problems to illustrate applications of the concepts; then we usually provide exercises, which you should perform as needed to reinforce your understanding. The exercises normally address only material in the preceding section, and are generally quite straightforward. They are drill, to familiarize you with the subject and concepts, and answers to the exercises are provided at the back of the book.

At the end of each chapter we have placed summary problems. These are intended to be more like real world or examination problems, where it is not indicated which section or sections they address. In some instances they may require the application of concepts from a number of sections or even chapters. You will find it a great advantage to have developed your familiarity with the concepts by doing drill exercises before tackling the end-of-chapter problems. Although answers to the exercises are given, answers to the end-of-chapter problems are not. One reason is that many problems in fluid mechanics require trial and error solution methods, and having answers reduces learning of such methods. Another is that as you progress in competence, you need to rely upon yourself more and learn ways to check yourself; real world problems do not come with accompanying answers.

In this text we stress the application of basic principles to the solution of practical engineering problems. *Only by working many problems can you truly understand the basic principles and how to apply them.* We feel this is very important! Because of this importance, we next include some suggestions that will aid you in solving problems.

1.4 APPROACH TO PROBLEM SOLVING

The following are four important steps to becoming a master of the assigned material:

1. Study the material of the next section(s) *before* it is covered in class. This way you will get so much more out of the review in class; also, you will be able to ask (and answer) perceptive questions!

2. Study the sample problems, and be sure you can work them yourself *without referring to the text,* i.e., "closed book".

3. Do enough of the drill Exercises, answers unseen, until you are confident of your familiarity with the basic material and procedures.

4. Do the (more challenging) homework Problems you have been assigned.

If you get stuck on any of these steps, that suggests you have not sufficiently mastered the previous step(s). Review those yourself before seeking help. Mastering the material by yourself will build your self-confidence, but of course you should always seek help if unable to master it alone.

In writing solutions to problems, for steps 3 and 4 above, the following substeps are often recommended, and are an excellent guide:

a. Thoroughly read and ponder the problem statement for a few moments before writing anything on paper.

b. Summarize information to be used, both that given and that obtained elsewhere; and summarize quantities to be found.

c. Draw a neat figure or figures, fully labeled, of the situation to be analyzed.

d. State all assumptions you consider necessary.

e. Reference all principles, equations, tables, etc., that you will use. (Remember all the available supporting resources, mentioned in Sec. 1.3.)

f. Solve the problem as far as possible algebraically before inserting numbers.

g. Check the dimensions of the various terms for consistency (per Sec. 1.5).

h. Insert numerical values for the variables, using a consistent (SI or BG) set of units (per Sec. 1.5). Evaluate a numerical answer, with units, and report it to an appropriate precision. (This should be no more precise, as a percentage, than that of the least precise inserted value; and however precise the inserted values may be, a common practical rule in engineering is to report results to three significant figures unless they begin with a "1", in which case give four figures.) Do not round off values in your calculator, only do so for presenting your answer.

i. Check your answer for reasonableness and accuracy by comparing it with expected results and by whatever other methods you can devise.

j. Check that any assumptions you made initially are satisfied or appropriate. Note any limitations that apply.

We suggest you do not attempt more advanced problems until you have mastered the less advanced ones. *Demonstrate* this mastery to yourself by achieving correct answers without referring to the text, i.e., "closed book".

To confirm that you have sufficiently mastered problem solving, practice working problems closed book with a time limit. This can be quite challenging, so doing this regularly can be helpful. You can reserve one or two of the homework problems in each assignment for this purpose.

Form a study group early on in the course with one or more study partners. It is very time effective to quiz one another about the categories that problems fall into, and about the procedures that should be used to solve them (without always doing all the calculations).

Not only do you need to learn and understand the material, but also you need to *know how and when to use it!* Seek and build understanding of *applications* for your knowledge, particularly to problems that are not straightforward. It is for non-straightforward problems that we need well-trained engineers.

Understanding is particularly demonstrated by successful application of the principles to situations different from those you have met before. So getting the correct answers to a few "plug and chug" exercises does not alone indicate understanding. Also know that *feeling* you are prepared is not reliable. You should *prove* it to yourself by (a) correctly solving problems closed book under a time limit, and (b) by correctly answering questions on the material.

Although the preceding emphasizes analysis, which can involve algebra, trial and error methods, graphical methods, and calculus, other problem solving methods such as computer and experimental techniques can be used, and should be mastered to a reasonable extent. Become familiar with the use of computers to solve problems by iterative procedures, to perform repetitive numerical evaluations, and to perform numerical integration, etc. Also, programmable calculators are becoming very powerful, with root finders to solve implicit equations and with many integration and graphing capabilities. Familiarity with these will greatly add to your effectiveness in fluid mechanics and as an engineer in general. Chapter 7 provides guidance on planning flow experiments and model tests. Take every opportunity to learn about practical issues in the laboratory and on field trips; never forget, as the title of this book reminds us, that all this theory and analysis is for application to the real world.

Problems in the real world of course are usually not like those in our textbooks. So next you will need to develop your abilities to *recognize* problems in our environment, and to clearly *define* (or formulate) them, before beginning analysis. Often you will find that various methods of solution may be used, and experience will help you select the most appropriate. In the real world the numerical results of analyzing a problem are not the ultimate goal; for those results then need to be interpreted in terms of the physical problem, and then recommendations need to be made for action.

Remain conscious of your goal, to become a capable and responsible engineer, and remain conscious of your path to that goal, which involves the many steps we have discussed here.

1.5 DIMENSIONS AND UNITS

To properly define a physical property or a fluid phenomenon, one must express the property or phenomenon in terms of some set of units. For example, the diameter of a pipe might be 16 centimeters and the average flow velocity 8 meters per second.[2] A different set of units might have been used, such as a diameter of 0.16 meter and a velocity of 800 centimeters per second. Or, the diameter and velocity might have been expressed in English (U.S. Customary) or other units. In this book we use two systems of units: the British Gravitational (BG) system when dealing with English units, and the SI (Système Internationale d'Unités) when dealing with metric units. The SI was adopted in 1960 at the Eleventh General International Conference on Weights and Measures, at which the United States was represented. As of 1995, nearly every major country in the world, except the United States, was using the SI; it appears likely that the United States will officialy adopt the SI within a few years. Because of the imminence of metrification in the United States, the need to be able to readily interact with the many users of SI units, and because English units have been used in the technical literature for so many years, it is essential that the engineer be familiar with both the systems, BG and SI, used in this book.

In fluid mechanics the basic dimensions are length (L), mass (M), time (T), force (F), and temperature (θ). In order to satisfy Newton's second law, $F = ma = MLT^{-2}$, where acceleration a is expressed by its basic dimensions as LT^{-2}, we note that units for only three of the first four of these dimensions may be assigned arbitrarily; the fourth unit must agree with the other three, and is therefore referred to as a ***derived unit***. In the two systems of units used in this book, the commonly used units for the five basic dimensions mentioned are:

---

[2] This book uses the American spelling *meter*, although the official spelling is *metre*.

Dimension	BG unit	SI unit
Length (L)	Foot (ft)	Meter (m)
Mass (M)	Slug[a]	Kilogram (kg)
Time (T)	Second (sec)	Second (s)
Force (F)	Pound (lb)	Newton (N)[b]
Temperature		
Absolute	Rankine (°R)	Kelvin (K)
Ordinary	Fahrenheit (°F)	Celsius (°C)

[a] Derived unit (lb·sec²/ft).
[b] Derived unit (kg·m/s²).

SI employs L, M, and T and derives F from MLT^{-2}. Force in SI is defined by the ***newton***, the force required to accelerate one kilogram of mass at a rate of one meter per second per second; that is,

$$1 \text{ N} = (1 \text{ kg})(1 \text{ m/s}^2)$$

On the other hand, the British Gravitational system, also sometimes known as the U.S. Customary system, employs L, F, and T, and derives M from $F/a = FL^{-1}T^2$. The BG unit of mass, the ***slug***, is therefore defined as that mass which accelerates at one foot per second per second when acted upon by a force of one pound; that is,

$$1 \text{ slug} = (1 \text{ lb})/(1 \text{ ft/sec}^2) = 1 \text{ lb·sec}^2/\text{ft}$$

or

$$1 \text{ lb} = (1 \text{ slug})(1 \text{ ft/sec}^2)$$

When working in the BG system, it often pays to keep mass expressed in basic units (lb·sec²/ft) for as long as possible.

We see that the definition of mass in the BG system depends on the definition of one ***pound***, which is the force of gravity acting on (or weight of) a platinum standard whose mass is 0.453 592 43 kg. ***Weight*** is the gravitational attraction force F between two bodies, of masses m_1 and m_2, given by Newton's Law of Gravitation as

$$F = G\frac{m_1 m_2}{r^2}$$

where G is the universal constant of gravitation and r is the distance between the centers of the two masses. If m_1 is the mass m of an object on the earth's surface and m_2 is the mass M of the earth then r is the radius of the earth, so that

$$F = m\left(\frac{GM}{r^2}\right)$$

and the weight of the object is

$$W = mg$$

where the gravitational acceleration $g = GM/r^2$. Clearly g varies slightly with altitude and latitude on earth, since the earth is not truly spherical, while in space and on other planets it is much different. Because the force (weight) depends on the value of g, which in turn varies with location, a system such as the BG system based on length (L), force (F), and time (T) is referred to as a ***gravitational system***. On the other hand, systems like the SI, which are based on length (L), mass (M), and time (T), are ***absolute*** because they are *independent* of the gravitational acceleration g.

A partial list of derived quantities encountered in fluid mechanics and their commonly used dimensions in terms of L, M, T, and F is as follows:

Quantity	Commonly used dimensions	BG unit	SI unit
Acceleration (a)	LT^{-2}	ft/sec^2	m/s^2
Area (A)	L^2	ft^2	m^2
Density (ρ)	ML^{-3}	slug/ft^3	kg/m^3
Energy, work or quantity of heat	FL	ft·lb	N·m = J
Flowrate (Q)	L^3T^{-1}	ft^3/sec (cfs)	m^3/s
Frequency	T^{-1}	cycle/sec (sec^{-1})	Hz (hertz, s^{-1})
Kinematic viscosity (ν)	L^2T^{-1}	ft^2/sec	m^2/s
Power	FLT^{-1}	ft·lb/sec	N·m/s = W
Pressure (p)	FL^{-2}	lb/in^2 (psi)	N/m^2 = Pa
Specific weight (γ)	FL^{-3}	lb/ft^3 (pcf)	N/m^3
Velocity (V)	LT^{-1}	ft/sec (fps)	m/s
Viscosity (μ)	FTL^{-2}	lb·sec/ft^2	N·s/m^2
Volume ($\forall$)	L^3	ft^3	m^3

Using the identity $F = MLT^{-2}$, all dimensions containing an F could have been expressed using an M instead, and vice versa. Other derived quantities will be dealt with when they are encountered in the text, and particularly in Chap. 7.

On the earth's surface the variation in g is small, and, by international agreement, standard gravity at sea level is 32.1740 ft/sec^2 or 9.806 65 m/s^2 (for problem solving we usually use 32.2 ft/sec^2 or 9.81 m/s^2). So variations in g are generally not considered in this text as long as we are analyzing problems on the earth's surface. Fluid problems for other locations, such as on the moon, where g is quite different from that on earth, can be handled by the methods presented in this text if proper consideration is given to the value of g.

For unit mass (1 slug or 1 kg) on the earth's surface, we note that

In BG units: $W = mg = (1 \text{ slug})(32.2 \text{ ft/sec}^2) = 32.2 \text{ lb};$

In SI units: $W = mg = (1 \text{ kg})(9.81 \text{ m/s}^2) = 9.81 \text{ N}.$

Other systems of units used elsewhere include the English Engineering (EE) system, the Absolute Metric (cgs) system, and the mks metric system. The EE system uses pound force (lbf) and pound mass (lbm), and the mks metric system uses kilogram force (kgf) and kilogram mass (kgm). As a result, both of these are said to be ***inconsistent systems***, because unit force does not cause unit mass to undergo unit acceleration; they require an additional proportionality constant or conversion factor. The SI and BG systems used in this book are ***consistent systems*** having conversion factors with a magnitude of one. Although the cgs metric system is both consistent and non-gravitational, it is little used for engineering applications because its unit of force, the dyne, is so small; 1 dyne = (1 g)(1 cm/s^2) = 10^{-5} N.

Do not be confused by popular usage of kilograms to measure weight (force). When European shoppers buy a kilo of sugar, say, in our terms they are buying sugar with a mass of 1 kg, in effect defining a force of 1 kg (1 kgf) = (1 kgm)(9.81 m/s^2), which is equivalent to 9.81 N. Because a 1-lb weight has a mass of about 0.4536 kg, the shoppers' conversion factor is 1.0/0.4536 = 2.205 lb/kgf. In engineering we are careful to distinguish between mass and weight, reserving kg for mass and using newtons for force in the SI system.

In this book we shall use the abbreviation kg for kilogram mass, and lb for pound force. The abbreviation lb for pound is taken from the Latin *libra*, plural *librae*, so the correct plural abbreviation is lb not lbs. The units second, minute, hour, day, and year are correctly abbreviated as s, min, h, d, and y in the SI system, and although in the BG system they should be abbreviated as sec, min, hr, day, and yr, it is common to use the SI abbreviations for both systems. There are many "nonstandard" or traditional abbreviations used by engineers, such as fps for ft/sec, gpm for gal/min, and cfs for ft^3/sec (also sometimes referred to as the second-foot and the cusec). The more common of these are included in the list just preceding this chapter. Acres, tons, and slugs are not abbreviated. When units are named after people, like the newton (N), joule (J), pascal (Pa), they are capitalized when abbreviated but not capitalized when spelled out. The abbreviation capital L for liter is a special case, used to avoid ambiguity. Also note that in the SI the unit for absolute temperature measurement is the Kelvin degree, which is abbreviated K *without* a degree (°) symbol.

When dealing with unusually large or very small numbers, a series of prefixes has been adopted for use with SI units. The most commonly used prefixes are given for convenient reference facing the inside front and back covers of this book. Hence Mg (megagram) represents 10^6 grams, mm (millimeter) represents 10^{-3} meters, and kN (kilonewton) represents 10^3 newtons for example. Note that multiples of 10^3 are preferred in engineering usage; other multiples like cm are to be avoided if possible. Also, in the SI it is conventional to separate sequences of digits into groups of three by spaces rather than by commas, as was done earlier for the mass of the standard pound. Thus ten cubic meters of water weigh 98 100 N, or 98.1 kN.

It is common to need to convert quantities given in BG units into SI units, and vice versa. Because time units are the same in both systems, we only need to convert units of length, and force or mass, from which all other units can then be derived. For length, by definition, one foot is *exactly*

0.3048 meters. For force, using $W = mg$ and definitions given earlier, 1 lb = (0.453 592 43 kg)(9.806 65 m/s^2), or about 4.448 N. For mass, 1 slug = (1 lb)/(1 ft/sec^2) is about equal to (4.448 N)/(0.3048 m/s^2) = 14.59 kg. Conversion factors for many other units, derived from these three basic ones, are given for convenience on the insides of the front cover (BG to SI) and back cover (SI to BG) of the book. Conversions of units *within* the BG system and *within* the SI are also included in the tables on the insides of the covers. On the facing pages are given some definitions, other useful conversions, and relations between the four principal temperature scales.

In the SI lengths are commonly expressed in millimeters (mm), centimeters (cm), meters (m), or kilometers (km), depending on the distance being measured. A meter is about 39 inches or 3.3 ft and a kilometer is approximately five-eighths of a mile. Areas are usually expressed in square centimeters (cm^2), square meters (m^2), or ***hectares*** (100 m × 100 m = 10^4 m^2), depending on the area being measured. The hectare, used for measuring large areas, is equivalent to about 2.5 acres. A newton is equivalent to approximately 0.225 lb. The SI unit of stress (or pressure), newton per square meter (N/m^2), is referred to as the ***pascal*** (Pa), and is equivalent to about 0.021 lb/ft^2 or 0.000 15 lb/in^2. In SI units energy, work or quantity of heat are ordinarily expressed in joules (J). A ***joule*** is equal to a newton-meter, i.e., J = N·m. The unit of power is the ***watt*** (W), which is equivalent to a joule per second, i.e., W = J/s = N·m/s.

When you have to work with less usual units, like centipoise (for viscosity) or ergs (for energy), it is best to convert them into SI or BG units as soon as possible.

Sample Problem 1.1 Bernoulli's equation for the flow of an ideal fluid, which is discussed in Chap. 5, may be written

$$\frac{p}{\gamma} + z + \frac{V^2}{2g} = \text{constant} \qquad (5.16)$$

where p = pressure, γ = specific weight, z = elevation, V = mean flow velocity, and g = acceleration of gravity. Demonstrate that this equation is dimensionally homogeneous, i.e., that all terms have the same dimensions.

Solution

Term 1: Dimensions of $\dfrac{p}{\gamma} = \dfrac{F/L^2}{F/L^3} = L$

Term 2: Dimensions of $z = L$

Term 3: Dimensions of $\dfrac{V^2}{2g} = \dfrac{(L/T)^2}{L/T^2} = L$

So all the terms have the same dimensions, L, which must also be the dimensions of the constant at the right-hand side of Eq. (5.16).

SAMPLE PROBLEM 1.2 Convert 200 Btu to (*a*) BG, (*b*) SI, and (*c*) cgs metric units of energy.

Solution

From inside the front cover:

$$1 \text{ Btu} = 778 \text{ ft}\cdot\text{lb}, \qquad 1 \text{ ft}\cdot\text{lb} = 1.356 \text{ N}\cdot\text{m} = 1.356 \text{ J}, \qquad 1 \text{ N} = 10^5 \text{ dyne.}$$

(*a*) For BG units:

$$200 \text{ Btu} = 200 \text{ Btu}\left(\frac{778 \text{ ft}\cdot\text{lb}}{1 \text{ Btu}}\right) = 155{,}600 \text{ ft}\cdot\text{lb.}$$

(*b*) For SI units:

$$155\,600 \text{ ft}\cdot\text{lb} = 155\,600 \text{ ft}\cdot\text{lb}\left(\frac{1.356 \text{ N}\cdot\text{m}}{1 \text{ ft}\cdot\text{lb}}\right)$$

$$= 210\,994 \text{ N}\cdot\text{m} = 211 \text{ kN}\cdot\text{m} = 211 \text{ kJ.}$$

(*c*) For cgs units:

$$210\,994 \text{ N}\cdot\text{m} = 211 \times 10^3 \text{ N}\cdot\text{m}\left(\frac{10^5 \text{ dyne}}{1 \text{ N}}\right)\left(\frac{10^2 \text{ cm}}{\text{m}}\right)$$

$$= 211 \times 10^{10} \text{ dyne}\cdot\text{cm} = 211 \times 10^{10} \text{ erg.}$$

EXERCISES

1.5.1 Demonstrate that Eq. (6.5) is dimensionally homogeneous.

1.5.2 Demonstrate that Eq. (10.57) is dimensionally homogeneous. Note that C_d is dimensionless.

1.5.3 Demonstrate that Eq. (11.2) is dimensionally homogeneous. Note that $\forall$ is a volume, h_L has the dimensions of length, and ν is kinematic viscosity.

1.5.4 Demonstrate that Eq. (12.4) is dimensionally homogeneous.

1.5.5 Demonstrate that Eq. (13.45) is dimensionally homogeneous. Note that k is dimensionless.

1.5.6 Using information from inside the cover of this book, determine the weight of a U.S. gallon of water in the following units: (*a*) pounds; (*b*) newtons; (*c*) dynes.

1.5.7 Using information from inside the cover of this book, determine the weight of one liter of water at 15°C in the following units: (*a*) pounds; (*b*) newtons; (*c*) dynes.

1.5.8 Using information from inside the cover of this book, convert 15 million U.S. gallons per day (mgd) into (*a*) BG and (*b*) SI units.

1.5.9 Using information from inside the cover of this book, convert 75 km/h into (*a*) SI and (*b*) BG units.

CHAPTER 2

Properties of Fluids

In this chapter we discuss a number of fundamental properties of fluids. An understanding of these properties is essential if one is to apply basic principles of fluid mechanics to the solution of practical problems.

2.1 DISTINCTION BETWEEN A SOLID AND A FLUID

The molecules of a solid are usually closer together than those of a fluid. The attractive forces between the molecules of a solid are so large that a solid tends to retain its shape. This is not the case for a fluid, where the attractive forces between the molecules are smaller. An ideal elastic solid will deform under load and, once the load is removed, will return to its original state. Some solids are plastic. These deform under the action of a sufficient load and deformation continues as long as a load is applied, providing rupture of the material does not occur. Deformation ceases when the load is removed, but the plastic solid does not return to its original state.

The intermolecular cohesive forces in a fluid are not great enough to hold the various elements of the fluid together. Hence a fluid will flow under the action of the slightest stress and flow will continue as long as the stress is present.

2.2 DISTINCTION BETWEEN A GAS AND A LIQUID

A fluid may be either a gas or a liquid. The molecules of a gas are much farther apart than those of a liquid. Hence a gas is very compressible, and when all external pressure is removed, it tends to expand indefinitely. A gas is therefore in equilibrium only when it is completely enclosed. A liquid is relatively incompressible, and if all pressure, except that of its own vapor pressure, is removed, the cohesion between molecules holds them together, so that the liquid does not expand indefinitely. Therefore a liquid may have a free surface, i.e., a surface from which all pressure is removed, except that of its own vapor.

A ***vapor*** is a gas whose temperature and pressure are such that it is very near the liquid phase. Thus steam is considered a vapor because its state is normally not far from that of water. A gas may be defined as a highly superheated vapor; that is, its state is far removed from the liquid phase. Thus air is considered a gas because its state is normally very far from that of liquid air.

The volume of a gas or vapor is greatly affected by changes in pressure or temperature or both. It is usually necessary, therefore, to take account of changes in volume and temperature in dealing with gases or vapors. Whenever significant temperature or phase changes are involved in dealing with vapors and gases, the subject is largely dependent on heat phenomena (***thermodynamics***). Thus fluid mechanics and thermodynamics are interrelated.

2.3 DENSITY, SPECIFIC WEIGHT, SPECIFIC VOLUME, AND SPECIFIC GRAVITY

The ***density*** ρ (rho)[1] of a fluid is its *mass* per unit volume, while the ***specific weight*** γ (gamma) is its *weight* per unit volume. In the British gravitational (BG) system (Sec. 1.5) density ρ will be in slugs per cubic foot (kg/m^3 in SI units), which may also be expressed as units of $lb \cdot sec^2/ft^4$ ($N \cdot s^2/m^4$ in SI units) (Sec. 1.5 and inside covers).

Specific weight γ represents the force exerted by gravity on a unit volume of fluid, and therefore must have the units of force per unit volume, such as pounds per cubic foot (N/m^3 in SI units).

Density and specific weight of a fluid are related as follows:

$$\rho = \frac{\gamma}{g} \qquad \text{or} \qquad \gamma = \rho g \tag{2.1}$$

Since the physical equations are dimensionally homogeneous, the dimensions of density are

$$\text{Dimensions of } \rho = \frac{\text{dimensions of } \gamma}{\text{dimensions of } g} = \frac{lb/ft^3}{ft/sec^2} = \frac{lb \cdot sec^2}{ft^4} = \frac{\text{mass}}{\text{volume}} = \frac{\text{slugs}}{ft^3}$$

[1] The names of Greek letters are given in the List of Symbols on page xix.

In SI units

$$\text{Dimensions of } \rho = \frac{\text{dimensions of } \gamma}{\text{dimensions of } g} = \frac{\text{N/m}^3}{\text{m/s}^2} = \frac{\text{N}\cdot\text{s}^2}{\text{m}^4} = \frac{\text{mass}}{\text{volume}} = \frac{\text{kg}}{\text{m}^3}$$

It should be noted that density ρ is absolute, since it depends on mass, which is independent of location. Specific weight γ, on the other hand, is not absolute, since it depends on the value of the gravitational acceleration g, which varies with location, primarily latitude and elevation above mean sea level.

Densities and specific weights of fluids vary with temperature. Commonly needed temperature variations of these quantities for water and air are provided in Appendix A. Densities and specific weights of common gases at standard atmospheric pressure and temperature are also listed there.

Specific volume v is the volume occupied by a unit mass of fluid.[2] It is commonly applied to gases, and is usually expressed in cubic feet per slug (m^3/kg in SI units). Specific volume is the reciprocal of density. Thus

$$v = \frac{1}{\rho} \tag{2.2}$$

Specific gravity s of a liquid is the ratio of its density to that of pure water at a standard temperature. Physicists use 39.2°F (4°C) as the standard, but engineers often use 60°F. In the metric system the density of water at 4°C is 1.00 g/cm^3 (or 1.00 g/mL),[3] equivalent to 1000 kg/m^3, and hence the specific gravity (which is dimensionless) of a liquid has the same numerical value as its density expressed in g/mL or Mg/m^3. Specific gravities and densities of various liquids at standard atmospheric pressure are given in Appendix A.

The specific gravity of a gas is the ratio of its density to that of either hydrogen or air at some specified temperature and pressure, but there is no general agreement on these standards, and so they must be explicitly stated in any given case.

Since the density of a fluid varies with temperature, specific gravities must be determined and specified at particular temperatures.

---

[2] Note that this book uses a "rounded" lower case v (vee), to help distinguish it from a capital V and from the Greek ν (nu).

[3] One cubic centimeter (cm^3) is equivalent to one milliliter (mL).

SAMPLE PROBLEM 2.1 The specific weight of water at ordinary pressure and temperature is 62.4 lb/ft³. The specific gravity of mercury is 13.56. Compute the density of water and the specific weight and density of mercury.

Solution

$$\rho_{water} = \frac{\gamma_{water}}{g} = \frac{62.4 \text{ lb/ft}^3}{32.2 \text{ ft/sec}^2} = 1.938 \text{ slugs/ft}^3 \qquad \textit{ANS}$$

$$\gamma_{mercury} = s_{mercury}\gamma_{water} = 13.56(62.4) = 846 \text{ lb/ft}^3 \qquad \textit{ANS}$$

$$\rho_{mercury} = s_{mercury}\rho_{water} = 13.56(1.938) = 26.3 \text{ slugs/ft}^3 \qquad \textit{ANS}$$

SAMPLE PROBLEM 2.2 The specific weight of water at ordinary pressure and temperature is 9.81 kN/m³. The specific gravity of mercury is 13.56. Compute the density of water and the specific weight and density of mercury.

Solution

$$\rho_{water} = \frac{9.81 \text{ kN/m}^3}{9.81 \text{ m/s}^2} = 1.00 \text{ Mg/m}^3 = 1.00 \text{ g/mL} \qquad \textit{ANS}$$

$$\gamma_{mercury} = s_{mercury}\gamma_{water} = 13.56(9.81) = 133.0 \text{ kN/m}^3 \qquad \textit{ANS}$$

$$\rho_{mercury} = s_{mercury}\rho_{water} = 13.56(1.00) = 13.56 \text{ Mg/m}^3 \qquad \textit{ANS}$$

EXERCISES

2.3.1 If a certain gasoline weighs 46 lb/ft³, what are the values of its density, specific volume, and specific gravity relative to water at 60°F? Use Appendix A.

2.3.2 A certain gas weighs 0.12 lb/ft³ at a certain temperature and pressure. What are the values of its density, specific volume, and specific gravity relative to air weighing 0.075 lb/ft³?

2.3.3 A certain gas weighs 18.0 N/m³ at a certain temperature and pressure. What are the values of its density, specific volume, and specific gravity relative to air weighing 12.0 N/m³?

2.3.4 The specific weight of glycerin is 78.7 lb/ft³. Compute its density and specific gravity. What is its specific weight in kN/m³?

2.3.5 If the specific weight of a liquid is 54 lb/ft³, what is its density?

2.3.6 If the specific weight of a liquid is 7.6 kN/m³, what is its density?

2.3.7 If the specific volume of a gas is 350 ft^3/slug, what is its specific weight in lb/ft^3?

2.3.8 If the specific volume of a gas is 0.80 m^3/kg, what is its specific weight in N/m^3?

2.3.9 Initially when 1000.00 mL of water at 10°C are poured into a glass cylinder, the height of the water column is 1000.0 mm. The water and its container are heated to 80°C. Assuming no evaporation, what then will be the depth of the water column if the coefficient of thermal expansion for the glass is 3.6×10^{-6} mm/mm per °C?

2.4 COMPRESSIBLE AND INCOMPRESSIBLE FLUIDS

Fluid mechanics deals with both incompressible and compressible fluids, that is, with fluids of either constant or variable density. Although there is no such thing in reality as an incompressible fluid, this term is applied where the change in density with pressure is so small as to be negligible. This is usually the case with liquids. Gases, too, may be considered incompressible when the pressure variation is small compared with the absolute pressure.

Liquids are ordinarily considered incompressible fluids, yet sound waves, which are really pressure waves, travel through them. This is evidence of the elasticity of liquids. In problems involving ***water hammer*** (Sec. 12.6) it is necessary to consider the compressibility of the liquid.

The flow of air in a ventilating system is a case where a gas may be treated as incompressible, for the pressure variation is so small that the change in density is of no importance. But for a gas or steam flowing at high velocity through a long pipeline, the drop in pressure may be so great that change in density cannot be ignored. For an airplane flying at speeds below 250 mph (100 m/s), the air may be considered to be of constant density. But as an object moving through the air approaches the velocity of sound, which is of the order of 700 mph (300 m/s), the pressure and density of the air adjacent to the body become materialy different from those of the air at some distance away, and the air must then be treated as a compressible fluid (Chap. 13).

2.5 COMPRESSIBILITY OF LIQUIDS

The compressibility (change in volume due to change in pressure) of a liquid is inversely proportional to its ***volume modulus of elasticity***, also known as the ***bulk modulus***. This modulus is defined as $E_v = -v\,dp/dv = -(v/dv)\,dp$, where v = specific volume and p = pressure. As v/dv is a dimensionless ratio, the units of E_v and p are identical. The bulk modulus is

Table 2.1
Bulk modulus of water E_v, psi[a]

Pressure, psia	Temperature, °F 32°	68°	120°	200°	300°
15	292,000	320,000	332,000	308,000	
1,500	300,000	330,000	342,000	319,000	248,000
4,500	317,000	348,000	362,000	338,000	271,000
15,000	380,000	410,000	426,000	405,000	350,000

[a] These values can be transformed to meganewtons per square meter by multiplying them by 0.006 89. The values in the first line are for conditions close to normal atmospheric pressure; for a more complete set of values at normal atmospheric pressure, see Table A.1 in Appendix A.

analogous to the modulus of elasticity for solids; however, for fluids it is defined on a volume basis rather than in terms of the familiar one-dimensional stress–strain relation for solid bodies.

In most engineering problems the bulk modulus at or near atmospheric pressure is the one of interest. The bulk modulus is a property of the fluid and for liquids is a function of temperature and pressure. In Table 2.1 are shown a few values of the bulk modulus for water. At any temperature it can be noted that the value of E_v increases continuously with pressure, but at any one pressure the value of E_v is a maximum at about 120°F (50°C). Thus water has a minimum compressibility at about 120°F (50°C).

Note that applied pressures, such as those in Table 2.1, are often specified in absolute terms, because atmospheric pressure varies. The units psia or kN/m^2 abs indicate absolute pressure, which is the actual pressure on the fluid, relative to absolute zero. The standard atmospheric pressure at sea level is about 14.7 psia or 101.3 kN/m^2 abs (1013 mbar abs) (see Sec. 2.9 and Table A.3). Bars and millibars were previously used in metric systems to express pressure; 1 mbar = 100 N/m^2. Most pressures are measured relative to the atmosphere, and are known as gage pressures. This is explained more fully in Sec. 3.4.

The volume modulus of mild steel is about 26,000,000 psi (170 000 MN/m^2). Taking a typical value for the volume modulus of cold water to be 320,000 psi (2200 MN/m^2), it is seen that water is about 80 times as compressible as steel. The compressibility of liquids covers a wide range. Mercury, for example, is approximately 8% as compressible as water, while the compressibility of nitric acid is nearly six times greater than that of water.

Table 2.1 shows that at any one temperature the bulk modulus of water does not vary a great deal for a moderate range in pressure. Rearranging the definition of E_v, as an approximation one may use for the case of a fixed mass of liquid at constant temperature

$$\frac{\Delta v}{v} \approx -\frac{\Delta p}{E_v} \tag{2.3a}$$

or

$$\frac{v_2 - v_1}{v_1} \approx -\frac{p_2 - p_1}{E_v} \tag{2.3b}$$

where E_v is the mean value of the modulus for the pressure range and the subscripts 1 and 2 refer to the before and after conditions.

Assuming E_v to have a value of 320,000 psi, it may be seen that increasing the pressure of water by 1000 psi will compress it only $\frac{1}{320}$, or 0.3%, of its original volume. Therefore it is seen that the usual assumption regarding water as incompressible is justified.

Sample Problem 2.3 At a depth of 8 km in the ocean the pressure is 81.8 MPa. Assume that the specific weight of sea water at the surface is 10.05 kN/m³ and that the average volume modulus is 2.34×10^9 N/m² for that pressure range. (*a*) What will be the change in specific volume between that at the surface and at that depth? (*b*) What will be the specific volume at that depth? (*c*) What will be the specific weight at that depth?

Solution

(*a*) Eq. (2.2): $v_1 = 1/\rho_1 = g/\gamma_1 = 9.81/10\,050 = 0.000\,976 \text{ m}^3/\text{kg}$

Eq. (2.3a): $\Delta v = -0.000\,976(81.8 \times 10^6 - 0)/(2.34 \times 10^9)$

$= -34.1 \times 10^{-6} \text{ m}^3/\text{kg}$ ***ANS***

(*b*) Eq. (2.3b): $v_2 = v_1 + \Delta v = 0.000\,942 \text{ m}^3/\text{kg}$ ***ANS***

(*c*) $\gamma_2 = g/v_2 = 9.81/0.000\,942 = 10\,410 \text{ N/m}^3$ ***ANS***

EXERCISES

2.5.1 Water in a hydraulic press is subjected to a pressure of 4500 psia at 68°F. If the initial pressure is 15 psia, approximately what will be the percentage decrease in specific volume? Use Table 2.1.

2.5.2 At a depth of 5 miles in the ocean the pressure is 11,930 psi. Assume that the specific weight at the surface is 64.00 lb/ft³ and that the average volume modulus is 340,000 psi for that pressure range. (*a*) What will be the change in specific volume between that at the surface and at that depth? (*b*) What will be the specific volume at that depth? (*c*) What will be the specific weight at that depth?

2.5.3 (*a*) What is the percentage change in the specific volume in Exer. 2.5.2? (*b*) What is the percentage change in the specific weight in Exer. 2.5.2?

2.5.4 To two significant figures what is the bulk modulus of water in MN/m^2 at 50°C under a pressure of 30 MN/m^2? Use Table 2.1.

2.5.5 At normal atmospheric conditions, approximately what pressure in psi must be applied to water to reduce its volume by 3%?

2.5.6 At normal atmospheric conditions, approximately what pressure in MPa must be applied to water to reduce its volume by 2%?

2.5.7 Water is contained in a long rigid cylinder whose inside diameter is 0.650 in. A plunger is used to apply pressure to the water. The length of the column of water is 20.00 in. What will be the length of the column of water if a force of 500 lb is applied to the plunger. Assume no leakage and zero friction between the plunger and the cylinder.

2.5.8 A rigid cylinder, inside diameter 15 mm, contains a column of water 500 mm long. What will the column length be if a force of 2 kN is applied to its end by a frictionless plunger? Assume no leakage.

2.6 SPECIFIC WEIGHT OF LIQUIDS

The specific weights of some common liquids at 68°F (20°C) and standard sea-level atmospheric pressure[4] with $g = 32.2$ ft/sec^2 (9.81 m/s^2) are given in Table 2.2. The specific weight of a *liquid* varies only slightly with pressure,

Table 2.2
Specific weights of common liquids at 68°F (20°C), 14.7 psia (1013 mbar abs) with $g = 32.2$ ft/sec^2 (9.81 m/s^2)

	lb/ft^3	kN/m^3
Carbon tetrachloride	99.4	15.6
Ethyl alcohol	49.3	7.76
Gasoline	42	6.6
Glycerin	78.7	12.3
Kerosene	50	7.9
Motor oil	54	8.5
Water	62.3	9.79
Seawater	63.9	10.03

[4] See Secs. 3.5 and 2.9.

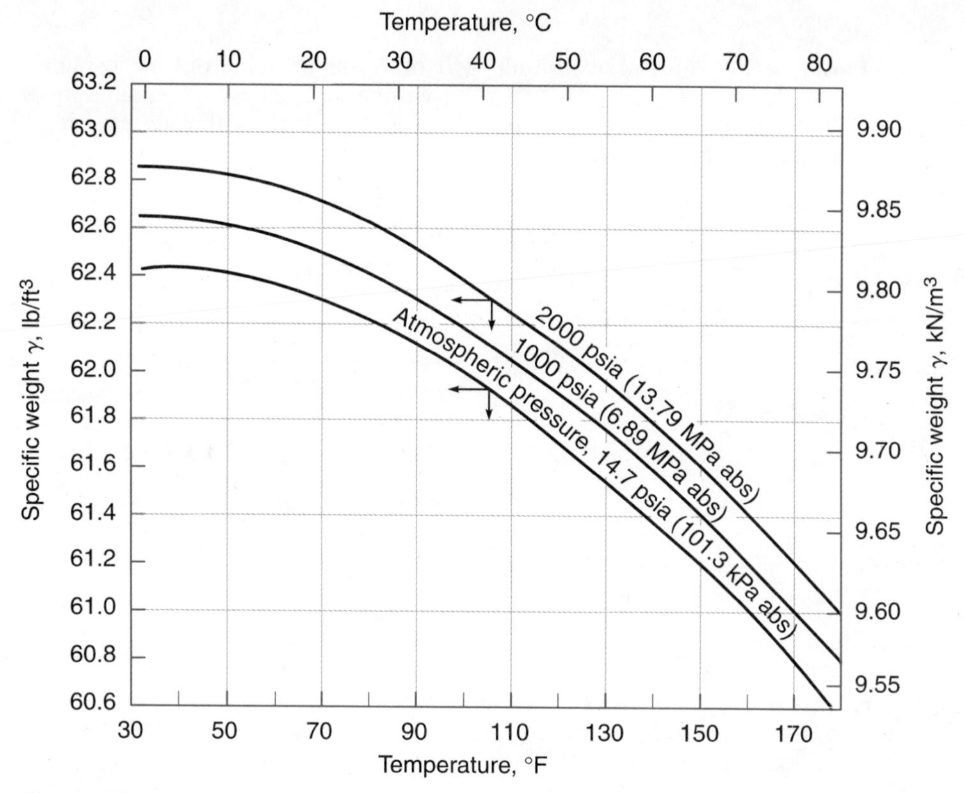

Figure 2.1
Specific weight γ of pure water as a function of temperature and pressure for condition where $g = 32.2$ ft/sec^2 (9.81 m/s^2).

depending on the bulk modulus of the liquid (Sec. 2.5); it also depends on temperature, and the variation may be considerable. Since specific weight γ is equal to ρg, the specific weight of a *fluid* depends on the local value of the acceleration of gravity in addition to the variations with temperature and pressure. The variation of the specific weight of water with temperature and pressure, where $g = 32.2$ ft/sec^2 (9.81 m/s^2), is shown in Fig. 2.1. The presence of dissolved air, salts in solution, and suspended matter will increase these values a very slight amount. Ocean water may ordinarily be assumed to weigh 64.0 lb/ft^3 (10.1 kN/m^3). Unless otherwise specified or implied by some specific temperature being given, the value to use for water in the problems in the text is $\gamma = 62.4$ lb/ft^3 (9.81 kN/m^3). Under extreme conditions the specific weight of water is quite different. For example, at 500°F (260°C) and 6000 psi (42 MN/m^2) the specific weight of water is 51 lb/ft^3 (8.0 kN/m^3).

SAMPLE PROBLEM 2.4 A vessel contains 85 L of water at 10°C and atmospheric pressure. If the water is heated to 70°C, what will be the percentage change in its volume? What weight of water must be removed to maintain the volume at its original value? Use Appendix A.

Solution

$$\forall_{10} = 85 \text{ L} = 0.085 \text{ m}^3$$

Table A.1: $\gamma_{10} = 9.804 \text{ kN/m}^3, \quad \gamma_{70} = 9.589 \text{ kN/m}^3$

$$\text{Weight of water} = \gamma\forall = \gamma_{10}\forall_{10} = \gamma_{70}\forall_{70}$$

i.e., $9.804(0.085) \text{ kN} = 9.589\forall_{70}; \quad \forall_{70} = 0.086\,91 \text{ m}^3$

$$\Delta\forall = \forall_{70} - \forall_{10} = 0.086\,91 - 0.085\,00 = 0.001\,91 \text{ m}^3$$

$$\Delta\forall/\forall_{10} = 0.001\,91/0.085 = 2.25\% \text{ increase} \qquad \textbf{\textit{ANS}}$$

$$\text{Must remove} \quad (0.001\,91 \text{ m}^3)(9589 \text{ N/m}^3) = 18.31 \text{ N} \qquad \textbf{\textit{ANS}}$$

EXERCISES

2.6.1 Approximately what is the specific weight of water in kN/m^3 at a temperature of 170°F under normal atmospheric conditions? What is the approximate specific weight at 170°F under a pressure of 2000 psia? Use Fig. 2.1.

2.6.2 A vessel contains 2.0 ft^3 of water at 40°F and atmospheric pressure. If the water is heated to 60°F, what will be the percentage change in its volume? What weight of water must be removed to maintain the volume at its original value? Use Appendix A.

2.6.3 A cylindrical tank (diameter = 10.00 m and depth = 6.00 m) contains water at 20°C and is brimful. If the water is heated to 50°C, how much water will spill over the edge of the tank? Assume the tank does not expand with the change in temperature. Use Appendix A.

2.7
PROPERTY RELATIONS FOR PERFECT GASES

The various properties of a gas, listed below, are related to one another (see, e.g., Appendix A, Tables A.2 and A.5). They differ for each gas. When the conditions of most real gases are far removed from the liquid phase, these relations closely approximate those of hypothetical ***perfect gases***. Perfect gases, are here (and often) defined to have constant specific heats[5] and to

[5] Specific heat and other thermodynamic properties of gases are discussed in Sec. 13.1.

obey the ***perfect-gas law***,

$$\frac{p}{\rho} = pv = RT \tag{2.4}$$

where p = absolute pressure
ρ = density (mass per unit volume)
v = specific volume ($1/\rho$)
R = a gas constant, the value of which depends upon the particular gas
T = absolute temperature in degrees Rankine or Kelvin[6]

For air, the value of R is 1715 ft·lb/(slug·°R) or 287 N·m/(kg·K) (Appendix A, Table A.5). Since $\gamma = \rho g$, Eq. (2.4) may also be written

$$\gamma = \frac{gp}{RT} \tag{2.5}$$

from which the specific weight of any gas at any temperature and pressure can be computed if R and g are known. Because Eqs. (2.4) and (2.5) relate the various gas properties at a particular state, they are known as ***equations of state*** and as ***property relations***.

In this book we shall assume that all gases are perfect. Perfect gases are also called ideal gases. Do not confuse a perfect (ideal) gas with an ideal fluid (Sec. 2.10).

Avogadro's law states that all gases at the same temperature and pressure under the action of a given value of g have the same number of molecules per unit of volume, from which it follows that the specific weight of a gas[7] is proportional to its molecular weight. Thus, if M denotes molecular weight, $\gamma_2/\gamma_1 = M_2/M_1$ and, from Eq. (2.5), $\gamma_2/\gamma_1 = R_1/R_2$ for the same temperature, pressure, and value of g. Hence

$$M_1R_1 = M_2R_2 = \text{constant} = R_0$$

R_0 is known as the ***universal gas constant***, and has a value of 49,709 ft·lb/(slug·°R) or 8312 N·m/(kg·K).

For real gases, the right-hand side of Eq. (2.4) is replaced by zRT, where z is a compressibility factor. Values of z and R may be found in texts on thermodynamics and in handbooks. However, for normally encountered monatomic and diatomic gases, z varies from unity by less than 3%, so a value of R for these may always be estimated by dividing R_0 by molecular

---

[6] Absolute temperature is measured above absolute zero. This occurs on the Fahrenheit scale at −459.67°F (0° Rankine) and on the Celsius scale at −273.15°C (0 Kelvin). Except for low temperature work, these values are usually taken as −460°F and −273°C. Remember that no degree symbol is used with Kelvin.

[7] The specific weight of air (molecular weight ≈ 29.0) at 68°F (20°C) and 14.7 psia (1013 mbar abs) with g = 32.2 ft/sec^2 (9.81 m/s^2) is 0.0752 lb/ft^3 (11.82 N/m^3).

weight. Thus water vapor in the atmosphere, because of its low partial pressure, may be treated as a perfect gas with $R = 49\,709/18 = 2760$ ft·lb/(slug·°R) [462 N·m/(kg·K)]. For steam at higher pressures this value is not applicable. In dealing with water vapor in the atmosphere Dalton's law of partial pressures states that the air and the water vapor each exert their own pressure as if the other were not present. Hence it is the partial pressure of each that is used in Eqs. (2.4) and (2.5) (see Sample Prob. 2.5).

As the pressure is increased and the temperature simultaneously lowered, a gas becomes a vapor, and as gases depart more and more from the gas phase and approach the liquid phase, the property relations become much more complicated than Eq. (2.4), and specific weight and other properties must then be obtained from vapor tables or charts. Such tables and charts exist for steam, ammonia, sulfur dioxide, freon, and other vapors in common engineering use.

Another fundamental equation for a perfect gas is

$$pv^n = p_1 v_1^n = \text{constant} \tag{2.6}$$

where p is absolute pressure, v $(=1/\rho)$ is specific volume, and n may have any nonnegative value from zero to infinity, depending upon the process to which the gas is subjected. Since this equation describes the change of the gas properties from one state to another for a particular process, it is known as a ***process equation***. If the process of change is at a constant temperature (isothermal), $n = 1$. If there is no heat transfer to or from the gas, the process is known as ***adiabatic***. A frictionless (and reversible) adiabatic process is called an ***isentropic*** process, and n is denoted by k, where $k = c_p/c_v$, the ratio of specific heat at constant pressure to that at constant volume.[8] For expansion with friction n is less than k, and for compression with friction n is greater than k. Values for k may be found in Appendix A, Table A.5, and in thermodynamics texts and in handbooks. For air and diatomic gases at usual temperatures, k may be taken as 1.4.

By combining Eqs. (2.4) and (2.6), it is possible to obtain other useful relations such as

$$\frac{T_2}{T_1} = \left(\frac{v_1}{v_2}\right)^{n-1} = \left(\frac{\rho_2}{\rho_1}\right)^{n-1} = \left(\frac{p_2}{p_1}\right)^{(n-1)/n} \tag{2.7}$$

SAMPLE PROBLEM 2.5 If an artificial atmosphere consists of 20% oxygen and 80% nitrogen by volume, at 14.7 psia and 60°F, what are (*a*) the specific weight and partial pressure of the oxygen and (*b*) the specific weight of the mixture?

[8] Specific heat and other thermodynamic properties of gases are discussed in Sec. 13.1.

Solution

Table A.5: R (oxygen) $= 1554$ ft^2/(sec^2·°R),

R (nitrogen) $= 1773$ ft^2/(sec^2·°R)

Eq. (2.5): 100% O_2: $\gamma = \dfrac{32.2(14.7\times 144)}{1554(460+60)} = 0.0843 \text{ lb/ft}^3$

Eq. (2.5): 100% N_2: $\gamma = \dfrac{32.2(14.7\times 144)}{1773(520)} = 0.0739 \text{ lb/ft}^3$

(*a*) Each ft^3 of mixture contains 0.2 ft^3 of O_2 and 0.8 ft^3 of N_2.

So for 20% O_2, $\gamma = 0.20(0.0843) = 0.016\,87 \text{ lb/ft}^3$ ***ANS***

From Eq. (2.5), for 20% O_2, $p = \dfrac{\gamma RT}{g} = \dfrac{0.016\,87(1554)520}{32.2}$

$= 423 \text{ lb/ft}^3 \text{ abs} = 2.94 \text{ psia}$ ***ANS***

Note that this $=$ 20%(14.7 psia).

(*b*) For 80% N_2, $\gamma = 0.80(0.0739) = 0.0591 \text{ lb/ft}^3$.

Mixture: $\gamma = 0.016\,87 + 0.0591 = 0.0760 \text{ lb/ft}^3$ ***ANS***

EXERCISES

2.7.1 A hydrogen-filled balloon of the type used in cosmic-ray studies is to be expanded to its full size, which is a 100-ft-diameter sphere, without stress in the wall at an altitude of 150,000 ft. If the pressure and temperature at this altitude are 0.14 psia and −67°F respectively, find the volume of hydrogen at 14.7 psia and 60°F that should be added on the ground. Neglect the balloon's weight.

2.7.2 If natural gas has a specific gravity of 0.6 relative to air at 14.7 psia and 68°F, what are its specific weight and specific volume at that same pressure and temperature. What is the value of R for the gas? Solve without using Table A.2.

2.7.3 A gas at 40°C under a pressure of 20 000 mbar abs has a specific weight of 241 N/m^3. What is the value of R for the gas? What gas might this be? Refer to Appendix A, Table A.5.

2.7.4 Calculate the density, specific weight and specific volume of air at 120°F and 50 psia.

2.7.5 Calculate the density, specific weight and specific volume of air at 50°C and 3400 mbar abs.

2.7.6 (*a*) If water vapor in the atmosphere has a partial pressure of 0.50 psia and the temperature is 90°F, what is its specific weight? (*b*) If the barometer reads 14.50 psia, what is the partial pressure of the (dry) air, and what is its specific weight? (*c*) What is the specific weight of the atmosphere (air plus the water vapor present)?

2.7.7 (*a*) If water vapor in the atmosphere has a partial pressure of 3500 Pa and the temperature is 30°C, what is its specific weight? (*b*) If the barometer reads 102 kPa abs, what is the partial pressure of the (dry) air, and what is its specific weight? (*c*) What is the specific weight of the atmosphere (air plus the water vapor present)?

2.7.8 If an artificial atmosphere consists of 20% oxygen and 80% nitrogen by volume, at 101.32 kN/m^2 abs and 20°C, what are (*a*) the specific weight and partial pressure of the oxygen, (*b*) the specific weight and partial pressure of the nitrogen, and (*c*) the specific weight of the mixture?

2.7.9 Prove that Eq. (2.7) follows from Eqs. (2.4) and (2.6).

2.8 COMPRESSIBILITY OF GASES

Differentiating Eq. (2.6) gives $npv^{n-1}\,dv + v^n\,dp = 0$. Inserting the value of dp from this into $E_v = -(v/dv)\,dp$ from Sec. 2.5 yields

$$E_v = np \tag{2.8}$$

So for an isothermal process of a gas $E_v = p$, and for an isentropic process $E_v = kp$.

Thus, at a pressure of 15 psia, the isothermal modulus of elasticity for a gas is 15 psi, and for air in an isentropic process it is 1.4×15 psi $= 21$ psi. Assuming from Table 2.1 a typical value of the modulus of elasticity of cold water to be 320,000 psi, it is seen that air at 15 psia is 320,000/15 = 21,000 times as compressible as cold water isothermally, or 320,000/21 = 15,000 times as compressible isentropically. This emphasizes the great difference between the compressibility of normal atmospheric air and that of water.

SAMPLE PROBLEM 2.6 (*a*) Calculate the density, specific weight, and specific volume of oxygen at 100°F and 15 psia (pounds per square inch absolute; see Sec. 2.7). (*b*) What would be the temperature and pressure of this gas if it were compressed isentropically to 40% of its original volume? (*c*) If the process described in (*b*) had been isothermal, and what would the temperature and pressure have been?

Solution

Table A.5 for oxygen (O_2): molecular weight $M = 32.0$, $k = 1.40$

(*a*) Sec. 2.7: $R \approx \dfrac{R_0}{M} = \dfrac{49{,}709}{32.0} = 1553 \text{ ft·lb/(slug·°R)}$ (as in Table A.5)

From Eq. (2.4): $\rho = \dfrac{p}{RT} = \dfrac{15 \times 144 \text{ lb/ft}^2}{[1553 \text{ ft·lb/(slug·°R)}][(460 + 100)\text{°R}]}$

$= 0.002\,48 \text{ slug/ft}^3$ ***ANS***

With $g = 32.2 \text{ ft/sec}^2$, $\gamma = \rho g = 0.0800 \text{ lb/ft}^3$ ***ANS***

Eq. (2.2): $v = \dfrac{1}{\rho} = \dfrac{1.0}{0.002\,48} = 403 \text{ ft}^3\text{/slug}$ ***ANS***

(*b*) Isentropic compression: $v_2 = 40\% v_1 = 0.4(403) = 161.1 \text{ ft}^3\text{/slug}$

$\rho_2 = 1/v_2 = 0.006\,21 \text{ slug/ft}^3$

Eq. (2.6) with $n = k$: $pv^k = (15 \times 144)(403)^{1.4} = (p_2 \times 144)(161.1)^{1.4}$

$p_2 = 54.1$ psia ***ANS***

From Eq. (2.4): $p_2 = 54.1 \times 144 \text{ psia} = \rho RT = 0.006\,21(1553)(460 + T_2)$

$T_2 = 348\text{°F}$ ***ANS***

(*c*) Isothermal compression: $T_2 = T_1 = 100\text{°F}$ ***ANS***

$pv =$ constant: $(15 \times 144)(403) = (p_2 \times 144)(0.4 \times 403)$

$p_2 = 37.5$ psia ***ANS***

Sample Problem 2.7 Calculate the density, specific weight, and specific volume of chlorine gas at 25°C and pressure of 600 kN/m² abs (kilonewtons per square meter absolute; see Sec. 2.7). Given the molecular weight of chlorine (Cl_2) = 71.

Solution

Sec. 2.7: $R = \dfrac{R_0}{M} = \dfrac{8312}{71} = 117.1 \text{ N·m/(kg·K)}$

From Eq. (2.4): $\rho = \dfrac{p}{RT} = \dfrac{600\,000 \text{ N/m}^2}{[117.1 \text{ N·m/(kg·K)}][(273 + 25)\text{K}]}$

$= 17.20 \text{ kg/m}^3$ ***ANS***

With $g = 9.81 \text{ m/s}^2$, $\gamma = \rho g = 168.7 \text{ N/m}^3$ ***ANS***

Eq. (2.2): $v = \dfrac{1}{\rho} = \dfrac{1}{17.20} = 0.0581 \text{ m}^3\text{/kg}$ ***ANS***

EXERCISES

2.8.1 Methane at 21 psia is compressed isothermally, and nitrogen at 15 psia is compressed isentropically. What is the modulus of elasticity of each gas? Which is the more compressible?

2.8.2 Methane at 120 kPa abs is compressed isothermally, and nitrogen at 90 kPa abs is compressed isentropically. What is the modulus of elasticity of each gas? Which is the more compressible?

2.8.3 Helium at 25 psia and 65°F is isentropically compressed to one-fifth of its original volume. What is its final pressure?

2.8.4 Helium at 165 kN/m^2 abs and 15°C is isentropically compressed to one-fifth of its original volume. What is its final pressure?

2.8.5 (*a*) If 10 m^3 of nitrogen at 30°C and 120 kPa is permitted to expand isothermally to 30 m^3, what is the resulting pressure? (*b*) What would the pressure and temperature have been if the process had been isentropic? The isentropic exponent k for nitrogen is 1.40.

2.9 STANDARD ATMOSPHERE

Standard atmospheres were first adopted in the 1920s in the United States and in Europe to satisfy a need for standardization of aircraft instruments and aircraft performance. As knowledge of the atmosphere increased, and man's activities in it rose to ever greater altitudes, such standards have been frequently extended and improved.

The International Civil Aviation Organization adopted its latest ***ICAO Standard Atmosphere*** in 1964, which extends up to 32 km (105,000 ft). The International Standards Organization adopted an ***ISO Standard Atmosphere*** to 50 km (164,000 ft) in 1973, which incorporates the ICAO standard. The United States has adopted the ***U.S. Standard Atmosphere,*** last revised in 1976. This incorporates the ICAO and ISO standards, and extends to at least 86 km (282,000 ft or 53.4 mi); for some quantities it extends as far as 1000 km (621 mi).

Variations of temperature and pressure in the U.S. Standard Atmosphere are presented graphically in Fig. 2.2. In the lowest 11.02 km (36,200 ft), called the ***troposphere***, the temperature decreases rapidly and linearly at a ***lapse rate*** of −6.489°C/km (−3.560°F/1000 ft). In the next layer, called the

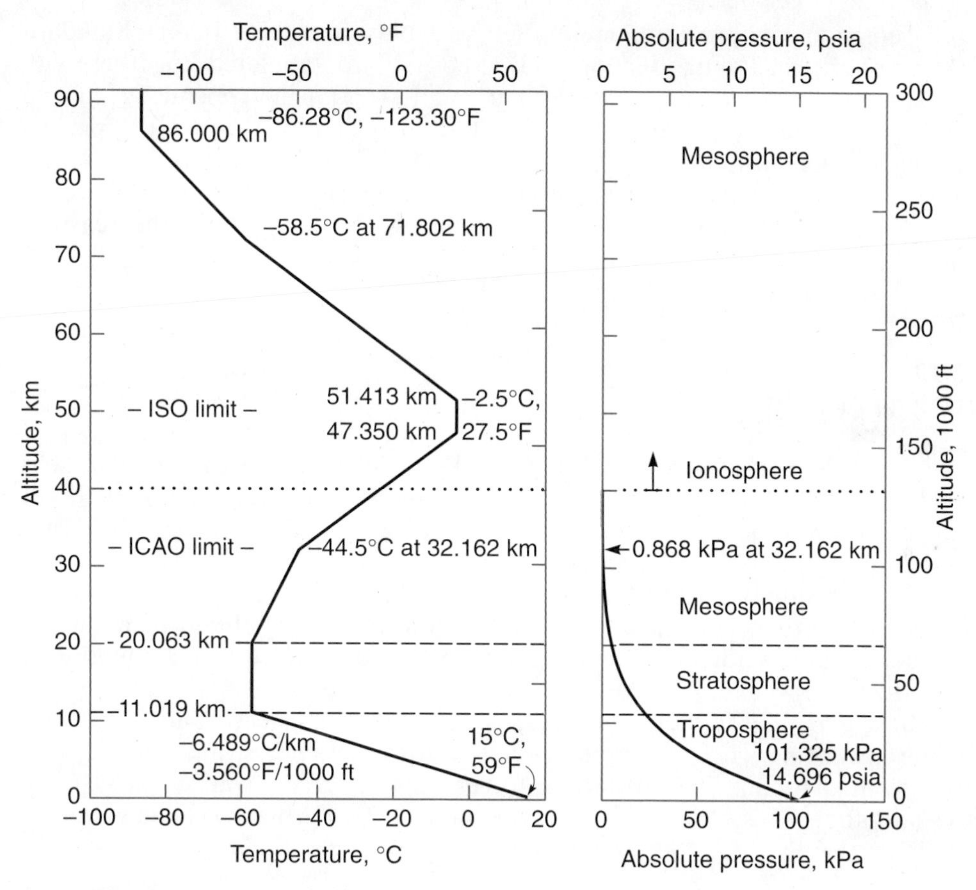

Figure 2.2
The U.S. Standard Atmosphere, temperature and pressure distributions.

stratosphere, about 9 km (30,000 ft) thick, the temperature remains constant at −56.5°C (−69.7°F). Next, in the ***mesosphere***, it increases, first slowly and then more rapidly, to a maximum of −2.5°C (27.5°F) at an altitude around 50 km (165,000 ft or 31 mi). Above this, in the ***ionosphere***, the temperature again decreases.

The standard absolute pressure behaves very differently from temperature (Fig. 2.2), decreasing quite rapidly and smoothly to almost zero at an altitude of 30 km (98,000 ft). The pressure profile was computed from the standard temperatures using methods of fluid statics (Sec. 3.2). The representation of the standard temperature profile by a number of linear functions of elevation (Fig. 2.2) greatly facilitates such computations (see Sample Prob. 3.1*d*).

Temperature, pressure, and other variables from the ICAO Standard Atmosphere, including density and viscosity, are tabulated together with gravitational acceleration out to 30 km and 100,000 ft in Appendix A, Table A.3. Such data are generally used in design calculations where the performance of high-altitude aircraft is of interest. The standard atmosphere serves as a good approximation of conditions to be expected in the atmosphere; of course the actual conditions vary somewhat with the weather, the seasons, and the latitude.

2.10 IDEAL FLUID

An ***ideal*** fluid may be defined as a fluid in which there is *no friction*; it is ***inviscid*** (its viscosity is zero). Thus the internal forces at any section within it are always normal to the section, even during motion. So these forces are purely pressure forces. Although such a fluid does not exist in reality, many fluids approximate frictionless flow at sufficient distances from solid boundaries, and so their behaviors can often be conveniently analyzed by assuming an ideal fluid. As noted in Sec. 2.7, take care to not confuse an ideal fluid with a perfect (ideal) gas.

In a ***real*** fluid, either liquid or gas, tangential or shearing forces always come into being whenever motion relative to a body takes place, thus giving rise to fluid friction, because these forces oppose the motion of one particle past another. These friction forces give rise to a fluid property called viscosity.

2.11 VISCOSITY

The ***viscosity*** of a fluid is a measure of its resistance to shear or angular deformation. Motor oil, for example, has high viscosity and resistance to shear, whereas gasoline has low viscosity. The friction forces in flowing fluid result from the cohesion and momentum interchange between molecules. The viscosities of typical fluids, and their dependence on temperature, are shown in Figs. 2.3 and 2.4. As the temperature increases, the viscosities of all liquids decrease, while the viscosities of all gases increase. This is because the force of cohesion, which diminishes with temperature, predominates with liquids, while with gases the predominating factor is the interchange of molecules between the layers of different velocities. Thus a rapidly moving molecule shifting into a slower-moving layer tends to speed up the latter. And a slow-moving molecule entering a faster-moving layer tends to slow down the faster-moving layer. This molecular interchange sets up a shear, or produces a friction force between adjacent layers. Increased molecular activity at higher temperatures causes the viscosity of gases to increase with temperature.

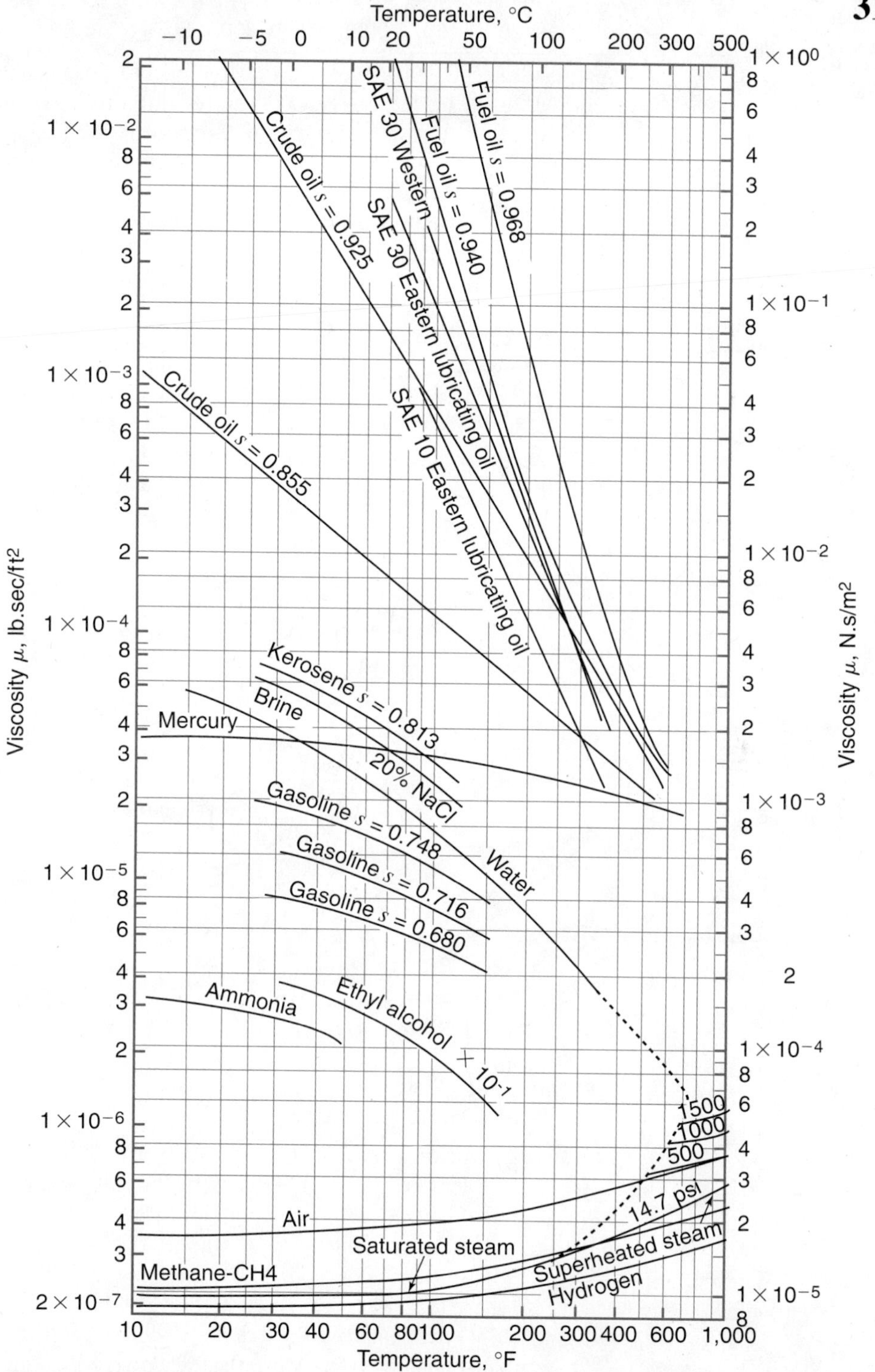

Figure 2.3
Absolute viscosity μ of fluids. (s = specific gravity at 60°F relative to water at 60°F.)

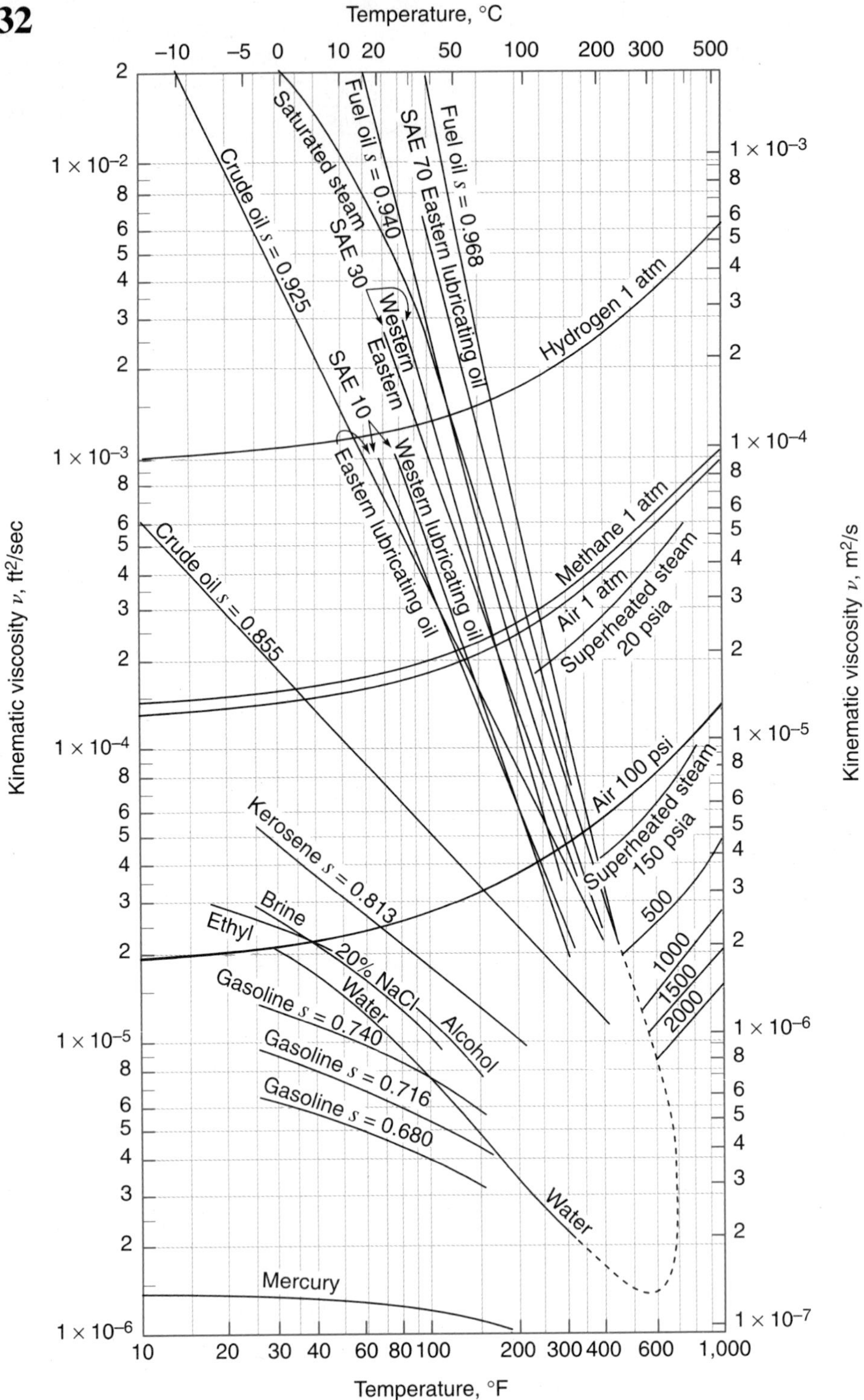

Figure 2.4
Kinematic viscosity ν of fluids. (s = specific gravity at 60°F relative to water at 60°F.)

Consider the classic case of two parallel plates (Fig. 2.5), sufficiently large that edge conditions may be neglected, placed a small distance Y apart, the space between being filled with the fluid. The lower surface is assumed to be stationary, while the upper one is moved parallel to it with a velocity U by the application of a force F corresponding to some area A of the moving plate.

At boundaries, particles of fluid adhere to the walls, and so their velocities are zero relative to the wall. This so-called ***no-slip condition*** occurs with all viscous fluids. Thus in Fig. 2.5 the fluid velocities must be zero where in contact with the plate at the lower boundary and U at the upper boundary. The form of the velocity variation with distance between these two extremes is called a ***velocity profile***. If the separation distance Y is not too great, the velocity U is not too high, and if there is no net flow of fluid through the space, the velocity profile will be linear, as shown in Fig. 2.5. The behavior is much as if the fluid were composed of a series of thin layers, each of which slips a little relative to the next. Experiments have shown that for a large class of fluids under such conditions

$$F \propto \frac{AU}{Y}$$

We see from similar triangles in Fig. 2.5 that U/Y can be replaced by the velocity gradient du/dy. If a constant of proportionality μ is now introduced, the shearing stress τ between any two thin sheets of fluid may be expressed by

$$\tau = \frac{F}{A} = \mu \frac{U}{Y} = \mu \frac{du}{dy} \tag{2.9}$$

Equation (2.9) is called ***Newton's equation of viscosity***, since it was first suggested by Sir Isaac Newton (1642–1727). Although better known for his formulation of the fundamental laws of motion and gravity and for the development of differential calculus, Newton, an English mathematician and natural philosopher, also made many pioneering studies in fluid mechanics. In transposed form Eq. (2.9) serves to define the proportionality constant

$$\mu = \frac{\tau}{du/dy} \tag{2.10}$$

which is called the ***coefficient of viscosity***, the ***absolute viscosity***, the ***dynamic viscosity*** (since it involves force), or simply the ***viscosity*** of the fluid.

We noted in Sec. 2.1 that the distinction between a solid and a fluid lies in the manner in which each can resist shearing stresses. A further distinction among various kinds of fluids and solids will be clarified by reference to Fig. 2.6. In the case of a solid, shear stress depends on the *magnitude* of the deformation; but Eq. (2.9) shows that in many fluids the shear stress is proportional to the *time rate* of (angular) deformation.

A fluid for which the constant of proportionality (i.e., the viscosity) does not change with rate of deformation is said to be a ***Newtonian fluid***, and can

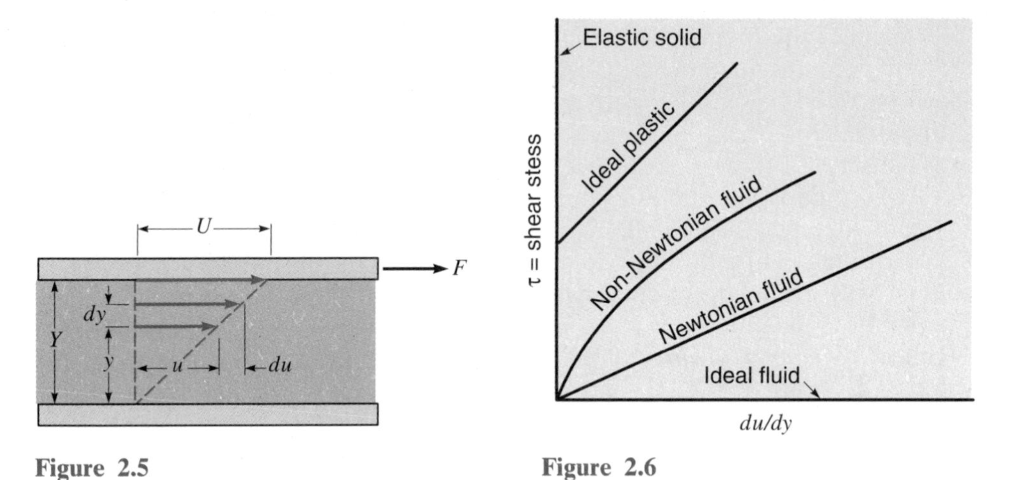

Figure 2.5

Figure 2.6

be represented by a straight line in Fig. 2.6. The slope of this line is determined by the viscosity. The ideal fluid, with no viscosity, is represented by the horizontal axis, while the true elastic solid is represented by the vertical axis. A plastic that sustains a certain amount of stress before suffering a plastic flow can be shown by a straight line intersecting the vertical axis at

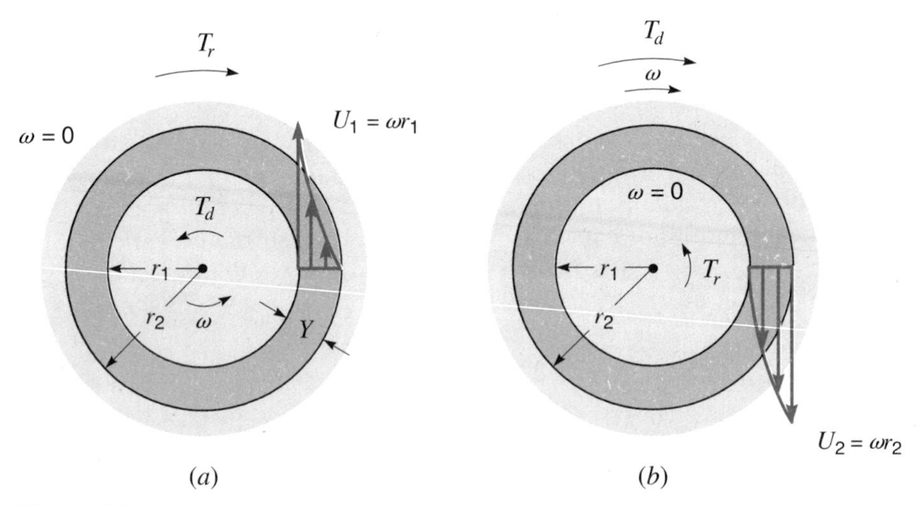

Figure 2.7
Velocity profile, rotating coaxial cylinders with gap completely filled with liquid. (*a*) Inner cylinder rotating. (*b*) Outer cylinder rotating. Z is the dimension at right angles to the plane of the sketch. Resisting torque = driving torque and $\tau \propto (du/dy)$.

$$\tau_1(2\pi r_1 Z)r_1 = \tau_2(2\pi r_2 Z)r_2, \qquad \left(\frac{du}{dy}\right)_1 = \left(\frac{du}{dy}\right)_2 \frac{r_2^2}{r_1^2}$$

the yield stress. There are certain non-Newtonian fluids[9] in which μ varies with the rate of deformation. These are relatively uncommon in engineering usage, so the remainder of this text will be restricted to the common fluids that under normal conditions obey Newton's law.

In the case of bulk fluid transport between the two parallel plates discussed previously, as could result from pressure-fed lubrication for example, the velocity profile becomes the sum of the previous linear profile plus a parabolic profile; the parabolic additions to (or subtractions from) the linear profile are zero at the walls or plates and maximum at the centerline.

In a ***journal bearing*** lubricating fluid fills the small annular space between a shaft and its surrounding support. This fluid layer is very similar to the layer between the two parallel plates, except it is curved. There is another more subtle difference, however. For coaxial cylinders (Fig. 2.7) with constant rotative speed ω, the resisting and driving torques must be equal. But because the radii at the inner and outer walls are different, it follows that the shear stresses and velocity gradients there must be different also (see Fig. 2.7 and equations below it). The shear stress and velocity gradient must vary continuously across the gap. However, as the gap distance $Y \to 0$, $du/dy \to U/Y$ = constant. So, when the gap is very small, the velocity profile can be assumed to be a straight line, and problems can be solved in a similar manner as for flat plates.

The dimensions of absolute viscosity are force per unit area divided by velocity gradient. In the British gravitational (BG) system the dimensions of absolute viscosity are as follows:

$$\text{Dimensions of } \mu = \frac{\text{dimensions of } \tau}{\text{dimensions of } du/dy} = \frac{\text{lb/ft}^2}{\text{fps/ft}} = \text{lb}\cdot\text{sec/ft}^2$$

In SI units

$$\text{Dimensions of } \mu = \frac{\text{N/m}^2}{\text{s}^{-1}} = \text{N}\cdot\text{s/m}^2$$

A widely used unit for viscosity in the metric system is the ***poise*** (P), named after Jean Louis Poiseuille (1799–1869). A French anatomist, Poiseuille was one of the first investigators of viscosity. The poise = 0.10 N·s/m². The ***centipoise*** (cP) (= 0.01 P = 1 mN·s/m²) is frequently a more convenient unit. It has a further advantage in that the viscosity of water at 68.4°F is 1 cP. Thus the value of the viscosity in centipoises is an indication of the viscosity of the fluid relative to that of water at 68.4°F.

In many problems involving viscosity the viscosity is divided by density. This ratio defines the ***kinematic viscosity*** ν, so called because force is not

[9] Typical non-Newtonian fluids include paints, printer's ink, gels and emulsions, sludges and slurries, and certain plastics. An excellent treatment of the subject is given by W. L. Wilkinson in *NonNewtonian Fluids,* Pergamon Press, New York, 1960.

involved, the only dimensions being length and time, as in kinematics (Sec. 1.1). Thus

$$\nu = \frac{\mu}{\rho} \tag{2.11}$$

Kinematic viscosity ν is usually measured in ft²/sec in the BG system, and is given in m²/s in the SI. Previously, in the metric system the common units were cm²/s, also called the ***stoke*** (St), after Sir George Stokes (1819–1903). The ***centistoke*** (cSt) (0.01 St = 10^{-6} m²/s) was often found a more convenient unit.

The absolute viscosity of all fluids is practically independent of pressure for the range that is ordinarily encountered in engineering work. For extremely high pressures, the values are somewhat higher than those shown in Fig. 2.3. The kinematic viscosity of gases varies with pressure because of changes in density.

The measurement of viscosity is described in Sec. 11.1.

SAMPLE PROBLEM 2.8 A 1-in-wide space between two horizontal plane surfaces is filled with SAE 30 western lubricating oil at 80°F. What force is required to drag a very thin plate of 4-ft² area through the oil at a velocity of 20 ft/min if the plate is 0.33 in from one surface?

Solution

Fig. 2.3: $\mu = 0.0063\ \text{lb}\cdot\text{sec/ft}^2$

Eq. (2.9): $\tau_1 = 0.0063 \times (20/60)/(0.33/12) = 0.0764\ \text{lb/ft}^2$

Eq. (2.9): $\tau_2 = 0.0063 \times (20/60)/(0.67/12) = 0.0394\ \text{lb/ft}^2$

From Eq. (2.9): $F_1 = \tau_1 A = 0.0764 \times 4 = 0.305\ \text{lb}$

From Eq. (2.9): $F_2 = \tau_2 A = 0.0394 \times 4 = 0.158\ \text{lb}$

$\text{Force} = F_1 + F_2 = 0.463\ \text{lb}$ ***ANS***

SAMPLE PROBLEM 2.9 In Fig. S2.9 oil of viscosity μ fills the small gap of thickness Y. (*a*) Neglecting fluid stress exerted on the circular underside, obtain an expression for the torque T required to rotate the truncated cone at constant speed ω. (*b*) What is the rate of heat generation, in joules per second, if the oil viscosity is 0.20 N·s/m², ω = 90 rpm, α = 45°, a = 45 mm, b = 60 mm, and Y = 0.2 mm?

Solution

(*a*) $U = \omega r$; for small gap Y, $\dfrac{du}{dy} = \dfrac{U}{Y} = \dfrac{\omega r}{Y}$

Eq. (2.9): $\tau = \mu \dfrac{du}{dy} = \dfrac{\mu \omega r}{Y}$; $dA = 2\pi r\, ds = \dfrac{2\pi r\, dy}{\cos \alpha}$

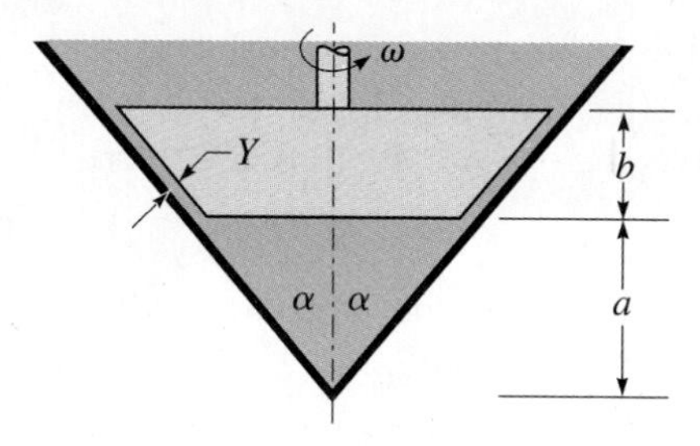

Figure S2.9

From Eq. (2.9): $dF = \tau\, dA = \dfrac{\mu\omega r}{Y}\left(\dfrac{2\pi r\, dy}{\cos\alpha}\right)$

$$dT = r\,dF = \frac{2\pi\mu\omega}{Y\cos\alpha} r^3\,dy; \qquad r = y\tan\alpha$$

$$dT = \frac{2\pi\mu\omega\tan^3\alpha}{Y\cos\alpha} y^3\,dy$$

$$T = \frac{2\pi\mu\omega\tan^3\alpha}{Y\cos\alpha}\int_a^{a+b} y^3\,dy; \qquad \frac{y^4}{4}\bigg|_a^{a+b} = \left[\frac{(a+b)^4}{4} - \frac{a^4}{4}\right]$$

$$T = \frac{2\pi\mu\omega\tan^3\alpha}{4Y\cos\alpha}[(a+b)^4 - a^4] \qquad \textit{\textbf{ANS}}$$

(*b*) $[(a+b)^4 - a^4] = (0.105\text{ m})^4 - (0.045\text{ m})^4 = 0.000\,117\,5\text{ m}^4$

$$\text{Heat generation rate} = \text{power} = T\omega = \frac{2\pi\mu\omega^2\tan^3\alpha}{4Y\cos\alpha}[(a+b)^4 - a^4]$$

$$= \frac{2\pi(0.20\text{ N·s/m}^2)(90\times 2\pi/60\text{ s}^{-1})^2(1)^3[0.000\,117\,5\text{ m}^4]}{4(2\times10^{-4}\text{ m})\cos 45^\circ}$$

$$= 23.2\text{ N·m/s} = 23.2\text{ J/s} \qquad \textit{\textbf{ANS}}$$

EXERCISES

2.11.1 A liquid has an absolute viscosity of 3.5×10^{-4} lb·sec/ft^2. It weighs 58 lb/ft^3. What are its absolute and kinematic viscosities in SI units?

2.11.2 (*a*) What is the ratio of the viscosity of water at a temperature of 70°F to that of water at 200°F? (*b*) What is the ratio of the viscosity of the crude oil in Fig. 2.3 (s = 0.925) to that of the gasoline (s = 0.680), both being at a temperature of 60°F? (*c*) In cooling from 300 to 80°F, what is the ratio of the change of the viscosity of the SAE 30 western oil to that of the SAE 30 eastern oil?

2.11.3 At 60°F what is the kinematic viscosity of the gasoline in Fig. 2.4, the specific gravity of which is 0.680? Give the answer in both BG and SI units.

2.11.4 To what temperature must the fuel oil with the higher specific gravity in Fig. 2.4 be heated in order that its kinematic viscosity may be reduced to three times that of water at 40°F?

2.11.5 Compare the ratio of the absolute viscosities of air and water at 70°F with the ratio of their kinematic viscosities at the same temperature and at 14.7 psia.

2.11.6 A flat plate 250 mm × 800 mm slides on oil (μ = 0.65 N·s/m^2) over a large plane surface. What force is required to drag the plate at 1.5 m/s, if the separating oil film is 0.5 mm thick?

2.11.7 A flat plate 1 ft × 2 ft slides on oil (μ = 0.02 lb·sec/ft^2) over a large plane surface. What force is required to drag the plate at 5 fps, if the separating oil film is 0.02 in thick?

2.11.8 A space of 20-mm width between two large plane surfaces is filled with SAE 30 western lubricating oil at 30°C. (*a*) What force is required to drag a very thin plate of 0.4-m^2 area between the surfaces at a speed of 0.2 m/s if this plate is equally spaced between the two surfaces? (*b*) If it is at a distance of 5 mm from one surface?

2.11.9 A journal bearing consists of an 80-mm shaft in an 80.4-mm sleeve 120 mm long, the clearance space (assumed to be uniform) being filled with SAE 30 western lubricating oil at 40°C. Calculate the rate at which heat is generated at the bearing when the shaft turns at 150 rpm. Express the answer in kN·m/s, J/s, Btu/hr, ft·lb/sec, and hp.

2.11.10 A hydraulic lift of the type commonly used for greasing automobiles consists of a 10.000-in-diameter ram that slides in a 10.006-in-diameter cylinder, the annular space being filled with oil having a kinematic viscosity of 0.0035 ft^2/sec and specific gravity of 0.85. If the rate of travel of the ram is 0.6 fps, find the frictional resistance when 10 ft of the ram is engaged in the cylinder.

2.11.11 A journal bearing consists of a 7.20-in shaft in a 7.21-in sleeve 8 in long, the clearance space (assumed to be uniform) being filled with SAE 30 eastern lubricating oil at 100°F. Calculate the rate at which heat is generated at the bearing when the shaft turns at 80 rpm. Express the answer in Btu/hr.

2.11.12 In using a rotating-cylinder viscometer, a bottom correction must be applied to account for the drag on the flat bottom of the inner cylinder.

Calculate the theoretical amount of this torque correction, neglecting centrifugal effects, for a cylinder of diameter d, rotated at a constant angular velocity ω, in a liquid of viscosity μ, with a clearance Δh between the bottom of the inner cylinder and the floor of the outer one.

2.11.13 Assuming a velocity distribution as shown in the diagram, which is a parabola having its vertex 12 in from the boundary, calculate the velocity gradients for $y = 0, 3, 6, 9$, and 12 in. Also calculate the shear stresses in lb/ft^2 at these points if the fluid viscosity is 500 cP.

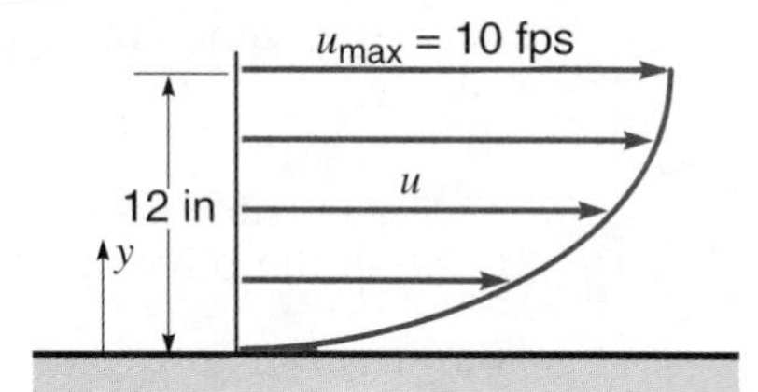

Figure X2.11.13

2.12 SURFACE TENSION

Capillarity

Liquids have cohesion and adhesion, both of which are forms of molecular attraction. ***Cohesion*** enables a liquid to resist tensile stress, while ***adhesion*** enables it to adhere to another body.[10] At the interface between a liquid and a gas, i.e., at the liquid surface, and at the interface between two ***immiscible*** (not mixable) liquids, the out of balance attraction force between molecules forms an imaginary film capable of resisting tension. This liquid property is known as ***surface tension***. Because this tension acts in a surface, we compare such forces by measuring the tension per unit length of surface. The surface tensions of various liquids cover a wide range, and they decrease slightly with increasing temperature. Values of the surface tension for water between the freezing and boiling points vary from 0.005 18 to 0.004 04 lb/ft (0.0756 to 0.0589 N/m); more typical values are given in Table A.1 of Appendix A.

[10] In 1877 Osborne Reynolds demonstrated that a $\frac{1}{4}$-in-diameter column of mercury could withstand a tensile stress of 3 atm for a time but that it would separate upon external jarring of the tube. Liquid tension (said to be as high as 400 atm) accounts for the rise of water in the very small channels of xylem tissue in tall trees. For practical engineering purposes, however, liquids are assumed to be incapable of resisting any direct tensile stress.

Values for other liquids are presented in Table A.4. ***Capillarity*** is the property of exerting forces on fluids by fine tubes or porous media; it is due to both cohesion and adhesion. When the cohesion is of less effect than the adhesion, the liquid will wet a solid surface with which it is in contact and rise at the point of contact; if cohesion predominates, the liquid surface will be depressed at the point of contact. For example, capillarity makes water rise in a glass tube, while mercury is depressed below the true level, as is shown by the insert in Fig. 2.8, which is drawn to scale and reproduced actual size. The curved liquid surface that develops in a tube is called a ***meniscus***.

Capillary rise in a tube is depicted in Fig. 2.9. From free-body considerations, equating the lifting force created by surface tension to the gravity force,

$$2\pi r\sigma \cos\theta = \pi r^2 h\gamma$$

so

$$h = \frac{2\sigma \cos\theta}{\gamma r} \tag{2.12}$$

where σ = surface tension in units of force per unit length
θ = wetting angle
γ = specific weight of liquid
r = radius of tube
h = capillary rise[11]

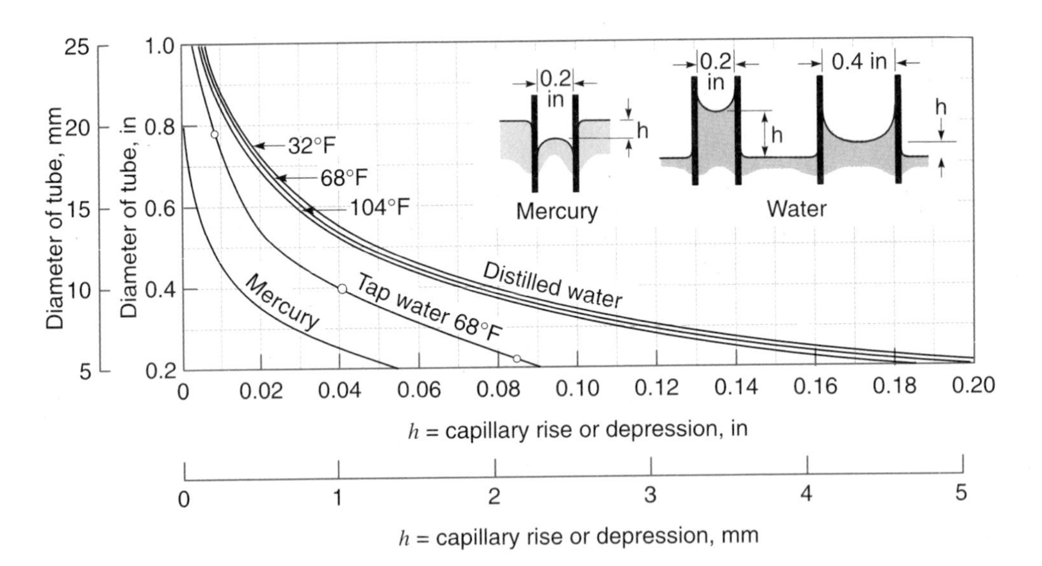

Figure 2.8
Capillarity in clean circular glass tubes.

[11] Measurements to a meniscus are usually taken to the point on the centerline.

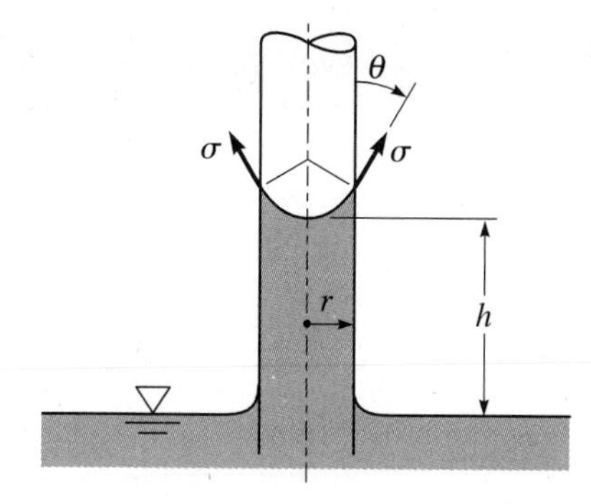

Figure 2.9
Capillary rise.

This expression can be used to compute the *approximate* capillary rise or depression in a tube. If the tube is clean, $\theta = 0°$ for water and about 140° for mercury. Note that the meniscus (Figs. 2.8 and 2.9) causes a small volume of liquid, near the tube walls, to be lifted in addition to the volume $\pi r^2 h$ used in Eq. (2.12). For larger tube diameters, with smaller capillary rise heights, this small additional volume can become a large fraction of $\pi r^2 h$. So Eq. (2.12) overestimates the amount of capillary rise or depression, particularly for larger diameter tubes. The curves of Fig. 2.8 are for water or mercury in contact with air. If mercury is in contact with water, the surface tension effect is slightly less than when in contact with air. For tube diameters larger than $\frac{1}{2}$ in (12 mm), capillary effects are negligible.

Surface tension effects are generally negligible in most engineering situations. However, they can be important in problems involving capillary rise, such as in the soil water zone. When small tubes are used to measure fluid properties, such as pressures, the readings must be taken aware of the surface tension effects; a true reading would be that which would occur if surface tension effects were zero. These effects can also be important in the formation of drops and bubbles, the breakup of liquid jets, and in hydraulic model studies where the model is small.

Sample Problem 2.10 Water at 10°C stands in a clean glass tube of 2-mm diameter at a height of 35 mm. What is the true static height?

Solution

Table A.1 at 10°C: $\gamma = 9804\ \text{N/m}^3,\ \sigma = 0.0742\ \text{N/m}.$

Sec. 2.12 for clean glass tube: $\theta = 0°.$

Eq. (2.12):

$$h = \frac{2\sigma}{\gamma r} = \frac{2(0.0742\ \text{N/m})}{(9804\ \text{N/m}^3)0.001\ \text{m}}$$

$$= 0.015\,14\ \text{m} = 15.14\ \text{mm}$$

Sec. 2.12: True static height $= 35.00 - 15.14 = 19.86$ mm ***ANS***

EXERCISES

2.12.1 Distilled water at 20°C stands in a glass tube of 6.0-mm diameter at a height of 22.0 mm. What is the true static height?

2.12.2 Tap water at 68°F stands in a glass tube of 0.27-in diameter at a height of 6.00 in. What is the true static height?

2.12.3 Use Eq. (2.12) to compute the capillary rise of water to be expected in a 0.30-in-diameter tube. Assume pure water at 68°F. Compare the result with Fig. 2.8.

2.12.4 Compute the capillary rise in mm of water at 5°C expected in a 1-mm-diameter tube.

2.12.5 Use Eq. (2.12) to compute the capillary depression of mercury ($\theta = 140°$) to be expected in a 0.06-in-diameter tube. At 68°F the surface tension of mercury is 0.0318 lb/ft.

2.13 VAPOR PRESSURE OF LIQUIDS

All liquids tend to evaporate or vaporize, which they do by projecting molecules into the space above their surfaces. If this is a confined space, the partial pressure exerted by the molecules increases until the rate at which molecules reenter the liquid is equal to the rate at which they leave. For this equilibrium condition, the vapor pressure is known as the ***saturation pressure***.

Molecular activity increases with increasing temperature and decreasing pressure, and so the saturation pressure does the same. At any given temperature, if the pressure on the liquid surface is lowered below the saturation pressure, a rapid rate of evaporation results, known as ***boiling***. Thus the saturation pressure may be known as the ***boiling pressure*** for a given temperature, and it is of practical importance for liquids.[12]

Table 2.3
Saturation vapor pressure of selected liquids at 68°F (20°C)

	psia	N/m² abs	mbar abs
Mercury	0.000 025	0.17	0.0017
Water	0.339	2 340	23.4
Kerosene	0.46	3 200	32
Carbon tetrachloride	1.76	12 100	121
Gasoline	8.0	55 000	550

[12] Values of the saturation pressure for water for temperatures from 32 to 705.4°F may be found in J. H. Keenan, *Thermodynamic Properties of Water including Vapor, Liquid and Solid States,* John Wiley & Sons, Inc., New York, 1969, and in other steam tables. There are similar vapor tables published for ammonia, carbon dioxide, sulfur dioxide, and other vapors of engineering interest.

The rapid vaporization and recondensation of liquid as it briefly flows through a region of low absolute pressure is called ***cavitation***. This phenomenon is often very damaging, and so must be avoided; it is described in more detail in Sec. 5.11.

The wide variation in saturation vapor pressure of various liquids is indicated in Table 2.3, and more are given in Appendix A, Table A.4. The very low vapor pressure of mercury makes it particularly suitable for use in barometers. Values for the vapor pressure of water at different temperatures are presented in Appendix A, Table A.1.

SAMPLE PROBLEM 2.11 At approximately what temperature will water boil if the elevation is 10,000 ft?

Solution

From Appendix A, Table A.3, the pressure of the standard atmosphere at 10,000-ft elevation is 10.11 psia. From Appendix A, Table A.1, the saturation pressure of water is 10.11 psia at about 193°F. Hence the water at 10,000 ft will boil at 193°F. ***ANS***

Compared with the boiling temperature of 212°F at sea level, this explains why it takes longer to cook at high elevations.

EXERCISES

2.13.1 At approximately what temperature will water boil in Mexico City (elevation 7400 ft)? Refer to Appendix A.

2.13.2 At what pressure in millibars absolute will 50°C water boil?

PROBLEMS

2.1 If the specific weight of a gas is 16.40 N/m^3, what is its specific volume in m^3/kg?

2.2 If a certain liquid weighs 7600 N/m^3, what are the values of its density, specific volume, and specific gravity relative to water at 15°C? Use Appendix A.

2.3 Find the change in volume of 12.00 lb of water at ordinary atmospheric pressure for the following conditions: (*a*) reducing the temperature 40°F from 160°F to 120°F; (*b*) reducing the temperature 40°F from 120°F to 80°F; (*c*) reducing the temperature 40°F from 80°F to 40°F. Calculate each and note the trend in the changes in volume.

2.4 Compare the change in volume of 2.00 m^3 of 80°C water for the following conditions: (*a*) a temperature decrease of 10°C from 80°C to 70°C with pressure constant; (*b*) a pressure increase of 10 MN/m^2 with temperature remaining constant at 80°C; (*c*) a temperature increase from 80°C to 90°C combined with a pressure decrease of 20 MN/m^2.

2.5 Repeat Exer. 2.6.3 for the case where the tank is made of a material that has a coefficient of thermal expansion of 4.8×10^{-6} mm/mm per °C.

2.6 A heavy closed steel chamber is filled with water at 50°F and atmospheric pressure. If the temperature of the water and the chamber is raised to 90°F, what will be the new pressure of the water? The coefficient of thermal expansion of the steel is 6.5×10^{-6} in/in per °F; assume the chamber is unaffected by the water pressure. Use Fig. 2.1.

2.7 (*a*) Calculate the density, specific weight and specific volume of oxygen at 20°C and 50 kN/m^2 abs. (*b*) If the oxygen is enclosed in a rigid container of constant volume, what will be the pressure if the temperature is reduced to −100°C?

2.8 If the specific weight of water vapor in the atmosphere is 0.1020 N/m^3 and that of the (dry) air is 11.62 N/m^3 when the temperature is 20°C, (*a*) what are the partial pressures of the water vapor and the dry air in kPa, (*b*) what is the specific weight of the atmosphere (air and water vapor), and (*c*) what is the barometric pressure in kPa and millibars?

2.9 When the ambient air is at 70°F, 14.7 psia, and contains 21% oxygen by volume, 4.5 lb of air are pumped into a scuba tank, capacity 0.75 ft^3. (*a*) What volume of ambient air was compressed? (*b*) When the filled tank has cooled to ambient conditions, what is the (gage) pressure of the air in the tank? (*c*) What is the partial pressure (psia) and specific weight of the ambient oxygen? (*d*) What weight of oxygen was put in the tank? (*e*) What is the partial pressure (psia) and specific weight of the oxygen in the tank?

2.10 (*a*) If 12 ft^3 of carbon dioxide at 60°F and 15 psia is compressed isothermally to 2 ft^3, what is the resulting pressure? (*b*) What would the pressure and temperature have been if the process had been isentropic? The isentropic exponent k for carbon dioxide is 1.28.

2.11 (*a*) If 350 L of carbon dioxide at 20°C and 120 kN/m^2 abs is compressed isothermally to 50 L, what is the resulting pressure? (*b*) What would the pressure and temperature have been if the process had been isentropic? The isentropic exponent k for carbon dioxide is 1.28.

2.12 The absolute viscosity of a certain gas is 0.0234 cP while its kinematic viscosity is 181 cSt, both measured at 1013 mbar abs and 100°C. Calculate its approximate molecular weight, and suggest what gas it may be.

2.13 A hydraulic lift of the type commonly used for greasing automobiles consists of a 300.00-mm-diameter ram that slides in a 300.15-mm-diameter cylinder, the annular space being filled with oil having a kinematic viscosity of 0.0004 m^2/s and specific gravity of 0.85. If the rate of travel of the ram is 0.2 m/s, find the frictional resistance when 2 m of the ram is engaged in the cylinder.

2.14 Repeat Exer. 2.11.11 for the case where the sleeve has a diameter of 7.68 in. Compute as accurately as possible the velocity gradient in the fluid at the shaft and sleeve.

2.15 A disk spins within an oil-filled enclosure, having 3-mm clearance from flat surfaces each side of the disk. The disk surface extends from radius 10 to 90 mm. What torque is required to drive the disk at 600 rpm if the oil's viscosity is 0.10 N·s/m^2?

2.16 It is desired to apply the general case of Sample Prob. 2.9 to the extreme cases of a journal bearing ($\alpha = 0$) and an end bearing ($\alpha = 90°$). But when $\alpha = 0$, $r = \tan\alpha = 0$, so $T = 0$; when $\alpha = 90°$, contact area $= \infty$ due to b, so $T = \infty$. Therefore devise an alternative general derivation that will also provide solutions to these two extreme cases.

2.17 Water at 40°F stands in a glass tube of 0.05-in diameter at a height of 7.32 in. Compute the true static height.

2.18 (*a*) Derive an expression for capillary rise (or depression) between two vertical parallel plates. (*b*) How much would you expect 15°C water to rise (in mm) if the clean glass plates are separated by 1.5 mm?

2.19 Water at 150°F is placed in a beaker within an airtight container. Air is gradually pumped out of the container. What reduction below standard atmospheric pressure of 14.7 psia must be achieved before the water boils?

CHAPTER 3

Fluid Statics

There are no shear stresses in fluids at rest; hence only normal pressure forces are present. Normal forces produced by static fluids are often very important. For example, they tend to overturn concrete dams, burst pressure vessels, and break lock gates on canals. Obviously, to design such facilities, we need to be able to compute the magnitudes and locations of normal pressure forces. Understanding them, we can develop instruments to measure pressures, and systems that transfer pressures, such as for automobile brakes and hoists.

Note that normal pressure forces alone can occur in a moving fluid if the fluid is moving in bulk without deformation, i.e., as if it were solid or rigid. For such an example, see Sec. 3.11. However, this is relatively rare.

The ***average pressure intensity*** p is defined as the force exerted on a unit area. If F represents the total normal pressure force on some finite area A, while dF represents the force on an infinitesimal area dA, the pressure is

$$p = \frac{dF}{dA} \tag{3.1}$$

If the pressure is uniform over the total area, then $p = F/A$. In the British gravitational (BG) system pressure is generally expressed in pounds per square inch (psi) or pounds per square foot (lb/ft^2 = psf), while in SI units the pascal (Pa = N/m^2) or kPa (kN/m^2) is commonly used. Previously bars and millibars were used in metric systems to express pressure; 1 mbar = 100 Pa.

3.1 PRESSURE AT A POINT THE SAME IN ALL DIRECTIONS

In a solid, because of the possibility of tangential stresses between adjacent particles, the stresses at a given point may be different in different directions.

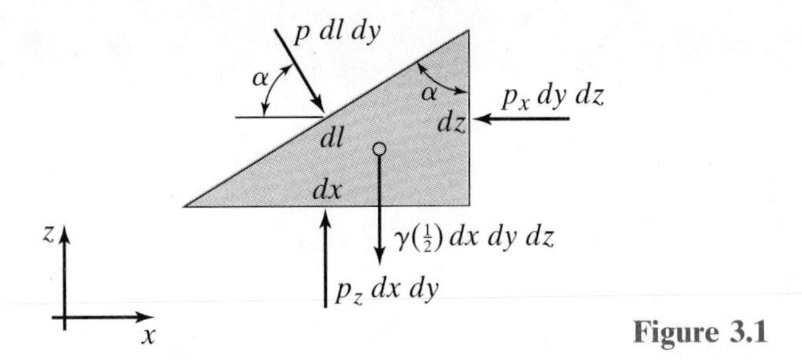

Figure 3.1

But no tangential stresses can exist in a fluid at rest, and the only forces between adjacent surfaces are pressure forces normal to the surfaces. Therefore the pressure at any point in a fluid at rest is the same in every direction.

This can be proved by reference to Fig. 3.1, showing a very small wedge-shaped element of fluid at rest whose thickness perpendicular to the plane of the paper is constant and equal to dy. Let p be the average pressure in any direction in the plane of the paper, let α be defined as shown, and let p_x and p_z be the average pressures in the horizontal and vertical directions.[1] The forces acting on the element of fluid, with the exception of those in the y direction on the two faces parallel to the plane of the paper, are shown in the diagram. For our purpose, forces in the y direction need not be considered because they cancel. Since the fluid is at rest, no tangential forces are involved. As this is a condition of equilibrium, the sum of the force components on the element in any direction must be equal to zero. Writing such an equation for the components in the x direction, $p\,dl\,dy\cos\alpha - p_x\,dy\,dz = 0$. Since $dz = dl\cos\alpha$, it follows that $p = p_x$. Similarly, summing forces in the z direction gives $p_z\,dx\,dy - p\,dl\,dy\sin\alpha - \frac{1}{2}\gamma\,dx\,dy\,dz = 0$. The third term is of higher order than the other two terms and so may be neglected. It follows from this that $p = p_z$. It can also be proved that $p = p_y$ by considering a three-dimensional case. The results are independent of α; hence the pressure at any point in a fluid at rest is the same in all directions.

3.2 VARIATION OF PRESSURE IN A STATIC FLUID

Consider the differential element of static fluid shown in Fig. 3.2. Since the element is very small, we can assume that the density of the fluid within the

[1] Note that the axes are arranged differently from those usually used in solid mechanics. They are chosen to retain a right-handed coordinate system, and to make z vertical, because z is traditionally used for elevation in fluid mechanics.

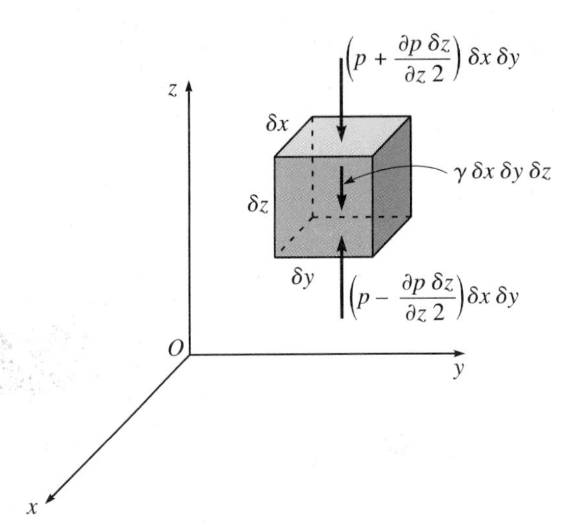

Figure 3.2

element is constant. Assume that the pressure at the center of the element is p and that the dimensions of the element are δx, δy and δz. The forces acting on the fluid element in the vertical direction are (*a*) the ***body force***, the action of gravity on the mass within the element, and (*b*) the ***surface forces***, transmitted from the surrounding fluid and acting at right angles against the top, bottom, and sides of the element. Because the fluid is at rest, the element is in equilibrium and the summation of forces acting on the element in any direction must be zero. If forces are summed in the horizontal direction, that is, x or y, the only forces acting are the pressure forces on the vertical faces of the element. To satisfy $\sum F_x = 0$ and $\sum F_y = 0$, the pressures on the opposite vertical faces must be equal. Thus $\partial p/\partial x = \partial p/\partial y = 0$ for the case of the fluid at rest.

Summing forces in the vertical direction and setting the sum equal to zero,

$$\sum F_z = \left(p - \frac{\partial p}{\partial z}\frac{\delta z}{2}\right)\delta x\,\delta y - \left(p + \frac{\partial p}{\partial z}\frac{\delta z}{2}\right)\delta x\,\delta y - \gamma\,\delta x\,\delta y\,\delta z = 0$$

This results in $\partial p/\partial z = -\gamma$, which, since p is independent of x and y, can be written as

$$\frac{dp}{dz} = -\gamma \tag{3.2}$$

This is the general expression that relates variation of pressure in a static fluid to vertical position. The minus sign indicates that as z gets larger (increasing elevation), the pressure gets smaller.

To evaluate pressure variation in a fluid at rest, one must integrate Eq. (3.2) between appropriately chosen limits. For incompressible fluids (γ = constant), Eq. (3.2) can be integrated directly. For compressible fluids, however, γ must be expressed algebraically as a function of z or p if one is to determine pressure accurately as a function of elevation. The variation of pressure in the earth's atmosphere is an important problem, and several approaches are illustrated in the following example.

Sample Problem 3.1 Compute the atmospheric pressure at elevation 20,000 ft, considering the atmosphere as a static fluid. Assume standard atmosphere at sea level. Use four methods; (*a*) air of constant density; (*b*) constant temperature between sea level and 20,000 ft; (*c*) isentropic conditions; (*d*) air temperature decreasing linearly with elevation at the standard lapse rate of 0.00356°F/ft.

Solution. From Appendix A, Table A.3 the conditions of the standard atmosphere at sea level are $T = 59.0°\text{F}$, $p = 14.70$ psia, $\gamma = 0.07648\ \text{lb/ft}^3$.

(a) Constant density

From Sec. 3.2: $$\frac{dp}{dz} = -\gamma; \quad dp = -\gamma\, dz; \quad \int_{p_1}^{p} dp = -\gamma \int_{z_1}^{z} dz$$

where subscript 1 indicates conditions at a reference elevation, sea level in this case,

so $$p - p_1 = -\gamma(z - z_1)$$

and $$p = 14.70 \times 144 - 0.07648(20{,}000) = 587\ \text{lb/ft}^2\ \text{abs} = 4.08\ \text{psia} \qquad \textit{ANS}$$

(b) Isothermal

From Sec. 2.7: $pv = \text{constant}$; hence $\dfrac{p}{\gamma} = \dfrac{p_1}{\gamma_1}$ if g is constant

Eq. (3.2): $$\frac{dp}{dz} = -\gamma, \quad \text{where} \quad \gamma = \frac{p\gamma_1}{p_1}$$

so $$\frac{dp}{p} = -\frac{\gamma_1}{p_1}\, dz$$

Integrating, $$\int_{p_1}^{p} \frac{dp}{p} = \ln\frac{p}{p_1} = -\frac{\gamma_1}{p_1}\int_{z_1}^{z} dz = -\frac{\gamma_1}{p_1}(z - z_1)$$

and $$\frac{p}{p_1} = \exp\left[-\left(\frac{\gamma_1}{p_1}\right)(z - z_1)\right]$$

Thus $$p = 14.70 \exp\left[-\frac{0.07648}{14.70 \times 144}(20{,}000)\right] = 7.14\ \text{psia} \qquad \textit{ANS}$$

(c) Isentropic

From Sec. 2.7: $pv^{1.4} = \dfrac{p}{\rho^{1.4}} = \text{constant}$; hence $\dfrac{p}{\gamma^{1.4}} = \text{constant} = \dfrac{p}{\gamma_1^{1.4}}$

Eq. (3.2): $\frac{dp}{dz} = -\gamma, \quad \text{where } \gamma = \gamma_1\left(\frac{p}{p_1}\right)^{1/1.4} = \gamma_1\left(\frac{p}{p_1}\right)^{0.714}$

so $$dp = -\gamma_1\left(\frac{p}{p_1}\right)^{0.714} dz$$

Integrating:

$$\int_{p_1}^{p} p^{-0.714}\, dp = -\gamma_1 p_1^{-0.714}\int_{z_1}^{z} dz$$

$$p^{0.286} - p_1^{0.286} = -0.286\gamma_1 p_1^{-0.714}(z - z_1)$$

$$p^{0.286} = (14.70 \times 144)^{0.286} - 0.286(0.07648)(14.70 \times 144)^{-0.714}(20{,}000)$$

$$p = 942 \text{ lb/ft}^2 \text{ abs} = 6.54 \text{ psia} \qquad \textbf{\textit{ANS}}$$

(d) Temperature decreasing linearly with elevation

For the standard lapse rate (Fig. 2.2): $T = a + bz$,

where $a = 59.00 + 459.67 = 518.67°\text{R}$ and $b = -0.003560°\text{R/ft}$

Eqs. (3.2) and (2.4): $\frac{dp}{dz} = -\rho g; \quad \rho = \frac{p}{RT}$

Combining to eliminate ρ, which varies, rearranging, and substituting for T,

$$\frac{dp}{p} = -\frac{g\, dz}{R(a + bz)}$$

Integrating: $$\int_1^2 \frac{dp}{p} = -\frac{g}{R}\int_1^2 \frac{dz}{a + bz}$$

$$\ln\left(\frac{p_2}{p_1}\right) = -\frac{g}{Rb}\ln\left(\frac{a + bz_2}{a + bz_1}\right) = \ln\left(\frac{a + bz_2}{a + bz_1}\right)^{-g/Rb}$$

i.e. $$\frac{p_2}{p_1} = \left(\frac{a + bz_2}{a + bz_1}\right)^{-g/Rb}$$

Here $$\frac{-g}{Rb} = \frac{-32.174}{1716(-0.003560)} = 5.27$$

and, from Table A.3: $p_1 = 14.696$ psia when $z_1 = 0$.

Thus $$\frac{p_2}{14.696} = \left(\frac{518.67 - 0.003560 \times 20{,}000}{518.67 + 0}\right)^{5.27} = 0.459$$

$$p_2 = 14.696(0.459) = 6.75 \text{ psia} \qquad \textbf{\textit{ANS}}$$

The latter approach corresponds to the standard atmosphere, described in Sec. 2.9 and in Table A.3 of Appendix A.

In Sample Problem 3.1*a* it was shown that, for the case of an incompressible fluid,

$$p - p_1 = -\gamma(z - z_1) \tag{3.3}$$

where p is the pressure at an elevation z. This expression is generally applicable to liquids, since they are only slightly compressible. Only where there are large changes in elevation, as in the ocean, need the compressibility of the liquid be considered, to arrive at an accurate determination of pressure variation. For small changes in elevation, Eq. (3.3) will give accurate results when applied to gases.

For the case of a liquid at rest, it is convenient to measure distances vertically downward from the free liquid surface. If h is the distance below the free liquid surface and if the pressure of air and vapor on the surface is arbitrarily taken to be zero, Eq. (3.3) can be written as

$$p = \gamma h \tag{3.4}$$

As there must always be some pressure on the surface of any liquid, the total pressure at any depth h is given by Eq. (3.4) plus the pressure on the surface. In many situations this surface pressure may be disregarded, as pointed out in Sec. 3.4.

From Eq. (3.4), it may be seen that all points in a connected body of constant density fluid at rest are under the same pressure if they are at the same depth below the liquid surface (Pascal's law). This indicates that a surface of equal pressure for a liquid at rest is a horizontal plane. Strictly speaking, it is a surface everywhere normal to the direction of gravity and is approximately a spherical surface concentric with the earth. For practical purposes, a limited portion of this surface may be considered a plane area.

EXERCISES

3.2.1 Neglecting the pressure upon the surface and the compressibility of water, what is the pressure in pounds per square inch at a depth of an ore deposit 12,000 ft below the surface of the ocean? The specific weight of ocean water under ordinary conditions is 64.0 lb/ft^3.

3.2.2 Neglecting the pressure upon the surface and the compressibility of water, what is the pressure in kPa at a depth of a wreck 4 km below the surface of the ocean? The specific weight of ocean water under ordinary conditions is 10.05 kN/m^3.

3.2.3 A pressure gage at elevation 7.5 m on the side of an industrial tank containing a liquid reads 63.0 kN/m^2. Another gage at elevation 4.0 m reads 90.2 kN/m^2. Compute the specific weight and density of the liquid.

3.2.4 A pressure gage at elevation 18.0 ft on the side of an industrial tank containing a liquid reads 11.4 psi. Another gage at elevation 12.0 ft reads 13.7 psi. Compute the specific weight, density, and specific gravity of the liquid.

3.3 PRESSURE EXPRESSED IN HEIGHT OF FLUID

Imagine an open tank of liquid upon whose surface there is no pressure, though in reality the minimum pressure upon any liquid surface is the pressure of its own vapor (Fig. 3.3). Disregarding this for the moment, by Eq. (3.4), the pressure at any depth h is $p = \gamma h$. If γ is assumed constant, there is a definite relation between p and h. That is, pressure (force per unit area) is equivalent to a height h of some fluid of constant specific weight γ. It is often more convenient to express pressure in terms of a height of a column of fluid rather than in pressure per unit area.

Even if the surface of the liquid is under some pressure, it is necessary only to convert this pressure into an equivalent height of the fluid in question and add this to the value of h shown in Fig. 3.3, to obtain the total pressure.

The preceding discussion has been applied to a liquid, but it is equally possible to use it for a gas or vapor by specifying some *constant* specific weight γ for the gas or vapor in question. Thus pressure p may be expressed in the height of a column of *any* fluid by the relation

$$h = \frac{p}{\gamma} \tag{3.5}$$

This relationship is true for any consistent system of units. If p is in pounds per square foot, γ must be in pounds per cubic foot, and then h will be in feet. In SI units, p may be expressed in kilopascals (kilonewtons per square meter),

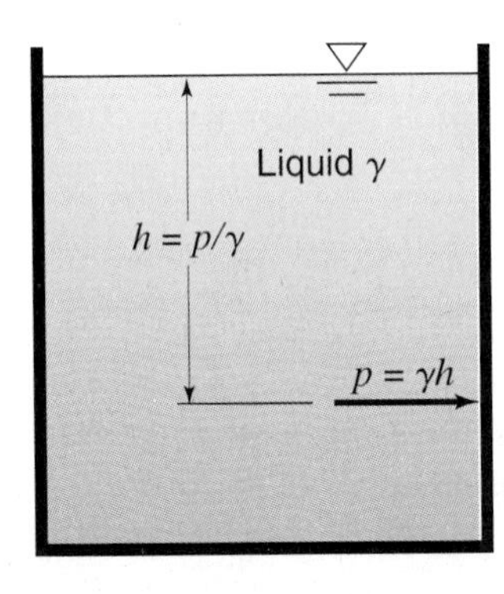

Figure 3.3

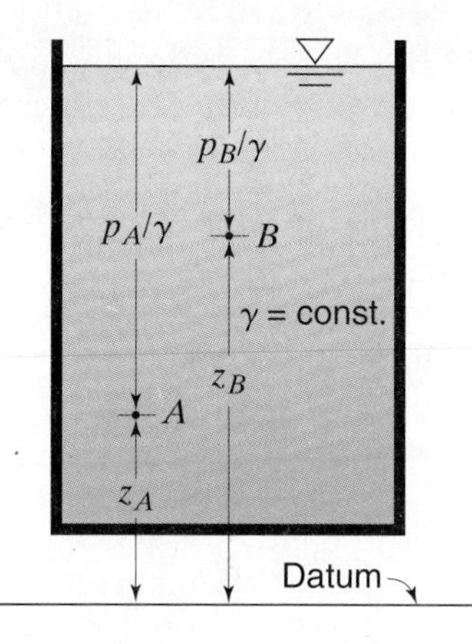

Figure 3.4

in which case if γ is expressed in kilonewtons/cubic meter, h will be in meters. When pressure is expressed in this fashion, it is commonly referred to as ***pressure head***. Because pressure is commonly expressed in pounds per square inch (or kPa in SI units), and since the value of γ for water is usually assumed to be 62.4 lb/ft^3 (9.81 kN/m^3), a convenient relationship is

$$h\ (\text{ft of } H_2O) = \frac{144 \times \text{psi}}{62.4} = 2.308 \times \text{psi}$$

or

$$h\ (\text{m of } H_2O) = \frac{\text{kPa}}{9.81} = 0.1020 \times \text{kPa}$$

It is convenient to express pressures occurring in one fluid in terms of height of another fluid, e.g., barometric pressure in millimeters of mercury.

Equation (3.3) may be expressed as follows:

$$\frac{p}{\gamma} + z = \frac{p_1}{\gamma} + z_1 = \text{constant} \tag{3.6}$$

This shows that for an incompressible fluid at rest, at any point in the fluid the sum of the elevation z and the pressure head p/γ is equal to the sum of these two quantities at any other point. The significance of this statement is that, in a fluid at rest, with an increase in elevation there is a decrease in pressure head, and vice versa. This concept is depicted in Fig. 3.4.

Sample Problem 3.2 An open tank contains water 1.40 m deep covered by a 2 m thick layer of oil (s = 0.855). What is the pressure head at the bottom of the tank, in terms of a water column?

Solution 1

From inside cover of book: $\gamma_w = 9.81\ \text{kN/m}^3$.

Sec. 2.3: $\gamma_o = 0.855(9.81) = 8.39\ \text{kN/m}^3$

Eq. (3.4) for interface: $p_i = \gamma_o h_o = (8.39)2 = 16.78\ \text{kN/m}^2$

Eq. (3.5) for water equivalent of oil:

$$h_{oe} = \frac{p_i}{\gamma_w} = \frac{16.78\ \text{kN/m}^2}{9.81\ \text{kN/m}^3} = 1.710\ \text{m of water}$$

So $h_{we} = h_w + h_{oe} = 1.40 + 1.710 = 3.11$ m of water ***ANS***

Solution 2
From Eq. (3.4) for bottom of tank:

$$p_b = \gamma_o h_o + \gamma_w h_w = (8.39)2 + 9.81(1.4) = 30.51\ \text{kN/m}^2$$

Eq. (3.5) for total water equivalent:

$$h_{we} = \frac{p_b}{\gamma_w} = \frac{30.51\ \text{kN/m}^2}{9.81\ \text{kN/m}^3} = 3.11\ \text{m of water}$$ ***ANS***

EXERCISES

3.3.1 An open tank contains 6.0 m of water covered with 2.5 m of oil ($\gamma = 8.0\ \text{kN/m}^3$). Find the gage pressure at the interface and at the bottom of the tank.

3.3.2 An open tank contains 9 ft of water covered with 2.5 ft of oil ($s = 0.86$). Find the gage pressure at the interface between the liquids and at the bottom of the tank.

3.3.3 If air had a constant specific weight of 0.0765 lb/ft^3 and were incompressible, what would be the height of air surrounding the earth to produce a pressure at the surface of 14.70 psia?

3.3.4 If the specific weight of a sludge can be expressed as $\gamma = 64.0 + 0.2h$, determine the pressure in psi at a depth of 12 ft below the surface. γ is in lb/ft^3, and h in ft below surface.

3.4 ABSOLUTE AND GAGE PRESSURES

If pressure is measured relative to absolute zero, it is called ***absolute*** pressure; when measured relative to atmospheric pressure as a base, it is called ***gage***

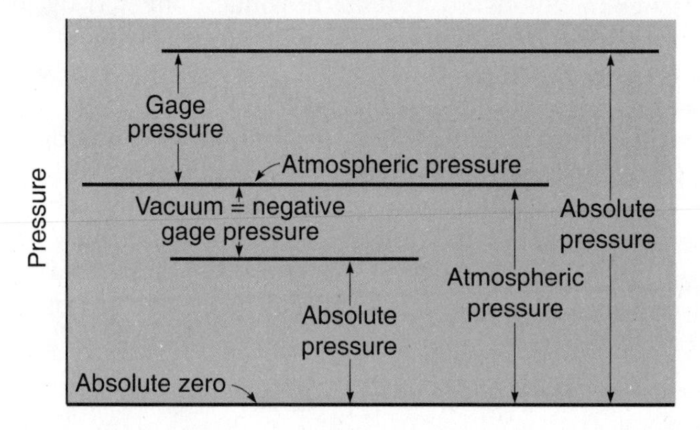

Figure 3.5

pressure. This is because practically all pressure gages register zero when open to the atmosphere, and hence measure the difference between the pressure of the fluid to which they are connected and that of the surrounding air.

If the pressure is below that of the atmosphere, it is designated as a ***vacuum***, and its gage value is the amount by which it is *below* that of the atmosphere. What is called a "high vacuum" is really a low absolute pressure. A perfect vacuum would correspond to absolute zero pressure.

All values of absolute pressure are positive, since a negative value would indicate tension, which is normally considered impossible in any fluid.[2] Gage pressures are positive if they are above that of the atmosphere and negative if they are vacuum (Fig. 3.5).

It can be seen from the foregoing discussion that the following relation holds:

$$p_{\text{abs}} = p_{\text{atm}} + p_{\text{gage}} \tag{3.7}$$

where p_{gage} may be positive or negative (vacuum).

The atmospheric pressure is also called the ***barometric*** pressure and varies with elevation above sea level. Also, at a given place it varies slightly from time to time because of changes in meteorological conditions.

In thermodynamics it is essential to use absolute pressure, because most thermal properties are functions of the actual pressure of the fluid, regardless of the atmospheric pressure. For example, the property relations for a perfect gas [Eq. (2.4)] are an equation in which absolute pressure must be used. In fact, absolute pressures must be employed in most problems involving gases and vapors.

[2] For an exception to this statement, see footnote 10 in Chap. 2.

The properties of liquids are usually not much affected by pressure, and hence gage pressures are commonly employed in problems dealing with liquids. Also it will usually be found that the atmospheric pressure appears on both sides of an equation, and hence cancels. Thus the value of atmospheric pressure is usually of no significance when dealing with liquids, and, for this reason as well, gage pressures are almost universally used with liquids. About the only case where the absolute pressure of a liquid needs to be considered is where conditions are such that the pressure may approach or equal the saturated vapor pressure. Throughout this text all numerical pressures will be understood to be gage pressures unless specifically given as absolute values.

EXERCISES

3.4.1 The absolute pressure on a gas is 38 psia and the atmospheric pressure is 925 mbar abs. Find the gage pressure in psi, kPa, and mbar.

3.4.2 If the atmospheric pressure is 860 mbar abs and a gage attached to a tank reads 370 mmHg vacuum, what is the absolute pressure within the tank?

3.4.3 If the atmospheric pressure is 14.10 psia and a gage attached to a tank reads 10.0 inHg vacuum, what is the absolute pressure within the tank?

3.4.4 If the atmospheric pressure is 970 mbar abs and a gage attached to a tank reads 220 mmHg vacuum, what is the absolute pressure within the tank?

3.4.5 A gage is connected to a tank in which the pressure of the fluid is 45 psi above atmospheric. If the absolute pressure of the fluid remains unchanged but the gage is in a chamber where the air pressure is reduced to a vacuum of 28 inHg, what reading in psi will then be observed?

3.4.6 A gage is connected to a tank in which the pressure of the fluid is 315 kPa above atmospheric. If the absolute pressure of the fluid remains unchanged but the gage is in a chamber where the air pressure is reduced to a vacuum of 668 mmHg, what reading in kPa will then be observed?

3.4.7 If the atmospheric pressure is 29.92 inHg, what will be the height of water in a water barometer if the temperature of the water is 70°F; 120°F? Be as precise as possible.

3.5 BAROMETER

The absolute pressure of the atmosphere is measured with a barometer. If a tube such as that in Fig. 3.6*a* has its open end immersed in a liquid that is exposed to atmospheric pressure, and if air is exhausted from the tube, the

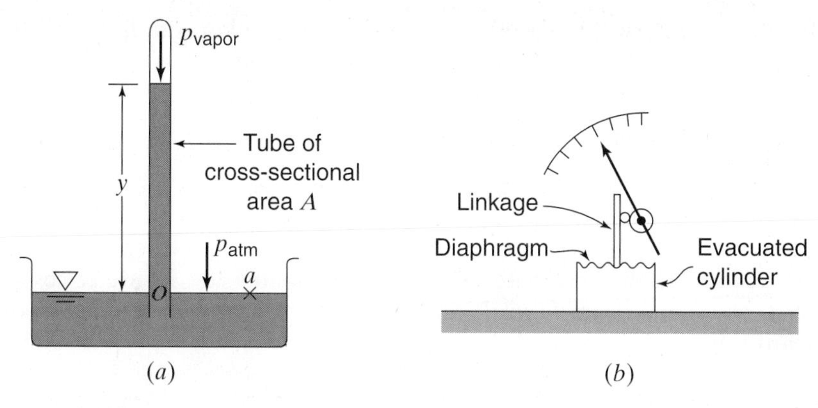

Figure 3.6
Types of barometers. (*a*) Mercury barometer. (*b*) Aneroid barometer.

liquid will rise in it. If the tube is long enough and if the air is completely removed, the only pressure on the surface of the liquid in the tube will be that of its own vapor pressure, and the liquid will have reached its maximum height.

From the concepts developed in Sec. 3.2, the pressure at O within the tube and at a at the surface of the liquid outside the tube must be the same; that is, $p_O = p_a$. From the conditions of static equilibrium of the liquid above O in the tube of cross-sectional area A (Fig. 3.6*a*),

$$p_{atm}A - p_{vapor}A - \gamma A y = 0$$

$$p_{atm} = \gamma y + p_{vapor} \tag{3.8}$$

If the vapor pressure on the surface of the liquid in the tube were negligible, then we would have

$$p_{atm} = \gamma y$$

The liquid employed for barometers of this type is usually mercury, because its density is sufficiently great to enable a reasonably short tube to be used, and also because its vapor pressure is negligibly small at ordinary temperatures. If some other liquid were used, the tube necessarily would be so high as to be inconvenient and its vapor pressure at ordinary temperatures would be appreciable; hence a nearly perfect vacuum at the top of the column would not be attainable. The height attained by the liquid would consequently be less than the true barometric height and would require a correction to the reading. When using a mercury barometer, to get as accurate a measurement of atmospheric pressure as possible, corrections for capillarity and vapor pressure should be made to the reading. An ***aneroid*** barometer measures the difference in pressure between the atmosphere and an evacuated cylinder by means of a sensitive elastic diaphragm and linkage system as indicated in Fig. 3.6*b*.

Since atmospheric pressure at sea level is so widely used, it is well to have in mind equivalent forms of expression. Application of Eq. (3.5) shows that standard sea-level atmospheric pressure may be expressed as follows:

$$14.696 \text{ psia (2116 psfa)} \quad \text{or} \quad 101.325 \text{ kPa abs (1013.25 mbar abs)}$$

$$29.92 \text{ inHg} \quad \text{or} \quad 760 \text{ mmHg}$$

$$33.91 \text{ ft of water} \quad \text{or} \quad 10.34 \text{ m of water.}$$

For convenience, these are listed on the pages facing the inside covers of the book. For most engineering work, they are generally rounded to three or four significant figures.

SAMPLE PROBLEM 3.3 What would be the reading on a barometer containing carbon tetrachloride at 68°F at a time when the atmospheric pressure was equivalent to 30.26 inHg?

Solution

$$30.26 \text{ inHg} \times \frac{14.696 \text{ psia}}{29.92 \text{ inHg}} = 14.86 \text{ psia}$$

Table A.4 for carbon tetrachloride at 68°F:

$$\rho = 3.08 \text{ slugs/ft}^3, \quad p_{\text{vapor}} = 1.90 \text{ psia.}$$

From Eq. (3.8):

$$y = \frac{p_{\text{atm}} - p_{\text{vapor}}}{\rho g}$$

$$= \frac{(14.86 - 1.90)144}{3.08(32.2)}$$

$$= 18.82 \text{ ft of carbon tetrachloride} \qquad \textbf{\textit{ANS}}$$

EXERCISES

3.5.1 If the atmospheric pressure were equivalent to 33.1 ft of water, what would be the reading on a barometer containing an alcohol ($s = 0.79$) if the vapor pressure of the alcohol at the temperature of observation were 2.12 psia?

3.5.2 If the atmospheric pressure is 890 mbar abs, what would be the reading in meters of a barometer containing water at 70°C?

3.6 MEASUREMENT OF PRESSURE

There are many ways by which pressure in a fluid may be measured. Some are discussed below.

Bourdon Gage

Pressures or vacuums are commonly measured by the bourdon gage of Fig. 3.7. In this gage a curved tube of elliptical cross section will change its curvature with changes in pressure within the tube. The moving end of the tube rotates a hand on a dial through a linkage system. A pressure and vacuum gage combined into one is known as a ***compound gage***, and is shown in Fig. 3.8. The pressure indicated by the gage is assumed to be that at its center. If the connecting piping is filled completely with fluid of the same density as that in A of Fig. 3.7 and if the pressure gage is graduated to read in pounds per square inch, as is customary, then

$$p_A \text{ (psi)} = \text{gage reading (psi)} + \frac{\gamma z}{144}$$

where γ is expressed in pounds per cubic foot and z in feet.

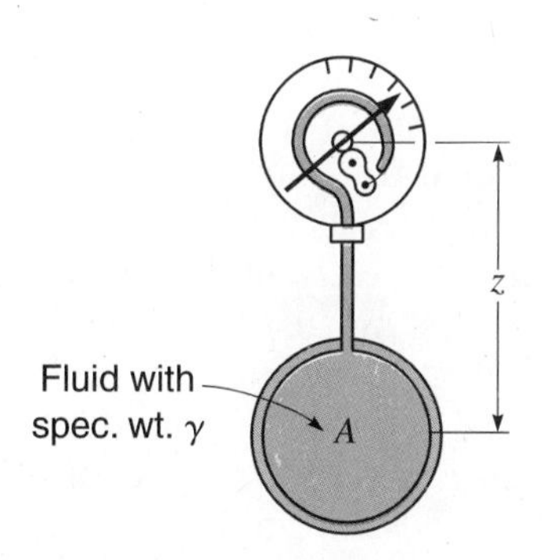

Figure 3.7
Bourdon gage.

Figure 3.8
Compound pressure and vacuum gage. Pressures in pounds per square inch, vacuums in inches of mercury.

A vacuum gage, or the negative-pressure portion of a compound gage, is traditionally graduated to read in in or mm of mercury. For vacuums,

$$\text{inHg vacuum at } A = \text{gage reading (inHg vacuum)} - \frac{\gamma z}{144}\left(\frac{29.92}{14.70}\right)$$

Here, once again, it is assumed that this fluid completely fills the connecting tube of Fig. 3.7. The elevation-correction terms, i.e., those containing z, may be positive or negative, depending on whether the gage is above or below the point at which the pressure determination is desired. The expressions given are for the situation depicted in Fig. 3.7. When measuring liquid pressures, the gage is usually set to measure the pressure at the centerline of the pipe. When measuring gas pressures, the elevation correction terms are generally negligible.

The above expressions, when written in SI units, require no conversion factors; however, care must be taken in dealing with decimal points when adding terms.

Pressure Transducer

A ***transducer*** is a device that transfers energy (in any form) from one system to another. A bourdon gage, for example, is a mechanical transducer in that it has an elastic element that converts energy from the pressure system to a displacement in the mechanical measuring system. An ***electrical pressure transducer*** converts the displacement of a mechanical system (usually a metal diaphragm) to an electric signal, either actively if it generates its own electrical output or passively if it requires an electrical input that it modifies as a function of the mechanical displacement. In one type of pressure transducer (Fig. 3.9) an electrical strain gage is attached to a diaphragm. As

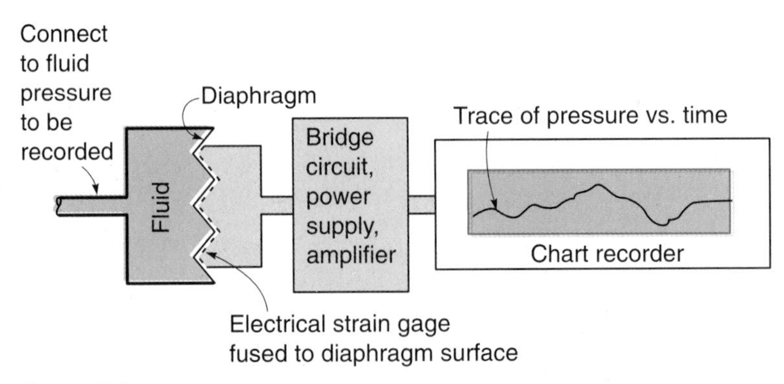

Figure 3.9
Schematic of an electrical strain-gage pressure transducer with a strip-chart recorder.

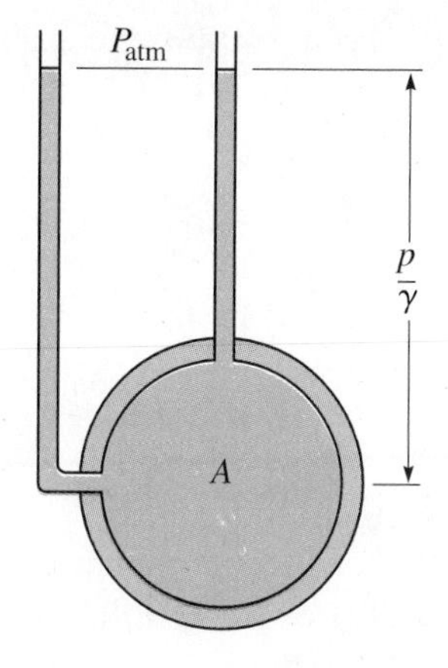

Figure 3.10
Piezometer (for measuring p/γ in liquids only).

the pressure changes, the deflection of the diaphragm changes. This, in turn, changes the electrical output, which, through proper calibration, can be related to pressure. Such a device when connected to a strip-chart recorder can be used to give a continuous record of pressure. In lieu of a strip-chart recorder, the data may be recorded at fixed time intervals on a tape or disk using a computer data acquisition system and/or it may be displayed on a panel in digital form.

Piezometer Column

A piezometer column is a simple device for measuring moderate pressures of liquids. It consists of a sufficiently long tube (Fig. 3.10) in which the liquid can freely rise without overflowing. The height of the liquid in the tube will give the value of the pressure head directly. To reduce capillary error the tube diameter should be at least 0.5 in (12 mm).

If the pressure of a *flowing* fluid is to be measured, special precautions should be taken in making the connection. The hole must be drilled absolutely normal to the interior surface of the wall, and the piezometer tube or the connection for any other pressure-measuring device must not project beyond the surface. All burrs and surface roughness near the hole must be removed, and it is well to round the edge of the hole slightly. Also, the hole should be small, preferably not larger than $\frac{1}{8}$ in (3 mm) diameter.

Simple Manometer

Since the open piezometer tube is cumbersome for use with liquids under high pressure and cannot be used with gases, the simple manometer or mercury U-tube of Fig. 3.11 is a convenient device for measuring pressures.

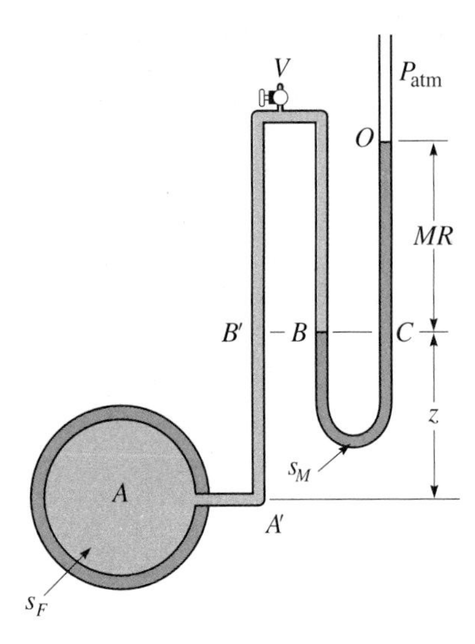

Figure 3.11
Open-end manometer (for measuring p/γ in liquids or gases).

To determine the ***gage pressure head*** at A, in terms of the liquid at A, one may write a ***gage equation*** based on the fundamental relations of hydrostatic pressures [Eq. (3.3)]. Although any units of pressure or pressure head may be used in the gage equation, it is generally advantageous to express all terms in *feet (or meters) of the fluid whose pressure is to be measured.* Let us define s_M as the specific gravity of the *manometer* (M) fluid (or gage fluid) and s_F as the specific gravity of the *fluid* (F) whose pressure is being measured. Also, let us identify a manometer reading by MR; in Fig. 3.11 this is the height OC. If y' is the height of a column of measured (F) fluid that would exert the same pressure at C as does the column of manometer fluid OC, height MR, then, from Eq. (3.4),

$$\text{gage pressure } p_C = \gamma_M MR = \gamma_F y'$$

and rearranging, making use of Sec. 2.3,

$$y' = (\gamma_M/\gamma_F)MR = (\rho_M/\rho_F)MR = (s_M/s_F)MR$$

Thus the gage pressure head at C, in terms of the fluid whose pressure is to be measured, as required, is $(s_M/s_F)MR$. This is also the head at B because the fluid in BC is in balance, while the head at A is greater than this by z, assuming the fluid in the connecting tube $A'B$ is of the same specifie weight as that of the fluid at A. For this simple case the head at A can be written down directly. But for more complicated gages it is helpful to commence the equation at the open end of the manometer with the pressure head there, then proceed through the entire tube to A, adding pressure head terms when descending and subtracting them when ascending, all in terms of equivalent columns of measured fluid (F), finally equating the result to the head at A. Portions of the same fluid with the same end elevations, like BC and $B'B$, may be omitted because they are in balance and so do not affect the pressure at A. Thus, for Fig. 3.11,

$$0 + \left(\frac{s_M}{s_F}\right)MR + z = \frac{p_A}{\gamma} \tag{3.9}$$ [3]

where γ is the specific weight of the liquid at A.

If the ***absolute pressure head*** at A is desired then the zero of the first term will be replaced by the atmospheric pressure head expressed in feet (or meters) of the fluid whose pressure is to be measured. For measuring the pressure in liquids, an air-relief valve V (Fig. 3.11) will provide a means for the escape of gas should any become trapped in tube $A'B$. If the fluid in A is a gas, the pressure head contribution from the distance z is generally negligible and can be neglected because of the relatively small specific gravity (or density) of the gas. If desired, the analysis of Eq. (3.9) could have been accomplished by expressing the terms in units of pressure rather than head. For example, proceeding from O to A on Fig. 3.11,

$$0 + \gamma_M MR + \gamma_F z = p_A$$

(see also footnote 3), and if this is divided through by γ_F, it is seen to be the same as Eq. (3.9).

In measuring a vacuum, for which the arrangement in Fig. 3.12 might be used, the resulting gage equation, subject to the same conditions as in the preceding case, is

$$0 - \left(\frac{s_M}{s_F}\right)MR + z = \frac{p_A}{\gamma} \tag{3.10}$$

Again, it would simplify the equation if one were measuring pressure in a gas, because the z term could be neglected. In measuring vacuums in liquids the arrangement in Fig. 3.13 is advantageous, since gas and vapors cannot

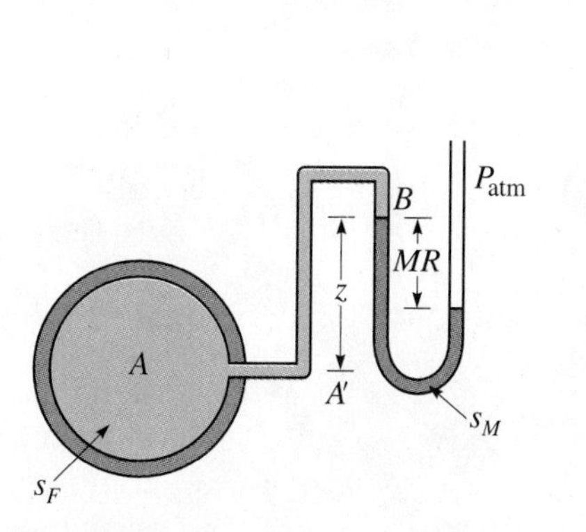

Figure 3.12
Negative-pressure manometer.

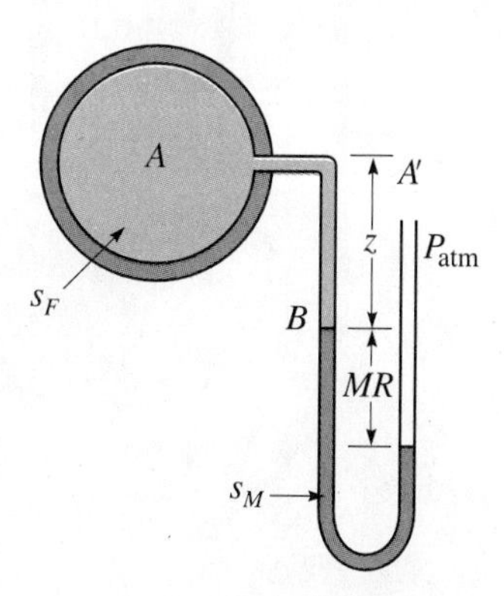

Figure 3.13
Negative-pressure manometer.

[3] In terms of gage pressure this equation can be expressed as: $0 + (s_M/s_F)\gamma MR + \gamma z = p_A$.

become trapped in the tube. For this case,

$$0 - \left(\frac{s_M}{s_F}\right)MR - z = \frac{p_A}{\gamma}$$

or

$$\frac{p_A}{\gamma} = -\left[z + \left(\frac{s_M}{s_F}\right)MR\right] \tag{3.11}$$

Although mercury is generally used as the measuring fluid in the simple manometer, other liquids (carbon tetrachloride for example) can be used. As the specific gravity of the measuring fluid approaches that of the fluid whose pressure is being measured, the reading becomes larger for a given pressure, thus increasing the accuracy of the instrument, provided the specific gravities are accurately known.

Differential Manometers

In many cases only the difference between two pressures is desired, and for this purpose differential manometers, such as shown in Fig. 3.14, may be

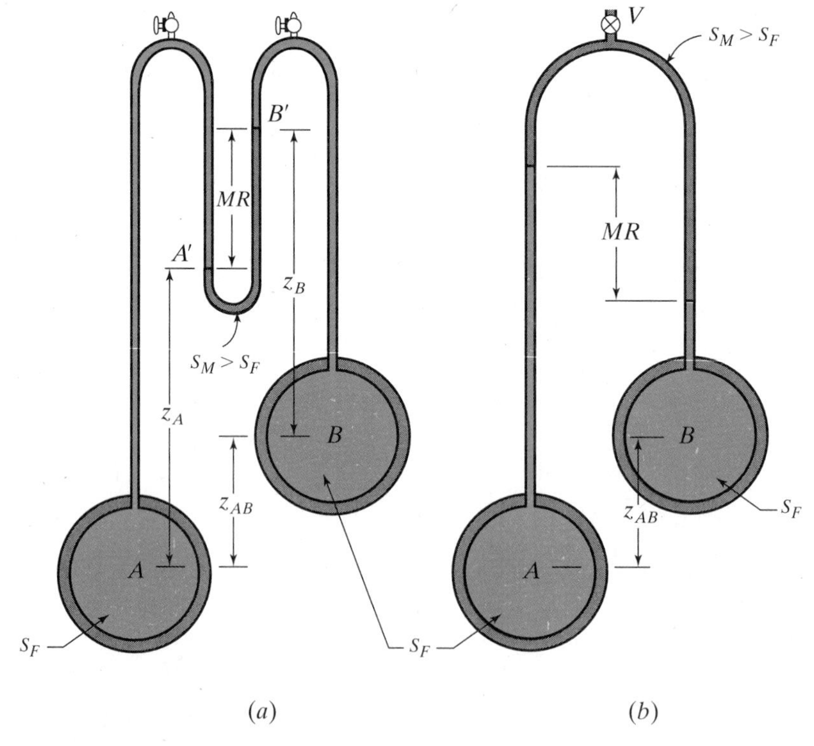

Figure 3.14
Differential manometers. (*a*) For measuring Δp in liquids or gases. (*b*) For measuring Δp in liquids only.

used. In Fig. 3.14*a* the measuring fluid is of greater density than that of the fluid whose pressure difference is involved. If the fluids in A and B (Fig. 3.14*a*) are of the same density, then, proceeding in a similar manner as before, through the manometer tubing from A to B, we obtain

$$\frac{p_A}{\gamma} - z_A - \left(\frac{s_M}{s_F}\right)MR + z_B = \frac{p_B}{\gamma}$$

So, by rearranging,

$$\frac{p_A}{\gamma} - \frac{p_B}{\gamma} = z_A - z_B + \left(\frac{s_M}{s_F}\right)MR$$

But, from Fig. 3.14*a*,

$$z_A + MR = z_B + z_{AB}$$

so

$$z_A - z_B = z_{AB} - MR$$

where z_{AB} is negative if B is below A, so that

$$\frac{p_A}{\gamma} - \frac{p_B}{\gamma} = z_{AB} - MR + \left(\frac{s_M}{s_F}\right)MR = z_{AB} + \left(\frac{s_M}{s_F} - 1\right)MR \tag{3.12}$$

Equation (3.12) is applicable only if the fluids in A and B have the same density. If these densities are different, the pressure head difference can be found by expressing all head components between A and B in terms of one or other of the fluids, as is done in Sample Prob. 3.4. It must be emphasized that by far the most common mistakes made in working differential-manometer problems are to omit the factor $(s_M/s_F - 1)$ for the gage difference MR, or to omit the -1 from this factor. The term $(s_M/s_F - 1)MR$ accounts for the *difference* in pressure heads due to the two columns of liquids (M) and (F) of height MR in the U-tube.

The differential manometer, when used with a heavy liquid such as mercury, is suitable for measuring large pressure differences. For a small pressure difference, a light fluid, such as oil, or even air, may be used, and in this case the manometer is arranged as in Fig. 3.14*b*. Of course, the manometer fluid must be one that will not mix with the fluid in A or B. By the same method of analysis as above, we can show for Fig. 3.14*b* that, for identical liquids in A and B,

$$\frac{p_A}{\gamma} - \frac{p_B}{\gamma} = z_{AB} + \left(1 - \frac{s_M}{s_F}\right)MR \tag{3.13}$$

where s_M/s_F, the ratio of the specific gravities (or densities or specific weights), has a value less than one. As the density of the manometer fluid approaches that of the fluid being measured, $(1 - s_M/s_F)$ approaches zero, and larger values of MR are obtained for small pressure differences, thus increasing the sensitivity of the gage. Once again, z_{AB} is negative if B is below A, and the equation must be modified if the densities of fluids A and B are different.

To determine pressure difference between liquids, it is often satisfactory to use air or some other gas as the measuring fluid (Fig. 3.14*b*). Air can then be pumped through valve *V* until the pressure is such as to bring the two liquid columns to a suitable level. Any change in pressure raises or lowers both liquid columns by the same amount so that the difference between them is constant. In this case the value of s_M/s_F may be considered to be zero, since the density of gas is so much less than that of a liquid.

Another way to obtain increased sensitivity is simply to incline the gage tube so that a vertical gage difference *MR* is transposed into a reading that is magnified by $1/\sin\alpha$, where α is the angle of inclination with the horizontal.

SAMPLE PROBLEM 3.4 Liquid *A* weighs 53.5 lb/ft³ (8.4 kN/m³). Liquid *B* weighs 78.8 lb/ft³ (12.4 kN/m³). Manometer liquid *M* is mercury. If the pressure at *B* is 30 psi (207 kPa), find the pressure at *A*. Express all pressure heads in terms of the liquid in bulb *B*.

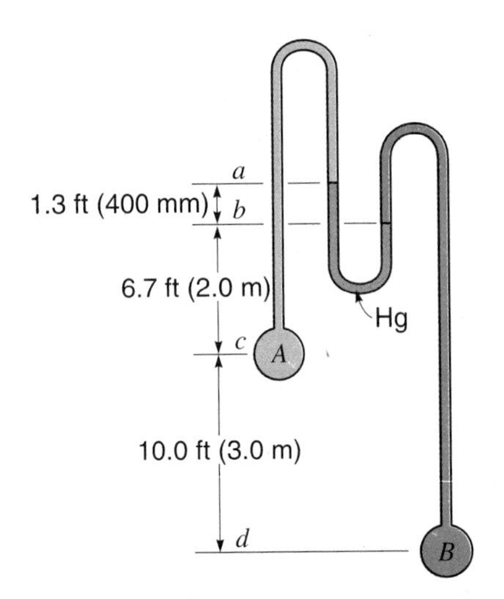

Figure S3.4

Solution

Proceeding from *A* to *B*:
$$\frac{p_A}{\gamma_B} - (\Delta z_{c-a})\frac{\gamma_A}{\gamma_B} + (\Delta z_{a-b})\frac{\gamma_M}{\gamma_B} + (\Delta z_{b-d})\frac{\gamma_B}{\gamma_B} = \frac{p_B}{\gamma_B}$$

BG units:
$$\frac{p_A}{\gamma_B} - 8.0\frac{53.5}{78.8} + 1.3\frac{13.56\times 62.4}{78.8} + 16.7 = \frac{p_B}{\gamma_B}$$

$$\frac{p_A}{\gamma_B} - 5.43 + 13.96 + 16.7 = \frac{30\times 144}{78.8} = 54.8 \text{ ft}$$

$$\frac{p_A}{\gamma_B} = 29.6 \text{ ft} \qquad p_A = 29.6\frac{78.8}{144} = 16.2 \text{ psi} \qquad \textbf{\textit{ANS}}$$

SI units: $$\frac{p_A}{\gamma_B} - 2.4\,\frac{8.4}{12.4} + 0.4\,\frac{13.56\times 9.81}{12.4} + 5.0 = \frac{p_B}{\gamma_B}$$

$$\frac{p_A}{\gamma_B} - 1.626 + 4.29 + 5.00 = \frac{207\ \text{kN/m}^2}{12.4\ \text{kN/m}^3} = 16.69\ \text{m}$$

$$\frac{p_A}{\gamma_B} = 9.03\ \text{m}, \quad p_A = 9.03\times 12.4 = 112.0\ \text{kN/m}^2 = 112.0\ \text{kPa} \qquad \textit{ANS}$$

EXERCISES

3.6.1 In Sample Prob. 3.4 suppose the atmospheric pressure at B is 1035 mbar abs. What would be the absolute pressure at A? Express in mbar abs and in mHg.

3.6.2 In the figure, originally there is a 6-in manometer reading. Atmospheric pressure is 14.70 psia. If the absolute pressure at A is doubled, what will be the manometer reading?

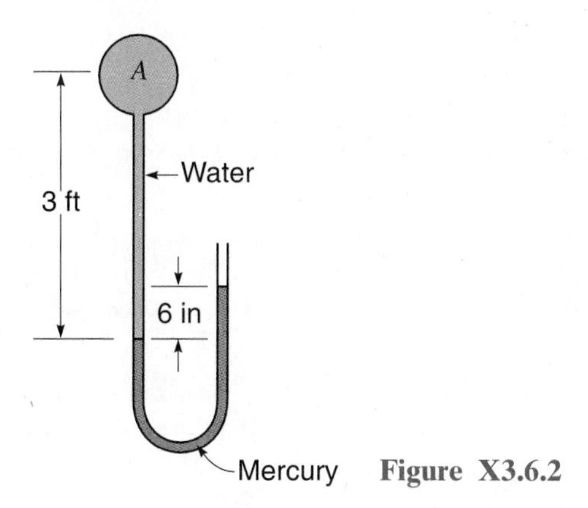

Figure X3.6.2

3.6.3 A mercury manometer (Fig. 3.11) is connected to a pipeline carrying water at 150°F and located in a room where the temperature is also 150°F. If the elevation of point B is 10 ft above A and the mercury manometer reading is 48 in, what is the pressure in the pipe in psi? Repeat, assuming all temperatures are 68°F. Be as precise as possible, and note the effect of temperature. Note that at 150°F the specific gravity of mercury is 13.45.

3.6.4 (*a*) Two vessels are connected to a differential manometer using mercury (s = 13.56), the connecting tubing being filled with water. The higher-pressure vessel is 5 ft lower in elevation than the other. Room temperature prevails. If the mercury reading is 4.0 in what is the pressure difference in feet of water, in psi? (*b*) If carbon tetrachloride (s = 1.59) were used instead of mercury, what would be the manometer reading for the same pressure difference?

3.6.5 (*a*) Two vessels are connected to a differential manometer using mercury ($s = 13.56$), the connecting tubing being filled with water. The higher-pressure vessel is 1.5 m lower in elevation than the other. Room temperature prevails. If the mercury reading is 100 mm, what is the pressure difference in m of water, in kPa? (*b*) If carbon tetrachloride ($s = 1.59$) were used instead of mercury, what would be the manometer reading for the same pressure difference?

3.6.6 Refer to the manometer of Fig. 3.14*b*. *A* and *B* are at the same elevation. Water is contained in *A* and rises in the tube to a level 68 in above *A*. Glycerin is contained in *B*. The inverted U-tube is filled with air at 15 psi and 70°F. Atmospheric pressure is 14.70 psia. Determine the difference in pressure between *A* and *B* if the manometer reading is 12 in. Express the answer in psi. What is the absolute pressure in *B* in inches of mercury, feet of glycerin?

3.6.7 Gas confined in a rigid container exerts a pressure of 25 psi when its temperature is 40°F. What pressure would the gas exert if the temperature were raised to 165°F? Barometric pressure remains constant at 29.0 inHg.

3.6.8 Gas confined in a rigid container exerts a pressure of 200 kPa when its temperature is 5°C. What pressure would the gas exert if the temperature were raised to 80°C? Barometric pressure remains constant at 29.0 inHg.

3.6.9 In the figure, atmospheric pressure is 14.70 psia; the gage reading at *A* is 3.5 psi; the vapor pressure of the alcohol is 1.4 psia. Compute *x* and *y*.

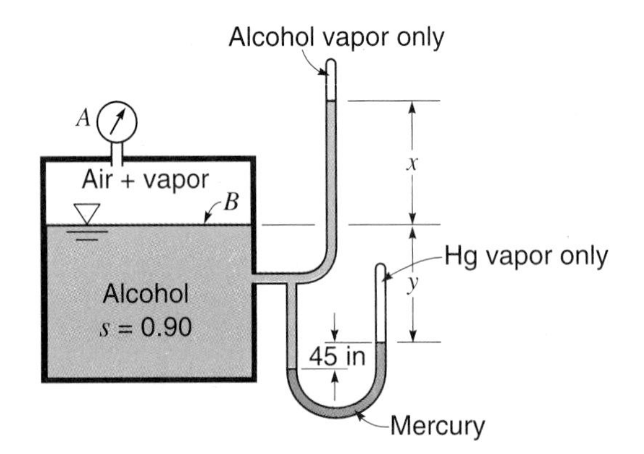

Figure X3.6.9

3.6.10 In Fig. X3.6.9 assume the following: atmospheric pressure = 900 mbar abs; vapor pressure of the alcohol = 130 mbar abs; $x = 3.50$ m and $y = 1.80$ m. Compute the reading on the pressure gage and on the manometer.

3.7 FORCE ON PLANE AREA

As noted previously in Sec. 3.1, no tangential force can exist within a fluid at rest. All forces are then normal to the surfaces in question. If the pressure is uniformly distributed over an area, the force is equal to the pressure times the area, and the point of application of the force is at the centroid of the area. For submerged horizontal areas, the pressure is uniform. In the case of compressible fluids (gases), the pressure variation with vertical distance is very small because of the low specific weight; hence, when computing the static fluid force exerted by a gas, p may usually be considered constant. Thus, for such cases,

$$F = \int p \, dA = p \int dA = pA \tag{3.14}$$

In the case of liquids the distribution of pressure is generally not uniform; hence further analysis is necessary. Consider a vertical plane whose upper edge lies in the free surface of a liquid (Fig. 3.15). Let this plane be perpendicular to the plane of the paper, so that MN is merely its trace. The

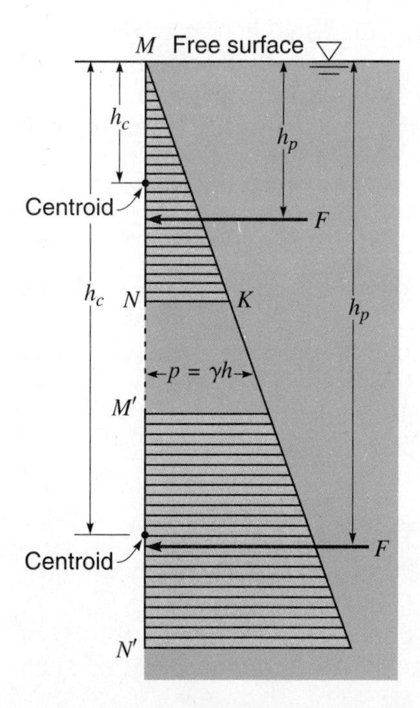

Figure 3.15

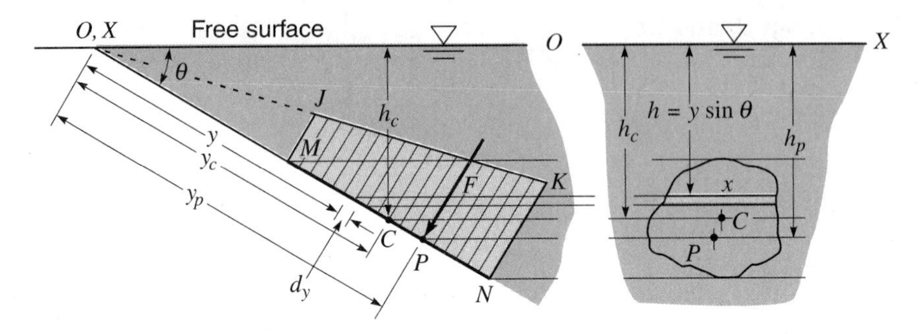

Figure 3.16

gage pressure will vary from zero at M to NK at N. Thus the total force on one side is the sum of the products of the elementary areas and the pressure upon them. From the pressure distribution, it is apparent that the resultant of this system of parallel forces must be applied at a point *below* the centroid of the area, since the centroid of an area is the point of application of the resultant of a system of *uniform* parallel forces.

If the plane is lowered to $M'N'$, the *proportionate* change of pressure from M' to N' is less than that from M to N. Hence the resultant pressure force will act nearer to the centroid of the plane surface, and the deeper the plane is submerged, the smaller the proportional pressure variation becomes and the closer the resultant moves to the centroid.

In Fig. 3.16 let MN be the trace of a plane area making an angle θ with the horizontal. To the right is drawn the projection of this area upon a vertical plane. The pressure distribution over the sloping area forms a ***pressure prism*** (MNKJ times width in Fig. 3.16), whose volume is equal to the total force F acting on the area. If the width x is contant then we can easily compute the volume of the pressure prism, using a mean pressure $= 0.5(MJ + NK)$, and so obtain F.

If x varies, we must integrate to find F. Let h be the variable depth to any point and y be the corresponding distance from OX, the intersection of the plane containing the area and the free surface.

Consider an element of area so chosen that the pressure is uniform over it. Such an element is a horizontal strip, width x, so $dA = x\, dy$. As $p = \gamma h$ and $h = y \sin \theta$, the force dF on a horizontal strip is

$$dF = p\, dA = \gamma h\, dA = \gamma y \sin \theta\, dA$$

Integrating,
$$F = \int dF = \gamma \sin \theta \int y\, dA = \gamma \sin \theta\, y_c A \tag{3.15}$$

where y_c is the distance from OX along the slope plane to the centroid C of the area A. If the vertical depth of the centroid is denoted by h_c then

$h_c = y_c \sin \theta$, and in general we have

$$F = \gamma h_c A \tag{3.16}$$

Thus the total force on any plane area submerged in a liquid is found by multiplying the specific weight of the liquid by the product of the area and the depth of its centroid. The value of F is independent of the angle of inclination of the plane so long as the depth of its centroid is unchanged.[4]

Since γh_c is the pressure at the centroid, we can also state that the total force on any plane area submerged in a liquid is the product of the area and the pressure at its centroid.

3.8 CENTER OF PRESSURE

The point of application of the resultant pressure force on a submerged area is called the ***center of pressure***. We need to know its location whenever we wish to work with the moment of this force.

The most general way of looking at the problem of forces on a submerged plane area is through the use of the recently discussed ***pressure prism concept*** (Sec. 3.7 and Fig. 3.16). The line of action of the resultant pressure force must pass through the centroid of the pressure prism (volume). As noted earlier, this concept is very convenient to apply for simple areas such as rectangles. For example, if the submerged area in Fig. 3.15 is of constant width then we know that the centroid of the pressure prism on area MN is $\frac{2}{3}MN$ below M.

If the shape of the area is not so regular, i.e., if the width x in Fig. 3.16 varies, then we must take moments and integrate. Taking OX in Fig. 3.16 as an axis of moments, the moment of an elementary force $dF = \gamma y \sin \theta \, dA$ is

$$y \, dF = \gamma \sin \theta \, y^2 \, dA$$

and if y_p denotes the distance to the center of pressure, using the concept that the moment of the resultant force equals the sum of the moments of the component forces,

$$y_p F = \gamma \sin \theta \int y^2 \, dA = \gamma \sin \theta \, I_O$$

where I_O is recognized as being the moment of inertia of the plane area about axis OX.

---

[4] For a plane submerged as in Fig. 3.16, it is obvious that Eq. (3.16) applies to one side only. As the pressure forces on the two sides are identical but opposite in direction, the net force on the plane is zero. In most practical cases where the thickness of the plane is not negligible, the pressures on the two sides are not the same.

If this last expression is divided by the value of F as given in Eq. (3.16), the result is

$$y_p = \frac{\gamma \sin \theta \, I_O}{\gamma \sin \theta \, y_c A} = \frac{I_O}{y_c A} \tag{3.17}$$

The product $y_c A$ is the static moment of area A about OX. Therefore Eq. (3.17) tells us that the distance from the center of pressure to the axis where the plane (extended) intersects the liquid surface is obtained by dividing the moment of inertia of the area A about the surface axis by its static moment about the same axis.

This may also be expressed in another form, by noting from the parallel axis theorem that

$$I_O = A y_c^2 + I_c$$

where I_c is the moment of inertia of an area about its centroidal axis. Thus

$$y_p = \frac{A y_c^2 + I_c}{y_c A} = y_c + \frac{I_c}{y_c A} \tag{3.18}$$

From this equation, it may again be seen that the location of the center of pressure P is independent of the angle θ; that is, the plane area may be rotated about axis OX without affecting the location of P. Also, it may be seen that P is always *below* the centroid C and that, as the depth of immersion is increased, y_c increases and P approaches C.

The lateral location of the center of pressure P may be determined by considering the area to be made up of a series of elemental horizontal strips. The center of pressure for each strip would be at the midpoint of the strip. Since the moment of the resultant force F must be equal to the moment of the distributed force system about any axis, say, the y axis,

$$X_p F = \int x_p p \, dA \tag{3.19}$$

where X_p is the lateral distance from the selected y axis to the center of pressure P of the resultant force F, and x_p is the lateral distance to the center of any elemental horizontal strip of area dA on which the pressure is p.

SAMPLE PROBLEM 3.5 Figure S3.5 shows a gate, 2 ft wide perpendicular to the sketch. It is pivoted at hinge H. The gate weighs 500 lb. Its center of gravity is

: Programmed computing aids (Appendix B) could help solve problems marked with this icon.

1.2 ft to the right of and 0.9 ft above H. For what values of water depth x above H will the gate remain closed? Neglect friction at the pivot and neglect thickness of the gate.

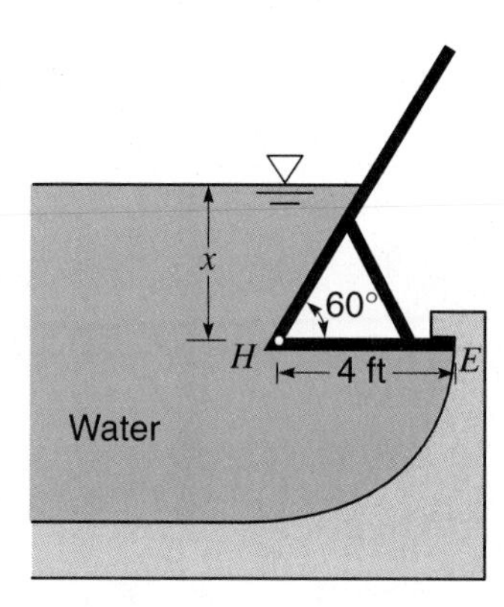

Figure S3.5

Solution

In addition to the reactive forces R_H at the hinge and R_E at end E, there are three forces acting on the gate: its weight W, the vertical hydrostatic force F_v upward on the rectangular bottom of the gate, and the slanting hydrostatic force F_s acting at right angles to the sloping rectangular portion of the gate. The magnitudes of the latter three forces are as follows:

Given: $$W = 500 \text{ lb}$$

Eq. (3.16): $$F_v = \gamma h_c A = \gamma(x)(4\times 2) = 8\gamma x$$

Eq. (3.16): $$F_s = \gamma h_c A = \gamma(x/2)\left(\frac{x}{\cos 30^\circ}\times 2\right) = 1.155\gamma x^2$$

A diagram showing these three forces is as follows:

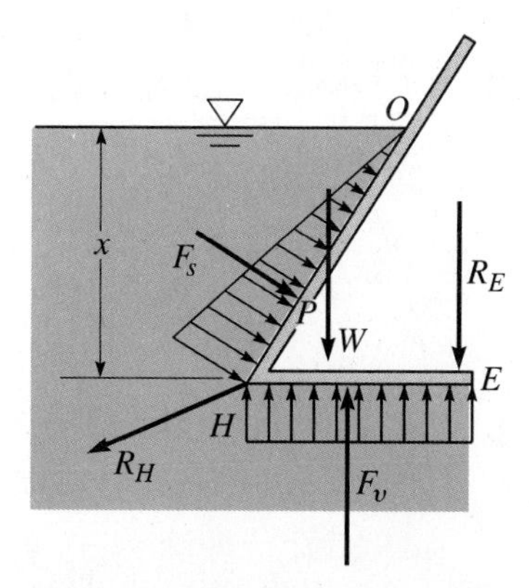

The moment arms of W and F_v with respect to H are 1.2 ft and 2.0 ft respectively. The moment arm of F_s gets larger as the water depth increases because the location of the center of pressure changes. The location of the center of pressure of F_s may be found from Eq. (3.18):

$$y_p = y_c + \frac{I_c}{y_c A}, \quad \text{where} \quad I_c = \frac{bh^3}{12}$$

with $h = x/\cos 30°$ and $y_c = 0.5h$. So

$$y_p = \frac{0.5x}{\cos 30°} + \frac{(1/12)2(x/\cos 30°)^3}{(0.5x/\cos 30°)[2(x/\cos 30°)]}$$

i.e., for F_s:

$$OP = y_p = 0.577x + \frac{2x}{12 \cos 30°} = 0.770x$$

Hence the moment arm of F_s with respect to H is $PH = x/\cos 30° - 0.770x = 0.385x$. [*Note:* In this case Eq. (3.18) need not be used to find the lever arm of F_s because the line of action of F_s for the triangular distributed load on the rectangular area is known to be at the third point between H and O, i.e., $HP = (1/3)(x/\cos 30°) = 0.385x$.]

When the gate is about to open (incipient rotation), $R_E = 0$ and the sum of the moments of all forces about H is zero, viz

$$\sum M = F_s(0.385x) + W(1.2) - F_v(2.0) = 0$$

i.e.,

$$1.155\gamma x^2(0.385x) + 500(1.2) - 8\gamma x(2) = 0$$

Substituting $\gamma = 62.4\ \text{lb/ft}^3$ gives

$$27.73x^3 + 600 - 998.4x = 0$$

Without an equation solver, we can solve this cubic equation by trials, seeking x values that make the left-hand side of the equation equal to zero. After two trial values, use linear interpolation to estimate the next trial value, as follows:

Trial x	*Left-hand side*	*Trial x*	*Left-hand side*	*Trial x*	*Left-hand side*
0.1	500.2	10.0	18,349	−10.0	−17,149
0.5	104.3	5.0	−925.3	−5.0	2125
0.6	6.95	6.0	600.0	−5.55	1400
0.61	−2.73	5.60	−120.6	−6.61	−810
		5.67	−5.58	−6.22	−136.3
				−6.28	1.15

Thus $x = 0.61$ ft or 5.67 ft or a negative (meaningless) root. Therefore, from inspection of the moment equation, the gate will remain closed when $0.61\ \text{ft} < x < 5.67$ ft. ***ANS***

Sample Problem 3.6 The cubic tank shown is half full of water. Find (*a*) the pressure on the bottom of the tank, (*b*) the force exerted by the fluids on a tank wall, and (*c*) the location of the center of pressure on a wall.

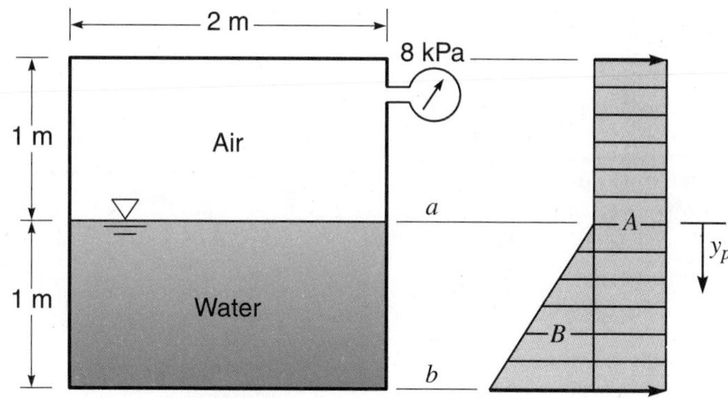

Figure S3.6

Solution

(*a*)
$$p_b = p_a + \gamma_w h_w = 8\text{ kN/m}^2 + (9.81\text{ kN/m}^3)(1\text{ m})$$
$$= 17.81\text{ kN/m}^2 = 17.81\text{ kPa} \qquad \textit{ANS}$$

(*b*) The force acting on the tank end is divided into two components, identified as A and B on the pressure distribution sketch. Component A has a *uniform* pressure distribution, due to the pressure of the confined air, which is transmitted through the water:

$$F_A = p_a A_a = (8\text{ kN/m}^2)(4\text{ m}^2) = 32\text{ kN}$$

For component B, the varying water pressure distribution on the lower half of the tank wall, the centroid C of the area of application is at

$h_c = y_c = 0.5$ m below the water surface, so, from Eq. (3.16),

$$F_B = \gamma_w h_c A_w = 9.81(0.5)2 = 9.81\text{ kN}$$

So the total force on the tank wall is

$$F = F_A + F_B = 32 + 9.81 = 41.8\text{ kN} \qquad \textit{ANS}$$

(*c*) The locations of the centers of pressure of the component forces are as follows.

$$(y_p)_A = 0\text{ m}$$

below the water surface, to the centroid of the square area for the uniform pressure.

$$(y_p)_B = \tfrac{2}{3}h_w = \tfrac{2}{3}(1\text{ m}) = 0.667\text{ m}$$

below the water surface for the varying pressure on the rectangular wetted wall area. This can also be found using Eq. (3.18) with $y_c = 0.5$ m, $I_c = bh^3/12 = 0.1667\text{ m}^4$, and $A = bh = 2\text{ m}^2$.

Taking moments: $\quad F(y_p) = F_A(y_p)_A + F_B(y_p)_B$

from which $\quad y_p = 0.1565$ m below the water surface $\qquad$ ***ANS***

SAMPLE PROBLEM 3.7 Water, oil, and air are present in the cylindrical storage tank as shown. (*a*) Find the pressure at the bottom of the tank. (*b*) Find the force exerted by the fluids on the end of the tank. (*c*) Determine the location of the center of pressure on the tank end.

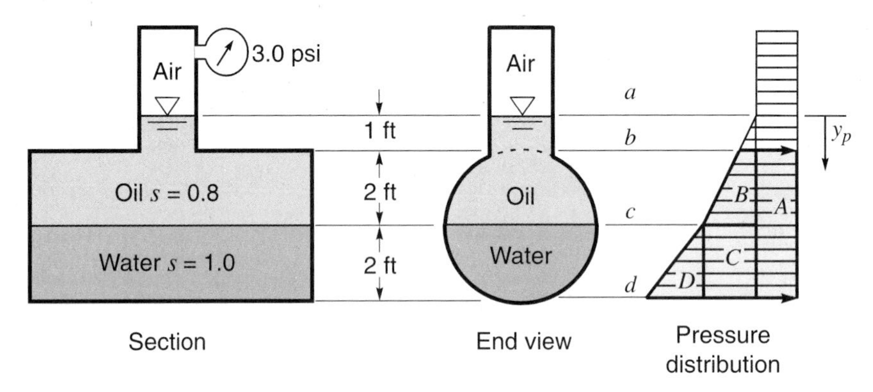

Figure S3.7

Solution

(*a*)
$$p_d = p_a + \gamma_{\text{oil}} h_{a-c} + \gamma_{\text{H}_2\text{O}} h_{c-d} = 3.0 + \frac{0.80 \times 62.4}{144} 3 + \frac{62.4}{144} 2$$
$$= 3.0 + 1.04 + 0.87 = 4.9 \text{ psi} \qquad \textbf{\textit{ANS}}$$

(*b*) The force acting on the tank end is divided into four components, identified as *A–D* on the pressure distribution sketch. Note that components *A* and *C* have *uniform* pressure distributions, due to air (*A*) or a different liquid (*C*) above the subject liquid.

As a preliminary, we note for the semicircular end areas ($r = 2$ ft) that

(i) $A = \pi r^2/2 = 6.28 \text{ ft}^2$;

(ii) from Appendix 1, Table A.7, the centroid is $4r/3\pi = 0.849$ ft from the center of the circular end.

For component *A*, the confined air exerts a uniform pressure of 3.0 psi, which is transmitted through the other fluids to the end of the tank.

$$F_A = pA = 3.0(144)\pi 2^2 = 5430 \text{ lb}$$

For component *B*, the varying oil pressure distribution acts on the upper half of the tank end, which has its centroid at

$$h_c = y_c = 3 - 0.849 = 2.15 \text{ ft below the free oil surface,}$$

so:
$$F_B = \gamma h_c A = 0.8 \times 62.4(2.15)\pi 2^2/2 = 675 \text{ lb}$$

The force F_C on the lower half of the end of the tank due to the uniform pressure produced by the 3-ft depth of oil above midheight is

$$F_C = pA = \gamma hA = (0.8 \times 62.4)3(\pi 2^2/2) = 941 \text{ lb}$$

For component *D*, the varying pressure distribution due to the water (only) on the lower half of the tank end has its centroid at

$$h_c = y_c = 0.849 \text{ ft below the top water surface,}$$

so: $$F_D = \gamma h_c A = 62.4(0.849)\pi 2^2/2 = 333 \text{ lb}$$

The total force F on the end of the tank is therefore

$$F = F_A + F_B + F_C + F_D = 7380 \text{ lb} \qquad \textbf{\textit{ANS}}$$

(*c*) As a preliminary to locating the center of pressure, we note that for the semicircular end area ($D = 4$ ft),
(i) from Table A.7: I about the center of the circle $= \pi D^4/128 = 6.28 \text{ ft}^4$, and
(ii) by the parallel axis theorem, $I_c = I - Ad^2 = 6.28 - 6.28(0.849)^2 = 1.756 \text{ ft}^4$.
The locations of the centers of pressure of the component forces are

$(y_p)_A = 3.0$ ft below the free oil surface, to the centroid of the circular area for uniform pressure;

Eq. (3.18): $(y_p)_B = y_c + \dfrac{I_c}{y_c A}$ below the free oil surface, for varying oil pressure on the upper semicircular area, and,

using $(y_c)_B$ and I_c from above: $$(y_p)_B = 2.15 + \frac{1.756}{2.15(\pi 2^2/2)} = 2.28 \text{ ft}$$

$(y_p)_C = y_c = 3 + 0.849 = 3.85$ ft below the free oil surface, to the *centroid* of the lower semicircular area, for uniform pressure on this area due to 3 ft of oil above the water;

Eq. (3.18): $(y_p)_D = y_c + \dfrac{I_c}{y_c A}$ below the water top surface, for the varying water pressure on the lower semicircular area, and,

using $(y_c)_D$ and I_c from above:

$$(y_p)_D = 0.849 + \frac{1.756}{0.849(\pi 2^2/2)} = 1.178 \text{ ft below the water top surface}$$

$$= 3 + 1.178 = 4.18 \text{ ft below the free oil surface}$$

Finally, $$Fy_p = F_A(y_p)_A + F_B(y_p)_B + F_C(y_p)_C + F_D(y_p)_D$$

$$y_p = 3.10 \text{ ft} \qquad \textbf{\textit{ANS}}$$

EXERCISES

3.8.1 If a triangle of height d and base b is vertical and submerged in liquid with its vertex at the liquid surface, derive an expression for the depth to its center of pressure.

3.8.2 Repeat Exer. 3.8.1 for the same triangle but with its vertex a distance a below the liquid surface.

3.8.3 If a triangle of height d and base b is vertical and submerged in a liquid with its base at the liquid surface, derive an expression for the depth to its center of pressure.

3.8.4 A circular area of diameter d is vertical and submerged in a liquid. Its upper edge is coincident with the liquid surface. Derive an expression for the depth to its center of pressure.

3.8.5 Refer to Sample Prob. 3.7. If the air pressure were 5 psi rather than 3 psi, compute the total force and determine the location of the center of pressure.

3.8.6 A rectangular plate submerged in water is 5 ft by 4 ft, the 5-ft side being horizontal and the 4-ft side being vertical. Determine the magnitude of the force on one side of the plate and the depth to its center of pressure if the top edge is (a) at the water surface; (b) 1 ft below the water surface; (c) 100 ft below the water surface.

3.8.7 Repeat Exer. 3.8.6 changing all dimensions from feet to meters, i.e., the plate is 5 m by 4 m.

3.8.8 A rectangular plate 5 ft by 4 ft is at an angle of 30° with the horizontal, and the 5-ft side is horizontal. Find the magnitude of the force on one side of the plate and the location of its center of pressure when the top edge is (a) at the water surface; (b) 1 ft below the water surface.

3.8.9 A rectangular area is 5 m by 6 m, with the 5 m side horizontal. It is placed with its centroid 4 m below a water surface and rotated about a horizontal axis in the plane area and through its centroid. Find the magnitude of the force on one side and the distance between the center of pressure and the centroid of the plane when the angle θ = 90, 60, 30, 0°.

3.8.10 A plane surface is circular and 5 ft in diameter. If it is vertical and the top edge is 2 ft below the water surface, find the magnitude of the force on one side and the depth to the center of pressure.

3.8.11 A plane surface is circular and 1.5 m in diameter. If it is vertical and the top edge is 1 m below the water surface, find the magnitude of the force on one side and the depth to the center of pressure.

3.8.12 Prove that for a plane area such that a straight line can be drawn through the midpoints of all horizontal elements, the center of pressure must lie on this line.

3.8.13 A vertical right-triangle of height d and base b submerged in liquid has its vertex at the liqiud surface. Find the distance from the vertical side to the center of pressure by (a) inspection; (b) calculus.

3.8.14 The right-triangular plate shown in Fig. X3.8.14 is submerged in a vertical plane with its base horizontal. Determine the depth and horizontal position of the center of pressure.

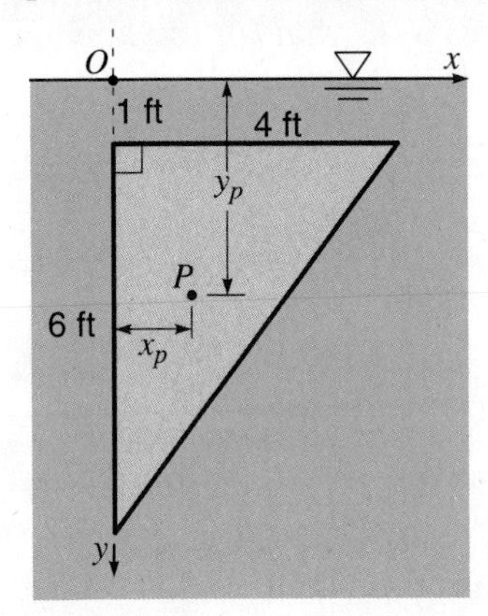

Figure X3.8.14

3.8.15 Repeat Exer. 3.8.14, but with the base 1.2 m long and submerged 0.3 m, and with the vertical edge 1.8 m long.

3.8.16 Figure X3.8.16 shows a cylindrical tank with 0.25-in-thick walls, containing water. What is the force on the bottom? What is the force on the annular surface *MM*? What is the weight of the water? Find the longitudinal (vertical) tensile stress in the sidewalls *BB* if (*a*) the tank is suspended from the top; (*b*) it is supported on the bottom. Neglect the weight of the tank.

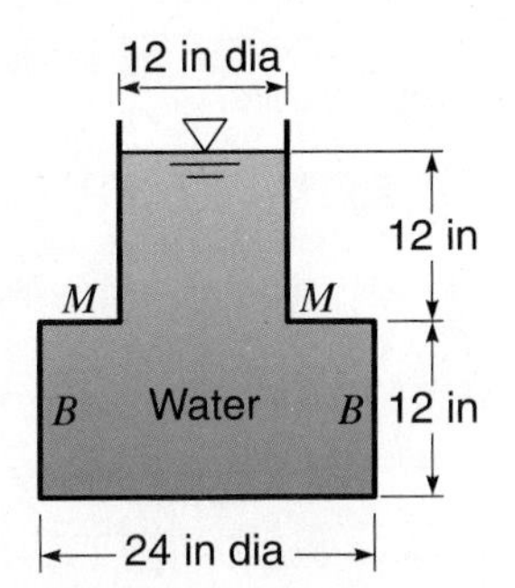

Figure X3.8.16

3.8.17 Find the magnitude and point of application of the force on the circular gate shown in Fig. X3.8.17.

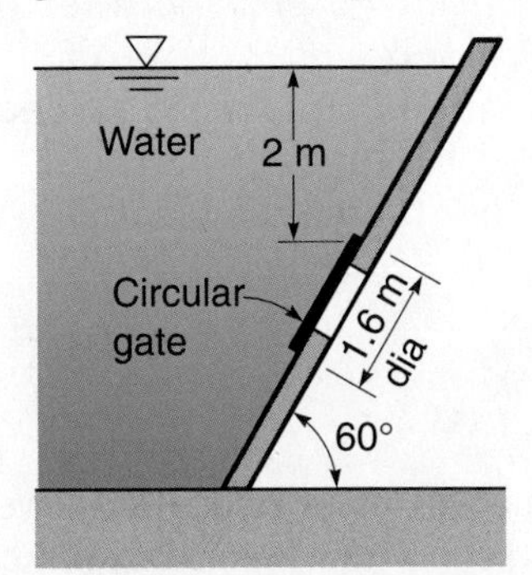

Figure X3.8.17

3.8.18 The gate MN in Fig. X3.8.18 rotates about an axis through N. If the width perpendicular to the plane of the figure is 3 ft, what torque applied to the shaft through N is required to hold the gate closed?

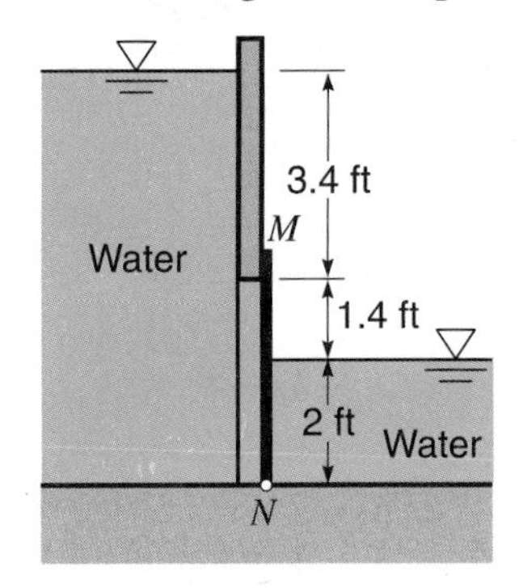

Figure X3.8.18

3.8.19 In Fig. X3.8.19 the rectangular flashboard MN shown in cross-section is pivoted at B. (a) What must be the maximum height of B above N if the flashboard is on the verge of tipping when the water surface rises to M? (b) If the flashboard is pivoted at the location determined in (a) and the water surface is 1 m below M, what are the reactions at B and N per m length of board perpendicular to the figure?

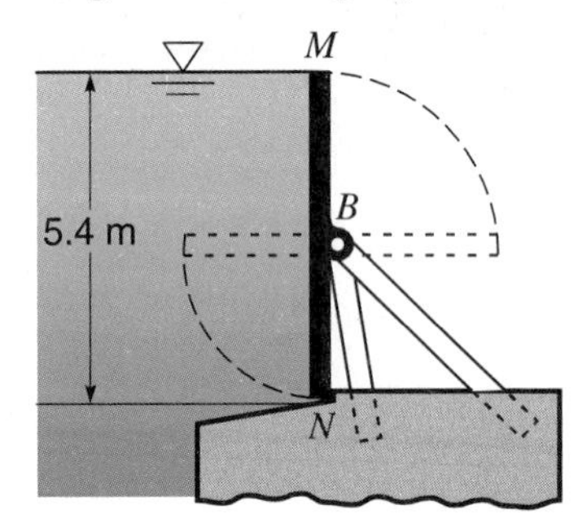

Figure X3.8.19

3.9 FORCE ON CURVED SURFACE

On any curved or warped surface such as MN in Fig. 3.17a, the force on the various elementary areas that make up the curved surface are different in direction and magnitude, so an algebraic summation is impossible. Hence Eq. (3.16) can be applied only to a plane area. But for nonplanar areas, component forces in certain directions can be found, and often without integration.

Horizontal Force on Curved Surface

Any irregular curved area MN (Fig. 3.17a) may be projected upon a vertical plane whose trace is $M'N'$ (Fig. 3.17b). The projecting elements, which are

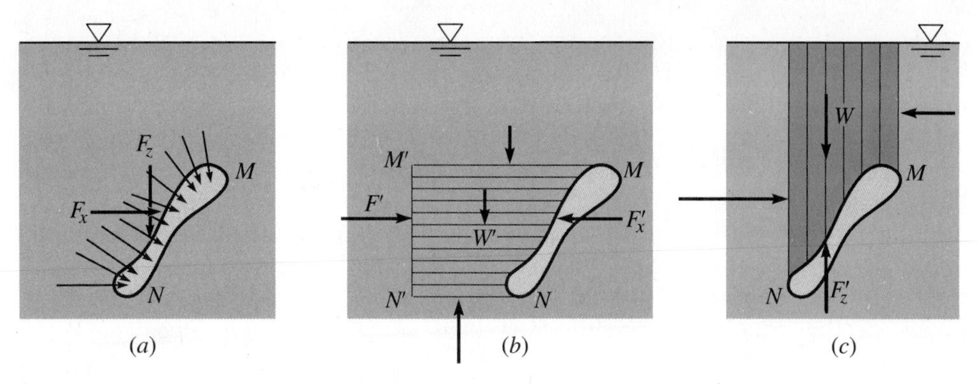

Figure 3.17
Hydrostatic forces on curved surfaces.

all horizontal, enclose a volume whose ends are the vertical plane $M'N'$ and the irregular area MN. This volume of liquid is in static equilibrium. Acting on the projected vertical area $M'N'$ is a force F'. The horizontal force component F_x' acts on the irregular end area MN and is equal and opposite to the F_x of Fig. 3.17*a*. Gravity W' is vertical, and the lateral forces on all the horizontal projection elements are normal to these elements and hence normal to F'. Thus the only horizontal forces on $MNN'M'$ are F' and F_x', and therefore

$$F' - F_x' = 0$$

and
$$F_x = F_x' = F' \tag{3.20}$$

Hence the horizontal force in any given direction on any area is equal to the force on the projection of that area upon a vertical plane normal to the given direction. The line of action of F_x must be the same as that of F'. Equation (3.20) is applicable to gases as well as liquids. In the case of a gas the horizontal force on a curved surface is given by the pressure multiplied by the projection of that area upon a vertical plane normal to the force.

Vertical Force on Curved Surface

The vertical force F_z on a curved or warped area, such as MN in Fig. 3.17*a*, can be found by considering the volume of liquid enclosed by the area and vertical elements extending to the level of the free surface (Fig. 3.17*c*). This volume of liquid is in static equilibrium. The only vertical forces on this volume of liquid are the force $F_G = p_G A$ due to any gas (at pressure p_G) above the liquid, the gravity force W downward, and F_z', the upward vertical force on the irregular area MN. The force F_z' (Fig. 3.17*c*) must be equal and opposite to the force F_z (Fig. 3.17*a*). Any other forces on the vertical elements are normal to the elements, and hence are horizontal. Therefore

$$F_z' - W - F_G = 0$$

and
$$F_z = F_z' = W + F_G \tag{3.21}$$

Hence the vertical force upon any area is equal to the weight of the volume of liquid above it, plus any superimposed gas pressure force. The line of action of F_z must be the resultant of F_G and W. F_G must pass through the centroid of the plan (surface projection) area, and W must pass through the center of gravity of the liquid volume. The portion of this volume above $M'M$ has a regular shape with volume equal to height times projected plan area, and has its centroid beneath the centroid of the plan area; the other portion, below $M'M$ and above the curved surface MN, may have a difficult shape and may require integration to find its volume and centroid. If only a gas is involved, the procedure is similar, but is much simplified because W is negligible.

For the case where the lower side of the surface is subjected to a force while the upper side is not, the vertical force component is the same in magnitude as that given by Eq. (3.21) but opposite in sense.

Resultant Force on Curved Surface

In general, there is no single resultant force on an irregular area, because the horizontal and vertical forces, as found in the foregoing discussion, may not be in the same plane. But in certain cases these two forces will lie in the same plane and then they can be combined into a single force.

SAMPLE PROBLEM 3.8 Find the horizontal and vertical components of the force exerted by the fluids on the horizontal cylinder in Fig. S3.8 if (*a*) the fluid to the left of the cylinder is a gas confined in a closed tank at a pressure of 35.0 kN/m^2; (*b*) the fluid to the left of the cylinder is water with a free surface at an elevation coincident with the uppermost part of the cylinder. Assume in both instances that atmospheric pressure occurs to the right of the cylinder.

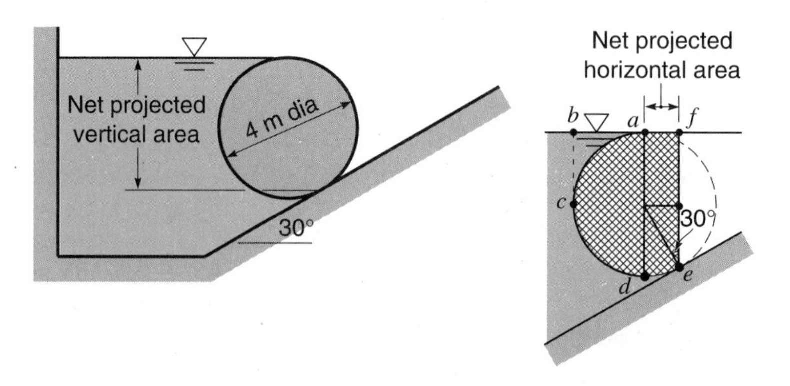

Figure S3.8

Solution

The net projection on a vertical plane of the portion of the cylindrical surface under consideration (see left-hand diagram) is, from the right-hand diagram, $ef = 2 + 2\cos 30° = 3.73$ m.

(*a*) For the gas,

$$F_x = pA_2 = 35.0 \text{ kN/m}^2(3.73 \text{ m}) = 130.5 \text{ kN/m to the right} \qquad \textbf{\textit{ANS}}$$

The vertical force of the gas on the surface *ac* is equal and opposite to that on the surface *cd*. Hence the net projection on a horizontal plane for the gas is $af = 2\sin 30° = 1$ m. Thus

$$F_z = pA_x = 35.0 \text{ kN/m}^2(1 \text{ m}) = 35.0 \text{ kN/m upward} \qquad \textbf{\textit{ANS}}$$

(*b*) For the fluid,

Eq. (3.16): $$F_x = \gamma h_c A = 9.81 \text{ kN/m}^2(\tfrac{1}{2}\times 3.73 \text{ m})(3.73 \text{ m}) = 68.3 \text{ kN/m to the right} \qquad \textbf{\textit{ANS}}$$

$$\begin{aligned}\text{Net } F_z &= \text{upward force on surface } cde - \text{downward force on surface } ca\\ &= \text{weight of volume } abcdefa - \text{weight of volume } abca\\ &= \text{weight of cross-hatched volume of liquid}\\ &= 9.81 \text{ kN/m}^3[\tfrac{210}{360}\pi 2^2 + \tfrac{1}{2}(1\times 2\cos 30°) + (1\times 2)] \text{ m}^2\\ &= 100.0 \text{ kN/m upward} \qquad \textbf{\textit{ANS}}\end{aligned}$$

EXERCISES

3.9.1 A vertical-thrust bearing for a large hydraulic gate is composed of a 10-in-radius bronze hemisphere mating into a steel hemispherical shell in the gate bottom. At what pressure must lubricant be supplied to the bearing so that a complete oil film is present if the vertical thrust on the bearing is 750,000 lb?

3.9.2 A tank with vertical ends contains water and is 6 m long normal to the plane of Fig. X3.9.2. The sketch shows a portion of its cross-section where *MN* is one-quarter of an ellipse with semiaxes *b* and *d*. If $b = 2.5$ m, $d = 4$ m, and $a = 1.0$ m, find, for the surface represented by *MN*, the magnitude and position of the line of action of (*a*) the horizontal component of force; (*b*) the vertical component of force; (*c*) the resultant force and its direction with the horizontal.

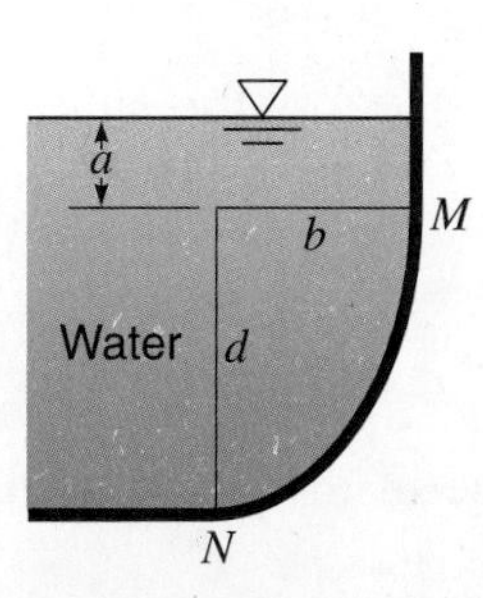

Figure X3.9.2

3.9.3 Find the answers called for in Exer. 3.9.2 if $a = 2$ ft, $b = 6$ ft, $d = 9$ ft, the tank is 12 ft long, and MN represents a parabola with vertex at N.

3.9.4 The cross-section of a tank is as shown in Fig. X3.9.4. BC is a cylindrical surface. If the tank contains water to a depth of 10 ft, determine the magnitude and location of the horizontal- and vertical-force components on the wall ABC.

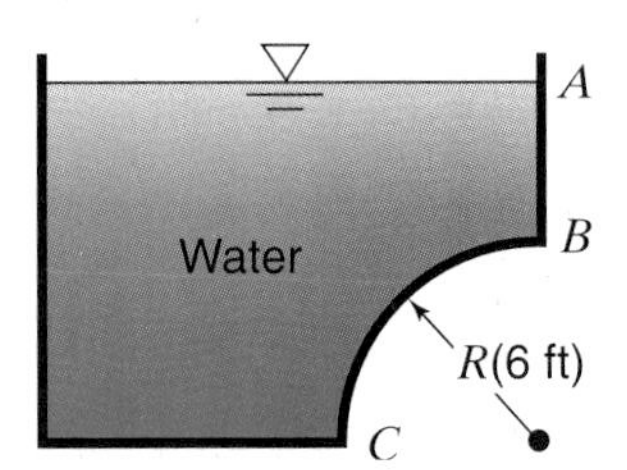

Figure X3.9.4

3.9.5 Find the answers called for in Exer. 3.9.4 if the radius of BC is 2 m and the water depth is 3.5 m.

3.9.6 Repeat Exer. 3.9.4 where the tank is closed and contains gas at a pressure of 10 psi.

3.9.7 Repeat Exer. 3.9.5 where the tank is closed and contains gas at a pressure of 75 kPa.

3.9.8 A spherical steel tank of 18 m diameter contains gas under a pressure of 380 kN/m^2. The tank consists of two half-spheres joined together with a weld. What will be the tensile force across the weld in kN/m? If the steel is 22.0 mm thick, what is the tensile stress in the steel? Express in kPa and in psi. Neglect the effects of cross-bracing and stiffeners.

3.9.9 The hemispherical body shown in Fig. X3.9.9 projects into a tank. Find the horizontal and vertical forces acting on the hemispherical projection for the following cases: (*a*) the tank is full of water with the free surface 5 ft above A; (*b*) the tank contains CCl_4 ($s = 1.59$) to the level of A overlain with water having its free surface 5 ft above A; (*c*) the tank is closed and contains only gas at a pressure of 6 psi; (*d*) the tank is closed and contains water to the level of A overlain with gas at a pressure of 2 psi. Assume the gas weighs 0.075 lb/ft^3.

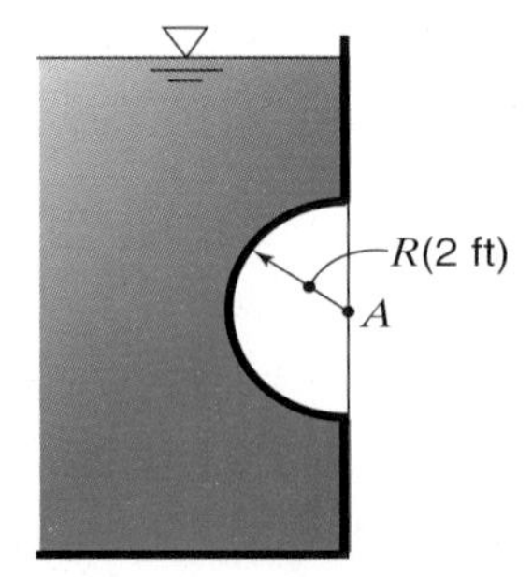

Figure X3.9.9

3.9.10 Determine the force required to hold the cone in the position shown in Fig. X3.9.10. Assume the cone is weightless.

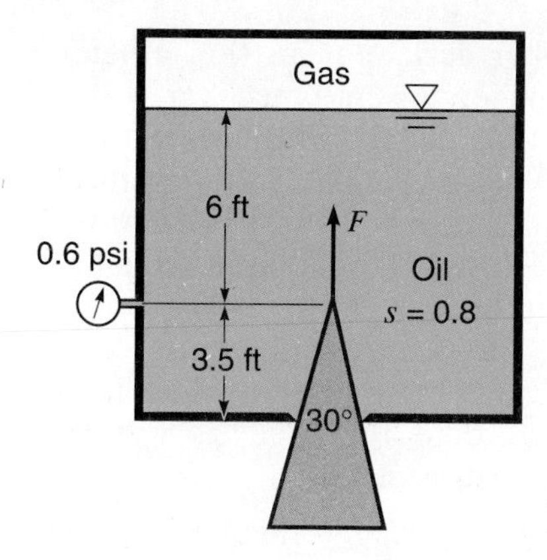

Figure X3.9.10

3.10 BUOYANCY AND STABILITY OF SUBMERGED AND FLOATING BODIES

Submerged Body

The body $DHCK$ immersed in the fluid in Fig. 3.18 is acted upon by gravity and the pressures of the surrounding fluid. On its upper surface the vertical component of the force is F_z and is equal to the weight of the volume of fluid $ABCHD$. In similar manner, the vertical component of force on the undersurface is F_z' and is equal to the weight of the volume of fluid $ABCKD$. The difference between these two volumes is the volume of the body $DHCK$.

Buoyancy**.** The buoyant force of a fluid is denoted by F_B, and it is vertically upward and equal to $F_z' - F_z$, which is equal to the weight of the volume of fluid $DHCK$. That is, *the* ***buoyant force*** *on any body is equal to the weight of fluid displaced*. This is probably the best known discovery of

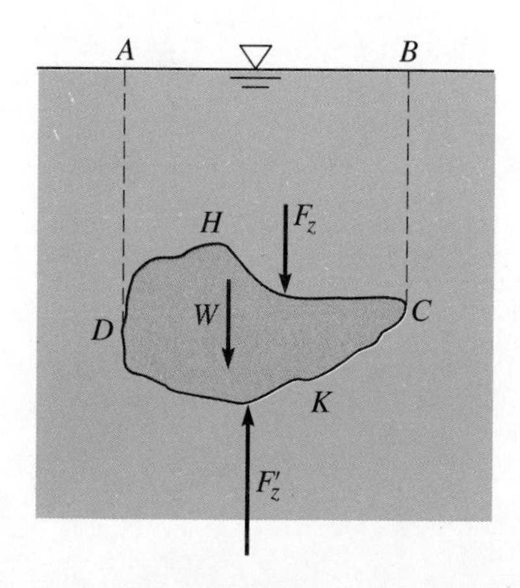

Figure 3.18

Archimedes (287–212 B.C.), a Greek philosopher acclaimed as the father of hydrostatics, and one of the earliest known pioneers of fluid mechanics.

If the body in Fig. 3.18 is in equilibrium, W is equal and opposite to F_B, which means that the densities of body and fluid are equal. If W is greater than F_B, the body will sink. If W is less than F_B, the body will rise until its density and that of the fluid are equal, as in the case of a balloon in the air or, in the case of a liquid with a free surface, the body will rise to the surface until the weight of the displaced liquid equals the weight of the body. If the body is less compressible than the fluid, there is a definite level at which it will reach equilibrium. If it is more compressible than the fluid, it will rise indefinitely, provided the fluid has no definite limit of height.

Stability. When a body in equilibrium is given a slight angular displacement (tilt or ***list***), a horizontal distance a then separates W and F_B, which in combination create moments that tend to rotate the body, as indicated in Fig. 3.19. If the moments tend to restore the body to its original position, the lesser of the two moments is called the ***righting moment*** (Fig. 3.19), and the body is said to be in ***stable equilibrium***. The stability of submerged or floating bodies depends on the relative positions of the buoyant force and the weight of the body. The buoyant force acts through the ***center of buoyancy*** B which corresponds to the center of gravity of the displaced fluid. *The criterion for stability of a fully submerged body* (balloon or submarine, etc.) *is that the center of buoyancy must be above the center of gravity of the body.* Inspection of Fig. 3.19 will confirm that if B where initially below G, the center of gravity, then the moment created by a tilt would tend to increase the displacement.

Floating Body

For a body in a liquid with a free surface, if its weight W is less than that of the same volume of liquid, it will rise and float on the surface as in Fig. 3.20,

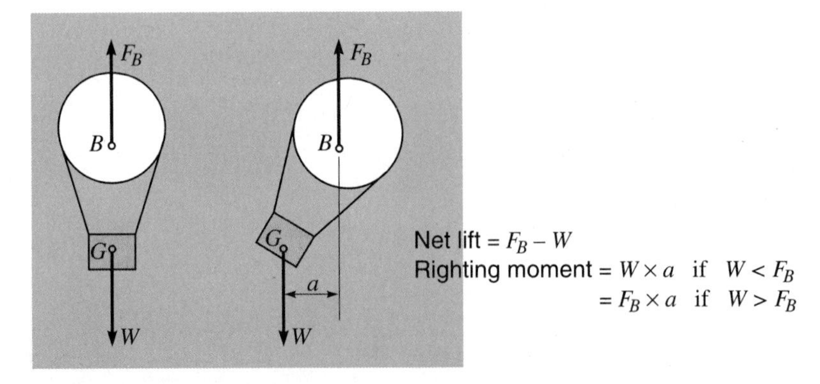

Figure 3.19
Submerged body (balloon).

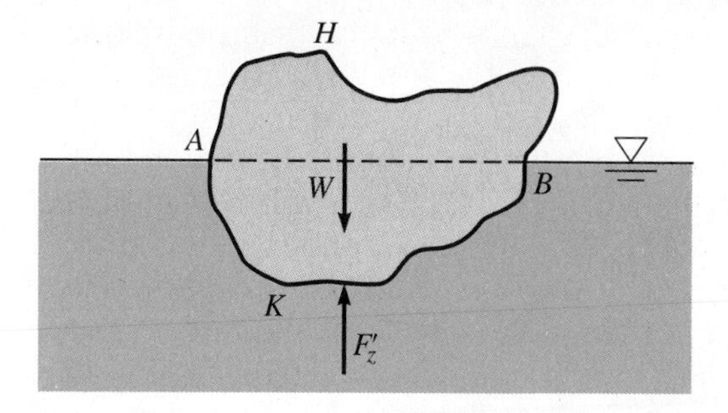

Figure 3.20

so that $W = F_B$. The body $AHBK$ is then acted upon by gravity and the pressures of the fluids in contact with it. The vertical component of force on the undersurface is F_z' and is equal to the weight of the volume of liquid AKB. This volume is the volume of liquid displaced by the body.

Buoyancy. The buoyant force F_B is vertically upward and equal to F_z'. So, just as for a fully submerged body, the buoyant force acting on a floating body is equal to the weight of liquid displaced. Thus a *floating body displaces a volume of liquid equivalent to its weight.* For equilibrium, the two forces W and F_B must be equal and opposite, and must lie in the same vertical line.

The atmospheric pressure is transmitted through the liquid to act equally on all surfaces of the body. As a result, it has zero net effect. Any buoyancy due to the weight of air displaced by the portion of the body above the liquid surface is usually negligible in comparison with the weight of liquid displaced.

A practical application of the buoyancy principle is the ***hydrometer,*** an instrument used to measure the specific gravity of liquids. It has a thin, uniform stem of constant cross-sectional area, say A. It is weighted to float upright as in Fig. 3.21*a*, with a reference mark at the water surface when floating in pure water ($s = 1.0$). When placed in a denser liquid of specific

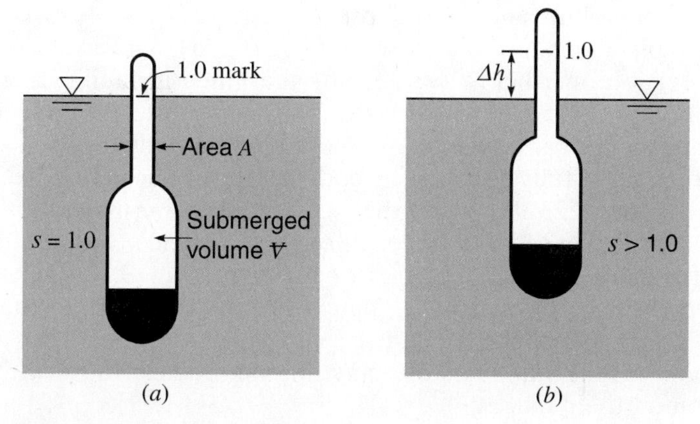

Figure 3.21
Hydrometer in (*a*) pure water; (*b*) a denser liquid.

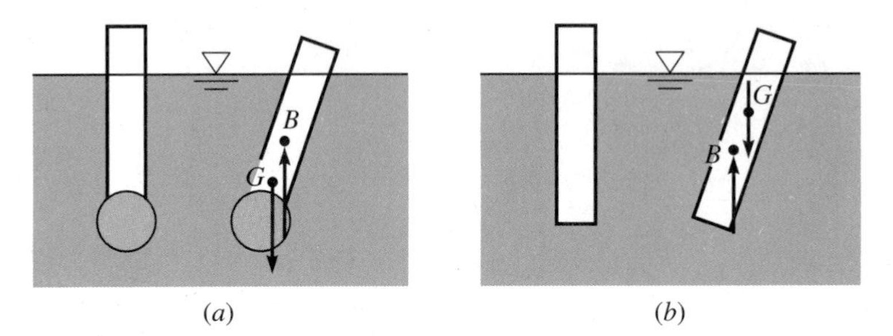

Figure 3.22
(*a*) Stable. (*b*) Unstable.

gravity s (Fig. 3.21*b*), the volume of liquid displaced must be smaller, so less is submerged and the reference mark is some height Δh above the water surface. If the submerged volume in pure water is $\forall$, then in the denser liquid it is $\forall - A\,\Delta h$, and the hydrometer's weight

$$W = \gamma_w \forall = (s\gamma_w)(\forall - A\,\Delta h)$$

from which

$$\Delta h = \frac{\forall}{A}\left(\frac{s-1}{s}\right) \tag{3.22}$$

Equation (3.22) enables the calculation of spacing for a specific gravity scale to be marked on the stem.

Stability. If a righting moment is developed when a floating body lists, the body will be stable regardless of whether the center of buoyancy is above or below the center of gravity. Examples of stable and unstable floating bodies are shown in Fig. 3.22. In these examples the stable body is the one where the center of buoyancy B is above the center of gravity G (Fig. 3.22*a*), and the unstable body has B below G (Fig. 3.22*b*). However, for floating bodies the location of B below G does not guarantee instability as it does for submerged bodies, discussed previously. This is because the position of the center of buoyancy B can move relative to a floating body as it tilts, due to its shape, whereas for a fully submerged body the position of B is fixed relative to the body. Figure 3.23 illustrates this point; it shows cross-sections through the hull of a ship that is stable even though B is below G. Because of the cross-sectional shape, as the ship tilts to the right (Fig. 3.23*b*) the center of gravity of the displaced water (i.e., B) moves to the right further than the line of action of the body weight W, and so it provides a righting moment $F_B \times a$. Clearly, therefore, the stabilities of many floating bodies (those with B below G) depend upon their shapes.

If liquid in the hull of a ship were unconstrained, the center of mass of the floating body would move toward the center of buoyancy when the ship

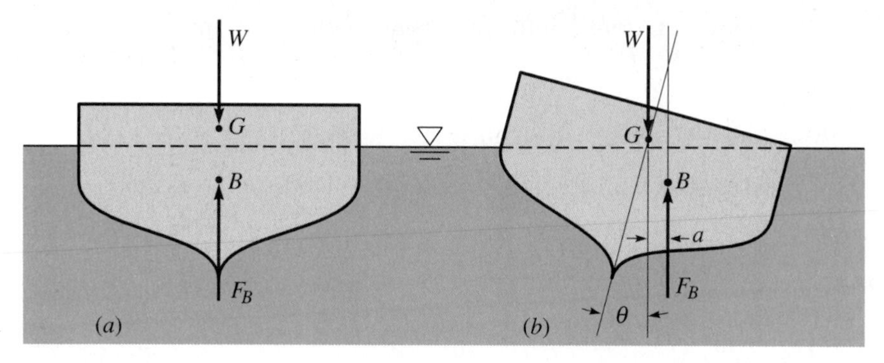

Figure 3.23

rolled, thus decreasing the righting couple and the stability. For this reason, liquid ballast or fuel oil in floating vessels should be stored in tanks or bulkheaded compartments.

Sample Problem 3.9 A pontoon, 15 ft long, 9 ft wide, and 4 ft high is built of uniform material, $\gamma = 45 \text{ lb/ft}^3$. (*a*) How much of it is submerged when floating in water? (*b*) If it is tilted about its long axis by an applied couple (no net force), to an angle of 12°, what will be the moment of the righting couple?

Solution

(*a*) Floating level, let d = the depth of submergence. Then

$$W = F_B; \quad 15(9)4(45) = 15(9)d(62.4); \quad d = 2.885 \text{ ft} \qquad \textit{ANS}$$

(*b*) At 12° tilt, let *AD* be the water line (see Fig. S3.9).

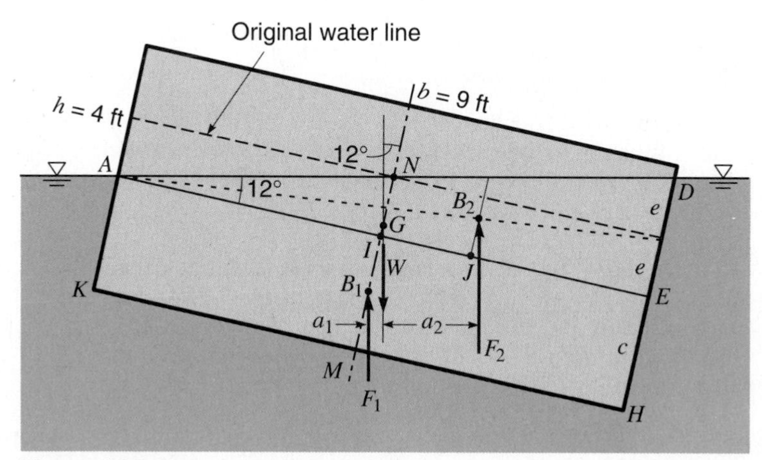

Figure S3.9

Divide the buoyancy force into two components B_1 and B_2, due to the rectangular block $AEHK$ and the triangular prism ADE of displaced water respectively.

$$DE = 2e = b \tan 12° = 9 \tan 12° = 1.913 \text{ ft}; \qquad NI = e = 0.957 \text{ ft}$$

As there is no net force, $MN = d = 2.885$ ft. Therefore

$$c = IM = MN - NI = 2.885 - 0.957 = 1.928 \text{ ft}$$

B_1 is at the centroid of the block $AEHK$, so

$$GB_1 = \tfrac{1}{2}(h - c) = \tfrac{1}{2}(4 - 1.928) = 1.036 \text{ ft}; \qquad a_1 = GB_1 \sin 12° = 0.215 \text{ ft}$$

$$F_1 = \gamma Lbc = 45(15)9(1.928) = 11{,}710 \text{ lb}$$

B_2 is at the centroid of the triangle ADE, so

$$JE = b/3, \qquad IJ = b/6 = 1.5 \text{ ft}, \qquad B_2J = \tfrac{2}{3}e = 0.638 \text{ ft}$$

G is at the centroid of the major rectangle, so $MG = h/2 = 2$ ft,

$$GI = MG - MI = MG - c = 2 - 1.928 = 0.0719 \text{ ft}$$

$$a_2 = IJ \cos 12° + (B_2J - GI) \sin 12° = 1.585 \text{ ft}$$

$$F_2 = \gamma Lbe = 45(15)9(0.957) = 5810 \text{ lb}$$

Counterclockwise moments about G:

$$\text{Righting moment} = F_2a_2 - F_1a_1 = 5810(1.585) - 11{,}710(0.215)$$

$$= 6690 \text{ lb·ft} \qquad \textit{ANS}$$

EXERCISES

3.10.1 An iceberg in the ocean floats with one-eighth of its volume above the surface. What is its specific gravity relative to ocean water, which weighs 64 lb/ft^3? What portion of its volume would be above the surface if the ice were floating in pure water?

3.10.2 A hydrometer (Fig. 3.22*a*) consists of a 6-mm-diameter cylinder of length 180 mm attached to a 20-mm-diameter weighted sphere. The cylinder has a mass of 0.6 g and the mass of the sphere is 6.4 g. At what level will this device float in liquids having specific gravities 0.8, 1.0, and 1.2? Is the scale spacing on the cylindrical stem uniform? Why or why not?

3.10.3 Determine the volume of an object that weighs 5.5 lb in water and 8 lb in oil ($s = 0.82$). What is the specific weight of the object?

3.10.4 A balloon weighs 160 lb and has a volume of 7200 ft^3. It is filled with helium, which weighs 0.0112 lb/ft^3 at the temperature and pressure of the air, which in turn weighs 0.0807 lb/ft^3. What load will the balloon support, or what force in a cable would be required to keep it from rising?

3.10.5 For the conditions shown in the Fig. X3.10.5, find the force F required to lift the concrete-block gate if the concrete weighs 23.6 kN/m^3. Neglect friction.

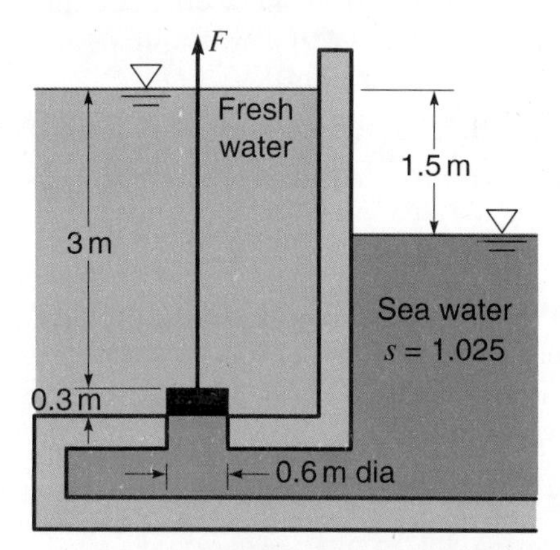

Figure X3.10.5

3.10.6 An 8-in-diameter solid cylinder of height 3 in weighing 3.4 lb is immersed in liquid ($\gamma = 52$ lb/ft^3) contained in a tall, upright metal cylinder having a diameter of 9 in (Fig. X3.10.6). Before immersion, the liquid was 3 in deep. At what level will the solid cylinder float? Find the distance z between the bottoms of the two cylinders.

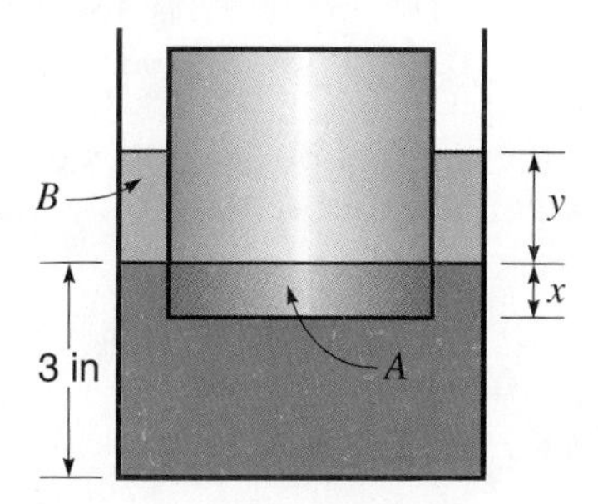

Figure X3.10.6

3.10.7 A cylindrical bucket 250 mm in diameter and 400 mm high weighing 20.0 N contains oil ($s = 0.80$) to a depth of 180 mm. (a) When placed to float in water, what will be the immersion depth to the bottom of the bucket? (b) What is the maximum volume of oil the bucket can hold and still float?

3.10.8 A metal block 1.5 ft square and 1 ft deep is floated on a body of liquid consisting of a 10-in-layer of water above a layer of mercury. The block weighs 120 lb/ft^3. What is the position of the bottom of the block? If a downward vertical force of 600 lb is applied to the center of this block, what is the new position of the bottom of the block? Assume that the tank containing the fluid is of infinite dimensions.

3.10.9 Two spheres, each 1.5 m in diameter, weigh 8 and 24 kN respectively. They are connected with a short rope and placed in water. (*a*) What is the tension in the rope and what portion of the lighter sphere's volume protrudes from the water? (*b*) What should be the weight of the heavier sphere in order for the lighter sphere to float halfway out of the water? Assume that the sphere volumes remain constant.

3.10.10 A solid, half-cylinder-shaped log, 1.50 ft radius and 10 ft long, floats in water with the flat face up (Fig. X3.10.10). (*a*) If the draft (immersion depth of the lowest point) is 0.90 ft, what is the uniform specific weight of the log? (*b*) The log tilts about its axis (zero net applied force) by less than 23°. Is it in stable equilibrium? Justify your answer with a sketch and logic. (*c*) If the log tilts by 20° (right side down; zero net applied force), what is the magnitude and sense of any moment that results?

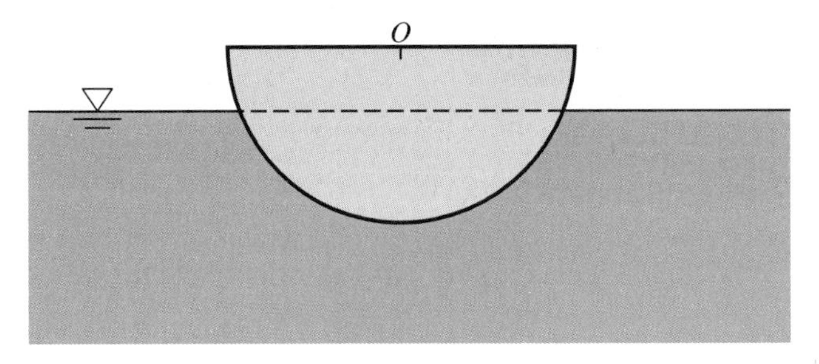

Figure X3.10.10

3.10.11 A solid, half-cylinder-shaped log, 0.48 m in radius and 2.5 m long, floats in water with the flat face up (see Fig. X3.10.10). (*a*) If the draft (immersion depth of the lowest point) is 0.30 m, what is the uniform specific weight of the log? (*b*) The log tilts about its axis (zero net applied force) by less than 22°. Is it in stable equilibrium? Justify your answer with a sketch and logic. (*c*) If the log tilts by 18° (left side down; zero net applied force), what is the magnitude and sense of any moment that results?

3.11 FLUID MASSES SUBJECTED TO ACCELERATION

Under certain conditions there may be no relative motion between the particles of a fluid mass yet the mass itself may be in motion. If a body of fluid in a tank is transported at a uniform velocity, the conditions are those of ordinary fluid statics. But if it is subjected to acceleration, special treatment is required. Consider the case of a liquid mass in an open tank moving horizontally with a linear acceleration a_x, as shown in Fig. 3.24*a*. A free-body

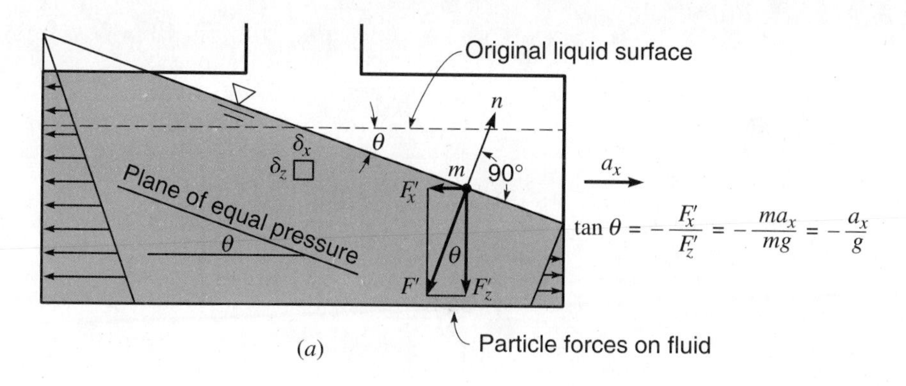

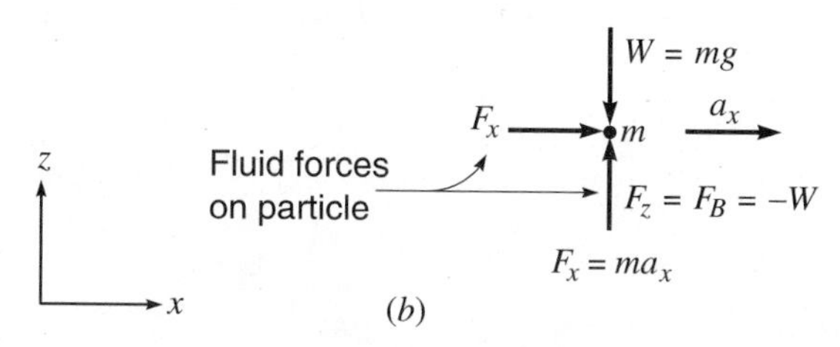

Figure 3.24
Liquid mass subjected to horizontal acceleration.

diagram (Fig. 3.24*b*) of a small particle (mass m) of liquid on the surface indicates that the forces exerted by the surrounding fluid on the particle are such that $F_z = F_B = -W$ and $F_x = ma_x$. F_z counterbalances W, so there is no acceleration in the z direction. F_x is the force required to produce acceleration a_x of the particle. Equal and opposite to these forces are F'_x and F'_z of Fig. 3.24*a*, the forces exerted by the particle on the surrounding fluid. The resultant of these forces is F'. The liquid surface must be at right angles to F', for if it were not, the particle would not maintain its fixed relative position in the liquid. Hence (Fig. 3.24*a*) $\tan\theta = -a_x/g$. The liquid surface and all other planes of equal hydrostatic pressure must be inclined at angle θ with the horizontal as shown in Fig. 3.24*a*.

Next consider the more general case where a fluid mass is subject to acceleration in both the x and z directions. Figure 3.25 shows a free-body diagram of an elemental cube of fluid, volume $\delta x\ \delta y\ \delta z$, at the center of which the pressure is p. Applying the equation of motion in the x-direction,

$$\sum F_x = ma_x$$

$$\left(p - \frac{\partial p}{\partial x}\frac{\delta x}{2}\right)\delta y\ \delta z - \left(p + \frac{\partial p}{\partial x}\frac{\delta x}{2}\right)\delta y\ \delta z = \rho\ \delta x\ \delta y\ \delta z\ a_x$$

which reduces to

$$\frac{\partial p}{\partial x} = -\rho a_x \qquad (3.23)$$

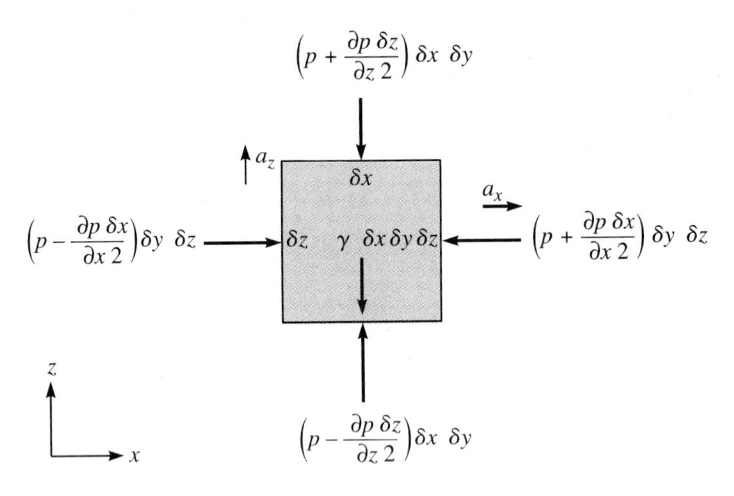

Figure 3.25

In the vertical direction $\qquad \sum F_z = ma_z$

$$\left(p - \frac{\partial p}{\partial z}\frac{\delta z}{2}\right) \delta x \; \delta y - \left(p + \frac{\partial p}{\partial z}\frac{\delta z}{2}\right) \delta x \; \delta y - \gamma \; \delta x \; \delta y \; \delta z = \rho \; \delta x \; \delta y \; \delta z \; a_z$$

where $\gamma = \rho g$ [Eq. (2.1)]. This yields

$$\frac{\partial p}{\partial z} = -\rho(a_z + g) \qquad (3.24)$$

Equations (3.23) and 3.24) can be employed to develop a generalization applicable to a fluid mass that is subject to acceleration in the x and z directions. The chain rule for the total differential of dp in terms of its partial drivatives is

$$dp = \frac{\partial p}{\partial x} dx + \frac{\partial p}{\partial z} dz$$

Substituting the expressions for $\partial p/\partial x$ and $\partial p/\partial z$ from Eqs. (3.23) and (3.24) gives

$$dp = -\rho(a_x) \, dx - \rho(a_z + g) \, dz \qquad (3.25)$$

Along a line of constant pressure, $dp = 0$. From Eq. (3.25), if $dp = 0$,

$$\frac{dz}{dx} = -\frac{a_x}{a_z + g} \qquad (3.26)$$

This then defines the slope $dz/dx = \tan\theta$ of a line of constant pressure within the accelerated fluid mass. The fluid surface is one such line.

From Eqs. (3.23) and (3.24), we may obtain the resultant of $\partial p/\partial x$ and

$\partial p/\partial z$, namely

$$\frac{\partial p}{\partial n} = -\rho\sqrt{a_x^2 + (a_z + g)^2} \tag{3.27}$$

where n is at right angles to the lines of equal pressure and in the direction of decreasing pressure (Fig. 3.24*a*). When $a_x = a_z = 0$, this equation reduces to $\partial p/\partial n = -\rho g = -\gamma$, which is essentially the same as the basic hydrostatic equation (3.2). Application of Eq. (3.27) indicates that, if fluid in a container is subjected to an upward acceleration, this increases pressures within the fluid; downward acceleration decreases them.

SAMPLE PROBLEM 3.10 At a particular instant an airplane is traveling upward at a velocity of 180 m/s in a direction that makes an angle of 40° with the horizontal. At this instant the airplane is losing speed at the rate of 4 m/s². Also it is moving on a concave-upward circular path having a radius of 2600 m. Determine for the given conditions the slope of the free liquid surface in the fuel tank of this vehicle.

Solution

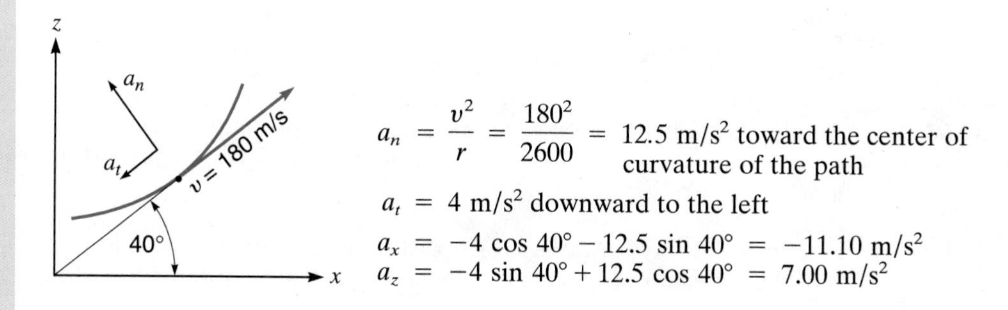

$$a_n = \frac{v^2}{r} = \frac{180^2}{2600} = 12.5 \text{ m/s}^2 \text{ toward the center of curvature of the path}$$

$$a_t = 4 \text{ m/s}^2 \text{ downward to the left}$$

$$a_x = -4\cos 40° - 12.5\sin 40° = -11.10 \text{ m/s}^2$$

$$a_z = -4\sin 40° + 12.5\cos 40° = 7.00 \text{ m/s}^2$$

Consider a liquid particle of mass m at the free surface. To achieve a_x and a_z, the forces exerted by the surrounding liquid on the liquid particle are in the directions shown:

m

$F_x = ma_x$

$F_z = ma_z$

$F_B = -W$

Forces exerted by the surrounding liquid on the liquid particle at the free surface

Equal and opposite to these forces are the forces F_x', F_z' and W exerted by the liquid particle on the surrounding liquid. Thus

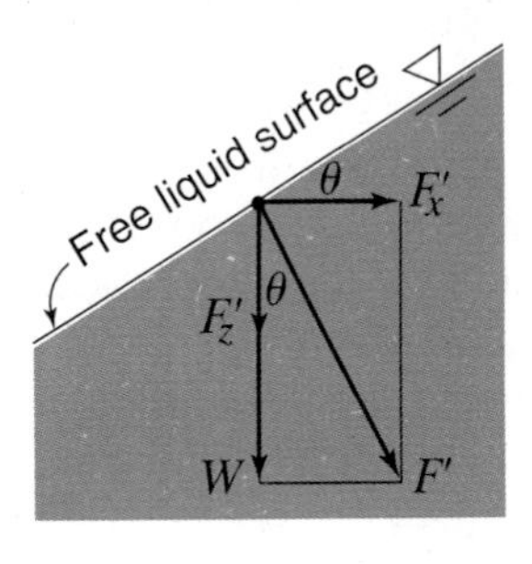

Forces exerted by the liquid particle at the free surface on the surrounding liquid

$$\tan\theta = -\frac{F_x'}{F_z' + W} = -\frac{ma_x}{ma_z + mg} = -\frac{a_x}{a_z + g} = -\frac{(-11.10)}{7.00 + 9.81}$$

$$\theta = \tan^{-1}(+0.660) = 33.4° \qquad \textit{ANS}$$

Or, alternatively,

Eq. (3.26): $\dfrac{dz}{dx} = -\left(\dfrac{-11.10}{7.00 + 9.81}\right) = +0.660$, slope of the free surface. ***ANS***

EXERCISES

3.11.1 What would be the hydrostatic gage pressure at a depth of 9 in in a bucket of oil ($s = 0.88$) that is in an elevator being accelerated upward at 12 ft/s^2?

3.11.2 What would be the hydrostatic gage pressure at a depth of 300 mm in a bucket of oil ($s = 0.86$) that is in an elevator being accelerated upward at 2.5 m/s^2?

3.11.3 A tank containing water to a depth of 5 ft is accelerated upward at 8 ft/s^2. Calculate the pressure on the bottom of the tank.

3.11.4 A tank containing water to a depth of 2.5 m is accelerated upward at 3.6 m/s^2. Calculate the pressure on the bottom of the tank.

3.11.5 Suppose the tank shown in Fig. 3.24 is rectangular and completely open at the top, It is 16 ft long, 6 ft wide, and 5 ft deep. If it is initially filled to the top, how much liquid will be spilled if it is given a horizontal acceleration $a_x = 0.25g$ in the direction of its length?

3.11.6 Suppose the tank of Fig. 3.24 is rectangular and completely open at the top. It is 18 m long, 5 m wide, and 4 m deep. If it is initially filled to the top, how much liquid will be spilled if it is given a horizontal acceleration $a_x = 0.4g$ in the direction of its length?

3.11.7 If the tank of Exer. 3.11.5 is closed at the top and is completely filled, what will be the pressure difference between the left-hand end at the top and the right-hand end at the top if the liquid has a specific weight of 50 lb/ft^3 and the horizontal acceleration is $a_x = 0.3g$? Sketch planes of equal pressure, indicating their magnitude; assume zero pressure in the upper right-hand corner.

3.11.8 If the tank of Exer. 3.11.6 is closed at the top and is completely filled, what will be the pressure difference between the left-hand end at the top and the right-hand end at the top if the liquid has a specific weight of 8.0 kN/m^3 and the horizontal acceleration is $a_x = 0.3g$? Sketch planes of equal pressure, indicating their magnitude; assume zero pressure in the upper right-hand corner.

PROBLEMS

3.1 Repeat Exer. 3.2.1, but consider the effects of compressibility (E_v = 300,000 psi). Neglect changes in density caused by temperature variations. (*Hint:* As a starting point, express Eq. (2.3) in terms of γ and integrate to determine γ as a function of z.)

3.2 On a certain day the barometric pressure at sea level is 30.0 inHg and the temperature is 60°F. The pressure gage on an airplane flying overhead indicates that the atmospheric pressure at that point is 9.7 psia and that the air temperature is 42°F. Calculate as accurately as you can the height of the airplane above sea level. Assume a linear decrease of temperature with elevation.

3.3 The tire of an airplane is inflated at sea level to 65 psi. Assuming the tire does not expand, what is the pressure within the tire at elevation 35,000 ft? Assume standard atmosphere. Express the answer in psi and psia.

3.4 The tire of an airplane is inflated at sea level to 380 kPa. Assuming the tire does not expand, what is the pressure within the tire at elevation (*a*) 10 000 m; (*b*) 20 000 m? Assume standard atmosphere. Express answers in both kPA and kPA abs.

3.5 What would be the manometer reading in Sample Prob. 3.4 if $p_B - p_A$ = 120 kPa?

3.6 The diameter of tube C in Fig. 3.11 is d_1, and that of tube B is d_2. Let z_0 be the elevation of the mercury above A when both mercury columns are at the same level. R is the distance the right-hand column of mercury rises above z_0 when the fluid in A is under pressure. Let γ' be the specific weight of the mercury (or any other measuring fluid), while γ is the specific weight of the fluid in A and the connecting tubing. Prove that

$$p_A = \gamma z_0 + \left[\gamma' + (\gamma' - \gamma)\left(\frac{d_1}{d_2}\right)^2\right]R = M + NR$$

where M and N are constants. It is seen that this equation involves only one variable, which is the reading R on the scale for column C. It also shows the significance of having d_2 large compared with d_1.

3.7 At a certain point the pressure in a pipeline containing gas ($\gamma = 0.05\ \text{lb/ft}^3$) is 5.6 in of water. The gas is not flowing. All temperatures are 60°F. What is the pressure in inches of water at another point in the line whose elevation is 650 ft greater than the first point? Make and state clearly any necessary assumptions.

3.8 A vertical semicircular area has its diameter in a liquid surface. Derive an expression for the depth to its center of pressure.

3.9 Repeat Exer. 3.8.6 for the case where a 2-ft-thick layer of oil ($s = 0.8$) is resting on the water, and replace "water surface" by "oil surface."

3.10 The Utah-shaped plate shown in Fig. P3.10 is submerged in oil ($s = 0.94$) and lies in a vertical plane. Find the magnitude and location of the hydrostatic force acting on one side of the plate.

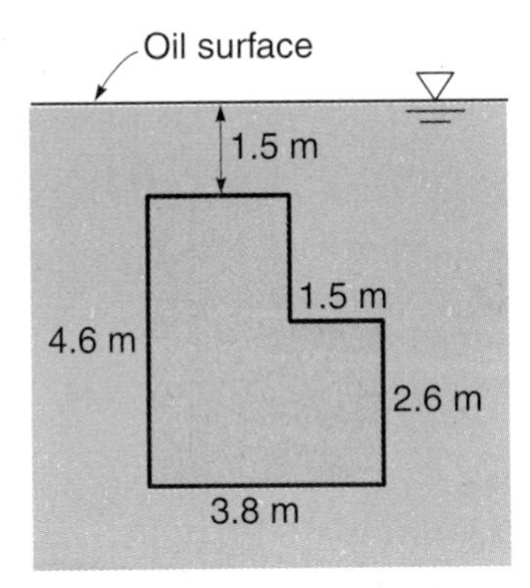

Figure P3.10

3.11 The common type of irrigation head gate shown in Fig. P3.11 is a plate that slides over the opening to a culvert. The coefficient of friction between the gate and its sliding ways is 0.6. Find the force required to slide open this 600-lb gate if it is set (*a*) vertically; (*b*) on a 2:1 slope ($n = 2$), as is common.

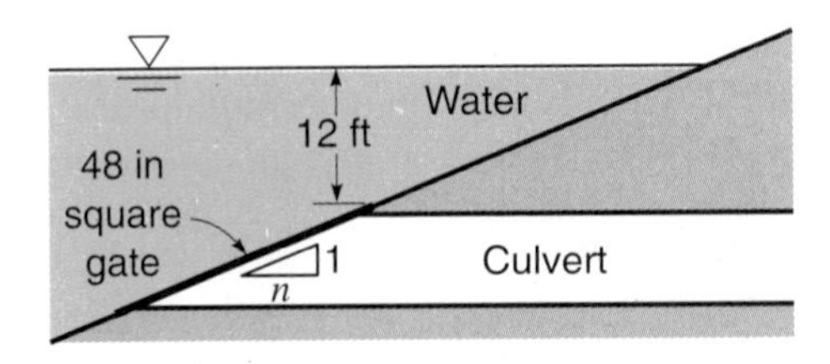

Figure P3.11

3.12 In the drainage of irrigated lands it is frequently desirable to install automatic flap gates to prevent a flood from backing up into the lateral drains from a river. Suppose a 50-in-square flap gate, weighing 1750 lb, is hinged 35 in above the center, as shown in Fig. P3.12, and the face is sloped 4° from the vertical. Find the depth to which water will rise behind the gate before it will open.

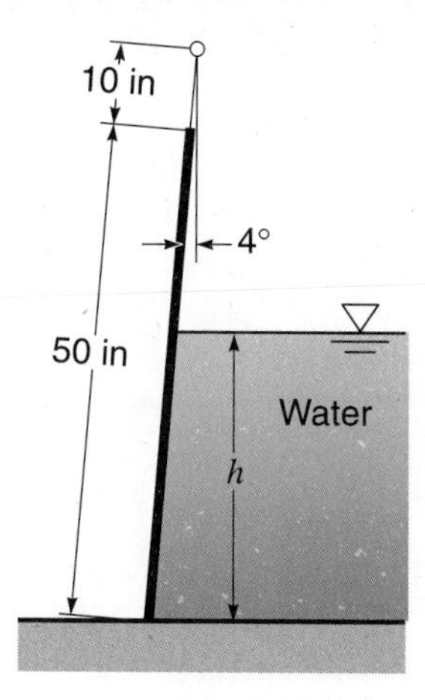

Figure P3.12

3.13 Find the minimum value of z for which the gate in Fig. P3.13 will rotate counterclockwise if the gate is (*a*) rectangular, 5 ft by 4 ft; (*b*) triangular, 4 ft base as axis, height 5 ft. Neglect friction in bearings.

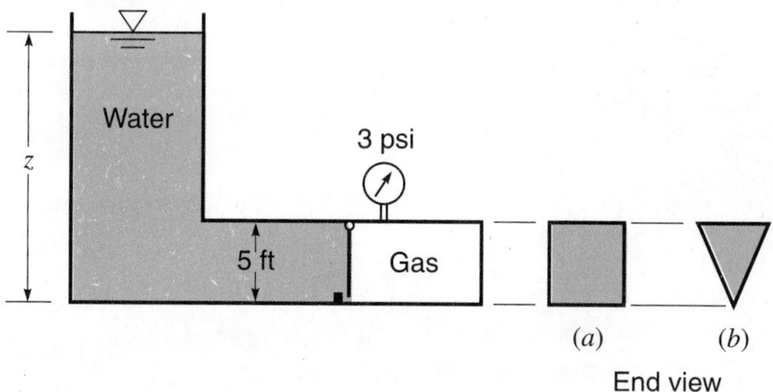

Figure P3.13

3.14 Referring to Fig. P3.14, what minimum value of b is necessary to keep the rectangular masonry wall from sliding if it weighs $160\,\text{lb/ft}^3$ and the coefficient of friction is 0.45? With this minimum b value, will it also be safe against overturning? Assume that water does not get underneath the block.

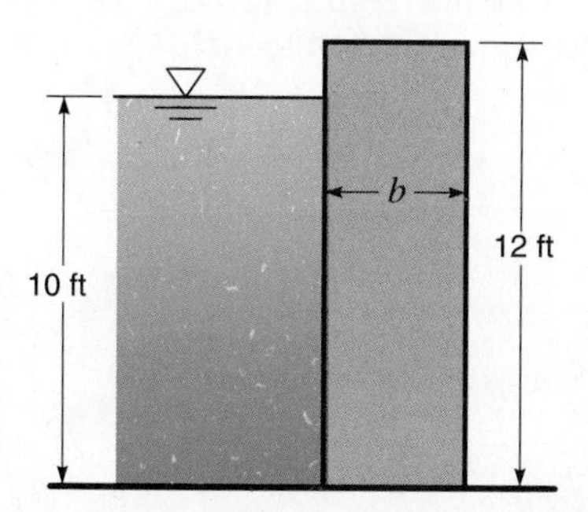

Figure P3.14

3.15 (*a*) Find horizontal and vertical forces per foot of width on the Tainter gate shown in Fig. P3.15. (*b*) Locate the horizintal force and indicate the line of action of the vertical force without actually computing its location. (*c*) Locate the vertical force (*hint:* consider the resultant).

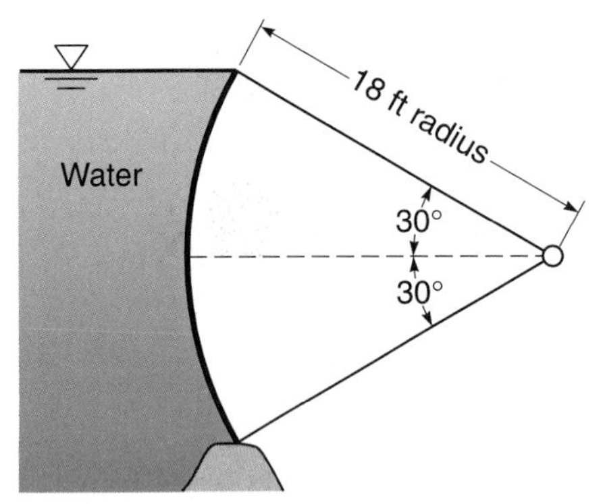

Figure P3.15

3.16 A tank has an irregular cross section as shown in Fig. P3.16. Determine as accurately as possible the magnitude and location of the horizontal- and vertical-force components on a one-foot length of the wall *ABCD* when the tank contains water to a depth of 8 ft. To determine areas, use a planimeter or count squares (1 ft grid); make a cardboard cutout, or take approximate moments of the squares, to locate the centroid.

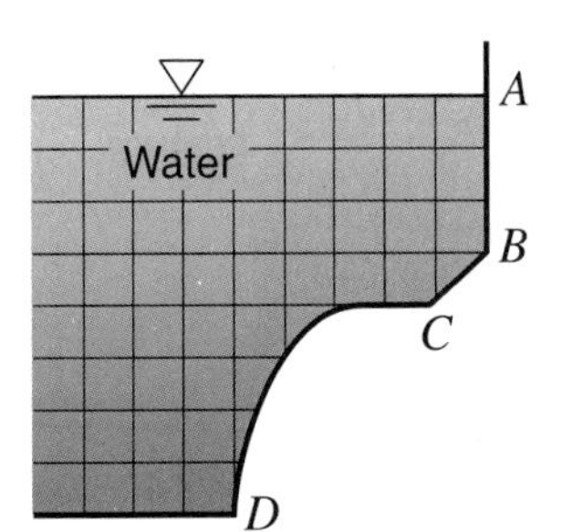

Figure P3.16

3.17 Repeat Exer. 3.9.4 where the tank is closed and contains 4 ft of water overlain with a gas that is under a pressure of 0.8 psi.

3.18 The cross-section of a gate is shown in Fig. P3.18. Its dimension normal to the plane of the paper is 8 m, and its shape is such that $x = 0.2y^2$. The gate is pivoted about *O*. Develop analytic expressions in terms of the water depth *y* upstream of the gate for the following: (*a*) horizontal force; (*b*) vertical force; (*c*) clockwise moment acting on the gate. Compute (*a*), (*b*), and (*c*) for the case where the water depth is 2.5 m.

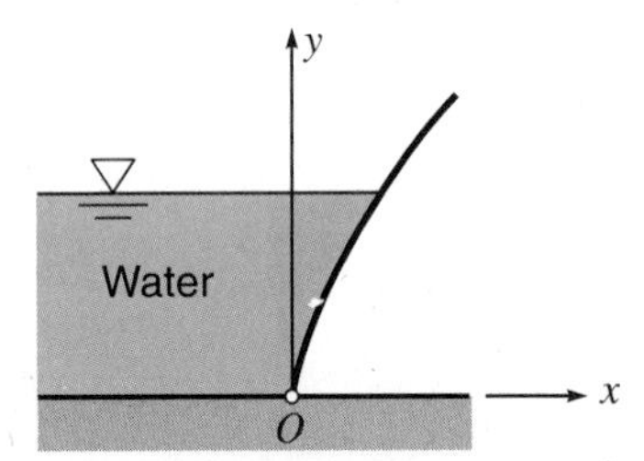

Figure P3.18

3.19 Find the approximate value of the maximum specific gravity of liquid for which the device of Exer. 3.10.2 will be stable.

3.20 A 2.0-ft^3 object weighing 650 lb is attached to a balloon of negligible weight and released in the ocean ($\gamma = 64\ lb/ft^3$). The balloon was originally inflated with 5.0 lb of air to a pressure of 20 psi. To what depth will the balloon sink? Assume that air temperature within the balloon stays constant at 50°F.

3.21 Work Prob. 3.20 with all data the same except assume the balloon was originally inflated with 5.0 lb of air to a pressure of 10 psi. In this latter case the balloon is more elastic because a lower pressure is obtained with the same amount of air.

3.22 A wooden pole weighing 2 lb/ft has a cross-sectional area of 6.7 in^2 and is supported as shown in Fig. P3.22. The hinge is frictionless. Find θ.

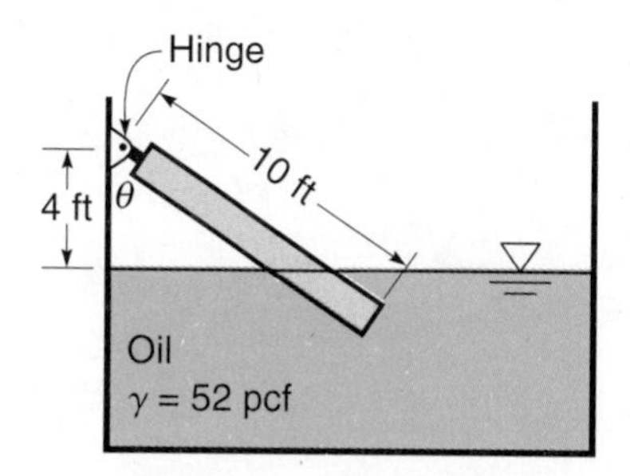

Figure P3.22

3.23 A rectangular block of uniform material and length $L = 3$ ft, width $b = 1$ ft, and depth $d = 0.20$ ft, is floating in a liquid. It assumes the position shown in Fig. P3.23 when a uniform vertical load of 1.30 lb/ft is applied at P. (*a*) Find the weight of the block. (*b*) If the load is suddenly removed, what is the righting moment before the block starts to move? (*Hint:* Refer also to Fig. 3.19.)

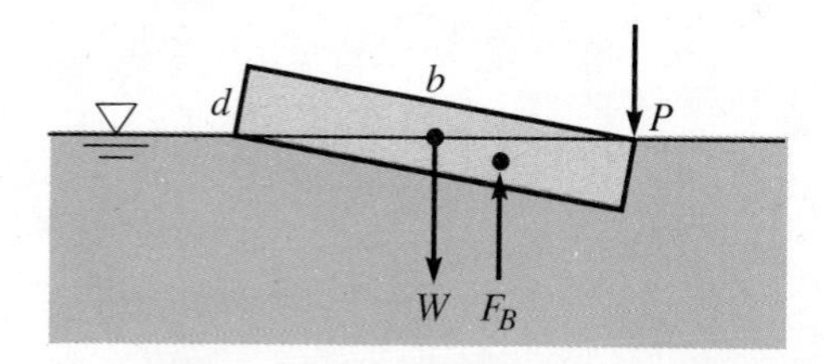

Figure P3.23

3.24 A rectangular block of uniform material and length $L = 800$ mm, width $b = 250$ mm, and depth $d = 50$ mm, is floating in a liquid. It assumes the position shown in Fig. P3.23 when a uniform vertical load of 20 N/m is applied at P. (*a*) Find the weight of the block. (*b*) If the load is suddenly removed, what is the righting moment before the block starts to move? (*Hint:* Refer also to Fig. 3.19.)

3.25 A solid block, 4 in wide by 4 in deep and 3 in high weighs 0.90 lb. It floats in liquid ($\gamma = 55$ lb/ft^3) inside a cubic container of side 5 in. Before immersion the liquid was 2 in deep. (*a*) At what level will the block float? Find the distance z from the bottom of the block to the bottom of the container. (*b*) If the block is tilted by a couple (no net force) to an angle of 15° so that two sides remain vertical, what will be the righting moment in in·lb?

3.26 At a particular instant an airplane is traveling upward at a velocity of 180 mph in a direction that makes an angle of 30° with the horizontal. At this instant the airplane is losing speed at the rate of 3.6 mph/sec. Also, it is moving on a concave-upward circular path having a radius of 5000 ft. Determine for the given conditions the slope of the free liquid surface in the fuel tank of this vehicle.

3.27 Refer to Sample Prob. 3.10. Suppose the velocity of the airplane is 220 m/s, with all other data unchanged. What then would be the slope of the liquid surface in the tank?

CHAPTER 4

Basics of Fluid Flow

In this chapter we shall deal with fluid velocities and accelerations and their distributions in space without consideration of any forces involved. As noted in Sec. 1.1, the subject that deals with velocities and flow paths without considering forces or energy is known as ***kinematics.***

Because only certain types of flow can be treated by the methods of kinematics, and because there are many different types of flow, we summarize these first to provide perspective. Some related concepts, most notably the control volume and the flow net, are also introduced.

4.1 TYPES OF FLOW

When speaking of fluid flow, we often refer to the flow of an ***ideal fluid*** (Sec. 2.10). Such a fluid is presumed to have no viscosity. This is an idealized situation that does not exist; however,, there are instances in engineering problems where the assumption of an ideal fluid is helpful. When we refer to the flow of a ***real fluid***, the effects of viscosity are introduced into the problem. This results in the development of shear stresses between neighboring fluid particles when they are moving at different velocities. In the case of an ideal fluid flowing in a straight conduit, all particles move in parallel lines with equal velocity (Fig.4.1*a*). In the flow of a real fluid the velocity adjacent to the wall will be zero; it will increase rapidly within a short distance from the wall and produce a velocity profile such as shown in Fig. 4.1*b*.

Flow may also be classified as that of an ***incompressible*** or ***compressible***

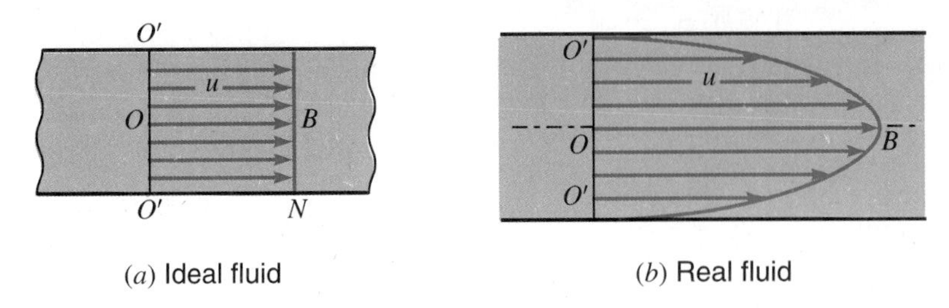

Figure 4.1
Typical velocity profiles. (*a*) Ideal fluid. (*b*) Real fluid

fluid. Since liquids are relatively incompressible, they are generally treated as wholly incompressible fluids. Under particular conditions where there is little pressure variation, the flow of gases may also be considered incompressible, though generally the effects of the compressibility of the gas should be considered. Some of the basic concepts governing the flow of compressible fluids are discussed in Chap. 13.

In addition to the flow of different types of fluids, i.e., real, ideal, incompressible, and compressible, there are various classifications of flow. Flow may be ***steady*** or ***unsteady*** with respect to time (see Sec. 4.3). It may be ***laminar*** or ***turbulent***, as discussed in the following section. Other classifications of flow include ***rotational*** or ***irrotational*** (Chap. 14), ***supercritical*** or ***subcritical*** (Chap, 10), etc. These and other common ways in which flow can be classified are listed in Table 4.1, in many cases with definitions.

4.2 LAMINAR AND TURBULENT FLOW

Whether laminar or turbulent flow occurs in a given situation, or how much of each occurs, is very important because of the strongly different effects these two different types of flow have on a variety of flow features, including on energy losses, velocity profiles, and mixing of transported materials.

That there are two distinctly different types of fluid flow was demonstrated in 1883 by Osborne Reynolds (see Sec. 7.4). He injected a fine, threadlike stream of colored liquid having the same density as water at the entrance to a large glass tube through which water was flowing from a tank. A valve at the discharge end permitted him to vary the flow. When the velocity in the tube was small, this colored liquid was visible as a straight line throughout the length of the tube, thus showing that the particles of water moved in parallel straight lines. As the velocity of the water was gradually increased by opening the valve further, there was a point at which the flow changed. The line would first become wavy, and then at a short distance from

Table 4.1
Classification of types of flow[a]

One-dimensional, two-dimensional *or* ***three-dimensional flow***
See Sec. 4.8 for discussion.

Real fluid flow or ***ideal fluid flow*** (also referred to as ***viscid*** and ***inviscid flow***)
Real fluid flow implies frictional (viscous) effects. Ideal fluid flow is hypothetical; it assumes no friction (i.e., viscosity of fluid = 0).

Incompressible fluid flow or ***compressible fluid flow***
Incompressible fluid flow assumes the fluid has constant density (ρ = constant). Though liquids are slightly compressible they are usually assumed to be incompressible. Gases are compressible; their density is a function of absolute pressure and absolute temperature [$\rho = f(p, T)$].

Steady or ***unsteady flow***
Steady flow means steady with respect to time. Thus all properties of the flow at every point remain constant with respect to time. In unsteady flow, the flow properties at a point change with time.

Pressure flow or ***gravity flow***
Pressure flow implies that flow occurs under pressure. Gases always flow in this manner. When a liquid flows with a free surface (for example, a partly full pipe) the flow is referred to as gravity flow, because gravity is the primary moving force. Liquids also flow under pressure (for example, a pipe flowing full).

Spatially constant or ***spatially variable flow***
Spatially constant flow occurs when the fluid density and the local average flow velocity are identical at all points in a flow field. If these quantities change along or across the flow lines, the flow is spatially variable. Examples of different types of spatially varied flow includes the local flow field around an object, flow through a gradual contraction in a pipeline, and the flow of water in a uniform gutter of constant slope receiving inflow over the length of the gutter.

Laminar or ***turbulent flow***
See Sec. 4.2 for a discussion of the difference between these two types of flow.

Established or ***unestablished flow***
This is discussed in Sec. 8.7.

Uniform or ***varied flow***
This classification is ordinarily used when dealing with open-channel (gravity) flow (Chap. 10). In uniform flow the cross section (shape and area) through which the flow occurs remains constant.

Subcritical or ***supercritical flow***
This classification is used with open-channel flow (Chap. 10)

Subsonic or ***supersonic flow***
This classification is used with compressible flow (Chap. 13).

Rotational or ***irrotational flow***
This is used in mathematical hydrodynamics (Chap. 14).

Other classifications of flow include ***converging*** or ***diverging***, ***disturbed***, ***isothermal*** (constant temperature), ***adiabatic*** (no heat transfer) and ***isentropic*** (frictionless adiabatic).

[a] Note that in a given situation these different types of flow may occur in combination. For example, flow of a liquid in a pipe is usually considered to be one-dimensional, incompressible, real fluid flow that may be steady or unsteady, laminar or turbulent. Such flow is commonly spatially constant and established.

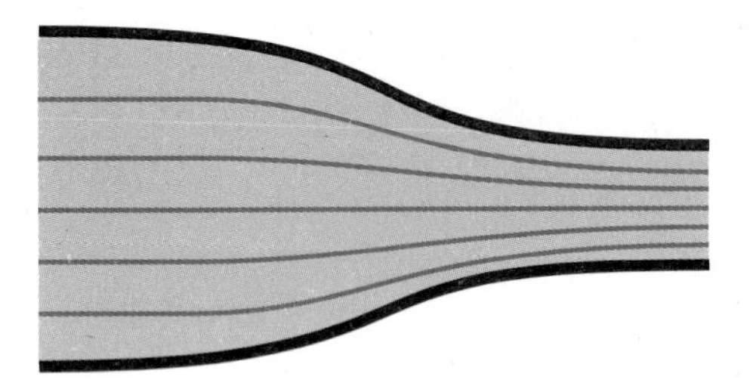

Figure 4.2

the entrance it would break into numerous vortices beyond which the color would be uniformly diffused so that no streamlines could be distinguished. Later observations have shown that in this latter type of flow the velocities are continuously subject to irregular fluctuations.

The first type is known as ***laminar, streamline***, or ***viscous*** flow. The significance of these terms is that the fluid appears to move by the sliding of laminations of infinitesimal thickness over adjacent layers, with relative motion of fluid particles occurring at a molecular scale; that the particles move in definite and observable paths or streamlines, as in Fig. 4.2; and also that the flow is characteristic of a viscous fluid or is one in which viscosity plays a significant part (Fig. 2.5 and Sec. 2.11).

The second type is known as ***turbulent*** flow, and is illustrated in Fig. 4.3, where (*a*) represents the irregular motion of a large number of particles during a very brief time interval, while (*b*) shows the erratic path followed by a single particle during a longer time interval. A distinguishing characteristic of turbulence is its irregularity, there being no definite frequency as in wave action, and no observable pattern as in the case of large swirls.

Large swirls and irregular movements of large bodies of fluid, which can be traced to obvious sources of disturbances, do not constitute turbulence, but may be described as ***disturbed flow***. By contrast, turbulence may be found in what appears to be a very smoothly flowing stream and one in which there is no apparent source of disturbance. Turbulent flow is characterized by fluctuations in velocity at all points of the flow field (Figs. 4.6 and 8.6*b*). These fluctuations arise because the fluid moves as a number of small, discrete particles or "packets" called ***eddies***, jostling each other around in a

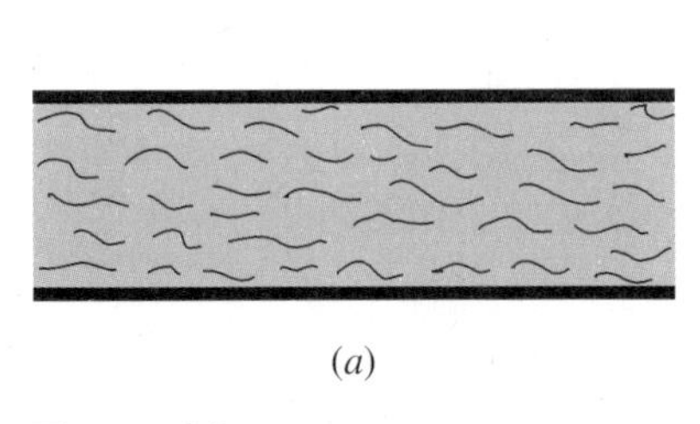

(*a*)

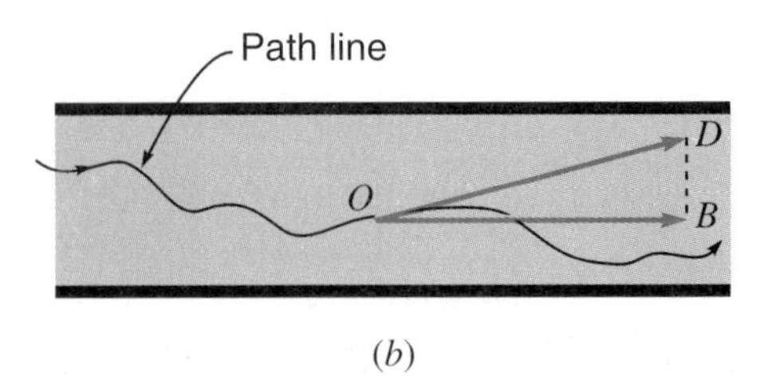

(*b*)

Figure 4.3
Turbulent flow

random manner. Although small, the smallest eddies are macroscopic in size, very much larger than the molecular sizes of the particles in laminar flow. The eddies interact with one another and with the general flow. They are the cause of the effective mixing action experienced with turbulent flow. Near boundaries eddies often rotate. They change shape and size with time as they move along with the flow. Each eddy dissipates its energy through viscous shear with its surroundings and eventually disappears. New eddies are continuously being formed. Large eddies (large-scale turbulence) have smaller eddies within them giving rise to small-scale turbulence. The resulting fluctuations in velocity are rapid and irregular and often can be detected only by a fast-acting probe such as a hot-wire or hot-film anemometer (Sec. 11.4).

At a certain instant the flow passing point O in Fig. 4.3*b* may be moving with the velocity OD. In turbulent flow OD will vary continuously both in direction and in magnitude. Fluctuations of velocity are accompanied by fluctuations in pressure, which is the reason why manometers or pressure gages attached to a pipe in which fluid is flowing usually show pulsations. In this type of flow an individual particle will follow a very irregular and erratic path, and no two particles may have identical or even similar motions. Thus a rigid mathematical treatment of turbulent flow is impossible, and instead statistical means of evaluation must be employed.

Criteria governing the conditions under which the flow will be laminar and those under which it will be turbulent are discussed in Sec. 8.2.

4.3 STEADY FLOW AND UNIFORM FLOW

A ***steady flow*** is one in which all conditions at any point in a stream remain constant with respect to *time,* but the conditions may be different at different points. A truly ***uniform flow*** is one in which the velocity is the same in both magnitude and direction at a given instant at every point in the fluid. Both of these definitions must be modified somewhat, since true steady flow is found only in laminar flow. In turbulent flow there are continual fluctuations in velocity and pressure at every point, as has been explained. But if the values fluctuate equally on both sides of a constant average value, the flow is called steady flow. However, a more exact definition for this case would be ***mean steady flow.***

Likewise, this strict definition of uniform flow can have little meaning for the flow of a real fluid where the velocity varies across a section, as in Fig. 4.1*b*. But when the size and shape of cross-section are constant along the length of channel under consideration, the flow is said to be ***uniform.***

Steady (or unsteady) and uniform (or nonuniform) flow can exist independently of each other, so that any of four combinations is possible. Thus the flow of liquid at a constant rate in a long straight pipe of constant diameter is ***steady uniform*** flow, the flow of liquid at a constant rate through

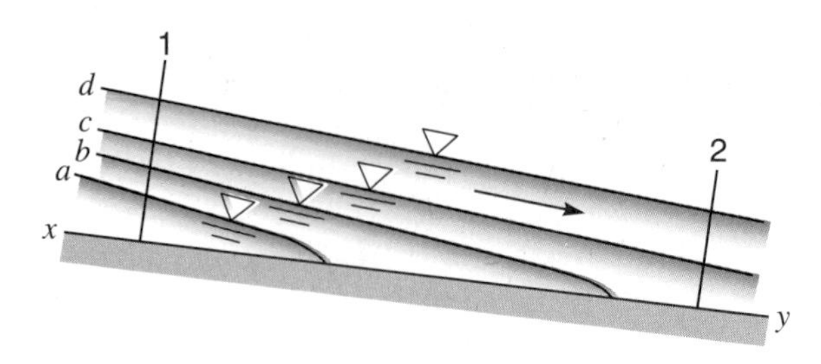

Figure 4.4
Unsteady flow in a canal

a conical pipe is ***steady nonuniform*** flow, while at a changing rate of flow these cases become ***unsteady uniform*** and ***unsteady nonuniform flow***, respectively.

Unsteady flow is a transient phenomenon, which may in time become either steady flow or zero flow. An example may be seen in Fig. 4.4, where (*a*) denotes the surface of a stream that has just been admitted to the bed of a canal by the sudden opening of a gate. After a time the water surface will be at (*b*), later at (*c*), and finally reaches equilibrium at (*d*). The unsteady flow has then become mean steady flow. Another example of transient phenomenon is when a valve is closed at the discharge end of a pipeline (Sec. 12.6), thus causing the velocity in the pipe to decrease to zero. In the meantime there will be fluctuations in both velocity and pressure within the pipe.

Unsteady flow may also include periodic motion such as that of waves on beaches, tidal motion in estuaries, and other oscillations. The difference between such cases and that of mean steady flow is that the deviations from the mean are very much greater and the time scale is also much longer.

EXERCISE

4.3.1 Classify the following cases of flow as to whether they are steady or unsteady, uniform or nonuniform: (*a*) water flowing from a tilted pail; (*b*) flow from a rotating lawn sprinkler; (*c*) flow through the hose leading to the sprinkler; (*d*) natural stream during dry-weather flow; (*e*) natural stream during flood; (*f*) flow in a city water-distribution main in a straight section of constant diameter and no side connections. (*Note*: There is room for legitimate argument in some of the above cases, which should stimulate independent thought.)

4.4 PATH LINES, STREAMLINES, AND STREAK LINES

A ***path line*** (Fig. 4.3*b*) is the trace made by a *single* particle over a *period* of time. If a camera were to take a time exposure of a flow in which a fluid particle was colored so it would register on the negative, the picture would show the course followed by the particle. This would be its path line. The path line shows the direction of the velocity of the particle at successive instants of time.

Streamlines show the mean direction of a *number* of particles at the *same* instant of time. If a camera were to take a very short time exposure of a flow in which there were a large number of particles, each particle would trace a short path, which would indicate its velocity during that brief interval. A series of curves drawn tangent to the means of the velocity vectors are streamlines.

Path lines and streamlines are identical in the steady flow of a fluid in which there are no fluctuating velocity components, in other words, for truly steady flow. Such flow may be either that of an ideal frictionless fluid or that of one so viscous and moving so slowly that no eddies are formed. This latter is the ***laminar*** type of flow, wherein the layers of fluid slide smoothly, one upon another. In turbulent flow, however, path lines and streamlines are not coincident, the path lines being very irregular while the streamlines are everywhere tangent to the local mean temporal velocity. The lines in Fig. 4.2 represent both path lines and streamlines if the flow is laminar; they represent only streamlines if the flow is turbulent.

In experimental fluid mechanics, a dye or other tracer is frequently injected into the flow to trace the motion of the fluid particles. If the flow is laminar, a ribbon of color results. This is called a ***streak line***, or ***filament line***. It is an instantaneous picture of the positions of all particles in the flow that have passed through a given point (namely, the point of injection). In utilizing fluid-tracer techniques it is important to choose a tracer with physical characteristics (especially density) the same as those of the fluid being observed. Thus the smoke rising from a cigarette, while giving the appearance of a streak line, does not properly represent the movement of the ambient air in the room because it is less dense than the air and therefore rises more rapidly.

4.5 FLOW RATE AND MEAN VELOCITY

The quantity of fluid flowing per unit time across any section is called the ***flow rate***. It may be expressed (1) in terms of ***volume flow rate (discharge)*** using BG units such as cubic feet per second (cfs), gallons per minute (gpm), million gallons per day (mgd), or (2) in terms of ***mass flow rate*** (slugs per second), or (3) ***weight flow rate*** (pounds per second). In SI units, cubic meters per second (m^3/s), kilograms per second (kg/s), and kilonewtons per second (kN/s) are fairly standard for expressing volume, mass, and weight flow rate respectively. In dealing with incompressible fluids, volume flow rate is commonly used, whereas weight flow rate or mass flow rate is more convenient with compressible fluids.

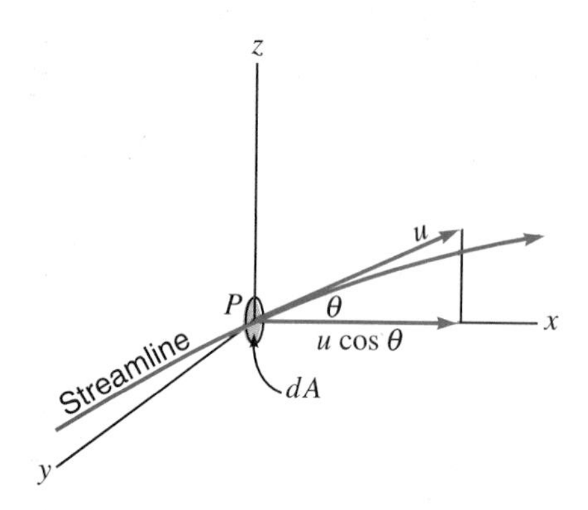

Figure 4.5

Figure 4.5 presents a streamline in steady flow lying in the xz plane. Element of area dA lies in the yz plane. The mean velocity at point P is u. The volume flow rate passing through the element of area dA is

$$dQ = \mathbf{u} \cdot d\mathbf{A} = (u \cos \theta)\, dA = u(\cos \theta\, dA) = u\, dA' \tag{4.1}$$

where dA' is the projection of dA on the plane normal to the direction of u. This indicates that the *volume flow rate is equal to the magnitude of the mean velocity multiplied by the flow area at right angles to the direction of the mean velocity*. The mass flow rate and the weight flow rate may be computed by multiplying the volume flow rate by the density and specific weight of the fluid respectively.

If the flow is turbulent, the ***instantaneous velocity component*** u_t along the streamline will fluctuate with time, even though the flow is nominally steady. A plot of u_t as a function of time is shown in Fig. 4.6. The average value of u_t over a period of time determines the time (temporal) mean value of velocity u at point P.

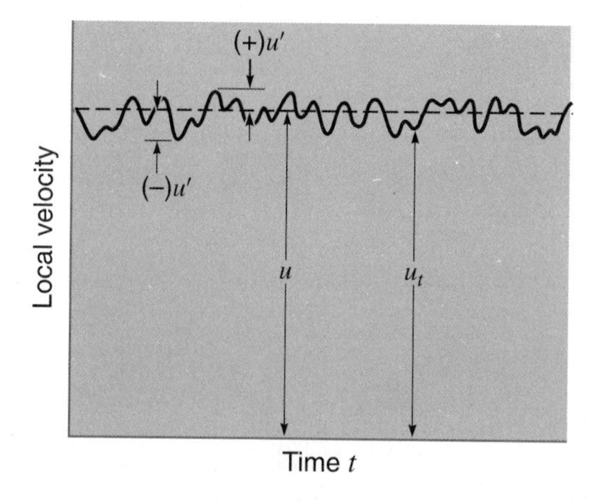

Figure 4.6
Fluctuating velocity at a point

The difference between u_t and u, which may be denoted by u', is called the ***turbulent fluctuation*** of this component; it may be either positive or negative. The ***time*** mean value of u' must be zero, as must the time means of all components transverse to the channel, such as BD in Fig. 4.3. Thus at any instant

$$u_t = u + u' \tag{4.2}$$

and u may be evaluated for any finite time t as $u = (1/t) \int_0^t u_t \, dt$.

In a real fluid the local time mean velocity u will vary across the section in some manner, such as that shown in Fig. 4.1*b* and hence the flow rate may be expressed as

$$Q = \int_A u \, dA = AV \tag{4.3}$$

or, for constant-density flow,

$$\dot{m} = \rho \int_A u \, dA = \rho AV = \rho Q \tag{4.4}$$

or

$$G = g\dot{m} = \gamma \int_A u \, dA = \gamma AV = \gamma Q \tag{4.5}$$

where u is the time mean velocity through an infinitesimal area dA, while V is the ***mean, or average, velocity*** over the entire sectional area A;[1] Q is the volume flow rate (cfs or m^3/s), $\dot{m}$ is the mass flow rate (slugs/s or kg/s),[2] and G is the weight flow rate (lb/s or kN/s).[3] If u is known as a function of A, the foregoing may be integrated. If only average values of V are known for different finite areas into which the total area may be divided, then

$$Q = A_a V_a + A_b V_b + \ldots + A_n V_n = AV \tag{4.6}$$

Similar expressions may be written for $\dot{m}$ and G. If the flow rate has been determined directly by some method, the mean velocity may be found by

$$V = \frac{Q}{A} = \frac{\dot{m}}{\rho A} = \frac{G}{\gamma A} \tag{4.7}$$

---

[1] Note that area A is defined by the surface at right-angles to the velocity vectors.

[2] Here, as used on m, and subsequently, the overdot represents the time derivative, as is standard practice.

[3] In Eqs. (4.4) and (4.5) the ρ and γ should be to the right of the integral sign if the density of the fluid varies across the flow.

SAMPLE PROBLEM 4.1 Air at 100°F and under a pressure of 40 psia flows in a 10-in-diameter ventilation duct at a mean velocity of 30 fps. Find the mass flow rate.

Solution. Table A.5 for air: $R = 1715$ ft·lb/(slug·°R)

From Eq. (2.4): $$\rho = \frac{p}{RT} = \frac{40(144)}{1715(460 + 100)} = 0.00600 \text{ slug/ft}^3$$

Eq. (4.4): $$\dot{m} = \rho AV = (0.00600)\frac{\pi}{4}\left(\frac{10}{12}\right)^2(30) = 0.0981 \text{ slug/sec} \qquad \textbf{\textit{ANS}}$$

EXERCISES

4.5.1 In the laminar flow of a fluid in a circular pipe the velocity profile is exactly a true parabola. The rate of discharge is then represented by the volume of a paraboloid. Prove that for this case the ratio of the mean velocity to the maximum velocity is 0.5.

4.5.2 A gas ($\gamma = 0.05$ lb/ft^3) flows at the rate of 1.0 lb/sec past section A through a long rectangular duct of uniform cross section 1.5 ft by 2 ft. At section B some distance along the duct the gas weighs 0.085 lb/ft^3. What is the average velocity of flow at sections A and B?

4.5.3 The velocity of a liquid ($s = 1.59$) in a 3-in pipeline is 1.8 fps. Calculate the rate of flow in cfs, gal/min, slugs/sec and lb/sec.

4.5.4 The velocity of a liquid ($s = 1.59$) in a 15-cm pipeline is 0.72 m/s. Calculate the rate of flow in L/s, m^3/s, kg/s, and kN/s.

4.5.5 Water flows at 4 gal/min through a small circular hole in the bottom of a large tank. Assuming the water in the tank approaches the hole radially, what is the velocity in the tank at 2, 4, and 8 in from the hole?

4.5.6 Water flows at 0.25 L/s through a small circular hole in the bottom of a large tank. Assuming the water in the tank approaches the hole radially, what is the velocity in the tank at 50, 100, and 200 mm from the hole?

4.6 SYSTEM AND CONTROL VOLUME

The concept of a free body diagram, as used in the statics of rigid bodies and in fluid statics (e.g., Fig. 3.1), is usually inadequate for the analysis of moving fluids. Instead, we frequently find the concepts of *system* and *control volume* to be useful in the analysis of fluid mechanics.

A ***fluid system*** refers to a *specific mass of fluid* within the boundaries defined by a closed surface. The shape of the system, and so the boundaries, may change with time, as when liquid flows through a constriction or when gas is compressed; as a fluid moves and deforms, so the system containing it moves and deforms. The size and shape of a system is entirely optional.

In contrast, a ***control volume*** refers to a *fixed region in space, which does not move or change shape* (Fig. 4.7). It is usually chosen as a region that fluid flows into and out of. Its closed boundaries are called the ***control surface***. Again, the size and shape of a control volume is entirely optional, although the boundaries are often chosen to coincide with some solid or other natural flow boundaries. Actually, the control surface may be in motion through space relative to an absolute frame of reference; this is acceptable provided the motion is limited to constant-velocity translation.

We shall now derive a general relationship between a system and a control volume that provides an important basis for the equations of continuity, energy, and momentum for moving fluids. This relationship is derived from what is commonly referred to as the ***control volume approach***, more formally known as the ***Reynolds transport theorem***. Addressing the motion of fluid as it moves through a given region, the control volume approach is also called the ***Eulerian approach***, in contrast to the ***Lagrangian approach*** of solid mechanics in which the motion of a particle is described by its position as a function of time.

Let X represent the total amount of some fluid property, such as mass, energy, or momentum, contained within specified boundaries at a specified time. The specified boundaries will be either those of a system, indicated by a subscript S, or those of a control volume, indicated by a subscript CV. Consider the general flow situation of Fig. 4.7. At time t, the boundaries of the system and the control volume were chosen to coincide, so $(X_S)_t = (X_{CV})_t$. At instant Δt later, the system has moved a little through the control volume and possibly slightly changed its shape; a small amount of new fluid $\Delta \forall_{CV}^{\text{in}}$ has entered the control volume, and another small amount of system fluid $\Delta \forall_{CV}^{\text{out}}$ has left the control volume, where $\forall$ represents volume. These

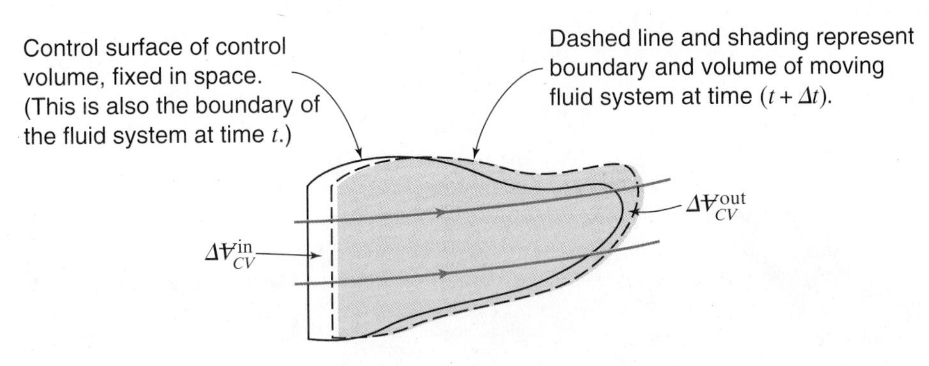

Figure 4.7
Fluid system, control volume, and differences

small volumes carry small amounts of property X with them, so that $\Delta X_{CV}^{\text{in}}$ enters and $\Delta X_{CV}^{\text{out}}$ leaves the control volume. Comparing X in the various volumes, we see that

$$(X_S)_{t+\Delta t} = (X_{CV})_{t+\Delta t} + \Delta X_{CV}^{\text{out}} - \Delta X_{CV}^{\text{in}}$$

Subtracting the equation for t from that for $t + \Delta t$, we obtain

$$(X_S)_{t+\Delta t} - (X_S)_t = (X_{CV})_{t+\Delta t} - (X_{CV})_t + \Delta X_{CV}^{\text{out}} - \Delta X_{CV}^{\text{in}}$$

or

$$\Delta X_S = \Delta X_{CV} + \Delta X_{CV}^{\text{out}} - \Delta X_{CV}^{\text{in}} \tag{4.8}$$

and dividing by Δt and letting $\Delta t \to 0$, we get

$$\frac{dX_S}{dt} = \frac{dX_{CV}}{dt} + \frac{dX_{CV}^{\text{out}}}{dt} - \frac{dX_{CV}^{\text{in}}}{dt} \tag{4.9}$$

These equations will be used in subsequent studies of continuity, energy, and momentum. The left-hand side of Eq. (4.9) is the rate of change of the total amount of any extensive property X within the moving system. The next term, dX_{CV}/dt, is the rate of change of the same property, but contained within the fixed control volume. The last two terms are the net rate of outflow of X passing through the control surface. So *Eq.* (4.9) *states that the difference between the rate of change of X within the system and that within the control volume is equal to the net rate of outflow from the control volume.*

4.7 EQUATION OF CONTINUITY

Although continuity is a strongly intuitive concept, it took early investigators a long time to formalize it. Pioneers in this effort include Hero (or Heron) of Alexandria (*circa* 100 A.D.), a Greek scientist, Leonardo da Vinci (1452–1519), the prodigious Italian artist and scientist, and Benedetto Castelli (1577–1643), a pupil of Galileo.

Let Fig. 4.8 represent a short length of a ***stream tube***, which may be

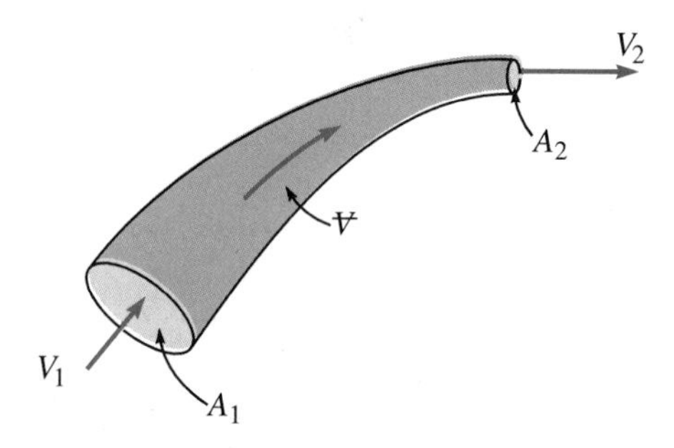

Figure 4.8
Portion of stream tube as control volume

assumed, for practical purposes, to be a bundle of streamlines. Since the stream tube is bounded on all sides by streamlines and since there can be no net velocity normal to a streamline, no fluid can leave or enter the stream tube except at the ends. The fixed volume between the two end sections is a control volume, of volume $\forall$ let us say. Using the relation of a system to a control volume developed in Sec. 4.6, and letting the general property X now be the mass m, Eq. (4.9) becomes

$$\frac{dm_S}{dt} = \frac{dm_{CV}}{dt} + \frac{dm_{CV}^{\text{out}}}{dt} - \frac{dm_{CV}^{\text{in}}}{dt} \tag{4.10}$$

But according to Newtonian physics (i.e., disregarding the possibility of converting mass to energy), the mass of a system must be conserved, so

$$\frac{dm_S}{dt} = 0 \tag{4.11}$$

Also, because the volume $\forall$ of the control volume is fixed, $m_{CV} = \forall\bar{\rho}_{CV}$ where $\bar{\rho}_{CV}$ is the mean density within the control volume, so

$$\frac{dm_{CV}}{dt} = \forall\frac{d\bar{\rho}_{CV}}{dt} = \forall\frac{\partial\bar{\rho}_{CV}}{\partial t} \tag{4.12}$$

since $\bar{\rho}_{CV}$ can vary only with time within the control volume. And, from Fig. 4.8, $\Delta m_{CV}^{\text{out}} = \rho_2\,\Delta\forall_2 = \rho_2 A_2 V_2\,\Delta t$, so that

$$\frac{dm_{CV}^{\text{out}}}{dt} = \rho_2 A_2 V_2 \tag{4.13}$$

and similarly,

$$\frac{dm_{CV}^{\text{in}}}{dt} = \rho_1 A_1 V_1 \tag{4.14}$$

Substituting Eqs. (4.11)–(4.14) into (4.10) and rearranging slightly, we obtain

$$\rho_1 A_1 V_1 - \rho_2 A_2 V_2 = \forall\frac{\partial\bar{\rho}_{CV}}{\partial t} \tag{4.15}$$

This is the general equation of continuity for flow through regions with fixed boundaries, in which $\partial\bar{\rho}_{CV}/\partial t$ is the time rate of change of the mean density of the fluid in $\forall$. The equation states that the net rate of mass inflow to the control volume is equal to the rate of increase of mass within the control volume.

For steady flow, $\partial\bar{\rho}_{CV}/\partial t = 0$ and

$$\rho_1 A_1 V_1 = \rho_2 A_2 V_2 = \dot{m} \tag{4.16a}$$

or

$$\gamma_1 A_1 V_1 = \gamma_2 A_2 V_2 = g\dot{m} = G \tag{4.16b}$$

These are the continuity equations that apply to steady, compressible or incompressible flow within fixed boundaries.

If the fluid is incompressible, ρ = constant; hence $\rho_1 = \rho_2$ and $\partial\rho/\partial t = 0$, and thus

$$A_1V_1 = A_2V_2 = Q \tag{4.17}$$

This is the continuity equation that applies to incompressible fluids for both steady and unsteady flow within fixed boundaries.[4]

Equations (4.16) and (4.17) are generally adequate for the analysis of flows in conduits with solid boundaries, but for the consideration of flow in space, as that of air around an airplane, for example, it is desirable to express the continuity equation in differential form, as indicated in Sec. 14.1. Or, for the case of unsteady flow of a liquid in an open channel (Fig. 4.4), the principle of conservation of mass indicates that the rate of flow past section 1 minus the rate of flow past section 2 is equal to the time rate of change of the volume of liquid $\forall$ contained in the channel between the two sections. Thus

$$Q_1 - Q_2 = d\forall/dt \tag{4.18}$$

EXERCISES

4.7.1 Gas flows at a steady rate in a 150-mm-diameter pipe that enlarges to a 200-mm-diameter pipe. (*a*) At a certain section of the 150-mm pipe the density of the gas is 175 kg/m^3 and the velocity is 18 m/s. At a certain section of the 200-mm pipe the velocity is 12 m/s. What must be the density of the gas at that section? (*b*) If these same data were given for the case of unsteady flow at a certain instant, could the problem be solved? Discuss.

4.7.2 Gas is flowing in a long 8-in-diameter pipe from *A* to *B*. At section *A* the flow is 0.62 lb/s while at the same instant at section *B* the flow is 0.68 lb/s. The distance between *A* and *B* is 750 ft. Find the mean value of the time rate of change of the specific weight of the gas between sections *A* and *B* at that instant.

4.7.3 Water flows in a river. At 8 A.M. the flow past bridge 1 is 2200 cfs. At the same instant the flow past bridge 2 is 1750 cfs. At what rate is water being stored in the river between the two bridges at this instant? Assume zero seepage and negligible evaporation.

4.7.4 Water flows in a river. At 9 A.M. the flow past bridge 1 is 58.4 m^3/s. At the same instant the flow past bridge 2 is 42.7 m^3/s. At what rate is water being stored in the river between the two bridges at this instant? Assume zero seepage and negligible evaporation.

---

[4] The continuity equations (4.16) and (4.17) are applicable to any stream tube in a flow system. Most commonly the continuity equation is applied to the stream tube that is coincident with the boundaries of the flow.

4.8 ONE-, TWO-, AND THREE-DIMENSIONAL FLOW

In true one-dimensional flow the velocity at all points has the same direction and (for an incompressible fluid) the same magnitude. Such a case is rarely of practical interest. However, the term ***one-dimensional method of analysis*** is applied to the flow between boundaries which are really three-dimensional, with the understanding that the "one dimension" is taken *along the central streamline* of the flow. Average values of velocity, pressure, and elevation across a section normal to this streamline are considered typical of the flow as a whole. Thus the equation of continuity in Sec. 4.7 is called the one-dimensional equation of continuity, even though it may be applied to flow in conduits that curve in space and in which the velocity varies across any section normal to the flow. It will be of increasing importance in the following chapters to recognize that, when high accuracy is desired, the equations derived by the one-dimensional method of analysis require refinement to account for the variation in conditions across the flow section.

If the flow is such that all streamlines are plane curves and are identical in a series of parallel planes, it is said to be ***two-dimensional***. In Fig. 4.9*a* the channel has a constant dimension perpendicular to the plane of the figure. Thus every cross section normal to the flow must be a rectangle of this constant width. ***Three-dimensional*** flow is illustrated in Fig. 4.9*b*, although in this particular case the flow is axially symmetric, which simplifies the analysis. A generalized three-dimensional flow, such as the flow of cool air from an air conditioning outlet into a room, is quite difficult to analyze. Such flows are often approximated as two-dimensional or as axially symmetric flow. This offers an advantage in that it is easier to draw diagrams describing the flow, and the mathematical treatment is much simpler.

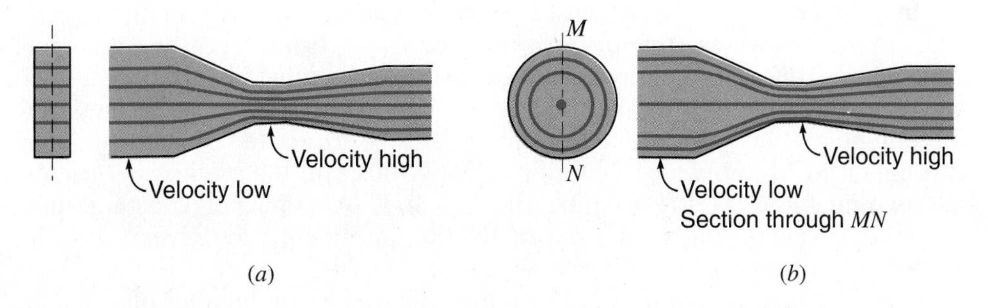

Figure 4.9
Two- and three-dimensional (axially symmetric) flow of an ideal fluid

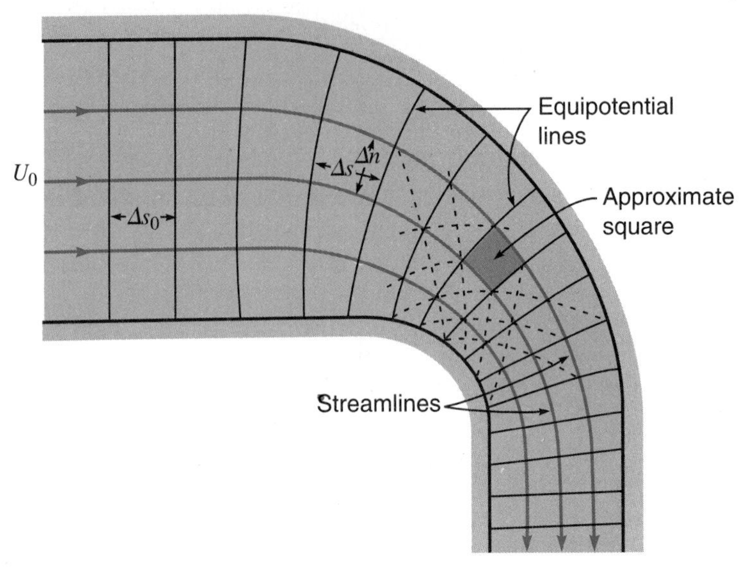

Figure 4.10
Flow net (two-dimensional flow)

4.9 THE FLOW NET

The streamlines and velocity distribution in the case of steady two-dimensional flow of an *ideal* fluid within any boundary configuration may be determined by a ***flow net***, such as is shown in Fig. 4.10. This is a network of streamlines and lines normal (perpendicular) to them so spaced that the distances between both sets of lines are inversely proportional to the local velocities. The streamlines show the mean direction of flow at any point. A fundamental property of the flow net is that it provides the one and only representation of the ideal flow within the given boundaries. It is also independent of the actual magnitude of the flow and, for the *ideal* fluid, is the same whether the flow is in one direction or the reverse.

In a number of simple cases it is possible to obtain mathematical expressions, known as ***stream functions*** (Sec. 14.4), from which one can plot streamlines. But even the most complex cases can be solved by plotting a flow net by a trial-and-error method. Although it is possible to construct nets for three-dimensional flow, treatment here will be restricted to the simpler two-dimensional net, which will more clearly illustrate the method. Consider the two-dimensional stream tube of Fig. 4.11. Assuming a constant unit thickness perpendicular to the paper, the continuity equation gives $V_1 \Delta n_1 = V_2 \Delta n_2$.

Consider next a region of uniform flow divided into a number of strips of equal width, separated by streamlines, as in Fig. 4.9*a*. Each strip represents a stream tube, and the flow is equally divided among the tubes. As the flow

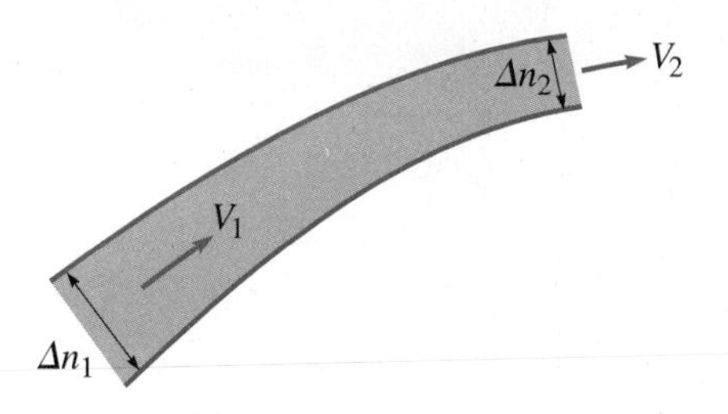

Figure 4.11

approaches a bend or obstruction, the streamline must curve so as to conform to the boundaries, but each stream tube still carries the same flow. Thus the spacing between all streamlines in the entire field is everywhere inversely proportional to the local velocities so that, for any section normal to the velocity,

$$V\,\Delta n = \text{constant} \tag{4.19}$$

To draw the streamlines, it is necessary to start by estimating not only the spacing between them but also their directions at all points. As an aid in the latter, we make use of normal, or ***equipotential***, lines. As an analogy consider the flow of heat through a homogeneous material enclosed between perfectly insulated boundaries. The heat might be considered to flow along the equivalent of streamlines. As there can be no flow of heat along a line of constant temperature, it follows that the heat flow must be everywhere perpendicular to isothermal lines. In like manner, streamlines must be everywhere perpendicular to equipotential lines. Because solid boundaries, across which there can be no flow, also represent streamlines, it follows that *equipotential lines must meet the boundaries everywhere at right-angles.*

If, as is usually most convenient, the equipotential lines are spaced the same distance apart as the streamlines in the region of uniform two-dimensional flow (as at the ends of Fig. 4.10), the net for that region is composed of perfect squares. In a region of deformed flow (as in the bend of Fig. 4.10) the quadrilaterals cannot remain square, but they will approach squares as the number of streamlines and equipotential lines are increased indefinitely by subdividing. It is frequently helpful, in regions where the deformation is marked, to introduce extra streamlines and equipotential lines spaced midway between the original ones.

In drawing a flow net, a beginner will do a lot of erasing, but with some practice a net can be sketched fairly easily to represent any boundary configuration. It is even possible to construct an approximate flow net for cases where one solid boundary does not exist and the fluid extends laterally indefinitely, as in the flow around an immersed object. Such a case reveals an advantage of the flow net that is not evident from Fig. 4.10. In the flow between confining solid boundaries it is always possible to determine the mean velocity across any section by dividing the total flow by the section area.

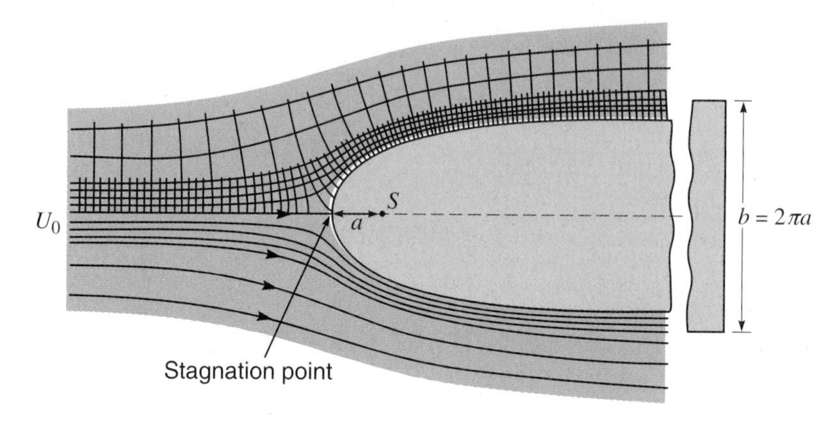

Figure 4.12
Two-dimensional flow of a frictionless fluid past a solid[5] whose surface is perpendicular to the plane of the paper. Streamlines or path lines for steady flow

For flow around an immersed object, as in Fig. 4.12, there is no fixed area by which to divide a definite flow, but the flow net in combination with Eq. (4.19) provides a good means of estimating velocities in the region around such an object. With increasing distance from the body's centerline the deflection of streamlines around it reduces, until it becomes negligible; this distance or deflection must be estimated in order to draw the flow net, and reasonable estimates will yield closely similar velocities near the body.

Where a channel is curved, the equipotential lines must diverge inasmuch as they radiate from centers of curvature. The distance between the associated streamlines must vary in the same way as that between the equipotential lines. Therefore, as in Fig. 4.10, the areas are smallest along the inner radius of the bend and increase toward the outside.

The accuracy of the final flow net can be checked by drawing diagonals, as indicated by a few dashed lines in Fig. 4.10. If the net is correct, these dashed lines also form a network of lines that cross each other at right angles and produce areas that approach squares in shape.

4.10 USE AND LIMITATIONS OF FLOW NET

Although the flow net is based on an ideal frictionless fluid, it may be applied to the flow of a real fluid within certain limits. Such limits are dictated by the extent to which the real fluid is affected by factors which the ideal-fluid theory neglects. The principal factor of this type is fluid friction.

[5] This surface shape can be generated as the boundary between the given flow field and that issuing from a source of strength $Q = bdU_0$ located at S, where d is the source length perpendicular to the figure (see Prob. 14.14).

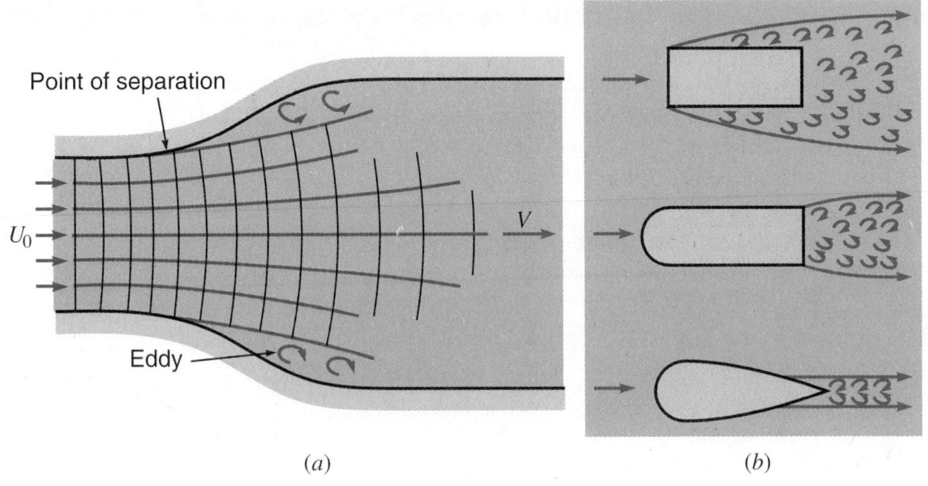

Figure 4.13
Separation in diverging flow. (*a*) Eddy formation in a diverging channel. (*b*) Turbulent wakes.

The viscosity effects of a real fluid are most pronounced at or near a solid boundary and diminish rapidly with distance from the boundary. Hence, for an airplane or a submerged submarine, the fluid may be considered as frictionless, except when very close to the object. The flow net always indicates a velocity next to a solid boundary, whereas a real fluid must have zero velocity adjacent to a wall. The region in which the velocity is so distorted, however, is confined to a relatively thin layer called the ***boundary layer*** (Secs. 8.7–8.9 and 9.2–9.4), outside of which the real fluid behaves very much like the ideal fluid.

The effect of the boundary friction is minimized when the streamlines are converging, but in a diverging flow there is a tendency for the streamlines not to follow the boundaries if the rate of divergence is too great. In a sharply diverging flow, such as is shown schematically in Fig. 4.13, there may be a ***separation*** of the boundary layer from the wall, resulting in ***eddies*** and even reverse flow in that region (Fig. 9.8). The flow is badly disturbed in such a case, and the flow net may be of limited value.

A practical application of the flow net may be seen in the flow around a body, as shown in Fig. 4.12. An example of this is the upstream portion of a bridge pier below the surface where surface wave action is not a factor. Except for a thin layer adjacent to the body, this diagram represents the flow in front of and around the sides of the body. The central streamline is seen to branch at the forward tip of the body to form two streamlines along the walls. At the forward tip the velocity must be zero; hence this point is called a ***stagnation point***. Other common applications are to flows over spillways, and to seepage flows through earth dams and through the ground under a concrete dam. In the first two of these cases the flow has a ***free surface*** at atmospheric pressure. To draw flow nets for free surface flows, one must make use of more advanced principles that are not covered in this text.

Considering the limitations of the flow net in diverging flow, it may be seen that, while the flow net gives a fairly accurate picture of the velocity distribution in the region near the upstream part of any solid body, it may give little information concerning the flow conditions near the rear because of the possibility of separation and eddies. The disturbed flow to the rear of a body is known as a ***turbulent wake*** (Fig. 4.13*b*). The space occupied by the wake may be greatly diminished by streamlining the body, i.e., giving the body a long slender tail, which tapers to a sharp edge for two-dimensional flow or to a point for three-dimensional flow.

SAMPLE PROBLEM 4.2 Figure 4.12 represents flow towards and around a bridge pier where $b = 5$ ft and $U_0 = 10$ fps. (*a*) Make a plot of the velocity along the flow centerline to the left of the solid, and along the boundary of the solid. (*b*) By what percentage does the maximum velocity along the boundary exceed the uniform velocity? (*c*) At what distance from the stagnation point does a velocity of 7.5 fps occur?

Solution

Eq. (4.19): $\quad V\Delta n = \text{const.} = U_0\,\Delta n_0$

So $\quad V = (\Delta n_0/\Delta n)10$ fps.

Use $b = 5$ ft to scale 1 ft distances along the centreline and around the boundary of the solid. On Fig. 4.12 measure the net "square" sizes, in both the flow (ΔL) and perpendicular (ΔW) directions, using three of four squares where appropriate and taking the average. Calculate Δn and V as shown in the table below:

Distance from stagnation pt, ft	−6	−5	−4	−3	−2	−1	0	1	2	3	4	5	6	7	8
Average ΔL, mm	0.98	1.02	1.06	1.06	1.17	1.25	—	0.88	0.70	0.67	0.66	0.73	0.75	0.78	0.84
Average ΔW, mm	0.88	0.91	0.94	1.00	1.30	1.80	—	0.95	0.80	0.74	0.75	0.76	0.79	0.90	0.93
$\Delta n = \frac{1}{2}(\Delta L + \Delta W)$, mm	0.93	0.97	1.00	1.03	1.24	1.53	—	0.92	0.75	0.71	0.71	0.74	0.77	0.84	0.88
$\Delta n_0/\Delta n = 0.93/\Delta n$	1.00	0.96	0.93	0.90	0.75	0.61	—	1.02	1.24	1.32	1.32	1.25	1.21	1.11	1.05
$V = 10(\Delta n_0/\Delta n)$, fps	10.0	9.6	9.3	9.0	7.5	6.1	0	10.2	12.4	13.2	13.2	12.5	12.1	11.1	10.5

(*a*)

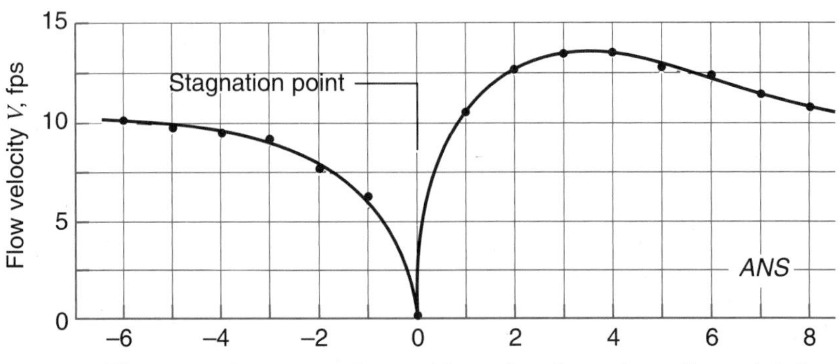

(*b*) $V_{max}/U_0 = 13.3/10.0 = 1.33$
Therefore V_{max} is 33% greater than U_0 ***ANS***
(*c*) From the above plot, a velocity of 7.5 fps occurs at about −1.9 ft and +0.4 ft from the stagnation point ***ANS***
Note that if the flow net had been perfectly constructed, the respective average ΔL and ΔW values would have been identical. Even so, the results obtained here are quite accurate, because the respective values of ΔL and ΔW were averaged. This problem can be solved analytically using principles of hydrodynamics, which yield $V_{max}/U_0 = 1.260$ (see, e.g., Prob. 14.14).

EXERCISES

4.10.1 An incompressible ideal fluid flows at 10 L/s through a circular 160-mm-diameter pipe into a conically converging nozzle like that of Sample Prob. 4.4 (diameter at *B* is 80 mm). Determine the average velocity of flow at sections *D* and *B*.

4.10.2 Figure X4.10.2 shows the flow net for two-dimensional flow from a rounded, long-slotted exit from a tank. If $U_0 = 2.5$ m/s, what is the approximate flow velocity at *A*?

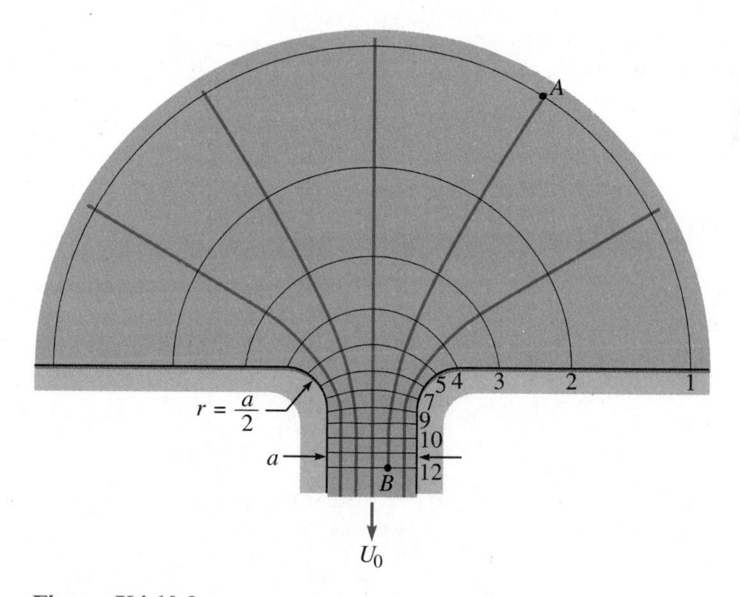

Figure X4.10.2

4.10.3 Given that U_0 in Fig. 4.10 is 5.0 fps, find approximately (*a*) the maximum velocity in the bend and (*b*) the uniform velocity in the downstream section.

4.10.4 Given that U_0 in Fig. 4.12 is 3.5 m/s, find approximately (*a*) the maximum and (*b*) the minimum velocity on the body surface.

4.11 FRAME OF REFERENCE IN FLOW PROBLEMS

In flow problems we are really concerned with the *relative* velocity between the fluid and the body. It makes no difference whether the body is at rest and the fluid flows past it or whether the fluid is at rest and the body moves through the fluid. There are thus two frames of reference. In one the observer (or the camera) is at rest with respect to the solid body. If the observer at rest with respect to a bridge pier views a steady flow past it or is on a ship moving at constant velocity through still water, the streamlines appear to him to be unchanging and therefore the flow is steady. But if he floats with the current past the pier or views a ship going by while he stands on the bank, the flow pattern that he observes is changing with time. Thus the flow is unsteady.

The same flow may therefore be either steady or unsteady according to the frame of reference. The case that is usually of more practical importance is steady, ideal flow, for which the streamlines and path lines are identical. In unsteady flow streamlines and path lines are entirely different from each other, and they also bear no resemblance to those of steady flow.

4.12 VELOCITY AND ACCELERATION IN STEADY FLOW

In a typical three-dimensional flow field the velocities may be everywhere different in magnitude and direction. Also, the velocity at any point in the field may change with time. Let us first consider the case where the flow is steady and thus independent of time. If the velocity of a fluid particle has components u, v,[6] and w parallel to the x, y, and z axes, then, for steady flow,

$$u_{st} = u(x, y, z) \tag{4.20a}$$

$$v_{st} = v(x, y, z) \tag{4.20b}$$

$$w_{st} = w(x, y, z) \tag{4.20c}$$

Applying the chain rule of partial differentiation, the acceleration of the fluid particle for steady flow can be expressed as

$$\mathbf{a}_{st} = \frac{d}{dt}\mathbf{V}(x, y, z) = \frac{\partial \mathbf{V}}{\partial x}\frac{dx}{dt} + \frac{\partial \mathbf{V}}{\partial y}\frac{dy}{dt} + \frac{\partial \mathbf{V}}{\partial z}\frac{dz}{dt} \tag{4.21}$$

where $$|\mathbf{V}| = (u^2 + v^2 + w^2)^{1/2}$$

[6] This text uses a rounded lower case v (vee) to help distinguish it from the capital V and from the Greek ν (nu) used for kinematic viscosity.

Noting that $dx/dt = u$, $dy/dt = v$, and $dz/dt = w$,

$$\mathbf{a}_{st} = u\frac{\partial \mathbf{V}}{\partial x} + v\frac{\partial \mathbf{V}}{\partial y} + w\frac{\partial \mathbf{V}}{\partial z} \tag{4.22}$$

This vector equation can be written as three scalar equations:

$$(a_x)_{st} = u\frac{\partial u}{\partial x} + v\frac{\partial u}{\partial y} + w\frac{\partial u}{\partial z} \tag{4.23a}$$

$$(a_y)_{st} = u\frac{\partial v}{\partial x} + v\frac{\partial v}{\partial y} + w\frac{\partial v}{\partial z} \tag{4.23b}$$

$$(a_z)_{st} = u\frac{\partial w}{\partial x} + v\frac{\partial w}{\partial y} + w\frac{\partial w}{\partial z} \tag{4.23c}$$

These equations show that even though the flow is steady, the fluid may possess an acceleration by virtue of a change in velocity with change in position. This type of acceleration is commonly referred to as ***convective acceleration***. With incompressible fluid flow, there is a convective acceleration wherever the effective flow area changes along the flow path. This is also true for compressible fluid flow, but, in addition, convective acceleration of a compressible fluid occurs wherever the density varies along the flow path irrespective of any changes in the effective flow area.

At times it is convenient to superimpose the coordinate system on the streamline pattern in such a way that the x axis is tangential to the streamline at a particular point of interest. In such a case we shall let s indicate distance along the streamline. Thus $\mathbf{V} = \mathbf{V}(s)$, and, since the perpendicular velocity components in Eq. (4.22) are zero, the acceleration of the fluid particle along the streamline at this point can conveniently be expressed as

$$\mathbf{a}_{st} = V\frac{\partial \mathbf{V}}{\partial s} \tag{4.24}$$

In the terminology of curvilinear motion, this is referred to as the ***tangential acceleration***. In uniform flow with ρ = constant this acceleration is zero.

At this point in our discussion we should recall that a particle moving steadily along a curved path has a ***normal acceleration*** a_n toward the center of curvature of the path. From mechanics,

$$a_n = \frac{V^2}{r} \tag{4.25}$$

where r is the radius of the path. A particle moving on a curved path will always have a normal acceleration, regardless of its behavior in the tangential direction.

SAMPLE PROBLEM 4.3 A two-dimensional flow field is given by $u = 2y$, $v = x$. (*a*) Sketch the flow field. (*b*) Derive a general expression for the velocity and acceleration (x and y are in units of length L; u and v are in units of L/T). (*c*) Find the acceleration in the flow field at point A ($x = 3.5$, $y = 1.2$).

Solution

(*a*) Velocity components u and v are plotted to scale, and streamlines are sketched tangentially to the resultant velocity vectors. This gives the following general picture of the flow field:

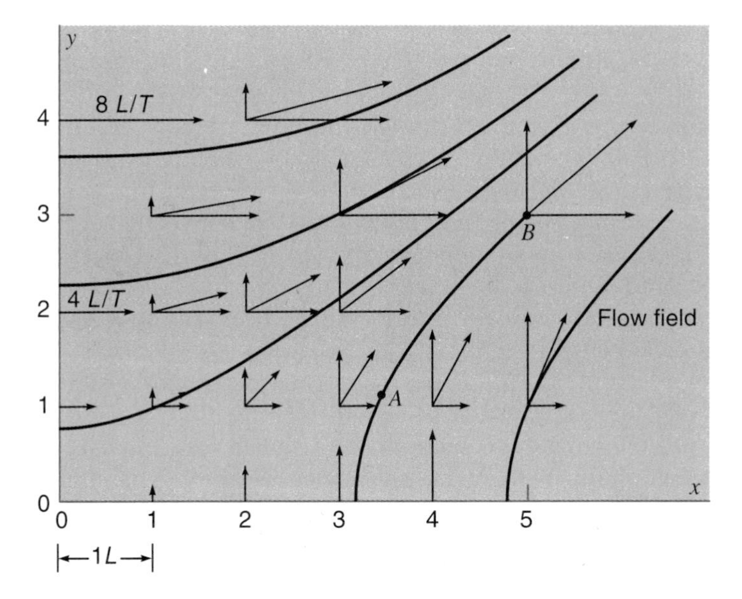

ANS

(*b*) $$V = (u^2 + v^2)^{1/2} = (4y^2 + x^2)^{1/2} \qquad \textbf{\textit{ANS}}$$

Eq. (4.23*a*): $$a_x = u\frac{\partial u}{\partial x} + v\frac{\partial u}{\partial y} = 2y(0) + x(2) = 2x$$

Eq. (4.23*b*): $$a_y = u\frac{\partial v}{\partial x} + v\frac{\partial v}{\partial y} = 2y(1) - x(0) = 2y$$

$$a = (a_x^2 + a_y^2)^{1/2} = (4x^2 + 4y^2)^{1/2} = 2(x^2 + y^2)^{1/2} \qquad \textbf{\textit{ANS}}$$

(*c*) At $A(3.5, 1.2)$,

$$(a_A)_x = 2x = 2(3.5) = 7.00\ L/T^2; \quad (a_A)_y = 2y = 2(1.2) = 2.40\ L/T^2$$

and $$a_A = [(a_A)_x^2 + (a_A)_y^2]^{1/2} = [(7.00)^2 + (2.40)^2]^{1/2} = 7.40\ L/T^2 \qquad \textbf{\textit{ANS}}$$

Rough check. Imagine a velocity vector at point A. This vector would have a magnitude approximately midway between that of the adjoining vectors, or $V_A \approx 4\ L/T$. The radius of curvature of the sketched streamline at A is roughly

$3L$. Thus $(a_A)_n \approx 4^2/3 \approx 5.3\ L/T^2$. The tangential acceleration of the particle at A may be approximated by noting that the velocity along the streamline increases from about 3.2 L/T, where it crosses the x axis, to about 7.8 L/T at B. The distance along the streamline between these two points is roughly 4 L. Hence a very approximate value of the tangential acceleration at A is

$$(a_A)_t = V\frac{\partial V}{\partial s} \approx 4\left(\frac{7.8 - 3.2}{4}\right) \approx 4.6\ L/T^2$$

Vector diagrams of these roughly computed normal and tangential acceleration components are plotted below for comparison with the true acceleration as given by the analytic expressions. In Chap. 14 we shall prove that the flow in this example is that of an incompressible fluid.

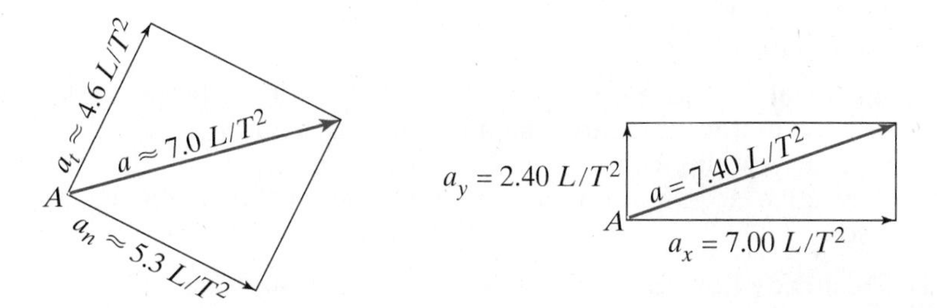

(*a*) Approximate vector diagram (*b*) True vector diagram

Acceleration at A

EXERCISES

4.12.1 A flow field is defined by $u = 2$, $v = 3$, $w = 4$. What is the velocity of flow? Specify units in terms of L and T.

4.12.2 The velocity along a streamline lying on the x axis is given by $u = 6 + x^{0.4}$. What is the convective acceleration at $x = 5$? Specify units in terms of L and T. Assuming the fluid is incompressible, is the flow converging or diverging?

4.12.3 A flow field is defined by $u = 2y$, $v = x$. Derive expressions for the x and y components of acceleration. Find the magnitude of the velocity and acceleration at the point (3, 1). Specify units in terms of L and T.

4.12.4 A flow field is defined by $u = 2x$, $v = y$. Derive expressions for the x and y components of acceleration. Find the magnitude of the velocity and acceleration at the point (3, 2). Specify units in terms of L and T.

4.12.5 A flow field is defined by $u = 2y$, $v = xy$. Derive expressions for the x and y components of acceleration. Find the magnitude of the velocity and acceleration at the point (2, 3). Specify units in terms of L and T.

4.12.6 A flow field is defined by $u = 3y$, $v = 2xy$, $w = 4z$. Derive expressions for the x, y, and z components of acceleration. Find the magnitude of the velocity and acceleration at the point (1, 2, 1). Specify units in terms of L and T.

4.12.7 The velocity along a circular streamline of radius 4.5 ft is 2.5 fps. Find the normal and tangential components of the acceleration if the flow is steady.

4.12.8 The velocity along a circular streamline of radius 2 m is 0.8 m/s. Find the normal and tangential components of the acceleration if the flow is steady.

4.12.9 A large tank contains an ideal liquid which flows out of the bottom of the tank through a 4.5-in-diameter hole. The rate of steady outflow is 6 cfs. Assume that the liquid approaches the center of the hole radially. Find the velocities and convective accelerations at points that are 2 and 3 ft from the center of the hole.

4.12.10 An ideal liquid flows out the bottom of a large tank through a 8-cm-diameter hole at a steady rate of 0.30 m^3/s. Assume the liquid approaches the center of the hole radially. Find the velocities and convective accelerations at points 0.5 and 1.0 m from the center of the hole.

4.12.11 The steady flow rate in each of the four stream tubes of Fig. 4.10 is 40 cfs per foot perpendicular to the plane of the figure. By scaling, the dimensions of the shaded "square" have been found to be 1.65 ft wide on the upstream face, 1.53 ft wide on the downstream face, and 1.67 ft along the flow line through its center; the radius of that flow line measures 11.1 ft. Find the normal, tangential, and resultant accelerations of a fluid particle at the center of the shaded area.

4.13 VELOCITY AND ACCELERATION IN UNSTEADY FLOW

If the flow is unsteady, then Eqs. (4.20a–c) take the form

$$u = u(x, y, z, t) \ldots \tag{4.26}$$

Following a similar procedure to that of the preceding section, we obtain the vector equation

$$\mathbf{a} = \left(u \frac{\partial \mathbf{V}}{\partial x} + u \frac{\partial \mathbf{V}}{\partial y} + w \frac{\partial \mathbf{V}}{\partial z}\right) + \frac{\partial \mathbf{V}}{\partial t} \tag{4.27}$$

and the following set of scalar equations

$$a_x = \left(u \frac{\partial u}{\partial x} + v \frac{\partial u}{\partial y} + w \frac{\partial u}{\partial z}\right) + \frac{\partial u}{\partial t} \tag{4.28a}$$

$$a_y = \left(u \frac{\partial v}{\partial x} + v \frac{\partial v}{\partial y} + w \frac{\partial v}{\partial z}\right) + \frac{\partial v}{\partial t} \tag{4.28b}$$

$$a_z = \left(u\frac{\partial w}{\partial x} + v\frac{\partial w}{\partial y} + w\frac{\partial w}{\partial z}\right) + \frac{\partial w}{\partial t} \tag{4.28c}$$

In the above set of equations the three terms in parentheses are recognized as the convective accelerations, while the $\partial \mathbf{V}/\partial t$, $\partial u/\partial t$, $\partial v/\partial t$, and $\partial w/\partial t$ terms represent the acceleration caused by the unsteadiness of the flow. This type of acceleration is commonly referred to as the ***local acceleration***.

If we let s represent distance along an instantaneous streamline, in the same manner as the previous section, we now have $\mathbf{V} = \mathbf{V}(s, t)$, and the tangential acceleration of a fluid particle along the streamline is given by

$$\mathbf{a} = V\frac{\partial \mathbf{V}}{\partial s} + \frac{\partial \mathbf{V}}{\partial t} \tag{4.29}$$

The first term on the right-hand side of this equation is the convective acceleration, which becomes zero in uniform flow (straight and parallel streamlines) with ρ = constant.

Sample Problem 4.4 The figure shows a cross-section through the centerline of a circular pipe with a conically converging nozzle. An incompressible ideal fluid flows through at $Q = (0.1 + 0.05t)$ cfs, where t is in sec. Find the average velocity and acceleration of the flow at points D and B when $t = 5$ sec.

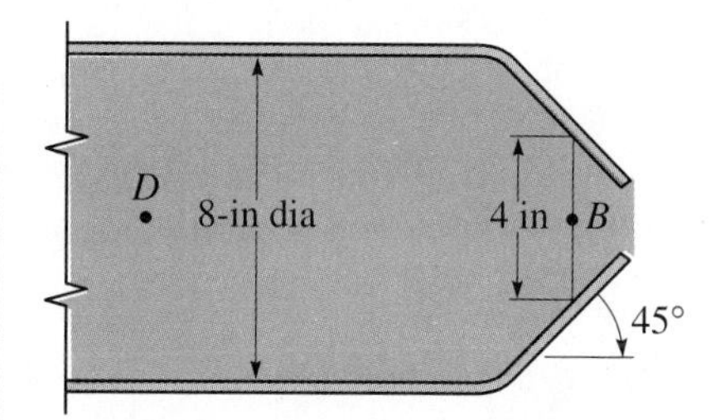

Solution

As a first step we sketch an approximate flow net to provide a general picture of the flow. We note that the flow is symmetric about the pipe axis (axisymmetric flow), so *the net is not a true two-dimensional flow net* (see Fig. 4.9).

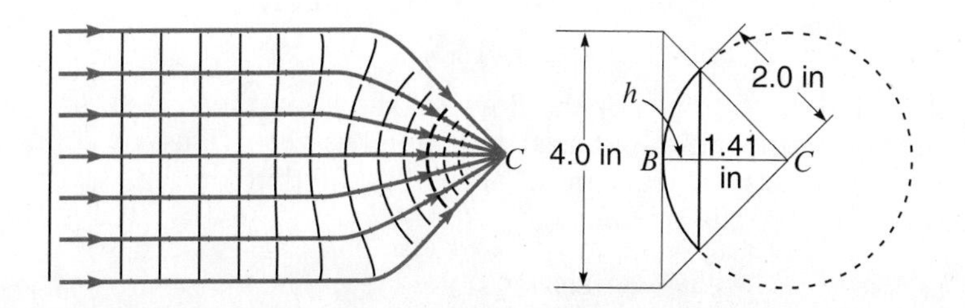

Since D and B are both on the pipe axis, $v = 0$ and $w = 0$ due to symmetry, so Eqs. (4.28) for these points reduce to

$$a_x = u\frac{\partial u}{\partial x} + \frac{\partial u}{\partial t}, \qquad a_y = 0, \qquad a_z = 0$$

At section D the streamlines are parallel and hence the area at right angles to the velocity vectors is a plane circle,

$$A_D = \frac{\pi}{4}\left(\frac{8}{12}\right)^2 = 0.349 \text{ ft}^2$$

So

$$u = \frac{Q}{A_D} = \frac{0.1 + 0.05t}{0.349} = \frac{2+t}{6.98}$$

and

$$\frac{\partial u}{\partial x} = 0, \qquad \frac{\partial u}{\partial t} = \frac{1}{6.98}$$

Thus at $t = 5$ s: $\quad V_D = u = \dfrac{2+5}{6.98} = 1.003$ fps $\qquad$ ***ANS***

and $\quad a_D = a_x = 1.003(0) + \dfrac{1}{6.98} = 0.1432$ ft/sec^2 $\qquad$ ***ANS***

At section B, however, the perpendicular flow area is the partial spherical surface through B, with center C and radius $r = 2$ in (see sketch). By table lookup, or by integration, this area is $2\pi rh$, where $h = r - r\cos 45° = 0.293r$. Thus $A_B = 2\pi r(0.293r) = 1.840r^2$.

On the centerline near B, $\quad u = \dfrac{Q}{A_B} = \dfrac{0.1 + 0.05t}{1.840r^2} = \dfrac{2+t}{36.8r^2}$

and since $x = \text{constant} - r$,

$$\frac{\partial u}{\partial x} = -\frac{\partial u}{\partial r} = -\left[\frac{-2(2+t)}{36.8r^3}\right] = \frac{2+t}{18.40r^3}$$

and

$$\frac{\partial u}{\partial t} = \frac{1}{36.8r^2}$$

Thus at $r = 2$ in and $t = 5$ sec:

$$V_B = u = \frac{2+5}{36.8(2/12)^2} = 6.85 \text{ fps} \qquad \textit{ANS}$$

and

$$a_B = a_x = 6.85\left[\frac{2+5}{18.40(2/12)^3}\right] + \frac{1}{36.8(2/12)^2}$$
$$= 563 \text{ (convective)} + 0.978 \text{ (local)}$$
$$= 564 \text{ ft/sec}^2 \qquad \textit{ANS}$$

It is should be noted that for the flow net shown in the sketch, the velocity at C is infinite because the flow area at that point is zero. This, of course, cannot occur; in the real case a jet somewhat similar to that of Fig. 11.13 will form downstream of the nozzle opening.

EXERCISES

4.13.1 A flow is defined by $u = 2(1+t)$, $v = 3(1+t)$, $w = 4(1+t)$. What is the velocity of flow at the point (3, 1, 4) at $t = 2$? What is the acceleration at that point at $t = 2$? Specify units in terms of L and T.

4.13.2 A two-dimensional flow field is given by $u = 3 + 2xy + 4t^2$, $v = xy^2 + 3t$. Find the velocity and acceleration of a particle of fluid at point (2, 1) at $t = 5$. Specify units in terms of L and T.

4.13.3 The flow velocity in fps along a circular streamline of radius 4 ft is $0.8 + 1.5t$. Find the normal and tangential components of the acceleration when $t = 1.3$ sec.

4.13.4 The flow velocity in m/s along a circular streamline of radius 1.6 m is $0.25 + 0.5t$. Find the normal and tangential components of the acceleration when $t = 1.4$ s.

PROBLEMS

4.1 The velocities in a circular conduit 8 in in diameter were measured at radii 0, 1.44, 2.60, and 3.48 in and were found to be 20.3, 19.7, 17.7, and 14.5 fps respectively. Find approximate values (graphically) of the volume flow rate and the mean velocity. Also determine the ratio of the mean velocity to the maximum velocity.

4.2 The velocities in a circular conduit 200 mm in diameter were measured at radii 0, 36, 65, and 87 mm and were found to be 7.0, 6.8, 6.1, and 5.0 m/s respectively. Find approximate values (graphically) of the volume flow rate and the mean velocity. Also determine the ratio of the mean velocity to the maximum velocity.

4.3 Carbon dioxide flows in a 2-in by 2-in duct at a pressure of 50 psi and a temperature of 80°F. If the atmospheric pressure is 13.4 psia and the velocity of flow is 15 fps, calculate the weight flow rate.

4.4 Nitrogen at 45°C and under a pressure of 2800 mbar abs flows in a 300-mm-diameter conduit at a mean velocity of 10 m/s. Find the mass flow rate.

4.5 A compressible fluid flows in an 18-in-diameter leaky pipe. Measurements are made simultaneously at two points A and B along the pipe that are 36,000 ft apart. Two sets of measurements are taken with an interval of exactly 40 min between them. The data are as follows:

Time	ρ_1, slug/ft^3	V_1, ft/sec	ρ_2, slug/ft^3	V_2, ft/sec
0	0.565	68	0.694	52
40 min	0.672	52	0.827	40

Assuming ρ varies linearly with respect to time and distance, compute the approximate average mass rate of leakage between A and B.

4.6 Refer to Fig. X4.10.2. If a is 3 in and U_0 is 10 fps, approximately how long will it take a particle to move from point A to point B on the same streamline? (*Note*: Between each pair of equipotential lines, measure Δs, and then compute the average velocity and time increment.)

4.7 Repeat Problem 4.6 using the following data: $a = 15$ cm and $U_0 = 0.5$ m/s. Find also the approximate velocity where the flow crosses equipotential line 3.

4.8 Make an approximate plot of the frictionless velocity (relative to U_0) along both the inner and the outer boundaries of Fig. 4.10. By what percent is the ideal maximum inner velocity greater than the ideal minimum outer velocity?

4.9 Consider the two-dimensional flow about a 2-in-diameter cylinder. Sketch the flow net for the ideal flow around one-quarter of the cylinder. Start with a uniform net of $\frac{1}{2}$-in squares, and fill in with $\frac{1}{4}$-in squares where desirable. (*Note*: It can be proved by classical hydrodynamics that the velocity tangent to the cylinder at a point 90° from the stagnation point is twice the uniform velocity.) From the flow net, determine the velocities (relative to U_0) along the center streamline from a point upstream where the velocity is uniform to the stagnation point, and then along the boundary of the cylinder from the stagnation point to the 90° point; plot them vs distance. By plotting a second curve on the same graph, compare the result thus obtained with the values given by the equation $V = 2U_0 \sin \theta$, where U_0 is the undisturbed stream velocity and θ is the angle subtended by the arc from the stagnation point to any point on the cylinder where V is desired.

4.10 Sketch the flow field defined by $u = 0$, $v = 3xy$, and derive expressions for the x and y components of acceleration. Find the acceleration at the point (2, 2). Specify units in terms of L and T.

4.11 Sketch the flow field defined by $u = 3y$, $v = 2$, and derive expressions for the x and y components of acceleration. Find the magnitude of the velocity and acceleration for the point having the coordinates (3, 4). Specify units in terms of L and T.

4.12 Sketch the flow field defined by $u = -2y$, $v = 3x$, and derive expressions for the x and y components of acceleration. As in Sample Prob. 4.3, find approximate values of the normal and tangential accelerations of the particle at the point (2, 3). Specify units in terms of L and T. Compare the value of $(a_n^2 + a_t^2)^{1/2}$ with the computed value $(a_x^2 + a_y^2)^{1/2}$.

4.13 Figure P4.13 shows to scale a two-dimensional stream tube. If the flow rate is 40 m^3/s per meter perpendicular to the plane of the sketch, determine approximate values of the normal and tangential acceleration of a fluid particle at A. What is the resultant acceleration of a particle at A?

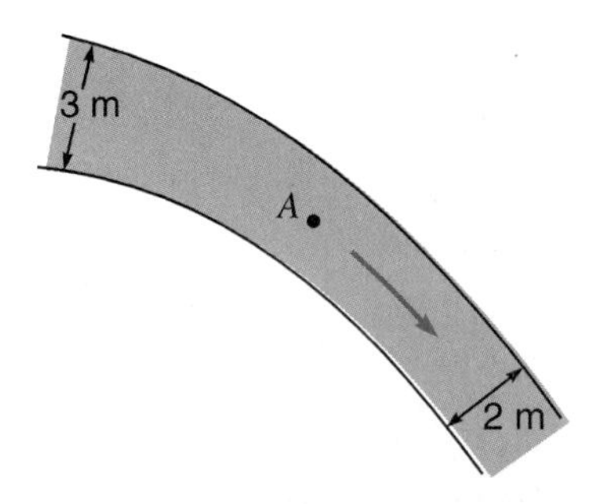

Figure P4.13

4.14 A large tank contains an ideal liquid which flows out of the bottom through a 4.5-in-diameter hole. The outflow rate $Q = 6 - 0.3t$, where Q is in cfs and t is in sec. Assume the liquid approaches the center of the hole radially. Find the local acceleration at a point 2 ft from the center of the hole at times $t = 8$ sec and 12 sec. What is the total acceleration at a point 3 ft from the center of the hole at $t = 12$ sec?

4.15 An ideal liquid flows out of the bottom of a large tank through an 8-cm-diameter hole. The outflow rate $Q = 0.25 - 0.01t^{0.5}$, where Q is in m^3/s and t is in s. Assume the liquid approaches the center of the hole radially. Find the local and convective accelerations at a point 0.4 m from the center of the hole at time = 15 s; what is the total acceleration?

4.16 Refer to the two-dimensional stream tube drawn to scale in Fig. P4.13. If the flow rate is $(20 - 3t)$ m^3/s per meter perpendicular to the plane of the sketch, with t in s, find approximate values of the normal, tangential, and total accelerations of a fluid particle at A when $t = 4$ s.

4.17 Assume that the streamlines for a two-dimensional flow of a frictionless incompressible fluid against a flat plate normal to the initial velocity may be represented by the equation xy = constant and that the flow is symmetrical about the plane through $x = 0$. A different streamline may be plotted for each value of the constant. Using a scale of 1 in = 6 units of distance, plot streamlines for values of the constant of 16, 64, and 128.

4.18 For the case in Prob. 4.17, it can be shown that the velocity components at any point are $u = ax$ and $v = -ay$, where a is a constant. Thus the actual velocity is $V = a\sqrt{x^2 + y^2} = ar$, where r is the radius to the origin. Let $a = \frac{1}{3}$; then if 1 in = 6 ft for the streamlines, 1 in = 2 fps for the velocity scale. Draw curves of equal velocity for values of 2, 4, 6, 8, and 10 fps. How does the velocity vary along the surface of the plate?

4.19 For three-dimensional flow with the y axis as the centerline, assume that the equation for the bounding streamline of a jet impinging vertically downward on a flat plate is $x^2y = 64$. (*a*) Plot the flow showing the centerline and bounding streamlines of the jet. (*b*) What is the approximate average velocity in the vertical jet at $y = 10$ if the average velocity in the vertical jet is 5.0 m/s at $y = 16$? (*c*) For the above conditions find the approximate velocity along the plate at $r = 12$, 24, and 36.

CHAPTER 5

Energy Considerations in Steady Flow

In this chapter we shall approach flow from the viewpoint of energy considerations. The first law of thermodynamics tells us that energy can be neither created nor destroyed. But it can, of course, be changed in form. It follows that all forms of energy are equivalent.

In Secs. 5.4–5.6 we shall derive flow equations based on such energy considerations, and in Sec. 5.7 we shall show how some of these equations can also be derived from Newton's second law.

The various forms of energy present in fluid flow are first introduced, in the next three sections.

5.1 KINETIC ENERGY OF A FLOWING FLUID

A body of mass m when moving at a velocity V possesses a kinetic energy, $\mathrm{KE} = \frac{1}{2}mV^2$. Thus if a fluid were flowing with all particles moving at the same velocity, its kinetic energy would also be $\frac{1}{2}mV^2$; this can be written as

$$\frac{\mathrm{KE}}{\mathrm{Weight}} = \frac{\frac{1}{2}mV^2}{\gamma \forall} = \frac{\frac{1}{2}(\gamma \forall)V^2}{\gamma g \forall} = \frac{V^2}{2g} \tag{5.1}$$

where $\forall$ represents the volume of the fluid mass. In BG units $V^2/2g$ is expressed in ft·lb/lb = ft and in SI units as N·m/N = m.

In most situations the velocities of the different fluid particles are not the same, so it is necessary to integrate all portions of the stream to obtain the true value of the kinetic energy. It is convenient to express the true value in terms of the mean velocity V and a factor α, known as the ***kinetic-energy correction factor***. Hence

$$\frac{\text{True KE}}{\text{Weight}} = \alpha \frac{V^2}{2g} \tag{5.2}$$

In order to obtain an expression for α, consider the case where the axial components of the velocity vary across a section (Fig. 4.1b). If u is the local axial velocity component at a point, the mass flow per unit of time through an elementary area dA is $\rho\, dQ = \rho u\, dA$. Thus the true flow of kinetic energy per unit of time across area dA is $\frac{1}{2}(\rho u\, dA)u^2 = \frac{1}{2}\rho u^3\, dA$. The weight rate of flow through dA is $\gamma\, dQ = \rho g u\, dA$. Thus, for the entire section,

$$\frac{\text{True KE/time}}{\text{Weight/time}} = \frac{\text{true KE}}{\text{weight}} = \frac{\frac{1}{2}\rho \int u^3\, dA}{\rho g \int u\, dA} = \frac{\int u^3\, dA}{2g \int u\, dA} \tag{5.3}$$

Comparing Eq. (5.3) with Eqs. (5.2) and (4.3), we get

$$\alpha = \frac{1}{V^2}\frac{\int u^3\, dA}{\int u\, dA} = \frac{1}{AV^3}\int u^3\, dA \tag{5.4}$$

As the average of cubes is always greater than the cube of the average, the value of α will always be more than 1. The greater the variation in velocity across the section, the larger will be the value of α. For laminar flow in a circular pipe, $\alpha = 2$ (see Sample Prob. 5.1); for turbulent flow in pipes, α ranges from 1.01 to 1.15, but it is usually between 1.03 and 1.06.

In some instances it is very desirable to use the proper value of α, but in most cases the error made in neglecting its divergence from 1.0 is negligible. As precise values of α are seldom known, it is customary in the case of turbulent flow to assume that the kinetic energy is $V^2/2g$ per unit weight of fluid, measured in units of ft·lb/lb = ft or N·m/N = m. In laminar flow the velocity is usually so small that the kinetic energy per unit weight of fluid is negligible.

SAMPLE PROBLEM 5.1 In laminar flow through a circular pipe the velocity profile is a parabola, the equation of which is $u = u_m[1-(r/r_0)^2]$, where u is the velocity at any radius r, u_m is the maximum velocity in the center of the pipe where $r = 0$, and r_0 is the radius to the wall of the pipe. Find α.

Solution

$$u = u_m\left[1-\left(\frac{r}{r_0}\right)^2\right], \qquad dA = 2\pi r\, dr$$

For Eq. (5.4):
$$\int u^3\, dA = 2\pi u_m^3 \int_0^{r_0}\left[1-\left(\frac{r}{r_0}\right)^2\right]^3 r\, dr$$
$$= 2\pi u_m^3 \int_0^{r_0}\left(r - 3\frac{r^3}{r_0^2} + 3\frac{r^5}{r_0^4} - \frac{r^7}{r_0^6}\right) dr$$
$$= 0.25\pi r_0^2 u_m^3$$

and $$Q = AV = \int u\,dA = 2\pi u_m \int_0^{r_0} \left[1 - \left(\frac{r}{r_0}\right)^2\right] r\,dr$$

$$= 2\pi u_m \int_0^{r_0} \left(r - \frac{r^3}{r_0^2}\right) dr = 0.5\pi u_m r_0^2$$

So $$V = \frac{Q}{A} = \frac{0.5\pi u_m r_0^2}{\pi r_0^2} = 0.5 u_m$$

Eq. (5.4): $$\alpha = \frac{1}{AV^3}\int u^3\,dA = \frac{0.25\pi r_0^2 u_m^3}{(\pi r_0^2)(0.5u_m)^3} = 2 \qquad \textbf{\textit{ANS}}$$

EXERCISES

5.1.1 Assume the velocity profile for turbulent flow in a circular pipe to be approximated by a parabola from the axis to a point very close to the wall where the local velocity is $u = 0.7u_m$, where u_m is the maximum velocity at the axis. The equation for this parabola is $u = u_m[1 - 0.3(r/r_0)^2]$. Find α.

5.1.2 Assume an open rectangular channel with the velocity at the surface twice that at the bottom and with the velocity varying as a straight line from top to bottom. Find α.

5.1.3 Find α for the case of a two-dimensional laminar flow, as between two flat plates, for which the velocity profile is parabolic.

5.2 POTENTIAL ENERGY

The potential energy of a particle of fluid depends on its elevation above an arbitrary datum plane. We are usually interested only in *differences* of elevation, and therefore the location of the datum plane used is determined solely by convenience. A fluid particle of weight W situated a distance z above datum possesses a potential energy of Wz. Thus its potential energy per unit weight is z, again measured in units of ft·lb/lb = ft or N·m/N = m.

5.3 INTERNAL ENERGY

Internal energy is stored energy that is associated with the molecular, or internal state of matter; it may be stored in many forms, including thermal, nuclear, chemical, and electrostatic. Here we shall only consider internal thermal energy, which is due to the motion of molecules and forces of attraction between them. This is described more fully in texts on thermodynamics. Experiments indicate that the internal thermal energy is primarily a function of temperature. For liquids and solids, the only exception occurs as they approach the vapor phase, when the internal thermal energy also depends on specific volume, or pressure. When a gas behaves as a perfect gas (Sec. 2.7), this also implies that the internal thermal energy is a function of temperature only. Internal thermal energy can be expressed in terms of energy i per unit of mass[1] or in terms of energy I per unit of weight. Note therefore that $i = gI$.

The zero of internal energy may be taken at any arbitrary temperature, since we are usually concerned only with differences. For a unit mass of substance at a constant volume, $\Delta i = c_v \Delta T$, where c_v is the specific heat at constant volume, whose units are ft·lb/(slug·°R) or N·m(kg·K) in SI units. Thus Δi is expressed in ft·lb/slug (N·m/kg in SI units). Internal energy I per unit of weight is usually expressed in ft·lb/lb = ft (N·m/N = m in SI units).[2]

5.4 GENERAL ENERGY EQUATION FOR STEADY FLOW OF ANY FLUID

Using the principles of Sec. 4.6, let us consider the energy of the fluid system and control volume defined within the stream tube of Fig. 5.1. The fixed

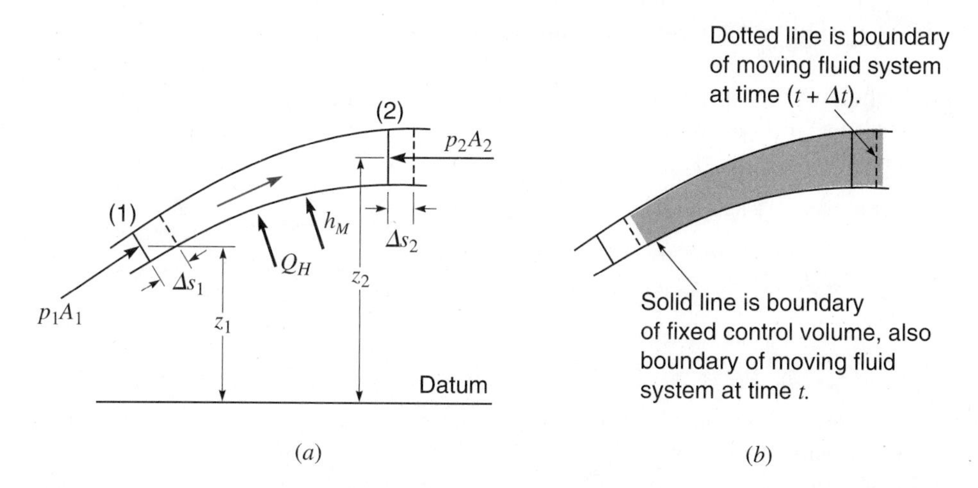

Figure 5.1

[1] In the technical literature internal energy per unit mass is commonly represented by the symbol u. In this text, however, we use i for internal energy per unit mass since u is used in several situations for velocity.

[2] In the BG system of measurement, internal energy I is sometimes expressed in Btu/lb; however, those units are rarely used today. Nevertheless, it is important to be familiar with such units when reading technical papers that were written a number of years ago. 1 Btu = 778 ft·lb.

control volume lies between sections 1 and 2, and the moving fluid system consists of the fluid mass contained at time t in the control volume. During a short time interval Δt we shall assume that the fluid moves a short distance Δs_1 at section 1 and Δs_2 at section 2. As we are restricting ourselves in these discussions to steady flow, $\gamma_1 A_1 \, \Delta s_1 = \gamma_2 A_2 \, \Delta s_2 = g \, \Delta m$, the weight of fluid entering and leaving the control volume during Δt. Recalling the analysis of Sec. 4.6, and letting the general property X now be the energy E, Eq. (4.8) becomes

$$\Delta E_S = \Delta E_{CV} + \Delta E_{CV}{}^{\text{out}} - \Delta E_{CV}{}^{\text{in}}$$

where, as before, subscript S denotes the moving fluid system and subscript CV denotes the fixed control volume. Because the flow is steady, conditions within the control volume do not change so $\Delta E_{CV} = 0$ and

$$\Delta E_S = \Delta E_{CV}{}^{\text{out}} - \Delta E_{CV}{}^{\text{in}} \tag{5.5}$$

Let us now apply the ***first law of thermodynamics*** to the fluid system. This law states that for steady flow, the external work done on any system plus the thermal energy transferred into or out of the system is equal to the change of energy of the system. In other words, for steady flow during time Δt,

$$\text{External work done} + \text{heat transferred} = \Delta E_S$$

Note that work, heat, and energy all have the same units, and thus are interchangeable under certain conditions.

External work may be done on the fluid system in various ways. It is done when the pressure forces acting on the boundaries move, in our case when $p_1 A_1$ and $p_2 A_2$ at the end sections move through Δs_1 and Δs_2, respectively. This work is referred to as ***flow work***. It may be expressed as

$$\begin{aligned}\text{Flow work} &= p_1 A_1 \, \Delta s_1 - p_2 A_2 \, \Delta s_2 \\ &= \frac{p_1}{\gamma_1}(\gamma A_1 \, \Delta s_1) - \frac{p_2}{\gamma_2}(\gamma_2 A_2 \, \Delta s_2) \\ &= \left(\frac{p_1}{\gamma_1} - \frac{p_2}{\gamma_2}\right) g \, \Delta m\end{aligned}$$

The minus signs in the second terms indicate that the force and displacement are in opposite directions.

In addition to flow work, if there is a machine between sections 1 and 2 then there will be ***shaft work***. During the short time interval Δt, we can write

$$\begin{aligned}\text{Shaft work} &= \frac{\text{weight}}{\text{time}} \times \frac{\text{energy}}{\text{weight}} \times \text{time} \\ &= \left(\gamma_1 A_1 \frac{ds_1}{dt}\right) h_M \, \Delta t = (\gamma_1 A_1 \, \Delta s_1) h_M \\ &= (\gamma_1 \, \Delta V\!\!\!\!-_1) h_M = (g \, \Delta m) h_M\end{aligned}$$

where h_M is the energy added to the flow by the machine per unit weight of flowing fluid. If the machine is a pump, h_M is positive; if the machine is a turbine, h_M is negative. Note that frictional shear stresses at the boundary of the fluid system also do work on the fluid within the system. These shear stresses are not external to the system and the work they do is converted to heat which tends to increase the temperature of the fluid within the system.

The heat transferred from an external source into the fluid system over a time interval Δt is

$$\text{Heat transferred} = \left(\gamma_1 A_1 \frac{ds_1}{dt}\right) Q_H \,\Delta t = (\gamma_1 A_1 \,\Delta s_1) Q_H = (g\,\Delta m) Q_H$$

where Q_H is the energy put into the flow by the external heat source per unit weight of flowing fluid. If the heat flow is out of the fluid, the value of Q_H is negative. So we find that the total energy added to (or removed from) the fluid system during time Δt is

$$\begin{aligned} \Delta E_S &= \text{external work done} + \text{heat transferred} \\ &= \text{flow work} + \text{shaft work} + \text{heat transferred} \\ &= \left(\frac{p_1}{\gamma_1} - \frac{p_2}{\gamma_2} + h_M + Q_H\right) g\,\Delta m \end{aligned} \tag{5.6}$$

To evaluate the right-hand side of Eq. (5.5), we first recall, as noted earlier, that for steady flow during time interval Δt, the weights of fluid entering the control volume at section 1 and leaving at section 2 are both equal to $g\,\Delta m$. From Secs. 5.1–5.3, we see that the energy (kinetic + potential + internal) carried across the boundary by $g\,\Delta m$ is

$$\Delta E = g\,\Delta m\left(z + \alpha \frac{V^2}{2g} + I\right)$$

Thus the change in energy of the control volume during Δt is

$$\Delta E_{CV}{}^{\text{out}} - \Delta E_{CV}{}^{\text{in}} = g\,\Delta m\left(z_2 + \alpha \frac{V_2^2}{2g} + I_2\right) - g\,\Delta m\left(z_1 + \alpha \frac{V_1^2}{2g} + I_1\right) \tag{5.7}$$

Substituting Eqs. (5.6) and (5.7) into Eq. (5.5), at the same time factoring out $g\,\Delta m$, we get

$$\frac{p_1}{\gamma_1} - \frac{p_2}{\gamma_2} + h_M + Q_H = \left(z_2 + \alpha_2 \frac{V_2^2}{2g} + I_2\right) - \left(z_1 + \alpha_1 \frac{V_1^2}{2g} + I_1\right)$$

or

$$\left(\frac{p_1}{\gamma_1} + z_1 + \alpha_1 \frac{V_1^2}{2g} + I_1\right) + h_M + Q_H = \left(\frac{p_2}{\gamma_2} + z_2 + \alpha_2 \frac{V_2^2}{2g} + I_2\right) \tag{5.8}$$

This equation applies to liquids, gases, and vapors, and to ideal fluids as well as to real fluids with friction. The only restriction is that it is for steady flow. The p/γ terms represent energy possessed by the fluid per unit weight of fluid by virtue of the pressure under which the fluid exists. Under proper circumstances, this pressure will be released and transformed to other forms of energy, i.e., kinetic, potential, or internal energy. Likewise, it is possible for these other forms of energy to be transformed into pressure energy.

In turbulent flow there are other forms of kinetic energy besides that of translation described in Sec. 5.1. These other forms are the rotational kinetic energy of eddies initiated by fluid friction (Sec. 4.2) and the kinetic energy of the turbulent fluctuations of velocity (Sec. 4.5). They are not represented by any specific terms in Eq. (5.8) because their effect appears indirectly. While the kinetic energy of translation can be converted into increases in p/γ or z, the kinetic energy due to eddies and turbulent fluctuations can never be transformed into anything but thermal energy. Thus they appear as an increase in the numerical value of I_2 over the value it would have if there were no friction.

The general energy equation (5.8) and the continuity equation are two important keys to the solution of many problems in fluid mechanics. For compressible fluids, it is necessary to have a third equation, which is the equation of state [Eq. (2.4)], which provides a relationship between density (or specific volume) and the absolute values of the pressure and temperature.

In many cases Eq. (5.8) is greatly simplified because certain quantities are equal and thus cancel each other, or are zero. Thus, if two points are at the same elevation, $z_1 - z_2 = 0$. If the conduit is well insulated or if the temperature of the fluid and that of its surroundings are practically the same, Q_H may be taken as zero. On the other hand, Q_H may be very large, as in the case of flow of water through a boiler tube. If there is no machine between sections 1 and 2 then the term h_M drops out. If there is a machine present, the rate of shaft work done by it or upon it may be determined by first solving Eq. (5.8) for h_M.

5.5 ENERGY EQUATIONS FOR STEADY FLOW OF INCOMPRESSIBLE FLUIDS, AND BERNOULLI'S THEOREM

For liquids, and even for gases and vapors where the change in pressure is very small, the fluid may be considered as incompressible for all practical purposes, and thus we may take $\gamma_1 = \gamma_2 = \gamma =$ constant. In turbulent flow the value of α is only a little more than unity (Sec. 5.1), and, as a simplifying assumption, it will be assumed equal to unity. If the flow is laminar, $V^2/2g$ is usually very small compared with the other terms in Eq. (5.8); hence little error is introduced if α is set equal to 1.0 rather than 2.0, its true value for laminar flow in circular pipes. Thus, for an incompressible fluid, Eq. (5.8) with

γ = constant and α = 1.0 becomes

$$\left(\frac{p_1}{\gamma}+z_1+\frac{V_1^2}{2g}\right)+h_M+Q_H = \left(\frac{p_2}{\gamma}+z_2+\frac{V_2^2}{2g}\right)+(I_2-I_1) \tag{5.9}$$

Fluid friction produces eddies and turbulence (Sec. 4.2), and these forms of kinetic energy are eventually transformed into thermal energy. If there is no heat transfer, the effect of friction is to produce an increase in temperature, so that I_2 becomes greater than I_1. Or if the flow is isothermal (T and I both constant), there must be a loss of heat Q_H from the system at a rate equal to the rate at which mechanical energy is being converted into thermal energy by friction.

A change in the internal energy of a fluid is accompanied by a change in temperature. If c is the specific heat[3] of the incompressible fluid then, on a mass basis,

$$\frac{\Delta(\text{internal energy})}{\text{Unit of mass}} = \Delta i = i_2 - i_1 = c(T_2 - T_1) \tag{5.10}$$

On a unit weight basis, the change of internal energy is equal to the heat added to or removed from the fluid plus the heat generated by fluid friction, i.e.,

$$\frac{\Delta(\text{internal energy})}{\text{Unit of weight}} = \Delta I = \frac{\Delta i}{g} = I_2 - I_1 = \frac{c}{g}(T_2 - T_1) = Q_H + h_L \tag{5.11}$$

where h_L is the fluid-friction energy loss per unit weight of fluid (ft·lb/lb = ft or N·m/N = J/N = m); h_L is commonly referred to as ***head loss***.

The occurrence of head loss follows directly from the ***second law of thermodynamics*** (the law of degradation of energy). This states that some forms of energy, such as kinetic and potential energies, which can be completely converted to other forms, are "superior" to other "inferior" forms, such as heat and internal energy, which can be only partially converted to the superior forms. Thus, while it is possible to completely convert a given amount of mechanical energy into heat, the opposite is only possible in part, resulting in the mechanical energy (head) loss that always occurs with viscous flow.

Equation (5.11) can be rewritten as

$$h_L = (I_2 - I_1) - Q_H = \frac{c}{g}(T_2 - T_1) - Q_H \tag{5.12}$$

---

[3] For water, c = 1 Btu/(mass of standard lb·°R) = 1 Btu(32.2 ft/s^2)/(lb·°R) = 32.2 Btu/(slug·°R). The Btu (British thermal unit) is defined on the pages facing the covers of this book, and the slug is defined in Sec. 1.5. In SI units, c for water = 1 cal/(g·K). These values can also be expressed as 25,000 ft·lb/(slug·°R) and 4187 N·m/(kg·K), equivalent to 25,000 ft^2/(s^2·°R) and 4187 m^2/(s^2·K) respectively. For the specific heats of other liquids, refer to Appendix A, Table A.4.

We see that if the loss of heat (Q_H negative) is greater than h_L then T_2 will be less than T_1. On the other hand, if there is any absorption of heat (Q_H positive), T_2 will be greater than the value which would have resulted from friction alone. Because friction in fluids generates much less heat than friction between solids, a large value of h_L produces only a very small rise in temperature if there is no heat transfer, or, stated another way, only a very small transfer of heat is required to keep incompressible flow isothermal.

If we substitute h_L for $(I_2 - I_1) - Q_H$ from Eq. (5.12), the general steady flow equation (5.9) for an incompressible fluid becomes

$$\left(\frac{p_1}{\gamma} + z_1 + \frac{V_1^2}{2g}\right) + h_M = \left(\frac{p_2}{\gamma} + z_2 + \frac{V_2^2}{2g}\right) + h_L \tag{5.13}$$

and if there is no machine between sections 1 and 2, the energy equation for an incompressible fluid becomes

$$\left(\frac{p_1}{\gamma} + z_1 + \frac{V_1^2}{2g}\right) = \left(\frac{p_2}{\gamma} + z_2 + \frac{V_2^2}{2g}\right) + h_L \tag{5.14}$$

The head loss h_L may be very large in some cases, and although for any real fluid it can never be zero, there are cases when it is so small that it may be neglected with small error.[4] In such special cases

$$\frac{p_1}{\gamma} + z_1 + \frac{V_1^2}{2g} = \frac{p_2}{\gamma} + z_2 + \frac{V_2^2}{2g} \tag{5.15}$$

from which it follows that

$$\frac{p}{\gamma} + z + \frac{V^2}{2g} = \text{constant} \tag{5.16}$$

The equation in either of these last two forms is known as Bernoulli's theorem, in honor of Daniel Bernoulli (1700–1782), the Swiss physicist who presented this theorem in 1738. Note that Bernoulli's theorem is for a *frictionless incompressible* fluid. However, it can be applied to real incompressible fluids with good results in situations where the frictional effects are very small.

---

[4] It is important to recognize that when frictional effects are very small, frictionless flow may be assumed. For example, the pressure around the nose of a streamlined body (Fig. 4.12) may be determined quite accurately by assuming frictionless flow; however, frictional effects must be considered if the shear stresses at the boundary are to be determined.

Sample Problem 5.2 Water flows at 10 m^3/s in a 1.5-m-diameter aqueduct; the head loss in a 1000 m length of this pipe is 20 m. Find the increase in water temperature assuming no heat enters or leaves the pipe.

Solution

Using Eq. (5.12):
$$20\text{ m} = h_L = \frac{c}{g}(T_2 - T_1)$$

From footnote 3: c for water $= 4187\text{ N·m/(kg·K)}$, and, rearranging,

$$\Delta T = T_2 - T_1 = \frac{gh_L}{c} = \frac{(9.81\text{ m/s}^2)(20\text{ m})}{4187[(\text{kg·m/s}^2)\text{·m}]/(\text{kg·K})}$$
$$= 0.0469\text{ K} \quad \textit{ANS}$$

Sample Problem 5.3 Glycerin (specific gravity 1.26) in a processing plant flows in a pipe at a rate of 700 L/s. At a point where the pipe diameter is 60 cm, the pressure is 300 kN/m^2. Find the pressure at a second point where the pipe diameter is 30 cm if the second point is 1.0 m lower than the first point. Neglect head loss.

Solution

$$\gamma_{\text{water}} = 9810\text{ N/m}^3 = 9.81\text{ kN/m}^3$$

$$V_1 = \frac{0.70\text{ m}^3/\text{s}}{\pi(0.3)^2\text{ m}^2} = 2.48\text{ m/s}, \quad V_2 = 4V_1 = 9.92\text{ m/s}$$

Eq. (5.15):
$$\frac{300}{1.26(9.81)} + 0 + \frac{(2.48)^2}{2(9.81)} = \frac{p_2}{1.26(9.81)} - 1.0 + \frac{(9.92)^2}{2(9.81)}$$

from which
$$p_2 = 254\text{ kN/m}^2 \quad \textit{ANS}$$

Sample Problem 5.4 Water is pumped through a pipeline to a treatment plant at a rate of 130 cfs. The 5-ft-diameter suction line is 3000 ft long, and the 3-ft-diameter discharge line is 1500 ft long. The pump adds energy to the water at a rate of 40 ft·lb/lb, and the total head loss is 56 ft. If the water pressure at the pipeline entrance is 50 psi, what is the pressure at the exit, which is 15 ft higher than the entrance?

Solution

$$V_1 = \frac{130}{(\pi/4)(5)^2} = 6.62\text{ fps}, \quad V_2 = \frac{130}{(\pi/4)(3)^2} = 18.39\text{ fps}$$

Eq. (5.13):
$$\left(\frac{50(144)}{62.4} + 0 + \frac{(6.62)^2}{2(32.2)}\right) + 40 = \left(\frac{p_2(144)}{62.4} + 15.0 + \frac{(18.39)^2}{2(32.2)}\right) + 56$$

from which
$$p_2 = 34.6\text{ psi} \quad \textit{ANS}$$

EXERCISES

5.5.1 A pipeline supplies water to a hydroelectric power plant, the elevation of which is 2200 ft below the level of the water at intake to the pipe. If 8% of this total, or 176 ft, is lost in friction in the pipe, what will be the value of ΔI in Btu/lb; and what will be the rise in temperature if there is no heat transfer?

5.5.2 A pipeline supplies water to a hydroelectric power plant, the elevation of which is 650 m below the level of the water at intake to the pipe. If 8% of this total, or 52 m, is lost in friction in the pipe, what will be the value of ΔI in J/N, and what will be the rise in temperature if there is no heat transfer?

5.5.3 A vertical pipe 5 ft in diameter and 64 ft long has a pressure head at the upper end of 20.2 ft of water. When the flow of water through it is such that the mean velocity is 15 fps, the friction loss is $h_L = 3.8$ ft. Find the pressure head at the lower end of the pipe when the flow is (*a*) downward; (*b*) upward.

5.5.4 A vertical pipe 1.8 m in diameter and 22 m long has a pressure head at the upper end of 5.6 m of water. When the flow of water through it is such that the mean velocity is 5 m/s, the friction loss is $h_L = 1.15$ m. Find the presssure head at the lower end of the pipe when the flow is (*a*) downward; (*b*) upward.

5.5.5 A conical pipe has diameters at the two ends of 1.2 and 4.2 ft and is 48 ft long. It is vertical, and the friction loss is $h_L = 7.6$ ft for flow of water in either direction when the velocity at the smaller section is 28 fps. If the smaller section is at the top and the pressure head there is 6.4 ft of water, find the pressure head at the lower end when the flow is (*a*) downward; (*b*) upward.

5.5.6 In Fig. X5.5.6 the pipe *AB* is of uniform diameter. The pressure at *A* is 25 psi and at *B* is 36 psi. In which direction is the flow, and what is the friction loss in feet of the fluid if the liquid has a specific weight of (*a*) 32 lb/ft^3; (*b*) 95 lb/ft^3?

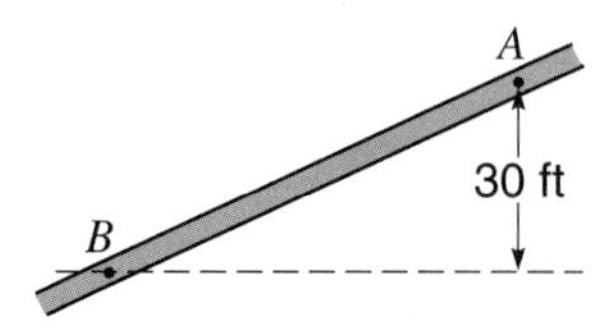

Figure X5.5.6

5.5.7 If the difference in elevation between *A* and *B* in Fig. X5.5.6 is 10 m and the pressures at *A* and *B* are 160 and 260 kN/m^2 respectively, find the direction of flow and the head loss in meters of liquid. Assume the liquid has a specific gravity of 0.88.

5.6 ENERGY EQUATION FOR STEADY FLOW OF COMPRESSIBLE FLUIDS

If sections 1 and 2 are so chosen that there is no machine between them, and if α is assumed as unity, Eq. (5.8) becomes

$$\left(\frac{p_1}{\gamma_1} + I_1 + z_1 + \frac{V_1^2}{2g}\right) + Q_H = \left(\frac{p_2}{\gamma_2} + I_2 + z_2 + \frac{V_2^2}{2g}\right) \tag{5.17}$$

For most compressible fluids, i.e., gases or vapors, the quantity p/γ is usually very large compared with $z_1 - z_2$ because of the small value of γ, and therefore the z terms are usually omitted. But $z_1 - z_2$ should not be ignored unless it is known to be negligible compared with the other quantities.

For gases and vapors, the p/γ and the I terms are usually combined into a single term called ***enthalpy***. Hence enthalpy represents a composite energy property possessed by a given mass (or weight) of gas or vapor. In thermodynamics enthalpy is commonly expressed in terms of energy per unit mass (h) rather than energy per unit weight. Thus[5]

$$h = i + \frac{p}{\rho} = gI + \frac{p}{\rho} \tag{5.18}$$

and so

$$I + \frac{p}{\gamma} = \frac{h}{g} \tag{5.19}$$

With these changes, Eq. (5.17) becomes

$$\frac{h_1}{g} + \frac{V_1^2}{2g} + Q_H = \frac{h_2}{g} + \frac{V_2^2}{2g} \tag{5.20}$$

This equation may be used for any gas or vapor and for any process. Some knowledge of thermodynamics is required to evaluate the enthalpies, and in the case of vapors it is necessary to use vapor tables or charts, because their properties cannot be expressed by any simple equations. Many more aspects of the flow of compressible fluids are discussed in Chap. 13.

SAMPLE PROBLEM 5.5 In an air conditioning system, air flows without heat gain or loss through a horizontal pipe of uniform diameter. At section 1 the pressure is 150 psia, the velocity is 80 fps, and the temperature is 70°F; at section 2 the pressure is 120 psia and the temperature is 50°F. Find (*a*) the change in kinetic energy of the air; (*b*) the head (mechanical energy) loss in Btu/lb; (*c*) the change in enthalpy; all between sections 1 and 2. Assume the air to be a perfect gas.

---

[5] Values of enthalpy h for vapors commonly used in engineering, such as steam, ammonia, freon, and others, may be obtained from vapor tables or charts. For a perfect gas and practically for real gases, $\Delta h = c_p \Delta T$, where c_p is specific heat at constant pressure. For air at usual pressures, c_p has a value of 6000 ft·lb/(slug·°R) [or 1003 N·m/(kg·K)]. These are equivalent to 6000 ft^2/(s^2·°R) [or 1003 m^2/(s^2·K)].

Solution

Eq. (2.4): $$\frac{p}{\rho T} = R = \text{constant}, \quad \text{so} \quad \frac{p_1}{\rho_1 T_1} = \frac{p_2}{\rho_2 T_2} \qquad (1)$$

From Eq. (4.16*a*): $$\frac{\dot{m}}{A} = \rho_1 V_1 = \rho_2 V_2 \qquad (2)$$

Multiplying (1) by (2) to eliminate ρ:

$$\frac{p_1 V_1}{T_1} = \frac{p_2 V_2}{T_2}$$

So $$V_2 = V_1\left(\frac{p_1 T_2}{p_2 T_1}\right) = 80\left(\frac{150}{120}\right)\frac{510}{530} = 96.2 \text{ fps}$$

(*a*) $$\Delta KE = \frac{V_2^2}{2g} - \frac{V_1^2}{2g} = \frac{96.2^2 - 80^2}{2(32.2)} = 44.4 \text{ ft·lb/lb increase} \qquad \textbf{\textit{ANS}}$$

(*b*) From Eq. (5.14) with $z_1 = z_2$:

$$h_L = \frac{p_1}{\rho_1 g} - \frac{p_2}{\rho_2 g} + \frac{V_1^2 - V_2^2}{2g}$$

Eq. (2.4): $$\frac{p}{\rho} = RT$$

Table A.5 for air: $R = 1715 \text{ ft}^2/(\text{s}^2\cdot°\text{R})$

So $$h_L = \frac{R}{g}(T_1 - T_2) + \frac{V_1^2 - V_2^2}{2g}$$

$$= \frac{1715}{32.2}(530 - 510) - 44.4 = 1021 \text{ ft}$$

$$h_L = \frac{1021 \text{ ft·lb/lb}}{778 \text{ ft·lb/Btu}} = 1.312 \text{ Btu/lb} \qquad \textbf{\textit{ANS}}$$

(*c*) From Eq. (5.20) with $Q_H = 0$:

$$\frac{h_1 - h_2}{g} = \frac{V_2^2 - V_1^2}{2g} = \Delta KE = 44.4 \text{ ft·lb/lb}$$

$$\Delta h = g(\Delta KE) = 32.2(44.4) = 1493 \text{ ft·lb/slug decrease} \qquad \textbf{\textit{ANS}}$$

EXERCISES

5.6.1 Gas, which may be assumed to be perfect, flows at a constant temperature through a uniform, horizontal pipe. At section 1 the pressure is 120 psia and the velocity is 60 fps; at section 2 the pressure is 100 psia and the velocity is 72 fps. Find (*a*) the change in enthalpy; (*b*) the gain or loss of heat per lb; both between sections 1 and 2. [*Hint*: Recall Eq. (2.4).]

5.6.2 Air flows isothermally (constant temperature) through a horizontal duct of constant cross section. At station 1 the pressure is 800 kPa abs and the velocity is 20 m/s; at station 2 the pressure is 1000 kPa abs and the velocity is 16 m/s. Find (*a*) the change in enthalpy; (*b*) the gain or loss of heat per newton; both between stations 1 and 2. Assume the air to be a perfect gas. [*Hint*: Recall Eq. (2.4).]

5.6.3 Oxygen flows without gain or loss of heat through a horizontal pipe of constant cross section. At section 1 the pressure is 170 psia, the velocity is 75 fps, and the temperature is 50°F; at section 2 the pressure is 125 psia, the velocity is 98 fps, and the temperature is 30°F. Find (*a*) the head (mechanical energy) loss in Btu/lb; (*b*) the change in enthalpy; both between sections 1 and 2. Assume the oxygen to be a perfect gas. [*Hint*: Recall Eq. (2.4).]

5.6.4 Air, which may be assumed to be a perfect gas, flows without heat gain or loss through a uniform horizontal pipe. At station 1 the pressure is 1135 kPa abs, the velocity is 25 m/s, and the temperature is 10°C; at station 2 the pressure is 830 kPa abs, the velocity is 33 m/s, and the temperature is 0°C. Find (*a*) the head (mechanical energy) loss in J/N; (*b*) the change in enthalpy; both between sections 1 and 2. [*Hint*: Recall Eq. (2.4).]

5.7 EULER'S EQUATION FOR STEADY MOTION OF AN IDEAL FLUID ALONG A STREAMLINE

Referring to Fig. 5.2, let us consider frictionless steady flow of a fluid along the streamline. We shall consider the forces acting on a small cylindrical element of the fluid in the direction of the streamline and apply Newton's second law, that is, $F = ma$. We recall from Secs. 4.12 and 4.13 that in steady flow the velocity does not vary at a point (local acceleration = 0), but that it may vary with position (convective acceleration $\neq 0$).

The forces tending to accelerate the fluid mass are the pressure forces on the two ends of the element

$$p\,dA - (p + dp)\,dA = -dp\,dA$$

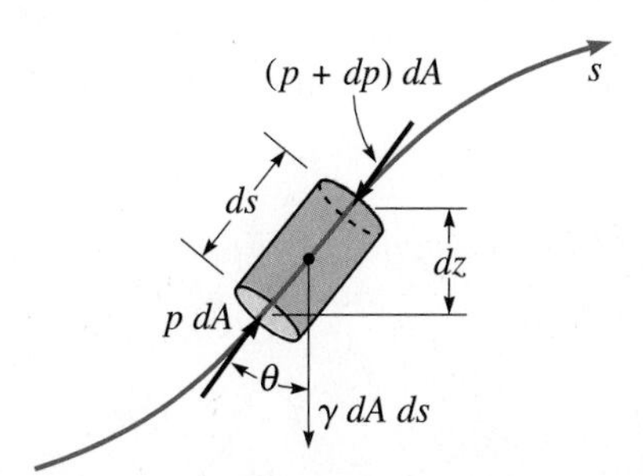

Figure 5.2
Element on streamtube (ideal fluid)

where dA is the cross-section of the element at right-angles to the streamline, and the weight component in the direction of motion

$$-\rho g\, ds\, dA \frac{dz}{ds} = -\rho g\, dA\, dz$$

The mass of the element is $\rho\, dA\, ds$, while we see from Eq. (4.24) that its acceleration for steady flow can be expressed as $V(dV/ds)$. Thus applying, $F = ma$, we get

$$-dp\, dA - \rho g\, dA\, dz = \rho\, ds\, dA\, V \frac{dV}{ds}$$

Dividing by $-\rho\, dA$,

$$\frac{dp}{\rho} + g\, dz + V\, dV = 0 \tag{5.21}$$

This equation is commonly referred to as the ***one-dimensional Euler***[6] ***equation***, because it was first derived by Leonhard Euler (1707–1783), a Swiss mathematician, in about 1750. It applies to both compressible and incompressible flow, since the variation of ρ over the elemental length ds is small. Equation (5.21) can also be expressed as

$$\frac{dp}{\gamma} + dz + d\frac{V^2}{2g} = 0 \tag{5.22}$$

For the case of a compressible ideal fluid, since $\gamma \neq$ constant, an equation relating γ (or ρ) to p and T must be introduced before integrating Eq. (5.22).

For the case of an incompressible fluid (γ = constant), Eq. (5.22) can be integrated to give

$$\int \frac{dp}{\gamma} + \int dz + \int d\frac{V^2}{2g} = \text{constant} \tag{5.23}$$

Thus

$$\frac{p}{\gamma} + z + \frac{V^2}{2g} = \text{constant} = \text{total head} = H \tag{5.24}$$

This is Bernoulli's equation [Eq. (5.16)] for steady flow of a frictionless incompressible fluid along a streamline. Thus we have developed the Bernoulli equation from two viewpoints: first from energy considerations and now from Newton's second law.

If there is no flow,

$$\frac{p}{\gamma} + z = \text{constant} \tag{5.25}$$

---

[6] Euler is pronounced (oi'lər), to rhyme with boiler.

This equation is identical to Eq. (3.6); it shows that for an incompressible fluid at rest, at any point in the fluid the summation of the pressure head p/γ plus the elevation z is equal to the sum of these two quantities at any other point.

EXERCISES

5.7.1 Assume the flow to be frictionless in the siphon shown in Fig. X5.7.1. Find the rate of discharge in cfs and the pressure head at B if the pipe has a uniform diameter of 4 in.

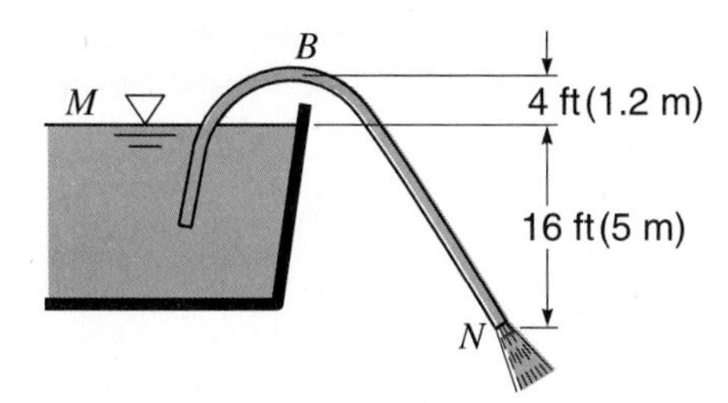

Figure X5.7.1

5.7.2 Referring to Fig. X5.7.1, assume the flow to be frictionless in the siphon. Find the rate of discharge in m^3/s and the pressure head at B if the pipe has a uniform diameter of 18 cm.

5.8 PRESSURE IN FLUID FLOW

Pressure in Conduits of Uniform Cross-Section

Let us now consider how pressure varies over a cross-section of flow in a uniform conduit. Figure 5.3 shows a small prism of a flowing fluid. Perpendicular to the motion and in the plane of the sketch, the forces acting on the faces of the prism are p_1A and p_2A as shown. Forces parallel to the direction of motion, namely the pressure and friction forces and the weight

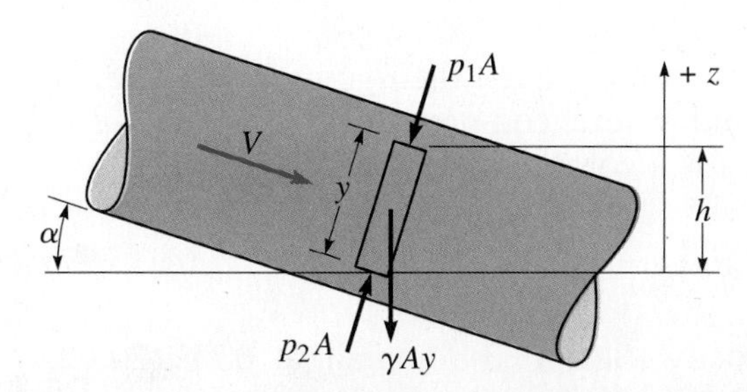

Figure 5.3

component, must balance out if the flow is steady and parallel. Summing the perpendicular forces, we get $p_1 A + \gamma A y \cos\alpha - p_2 A = 0$, where y is the dimension of the prism as shown, and A is its cross-sectional area. From this, we get

$$p_2 - p_1 = \gamma y \cos\alpha = \gamma h = \gamma(z_1 - z_2) = -\gamma(\Delta z)$$

which is similar to Eq. (3.3). Therefore in any plane perpendicular to the direction of flow the pressure varies according to the hydrostatic law if the flow is parallel and steady. The average pressure is then the pressure at the center of gravity of such an area. The pressure is lowest near the top of the pipe, and cavitation (Sec. 5.11), if it were to occur, would appear there first. On a horizontal diameter through the pipe the pressure is everywhere the same. Since the velocity is higher near the center than near the walls, it follows that the local energy head is also higher near the center. This emphasizes the fact that a flow equation applies along the same streamline, but not between two streamlines, any more than between two streams in two separate channels.

Static Pressure

In a flowing fluid, the fluid pressure p is called the ***static pressure*** because it is the pressure that would be measured by an instrument that is static with respect to the fluid, i.e., moving with the fluid. It is measured by piezometer tubes (Sec. 3.6) and other devices that attempt to minimize disturbance to the flow (see Sec. 11.2).

Stagnation Pressure

The center streamline in Fig. 4.12 shows that the velocity becomes zero at the stagnation point. If p/γ denotes the static-pressure head at some distance away where the velocity is V, while p_0/γ denotes the pressure head at the stagnation point, then, applying Eq. (5.15) to these two points, $p/\gamma + 0 + V^2/2g = p_0/\gamma + 0 + 0$, or the ***stagnation pressure*** is

$$p_0 = p + \gamma \frac{V^2}{2g} = p + \rho \frac{V^2}{2} \tag{5.26}$$

The quantity $\gamma V^2/2g$, or $\rho V^2/2$, is sometimes called the ***dynamic pressure.***

Equation (5.26) applies to a fluid where compressibility may be disregarded. In Sec. 13.5 it is shown that for a compressible fluid,

$$p_0 = p + \rho \frac{V^2}{2}\left(1 + \frac{V^2}{4c^2} + \ldots\right) \tag{5.27}$$

where c is the sonic (acoustic) velocity (Sec. 13.3). For air at 68°F (20°C),

$c \approx 1130$ fps (345 m/s). If $V = 226$ fps (69 m/s) the error in neglecting the compressibility factor, which is the quantity in parentheses, is only one percent. But for higher values of V, the effect becomes much more important. Equation (5.27) is, however, restricted to values of V/c less than one.

EXERCISES

5.8.1 Find the stagnation pressure on the nose of a submarine moving at 10.5 knots in seawater ($\gamma = 64$ lb/ft^3) when it is 55 ft below the surface.

5.8.2 Find the stagnation pressure on the nose of a submarine moving at 8 m/s in seawater ($\gamma = 10\,050$ N/m^3) when it is 25 m below the surface.

5.8.3 Find the stagnation pressure on the nose of a fish swimming at 25 fps in fresh water ($\gamma = 62.4$ lb/ft^3) when it is 5 ft below the surface.

5.9 HEAD

In Eq. (5.14) each term has the dimensions of *length*. Thus p/γ, called the ***pressure head***, represents the energy per unit weight stored in the fluid by virtue of the pressure under which the fluid exists; z, called the ***elevation head***, represents the potential energy per pound of fluid, and $V^2/2g$, called the ***velocity head***, represents the kinetic energy per pound of fluid. The sum of these three terms is called the ***total head*** usually denoted by H, so that

$$H = \frac{p}{\gamma} + z + \frac{V^2}{2g} \tag{5.28}$$

Each term in this equation, although ordinarily expressed in feet (or meters), represents *foot pounds of energy per pound of fluid flowing* (newton meters of energy per newton of fluid flowing in SI units). Note also that the sum of the middle two terms above, $(p/\gamma + z)$, is called the ***piezometric head*** or the ***static head*** (see Sec. 5.12).

For a frictionless incompressible fluid with no machine between 1 and 2, $H_1 = H_2$, but for a real fluid,

$$H_1 = H_2 + h_L \tag{5.29}$$

which is merely a brief way of writing Eq. (5.14). For a real fluid, it is obvious that if there is no input of energy head h_M by a machine between sections 1 and 2, the total head must decrease in the direction of flow.

If there is a machine between sections 1 and 2 then

$$H_1 + h_M = H_2 + h_L \tag{5.30}$$

If the machine is a pump, $h_M = h_p$, where h_p is the energy head put into the flow by the pump. If the machine is a turbine, $h_M = -h_t$, where h_t is the energy head extracted from the flow by the turbine.

5.10 POWER CONSIDERATIONS IN FLUID FLOW

In deriving Eq. (5.14), the term $g\,\Delta m$ representing weight of fluid was factored out; thus every term of the equation represents energy per unit weight (i.e., energy head). If the energy head is multiplied by the weight rate of flow, the resulting product represents power, or rate of energy transfer, since

$$\text{Power} = \frac{\text{energy}}{\text{time}} = \frac{\text{energy}}{\text{weight}} \times \frac{\text{weight}}{\text{time}} = h \times G = h \times g\dot{m} = h\gamma Q \qquad (5.31)$$

In BG units,

$$\text{Horsepower} = \frac{\gamma Q h}{550} \qquad (5.32)$$

while in SI units,

$$\text{Kilowatts} = \frac{\gamma Q h}{1000} \qquad (5.33)$$

where γ = the unit weight of fluid, lb/ft^3 (N/m^3 in SI units)
Q = the rate of flow, ft^3/sec (m^3/s in SI units)
h = the energy head, ft (m in SI units)

Note: 1 hp = 550 ft·lb/sec = 0.745 700 kW.

In these equations h may be any head for which the corresponding power is desired. For example, to find the power extracted from the flow by a turbine (i.e., the rate at which shaft work is done on a turbine, see Sec. 5.4), substitute h_t for h; to find the power of a jet, substitute $V_j^2/2g$ for h, where V_j is the jet velocity; and to find the power lost because of fluid friction, substitute h_L for h.

When power is transmitted through a process or machine, some power is lost in the process due to friction. The ***efficiency*** of the transmission is the fraction of the power input that appears in the output, i.e.

$$\text{Efficiency } \eta = \frac{\text{power output}}{\text{power input}} \qquad (5.34)$$

The efficiency of pumps and turbines is discussed in more detail in Secs. 15.3 and 16.9, respectively.

With respect to power, it may be recalled from mechanics that the power developed when a force F acts on a translating body, or when a torque T acts on a rotating body, is given by

$$\text{Power} = Fu = T\omega \tag{5.35}$$

where u is linear velocity in feet per second (or meters per second) and ω is angular velocity in radians per second. The force F represents the component force in the direction of the velocity u. These equations will be referred to in Chap. 6, where the dynamic forces exerted by moving fluids are discussed, and again in Chaps. 15 and 16, in the discussion of turbomachinery.

Sample Problem 5.6 Find the rate of energy loss due to pipe friction for the pipe of Sample Prob. 5.2.

Solution

Eq. (5.33):

$$\text{Rate of energy loss} = \frac{\gamma Q h}{1000}, \qquad \text{where } h = h_L$$

$$= \frac{(9810\ \text{N/m}^3)(10\ \text{m}^3/\text{s})(20\ \text{m})}{1000}$$

$$= 1962\ \text{kW} \qquad \textbf{\textit{ANS}}$$

Sample Problem 5.7 A liquid with a specific gravity of 1.26 is being pumped in a pipeline from A to B. At A the pipe diameter is 24 in (60 cm) and the pressure is 45 psi (300 kN/m^2). At B the pipe diameter is 12 in (30 cm) and the pressure is 50 psi (330 kN/m^2). Point B is 3 ft (1.0 m) lower than A. Find the flow rate if the pump puts 22 hp (16 kW) into the flow. Neglect head loss.

Solution (BG units)

Eq. (5.32):

$$\text{HP} = \frac{(1.26 \times 62.4)Qh_p}{550}$$

Rearranging:

$$h_p = \frac{153.9}{Q}$$

Eq. (5.14) with elevation A as datum, and with $h_L = 0$ (given), using $V = Q/A$, gives

$$\frac{45(144)}{1.26(62.4)} + 0 + \frac{(Q/\pi)^2}{2(32.2)} + \frac{153.9}{Q} = \frac{50(144)}{1.26(62.4)} - 3 + \frac{[Q/(0.25\pi)]^2}{2(32.2)}$$

i.e.,

$$\frac{Q^2}{42.37} + 6.157 - \frac{153.9}{Q} = 0$$

Without an equation solver, we can solve this cubic equation by trials (see

---

■: Programmed computing aids (Appendix B) could help solve problems marked with this icon.

Sample Prob. 3.5) as follows:

Trial Q:	10.0	20.0	17.0	14.0	14.15
Left side:	−22.26	7.903	3.925	−0.210	0.006

Thus $$Q = 14.15 \text{ cfs} \qquad \textbf{\textit{ANS}}$$

Note: The other two roots of this equation are imaginary.

Solution (SI Units)

Eq. (5.33): $$\text{kW} = 16 = \frac{(1.26\times 9810)Qh_p}{1000}$$

Rearranging: $$h_p = \frac{1.294}{Q}$$

Eq. (5.14) with elevation A as datum, and with $h_L = 0$ (given), using $V = Q/A$, gives

$$\frac{300\times 10^3}{1.26(9810)} + 0 + \left[\frac{Q}{\pi(0.3)^2}\right]^2 \frac{1}{2(9.81)} + \frac{1.294}{Q} = \frac{330\times 10^3}{1.26(9810)} - 1.0 + \left[\frac{Q}{\pi(0.15)^2}\right]^2 \frac{1}{2(9.81)}$$

By trial, $$Q = 0.418 \text{ m}^3/\text{s} \qquad \textbf{\textit{ANS}}$$

Note: The other two roots of this cubic equation in Q are imaginary.

EXERCISES

5.10.1 Water entering a pump through an 8-in-diameter pipe at 4 psi has a flow rate of 4 cfs. It leaves the pump through a 4-in-diameter pipe at 15 psi. Assuming that the suction and discharge sides of the pump are at the same elevation, find the horsepower delivered to the water by the pump.

5.10.2 After entering a pump through an 18-cm-diameter pipe at 35 kN/m^2, oil ($s = 0.82$) leaves the pump through a 12-cm-diameter pipe at 120 kN/m^2. The suction and discharge sides of the pump are at the same elevation. Find the rate at which energy is delivered to the oil by the pump if the flow rate is 70 L/s.

5.10.3 Water from a reservoir is being supplied to a powerhouse that is located 975 ft below the reservoir surface. Discharging through a nozzle, the water has a jet velocity 245 fps and jet diameter of 6 in. Find the horsepower lost to friction between the reservoir and the jet, and find the horsepower of the jet.

5.10.4 Water from a reservoir is being supplied to a powerhouse that is located 325 m below the surface of the reservoir. Discharging through a nozzle, the water has a jet velocity of 75 m/s and jet diameter of 25 cm. Find the kW lost to friction between the reservoir and the jet, and find the power of the jet in kW.

5.10.5 A turbine, located 765 ft below the water surface of the intake, carries a flow of 125 cfs. The pipeline leading to it has a friction loss of 30 ft. Find the horsepower delivered if the turbine efficiency is 90%.

5.11 CAVITATION

The rapid vaporization and recondensation of liquid as it briefly flows through a region of low absolute pressure is called ***cavitation***, as first noted in Sec. 2.13. This phenomenon is not possible in gas flow, because a gas does not change state at low pressure, whereas a liquid will change to a gas (vapor) if the pressure is low enough. The possibility of cavitation occurring in liquid flows must be investigated, because it can cause serious damage.

The dangerous, temporary low-pressure conditions associated with cavitation result from temporary high velocities, in accordance with Bernoulli's theorem [Eq. (5.16)]. In view of that theorem, at a given location (elevation z = constant) in a liquid flow where no energy is added or removed, if the velocity head increases, there must be a corresponding decrease in the pressure head. However, so long as there is some liquid present to evaporate, there is a minimum absolute pressure possible, namely, the vapor pressure of the liquid. The vapor pressure depends upon the liquid and its temperature. If the conditions are such that a calculation results in a lower absolute pressure than the vapor pressure, this simply means that the assumptions upon which the calculations are based no longer apply. Thus the critical condition for cavitation is

$$\left(\frac{p_{\text{crit}}}{\gamma}\right)_{\text{abs}} = \frac{p_v}{\gamma}$$

But

$$\left(\frac{p_{\text{crit}}}{\gamma}\right)_{\text{abs}} = \frac{p_{\text{atm}}}{\gamma} + \left(\frac{p_{\text{crit}}}{\gamma}\right)_{\text{gage}}$$

so that

$$\left(\frac{p_{\text{crit}}}{\gamma}\right)_{\text{gage}} = -\left(\frac{p_{\text{atm}}}{\gamma} - \frac{p_v}{\gamma}\right) \tag{5.36}$$

where p_{atm}, p_v, and p_{crit} represent the atmospheric pressure, the vapor pressure, and the critical (or minimum) possible pressure, respectively, in liquid flow. Equation (5.36) shows that the gage pressure head in a flowing liquid can be negative, but no more negative than $p_{\text{atm}}/\gamma - p_v/\gamma$.

If at any point the local velocity is so high that the pressure in a liquid is reduced to its vapor pressure, the liquid will then vaporize (or boil) at that point, and bubbles of vapor will form. As the fluid flows into a region of higher pressure, the bubbles of vapor will suddenly condense; in other words, they may be said to ***collapse*** or ***implode***. This action may produce very high dynamic pressure upon the adjacent solid walls, and since this action is continuous and has a high frequency, the material in that zone may be damaged.

Turbine runners, pump impellers, and ship screw propellers are often severely and quickly damaged by such action, because holes are rapidly produced in the metal (see Fig. 16.15). Similar damage can occur immediately downstream of partly open valves. Overflow spillways and other types of hydraulic structures built of concrete are also subject to damage from cavitation. The damaging action is commonly referred to as ***pitting.*** Not only is cavitation destructive, but it may cause a drop in efficiency of the machine or propeller or other device, and it may produce undesirable cavitation noise and vibration.

In order to avoid cavitation, it is necessary that the absolute pressure at every point be above the vapor pressure. There are various ways to ensure this. One way is to raise the general pressure level, by placing the device below the intake level so that the liquid flows to it by gravity rather than being drawn up by suction. Another way is to design the machine so that there are no local velocities high enough to produce such a low pressure. A third way is to admit atmospheric air into the low-pressure zone; this is often done downstream of partly open valves and on overflow spillways (see Sec. 11.13).

Figure 5.4 shows photographs of blades for an axial-flow pump set up in a transparent-lucite working section where the pressure level can be varied. For *a*, *b*, and *c*, there was the same water velocity on the same vane but with decreasing absolute pressures. This resulted in the formation of a vapor

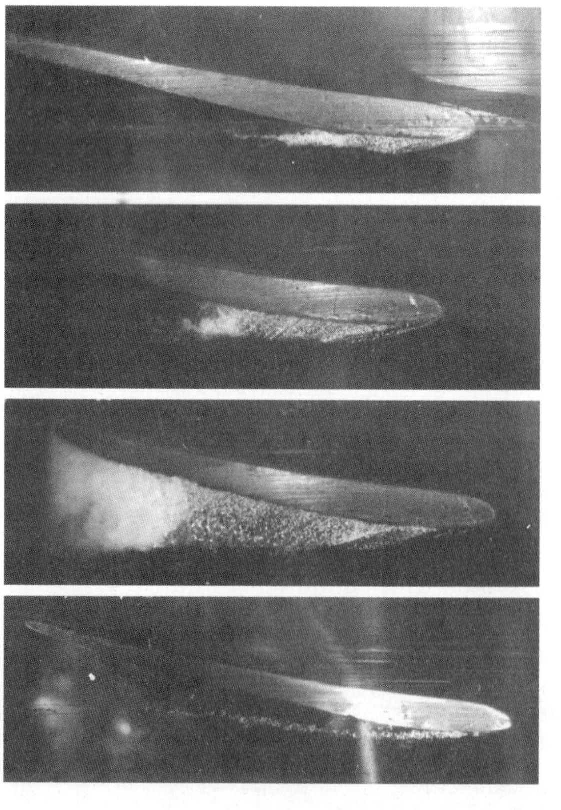

Figure 5.4
Cavitation phenomena: flow around an axial-flow pump blade illustrating the effect of reducing absolute pressure in (*a*), (*b*), and (*c*), and the effect of a slight change of shape in (*d*). (*Photographs by Hydrodynamics Laboratory, California Institute of Technology*)

pocket of increasing size. For *d*, the stream flow and the pressure were the same as for *b*, but the nose of the blade was slightly different in shape, which gave a different type of bubble formation. This shows the effect of a slight change in design.

Sample Problem 5.8 A liquid ($s = 0.86$) with a vapor pressure of 3.8 psia flows through the horizontal constriction in Fig. S5.8. Atmospheric pressure is 26.8 inHg. Find the maximum theoretical flow rate (i.e., at what Q does cavitation occur?). Neglect head loss.

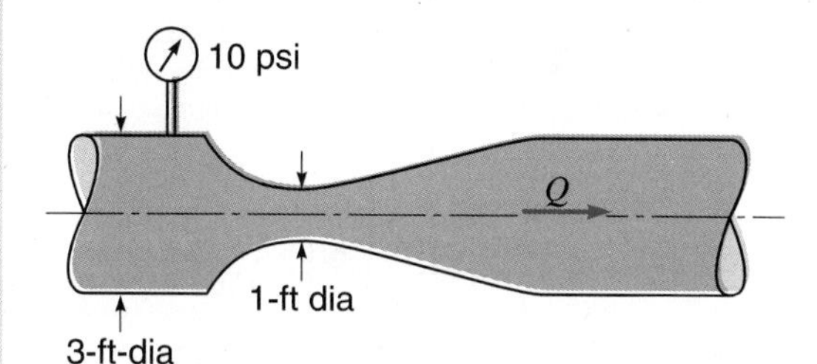

Figure S5.8

Solution

Since the standard atmosphere is equivalent to 29.92 inHg and 14.70 psia (Sec. 3.5),

$$p_{atm} = \frac{26.8}{29.92}(14.70) = 13.16 \text{ psia}$$

Eq. (5.36):
$$\left(\frac{p_{crit}}{\gamma}\right)_{gage} = -\left[\frac{13.16 - 3.8}{0.86(62.4)}\right]144 = -25.1 \text{ ft}$$

$$V_1 = \frac{Q}{A_1} = \frac{4Q}{\pi D_1^2} = \frac{4Q}{\pi 3^2} = \frac{Q}{7.07}; \quad V_2 = \frac{4Q}{\pi 1^2} = \frac{Q}{0.785}$$

Eq. (5.15):
$$\frac{10(144)}{0.86(62.4)} + 0 + \left(\frac{Q}{7.07}\right)^2 \frac{1}{2(32.2)} = -25.1 + 0 + \left(\frac{Q}{0.785}\right)^2 \frac{1}{2(32.2)}$$

$$Q = 45.7 \text{ cfs} \qquad \textit{ANS}$$

EXERCISES

5.11.1 Water at a temperature of 160°F flows through the horizontal constriction of Fig. S5.8 when the atmospheric pressure is 27.9 inHg. Neglecting head loss, find the flow rate at which cavitation begins.

5.11.2 Water at a temperature of 50°C flows through a horizontal constriction similar to that in Fig. S5.8 when the atmospheric pressure is 709 mmHg. The gage reading is 30 kN/m², $d_1 = 0.5$ m, and $d_2 = 0.2$ m. Neglecting head loss, find the flow rate at which cavitation begins.

5.11.3 Water at a temperature of 80°F flows through the horizontal constriction of Fig. S5.8 at a rate of 65 cfs when the atmospheric pressure is 27.9 inHg. Neglecting head loss, find the largest throat constriction diameter d_2 which will cause cavitation.

5.11.4 Water at a temperature of 50°C flows at a rate of 2 m^3/s through a horizontal constriction similar to that in Fig. S5.8 when the atmospheric pressure is 750 mmHg. The gage reading is 30 kN/m^2 and $d_1 = 0.5$ m. Neglecting head loss, find the largest throat constriction diameter d_2 that will cause cavitation.

5.11.5 Referring to Fig. S5.8, find the maximum theoretical flow rate for water at 70°C. Neglect head loss. The diameters are 80 cm and 40 cm respectively, the upstream pressure is 20 kN/m^2, and the atmospheric pressure is 750 mmHg.

5.12 DEFINITION OF HYDRAULIC GRADE LINE AND ENERGY LINE

When dealing with flow problems involving liquids, the concept of energy line and hydraulic grade line is usually advantageous. Even with gas flow, the concepts can be useful. The term $p/\gamma + z$ is referred to as the ***piezometric head***, because it represents the level to which liquid will rise in a piezometer tube (Sec. 3.6). The ***piezometric head line**, or **hydraulic grade line*** (HGL), is a line drawn through the tops of the pizeometer columns. A pitot tube (Sec. 11.3), a small open tube with its open end pointing upstream, will intercept the kinetic energy of the flow and hence indicate the ***total energy head***, $p/\gamma + z + u^2/2g$. Referring to Fig. 5.5, which depicts the flow of an ideal fluid,

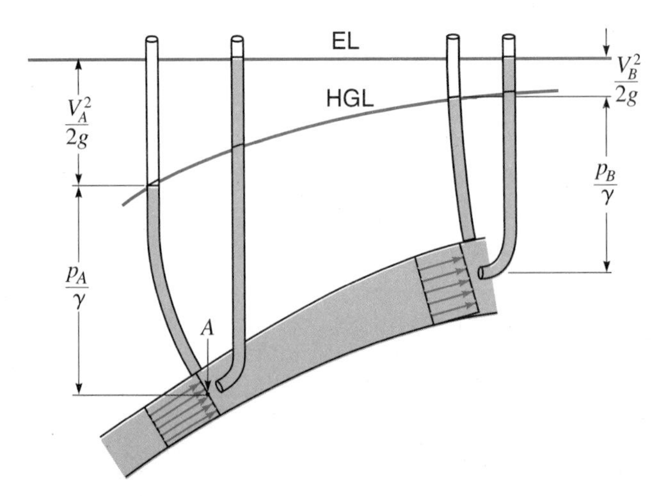

Figure 5.5
Ideal fluid

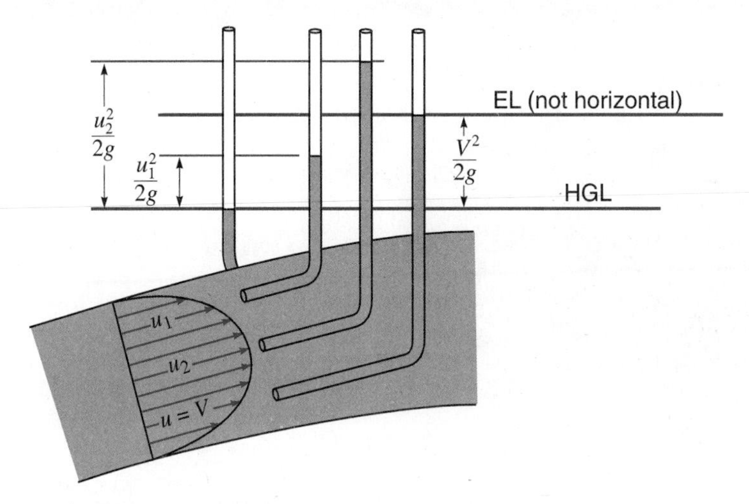

Figure 5.6
Real fluid

the vertical distance from point A on the stream tube to the level of the piezometric head at that point represents the ***static pressure head*** in the flow at point A (see also Sec. 5.8). The vertical distance from the liquid level in the piezometer tube to that in the pitot tube is $V^2/2g$. In Fig. 5.5 the line sketched through the pitot-tube liquid levels is known as the ***energy line*** (EL). For flow of an ideal fluid, the energy line is horizontal, since there is no head loss; for a real fluid, the energy line slopes downward in the direction of flow because of head loss due to friction.

Because the local velocity u usually varies across a flow cross-section, as depicted in Fig. 5.6, the reading obtained by a pitot tube will depend on the precise location of its submerged open end. For a pitot tube to indicate the true level of the energy line, it must be placed in the flow at a point where $u^2/2g = \alpha(V^2/2g)$, or, in other words, where $u = \sqrt{\alpha}V$. If α (Sec. 5.1) is assumed to have a value of 1.0 then, to indicate the true energy line, the tube must be placed in the flow at a point where $u = V$. One rarely knows ahead of time where in the flow $u = V$; hence the correct positioning of a pitot tube, in order that it indicate the true position of the energy line, is generally unknown.

Sample Problem 5.9 Water flows in a wide open channel as shown in Fig. S5.9. Two pitot tubes are connected to a differential manometer containing a liquid ($s = 0.82$). Find u_A and u_B.

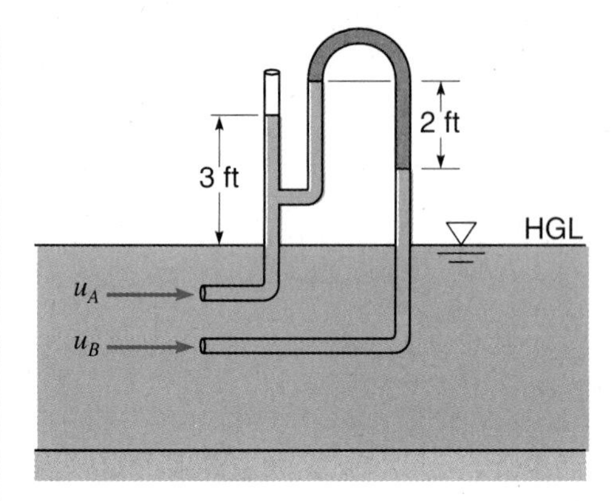

Figure S5.9

Solution

The water surface in the channel is the HGL, and the water surface in the pitot tube is at the EL, so the difference is

$$\frac{u_A^2}{2g} = 3 \text{ ft}$$

from which $\qquad u_A = \sqrt{2(32.2)3} = 13.90 \text{ fps} \qquad$ ***ANS***

From Sec. 3.6, Fig. 3.14*b*, and Eq. (3.13) for the manometer,

$$\frac{p_A}{\gamma} - \frac{p_B}{\gamma} = z_B - z_A + \left(1 - \frac{s_M}{s_F}\right)MR$$

The tip of piezometer A is a stagnation point (Sec. 4.10) where $V = 0$, so, considering Eq. (5.15) for the approaching streamline, with y_A as the depth of point A, we obtain

$$y_A + z_A + \frac{u_A^2}{2g} = \frac{p_A}{\gamma} + z_A + 0, \qquad \text{i.e.,} \qquad \frac{p_A}{\gamma} = \frac{u_A^2}{2g} + y_A$$

and, subtracting a similar equation for point B, we obtain

$$\frac{p_A}{\gamma} - \frac{p_B}{\gamma} = \frac{u_A^2}{2g} - \frac{u_B^2}{2g} + y_A - y_B$$

Substituting for $\Delta p/\gamma$ from the manometer equation written previously,

$$z_B - z_A + \left(1 - \frac{s_M}{s_F}\right)MR = \frac{u_A^2}{2g} - \frac{u_B^2}{2g} + y_A - y_B$$

and, noting that $y_A + z_A = y_B + z_B$ = the elevation of the HGL, this simplifies to

$$\frac{u_A^2}{2g} - \frac{u_B^2}{2g} = \left(1 - \frac{s_M}{s_F}\right) MR$$

i.e., $$3 - \frac{u_B^2}{2g} = (1 - 0.82)2 = 0.360 \text{ ft}$$

from which $u_B = \sqrt{(2(32.2)(3 - 0.360)} = 13.04$ fps ***ANS***

5.13 LOSS OF HEAD AT SUBMERGED DISCHARGE

When a fluid with a velocity V is discharged from the end of a pipe into a tank or reservoir that is so large that the velocity within it is negligible, the entire kinetic energy of the flow is dissipated. That this is so can be seen by examining Fig. 5.7. In the pipe up to point (a) the kinetic energy of the flowing fluid per unit weight of fluid is $V^2/2g$, but at point (b) in the tank the velocity is zero and hence the kinetic energy per unit weight of fluid is also zero. Thus the loss of head in this case, with ***submerged discharge***, is $V^2/2g$. The loss occurs after the fluid leaves the end of the pipe. This is a situation where fast moving fluid impinges on stationary fluid. It is an impact situation not unlike the case where a fast moving mudball collides with an immovable wall. The loss of head at submerged discharge is $V^2/2g$, irrespective of whether the fluid is ideal or real, compressible or incompressible. A more detailed discussion of this topic is presented in Sec. 8.19.

5.14 APPLICATION OF HYDRAULIC GRADE LINE AND ENERGY LINE

Familiarity with the concept of the energy line and hydraulic grade line is useful in the solution of flow problems involving incompressible fluids. If a

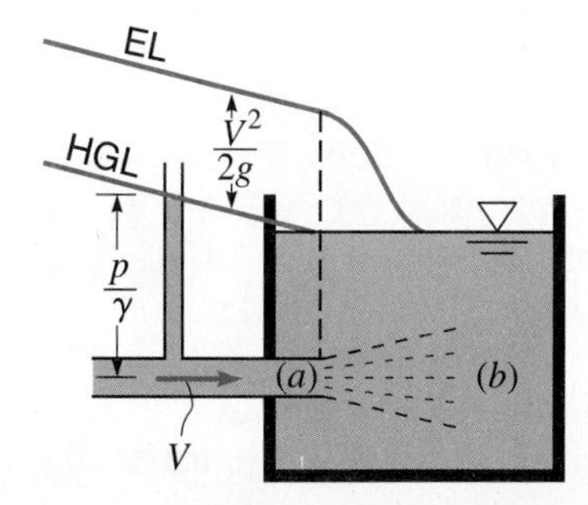

Figure 5.7

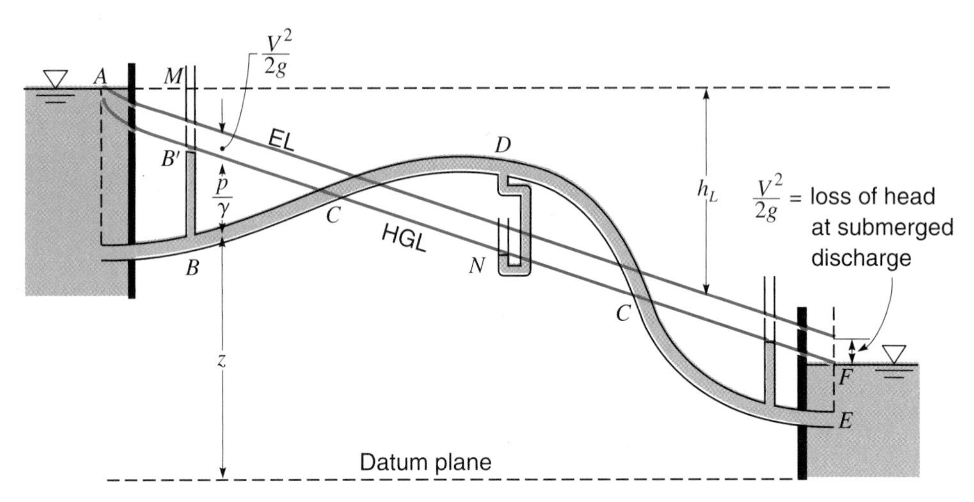

Figure 5.8
Hydraulic and energy grade lines

piezometer tube is erected at B in Fig. 5.8, the liquid will rise in it to a height BB' equal to the pressure head existing at that point. If the end of the pipe at E were closed so that no flow would take place, the height of this column would then be BM. The drop from M to B' when flow occurs is due to two factors, one of these being that a portion of the pressure head has been converted into the velocity head which the liquid has at B, and the other that there has been a loss of head due to fluid friction between A and B.

As noted in Sec. 5.12, if a series of piezometers were erected along the pipe, the liquid would rise in them to various levels along what is called the ***hydraulic grade line*** (Figs. 5.5 and 5.8). It may be observed that the hydraulic grade line represents what would be the free surface if one could exist and maintain the same conditions of flow.

The hydraulic grade line indicates the pressure along the pipe, since at any point the vertical distance from the pipe to the hydraulic grade line is the pressure head at that point, assuming the profile to be drawn to scale. At C this distance is zero, thus indicating that the absolute pressure within the pipe at that point is atmospheric. At D the pipe is above the hydraulic grade line, indicating that there the pressure head is $-DN$, or a vacuum of DN ft (or m) of liquid.

If the profile of a pipeline is drawn to scale, not only does the hydraulic grade line enable the pressure head to be determined at any point by measurement on the diagram, but it shows by mere inspection the variation of the pressure in the entire length of the pipe. The hydraulic grade line is a straight line only if the pipe is straight and of uniform diameter. But for the gradual curvatures that are often found in long pipelines, the deviation from a straight line will be small. Of course, if there are local losses of head, aside

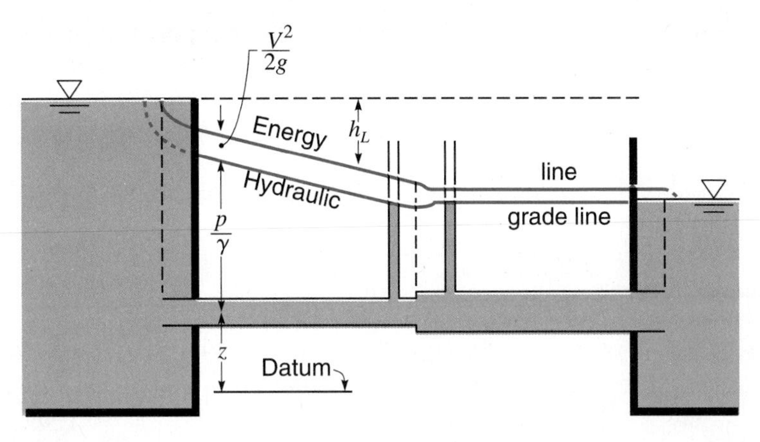

Figure 5.9
(*Plotted to scale*)

from those due to normal pipe friction, there may be abrupt drops in the hydraulic grade line. Changes in diameter with resultant changes in velocity will also cause abrupt changes in the hydraulic grade line.

If the velocity head is constant, as in Fig. 5.8, the drop in the hydraulic grade line between any two points is the value of the loss of head between those two points, and the slope of the hydraulic grade line is then a measure of the rate of loss. Thus in Fig. 5.9 the rate of loss in the larger pipe is much less than in the smaller pipe. If the velocity changes, the hydraulic grade line might actually rise in the direction of flow, as shown in Figs 5.9 and 5.10.

The vertical distance from the level of the surface at A in Fig. 5.8 down to the hydraulic grade line represents $V^2/2g + h_L$ from A to any point in

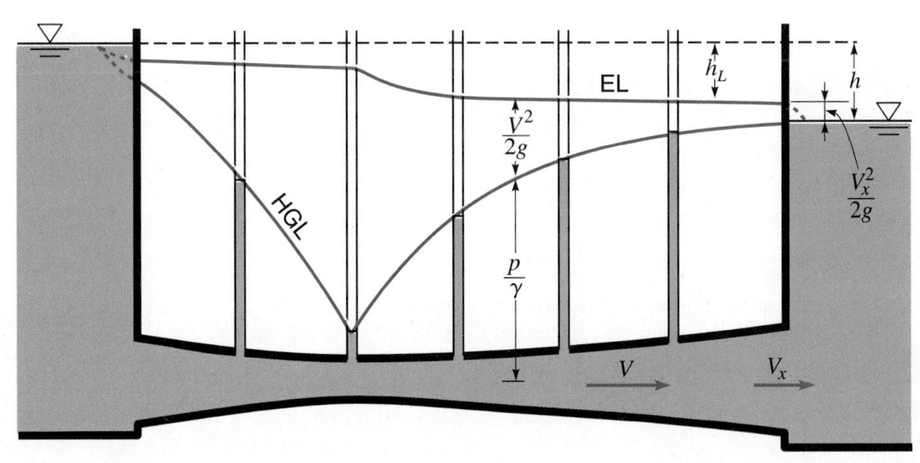

Figure 5.10
(*Plotted to scale*)

question. Hence the position of the grade line is independent of the position of the pipe. Therefore it is not necessary to compute pressure heads at various points in the pipe to plot the hydraulic grade line. Instead, values of $V^2/2g + h_L$ from A to various points can be set off below the horizontal line through A, and this procedure is often more convenient. If the pipe is of uniform diameter, it is necessary to locate only a few points, and often only two are required.

If Fig. 5.8 represents to scale the profile of a pipe of uniform diameter, the hydraulic grade line can be drawn as follows. At the intake to the pipe there will be a drop below the surface at A, which should be set off equal to $V^2/2g$ plus a local entrance loss. (This latter is explained in Sec. 8.18.) At E the pressure is EF, and hence the grade line must end at the surface at F. If the pipe discharged freely into the air at E, the line would pass through E. The location of other points, such as B' and N, may be computed if desired. In the case of a *long* pipe of uniform diameter the error is very small if the hydraulic grade line is drawn as a straight line from A to F if the discharge is submerged, or from A to E if there is a free discharge into the atmosphere.

If values of h_L are set off below the horizontal line through A, the resulting line represents values of the total energy head H measured above any arbitrary datum plane inasmuch as the line is above the hydraulic grade line a distance equal to $V^2/2g$. This line is the ***energy grade line***, usually referred to as simply the ***energy line*** (see also Sec. 5.12). It shows the rate at which the energy decreases, and it must always drop downward in the direction of flow unless there is an energy input from a pump. The energy line is also independent of the position of the pipeline.

Energy lines are shown in Figs. 5.8–5.10. The last one, plotted to scale, shows that the chief loss of head is in the diverging portion and just beyond the section of minimum diameter. In all three of these cases there is a submerged discharge and hence the velocity head is lost at discharge. Note that in Fig. 5.10 the loss is greatly reduced by the conical diffuser (diverging pipe), which results in an enlarged discharge area and hence a reduced velocity at discharge. The large pressure changes that occur in converging–diverging pipes similar to Fig. 5.10 provide a very convenient means of measuring flow rates, as described in Sec. 11.7.

EXERCISES

5.14.1 A pump, having an efficiency of 90%, lifts water to a height of 465 ft at the rate of 250 cfs. The friction loss in the pipe is 35 ft. What is the required horsepower? Also provide a sketch of the energy line and the hydraulic grade line of this system.

5.14.2 A pump, having an efficiency of 90%, lifts water to a height of 155 m at the rate of 7.5 m^3/s. The friction loss in the pipe is 13 m. What is the required pump power in kW? Also provide a sketch of the energy line and the hydraulic grade line of this system.

5.14.3 Assume there is friction loss in the siphon of Fig. X5.7.1. The loss between the intake and B is 0.7 m and between B and N is 1.1 m. What is the rate of discharge and pressure head at B when the diameter is 18 cm?

5.14.4 Assume there is friction loss in the siphon of Fig. X5.7.1. The loss between the intake and B is 3 ft and that between B and N is 4 ft. What is the rate of discharge and pressure head at B when the diameter is 8 in?

5.14.5 Refer to Fig. X5.7.1. Find the maximum difference in elevation between M and N instead of the 5 m shown in the sketch. Assume friction is neglected and the minimum pressure allowable in the siphon is a vacuum of -9.8 m of water.

5.14.6 Refer to Fig. X5.7.1. Find the maximum difference in elevation between M and N instead of the 16 ft shown in the sketch. Assume friction is neglected and the minimum pressure allowable in the siphon is a vacuum of -32.8 ft of water.

5.14.7 In Fig. X5.14.7 let $a = 25$ ft, $b = 60$ ft, $c = 40$ ft, and $d = 2$ ft. All the losses of energy are to be ignored when the stream discharging into the air at E has a diameter of 4 in. What are pressure heads at B, C, and D if the diameter of the vertical pipe is 5 in?

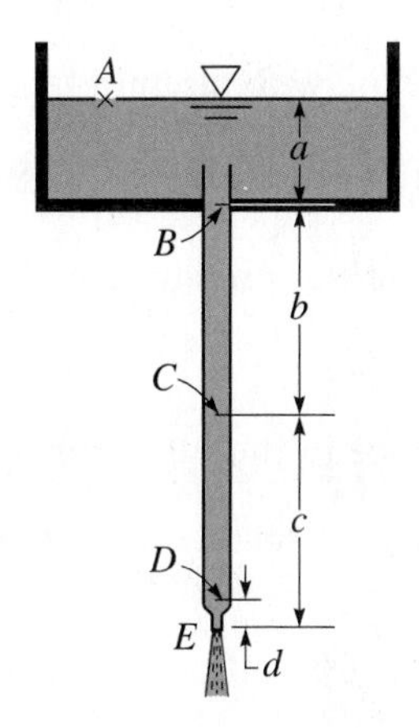

Figure X5.14.7

5.14.8 In Fig. X5.14.7 let $a = 7.5$ m, $b = c = 15$ m, and $d = 30$ cm. All the losses of energy are to be ignored when the stream discharging into the air at E has a diameter of 8 cm. What are pressure heads at B, C, and D if the diameter of the vertical pipe is 12 cm?

5.15 METHOD OF SOLUTION OF FLOW PROBLEMS

For the solutions of problems of liquid flow, there are two fundamental equations: the equation of continuity (4.17) and the energy equation in one of the forms from Eqs. (5.8)–(5.15). The following procedure may be employed:

1. Choose a datum plane through any convenient (lower) point.
2. Note at what sections the velocity is known or is to be assumed. If at any point the section area is great compared with its value elsewhere, the velocity head is so small that it may be disregarded.
3. Note at what points the pressure is known or is to be assumed. In a body of liquid at rest with a free surface the pressure is known at every point within the body. The pressure in a jet is the same as that of the medium surrounding the jet.
4. Note whether or not there is any point where all three terms, pressure, elevation, and velocity, are known.
5. Note whether or not there is any point where there is only one unknown quantity.

It is generally possible to write an energy equation that will fulfill conditions 4 and 5. If there are two unknowns in the equation then the continuity equation must be used also. The application of these principles is shown in the following illustrative examples.

SAMPLE PROBLEM 5.10 In a fire fighting system, a pipeline with a pump leads to a nozzle as shown in Fig. S5.10. Find the flow rate when the pump develops a head of 80 ft. Assume that the head loss in the 6-in-diameter pipe may be expressed by $h_L = 5V_6^2/2g$, while the head loss in the 4-in-diameter pipe is $h_L = 12V_4^2/2g$. Sketch the energy line and hydraulic grade line, and find the pressure head at the suction side of the pump.

Solution

Select the datum as the elevation of the water surface in the reservoir. If V_3 is the jet velocity, note from continuity that

$$V_6 = (\tfrac{3}{6})^2 V_3 = 0.25V_3, \qquad V_4 = (\tfrac{3}{4})^2 V_3 = 0.563V_3$$

Writing an energy equation from the surface of the reservoir (point 1) to the jet (point 3),

$$\left(\frac{p_1}{\gamma} + z_1 + \frac{V_1^2}{2g}\right) - h_{L_6} + h_p - h_{L_4} = \frac{p_3}{\gamma} + z_3 + \frac{V_3^2}{2g}$$

$$0 + 0 + 0 - 5\frac{V_6^2}{2g} + 80 - 12\frac{V_4^2}{2g} = 0 + 10 + \frac{V_3^2}{2g}$$

Express all velocities in terms of V_3:

$$-\frac{5(0.25V_3)^3}{2g} + 80 - 12\frac{(0.563V_3)^2}{2g} = 10 + \frac{V_3^2}{2g}$$

$$V_3 = 29.7 \text{ fps}$$

$$Q = A_3V_3 = \frac{\pi}{4}\left(\frac{3}{12}\right)^2 29.7 = 1.458 \text{ cfs} \qquad \textbf{\textit{ANS}}$$

Head loss in suction pipe:

$$h_L = 5\frac{V_6^2}{2g} = \frac{5(0.25V_3)^2}{2g} = \frac{0.312V_3^2}{2g}$$
$$= 4.28 \text{ ft}$$

Head loss in discharge pipe

$$h_L = 12\frac{V_4^2}{2g} = \frac{12(0.563V_3)^2}{2g} = 52.0 \text{ ft}$$

$$\frac{V_3^2}{2g} = 13.70 \text{ ft}, \quad \frac{V_4^2}{2g} = 4.33 \text{ ft}, \quad \frac{V_6^2}{2g} = 0.856 \text{ ft}$$

ANS

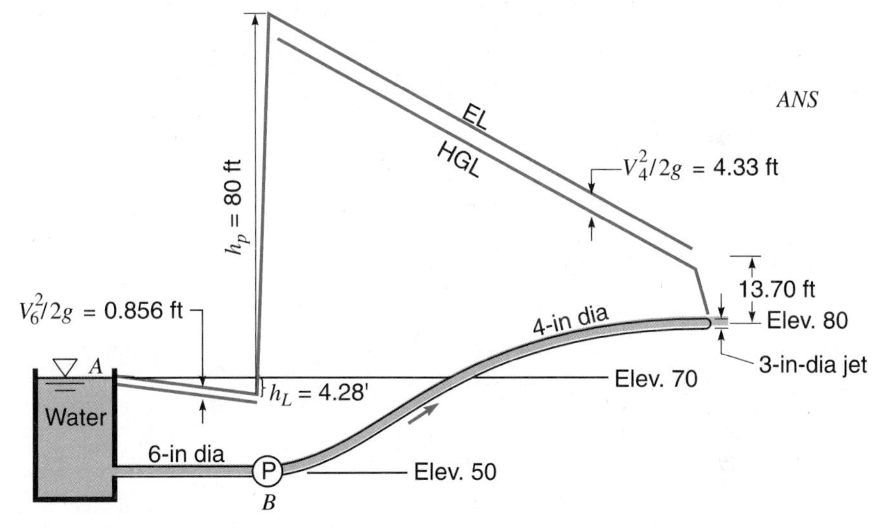

Figure S5.10

The energy line and hydraulic grade line are drawn on the figure to scale. Inspection of the figure shows that the pressure head on the suction side of the pump is

$$p_B/\gamma = 70 - 50 - 4.28 - 0.856 = 14.86 \text{ ft} \qquad \textit{ANS}$$

Likewise, the pressure head at any point in the pipe may be found if the figure is to scale.

SAMPLE PROBLEM 5.11. Given the two-dimensional flow shown in Fig. S5.11. Determine the flow rate per m width of channel. Assume no head loss.

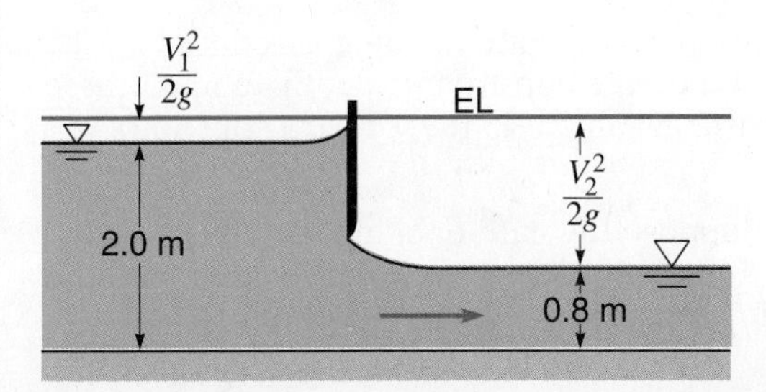

Figure S5.11

Solution

Select the datum as the (effectively horizontal) channel bed. The hydraulic grade line is represented by the water surface in the region where the streamlines are parallel. The energy line is a distance $V^2/2g$ above the water surface, assuming $\alpha = 1.0$. If there is a no head loss, the energy line is horizontal. Writing the energy equation (5.15) from section 1 to 2, we have

$$0 + 2.0 + \frac{V_1^2}{2g} = 0 + 0.8 + \frac{V_2^2}{2g} \qquad (1)$$

Note that this applies either (*a*) between points on the water surface, with $p_1 = p_2 = 0$, $z_1 = 2.0$, and $z_2 = 0.8$, or (*b*) between points on the bed, with $z_1 = z_2 = 0$, $p_1/\gamma = 2.0$, and $p_2/\gamma = 0.8$.

But from the continuity equation (4.17), for 1 m of channel width perpendicular to the figure,

$$(2 \times 1)V_1 = (0.8 \times 1)V_2 \qquad (2)$$

Substituting Eq. (1) into Eq. (2), and using $g = 9.81\ \text{m/s}^2$, we obtain

$$V_1 = 2.12\ \text{m/s}, \quad V_2 = 5.29\ \text{m/s}, \quad \frac{V_1^2}{2g} = 0.229\ \text{m}, \quad \frac{V_2^2}{2g} = 1.429\ \text{m}$$

and $Q = A_1V_1 = (2 \times 1)2.12 = 4.24\ \text{m}^3/\text{s}$ (for 1 m of channel width) ***ANS***

EXERCISES

5.15.1 Refer to Sample Problem 5.11. If the depths upstream and downstream of the gate were 8.0 ft and 4.0 ft respectively, find the flow rate per foot of channel width.

5.15.2 Refer to Sample Problem 5.11. If the depths upstream and downstream of the gate were 2.0 m and 0.6 m respectively, find the flow rate per meter of channel width.

5.15.3 Refer to Sample Problem 5.11. Suppose the gate opening is reduced so the depth downstream is 2.0 ft. Find the upstream depth under these conditions if the flow rate remains constant at 45 ft^3/sec per ft of width.

5.15.4 Refer to Sample Problem 5.11. Suppose the gate opening is reduced so the depth downstream is 0.7 m. Find the upstream depth under these conditions if the flow rate remains constant at 4.24 m^3/s per m of width.

5.16 JET TRAJECTORY

A free liquid jet in air will describe a ***trajectory***, or path under the action of gravity, with a vertical velocity component that is continually changing. The trajectory is a streamline, and consequently, if air friction is neglected, Bernoulli's theorem may be applied to it, with all the pressure terms zero. Thus the sum of the elevation and velocity head must be the same for all points of the curve. The energy grade line is a horizontal line at distance $V_0^2/2g$ above the nozzle, where V_0 is the initial velocity of the jet as it leaves the nozzle (Fig. 5.11).

The equation for the trajectory may be obtained by applying Newton's equations of uniformly accelerated motion to a particle of the liquid passing from the nozzle to point P, whose coordinates are (x, z), in time t. Then $x = V_{x0}t$ and $z = V_{z0}t - \frac{1}{2}gt^2$. Evaluating t from the first equation and substituting it in the second gives

$$z = \frac{V_{z0}}{V_{x0}} x - \frac{g}{2V_{x0}^2} x^2 \tag{5.37}$$

By setting $dz/dx = 0$, we find that z_{max} occurs when $x = V_{x0}V_{z0}/g$. Substituting this value for x in Eq. (5.37) gives $z_{max} = V_{z0}^2/2g$. Thus Eq. (5.37) is that of an inverted parabola having its vertex at $x_0 = V_{x0}V_{z0}/g$ and $z_0 = V_{z0}^2/2g$. Since the velocity at the top of the trajectory is horizontal and equal to V_{x0}, the distance from this point to the energy line is evidently $V_{x0}^2/2g$. This may be obtained in another way by considering that $V_0^2 = V_{x0}^2 + V_{z0}^2$. Dividing each term by $2g$ gives the relations shown in Fig. 5.11.

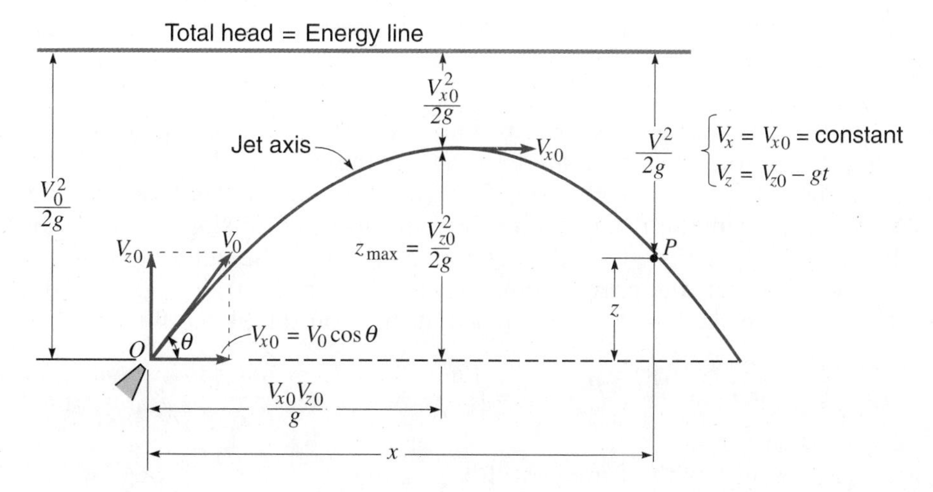

Figure 5.11
Jet trajectory

If the jet is initially horizontal, as in the flow from a vertical orifice, $V_{x0} = V_0$ and $V_{z0} = 0$. Equation (5.37) is then readily reduced to an expression for the initial jet velocity in terms of the coordinates from the vena contracta (Fig. 11.14) to any point of the trajectory, z now being positive downward:

$$V_0 = x\sqrt{\frac{g}{2z}} \tag{5.38}$$

SAMPLE PROBLEM 5.12 If a water jet is inclined upward 30° from the horizontal, what must be its initial velocity to reach over a 10 ft wall at a horizontal distance of 60 ft, neglecting friction?

Solution

$$V_{x0} = V_0 \cos 30° = 0.866V_0$$

$$V_{z0} = V_0 \sin 30° = 0.5V_0$$

From Newton's laws,

$$x = 0.866\ V_0 t = 60 \tag{1}$$

$$z = 0.5V_0 t - 0.5gt^2 = 10 \tag{2}$$

From Eq. (1), $t = 69.3/V_0$. Substituting this into Eq. (2),

$$0.5V_0\frac{69.3}{V_0} - \frac{32.2}{2}\left(\frac{69.3}{V_0}\right)^2 = 10$$

from which $V_0^2 = 3140$, or $V_0 = 56.0$ fps ***ANS***

When a fluid is discharged into a second fluid with a similar density, a plume of the first fluid forms. We are familiar with smoke and steam plumes; similar plumes form when treated sewage effluent is discharged under the ocean from outfall sewers (see Exer. 5.16.3). Such plumes rise because $s_1 < s_2$. To a first approximation, neglecting fluid friction and mixing, if the second fluid is not moving then the path of the plume may be computed as a jet trajectory. However, the gravitational acceleration in the trajectory equations must then be replaced by the force per unit mass on the plume fluid, which is

$$g' = \frac{(\rho_1 - \rho_2)g\forall}{\rho_1\forall} = \frac{(s_1 - s_2)g}{s_1} = g\left(1 - \frac{s_2}{s_1}\right) \tag{5.39}$$

For a rising plume, both g' and z will be negative.

EXERCISES

5.16.1 It is required to throw a fire stream so as to reach the window in the wall shown in Fig. X5.16.1 ($h = 14$ m). Assuming a jet velocity of 25 m/s and neglecting air friction, find the angle (or angles) of inclination which will achieve this result, given $d = 23$ m and $a = 2$ m.

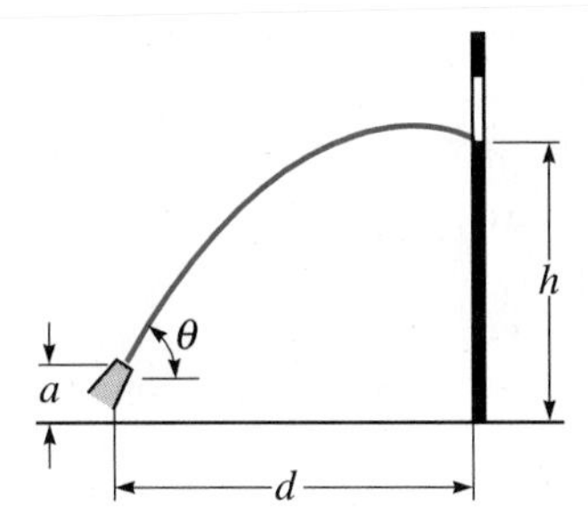

Figure X5.16.1

5.16.2 A jet issues horizontally from an orifice in the vertical wall of a large tank (Fig. X5.16.2). Neglecting air resistance, determine the velocity of the jet for the following variety of trajectories: (*a*) $x = 1.0$ m, $y = 1.0$ m; (*b*) $x = 2.0$ m, $y = 2.0$ m; (*c*) $x = 3.0$ m, $y = 3.0$ m; (*d*) $x = 4.0$ m, $y = 4.0$ m. Express the answers in m/s.

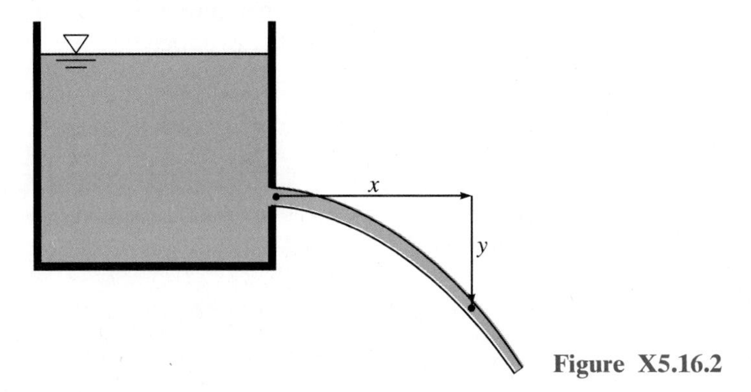

Figure X5.16.2

5.16.3 Freshwater sewage effluent is discharged from a horizontal outfall pipe on the floor of the ocean at a point where the depth is 120 ft. When the ocean is still, the jet is observed to rise to the surface at a point 95 ft horizontally from the end of the pipe. Assuming the ocean water to have a specific gravity of 1.03 and neglecting fluid friction and mixing of the jet with the ocean water, find the velocity at the end of the outfall. [*Note*: In this case the jet is submerged, and it is no longer possible to neglect the density of the surrounding medium; hence the value of g in Eqs. (5.37) and (5.38) must be adjusted accordingly.]

5.17 FLOW IN A CURVED PATH

The energy equations previously developed apply fundamentally to flow along a streamline or along a stream of large cross-section if certain average values are used. Now conditions will be investigated in a direction normal to a streamline. Figure 5.12 shows *an element of fluid moving in a horizontal plane*[7] with a velocity V along a curved path of radius r. The element has a linear dimension dr in the plane of the paper and an area dA normal to the plane of the paper. The mass of this fluid element is $\rho\, dA\, dr$, and the normal component of acceleration is V^2/r. Thus the centripetal force acting upon the element toward the center of curvature is $\rho\, dA\, dr\, V^2/r$. As the radius increases from r to $r + dr$, the pressure will change from p to $p + dp$. Thus the resultant force in the direction of the center of curvature is $dp\, dA$. Equating these two forces and dividing by dA,

$$dp = \rho \frac{V^2}{r}\, dr \tag{5.40}$$

When horizontal flow is in a straight line for which r is infinity, the value of dp is zero. That is, no difference in pressure can exist in the horizontal direction transverse to horizontal flow in a straight line.

As dp is positive if dr is positive, the equation shows that pressure increases from the concave to the convex side of the stream, but the exact way in which it increases depends upon the way in which V varies with the radius. In the next two sections two important practical cases will be presented in which V varies in two different ways.

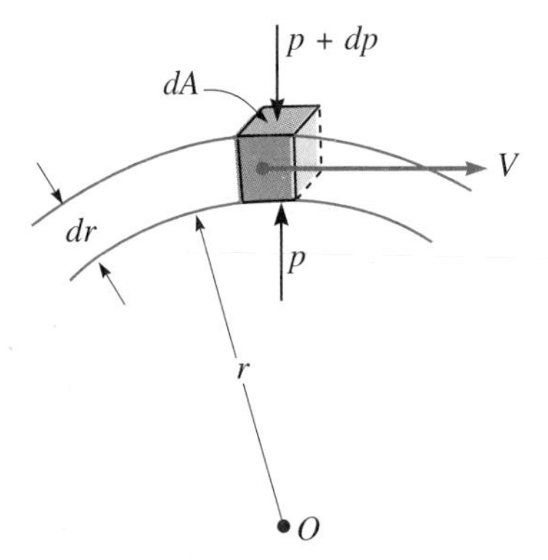

Figure 5.12

[7] A more generalized analysis of flow along a curved path in a vertical or inclined plane leads to a result that includes z-terms.

EXERCISES

5.17.1 Figure X5.17.1 shows a two-dimensional ideal flow in a vertical plane. Data are as follows: $r = 12$ ft, $b = 5$ ft, $\gamma = 62.4$ lb/ft^3, $V = 24$ fps. If the pressure at A is 6 psi, find the pressure at B.

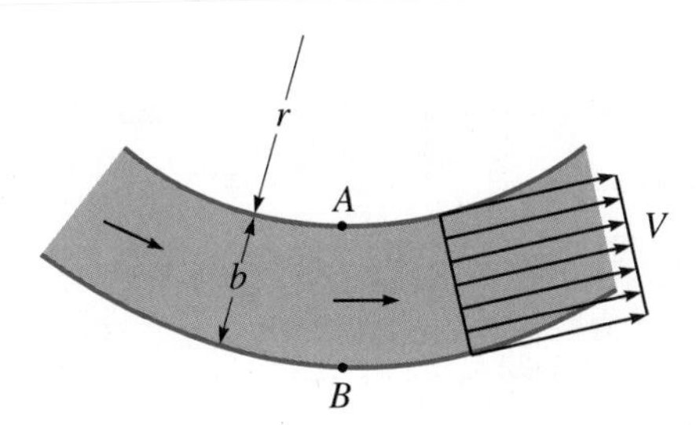

Figure X5.17.1

5.17.2 Refer to Fig. X5.17.1. Flow occurs in a vertical plane. Data are as follows: $r = 7$ m, $b = 3$ m, $\gamma = 9.81$ kN/m^3, $V = 5$ m/s. Find the pressure at A if the pressure at B is 150 kN/m^2.

5.18 FORCED OR ROTATIONAL VORTEX

In theory, a fluid may be made to rotate as a solid body without relative motion between particles, either by the rotation of a containing vessel or by stirring the contained fluid, so as to force it to rotate. Thus an ***external torque*** is applied. A common example is the rotation of liquid within a centrifugal pump or of gas in a centrifugal compressor.

Cylindrical Forced Vortex

If the entire body of fluid rotates as a solid then V varies directly with r; that is, $V = r\omega$, where ω is the imposed angular velocity. Inserting this value in Eq. (5.40), for the case of rotation about a vertical axis in any horizontal plane we have

$$dp = \rho\omega^2 r\, dr = \frac{\gamma}{g}\omega^2 r\, dr$$

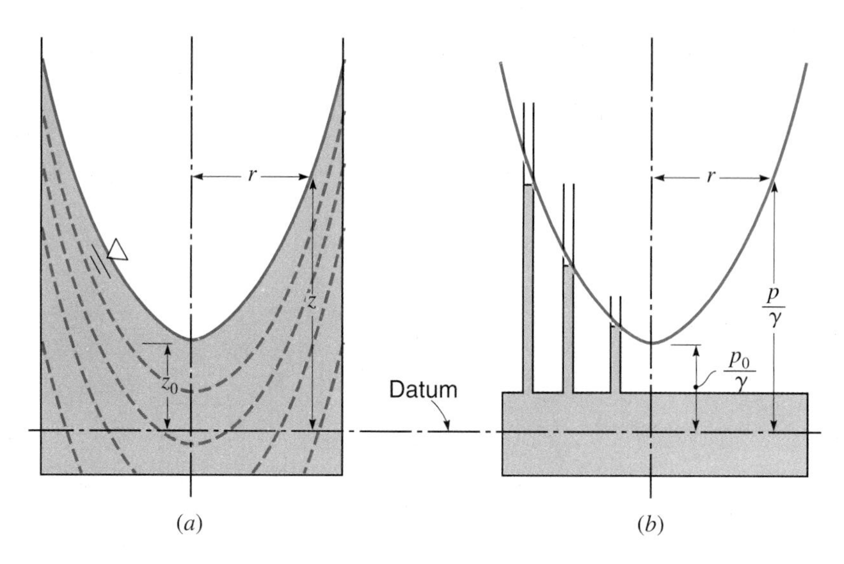

Figure 5.13
Forced vortex. (*a*) Open vessel. (*b*) Closed vessel.

Between any two radii r_1 and r_2, this integrates to give

$$\frac{p_2}{\gamma} - \frac{p_1}{\gamma} = \frac{\omega^2}{2g}(r_2^2 - r_1^2) \tag{5.41}$$

If p_0 is the value of the pressure when $r_1 = 0$, this becomes

$$\frac{p}{\gamma} = \frac{\omega^2}{2g} r^2 + \frac{p_0}{\gamma} \tag{5.42}$$

which is seen to be the equation of a parabola. In Fig. 5.13*a* it is seen that if the fluid is a liquid then the pressure head p/γ at any point is equal to z, the depth of the point below the free surface. Hence the preceding equations may also be written as

$$z_2 - z_1 = \frac{\omega^2}{2g}(r_2^2 - r_1^2) \tag{5.43}$$

and

$$z = \frac{\omega^2}{2g} r^2 + z_0 \tag{5.44}$$

where z_0 is the elevation when $r_1 = 0$. Equations (5.43) and (5.44) are the equations of the free surface, if one exists, or in any case are the equations for any surface of equal pressure; these are a series of paraboloids as shown by the dotted lines in Fig. 5.13*a*.

For the open vessel shown in Fig. 5.13*a*, the pressure head at any point is equal to its depth below the free surface. If the liquid is confined within a vessel, as shown in Fig. 5.13*b*, the pressure along any radius will vary in just the same way as if there were a free surface. Hence the two are equivalent.

In the preceding discussion the axis of the vessel was assumed to be vertical; however, the axis might be inclined. Since pressure varies with elevation as well as radius, a more general equation applicable to fluid in a closed tank with an inclined axis is

$$\frac{p_2}{\gamma} - \frac{p_1}{\gamma} + z_2 - z_1 = \frac{\omega^2}{2g}(r_2^2 - r_1^2) \tag{5.45}$$

Equation (5.41) is the special case where $z_1 = z_2$ (closed tank with vertical axis), and Eq. (5.43) is the special case where $p_1 = p_2$ (open tank with vertical axis). It should be noted that Eqs. (5.41)–(5.45) are not energy equations, since they represent conditions across streamlines rather than along a streamline.

Spiral Forced Vortex

So far the discussion has been confined to the rotation of all particles in concentric circles. Suppose that there is now superimposed a flow with a velocity having radial components either outward or inward. If the height of the walls of the open vessel in Fig. 5.13*a* were less than that of the liquid surface, and if liquid were supplied to the center at the proper rate by some means, then it is obvious that liquid would flow outward. If, on the other hand, liquid flowed into the tank over the rim from some source at a higher elevation and were drawn out at the center, the flow would be inward. The combination of this approximately radial flow with the circular flow would result in path lines that were some form of spirals.

If the closed vessel in Fig. 5.13*b* is arranged with suitable openings near the center and also around the periphery, and if it is provided with vanes, as shown in Fig. 5.14, it becomes either a centrifugal pump impeller or a turbine runner, as the case may be. These vanes constrain the flow of the liquid and determine both its relative magnitude and its direction. If the area of the

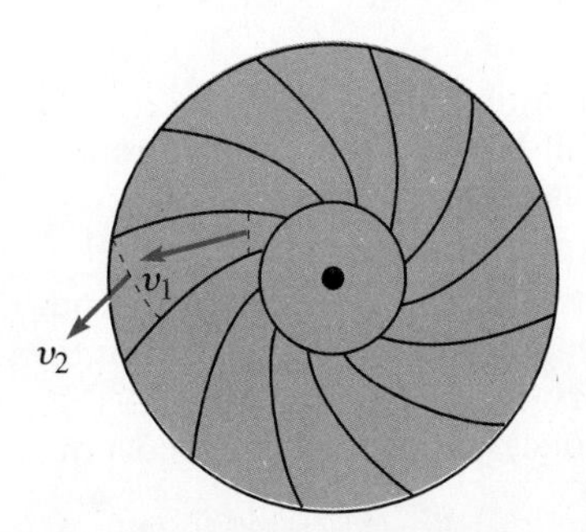

Figure 5.14
Flow through a rotor

passages normal to the direction of flow is a, the equation of continuity fixes the relative velocities, since

$$Q = a_1 v_1 = a_2 v_2 = \text{constant}$$

This relative flow is the flow as it would appear to an observer or a camera, revolving with the rotor. Neglecting friction losses and assuming a vertical axis of rotation, the pressure difference due to this superimposed flow alone is found by the energy equation to be $p_2/\gamma - p_1/\gamma = (v_1^2 - v_2^2)/2g$.

Hence, for the case of rotation with flow (i.e., spiral forced vortex), the total pressure difference between two points is found by adding together the pressure differences due to the two flows considered separately. That is, for the case of a vertical axis,

$$\frac{p_2}{\gamma} - \frac{p_1}{\gamma} = \frac{\omega^2}{2g}(r_2^2 - r_1^2) + \frac{v_1^2 - v_2^2}{2g} \tag{5.46}$$

Of course, friction losses will modify this result to some extent. If the axis is inclined, z-terms must be added to the equation. It is seen that Eq. (5.41) is a special case of Eq. (5.46) when $v_1 = v_2$ either when both are finite or when both are zero.

For a forced vortex with spiral flow, energy is put into the fluid in the case of a pump and extracted from it in the case of a turbine. In the limiting case of zero flow, when all path lines become concentric circles (i.e., a cylindrical forced vortex), energy input from some external source is still necessary for any real fluid in order to maintain the rotation.

EXERCISES

5.18.1 A closed vessel 16 in in diameter completely filled with fluid is rotated at 1800 rpm. What will be the pressure difference between the circumference and the axis of rotation in feet of the fluid and in pounds per square inch if the fluid is (*a*) air with a specific weight of 0.076 lb/ft^3; (*b*) water a 70°F; (*c*) oil with a specific weight of 46 lb/ft^3?

5.18.2 A closed vessel 120 cm in diameter is completely filled with oil (γ = 8.3 kN/m^3) and is rotated at 400 rpm. What will be the pressure difference between the circumference and axis of rotation? Express answer in Pa.

5.18.3 An open cylindrical vessel partially filled with water is 2 ft in diameter and rotates about its axis, which is vertical. How many revolutions per minute would cause the water surface at the periphery to be 5 ft higher than the water surface at the axis? What would be the necessary speed for the same conditions if the fluid were mercury?

5.19 FREE OR IRROTATIONAL VORTEX

In the free vortex there is no expenditure of energy whatever from an outside source, and the fluid rotates by virtue of some rotation previously imparted to it or because of some internal action. Some examples are a whirlpool in a river, the rotary flow that often arises in a shallow vessel when liquid flows out through a hole in the bottom (as is often seen when water empties from a bathtub), and the flow in a centrifugal-pump casing just outside the impeller or that in a turbine casing as the water approaches the guide vanes.

As no energy is imparted to the fluid, it follows that, neglecting friction, H is constant throughout; that is, $p/\gamma + z + V^2/2g = \text{constant}$.

Cylindrical Free Vortex

The angular momentum with respect to the center of rotation of a particle of mass m moving along a circular path of radius r at a velocity V_u is $mV_u r$, where V_u is the velocity along the circular path (i.e., tangential velocity).[8] Newton's second law states that, for the case of rotation, the torque is equal to the time rate of change of angular momentum. Hence torque $= d(mV_u r)/dt$. In the case of a free vortex (frictionless fluid) there is no torque applied; therefore $mVr = \text{constant}$, and thus $V_u r = C$, where the value of C is determined by knowing the value of V at some radius r. Assuming a vertical axis of rotation and inserting $V_u = C/r$ in Eq. (5.40), we obtain

$$dp = \rho \frac{C^2}{r^2} \frac{dr}{r} = \frac{\gamma}{g} \frac{C^2}{r^3} dr$$

Between any two radii r_1 and r_2, this integrates to give

$$\frac{p_2}{\gamma} - \frac{p_1}{\gamma} = \frac{C^2}{2g}\left(\frac{1}{r_1^2} - \frac{1}{r_2^2}\right) = \frac{V_{u1}^2}{2g}\left[1 - \left(\frac{r_1}{r_2}\right)^2\right] \tag{5.47}$$

If there is a free surface, the pressure head p/γ at any point is equal to the depth below the surface. Also, at any radius the pressure varies in a vertical direction according to the hydrostatic law. Hence this equation is merely a special case where $z_1 = z_2$.

As $H = p/\gamma + z + V_u^2/2g = \text{constant}$, it follows that at any radius r

$$\frac{p}{\gamma} + z = H - \frac{V_u^2}{2g} = H - \frac{C^2}{2gr^2} = H - \frac{V_{u1}^2}{2g}\left(\frac{r_1}{r}\right)^2 \tag{5.48}$$

[8] In this chapter V_u is used to represent the tangential component of velocity. Other symbols commonly used to represent tangential velocity are V_t and V_θ.

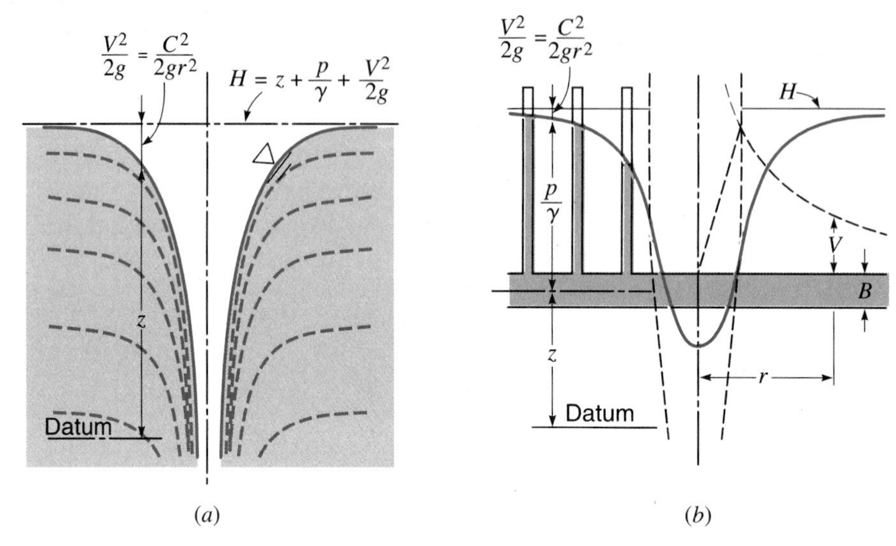

Figure 5.15
Free vortex. (*a*) Free surface. (*b*) Fluid enclosed

Assuming the axis to be vertical, the pressure along the radius can be found from this equation by taking z constant; and for any constant pressure p, values of z determining a surface of equal pressure, can be found. If p is zero, the values of z determine the free surface (Fig. 5.15*a*), if one exists.

Equation (5.48) shows that H is the asymptote approached by $p/\gamma + z$ as r approaches infinity and V_u approaches zero. On the other hand, as r approaches zero, V_u approaches infinity, and $p/\gamma + z$ approaches minus infinity. Since this is physically impossible, the free vortex cannot extend to the axis of rotation. In reality, since high velocities are attained as the axis is approached, the friction losses, which vary as the square of the velocity, become of increasing importance and are no longer negligible. Hence the assumption that H is constant no longer holds. The core of the vortex tends to rotate as a solid body, as in the central part of Fig. 5.15*b*.

Spiral Free Vortex

If a radial flow is superimposed upon the concentric flow previously described, the path lines will then be spirals. If the flow passes out through a circular hole in the bottom of a shallow vessel, the surface of the liquid takes the form shown in Fig. 5.15*a*, with an air ***core*** sucked down the hole. If an outlet symmetrical with the axis is provided in the arrangement shown in Fig. 5.15*b*, we might have a flow component either radially inward or radially outward. If the two plates shown are a constant distance B apart, the radial flow component with a velocity V_r is then across a series of concentric cylindrical surfaces whose area is $2\pi rB$. Thus

$$Q = 2\pi rBV_r = \text{constant}$$

from which it is seen that rV_r = constant. Therefore the radial velocity varies in the same way with r that the circumferential velocity did in the preceding discussion of the free cylindrical vortex. The pressure variation in a spiral free vortex (Fig. 5.15*b*) is given by

$$\frac{p_2}{\gamma} - \frac{p_1}{\gamma} = \frac{V_1^2}{2g} - \frac{V_2^2}{2g} \tag{5.49}$$

where $V = \sqrt{V_r^2 + V_u^2}$, the velocity of flow.

Sample Problem 5.13 A centrifugal pump with a 12-in-diameter impeller is surrounded by a casing that has a constant height of 1.5 in between sections *a* and *b* and that then enlarges into a volute at *c* (Fig. S5.13). Water leaves the impeller with a velocity of 60 fps at an angle of 15° with the tangent. (*a*) At what rate is water flowing through the pump? (*b*) Neglecting friction, what will be the magnitude and direction of the velocity at *b* and what will be the gain in pressure head from *a* to *b*?

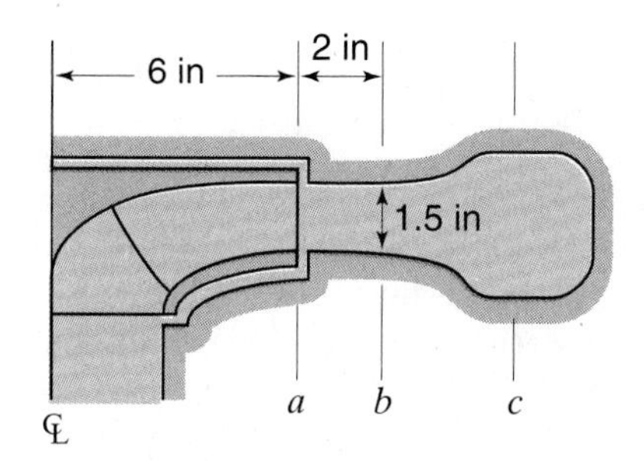

Figure S5.13

Solution

(*a*) Flow through the pump $Q = A_a(V_r)_a = 2\pi r_a B(V_r)_a$, where $(V_r)_a$ = 60 sin 15° = 15.53 fps

$$Q = 2\pi(6/12)(1.5/12)15.53 = 6.10 \text{ cfs} \qquad \textit{ANS}$$

(*b*) From continuity: $Q = 2\pi r_a B(V_r)_a = 2\pi r_b B(V_r)_b$,

hence
$$\frac{(V_r)_b}{(V_r)_a} = \frac{r_a}{r_b} = \frac{6}{8}$$

Since torque = 0 in the space between a and b, angular momentum must be conserved. Thus

$$m(V_u)_a r_a = m(V_u)_b r_b$$

so

$$\frac{(V_u)_b}{(V_u)_a} = \frac{r_a}{r_b} = \frac{6}{8}$$

The region between a and b is a spiral free vortex.

Since V_u and V_r are both seen to decrease in the same proportion as flow moves from a to b, the angle α does not change and

$$V_b/V_a = \tfrac{6}{8}; \quad V_b = (\tfrac{6}{8})60 = 45 \text{ fps} \quad \text{at } 15° \text{ with the tangent} \qquad \textbf{\textit{ANS}}$$

Finally, writing the energy equation along the flow lines gives

$$\frac{p_b}{\gamma} - \frac{p_a}{\gamma} = \frac{V_a^2}{2g} - \frac{V_b^2}{2g} = \frac{60^2 - 45^2}{2g} = 24.5 \text{ ft} \qquad \textbf{\textit{ANS}}$$

EXERCISES

5.19.1 Refer to Sample Prob. 5.13. If the impeller diameter is 22 cm, the casing height is 4 cm between a and b, and water leaves the impeller with a velocity of 18 m/s at an angle of 16° with the tangent, find the flow rate, the magnitude and direction of the velocity at b (where r = 16 cm), and the pressure increase from a to b. Neglect friction.

5.19.2 Refer to Sample Prob. 5.13. If the impeller diameter is 10 in, the casing height is 1.8 in between a and b, and water leaves the impeller with a velocity of 50 fps at an angle of 16° with the tangent, find the flow rate, the magnitude and direction of the velocity at b (where r = 7 in), and the pressure increase from a to b. Neglect friction.

PROBLEMS

5.1 Assume the seventh-root law [Eq. (8.38)] for a turbulent-velocity distribution between two smooth flat plates. Find α.

5.2 Assume the seventh-root law [Eq. (8.38)] for a turbulent-velocity distribution in smooth pipe flow. Find α.

5.3 A pipeline supplies water to a hydroelectric plant from a reservoir in which the water temperature is 58.8°F. (a) Suppose that in the length of the pipe there is a total loss of heat to the surrounding air of 0.28 Btu/lb of water and the temperature of the water at the power house is 58.7°F. What is the friction loss per pound of water? (b) With the same flow as in (a), what will be the temperature of the water at the power house if there is absorption of heat from hot sunshine at the rate of 3.2 Btu/lb of water?

5.4 Water is flowing at 15 m^3/s through a long pipe. The temperature of the water rises 0.23°C when heat is transferred to the water at the rate of 5600 kN·m/s. Find the head loss in the pipe.

5.5 Assume frictionless flow in a long, horizontal, conical pipe, where the diameter is 3.8 ft at one end and 2.2 ft at the other. The pressure head at the smaller end is 12.5 ft of water. If water flows through this cone at the rate of 120 cfs, find the velocities at the two ends and the pressure head at the larger end.

5.6 Water flows through a long, horizontal, conical diffuser at the rate of 4.5 m^3/s. The diameter of the diffuser changes from 1.2 m to 1.8 m. The pressure at the smaller end is 7.5 kN/m^2. Find the pressure at the downstream end of the diffuser, assuming frictionless flow. Assume also, that the angle of the cone is so small that separation of the flow from the walls of the diffuser does not occur.

5.7 In Fig. P5.7 water is admitted at the center at a rate of 3 cfs and is discharged into the air around the periphery. The upper circular plate in the figure is horizontal and is fixed in position, while the lower annular plate is free to move vertically and is not supported by the pipe in the center. The annular plate weights 6 lb, and the weight of the water on it should be considered. (*a*) If the distance between the two plates is to be maintained at 2 in, what is the total weight W that can be supported? (*b*) What is the pressure head where the radius is 4 in, and what is it at a radius of 8 in?

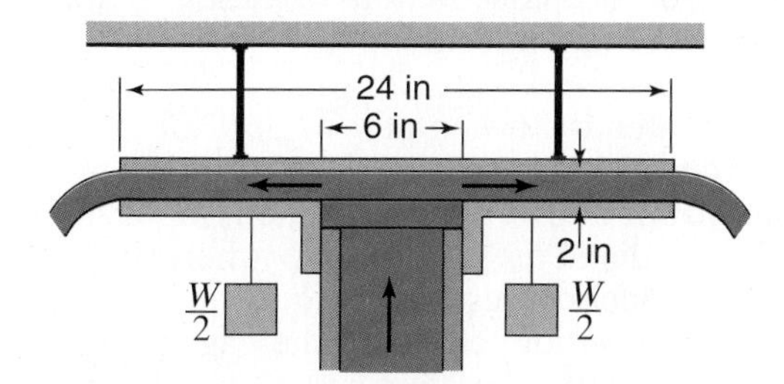

Figure P5.7

5.8 Plot the stagnation pressure (psia) on an object as it passes through air at sea level (standard atmosphere) as a function of velocity. Repeat for movement through air at 10,000 ft elevation. Let V vary from zero to c using 0, 25, 50, 75, and 100% of c.

5.9 Wind blows at a velocity of 20 m/s against the side of a pole at an elevation of 1000 m above sea level. What is the stagnation pressure assuming standard atmospheric conditions? Express answer as a gage pressure and as an absolute pressure in kN/m^2, Pa, and mmHg.

5.10 Find the stagnation pressure on a tree trunk at an elevation of 2000 m if the wind speed is 15 m/s.

5.11 A pump, with an efficiency of 90%, circulates water at the rate of 2500 gpm in a closed circuit that holds 8500 gal. The net head developed by the pump is 360 ft. What is the change in water temperature after one hour, assuming that the bearing friction is negligible and that there is no heat loss from the system?

5.12 A pump, with an efficiency of 90%, circulates water at the rate of 150 L/s in a closed circuit that holds 50 m^3. The net head developed by the pump is 110 m. What is the change in water temperature after one hour, assuming that the bearing friction is negligible and that there is no heat loss from the system?

5.13 In Fig. 5.10 neglect all head losses except at discharge, and assume water is flowing. If h = 15 ft and the water surface in the lower reservoir is 12 ft higher than the constriction, find the highest permissible temperature of the water in order that there be no cavitation. The diameter of the constriction is 80% of the diameter of the pipe where it joins the downstream tank. Atmospheric pressure is 13.6 psia.

5.14 In Fig. 5.10 neglect all head losses except at discharge, and assume water is flowing. If h = 4 m and the water surface in the lower reservoir is 5.5 m higher than the constriction, find the highest permissible temperature of the water in order that there be no cavitation. The diameter of the constriction is 70% of the diameter of the pipe where it joins the downstream tank. Atmospheric pressure is 97 kN/m^2 abs.

5.15 Redo Prob. 5.13, but this time let the water temperature be 50°F. Find the minimum permissible diameter of the constriction in order to not have cavitation. Express the answer as a fraction of the outlet diameter.

5.16 Referring to Fig. 5.10, assume water is flowing and neglect all head losses except at discharge. Find the flow rate if h = 8 ft. Assuming that the water surface in the lower reservoir is 10 ft above the constriction, the diameter of the constriction is two-thirds the diameter of the pipe where it joins the downstream tank, and the atmospheric pressure is equal to the standard atmospheric pressure at 10,000 ft elevation, calculate the gage pressure and the absolute pressure in the constriction. The diameter of the constriction is 14 in.

5.17 Referring to Fig. 5.10, assume water is flowing and neglect all head losses except at discharge. Find the flow rate if h = 2 m. Assuming that the water surface in the lower reservoir is 4 m above the constriction, the diameter of the constriction is two-thirds the diameter of the pipe where it joins the downstream tank, and the atmospheric pressure is equal to the standard atmospheric pressure at 3000 m elevation, calculate the gage pressure and the absolute pressure in the constriction. The diameter of the constriction is 36 cm.

5.18 Repeat Prob. 5.16, assuming head losses are as follows: 6 in in the converging section and 30 in in the diverging section.

5.19 Repeat Prob. 5.17, assuming head losses are as follows: 0.15 m in the converging section and 0.8 m in the diverging section.

5.20 In Fig. P5.20 friction loss between A and B is neglected while between B and C it is $0.1(V_B^2/2g)$. Find the pressure heads at A and C if the liquid is flowing from A to C at the rate of 300 L/s.

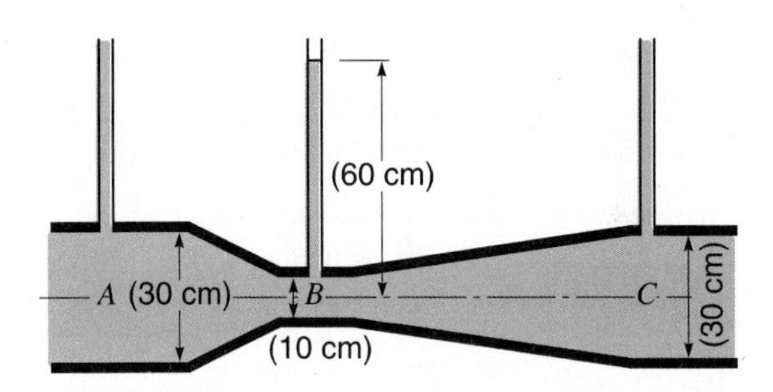

Figure P5.20

5.21 In Fig. P5.21 assume the tube flows full. At B, the diameter of the tube is 3 in and the diameter of the water jet discharging into the air at C is 4.5 in. (a) If all friction losses are neglected, what are the velocity and the pressure head at B if $h = 10$ ft. (b) What is the rate of discharge in cfs? And what would it be if the tube were cut off at B?

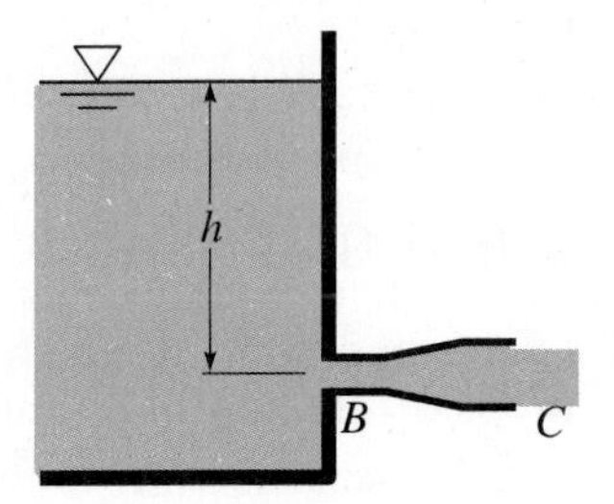

Figure P5.21

5.22 Referring to Fig. P5.21, assume the tube flows fully and all friction losses are neglected. The diameter at B is 6 cm and the diameter of the jet discharging into the air is 8 cm. If $h = 5$ m, what is the flow rate? What is the pressure head at B? What would be the flow rate if the tube were cut off at B?

5.23 In Fig. P5.23 friction losses in the pipe are $V^2/2g$ with the barometer pressure at 12.50 psia. The liquid in the suction pipe has a velocity of 7 fps. What would be the maximum allowable value of z if the liquid were (a) water at 70°F; (b) gasoline at a vapor pressure of 9 psia with a specific weight of 47 lb/ft^3?

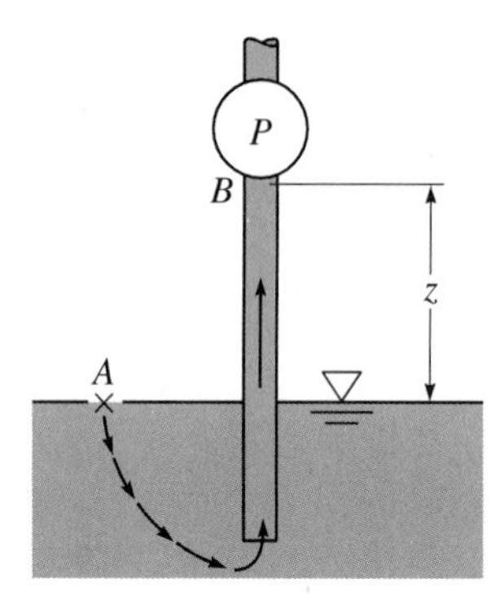

Figure P5.23

5.24 In Fig. P5.23 friction losses in the pipe are $V^2/2g$ with the barometer pressure at 90 kN/m^2. The liquid in the suction pipe has a velocity of 1.8 m/s. What would be the maximum allowable value of z if the liquid were (*a*) water at 20°C; (*b*) gasoline at a vapor pressure of 49 kN/m^2 abs, with a specific weight of 8 kN/m^3?

5.25 A discharge pressure gage reading, taken at a point of 6.5 ft above the center line of a pump, is 25 psi. A suction pressure gage reading, taken 2.5 ft below the center line, indicates a vacuum of 12 inHg when gasoline ($s = 0.75$) is pumped at the rate of 1.5 cfs. The diameters of the suction and discharge pipes of the pump are 8 and 6 in, respectively. What is the power delivered to the fluid? Sketch the energy line and the hydraulic grade line.

5.26 For this problem, use the same data as in Sample Prob. 5.10, except that, instead of the pump developing 80 ft of head, it delivers 110 hp to the water. Find the new flow rate. Plot the energy line and hydraulic grade line. Calculate the pressure on the suction side of the pump.

5.27 In Fig. 4.12 the velocity of the undisturbed field is 22 fps and the velocities very near the surface at radii from the "source" making angles with the axis of 0, 60, 120, and 150° are 0, 17.5, 23.7, and 21.9 fps, respectively. What will be the elevation of the surface of a liquid relative to that of the free surface of the undisturbed field? (This problem illustrates the way in which the water surface drops alongside a bridge pier or past the side of a moving ship.)

5.28 In Fig. 4.12 the velocity of the undisturbed field is 6 m/s and the velocities very near the surface at radii from the "source" making angles with the axis of 0, 60, 120, and 150° are 0, 4.8, 6.5, and 6.0 m/s respectively. What will be the elevation of the surface of a liquid relative to that of the free surface of the undisturbed field? (This problem illustrates the way in which the water surface drops alongside a bridge pier or past the side of a moving ship.)

5.29 If the body shown in Fig. 4.12 is not two-dimensional but is a solid of revolution about a horizontal axis, the flow will be three-dimensional and the streamlines will be differently spaced. Also, the distance between the stagnation point and the "source" will be $d/4$, where d is the diameter at a great distance from the stagnation point. At points very near the surface at radii from the source making angles with the axis of 0, 60, 120, and 150°, the velocities are 0, 14.0, 21.3, and 19.8 fps respectively when the velocity of the undisturbed field is 19 fps. If the body is an airship and the atmospheric pressure in the undisturbed field is 14 psia, what will be the pressures at these points, for air temperature of 53.9°F?

5.30 In Prob. 5.29 assume the body is a submarine with diameters at the four points of 0, 8.24, 14.28, and 15.90 ft respectively. If the submarine is submerged in the ocean ($\gamma = 64.1 \text{ lb/ft}^3$) with its axis 50 ft below the surface, find the pressures in pounds per square inch at these points along the top and along the bottom.

5.31 By manipulation of Eq. (5.37), demonstrate that it represents a standard parabola of the form $z - z_0 = a(x - x_0)^2$, where a is a constant and x_0 and z_0 are the coordinates of the vertex.

5.32 Find the maximum ideal horizontal range of a jet having an initial velocity of 85 fps. At what angle of inclination is this obtained?

5.33 Repeat Exer. 5.17.1. Let $V = Q/A = 24$ fps, but assume a parabolic velocity profile.

5.34 Using Fig. X5.17.1, which depicts a two-dimensional flow in a vertical plane, find the pressure at B if the pressure at A is 36 kN/m^2. Data are as follows: $r = 3$ m, $b = 1.2$ m, $\gamma = 9.81 \text{ kN/m}^3$, $V = Q/A = 6$ m/s. Assume a parabolic velocity profile.

5.35 In Fig. 5.14 suppose that the vanes are all straight and radial, that $r_1 = 0.3$ ft, $r_2 = 0.9$ ft, and that the height perpendicular to the plane of the figure is constant at $b = 0.25$ ft. Then $a = 2\pi rb$. If the speed is 1000 rpm and the flow of liquid is 9.6 cfs, find the difference in the pressure head between the outer and the inner circumferences, neglecting friction losses. Does it make any difference whether the flow is outward or inward?

5.36 In Fig. 5.14 suppose that the vanes are all straight and radial, that $r_1 = 10$ cm, $r_2 = 20$ cm, and that the height perpendicular to the plane of the figure is constant at $b = 8$ cm. Then $a = 2\pi rb$. If the speed is 1000 rpm and the flow of liquid is 0.3 m^3/s, find the difference in the pressure head between the outer and the inner circumferences, neglecting friction losses. Does it make any difference whether the flow is outward or inward?

5.37 An air duct of 2.5 ft by 2.5 ft square cross-section turns a bend of radius 5 ft as measured to the center line of the duct. If the measured pressure difference between the inside and outside walls of the bend is 1.5 in of water, estimate the rate of air flow in the duct. Assume standard sea-level conditions in the duct and assume ideal flow around the bend.

5.38 An air duct of 1.2 m by 1.2 m square cross-section turns a bend of radius 2.4 m as measured to the center line of the duct. If the measured pressure difference between the inside and outside walls of the bend is 5 cm of water, estimate the rate of air flow in the duct. Assume standard sea-level conditions in the duct and assume ideal flow around the bend.

5.39 Assume ideal fluid. The pressure at section 1 in the Fig. P5.39 is 10 psi, $V_1 = 15$ fps, $V_2 = 50$ fps, and $\gamma = 60\ \text{lb/ft}^3$. (*a*) Determine the reading on the manometer. (*b*) If the downstream piezometer were replaced with a pitot tube, what would be the manometer reading? Comment on the practicality of these arrangements.

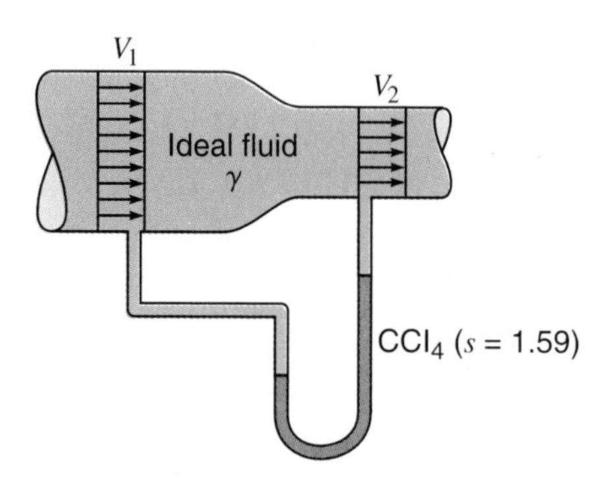

Figure P5.39

5.40 Refer to Fig. P5.39. Assume an ideal fluid with $\rho = 900\ \text{kg/m}^3$. The pressure at section 1 is $100\ \text{kN/m}^2$; $V_1 = 10$ m/s, $V_2 = 20$ m/s. (*a*) Determine the reading on the manometer. (*b*) If the downstream piezometer were replaced with a pitot tube, what would be the manometer reading? Comment on the practicality of these arrangements.

CHAPTER 6

Momentum and Forces in Fluid Flow

Previously, two important fundamental concepts of fluid mechanics were presented: the continuity equations and the energy equation. In this chapter a third basic concept, the ***impulse–momentum principle***, will be developed. This concept is of particular importance in flow problems where the determination of forces is involved. Such forces occur whenever the velocity of a stream of fluid is changed in either direction or magnitude. By the law of action and reaction, an equal and opposite force is exerted by the fluid upon the body producing the change. After developing the impulse–momentum principle, its application to a number of important engineering problems is discussed.

6.1 DEVELOPMENT OF THE IMPULSE–MOMENTUM PRINCIPLE

The impulse–momentum principle will be derived from Newton's second law. The flow may be compressible or incompressible, real (with friction) or ideal (frictionless), steady or unsteady, and the equation need not be applied along a streamline. In Chap. 5 when applying the energy equation to real fluids we found that the energy loss must be computed. This difficulty is not encountered in momentum analysis.

Newton's second law may be expressed as

$$\sum \mathbf{F} = \frac{d(m\mathbf{V})_S}{dt} \qquad (6.1)$$

This states that the sum of the external forces **F** on a body of fluid or system

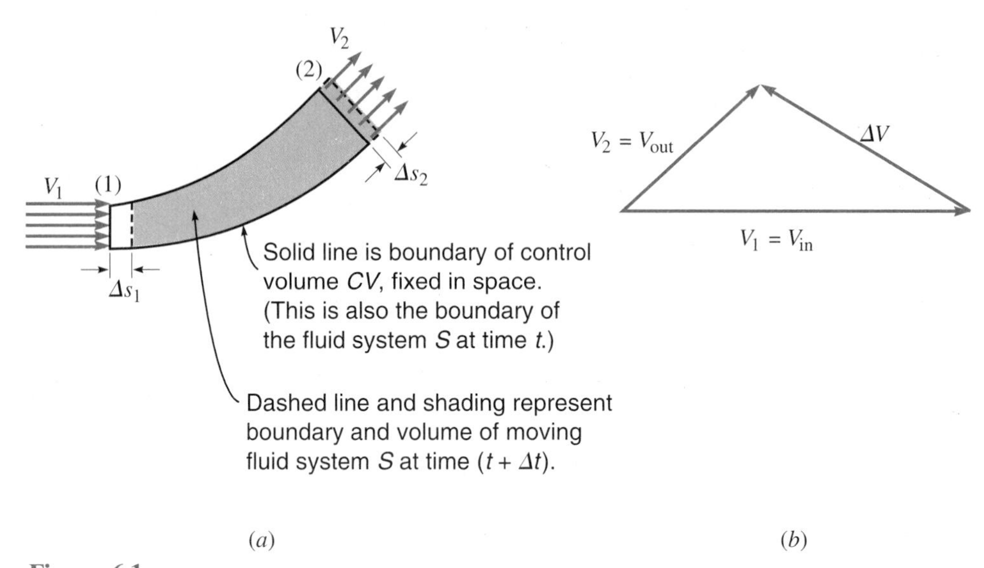

Figure 6.1
(a) Control volume for steady flow with control surface cutting a constant-velocity stream at right angles. (b) Velocity relations.

S is equal to the rate of change of linear momentum $m\mathbf{V}$ of that body or system. The boldface symbols $\mathbf{F}$ and $\mathbf{V}$ represent vectors, and so the change in momentum must be in the same direction as the force. Because Eq. (6.1) can also be expressed as $\sum(\mathbf{F})\,dt = d(m\mathbf{V})_S$, i.e., impulse equals change of momentum, the terminology ***impulse–momentum principle*** is used.

Using the principles of Sec. 4.6, let us consider the linear momentum of the ***fluid system*** and ***control volume*** defined within the stream tube of Fig. 6.1a, just as we did for energy in Sec. 5.4. The fixed control volume lies between sections 1 and 2, and the moving fluid system consists of the fluid mass contained at time t in the control volume. During a short time interval Δt, we shall assume that the fluid moves a short distance Δs_1 at section 1 and Δs_2 at section 2. Recalling the analysis of Sec. 4.6, and letting the general property X now be the momentum $m\mathbf{V}$, Eq. (4.9) becomes

$$\frac{d(m\mathbf{V})_S}{dt} = \frac{d(m\mathbf{V})_{CV}}{dt} + \frac{d(m\mathbf{V})_{CV}^{\text{out}}}{dt} - \frac{d(m\mathbf{V})_{CV}^{\text{in}}}{dt} \tag{6.2}$$

where, as before, subscript S denotes the moving fluid system and subscript CV denotes the fixed control volume. So, setting this equal to Eq. (6.1),

Unsteady flow:
$$\sum \mathbf{F} = \frac{d(m\mathbf{V})_{CV}}{dt} + \frac{d(m\mathbf{V})_{CV}^{\text{out}}}{dt} - \frac{d(m\mathbf{V})_{CV}^{\text{in}}}{dt} \tag{6.3}$$

On the right-hand side of this equation, the first term represents the rate of change or accumulation of momentum within the fixed control volume, whereas the second and third terms respectively represent the rates at which momentum enters and leaves the control volume. The entire Eq. (6.3) states that the resultant force acting on a fluid mass is equal to the rate of change of momentum of the fluid mass. It is perfectly general. It applies to compressible or incompressible, real or ideal, and steady or unsteady flow.

In the case of steady flow, conditions within the control volume do not change, so $d(m\mathbf{V})_{CV}/dt = 0$, and the equation becomes

$$\text{Steady flow:} \qquad \sum \mathbf{F} = \frac{d(m\mathbf{V})_{CV}{}^{\text{out}}}{dt} - \frac{d(m\mathbf{V})_{CV}{}^{\text{in}}}{dt} \qquad (6.4)$$

Thus, for steady flow the net force on the fluid mass is equal to the net rate of *outflow* of momentum across the control surface.

Since Eqs. (6.1)–(6.4) are vectorial equations, they can also be expressed as scalar equations in terms of forces and velocities in the x, y, and z directions, respectively.

It is advantageous to select a control volume so that the control surface is normal to the velocity where it cuts the flow. Consider such a situation in Fig. 6.1*a*. Also, let the velocity be constant where it cuts across the control surface, and let us restrict ourselves to steady flow so that Eq. (6.4) is applicable. Since

$$\frac{d(m\mathbf{V})_1}{dt} = \frac{dm_1}{dt}\mathbf{V}_1 = \dot{m}_1\mathbf{V}_1 = \rho_1 Q_1 \mathbf{V}_1$$

and the same relations hold for section 2, Eq. (6.4) may be written

$$\text{Steady flow:} \quad \sum \mathbf{F} = \dot{m}_2\mathbf{V}_2 - \dot{m}_1\mathbf{V}_1 = \rho_2 Q_2 \mathbf{V}_2 - \rho_1 Q_1 \mathbf{V}_1 \qquad (6.5)$$

But since the flow we are considering is steady, from continuity, $\dot{m}_1 = \dot{m}_2 = \dot{m} = \rho_1 Q_1 = \rho_2 Q_2 = \rho Q$. Also, using the vector relations of Fig. 6.1*b*, it is convenient to write $\mathbf{\Delta V} = \mathbf{V}_2 - \mathbf{V}_1 = \mathbf{V}_{\text{out}} - \mathbf{V}_{\text{in}}$. Using these, Eq. (6.5) becomes

$$\text{Steady flow:} \qquad \sum \mathbf{F} = \dot{m}(\mathbf{\Delta V}) = \rho Q(\mathbf{\Delta V}) \qquad (6.6)$$

The direction of $\sum \mathbf{F}$ must be the same as that of $\mathbf{\Delta V}$. Note that the $\sum \mathbf{F}$ represents the vectorial summation of *all* forces acting *on the fluid mass,* including gravity forces, shear forces, and pressure forces including those exerted by fluid surrounding the fluid mass under consideration as well as the pressure forces exerted by the solid boundaries in contact with the fluid mass. Often the force sought is just *one* of these many forces. Frequently it is not even one of them, but instead it is *opposite* to one of them, being the force of the liquid acting on a boundary. The right-hand side of Eq. (6.6) represents the change in momentum per unit time.

Since Eq. (6.6) is vectorial, it can be expressed by the following scalar equations:

Steady flow:

$$\sum F_x = \dot{m}(\Delta V_x) = \rho Q(\Delta V_x) = \rho Q(V_{2x} - V_{1x}) \tag{6.7a}$$

$$\sum F_y = \dot{m}(\Delta V_y) = \rho Q(\Delta V_y) = \rho Q(V_{2y} - V_{1y}) \tag{6.7b}$$

$$\sum F_z = \dot{m}(\Delta V_z) = \rho Q(\Delta V_z) = \rho Q(V_{2z} - V_{1z}) \tag{6.7c}$$

In Sec. 6.4 and succeeding sections these equations will be applied to several situations that are commonly encountered in engineering practice. If the flow in a single stream tube splits up into several streamtubes, the ρQV values of each stream tube are computed separately and then substituted into Eqs. (6.5)–(6.7). (See Sample Problem 6.2.) The great advantage of the impulse–momentum principle is that we need not be concerned with the details of what is occurring within the flow; only the conditions at the end sections of the control volume govern the analysis.

6.2 NAVIER–STOKES EQUATIONS

A set of differential equations that describe the motion of a real fluid can be derived for the general case by considering the forces acting on a small element or control volume of fluid like Fig. 3.2. The forces include gravitational, viscous (frictional), and pressure forces. Newton's equation of viscosity [Eq. (2.9)] for one-dimensional flow must be generalized to three-dimensional flow before it can be incorporated.

The full derivation of these equations is lengthy and involved, and beyond the scope of this text. However, for an incompressible fluid with constant viscosity, in rectangular coordinates with z increasing vertically upwards, the result is

$$-\frac{\partial p}{\partial x} + \mu\left(\frac{\partial^2 u}{\partial x^2} + \frac{\partial^2 u}{\partial y^2} + \frac{\partial^2 u}{\partial z^2}\right) = \rho\left[\frac{\partial u}{\partial t} + u\frac{\partial u}{\partial x} + v\frac{\partial u}{\partial y} + w\frac{\partial u}{\partial z}\right] \tag{6.8a}$$

$$-\frac{\partial p}{\partial y} + \mu\left(\frac{\partial^2 v}{\partial x^2} + \frac{\partial^2 v}{\partial y^2} + \frac{\partial^2 v}{\partial z^2}\right) = \rho\left[\frac{\partial v}{\partial t} + u\frac{\partial v}{\partial x} + v\frac{\partial v}{\partial y} + w\frac{\partial v}{\partial z}\right] \tag{6.8b}$$

$$-\rho g - \frac{\partial p}{\partial z} + \mu\left(\frac{\partial^2 w}{\partial x^2} + \frac{\partial^2 w}{\partial y^2} + \frac{\partial^2 w}{\partial z^2}\right) = \rho\left[\frac{\partial w}{\partial t} + u\frac{\partial w}{\partial x} + v\frac{\partial w}{\partial y} + w\frac{\partial w}{\partial z}\right] \tag{6.8c}$$

These fundamental general equations of motion are known as the ***Navier–Stokes equations***. They are named after the French scientist, C. L. M. H. Navier (1785–1836), who today we would describe as a civil engineer, and the

English physicist, Sir George Stokes (1819–1903), both of whom first derived them. They are second-order nonlinear partial differential equations that have not been analytically solved in general, although analytical and numerical solutions have been obtained for certain specific situations. Their complete derivation, for rectangular, cylindrical, and spherical coordinates, can be found in advanced fluid mechanics texts.

The Navier–Stokes equations are in fact just a differential form of the linear momentum principle. Thus, on the left-hand sides of Eqs. (6.8), we have the body force per unit volume (the term in g) and the surface force per unit volume (pressure force represented by terms in p, and viscous force represented by terms with μ and parentheses). These are equal to the time rate of change of momentum on the right-hand side, represented in the square brackets as the local acceleration (the derivatives with respect to time) and the convective acceleration (the other terms).

When Eqs. (6.8) are written in terms of the normal stresses σ and the shear stresses τ, they are known as the ***Cauchy equations***. For an ideal fluid ($\mu = 0$), they reduce to a set of three-dimensional equations known as the ***Euler equations*** of motion, which are the same as Eqs. (6.8) but with the terms in μ and parentheses eliminated.

6.3 MOMENTUM CORRECTION FACTOR

If the velocity is not uniform over a section, the momentum per unit time transferred across that section is greater than that computed by using the mean velocity. Thus the rate of momentum transfer (***momentum flux***) across an elementary area dA, where the local velocity is u, is $\dot{m}u = (\rho u\, dA)u = \rho u^2\, dA$, and the rate of momentum transfer across the entire section is $\rho \int_A u^2\, dA$, while that computed by using the mean velocity is $\rho QV = \rho AV^2$. Hence the correction factor by which ρQV should be multiplied to obtain the true momentum per unit time is

$$\beta = \frac{1}{AV^2}\int_A u^2\, dA \tag{6.9}$$

For laminar flow in a circular pipe, $\beta = \frac{4}{3}$, but for turbulent flow in circular pipes, it usually ranges from 1.005 to 1.05, as shown by Eq. (8.34*b*). For open-channel flow, it may be greater. Unless otherwise specified, the value of β in the ensuing discussion will be taken as 1.0.

EXERCISES

6.3.1 For laminar flow as in Sample Prob. 5.1, find β.

6.3.2 For the turbulent-flow case as approximated in Exer. 5.1.1, find β.

6.4 APPLICATIONS OF THE MOMENTUM PRINCIPLE

A common application of the impulse–momentum principle is that of finding forces exerted by flowing fluid on structures open to the atmosphere, like gates (e.g., Fig. 11.34*a*) and overflow spillways (e.g., Fig. 11.32).

It is very important to remember that the momentum principle only deals with forces that act on the fluid mass in a designated control volume; fluid forces acting on a structure are equal and opposite to boundary pressure forces acting on the fluid. To avoid confusion over signs, students are strongly advised to first solve for the magnitude and direction of the reaction force of the structure on the fluid, and only in the last step to find the equal and opposite force of the fluid on the structure.

The force of a fluid on a structure is usually distributed as varying pressure forces over the surface. We are normally interested in the resultant of this distribution, and we usually only consider pressures that differ from atmospheric. If the total resultant force is required, this is usually obtained by writing equations like (6.7) in order to first find the perpendicular components of the required force.

First, a control volume must be established, and, as noted in Sec. 6.1, this should be done by cutting the flow normal to the velocity, along a boundary where the velocity is constant. Applications to this type of problem are best further discussed with a sample problem.

SAMPLE PROBLEM 6.1 The water passage shown in Fig. S6.1 is 10 ft (3 m) wide normal to the plane of the figure. Determine the horizontal force acting on the shaded structure. Assume ideal flow.

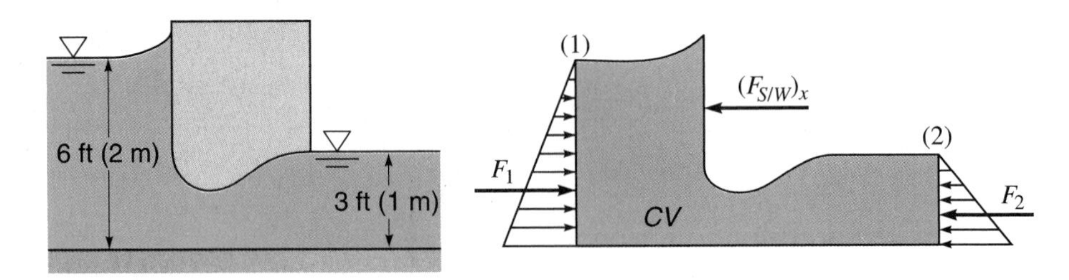

Figure S6.1

Solution (BG units)

In free-surface flow such as this where the streamlines are parallel, the water surface is coincident with the hydraulic grade line. Writing an energy equation from the upstream section to the down-stream section,

$$6+\frac{V_1^2}{2g} = 3+\frac{V_2^2}{2g} \qquad (1)$$

From continuity,

$$6(10)V_1 = 3(10)V_2 \qquad (2)$$

Substituting Eq. (2) into Eq. (1) yields

$$V_1 = 8.02\text{ fps}, \qquad V_2 = 16.05\text{ fps}$$

$$Q = A_1V_1 = A_2V_2 = 481\text{ cfs}$$

Next take a free-body diagram of the control volume of water shown in the figure and apply the impulse-momentum equation (6.7*a*),

$$F_1 - F_2 - (F_{S/W})_x = \rho Q(V_2 - V_1)$$

where $(F_{S/W})_x$ represents the force of the structure on the water in the horizontal direction.

From Eq. (3.16), we have $F_1 = \gamma h_{c1}A_1$ and $F_2 = \gamma h_{c2}A_2$. Hence

$$62.4(3)(10\times 6) - 62.4(1.5)(10\times 3) - (F_{S/W})_x = 1.94(481)(16.05 - 8.02)$$

and

$$(F_{S/W})_x = +936\text{ lb} = 936\text{ lb}\leftarrow$$

The positive sign means that the assumed direction is correct. The force of the water on the structure is equal and opposite, namely,

$$(F_{W/S})_x = 936\text{ lb}\rightarrow \qquad \textbf{\textit{ANS}}$$

Note that the momentum principle will not permit one to obtain the vertical component of the force of the water on the shaded structure, because the pressure distribution along the bottom of the channel is unknown. The pressure distribution along the boundary of the structure and along the bottom of the channel can be estimated by sketching a flow net and applying Bernoulli's principle. The horizontal and vertical components of the force can be found by computing the integrated effect of the pressure-distribution diagram.

Solution (SI units)

Energy:

$$2 + \frac{V_1^2}{2(9.81)} = 1 + \frac{V_2^2}{2(9.81)} \qquad (3)$$

Continuity:

$$2(3)V_1 = 1(3)V_2 \qquad (4)$$

Substituting Eq. (4) into Eq. (3) yields

$$V_1 = 2.56\text{ m/s}, \qquad V_2 = 5.11\text{ m/s}$$

$$Q = A_1V_1 = A_2V_2 = 15.34\text{ m}^2\text{/s}$$

Applying the impulse–momentum equation (6.7*a*) to the free-body diagram,

$$F_1 - F_2 - (F_{S/W})_x = \rho Q(V_2 - V_1)$$

$$9.81(1)(2)(3) - 9.81(0.5)(1)(3) - (F_{S/W})_x = 1.0(15.34)(5.11 - 2.56)$$

$$(F_{S/W})_x = +4.91 \text{ kN} = 4.91 \text{ kN} \leftarrow$$

So $\qquad (F_{W/S})_x = 4.91 \text{ kN} \rightarrow \qquad$ ***ANS***

There are numerous other fluid-flow situations where the impulse–momentum principle is useful. In the following sections we shall apply it to find forces exerted on pressure conduits such as at bends and nozzles, to forces exerted by jets on stationary and moving vanes or blades, and subsequently to rotating machines like pumps, turbines, propellers, and windmills. The momentum principle is also used to develop an expression for the head loss in a pipe expansion (Sec. 8.21) and for the conjugate depths of a hydraulic jump (Sec. 10.18). It is also employed in the development of the relationships in a shock wave (Sec. 13.10).

EXERCISES

6.4.1 A cylindrical drum of radius 2 ft is securely held in position in an open channel of rectangular section. The channel is 10 ft wide, and the flow rate is 240 cfs. Water flows beneath the drum as shown in Fig. X6.4.1. Determine the horizontal thrust on the cylinder using impulse–momentum. Neglect fluid friction.

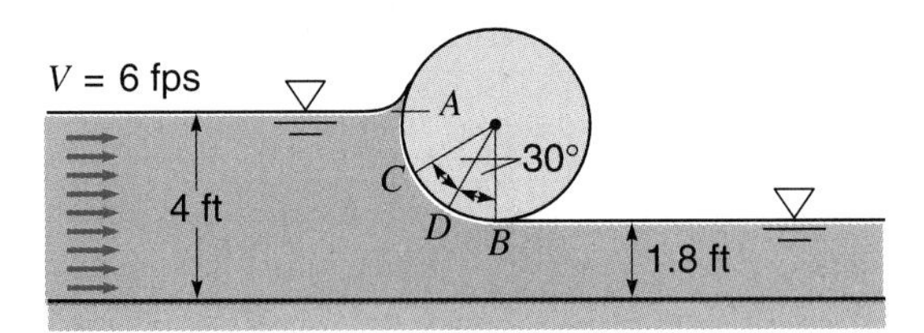

Figure X6.4.1

6.4.2 Find the horizontal thrust of the water on each meter of width of the sluice gate shown in Fig. X6.4.2, given $y_1 = 2.2$ m, $y_2 = 0.4$ m, and $y_3 = 0.5$ m. Neglect friction.

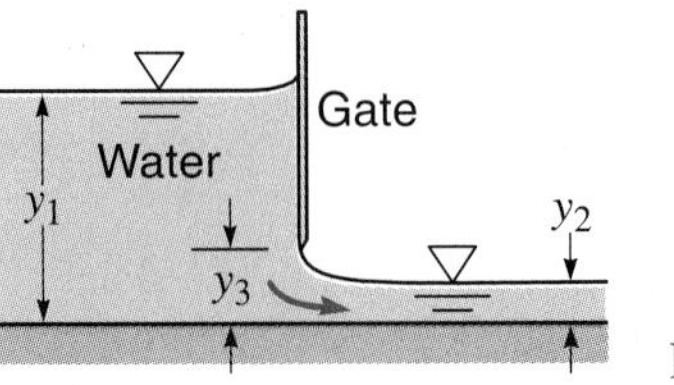

Figure X6.4.2

6.4.3 Refer to Fig. X6.4.2. Find the horizontal thrust of the water on each foot of width of the sluice gate, given $y_1 = 7$ ft, $y_2 = 1.2$ ft, and $y_3 = 1.4$ ft. Neglect friction.

6.4.4 Flow occurs over a spillway of constant section as shown in Fig. X6.4.4. Assuming ideal flow, determine the resultant horizontal force on the spillway per foot of spillway width (perpendicular to the spillway section), given that $y_1 = 4.2$ ft and $y_2 = 0.7$ ft.

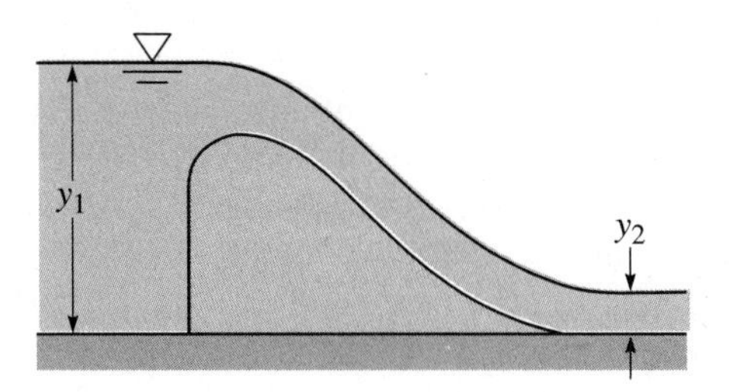

Figure X6.4.4

6.4.5 Flow occurs over the spillway of constant section as shown in Fig. X6.4.4. Given that $y_1 = 4.2$ m and $y_2 = 0.7$ m, determine the horizontal force on the spillway per meter of spillway width (perpendicular to the spillway section). Assume ideal flow.

6.5 FORCE EXERTED ON PRESSURE CONDUITS

Consider the case of horizontal flow to the right through the reducer of Fig. 6.2*a*. A free-body diagram of the forces acting on the fluid mass contained in the reducer (the control volume, *CV*) is shown in Fig. 6.2*b*. We shall apply Eq. (6.7*a*) to this fluid mass to examine the forces that are acting in the x direction. The forces p_1A_1 and p_2A_2 represent pressure forces exerted by fluid located just upstream and just downstream of the control volume. The force $(F_{R/F})_x$ represents the force exerted *by the reducer on the fluid* (R/F) in the x direction. Neglecting shear forces at the boundary of the reducer, the force $(F_{R/F})_x$ is the resultant (integrated) effect of the normal pressure forces that are exerted on the fluid by the wall of the reducer. The intensity of pressure at the wall will decrease as the diameter decreases because of the increase in velocity head, in accordance with Bernoulli's theorem (Eq. 5.16). A typical pressure diagram is shown in Fig. 6.3.

The effect that atmospheric pressure has on such analyses can be confusing. Gage pressures are shown in Fig. 6.3. If absolute pressures were

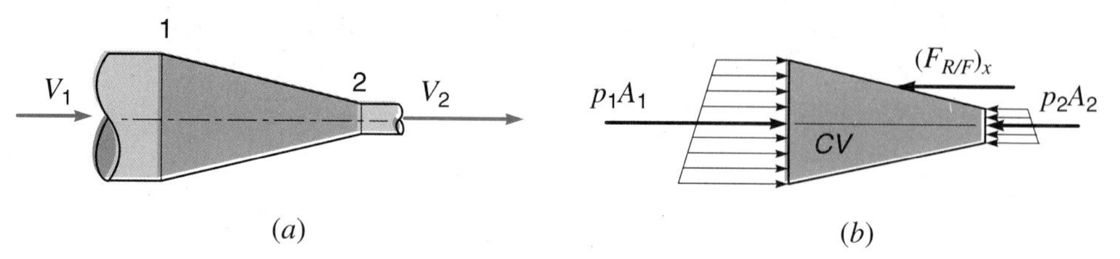

Figure 6.2

used, all the pressures shown would have to be increased by a constant amount, p_{atm} (about 14.70 psi or 101.3 kPa at sea level). This would increase $p_1A_1 - p_2A_2$, $F_{R/F}$, and the equal and opposite force exerted *by the fluid on the reducer*, $F_{F/R}$. However, this increased force on the inside of the reducer would be exactly balanced by the force of the atmosphere on the outside. Therefore the *net* force on the reducer, which is produced by the fluid flow and which tends to move the reducer, is unaffected by the atmospheric pressure. It is this *net* force in which we are interested, and it can most easily be obtained by excluding atmospheric pressure, i.e., by using gage pressures. It is therefore customary to use *gage* pressures for p_1 and p_2.

Applying Eq. (6.7*a*) and assuming the fluid to be ideal with $(F_{R/F})_x$ directed as shown, since the entry and exit velocities are parallel to the x direction, we get

$$\sum F_x = p_1A_1 - p_2A_2 - (F_{R/F})_x = \rho Q(V_2 - V_1) \tag{6.10}$$

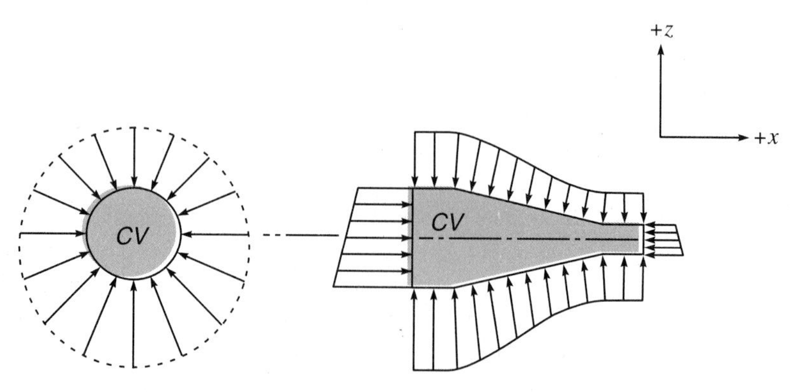

Figure 6.3
Gage pressure distribution on the fluid in a reducer.

In Eq. (6.10) each term can be evaluated independently from the given flow data, except $(F_{R/F})_x$, which is the quantity we wish to find. Rewriting Eq. (6.10), the result is

$$(F_{R/F})_x = p_1A_1 - p_2A_2 - \rho Q(V_2 - V_1) \tag{6.11}$$

This gives the value of the net force exerted by the *reducer on the fluid* in the x direction. This force acts to the left as assumed in Fig. 6.2*b* and as applied in Eq. (6.10). The force of the *fluid on the reducer* is, of course, equal and opposite to that of the reducer on the fluid. If the flow were to the left in Fig. 6.2, a similar analysis would apply, but it is necessary to be consistent in regard to plus and minus signs. The usual convention is to consider the direction in which the flow is occurring as the positive direction.

Consideration of the weight of fluid between sections 1 and 2 in Fig. 6.2 results in the conclusion that pressures are larger on the bottom half of the pipe than on the upper half. It should be noted that it is the conditions at the end sections of the control volume that govern the analysis. What occurs within the flow between sections 1 and 2 is unimportant so far as the determination of forces is concerned. Figure 6.3 gives a schematic representation of the pressure distribution on the fluid within the reducer. The integrated effect of the pressures exerted by the reducer itself is equivalent in the x direction to $(F_{R/F})_x$ and in the z direction to the weight of fluid between sections 1 and 2.

If the fluid undergoes a change in both direction and velocity, as in the reducing pipe bend in Fig. 6.4, the procedure is similar to that of the

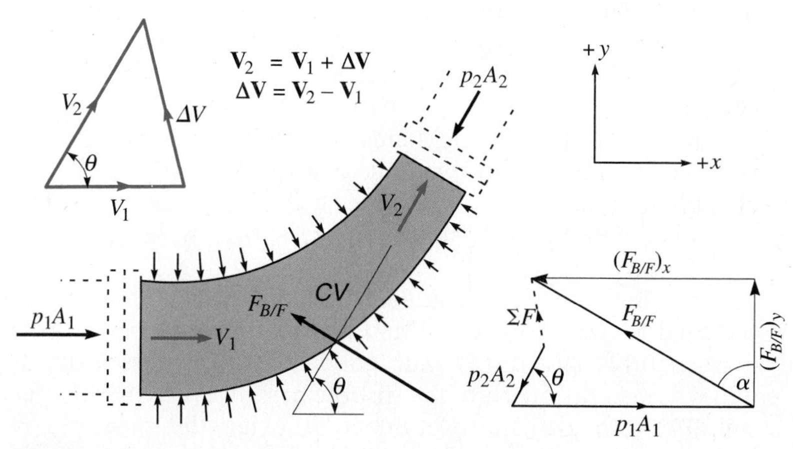

Figure 6.4
Forces on the fluid in a reducing bend.

preceding case, except that it is convenient to deal with components. Assuming the flow is in a horizontal plane so that the weight can be neglected, applying Eq. (6.7*a*) by summing up forces acting on the fluid in the x direction, and equating them to the change in fluid momentum in the x direction (where $V_{2x} = V_2 \cos \theta$) gives

$$\sum F_x = p_1 A_1 - p_2 A_2 \cos \theta - (F_{B/F})_x = \rho Q (V_2 \cos \theta - V_1) \qquad (6.12)$$

which, when rewritten for the force we wish to find, becomes

$$(F_{B/F})_x = p_1 A_1 - p_2 A_2 \cos \theta - \rho Q (V_2 \cos \theta - V_1) \qquad (6.13)$$

Similarly, in the y direction, where $V_{2y} = V_2 \sin \theta$ and $V_{1y} = 0$,

$$\sum F_y = 0 - p_2 A_2 \sin \theta + (F_{B/F})_y = \rho Q (V_2 \sin \theta - 0) \qquad (6.14)$$

which, when rewritten, becomes

$$(F_{B/F})_y = p_2 A_2 \sin \theta + \rho Q V_2 \sin \theta \qquad (6.15)$$

In a specific case, if the numerical values of $(F_{B/F})_x$ and $(F_{B/F})_y$ as determined by these equations are positive then the assumed directions are correct. A negative value for either one merely indicates that that component is in the direction opposite from that assumed.

Note that $\sum \mathbf{F} = \rho Q \, \Delta \mathbf{V}$ is the resultant of *all* the forces acting on the fluid, which *includes* the pressure forces on the two ends and the force $\mathbf{F}_{B/F}$ exerted by the bend on the fluid. The directions of $\sum \mathbf{F}$ and $\Delta \mathbf{V}$ must be the same (see Fig. 6.4). The value of $F_{B/F}$ is $\sqrt{(F_{B/F})_x^2 + (F_{B/F})_y^2}$, and its direction α may be obtained from the force diagram shown in Fig. 6.4.

The total force $\mathbf{F}_{F/B}$ exerted by the fluid on the bend is equal in magnitude but opposite in direction to the force $\mathbf{F}_{B/F}$ of the bend on the fluid. The force of the fluid on the bend tends to move the portion of the pipe under consideration. Hence, to prevent damage where such changes in velocity or alignment occur, a large pipe will usually be "anchored" by attaching it to a concrete block of sufficient size and/or weight to provide the necessary resistance.

If the flow in Fig. 6.4 had been in a vertical plane, i.e., y was vertical, the weight of the fluid between sections 1 and 2 would have to be estimated and included in Eqs. (6.14) and (6.15). The effect of shear stresses due to fluid friction could be introduced into the problem; however, these effects are usually small. If there are multiple inlets or exits, the principle remains the same: $\sum \mathbf{F} = \sum (\rho Q \mathbf{V})_{\text{out}} - \sum (\rho Q \mathbf{V})_{\text{in}}$ (see the following sample problem).

Sample Problem 6.2 Determine the magnitude and direction of the resultant force exerted on this double nozzle by water flowing through it as shown in Fig. S6.2. Both nozzle jets have a velocity of 12 m/s. The axes of the pipe and both nozzles lie in a horizontal plane. $\gamma = 9.81\ \mathrm{kN/m^3}$. Neglect friction

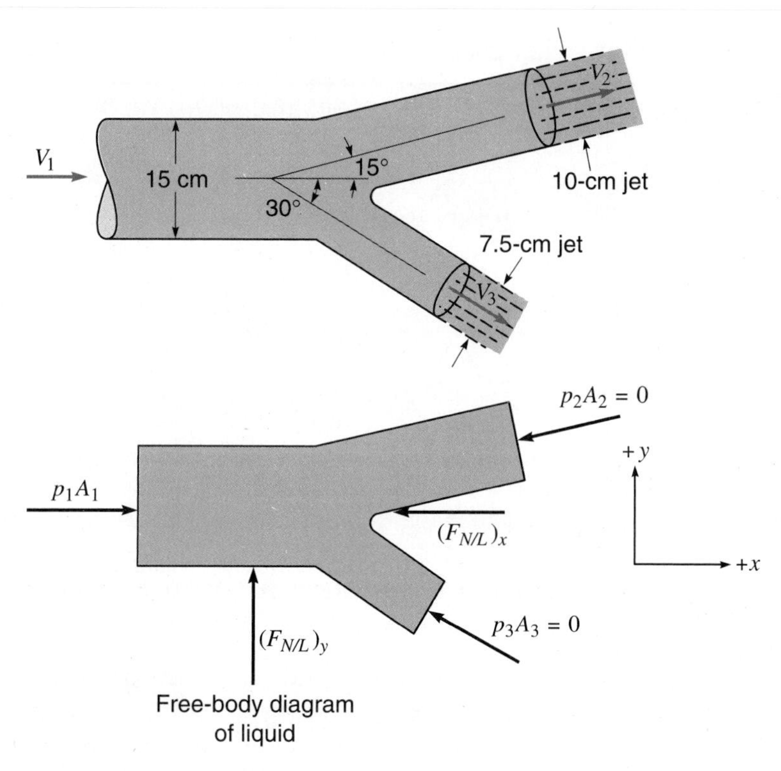

Figure S6.2

Solution

Continuity:

$$A_1V_1 = A_2V_2 + A_3V_3$$

$$15^2V_1 = 10^2(12) + 7.5^2(12), \quad V_1 = 8.33\ \mathrm{m/s}$$

$$Q_1 = \frac{\pi}{4}(0.15)^2 8.33 = 0.1473\ \mathrm{m^3/s}, \quad Q_2 = 0.0942\ \mathrm{m^3/s}, \quad Q_3 = 0.0530\ \mathrm{m^3/s}$$

Jets 2 and 3 are "free," i.e., in the atmosphere, so $p_2 = p_3 = 0$.

Energy equation:
$$\frac{p_1}{\gamma} + \frac{8.33^2}{2(9.81)} = 0 + \frac{12^2}{2(9.81)}$$

$$\frac{p_1}{\gamma} = 3.80\ \mathrm{m}, \quad p_1 = 37.3\ \mathrm{kN/m^2}, \quad p_1A_1 = 0.659\ \mathrm{kN}$$

$$\sum F_x = p_1A_1 - (F_{N/L})_x = (\rho Q_2 V_{2x} + \rho Q_3 V_{3x}) - \rho Q_1 V_{1x}$$

$$\rho = \frac{\gamma}{g} = \frac{9.81 \text{ kN/m}^3}{9.81 \text{ m/s}^2} = 1.0 \frac{\text{kN}\cdot\text{s}^2}{\text{m}^4} = 10^3 \frac{\text{kg}}{\text{m}^3}$$

$$V_{2x} = V_2 \cos 15° = 12(0.966) = 11.6 \text{ m/s}$$

$$V_{3x} = V_3 \cos 30° = 12(0.866) = 10.4 \text{ m/s}, \quad V_{1x} = V_1 = 8.33 \text{ m/s}$$

$$0.659 - (F_{N/L})_x = 10^3(0.0942)(11.6) + 10^3(0.0530)(10.4) - 10^3(0.1473)8.33$$
$$= 0.417 \text{ kN}$$

$$(F_{N/L})_x = 0.659 - 0.417 = 0.242 \text{ kN} \leftarrow$$

$$\sum F_y = (F_{N/L})_y = (\rho Q_2 V_{2y} + \rho Q_3 V_{3y}) - \rho Q_1 V_{1y}$$

$$V_{2y} = V_2 \sin 15° = 12(0.259) = 3.11 \text{ m/s}$$

$$V_{3y} = -V_3 \sin 30° = -12(0.50) = -6.00 \text{ m/s}, \quad V_{1y} = 0$$

$$(F_{N/L})_y = 10^3(0.0942)(3.11) + 10^3(0.0530)(-6.00) - 10^3(0.1473)(0)$$
$$= 0.291 - 0.318 - 0 = -0.027 \text{ kN} \uparrow = 0.027 \text{ kN} \downarrow$$

The minus sign indicates that the assumed direction of $(F_{N/L})_y$ was wrong. Therefore $(F_{N/L})_y$ acts in the negative y direction. $F_{L/N}$ is equal and opposite to $F_{N/L}$.

$$(F_{L/N})_x = 0.242 \text{ kN} \rightarrow \quad \text{(in the positive } x \text{ direction)}$$
$$(F_{L/N})_y = 0.027 \text{ kN} \uparrow \quad \text{(in the positive } y \text{ direction)}$$
$$F_{L/N} = 0.243 \text{ kN at } 5.90° \quad \textit{ANS}$$

EXERCISES

6.5.1 A nozzle that discharges a 4-in-diameter water jet into the air is on the right end of a horizontal 10-in-diameter pipe. In the pipe the water has a velocity of 15 fps and a gage pressure of 65 psi. Find the magnitude and direction of the resultant axial force exerted upon the nozzle, and the head loss in the nozzle.

6.5.2 A nozzle that discharges a 6-cm-diameter water jet into the air is on the right end of a horizontal 12-cm-diameter pipe. In the pipe the water has a velocity of 4 m/s and a gage pressure of 400 kN/m^2. Find the magnitude and direction of the resultant axial force exerted upon the nozzle, and the head loss in the nozzle.

6.5.3 A diverging nozzle that discharges an 8-in-diameter water jet into the air is on the right end of a horizontal 6-in-diameter pipe. If the velocity in the pipe is 12 fps, find the magnitude and direction of the resultant axial force exerted upon the nozzle. Neglect fluid friction.

6.5.4 Water under a gage pressure of 60 psi flows with a velocity of 12 fps through a right-angled bend that has a uniform diameter of 10 in. The bend lies in a horizontal plane, and water enters from the west and leaves towards the north. Assuming no drop in pressure, what is the magnitude and direction of the resultant force acting on the bend?

6.5.5 Water under a gage pressure of 400 kN/m^2 flows with a velocity of 4 m/s through a right-angled bend that has a uniform diameter of 26 cm. The bend lies in a horizontal plane, and water enters from the west and leaves towards the north. Assuming no drop in pressure, what is the magnitude and direction of the resultant force acting on the bend?

6.5.6 Refer to the double nozzle of Prob. 6.9, in which both water jets have a velocity of 35 fps. What angle should the 6-in jet make with the axis of the 8-in-diameter pipe so that the resultant force is along the axis of the 8-in-diameter pipe?

6.6 FORCE EXERTED ON A STATIONARY VANE OR BLADE

A procedure similar to that of Sec. 6.5 may be employed to find the force exerted on a stationary vane or blade. The main difference is that with a vane or blade the fluid is in contact with the atmosphere; hence the gage pressures p in the jet are zero and the pA forces disappear. A jet in contact with the atmosphere in this way is referred to as a ***free jet***. Another difference is that in many types of fluid machinery where vanes or blades are used the velocities are often so high that the neglect of friction may introduce a sizeable error. In such cases, for accurate results, friction should be considered. This is usually handled by prescribing a reduction in the velocity of the flow between its arrival and departure points on the blade. The following sample problem illustrates these points.

Sample Problem 6.3 A free jet of water with an initial diameter of 2 in strikes the vane shown in Fig. S6.3. Suppose that $\theta = 30°$ and $V_1 = 100$ fps. Owing to friction losses assume that $V_2 = 95$ fps. Flow occurs in a horizontal plane. Find the resultant force on the blade.

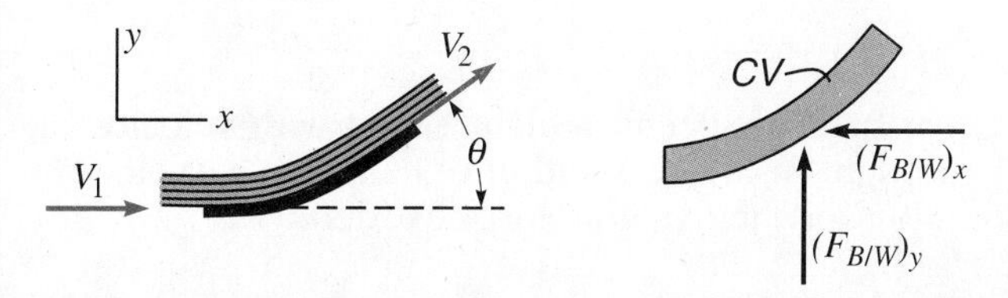

Figure S6.3

Solution

Take a free body diagram of the element of fluid (control volume) in contact with the blade. Assume the forces acting on the element are as shown in the sketch. The forces $(F_{B/W})_x$ and $(F_{B/W})_y$ represent the components (directions assumed) of the net force of the blade on the water (B/W) in the x and y directions. This net force includes shear stresses tangential to the blade and pressure forces normal to the blade.

Applying Eq. (6.7*a*) along the x axis and noting that $A = \pi(2/12)^2/4 = 0.0218\ \text{ft}^2$,

$$-(F_{B/W})_x = \rho Q(V_{2x} - V_{1x}) = 1.94(0.0218 \times 100)(0.866 \times 95 - 100)$$

$$= 4.23(-17.7) = -75.0\ \text{lb}$$

Hence $\quad (F_{B/W})_x = +75.0\ \text{lb} = 75.0\ \text{lb} \leftarrow$

The plus sign indicates that the assumed direction of $(F_{B/W})_x$ was correct.

Applying Eq. (6.7*b*) along the y axis,

$$+(F_{B/W})_y = \rho Q(V_{2y} - V_{1y}) = 4.23(0.50 \times 95 - 0) = 201\ \text{lb} = 201\ \text{lb} \uparrow$$

The resultant force of the blade on the control volume is the sum of these two components. The force of the fluid on the blade is equal and opposite to this. The resultant force on the blade is 214 lb at an angle of 69.5° ↘ ***ANS***

Note that if friction where neglected (i.e., $V_2 = V_1 = 100$ fps), the forces would have been calculated as $(F_{B/W})_x = 56.7$ lb and $(F_{B/W})_y = 212$ lb. When the angle of deflection θ from the initial direction of the jet is less than 90°, we find that friction increases the value of $(F_{B/W})_x$ over the value it would have if there were no friction. When θ is greater than 90°, friction decreases the value of F_x. On the other hand, friction decreases of value of F_y for any value of angle θ.

If the flow had been in a vertical plane, the effect on V_2 of the higher elevation at exit from the blade would have to be considered, and the weight of the liquid on the blade would have to be estimated and added to $\rho Q(\Delta V_z)$ to get the total value of $(F_{B/W})_z$.

EXERCISES

6.6.1 A jet containing any type of fluid of specific weight γ and with velocity V and area A is deflected through an angle θ without changing the velocity magnitude. Derive an equation for the dynamic force exerted.

6.6.2 Refer to Fig. S6.3. Assume that friction is negligible, that $\theta = 115°$, and that the water jet has a velocity of 95 fps and a diameter of 1 in. Find (*a*) the component of the force in the direction of the jet; (*b*) the force component normal to the jet; (*c*) the magnitude and direction of the resultant force exerted on the vane.

6.6.3 Refer to Fig. S6.3. Assume that friction is negligible, that $\theta = 115°$, and that the water jet has a velocity of 25 m/s and a jet diameter of 4 cm. Find (*a*) the component of the force in the direction of the jet; (*b*) the force component normal to the jet; (*c*) the magnitude and direction of the resultant force exerted on the vane.

6.6.4 Solve Exer. 6.6.2, assuming that friction reduces V_2 to 70 fps.

6.6.5 Solve Exer. 6.6.3, assuming that friction reduces V_2 to 20 m/s.

6.6.6 Calculate the resultant force on a large flat plate if the jet in Exer. 6.6.2 were to strike it normally.

6.6.7 Suppose the jet in Exer. 6.6.3 were to strike a large flat plate normally. What would be the resultant force on the plate?

6.6.8 In Exer. 6.6.6 assume the center of the jet is coincident with the center of the circular plate. find (*a*) the stagnation pressure and (*b*) the average pressure on the plate if the area of the plate is 22 times the area of the jet.

6.6.9 In Exer. 6.6.7 assume the center of the jet is coincident with the center of the circular plate. Find (*a*) the stagnation pressure and (*b*) the average pressure on the plate if the area of the plate is 25 times the area of the jet.

6.7 MOVING VANES: RELATION BETWEEN ABSOLUTE AND RELATIVE VELOCITIES

In much of the work that follows it will be necessary to deal with both absolute and relative velocities of the fluid. The absolute velocity $\mathbf{V}$ of a body (Fig. 6.5) is its velocity relative to the earth. The relative velocity $\boldsymbol{v}$ of a body is its velocity relative to a second body,[1] which may in turn be in motion with absolute velocity $\mathbf{u}$ relative to the earth.

---

[1] This text uses a rounded lower case v (vee) to help distinguish it from the capital V and from the Greek ν (nu) used for kinematic viscosity.

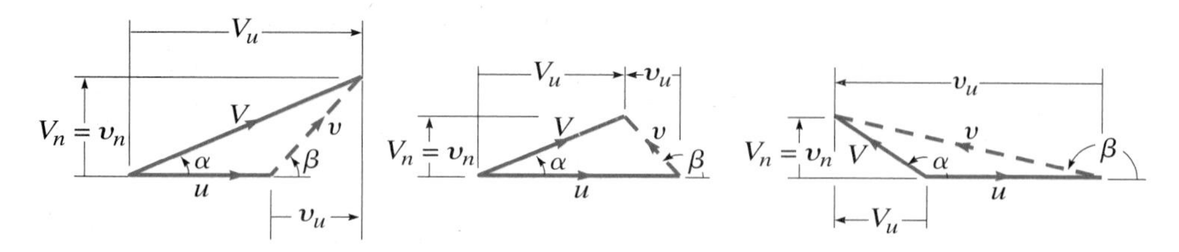

Figure 6.5
Relative (dashed) and absolute velocity relations, with their components.

The absolute velocity $\mathbf{V}$ of the first body is the vector sum of its velocity $\boldsymbol{v}$ relative to the second body and the absolute velocity $\mathbf{u}$ of the latter. The relation of the three is thus

$$\mathbf{V} = \mathbf{u} + \boldsymbol{v} \tag{6.16}$$

Because the directions of these three velocities may vary, we have represented them by vectors and by the boldface variables $\mathbf{V}$, $\mathbf{u}$, and $\boldsymbol{v}$ in equations. The relative velocities $\boldsymbol{v}$ are shown dashed in Fig. 6.5, which presents three different possible forms of the relations.

Let us define α and β as the angles made by the absolute and relative velocities of a fluid, respectively, with the positive direction of the linear velocity $\mathbf{u}$ of some point on a solid body. It is seen from Fig. 6.5 that, whatever the shape of the velocity vector triangle, the velocity components parallel and normal to $\mathbf{u}$ are always given by

$\|$ to $\mathbf{u}$: $$V_u = u + v_u = V \cos \alpha = u + v \cos \beta \tag{6.17}$$

$\perp$ to $\mathbf{u}$: $$V_n = v_n = V \sin \alpha = v \sin \beta \tag{6.18}$$

Here the subscript u indicates components parallel to $\mathbf{u}$, and the subscript n indicates components normal to $\mathbf{u}$. For *rotating* blades (Secs. 6.11–6.14), we shall see that u corresponds to the tangential direction and n corresponds to the radial direction.

6.8 FORCE EXERTED BY A JET ON A MOVING VANE OR BLADE

Single Blade, Moving Parallel to Jet

We can determine the force exerted by a stream on a single moving object by an equation very similar to Eq. (6.6), *provided the flow is steady and the body has a motion of translation along the same line as the initial stream,* i.e., provided $\mathbf{u}$ is parallel to $\mathbf{V}_1$. If the latter condition is not fulfilled, the case becomes a complex one of unsteady flow.

There are two principal differences between the action upon a stationary and a moving object. The first is that the amount of fluid that strikes a single moving object in any time interval Δt must be different from that which strikes a stationary object, and so this changes the rate of momentum transfer. The second difference is that for a moving object we must consider both relative and absolute velocities, which makes the determination of the required $\mathbf{\Delta V}$ more difficult.

Let us consider the first issue, regarding the rate of momentum transfer. If the cross-sectional area of a stream is A_1 and its velocity is V_1 then the rate at which fluid is emitted from the nozzle in terms of volume is $Q = A_1 V_1$ and the rate in terms of mass is $\dot{m} = \rho Q = \rho A_1 V_1$. But for the case of a single object, the amount of fluid that strikes the body per unit time will be less than this if the body is moving away from the nozzle and more than this if it is moving toward the nozzle. As an extreme case, suppose the object is moving away from the jet (along the same axis) and with the same or higher velocity, magnitude u. It is clear that none of the fluid will then act upon the body. But if it is moving with a velocity u less than that of the jet, the amount of fluid that strikes the body per unit time will be proportional to the difference between the two velocities, i.e., to $V_1 - u = v_1$. Accordingly, the rate at which fluid strikes the moving body will be, on a volume basis,

$$Q' = A_1(V_1 - u) = A_1 v_1 \tag{6.19}$$

and, on a mass basis,

$$\dot{m}' = \rho Q' = \rho A_1(V_1 - u) = \rho A_1 v_1 \tag{6.20}$$

The difference between Q and Q' may also be seen by considering Fig. 6.6, where the fluid issues from a nozzle at the rate $Q = A_1 V_1$ per unit time. But in this unit of time the object will have moved away from the nozzle a distance u, and the volume of fluid between the two will have been increased by the amount $A_1 u$. Since V_1 is greater than u, the difference, equal to $A_1 V_1 - A_1 u = A_1(V_1 - u) = A_1 v_1$, must be the amount that struck the object within unit time.

Let us now consider the second issue, regarding the relative velocities. The $\mathbf{\Delta V}$ in the impulse–momentum equation (6.6) is the difference of the absolute velocities, $\mathbf{V}_{out} - \mathbf{V}_{in}$. In Fig. 6.6 this is the same as $\mathbf{V}_2 - \mathbf{V}_1$, and $\mathbf{\Delta V}$ is shown on a velocity triangle. Usually $\mathbf{V}_2$ is unknown, and therefore so is the required $\mathbf{\Delta V}$. These can be obtained by first solving the relative velocity triangle (Fig. 6.5) for exit, and by next solving the absolute velocity triangle in Fig. 6.6. However, much of this work can often be avoided, because for the special case we are considering here in which u is constant for the entire vane or body, we have

$$\begin{aligned}
\mathbf{\Delta V} &= \mathbf{V}_2 - \mathbf{V}_1 \\
&= (\mathbf{u} + \boldsymbol{v}_2) - (\mathbf{u} + \boldsymbol{v}_1) \\
&= \boldsymbol{v}_2 - \boldsymbol{v}_1 \\
&= \mathbf{\Delta} \boldsymbol{v}
\end{aligned}$$

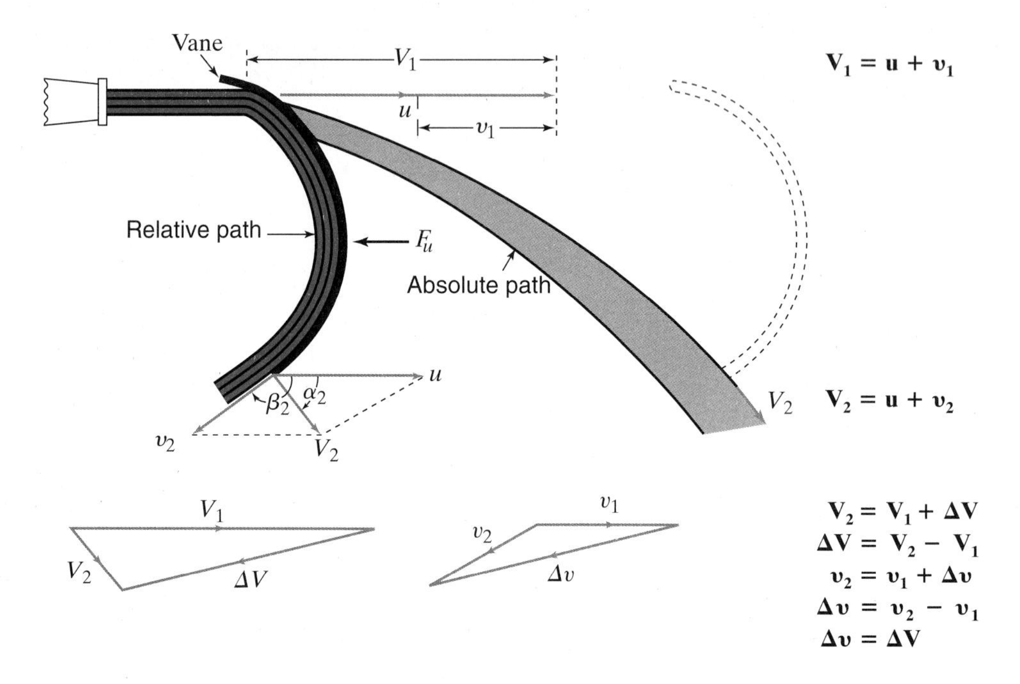

Figure 6.6
Jet acting on a vane in translation.

Thus either $\mathbf{\Delta V}$ or $\mathbf{\Delta v}$ may be used in this case, and $\mathbf{\Delta v}$ is often easier to obtain.

Noting the outcomes of the two issues just discussed, for the present case of a single moving vane, Eq. (6.6) therefore becomes

$$\sum \mathbf{F} = \dot{m}'(\mathbf{\Delta V}) = \dot{m}'(\mathbf{\Delta v}) = \rho Q'(\mathbf{\Delta V}) = \rho Q'(\mathbf{\Delta v}) \tag{6.21}$$

Usually we are most interested in the force exerted in the direction of the jet, which we previously named the u or the x direction. From Fig. 6.6 and Eq. (6.21), noting that $p_1 = p_2 = 0$ for a free jet, we obtain

$$\sum F_u = 0 - (F_{B/W})_u = \rho Q'(\Delta V_u) = \rho Q'(\Delta v_u) \tag{6.22}$$

where the minus sign confirms that the u component of the force of the blade on the water, $(F_{B/W})_{u'}$, acts to the left in Fig. 6.6. Equal and opposite to this force is the u component of the force of the water jet on the blade, $(F_{W/B})_u$, which acts to the right in Fig. 6.6, i.e., in the direction of the blade movement.

Recalling that $\rho = \gamma/g$ and $Q' = A_1 v_1$, another convenient form of these last two equations, which can be used for either force, is

Jet on single blade:

$$F_u = \frac{\gamma A_1 v_1}{g} \Delta v_u \tag{6.23}$$

In Fig. 6.6, by the time a particle of fluid that strikes the moving vane at the instant it is in the position shown by the solid line has reached the point of outflow from the vane, the latter will have reached the position shown by the dashed outline. Thus two paths may be traced for the fluid: one relative to the moving vane, which is as it would appear to an observer (or a camera) moving with the vane, and the other relative to the earth, termed the ***absolute path***, as it would appear to an observer (or a camera) stationary with respect to the earth.

A study of Fig. 6.6 shows that the direction of the relative velocity at outflow from the vane is determined by the shape of the latter, but the relative velocity at entrance, just *before* the fluid strikes the vane, is determined solely by the relation between V_1 and u. Just *after* the fluid strikes the vane, its relative velocity must be tangent to the vane surface. To avoid excess energy loss, these two directions should agree; otherwise there will be an abrupt change in velocity and direction of flow at this point, known as ***shock***.

Series of Rotating Blades

A stream of fluid impinging on a single moving vane, as just discussed, occurs rarely. More commonly, the jet is directed on a *series* of closely spaced vanes, as with a Pelton wheel (Figs. 16.1 and 16.3). In such cases, whatever flow does not impinge on the first bucket will strike the second one, and so on around the circle. For this reason, for a series of vanes the full momentum transfer rate $\dot{m} = \rho Q = \rho A_1 V_1$ is used instead of the rate give by Eq. (6.20). Thus the u component of the force exerted by the fluid on a series of vanes can be expressed as

Jet on series of blades:

$$F_u = \dot{m} \Delta V_u = \frac{\gamma A_1 V_1}{g} \Delta V_u = \frac{\gamma A_1 V_1}{g} \Delta v_u \tag{6.24}$$

The force of the series of vanes on the fluid is again in the direction of ΔV_u, that is, to the left in Fig. 6.6.

SAMPLE PROBLEM 6.4 A 2-in-diameter water jet with a velocity of 100 fps impinges on a single vane moving in the same direction (thus $F_x = F_u$) at a velocity of 60 fps. If $\beta_2 = 150°$ and friction losses over the vane are such that $v_2 = 0.9v_1$, compute the net force exerted by the water on the vane.

Solution

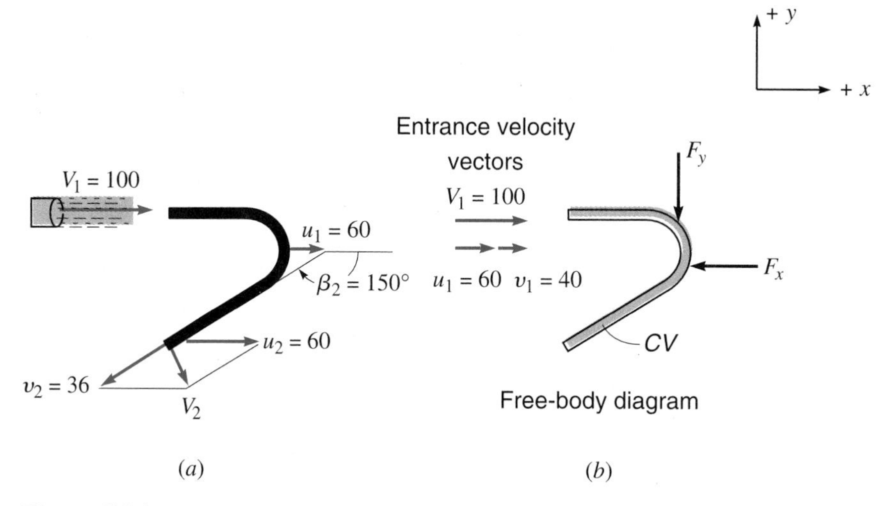

Figure S6.4

The velocity vector diagrams at entrance and exit to the vane are shown in Fig. S6.4. Since $v_2 = 0.9(40) = 36$ fps,

Eq. (6.17):

$$V_{2u} = V_2 \cos \alpha_2 = u + v_2 \cos \beta_2 = 60 + 36 \cos 150° = 28.8 \text{ fps} \qquad (1)$$

Eq. (6.18): $V_{2n} = V_2 \sin \alpha_2 = v_2 \sin \beta_2 = -36 \sin 150° = -18 \text{ fps} \qquad (2)$

$$Q' = A_1 v_i = A_i(V_1 - u) = \frac{\pi}{4}\left(\frac{2}{12}\right)^2 (100 - 60) = 0.873 \text{ cfs}$$

Eq. (6.22):

$$-F_x = \rho Q'(V_2 \cos \alpha_2 - V_1) = 1.938(0.873)(28.8 - 100) = -120.4 \text{ lb}$$

So $F_x = 120.4$ lb. The force of the vane on the water is to the left, as assumed; hence the force of water on the vane is 120.4 lb to the right.

$$-F_y = \rho Q'(V_2 \sin \alpha_2 - 0) = 1.938(0.873)(-18) = -30.4 \text{ lb}$$

Thus $F_y = 30.4$ lb in the direction shown. The force of water on the vane is equal and opposite and thus 30.4 lb upward.

Therefore the net

$$F_{W/B} = 124.2 \text{ lb at } 14.19° \nearrow \quad \textit{ANS}$$

If needed, (1) and (2) may be solved simultaneously to yield $V_2 = 34.0$ fps, $\alpha_2 = 32.0°$.

Note that if the blade were one of a *series* of blades,

$$Q = A_1 V_1 = \frac{\pi}{4}\left(\frac{2}{12}\right)^2 (100) = 2.18 \text{ cfs}$$

Eq. (6.24): $-F_x = \rho Q(V_2 \cos \alpha_2 - V_1) = 1.938(2.18)(28.8 - 100) = -301$ lb

Also, for the case of a series of blades, energy considerations can be used for the solution. The horsepower of the original jet is

Eq. (5.32): $$\text{HP}_{\text{in}} = \frac{\gamma Q(V_1^2/2g)}{550} = \frac{62.4(2.18)(100)^2}{550(2)32.2} = 38.4$$

The horsepower of the water as it leaves the system is

$$\text{HP}_{\text{out}} = \frac{\gamma Q(V_2^2/2g)}{550} = 4.44$$

The horsepower transferred to the blades (i.e., out of the fluid) is

$$\text{HP}_{\text{transfer}} = \frac{Fu}{550} = \frac{(301)(60)}{550} = 32.8$$

An equation for conservation of energy expressed in terms of power is

$$\text{HP}_{\text{in}} - \text{HP}_{\text{out}} - \text{HP}_{\text{transfer}} - \text{HP}_{\text{friction loss}} = 0$$

Thus $$38.4 - 4.4 - 32.8 = \text{HP}_{\text{friction loss}}$$

Therefore $$\text{HP}_{\text{friction loss}} = 1.2$$

This may be verified by computing

$$\frac{\gamma Q(v_1^2/2g) - \gamma Q(v_2^2/2g)}{550} = \frac{62.4(2.18)[(40)^2 - (36)^2]}{550(2)32.2} = 1.2 \text{ HP}$$

It should be noticed that the horsepower loss due to friction is small. Commonly, in problems of this type with free jets, an assumption that $v_1 = v_2$ in magnitude will give reasonably good results.

EXERCISES

6.8.1 If a jet of fluid strikes a single body moving in the same direction with a velocity u, flows over it without friction loss, and leaves with a relative velocity in the direction of β_2, prove that $F_u = (\gamma A_1/g)(1 - \cos\ \beta_2)(V_1 - u)^2$.

6.8.2 A jet of water strikes a single vane, which reverses it through 180° without friction loss. If the jet has an area of 3.5 in² and a velocity of 175 fps, find the force exerted if the vane moves (*a*) in the same direction as the jet with a velocity of 75 fps; (*b*) in a direction opposite to that of the jet with a velocity of 75 fps.

6.8.3 A jet of water strikes a single vane, which reverses it through 180° without friction loss. If the jet has an area of 25 cm² and a velocity of 55 m/s, find the force exerted if the vane moves (*a*) in the same direction as the jet with a velocity of 20 m/s; (*b*) in a direction opposite to that of the jet with a velocity of 20 m/s.

6.8.4 What would be the resultant force components on the single vane of Sample Prob. 6.4 if it were traveling to the left toward the nozzle at 15 fps?

6.8.5 A series of vanes is acted on by a 4-in water jet having a velocity of 105 fps, with $\alpha_1 = \beta_1 = 0$. Neglect friction and find the required blade angle β_2 in order that the resultant force acting on the vane in the direction of the jet is 200 lb. Solve using vane velocities of 0, 15, 45, and 75 fps. Also find the maximum possible vane velocity.

6.9 REACTION OF A JET

Consider a jet issuing steadily from a tank (Fig. 6.7). The tank is large enough so that the velocities within it may be neglected. Let the area of the jet be A_2 and its velocity V_2, and assume an ideal fluid, $V_2 = \sqrt{2gh}$. In this case, with the jet flowing to the right, the impulse–momentum principle tells us that a force equal to $\rho Q_2 V_2$ is exerted to the left on the tank. That this is so may be seen by applying Eq. (6.7*a*) to the free-body diagram (Fig. 6.7*b*) of the liquid in the tank. In Fig. 6.7*b* the heavy vectors represent the two resultant component forces of the tank on the liquid, while the distributed load represents the force of the liquid on the tank. The figure shows the distributed load in the vertical plane through the centerline of the jet.

Applying Eq. (6.7*a*) to the liquid, we get

$$\sum F_X = (F_{T/L})_x = \rho Q_2(V_2 - 0) = \rho A_2 V_2^2 = \rho A_2(2gh) = 2\gamma h A_2 \qquad (6.25)$$

This is the net force of the tank on the ideal liquid in the x direction; it acts to the right, and causes the change of velocity of the flowing liquid from zero to V_2. Equal and opposite to this force is the force of the liquid on the tank, often referred to as the ***jet reaction***. If the tank were supported on frictionless rollers, it would be moved to the left by this force. The net force $\rho Q_2 V_2$ is

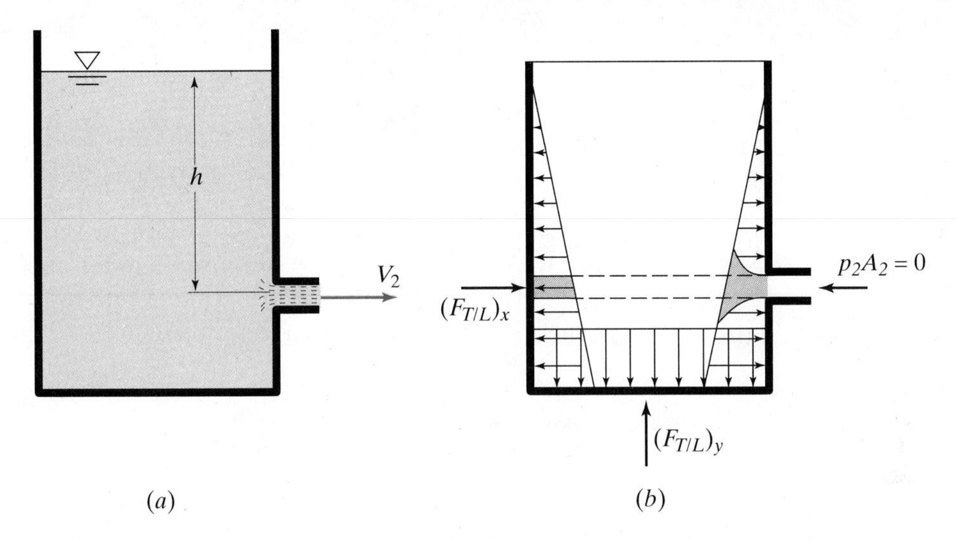

Figure 6.7

equal to the difference in the magnitude of the pressure forces on the two ends of the tank. On the left end of the tank a normal hydrostatic pressure exists, while on the right end of the tank there is a lowering of the pressure near the orifice because of the increase in velocity within the tank in that region (see Fig. 6.7*b*). By observing the last term of Eq. (6.25) we see that this net force is equal to *twice* the hydrostatic force on A_2. Thus the net force (shaded area at the right end of the tank) is equal to twice the hydrostatic force on A_2 (shaded area at the left end of the tank).

Refer now to Fig. 6.8, where a jet of the same liquid of cross-sectional

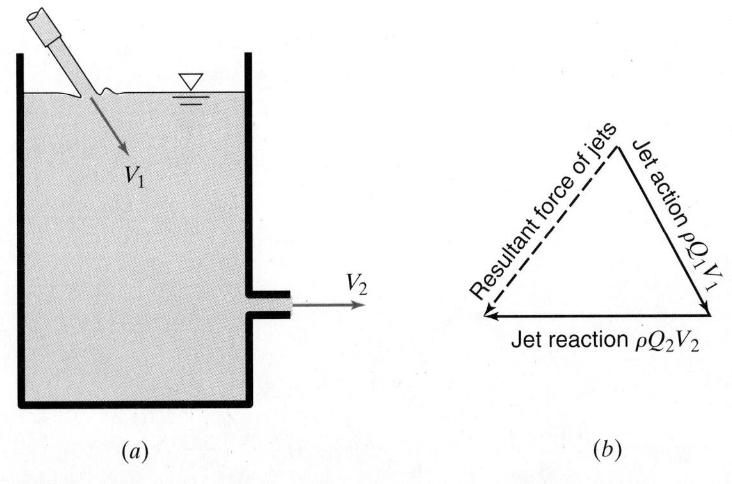

Figure 6.8

area A_1 is discharged into the tank with a velocity V_1. In this case a force $\mathbf{F} = \rho Q_1 \mathbf{V}_1$ is exerted by the jet on the liquid, which, in turn, transmits the force to the tank. This is referred to as ***jet action***.

The resultant force on the tank caused by one jet entering the tank at section 1 and the other jet leaving the tank at section 2 is the vector sum of $\rho Q_1 \mathbf{V}_1$ and $\rho Q_2 \mathbf{V}_2$, where the first vector (jet action) acts in the direction of V_1 (downward to the right in Fig. 6.8) and the second vector (jet reaction) acts in the direction opposite to that of V_2. Thus a jet entering a system acts on the system in the direction in which the jet is traveling, while a jet leaving a system acts on the system in the direction opposite to that in which the jet is traveling.

SAMPLE PROBLEM 6.5 Figure S6.5*a* shows a curved pipe section 40 ft long that is attached to the straight pipe section as shown. Determine the resultant force on the curved pipe, and find the horizontal component of the jet reaction. All significant data are given in the figure. Assume an ideal liquid with $\gamma = 55 \text{ lb/ft}^3$.

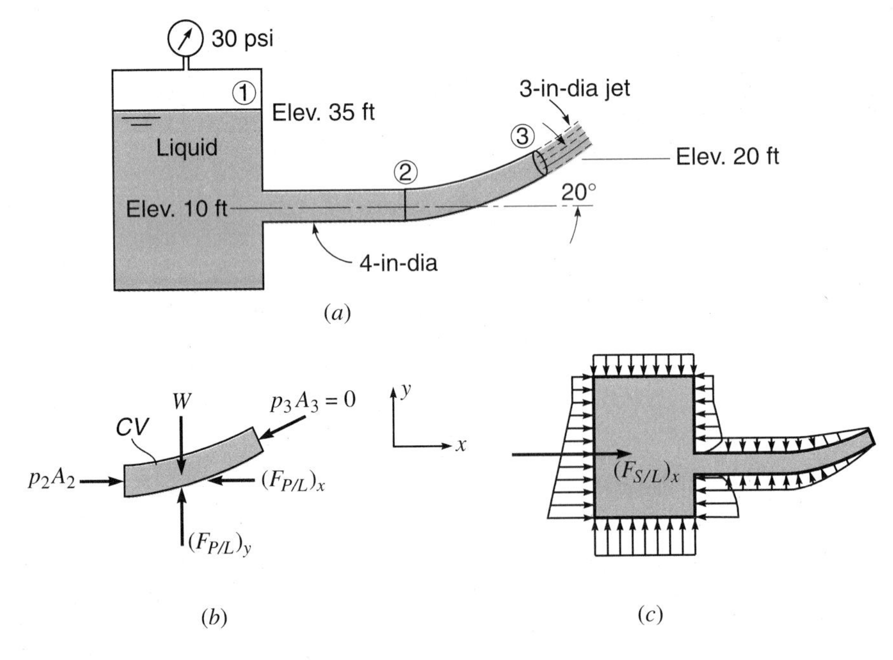

Figure S6.5

Solution

Energy Eq. (5.15) between points 1 and 3 in Fig. S6.5*a* gives

$$\frac{30(144)}{55} + 35 + 0 = 0 + 20 + \frac{V_3^2}{2(32.2)}$$

from which the jet velocity $V_3 = 77.6$ fps.

$$A_3 = \frac{\pi}{4}\left(\frac{3}{12}\right)^2 = 0.0491 \text{ ft}^2, \quad \text{so} \quad Q = A_3V_3 = 3.81 \text{ cfs}$$

$$A_2 = \frac{\pi}{4}\left(\frac{4}{12}\right)^2 = 0.0873 \text{ ft}^2, \quad \text{so} \quad V_2 = \frac{Q}{A_2} = 43.6 \text{ fps}$$

Energy Eq. (5.15) between points 2 and 3 gives

$$\frac{p_2(144)}{55} + 10 + \frac{(43.6)^2}{2(32.2)} = 0 + 20 + \frac{(77.6)^2}{2(32.2)}$$

from which $$p_2 = 28.3 \text{ psi.}$$

The free-body diagram of the forces acting on the liquid contained in the curved pipe between points 2 and 3 (the control volume) is shown in Fig. S6.5*b*. Applying Eq. (6.7*a*),

$$\sum F_x = p_2A_2 - p_3A_3 \cos 20° - (F_{P/L})_x = \rho Q(V_3 \cos 20° - V_2)$$

where $(F_{P/L})_x$ represents the force of the curved pipe on the liquid in the x direction. Since section 3 is a jet in contact with the atmosphere, $p_3 = 0$. The liquid density $\rho = 1.938(55/62.4) = 1.708$ slug/ft^3. Thus

$$28.3\left(\frac{\pi}{4}\times 4^2\right) - 0 - (F_{P/L})_x = (1.708)3.81(77.6 \cos 20° - 43.6)$$

$$356 - (F_{P/L})_x = 191$$

$$(F_{P/L})_x = +165 \text{ lb} = 165 \text{ lb} \leftarrow$$

The plus sign indicates that the assumed direction is correct. In the y direction the p_2A_2 force has no component. Estimating the weight of liquid W as 150 lb,

$$\sum F_y = 0 - 0 + (F_{P/L})_y - 150 = (1.708)3.81(77.6 \sin 20° - 0) = 173 \text{ lb}$$

$$(F_{P/L})_y = 173 + 150 = +323 \text{ lb} = 323 \text{ lb} \uparrow$$

The resultant force of liquid on the curved pipe is equal and opposite to that of the curved pipe on liquid. The resultant force of liquid on the curved pipe is $[(165)^2 + (323)^2]^{1/2} = 363$ lb downward and to the right at an angle of 62.9° with the horizontal. ***ANS***

The horizontal jet reaction is best found by taking a free-body diagram of the liquid in the entire system (S) as shown in Fig. S6.5*c*:

$$(F_{S/L})_x = \rho Q(V_3 \cos 20° - 0) = +475 \text{ lb} = 475 \text{ lb} \rightarrow$$

where $(F_{S/L})_x$ represents the force of the system on the liquid in the x direction. $(F_{S/L})_x$ is equivalent to the integrated effect of the x components of the pressure vectors shown in Fig. S6.5*c*. Equal and opposite to $(F_{S/L})_x$ is the force of the liquid on the system, i.e., the jet reaction. Hence the horizontal jet reaction is a 475-lb force to the left. ***ANS***

In summary, therefore, there is a 165-lb force to the right tending to separate the curved pipe section from the straight pipe section, while at the same time there is a 475-lb force tending to move the entire system to the left.

EXERCISES

6.9.1 Find the thrust developed when water is pumped in through an 8-in-diameter pipe in the bow of a boat at $v = 7$ fps and emitted through a 4-in diameter pipe in the stern of the boat.

6.9.2 Find the thrust developed when water is pumped in through a 20-cm-diameter pipe in the bow of a boat at $v = 2$ m/s and emitted through a 10-cm-diameter pipe in the stern of the boat.

6.10 JET PROPULSION

In Sec. 6.9 an expression was derived for the reaction of a jet from a stationary tank. Assume now that the tank in Fig. 6.7 is moving to the left with a velocity u. If the orifice is small compared with the size of the tank, the relative velocity within the latter may be disregarded, as may also any change in h for a short interval of time. Thus the absolute velocity of the fluid within the tank is $V_1 = u$ to the left. If the jet issues from the orifice with a relative velocity v_2, taking velocities to the right as positive, the absolute velocity of the jet will be $V_2 = v_2 - u$. Hence

$$\Delta V = V_2 - V_1 = (v_2 - u) - (-u) = v_2$$

The same result is obtained by referring to Fig. 6.7 for the case of a stationary tank (i.e., $u = 0$). In this instance $\Delta V = V_2 - 0 = v_2$. Thus the force of reaction is independent of the velocity of the tank, and Eq. (6.25) applies for either rest or motion.

Rocket

Both the fuel and the oxygen for combustion are contained within a rocket, which is analogous to the tank of Fig. 6.7. The only difference is that the exit pressure p_0 of the gases leaving the orifice or nozzle at section 2 may exceed the atmospheric pressure p_a. If A_2 equals the area of the jet, the rocket thrust is

$$F = \rho A_2 v_2^2 + (p_0 - p_a) A_2 \tag{6.26}$$

where v_2 is the velocity at which the jet issues from the rocket. The thrust F is independent of the speed of the rocket.

Jet Engine

By ***jet engine*** is meant a device that carries only its fuel, and takes in the air for combustion from the atmosphere. It is analogous to the tank of Fig. 6.8, including the intake of fluid at section 1, except that the velocity of the air received is usually in the same straight line as the velocity of the exit jet at section 2. There are three forms of jet engines, but the equation is the same for all three. The ***ram jet*** must be brought up to a high speed by rockets or some other means, and then it scoops in the air from in front and compresses it by virtue of the stagnation pressure due to its speed. The ***turbojet*** can take off from the ground, since in it the air is compressed by a compressor driven by a gas turbine, the exhaust from which supplies the jet propulsion. Then there is a ***pulsating*** machine, which scoops in air in cycles. The inlet is then closed, the fuel–air mixture is exploded; a jet then gives the device a spurt; and the process is repeated.

The thrust of a jet engine is

$$F = (\dot{m}_a + \dot{m}_f)v_2 - \dot{m}_a u \tag{6.27}$$

where $\dot{m}_a$ = mass of air taken in per second
$\dot{m}_f$ = mass of fuel consumed per second
v_2 = velocity of exhaust with respect to the engine
u = velocity of flight = velocity of air entry with respect to the engine

The thrust will vary with the speed of flight. Usually $p_0 = p_a$, and so the second term of Eq. (6.26) is not included in Eq. (6.27).

EXERCISES

6.10.1 Find the thrust of a turbojet whose speed is 700 fps and whose air intake rate is 40 lb/sec. The air/fuel ratio is 30:1 and the exhaust velocity is 1800 fps.

6.10.2 Find the thrust of a turbojet whose speed is 300 m/s and whose air intake rate is 10 kg/s. The air/fuel ratio is 22:1 and the exhaust velocity is 500 m/s.

6.11 ROTATING MACHINES: CONTINUITY, VELOCITIES, TORQUE[2]

In this section we shall consider only rotating *hydraulic* machines, designed for incompressible liquids like water. Further, we shall consider here only those machines that flow full of liquid, so that pressures may vary within them. Such machines fall into two categories, ***pumps*** and ***turbines***, which are discussed further in Chaps. 15 and 16.

These machines work when liquid flows through a rotating element known as a ***rotor***. The rotor has a number of blades or vanes, which guide the flow through and which make possible the energy exchange between liquid and machine. In pumps the rotor is called an ***impeller***, and in turbines it is called a ***runner***.

Different machines have different flow patterns. Figures 6.9–6.11 show two-dimensional flow in planes normal to the axis of rotation. This is known as ***radial flow***. The streamlines and velocities lie in the plane of the paper and so are readily represented. Do not misinterpret the term 'radial flow' to mean that the flow is confined along radial lines like the spokes of a wheel, it is not.

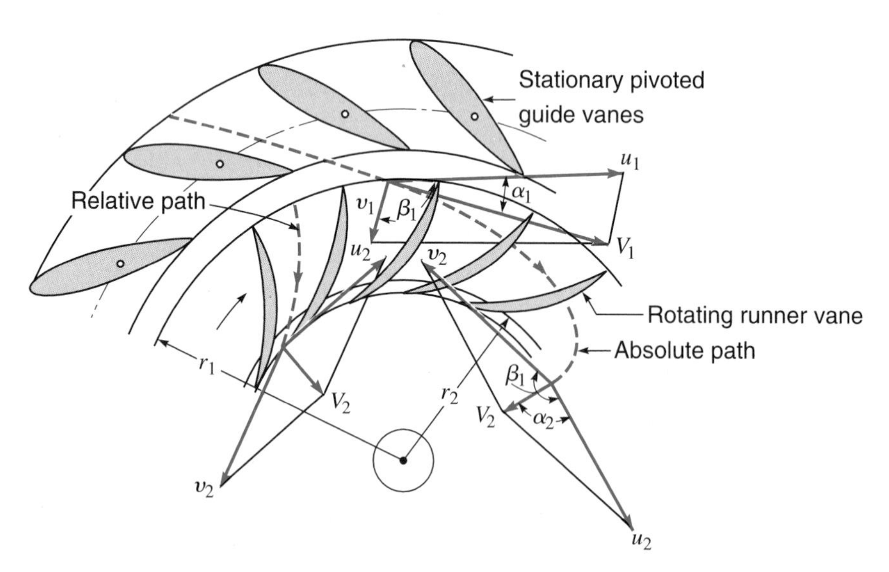

Figure 6.9
Radial-flow hydraulic turbine. (Flow is inward.)

[2] This and the remaining sections of this chapter will be of greater interest to those concerned with applications of the impulse–momentum principle to rotating machines for liquids and gases.

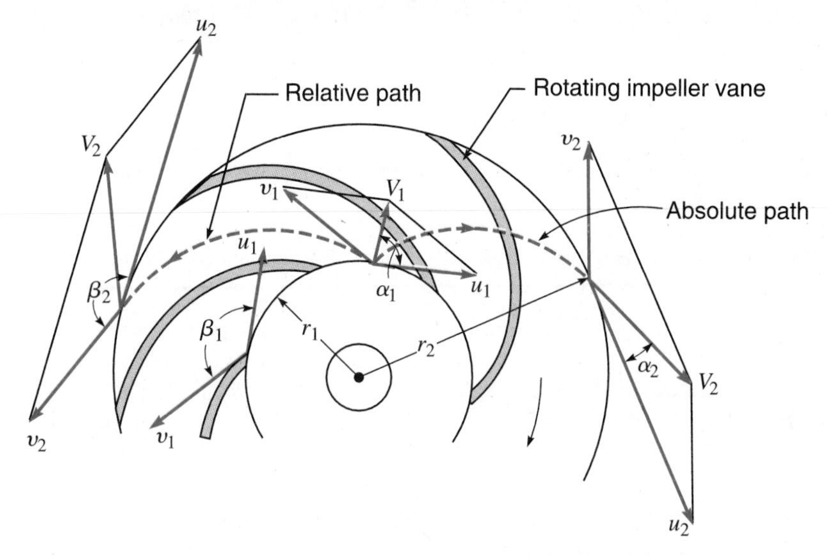

Figure 6.10
Centrifugal-pump impeller with radial flow. (Flow is outward.)

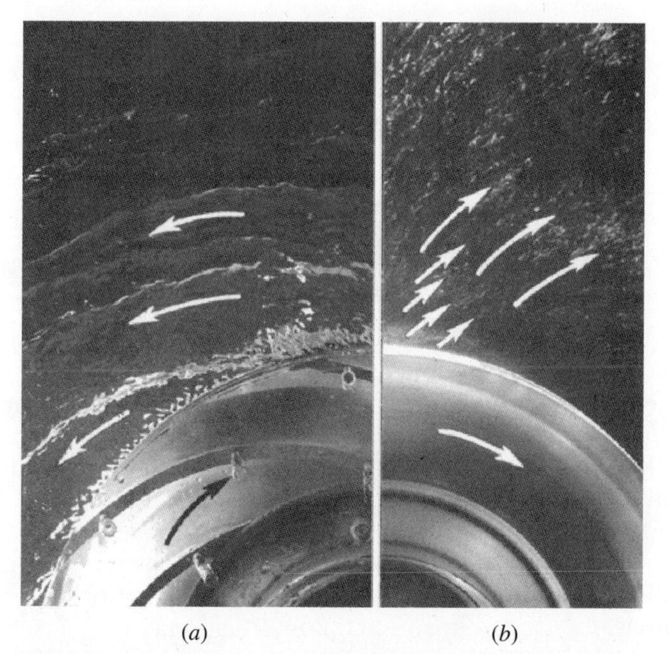

Figure 6.11
Radial-flow-pump impeller rotating at 200 rpm.
(*a*) Instantaneous photo showing relative flow. (*b*) Time exposure showing absolute flow. (*Photographs from Hydrodynamics Laboratory, California Institute of Technology.*)

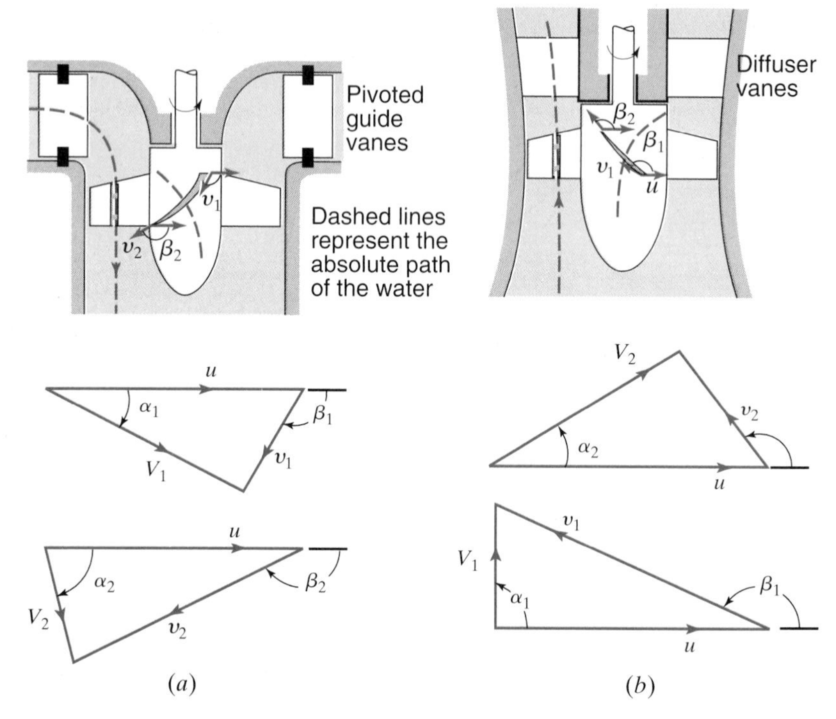

Figure 6.12
(*a*) Axial-flow hydraulic turbine. (*b*) Axial-flow pump.

With radial flow, although the flow paths remain in the normal plane, they are curved, as indicated in Figs. 6.9–6.11. Radial flow is usually inward in turbines for practical reasons of construction; in pumps it is outward due to centrifugal action, and such pumps are called ***centrifugal pumps***.

A different flow pattern is shown in Fig. 6.12, both for a turbine and a pump, in which fluid particles remain at constant distances from the rotation axis. This is known as ***axial flow***, which occurs for example through fans and ship's propellers. The streamlines are spirals on coaxial cylinders.

Mixed flow is intermediate between the two extremes just described, and its velocities have radial, axial, and tangential components. A streamline is a conical spiral with a varying radius from the axis of rotation. Needless to say, this is a complicated three-dimensional flow situation.

Although correct and accurate-looking equations for velocities and torque, etc., will be developed in the following paragraphs and sections, we must remain aware that it is difficult to determine the numerical values to be used in them. Thus fluid particles in different streamlines may flow with different velocities, and it is necessary to estimate what the average velocity may be. Also, it is known that the average direction of a stream is often different from that of the vane that it is supposed to follow, but as yet there is no exact knowledge of the amount of deviation in every case. Thus even the average velocity is not given precisely by dividing the flow by the cross-sectional area of a rotor passage. Furthermore, the entrance or exit edges of vanes are not always parallel to the axis of rotation, and thus the radii will be different for different streamlines.

Despite these defects, the idealized theory is useful. It indicates the shape or nature of the performance curves of a given machine; it shows the influence of each separate factor; and it suggests the direction in which changes in design should be made in order to alter the characteristics that have been found by test of an existing machine.

Continuity

For radial-flow machines, an approximate analysis of the flow behavior can be made by assuming that all elements of the vanes are parallel to the rotation axis and that water enters and leaves the vanes smoothly. An example of such an analysis is presented in Sample Problem 6.6. A key aspect of the analysis involves use of the principle of continuity of flow in the radial direction. Namely, $Q = (A_c V_r)_1 = (A_c V_r)_2$, where A_{c1} and A_{c2} represent circumferential flow areas, and V_{r_1} and V_{r_2} represent the radial components of the velocity of the water at r_1 and r_2. Because the vanes occupy some of the space, the circumferential flow area A_c should be found from the total circumferential area multiplied by a reduction factor m, which can be calculated from the blade thickness, the number of blades, and the circumference. Thus $A_c = m2\pi rz$, where z is the width of the flow passage between the sides of the turbine, and $m = 1 - nt/2\pi r$, where n is the number of blades and t is the blade thickness. Often m_1 and m_2 are assumed to have a value of 1.0, although in fact m must be less than 1.0, perhaps about 0.8–0.9. Usually the passage widths z_1 and z_2 are equal. With such an approach, therefore, given Q and the machine dimensions, we can calculate an average value of the radial velocity component V_r at any radius in a radial-flow machine.

For axial-flow machines, it would appear that a similar analysis could be applied to the flow area, now annular in shape. However, because the blade speed varies with radius, velocity variations are more complex, making analyses less reliable.

Velocity Triangles for Radial Flow

We can continue our theoretical analysis, and solve velocity triangles like those on Figs. 6.9 and 6.10, if, in addition to the flow rate and machine dimensions discussed just above, we know the blade angles β and the speed of the rotor ω. Before doing so, however, it is important that we clearly understand the various motions, absolute and relative flow paths, and velocity triangles represented on Figs. 6.9 and 6.10.

First, we note that the tangential blade speed u is given by $r\omega$. The radial component of the flow velocity, V_r, obtained as described above, is perpendicular to u. We see that we can use the relations of Sec. 6.7 and Fig. 6.5 to solve the velocity triangles, noting that the subscript n (for normal) in Sec. 6.7 has now become the subscript r, for radial. Also, we note from Sec. 6.7 that $v_r = V_r$.

For each radius of interest, we can proceed as follows. Find $v_u = v \cos \beta$ from $V_r/\tan \beta$. Next find v from $v_u/\cos \beta$ or from $v_r/\sin \beta = V_r/\sin \beta$ or from $\sqrt{v_u^2 + v_r^2}$. Then find $V_u = V \cos \alpha$ from $u + v_u$ [Eq. (6.17)]. Next find α from $\tan^{-1} (V_r/V_u)$, and last find V from $V_u/\cos \alpha$ or from $\sqrt{V_u^2 + V_r^2}$. An example of such calculations is given in Sample Prob. 6.6. When making repetitive calculations of this type, it is very helpful to tabulate (or do so by using a spreadsheet).

Torque

Because the radius usually varies as fluid flows through a rotor, it is desirable to compute ***torque*** rather than force. The resultant torque is the sum of the torques produced by all the elementary forces, but it has been shown that this sum may be considered as equivalent to two single forces, one concentrated at the entrance to and the other at the exit from any device. For steady flow, these equivalent forces have been shown to be $\rho Q V_1$ and $\rho Q V_2$ [see, for example, Eq. (6.6)]. Referring to Figs. 6.9 and 6.10 and taking moments, the torque produced is

$$T = \rho Q(r_1 V_1 \cos \alpha_1 - r_2 V_2 \cos \alpha_2) \tag{6.28}$$

This equation is a statement that torque equals the time rate of change of moment of momentum. As before in Sec. 6.7, α and β are defined as the angles made by the absolute and relative velocities of the fluid, respectively, with the positive direction of the linear velocity u of a point on the moving body (tip or root of the blade).

If T as given by this equation is positive, it is the value of the torque exerted by the fluid on the runner of a turbine. The torque output from the shaft of the turbine is less than this because of mechanical friction. If the value of T is negative, it represents the torque exerted on the fluid by the impeller of a pump or compressor or fan. The torque input to the shaft of such a machine is greater than this because of mechanical friction.

It is immaterial in the use of Eq. (6.28) and subsequent equations whether the fluid flows radially inward, as in Fig. 6.9, or radially outward, as in Figs. 6.10 and 6.11, or remains at a constant distance from the axis, as in Fig. 6.12. In any case, r_1 is the radius at entrance and r_2 is that at exit.

Sample Problem 6.6 A radial-flow turbine runner has 18 blades each 0.2 in thick, with $r_1 = 10$ in, $\beta_1 = 65°$, $r_2 = 6$ in, and $\beta_2 = 122°$. The width of the flow passage between the two sides of the turbine is 4 in. When rotating at 180 rpm, the water flow rate is observed to be 7.5 cfs. (*a*) For both entrance and exit conditions, tabulate values of m, A_c, and V_r, showing the method of calculation. (*b*) For the same conditions, tabulate values of v_u, v, u, V_u, α, and V, and draw and label velocity triangles. (*c*) Find the torque exerted by the water, and the horsepower delivered to the shaft. Assume that water enters and leaves the blades without shock.

Solution

(a) Continuity

$$m = 1 - \frac{nt}{2\pi r} = 1 - \frac{18(0.2)}{2\pi r}$$

$$A_c = m2\pi rz = m2\pi r(\tfrac{4}{12}); \quad V_r = v_r = \frac{Q}{A} = \frac{Q}{mA_c}$$

Substituting for r-values and evaluating,

Point	r, ft	m	A_c, ft^2	$V_r = v_r$, fps	
1	0.833	0.943	1.645	4.56	*ANS*
2	0.5	0.905	0.947	7.92	*ANS*

(b) Velocity triangles

$$\omega = \frac{2\pi n}{60} = 180\left(\frac{2\pi}{60}\right) = 18.85 \text{ rad/sec}$$

$$v_u = \frac{V_r}{\tan\beta}; \quad v = \frac{v_u}{\cos\beta}; \quad u = r\omega; \quad V_u = u + v_u;$$

$$\alpha = \tan^{-1}\left(\frac{V_r}{V_u}\right); \quad V = \frac{V_u}{\cos\alpha}$$

Point	r, ft	β	v_u, fps	v, fps	u, fps	
1	0.833	65°	2.13	5.03	15.71	*ANS*
2	0.5	122°	−4.95	9.34	9.43	*ANS*

Point	r, ft	V_u, fps	α	V, fps	
1	0.833	17.83	14.3°	18.41	*ANS*
2	0.5	4.48	60.5°	9.10	*ANS*

Entrance/outer periphery (Point 1):

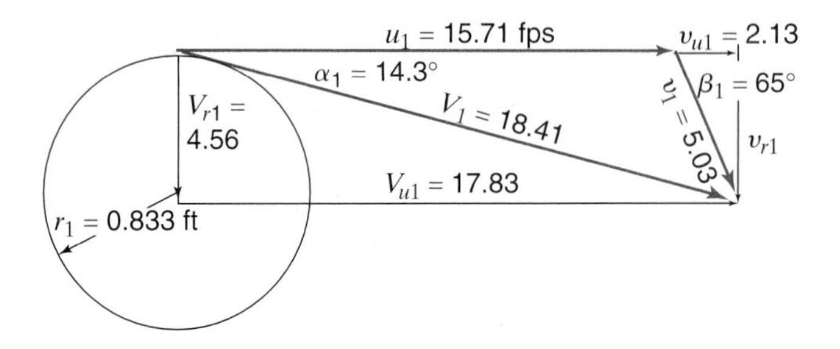

Exit/inner periphery (Point 2):

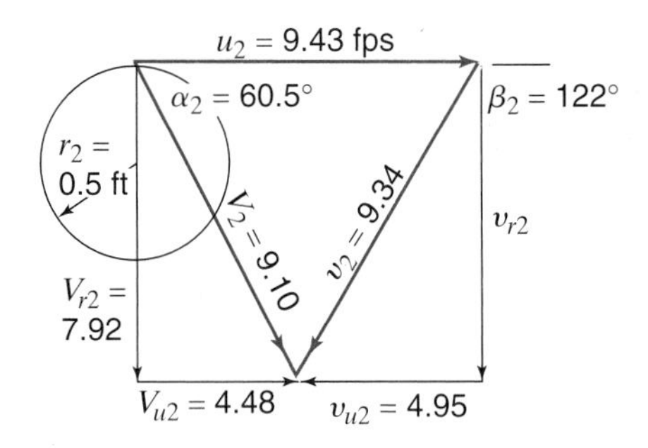

Figure S6.6

(c) Torque

Eq. (6.28): $T = \rho Q|\Delta(rV_u)| = \rho Q(r_1 V_{u1} - r_2 V_{u2})$

$= (62.4/32.2)7.5(0.83 \times 17.83 - 0.5 \times 4.48)$

$= 183.4$ ft·lb ***ANS***

Eqs. (5.35) and (5.32): $\text{HP} = T\omega/550 = 183.4(18.85)/550 = 6.28$ HP ***ANS***

EXERCISES

6.11.1 The absolute velocity of a jet of steam impinging upon the blades of a steam turbine is 3800 fps, and that leaving is 2600 fps. $\alpha_1 = 20°$, $\alpha_2 = 150°$, $u_1 = u_2 = 500$ fps, and $r_1 = r_2 = 0.5$ ft. Find the torque exerted on the rotor and the power delivered to it if the steam flows at 0.4 lb/sec.

6.11.2 The absolute velocity of a jet of steam impinging upon the blades of a steam turbine is 1200 m/s, and that leaving is 950 m/s. $\alpha_1 = 20°$, $\alpha_2 = 150°$, $u_1 = u_2 = 180$ m/s, and $r_1 = r_2 = 12$ cm. Find the torque exerted on the rotor and the power delivered to it if the steam flows at 2 N/s.

6.12 HEAD EQUIVALENT OF MECHANICAL WORK

If Eq. (6.28) for torque is multiplied by angular velocity ω, the product represents the rate at which mechanical energy is delivered *by* the fluid to a turbine or at which mechanical energy is delivered *to* the fluid by a pump. From Eqs. (5.31) and (5.35), power $= \gamma QH = T\omega$. Replacing H by a specific value h_M (for machine) and noting that, when Eq. (6.28) is multiplied by ω, $r_1\omega = u_1$ and $r_2\omega = u_2$, we have $\gamma Q h_M = T\omega = \rho Q(u_1 V_1 \cos \alpha_1 - u_2 V_2 \cos \alpha_2)$, or

$$h_M = \frac{u_1 V_1 \cos \alpha_1 - u_2 V_2 \cos \alpha_2}{g} \tag{6.29}$$

which is the ***head utilized*** by a turbine (h_t) or, when h_M is negative, the ***head imparted*** to the fluid by the impeller of a pump (h_p).

If the value of h_M, as determined by Eq. (6.29), is positive, it is the mechanical work done by the fluid on the vanes of a turbine runner per unit weight of fluid. If the value is negative, it is the mechanical work done on the fluid by the impeller of a pump or similar device per unit weight of fluid. Obviously, the work done by or on the fluid is equal to the loss or gain of energy, respectively, of the fluid.

Sample Problem 6.7 For the same turbine as Sample Prob. 6.6, find the head converted into mechanical work.

Solution

Eq. (6.29): $$h_M = \left|\frac{\Delta(uV_u)}{g}\right| = \frac{u_1 V_{u1} - u_2 V_{u2}}{g}$$

$$= \frac{15.708 \times 17.833 - 9.425 \times 4.477}{32.2} = 7.39 \text{ ft} \qquad \textbf{\textit{ANS}}$$

Note: $\text{HP} = \gamma Q h_M/550 = 62.4(7.5)7.39/550 = 6.29 \text{ HP}$

This agrees well with Part (*c*) of Sample Prob. 6.6.

6.13 FLOW THROUGH A ROTATING CHANNEL

The passageway between the vanes of a turbine or pump is a channel that rotates as the flow passes through. The usual energy equation (5.14) may be written between entrance to and exit from such a passage that is itself rotating about an axis, but in addition to the friction loss h_L there is an additional loss h_M, due to the fact that the fluid is delivering mechanical work and losing energy thereby. (If the passage is that of a pump, the numerical value of h_M will be negative.) Thus

$$H_1 - H_2 = h_L + h_M$$

or
$$\left(\frac{p_1}{\gamma} + z_1 + \frac{V_1^2}{2g}\right) - \left(\frac{p_2}{\gamma} + z_2 + \frac{V_2^2}{2g}\right) = h_L + \frac{u_1 V_1 \cos \alpha_1 - u_2 V_2 \cos \alpha_2}{g} \tag{6.30}$$

We recall that h_M was derived from Eq. (6.28) for torque, which is based on the impulse–momentum principle.

Wishing to eliminate angles from this equation, we note from Eq. (6.17) that

$$V \cos \alpha = u + v \cos \beta$$

Also, by trigonometry on the relative velocity triangle (Fig. 6.13), noting that β is an external angle and that $\cos (180° - \beta) = -\cos \beta$, we obtain

$$V^2 = v^2 + u^2 + 2vu \cos \beta$$

Inserting these values into Eq. (6.30) to eliminate α and β, it reduces to

$$\left(\frac{p_1}{\gamma} + z_1 + \frac{v_1^2 - u_1^2}{2g}\right) - \left(\frac{p_2}{\gamma} + z_2 + \frac{v_2^2 - u_2^2}{2g}\right) = h_L \tag{6.31}$$

This equation is sometimes called the ***equation of relative velocities***, because the absolute velocities of the energy equation are replaced by relative velocities.

If there is no flow, both v_1 and v_2 become zero and Eq. (6.31) reduces to that of a forced vortex [Eq. (5.45)], since $u = \omega r$. If there is no rotation, both u_1 and u_2 become zero, the relative velocities become absolute velocities, and the equation becomes the usual energy equation. The frame of reference having been changed, the mechanical work done does not appear as a separate term in Eq. (6.31).

EXERCISE

6.13.1 Develop Eq. (6.31) by making the substitutions indicated in the text.

6.14 REACTION WITH ROTATION

The force of reaction of a jet from a stationary body is given in Sec. 6.9 and that from a body in translation in Sec. 6.10. Since Sec. 6.13 develops the equation for the flow through a channel in rotation, we are now ready to consider the force of reaction of a fluid discharged from a rotating body.

A familiar object to illustrate this subject is the rotating lawn sprinkler. In Fig. 6.13 assume that the cross-sectional area of the arms is so large relative to the area of the jets that fluid-friction loss in the arms may be neglected. Water enters at the center, where $r_1 = 0$, so that $u_1 = 0$ in Eq. (6.31). With the sprinkler arms lying in a horizontal plane, $z_1 - z_2 = 0$, and for the jets discharging into the air, p_2 is atmospheric pressure and will be regarded as zero. Since friction is neglected, $h_L = 0$, and if we let $h = p_1/\gamma + v_1^2/2g$ then Eq. (6.31) applied to Fig. 6.13 becomes

$$h - \left(\frac{v_2^2 - u_2^2}{2g}\right) = 0$$

or

$$v_2 = \sqrt{2gh + u_2^2} \tag{6.32}$$

where h is the sum of the pressure head and velocity head at entry to the sprinkler.

If a_2 denotes the sum of the areas of all the jets (two in Fig. 6.13), then $Q = a_2 v_2$. This, with Eq. (6.32), shows that the discharge increases with the rotative speed, since $u_2 = r_2\omega$.

From Eq. (6.17) for the relative velocity triangle in Fig. 6.13, the tangential component of the absolute velocity of discharge is $V_{u_2} = u_2 + v_2 \cos \beta_2$, and hence the tangential component of the force of reaction is

$$F_u = \frac{\gamma Q}{g}\,\Delta V_u = \frac{\gamma a_2 v_2}{g}(0 - V_{u_2}) = -\frac{\gamma a_2 v_2}{g}(u_2 + v_2 \cos \beta_2)$$

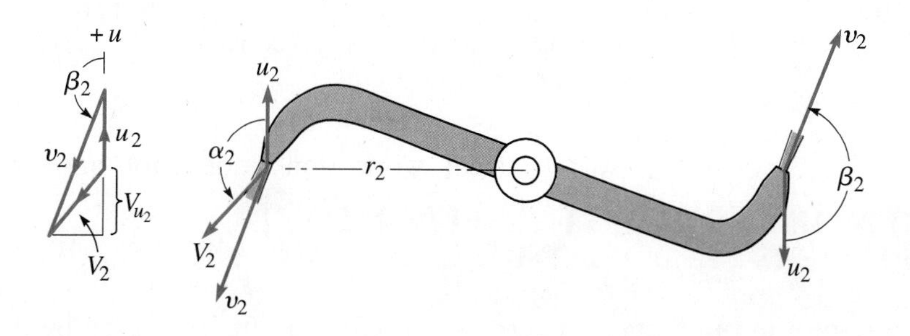

Figure 6.13

As the radius is a factor in any rotating body, it is usually better to compute torque rather than a force. In this case the torque is

$$T = F_u r_2 = -\frac{\gamma a_2 v_2}{g} r_2(u_2 + v_2 \cos \beta_2) \tag{6.33}$$

The ideal maximum speed, or ***runaway speed***, is when $T = 0$, and this will be the case when $u_2 = -v_2 \cos \beta_2$ and when $V_2 \cos \alpha_2 = 0$ or $\alpha_2 = 90°$. Because of mechanical friction, this condition will never be realized. Of the total power supplied to the sprinkler, the greater part is lost in the kinetic energy of the jets. The total power *developed* by the sprinkler is used in overcoming friction in the bearings and air resistance. If there were more arms, with larger orifices, so as to discharge more water, there could be a surplus of power, which would be useful power delivered. A primitive turbine constructed in this manner was known as Barker's mill.

EXERCISES

6.14.1 The flow from a lawn sprinkler such as Fig. 6.13 is 140 L/min, $\beta_2 = 180°$, and the total area of the jets is 120 mm^2. The jets are located 20 cm from the center of rotation. Determine the speed of rotation if there is no friction.

6.14.2 A lawn sprinkler like that of Fig. 6.13 with $\beta_2 = 156°$ has a total jet area of 0.00086 ft^2 at a radius of 16 in. Compute the discharge rate, the torque exerted by the water, and the power developed, when $h = 120$ ft and the sprinkler is prevented from rotating.

6.14.3 Repeat Exer. 6.14.2 with the following changes: total jet area $= 0.75$ cm^2, radius $= 36$ cm, $h = 45$ m.

6.14.4 How fast would the sprinkler of Exer. 6.14.2 rotate if there were no mechanical friction or air resistance (i.e., consider the case where $T = 0$)? This is known as ***runaway speed***.

6.14.5 How fast would the sprinkler of Exer. 6.14.3 rotate if there were no mechanical friction or air resistance (i.e., consider the case where $T = 0$)? This is known as ***runaway speed***.

6.15 MOMENTUM PRINCIPLE APPLIED TO PROPELLERS AND WINDMILLS

In the case of a fan in a duct, the cross-section of the fluid affected by the fan is the same upstream as it is downstream, and the principal effect of a fan is to increase the pressure in the duct. In the case of a propeller revolving in free air, however, this is not so. The pressure must necessarily be the same at a distance either upstream or downstream from the propeller. How, then, may the revolving blades be considered to do work on the air? This situation may be analyzed by consideration of the ***slipstream***, or ***propeller race***, which is nothing more than the body of air affected by the propeller (Fig. 6.14). The flow is undisturbed at section 1 upstream from the propeller, and is accelerated as it approaches the propeller. Additional increase in velocity occurs downstream of the propeller until it reaches a value of V_4 at section 4. It is customary to replace the propeller in simple slipstream theory with a stationary ***actuating disk*** across which the pressure is made to rise, as shown in the pressure profile below the slipstream of Fig. 6.14 and also in Fig. 6.15. We thereby neglect the rotational effect of the propeller, together with the helical path of vortices shed from the blade tips (Sec. 9.8). The thrust force

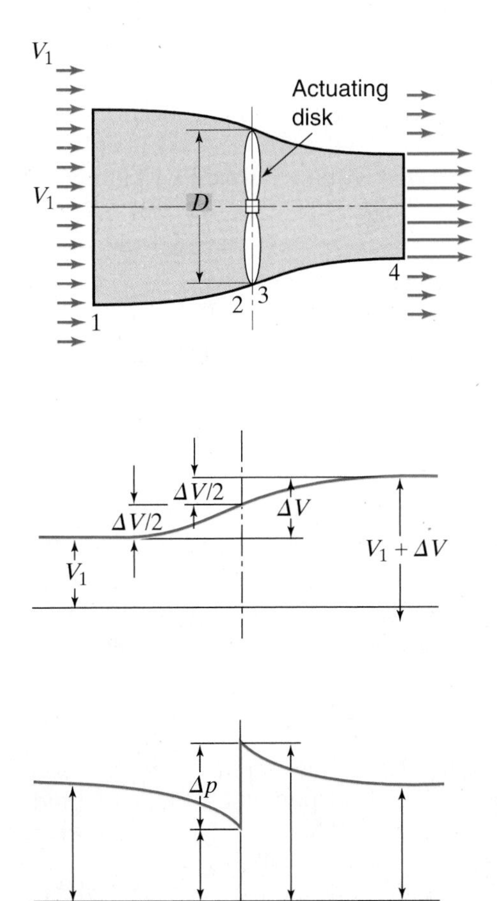

Figure 6.14
Slipstream of propeller in free fluid. V_1 represents the velocity of the undisturbed fluid relative to the propeller; p_0 represents the undisturbed pressure in the fluid.

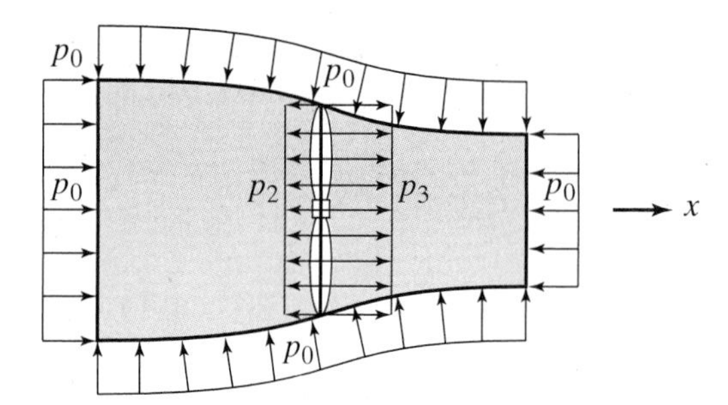

Figure 6.15
Forces acting on the fluid within the slipstream of Fig. 6.14.

F_T will be given by the pressure change at the disk times the area of the disk,

$$F_T = \frac{\pi D^2}{4}(p_3 - p_2) = A(\Delta p) \tag{6.34}$$

where D and A represent the diameter and area of the actuating disk, and p_2 and p_3 represent the pressures just upstream and downstream of the propeller, as indicated in Figs. 6.14 and 6.15. It should be noted that the pressures exerted on the boundary of the slipstream between sections 1 and 4 balance one another out and need not be considered.

By Newton's second law, the force F_T must equal the rate of change of momentum of the fluid upon which it acts. If we let Q be the rate of flow through the slipstream,

$$F_T = \rho Q(\Delta V) = \rho AV(V_4 - V_1) \tag{6.35}$$

where V represents the mean velocity through the actuating disk, and V_1 and V_4 are the velocities in the slipstream at sections 1 and 4 of Fig. 6.14, where the pressures correspond to the normal undisturbed pressure p_0 in the flow field.

The propeller we are considering could be a stationary one like a fan or a moving one such as the propeller of a moving aircraft or ship. If it were a fan, V_1 would generally be equal to zero, and the slipstream upstream of the fan would have a much larger diameter than that shown in Fig. 6.14. If we were dealing with a propeller of a moving aircraft or ship, the craft would be moving to the left with a velocity V_1 through a stationary fluid, in which case Fig. 6.14 shows velocities relative to the craft.

Writing the Bernoulli equation (5.15) from a point upstream where the velocity is V_1 to a point downstream where it is $V_4 = V_1 + \Delta V$, recognizing that the pressure terms at these points cancel and (assuming an ideal fluid) that the disk adds $\Delta p/\gamma$ units of energy to the fluid per unit weight of fluid, we get

$$\frac{V_1^2}{2g} + \frac{\Delta p}{\gamma} = \frac{(V_1 + \Delta V)^2}{2g} \tag{6.36}$$

Equating Eqs. (6.34) and (6.35) and solving for Q in terms of Δp, and then substituting into this expression for Q, the expression for Δp that results from solving Eq. (6.36) gives

$$Q = A\left(V_1 + \frac{\Delta V}{2}\right) \tag{6.37}$$

As $\Delta V = V_4 - V_1$, this may be expressed as

$$Q = A\left(V_1 + \frac{V_4 - V_1}{2}\right) = A\left(\frac{V_1 + V_4}{2}\right) = AV$$

This shows that the velocity V at the disk is the average of the upstream and downstream velocities. It also shows that one-half of ΔV occurs upstream of the propeller, while the other half of ΔV occurs downstream.

Solving Eq. (6.36) for ΔV and substituting F_T/A for Δp from Eq. (6.34) gives

$$\Delta V = -V_1 + \sqrt{V_1^2 + \frac{2F_T}{A\rho}} \tag{6.38}$$

We may use the slipstream analysis to determine the maximum possible efficiency of a propeller. The power output P_{out} is given by

$$P_{out} = F_T V_1 = (\rho Q\ \Delta V)V_1$$

The power input P_{in} is that required to increase the velocity of the fluid in the slipstream from V_1 to V_4. Applying Eq. (5.31), we get

$$P_{in} = \gamma Q\left(\frac{V_4^2}{2g} - \frac{V_1^2}{2g}\right) = \tfrac{1}{2}\rho Q(V_4^2 - V_1^2)$$

$$= \rho Q\left(\frac{V_4 + V_1}{2}\right)(V_4 - V_1) = (\rho Q V)(\Delta V)$$

The efficiency η is given by the ratio of the power output to the power input. Thus

$$\eta = \frac{P_{out}}{P_{in}} = \frac{(\rho Q\ \Delta V)V_1}{(\rho Q V)\ \Delta V} = \frac{V_1}{V} = \frac{V_1}{V_1 + \frac{1}{2}\Delta V} = \frac{1}{1 + \frac{1}{2}(\Delta V/V_1)} \tag{6.39}$$

The efficiency is seen to be a function of the ratio $\Delta V/V_1$. The efficiency approaches 100% as ΔV approaches zero, but if $\Delta V = 0$, the propeller produces no force. The actual maximum efficiency of aircraft propellers is about 85%. However, the efficiency of an airplane propeller drops rapidly at speeds in excess of 400 mph (600 km/h) because of compressibility effects. For ships, propeller efficiencies of only 60 or 70% are attainable.

A windmill is essentially the opposite of a propeller in that the function of a windmill is to extract energy from the wind. The slipstream for a windmill expands as it passes the actuated disk, and the pressure drops, as does the velocity. By a procedure similar to the one for a propeller, it can be shown that the maximum theoretical efficiency of a windmill is 59.3%. Because of friction and other losses, the actual efficiency of windmills rarely exceeds 40%.

SAMPLE PROBLEM 6.8 Find the thrust and efficiency of two 6.5-ft-diameter propellers through which flows a total of 20,000 cfs of air (0.072 lb/ft^3). The propellers are attached to an airplane moving at 150 mph through still air. Find also the pressure rise across the propellers and the horsepower input to each propeller. Neglect eddy losses.

Solution

The velocity of air relative to the airplane is

$$V_1 = 150 \text{ mph} = 150\left(\tfrac{44}{30}\right) = 220 \text{ fps}$$

The velocity of air through the actuating disk is

$$V = V_1 + \frac{\Delta V}{2} = \frac{Q}{A} = \frac{20{,}000/2}{(\pi/4)(6.5)^2} = 301 \text{ fps}$$

Thus
$$\Delta V = 2(301 - 220) = 162.7 \text{ fps}$$

Eq. (6.35):
$$F_T = \rho Q\, \Delta V = \frac{0.072}{32.2}(20{,}000)(162.7)$$

$$= 7280 \text{ lb} \quad \text{(total thrust of both propellers)} \qquad \textit{ANS}$$

Eq. (6.39):
$$\eta = \frac{1}{1 + \Delta V/2V_1} = \frac{1}{1 + 162/440} = 0.730 = 73\% \qquad \textit{ANS}$$

F_T on one propeller $= 7280/2 = 3640$ lb. But $F_T = (\Delta p)(A)$, thus $3640 = \Delta p(\pi/4)(6.5)^2$,

$$\Delta p = 109.6 \text{ psf} = 0.761 \text{ psi} \qquad \textit{ANS}$$

$$\text{hp/propeller} = \frac{\gamma Q(\Delta p/\gamma)}{550} = \frac{Q(\Delta p)}{550} = \frac{10{,}000(109.6)}{550} = 1994 \qquad \textit{ANS}$$

Check:
$$\text{hp/propeller} = \frac{F_T(V_1 + \Delta V/2)}{550} = \frac{3640(301)}{550} = 1994$$

EXERCISES

6.15.1 An 18-in-diameter household fan drives air (γ = 0.076 lb/ft^3) at a rate of 1.80 lb/sec. (*a*) Find the thrust exerted by the fan. (*b*) What is the pressure difference on the two sides of the fan? (*c*) Find the required horsepower to drive the fan. Neglect losses.

6.15.2 A 1.8-m-diameter fan drives air (γ = 12 N/m^3) at a rate of 50 N/s. (*a*) Find the thrust exerted by the fan. (*b*) What is the pressure difference on the two sides of the fan? (*c*) Find the required power to drive the fan. Neglect losses.

6.15.3 A 12-in electric fan is placed on a frictionless mount and is observed to exert a thrust of 0.8 lb. (*a*) Find the approximate velocity of the slipstream of standard air (sea level) that it produces. (*b*) If 45% of the power supplied to the blades is lost in eddies and friction and if the driving motor has an efficiency of 60%, find the required electrical input in watts.

6.15.4 A 30-cm electric fan is placed on a frictionless mount and is observed to exert a thrust of 2.5 N. (*a*) Find the approximate velocity of the slipstream of standard air (sea level) that it produces. (*b*) If 45% of the power supplied to the blades is lost in eddies and friction and if the driving motor has an efficiency of 60%, find the required electrical input in watts.

6.15.5 A fan sucks air from outside to inside a building through a 20-in-diameter duct. The density of the air is 0.0022 slug/ft^3. (*a*) If the pressure difference across the two sides of the fan is 4.0 in of water, determine the flow rate of the air in cubic feet per second (*b*) What thrust must the fan support be designed to withstand?

PROBLEMS

6.1 For laminar flow between two stationary parallel plates such as to give two-dimensional flow, find (*a*) the ratio of mean velocity to maximum velocity; (*b*) α; (*c*) β. The velocity profile is parabolic as in Exer. 6.3.1.

6.2 In Sample Prob. 6.1 suppose the passage narrowed down to a width of 9 ft at the second section. With the same depths, find the flow rate and the horizontal force on the structure.

6.3 In Sample Prob. 6.1 suppose the passage narrowed down to a width of 2.0 m at the second section. With the same depths, find the flow rate and the horizontal force on the structure.

6.4 A hydraulic jump (Sec. 10.18) occurs in a "diamond-shaped" transparent closed conduit as shown in Fig. P6.4. The conduit is horizontal, and the water depth just upstream of the jump is 2.0 ft. The conduit is completely full of water downstream of the jump. Pressure-gage readings are as shown in the figure. (*a*) Compute the flow rate. Note that, because of turbulence in the jump, there is a substantial energy loss. Hence ideal flow cannot be assumed. Shear forces along the boundary may be neglected however. (*b*) Determine the horsepower loss in the jump.

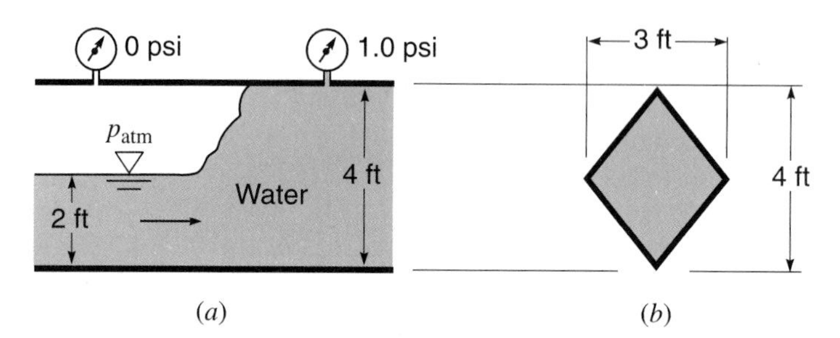

Figure P6.4

6.5 The diameters in Fig. 6.2 are 42 in and 30 in. At the larger end the pressure is 95 psi and the velocity is 10 fps. Neglecting friction, find the resultant force on the conical reducer if water flows (*a*) to the right; (*b*) to the left. (*c*) What would happen to the two previous answers if friction were not neglected?

6.6 The diameters in Fig. 6.2 are 80 cm and 50 cm. At the larger end the pressure is 650 kN/m^2 and the velocity is 3 m/s. Neglecting friction, find the resultant force on the conical reducer if water flows (*a*) to the right; (*b*) to the left. (*c*) What would happen to the two previous answers if friction were not neglected?

6.7 A reducing right-angled bend has the same orientation as in Exer. 6.5.4. Water enters with a velocity of 6 fps and a pressure of 4 psi. The diameter at the entrance is 22 in and that at the exit is 20 in. Neglecting any friction loss, find the magnitude and direction of the resultant force on the bend.

6.8 A reducing right-angled bend has the same orientation as in Exer. 6.5.4. Water enters with a velocity of 3 m/s and a pressure of 30 kN/m^2. The diameter at the entrance is 50 cm and that at the exit is 40 cm. Neglecting any friction loss, find the magnitude and direction of the resultant force on the bend.

6.9 Referring to Fig. P6.9, both nozzle jets are in the atmosphere and have a velocity of 35 fps. the liquid has a specific weight of 62.4 lb/ft^3. The axes of

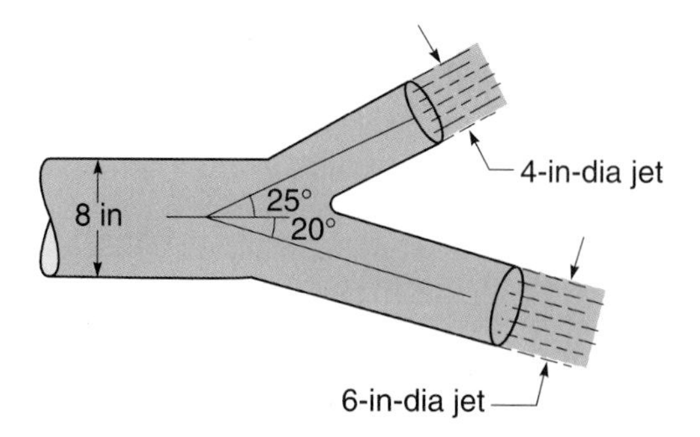

Figure P6.9

the pipe and both nozzles all lie in a horizontal plane. Find the magnitude and direction of the resultant force on this double nozzle while neglecting friction.

6.10 Assuming ideal flow, determine the total pull on the bolts in Fig. P6.10, where $y = 6$ ft, $d_1 = 2$ in, $d_2 = 4$ in, $d_3 = 1$ in and γ_M of the manometer fluid (oil) is 52 pcf.

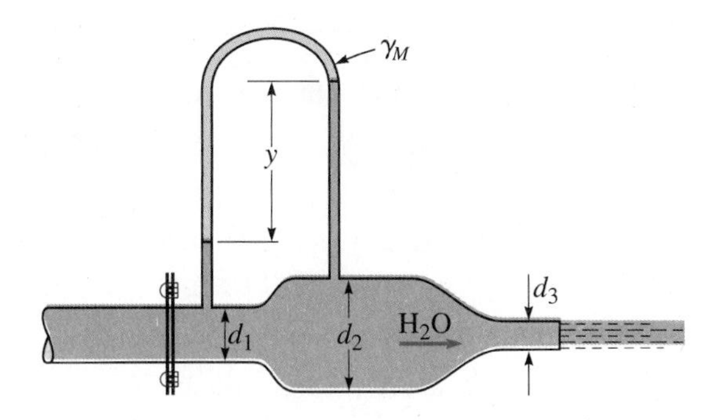

Figure P6.10

6.11 Repeat Prob. 6.10 for the case where $y = 180$ cm, $d_1 = 5$ cm, $d_2 = 10$ cm, $d_3 = 2.5$ cm, and the manometer liquid has a specific gravity of 0.80.

6.12 If a jet of any fluid has its velocity V_2 reduced to $0.7V_1$ by friction while it is deflected through an angle θ, derive an equation for the dynamic force exerted in terms of $\dot{m}$, V_1, and θ.

6.13 Plotted to scale in Fig. P6.13 are streamlines in the plane of the center of a free jet impinging vertically on a horizontal circular plate. By scaling off the pertinent dimensions, determine the velocity of the water as it leaves the plate and the total resultant force exerted by the water on the plate. The jet diameter is 28 cm and the stagnation pressure at point S is 5.5 kN/m^2.

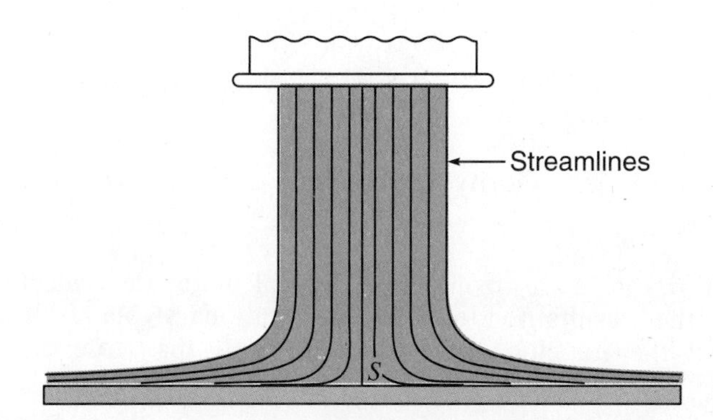

Figure P6.13

6.14 Repeat Prob. 6.13 with a jet diameter of 4.2 in and a jet velocity of 23 fps.

6.15 A horizontal jet of water issues from an orifice in the side of a tank under a head h_1 and strikes a large plate a short distance away that covers the end of a horizontal tube in the side of a second tank. The second tank contains oil of specific weight 50 lb/ft^3 at rest. The height of the oil above the tube is h_2. The jet diameter is three-fourths of the inside diameter of the tube. The jet and the tube are the same elevation. (*a*) If the impact of the water is just sufficient to hold the plate in place, find the relation between h_1 and h_2. Neglect the weight of the plate and assume ideal flow. (*b*) Consider the effect of the weight of the plate. Find h_2 if $h_1 = 9$ ft, the weight of the plate = 45 lb, the jet diameter = 1.4 in, and the coefficient of friction between the plate and tube = 0.5. (*c*) Repeat part (*b*) with the plate weighing only 5 lb.

6.16 Repeat Prob. 6.15 with the following data: (*a*) oil specific weight = 7500 N/m^3; (*b*) $h_1 = 2.8$ m, $W = 180$ N, jet diameter = 3.5 cm, μ = 0.5; (*c*) $W = 20$ N.

6.17 At section P a 5-in-diameter water jet with a velocity of 28 fps is directed vertically upward against the cone shown in Fig. P6.17. Neglecting friction and assuming the streamlines at Q are parallel, find the weight of the cone if $a = 1.5$ ft, $b = 0.6$ ft, and $c = 4$ ft.

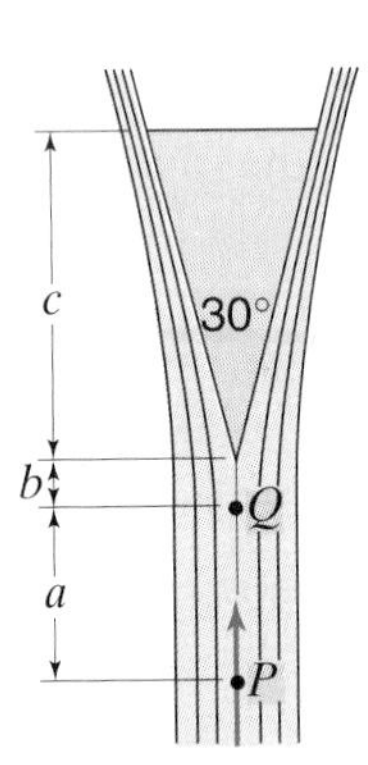

Figure P6.17

6.18 Repeat Prob. 6.17 if the jet diameter is 0.6 m, and the jet velocity is 22 m/s at P, and $a = 1.6$ m, $b = 0.6$ m, and $c = 3.5$ m.

6.19 Repeat Prob. 6.17 with jet velocity 19 fps, and $a = 0$, $b = 2$ ft, and $c = 4$ ft.

6.20 Assuming ideal flow in a horizontal plane, calculate the magnitude and direction of the resultant force on the stationary blade in Fig. P6.20, knowing that the velocity of the water jet is 40 fps. Note that the jet is divided by the splitter so that one-third of the water is diverted toward A.

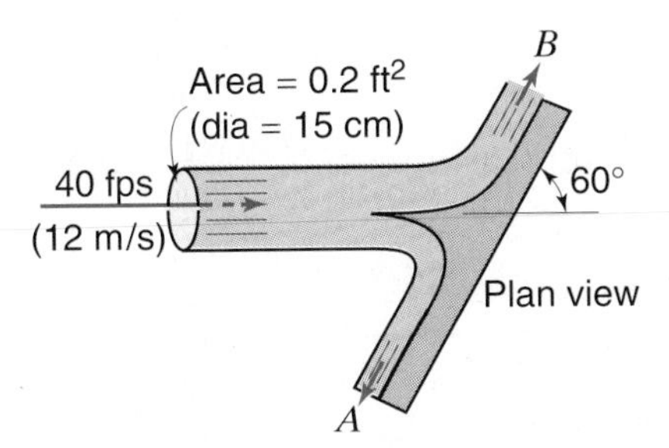

Figure P6.20

6.21 Refer to Fig. P6.20. Assuming ideal flow in the horizontal plane, calculate the magnitude and direction of the resultant force on the stationary blade, knowing that the velocity of the water jet is 12 m/s. Note that the jet is divided by the splitter so that one-third of the water is diverted toward A.

6.22 A locomotive tender running at 25 mph scoops up water from a trough between the rails, as shown in the figure. The scoop delivers water at a height $h = 7.5$ ft above its original level and in the direction of motion. The area of the stream of water at entrance is 40 in^2. The water is everywhere under atmospheric pressure. Neglecting all losses, what is the absolute velocity of the water as it leaves the scoop? What is the force acting on the tender caused by this? What is the minimum speed at which water will be delivered to the height h above the original level?

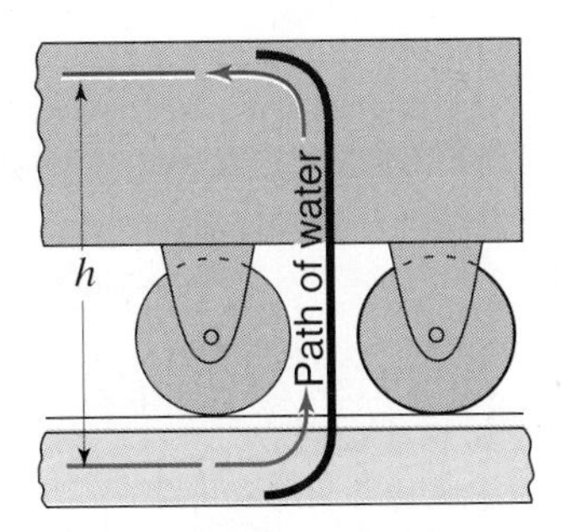

Figure P6.22

6.23 Repeat Prob. 6.22 for the following data: locomotive speed = 12 m/s, $h = 2.3$ m, stream area = 300 cm^2.

6.24 A 4-in-diameter water jet has a velocity of 130 fps. It strikes a single vane, which has an angle $\beta_2 = 90°$ and which is moving in the same direction as the jet with a velocity u. When u has values of 0, 45, 65, 85, 105, and 130 fps, find the values of (a) $\dot{m}'g$; (b) $V_2 \cos \alpha_2$; (c) ΔV_u; (d) Δv_u; (e) F_u. Assume $v_2 = 0.9v_1$. Present answers in a neat tabular form or on a spreadsheet.

6.25 Assume all the data in Prob. 6.24 are the same except that $\beta_2 = 180°$. Find the values of (*a*) $\dot{m}'g$; (*b*) v_2; (*c*) V_2; (*d*) ΔV; (*e*) Δv; (*f*) F_u. Assume $v_2 = 0.8v_1$. Present answers in a neat tabular form or on a spreadsheet.

6.26 Repeat Prob. 6.25, except that $v_2 = 0.7v_1$.

6.27 Assume that all the data are the same as in Prob. 6.24, except that $\beta_2 = 145°$ and $v_2 = 0.7v_1$. Find the values of (*a*) $v_2 \cos \beta_2$; (*b*) $V_2 \cos \alpha_2$; (*c*) ΔV_u; (*d*) Δv_u; (*e*) F_u. Present answers in a neat tabular form or on a spreadsheet.

6.28 For the conditions of Prob. 6.20, compute the magnitude and direction of the resultant force on the single blade if it is moving to the right at a velocity of 15 fps.

6.29 For the conditions of Prob. 6.21, compute the magnitude and direction of the resultant force on the single blade if it is moving to the right at a velocity of 5 m/s.

6.30 A 3-in-diameter air jet impinges on a series of blades, entering smoothly, and having the absolute velocities (fps) shown in Fig. P6.30. Assume $\gamma = 0.075 \text{ lb/ft}^3$, that the pressure is the same on both sides, and neglect friction. (*a*) What is the velocity of the blades and the power being transmitted to them? (*b*) Determine the necessary blade angles at entrance and exit.

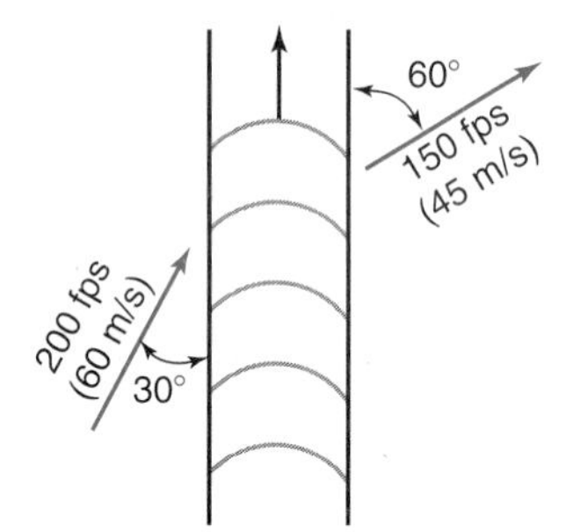

Figure P6.30

6.31 A 6-cm-diameter air jet impinges on a series of blades, entering smoothly, and having the absolute velocities (m/s) shown in Fig. P6.30. Assume $\gamma = 11 \text{ N/m}^3$, that the pressure is the same on both sides, and neglect friction. (*a*) What is the velocity of the blades and the power being transmitted to them? (*b*) Determine the necessary blade angles at entrance and exit.

6.32 Suppose the blade of Prob. 6.20 is one of a series of blades that are moving to the right at 10 fps. (*a*) Determine the resultant horizontal force on the blade system, and (*b*) compute the power transferred to the blades. (*c*) Compute the power of the jet and of the water leaving the blade system to verify an energy balance (Sample Prob. 6.4).

6.33 Suppose the blade of Prob. 6.21 is one of a series of blades that are moving to the right at 4 m/s. (*a*) Determine the resultant horizontal force on the blade system, and (*b*) compute the power transferred to the blades.

(c) Compute the power of the jet and of the water leaving the blade system to verify an energy balance (Sample Prob. 6.4).

6.34 An ideal liquid ($\gamma = 62.4\ \text{lb/ft}^3$) flows from a 2-ft-diameter tank as shown in Fig. P6.34. The jet diameter is 3 in. If the static coefficient of friction between the tank and floor is 0.56, determine the minimum value of h at which the tank will commence to move to the left. The tank itself weighs 100 lb.

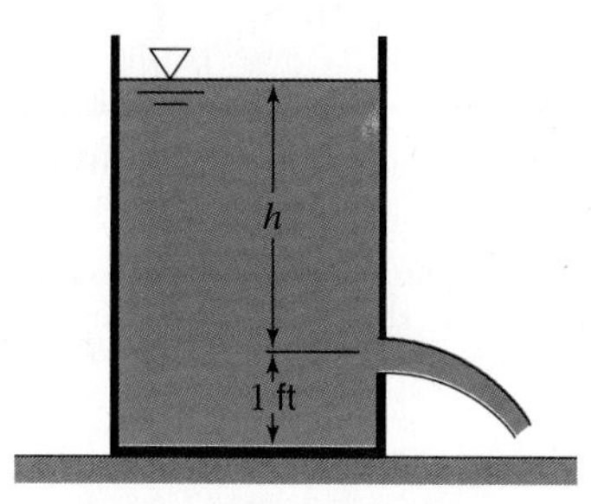

Figure P6.34

6.35 Find the magnitude and direction of the resultant force of the fluid on the compressor shown in Fig. P6.35. Air ($\gamma = 0.075\ \text{lb/ft}^3$) enters at A through a 4-ft^2 area at the velocity of 15 fps. Air is discharged at B through a 3-ft^2 area at a velocity of 17 fps.

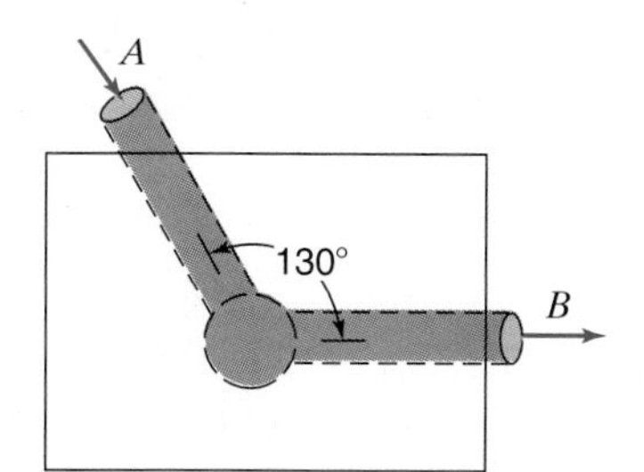

Figure P6.35

6.36 Repeat Prob. 6.35 for the case where a gas ($\gamma = 12.1\ \text{N/m}^3$) enters at A through a 60-cm-diameter pipe at 5 m/s and leaves at B through a 50-cm-diameter pipe at 7 m/s.

6.37 A rocket has a propellant flow rate of 21.6 lb/sec through a nozzle with a throat area of 9.3 in^2. The nozzle is designed to expand the gases down to 14.7 psia at exit. The exit area of the nozzle is 48.5 in^2 and the exhaust velocity is 6370 fps. Find the rocket's thrust (a) at sea level and (b) at an elevation of 20,000 ft where the barometer pressure is 6.75 psia. (c) Find the specific weight of the exhaust gas.

6.38 A radial-flow turbine has the following dimensions: $r_1 = 0.5$ m, $r_2 = 0.3$ m, $\beta_1 = 74°$, and $\beta_2 = 126°$. The width of the flow passage between the two sides of the turbine is 0.2 m. When operating at 160 rpm, the flow rate through the turbine is 1.5 m^3/s. Find (a) the torque exerted by the water; (b) the power delivered to the shaft; (c) the head converted into mechanical work.

6.39 A radial-flow turbine has the following dimensions: $r_1 = 3.2$ ft, $r_2 = 1.6$ ft, $\beta_1 = 76°$, and $\beta_2 = 135°$. The width of the flow passage between the two sides of the turbine is 0.65 ft. When operating at 150 rpm, the flow rate through the turbine is 60 cfs. Find (*a*) the torque exerted by the water; (*b*) the horsepower delivered to the shaft; (*c*) the head converted into mechanical work.

6.40 A paddle wheel with vanes that are all straight and radial is to be used as a crude centrifugal pump for water (Fig. P6.40): $r_1 = 3$ in, $r_2 = 9$ in, and the height z perpendicular to the plane of the figure is 0.2 ft. If the speed is 1200 rpm and the flow is 3380 gpm, (*a*) at the centerline elevation find the difference in pressure (psi) between the inner and outer circumferences, neglecting friction losses, (*b*) find which of these two points has the higher pressure, (*c*) compute the torque required to drive the pump, (*d*) calculate the horsepower requirement, and (*e*) verify that the horsepower requirement is equal to the difference between the horsepower of the outflow minus the horsepower of the inflow.

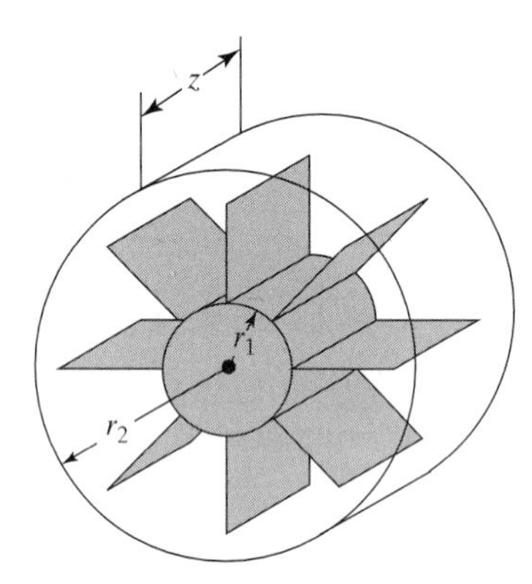

Figure P6.40

6.41 Repeat Prob. 6.40, where the data are given in SI units as follows: $r_1 = 6.5$ cm, $r_2 = 21.5$ cm, the height z perpendicular to the plane of figure = 4.8 cm. The speed is 1200 rpm and the flow is 150 L/s. Express pressure difference in kN/m^2 and power in kW.

6.42 Consider a lawn sprinkler such as that in Fig. 6.13 with $\beta_2 = 160°$, with the total area of the jets at a radius of 15 in being 0.0008 ft^2. When $h = 144$ ft, compute the rate of discharge, the torque exerted by the water, and the power developed if the rotative speed of the sprinkler is 400 rpm. Neglect fluid friction, but note that the calculated torque is that required to overcome mechanical friction and air resistance.

6.43 Repeat Prob. 6.42 with the following changes: radius = 40 cm, total jet area = 0.7 cm^2, $h = 42$ m.

6.44 For a lawn sprinkler like that of Fig. 6.13, develop an expression for the runaway speed ω in terms of h, r, and β_2. This would occur if there were no mechanical friction or air resistance, i.e., zero torque.

6.45 At what approximate speed will the sprinkler of Prob. 6.42 develop maximum horsepower?

6.46 Apply the momentum and energy principles to the case of a windmill (essentially the opposite of a propeller), to determine the maximum theoretical efficiency based on an input energy available from the wind velocity in a stream tube having a cross-section equivalent to that of the windmill blade circle.

CHAPTER 7

Similitude and Dimensional Analysis

7.1 DEFINITION AND USES OF SIMILITUDE

It is usually impossible to determine all the essential facts for a given fluid flow by pure theory, and hence dependence must often be placed upon experimental investigations. The number of tests to be made can be greatly reduced by the systematic use of dimensional analysis and the laws of similitude or similarity. For these permit test data to be applied to cases other than those observed.

Thus the similarity laws enable us to make experiments with a convenient fluid such as water or air, for example, and then apply the results to a fluid which is less convenient to work with, such as hydrogen, steam, or oil. Also, in both hydraulics and aeronautics, valuable results can be obtained at a minimum cost by tests made with small-scale models of the full-size apparatus. The laws of similitude make it possible to predict the performance of the ***prototype***, which means the full-size device, from tests made with the model. It is not necessary that the same fluid be used for the model and its prototype. Neither is the model necessarily smaller than its prototype. Thus the flow in a carburetor might be studied in a very large model. And the flow of water at the entrance to a small centrifugal-pump runner might be investigated by the flow of air at the entrance to a large model of the runner.

A few examples where models may be used are ships in towing basins, airplanes in wind tunnels, hydraulic turbines, centrifugal pumps, spillways of dams, river channels, and the study of such phenomena as the action of waves and tides on beaches, soil erosion and the transportation of sediment.

It should be emphasized that the model need not necessarily be different in size from its prototype. In fact, it may be the same device, the variables in this case being the velocity and the physical properties of the fluid.

7.2 GEOMETRIC SIMILARITY

One of the desirable features in model studies is that there be ***geometric similarity***, which means that the model and its prototype be identical in shape but differ only in *size*. The important consideration is that the flow patterns be geometrically similar. If subscripts p and m denote prototype and model, respectively, we shall define the ***scale ratio***[1] to be

$$L_r = \frac{L_p}{L_m} \tag{7.1}$$

the ratio of the linear dimensions of the prototype to, or divided by, the corresponding dimensions in the model. It follows that areas vary as L_r^2 and volumes as L_r^3. Complete geometric similarity is not always easy to attain. For example, the surface roughness of a small model may not be reduced in proportion unless it is possible to make its surface very much smoother than that of the prototype. Similarly, in the study of sediment transportation, it may not be possible to scale down the bed materials without having material so fine as to be impractical. Fine powder, because of cohesive forces between the particles, does not simulate the behavior of sand. Again, in the case of a river the horizontal scale is usually limited by the available floor space, and this same scale used for the vertical dimensions may produce a stream so shallow that capillarity has an appreciable effect and also the bed slope may be such that the flow is laminar. In such cases it is necessary to use a distorted model, which means that the vertical scale is larger than the horizontal scale. Then, if the horizontal scale ratio is denoted by L_r and the vertical scale ratio by $L_{r'}$, the cross section area ratio is $L_r L_{r'}$.

7.3 KINEMATIC SIMILARITY

Kinematic similarity implies that, *in addition* to geometric similarity, the ratio of the *velocities* at all corresponding points in the flows are the same. The ***velocity ratio*** is

$$V_r = \frac{V_p}{V_m} \tag{7.2}$$

[1] The reciprocal of the scale ratio will be referred to here as the ***model ratio***, or model scale, $\lambda = L_m/L_p$. Thus a model ratio of 1:20 or $\lambda = 0.05$ corresponds to a scale ratio of 20:1 or $L_r = 20$ (the prototype is 20 times larger than the model).

and this is a constant for kinematic similarity. Its value in terms of L_r will be determined by dynamic considerations, as explained in the following section.

As time T is dimensionally L/V, the time scale is

$$T_r = \frac{L_r}{V_r} \tag{7.3}$$

and in a similar manner the acceleration scale is

$$a_r = \frac{L_r}{T_r^2} = \frac{V_r^2}{L_r} \tag{7.4}$$

7.4 DYNAMIC SIMILARITY

Two systems have ***dynamic similarity*** if, *in addition to* kinematic similarity, corresponding forces are in the same ratio in both. The ***force ratio*** is

$$F_r = \frac{F_p}{F_m} \tag{7.5}$$

which must be constant for dynamic similarity.

Forces that may act on a fluid element include those due to gravity (F_G), pressure (F_P), viscosity (F_V), and elasticity (F_E). Also, if the element of fluid is at a liquid–gas interface, there are forces due to surface tension (F_T). If the sum of forces on a fluid element does not add up to zero, the element will accelerate in accordance with Newton's law. Such an unbalanced force system can be transformed into a balanced system by adding an inertia force F_I that is equal and opposite to the resultant of the acting forces. Thus, generally,

$$\sum \mathbf{F} = \mathbf{F}_G + \mathbf{F}_P + \mathbf{F}_V + \mathbf{F}_E + \mathbf{F}_T = \text{resultant}$$

and

$$\mathbf{F}_I = -\text{resultant}$$

Thus

$$\mathbf{F}_G + \mathbf{F}_P + \mathbf{F}_V + \mathbf{F}_E + \mathbf{F}_T + \mathbf{F}_I = 0$$

These forces may be expressed in the simplest terms as follows:

Gravity: $F_G = mg = \rho L^3 g$

Pressure: $F_P = (\Delta p)A = (\Delta p)L^2$

Viscosity: $F_V = \mu\left(\frac{du}{dy}\right)A = \mu\left(\frac{V}{L}\right)L^2 = \mu VL$

Elasticity: $F_E = E_v A = E_v L^2$

Surface tension: $F_T = \sigma L$

Inertia: $F_I = ma = \rho L^3 \frac{L}{T^2} = \rho L^4 T^{-2} = \rho V^2 L^2$

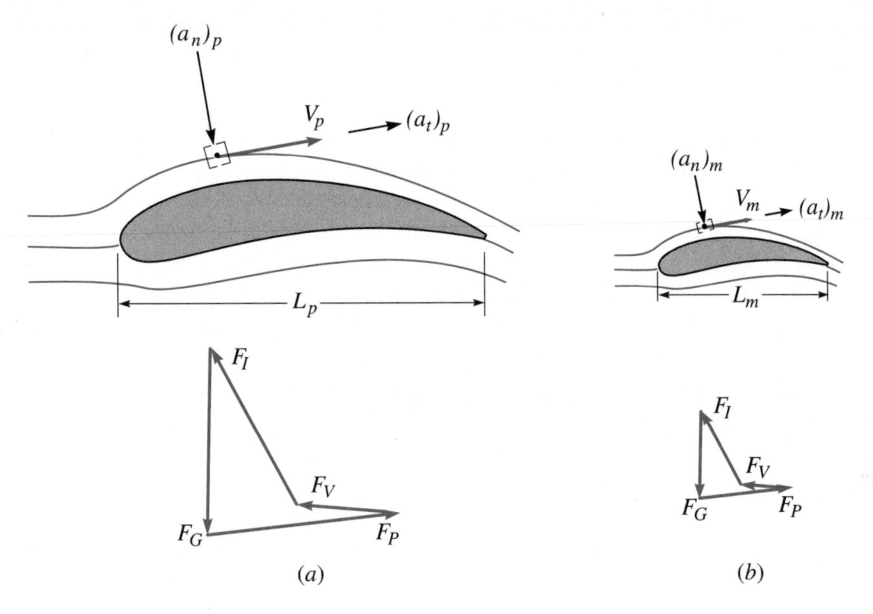

Figure 7.1
(*a*) Prototype. (*b*) Model. $L_r = L_p/L_m$; $V_r = V_p/V_m$; $F_r = F_p/F_m$.

In many flow problems some of these forces are either absent or insignificant. In Fig. 7.1 are depicted two geometrically similar flow systems. Let it be assumed that they also possess kinematic similarity and that the forces acting on any fluid element are F_G, F_P, F_V, and F_I. Then dynamic similarity will be achieved if

$$\frac{F_{G_p}}{F_{G_m}} = \frac{F_{P_p}}{F_{P_m}} = \frac{F_{V_p}}{F_{V_m}} = \frac{F_{I_p}}{F_{I_m}} = F_r$$

where subscripts p and m refer to prototype and model as before. These relations can also be expressed as

$$\left(\frac{F_I}{F_G}\right)_p = \left(\frac{F_I}{F_G}\right)_m, \quad \left(\frac{F_I}{F_P}\right)_p = \left(\frac{F_I}{F_P}\right)_m, \quad \left(\frac{F_I}{F_V}\right)_p = \left(\frac{F_I}{F_V}\right)_m$$

Each of the quantities is dimensionless. With four forces acting, there are three independent expressions that must be satisfied; with three forces, there are two independent expressions; and so on. The significance of the dimensionless ratios is discussed in the following subsections.

Reynolds Number[2]

In the flow of a fluid through a completely fllled conduit, gravity does not affect the flow pattern. Also, since there are no free liquid surfaces, it is obvious that capillarity is of no practical importance. Therefore the significant forces are inertia and fluid friction due to viscosity. The same is true of an airplane traveling at speeds below that at which compressibility of the air is appreciable. Also, for a submarine submerged far enough so as not to produce waves on the surface, the only forces involved are those of friction and inertia.

Considering the ratio of *inertia forces to viscous forces,* the parameter obtained is called the ***Reynolds number***, or **R**, in honor of Osborne Reynolds (1842–1912), the English physicist and professor who presented this in a publication of his experimental work in 1882. But it was Lord Rayleigh (1842–1919), another English physicist and a Nobel Laureate, who 10 years later developed the theory of dynamic similarity. The ratio of these two forces is

$$\mathbf{R} = \frac{F_I}{F_V} = \frac{L^2V^2\rho}{LV\mu} = \frac{LV\rho}{\mu} = \frac{LV}{\nu} \tag{7.6}$$

For any consistent system of units, **R** is a dimensionless number. The linear dimension L may be any length that is significant in the flow pattern. Thus, for a pipe completely filled, it might be either the diameter or the radius, and the numerical value of **R** will vary accordingly. General usage prescribes L as the pipe diameter. Thus, for a pipe flowing full, $\mathbf{R} = DV\rho/\mu = DV/\nu$, where D is the diameter of the pipe.

If two systems, such as a model and its prototype, or two pipelines with different fluids, are to be dynamically equivalent so far as inertia and viscous forces are concerned, they must both have the same value of **R**. Thus, for such a case, dynamic similarity is achieved when

$$\left(\frac{LV}{\nu}\right)_m = \mathbf{R}_m = \mathbf{R}_p = \left(\frac{LV}{\nu}\right)_p \tag{7.7}$$

For the same fluid in model and prototype, Eq. (7.7) shows that for dynamic similarity, a high velocity must be used with a model of small linear dimensions. The fluid used in the model need not be the same as that in the prototype, provided L and V are so chosen as to give the same value of **R** in model and prototype.

[2] It is now standard practice to represent Reynolds number, Froude number, etc., by bold face **R**, **F**, etc. For handwriting on the blackboard and overhead projector, it is suggested that $\mathbb{R}$ and $\mathbb{F}$ etc. with double lines on the left be used. The symbols *Re* and N_R are also sometimes used for Reynolds number, and *Fr* and N_F for Froude number, etc.

Sample Problem 7.1 If the Reynolds numbers of a model and its prototypes are the same, find an expression for V_r, T_r, and a_r.

Solution

$$\mathbf{R} = \frac{L_m V_m}{\nu_m} = \frac{L_p V_p}{\nu_p}$$

$$V_r = \frac{V_p}{V_m} = \frac{L_m \nu_p}{L_p \nu_m} = \frac{\nu_r}{L_r} = \left(\frac{\nu}{L}\right)_r \qquad \textit{ANS}$$

$$T_r = \frac{L_r}{V_r} = \left(\frac{L^2}{\nu}\right)_r \qquad \textit{ANS}$$

$$a_r = \frac{V_r}{T_r} = \left(\frac{\nu^2}{L^3}\right)_r \qquad \textit{ANS}$$

Froude Number[3]

Considering *inertia and gravity forces* alone, a ratio is obtained called a ***Froude number,*** *or* **F**, in honor of William Froude (1810–1879), a British naval architect who experimented with flat plates towed lengthwise through water in order to estimate the resistance of ships due to wave action. The ratio of inertia forces to gravity forces is

$$\frac{\rho L^2 V^2}{\rho g L^3} = \frac{V^2}{gL}$$

Although this is sometimes defined as a Froude number, it is more common to use the square root so as to have V in the first power, as in the Reynolds number. Thus a Froude number is

$$\mathbf{F} = \frac{V}{\sqrt{gL}} \tag{7.8}$$

Systems involving gravity and inertia forces include the wave action set up by a ship, the flow of water in open channels, the forces of a stream on a bridge pier, the flow over a spillway, the flow of a jet from an orifice, and other cases where gravity is the dominant factor.

For the computation of **F**, the length L must be some linear dimension that is significant in the flow pattern. For a ship, it is commonly taken as the length at the waterline. For an open channel, it is taken as the depth of flow. For situations where inertia and gravity forces predominate, dynamic similarity is achieved when

$$\left(\frac{V}{\sqrt{gL}}\right)_m = \mathbf{F}_m = \mathbf{F}_p = \left(\frac{V}{\sqrt{gL}}\right)_p \tag{7.9}$$

---

[3] Froude is pronounced ($\overline{\text{froo}}$d), to rhyme with brood.

In some flow situations fluid friction is a factor as well as gravity and inertia. In such cases, to achieve dynamic similarity, the Reynolds number and the Froude number criteria must both be satisfied simultaneously. A comparison of Eqs. (7.7) and (7.9) shows that the two cannot be satisfied at the same time with fluids of the same viscosity. The only way to satisfy Eqs. (7.7) and (7.9) for both model and prototype is to use fluids of different viscosities. Simultaneous solution of Eqs (7.7) and (7.9) yields $(\nu_m/\nu_p) = (L_m/L_p)^{3/2}$. Sometimes a fluid with the proper viscosity can be found, but usually this is either impractical or impossible. Consequently, the usual technique is to operate the model so as to satisfy one of the dimensionless numbers and then correct the results with experimental data dependent on the other dimensionless number. In the case of a ship, the towing of a model will give the total resistance, from which must be subtracted the empirically computed skin friction to determine the wave-making resistance of the model. Applying the Froude number criterion then permits one to determine the wave-making resistance of the full-size ship. A computed skin friction for the ship is then added to this value to give the total ship resistance. The details of such calculations are deferred to Chap. 9.

When water flows in an open channel, fluid friction as well as gravity and inertia may be a factor. However, for the flow of water in an open channel, there is often fully developed turbulence, in which case the hydraulic friction is independent of the Reynolds number (Sec. 8.11). Thus, for this case, identical Froude numbers will give dynamic similarity. When a high-viscosity liquid flows in an open channel or when water flows at a relatively low Reynolds number, the effect of Reynolds number cannot be neglected.

The scale ratios for Froude number similarity will now be discussed. From Eq. (7.8), V varies as $\sqrt{gL}$, and if g is considered to be the same in prototype and model, as is usually the case, then, from Eq. (7.2),

$$V_r = \frac{V_p}{V_m} = \sqrt{\frac{L_r}{1}} \qquad \text{(for the same } \mathbf{F} \text{ and same } g\text{)}$$

and, from Eq. (7.3), the ratio of time for prototype to model is

$$T_r = \frac{T_p}{T_m} = \frac{L_r}{V_r} = \sqrt{\frac{L_r}{1}} \qquad \text{(for the same } \mathbf{F} \text{ and same } g\text{)}$$

and $$a_r = \frac{V_r}{T_r} = 1 \qquad \text{(for the same } \mathbf{F} \text{ and same } g\text{)}$$

A knowledge of the time scale is useful in the study of cyclic phenomena such as waves and tides.

Since the velocity varies as $\sqrt{L_r}$ and the cross-sectional area as L_r^2, it follows that

$$Q_r = \frac{Q_p}{Q_m} = \frac{L_r^{5/2}}{1} \qquad \text{(for the same } \mathbf{F} \text{ and same } g\text{)}$$

As mentioned in Sec. 7.2, for river models it is usually necessary to use an enlarged vertical scale to provide the model with adequate depth.[4] Application of the Froude number indicates that in this case V varies as $\sqrt{L_{r'}}$, where $L_{r'}$ is the scale ratio in the vertical direction. Thus

$$\frac{Q_p}{Q_m} = \frac{L_r L_{r'}^{3/2}}{1} \qquad \text{(for the same } \mathbf{F} \text{ and same } g\text{)}$$

Mach Number

When compressibility is important, it is necessary to consider the ratio of the *inertia to the elastic forces.* The ***Mach number*** **M** is defined as the square root of this ratio. It is named after Ernst Mach (1838–1916), an Austrian physicist and philosopher who investigated the shock waves of supersonic projectiles in the 1880s. Thus

$$\mathbf{M} = \left(\frac{\rho V^2 L^2}{E_v L^2}\right)^{1/2} = \frac{V}{\sqrt{E_v/\rho}} = \frac{V}{c} \tag{7.10}$$

where c is the sonic velocity (or celerity) in the medium in question (see Sec. 13.3). Thus the Mach number is the ratio of the fluid velocity (or the velocity of the body through a stationary fluid) to that of a sound wave in the same medium. If **M** is less than 1, the flow is ***subsonic***; if it is equal to 1, it is ***sonic***; if it is greater than 1, the flow is called ***supersonic***; and for extremely high values of **M**, the flow is called ***hypersonic.*** The Mach number squared is equal to the Cauchy number.

Weber Number

In a few cases of flow, surface tension may be important, but normally it is negligible. The ratio of *inertia forces to surface tension* is $\rho V^2 L^2/\sigma L$, the square root of which is known as the ***Weber number***:

$$\mathbf{W} = \frac{V}{\sqrt{\sigma/\rho L}} \tag{7.11}$$

It is named in honor of Moritz Weber (1871–1951), who developed the

[4] Enlarged vertical scales are also commonly employed in models of large water bodies such as lakes, reservoirs, estuaries, and bays.

modern laws of similitude. An illustration of its application is at the leading edge of a very thin sheet of liquid flowing over a surface.

A dimensionless quantity related to the ratio of the *inertia forces to the pressure forces* is known as the ***Euler number***. It is named after the Swiss mathematician Leonhard Euler (1707–1783), renowned for his prolific work in pure mathematics. It is expressed in a variety of ways, one form being

$$\mathbf{E} = \frac{V}{\sqrt{2(\Delta p/\rho)}} = \frac{V}{\sqrt{2g(\Delta p/\gamma)}} \tag{7.12}$$

If only pressure and inertia influence the flow, the Euler number for any boundary form will remain constant. If other parameters (viscosity, gravity, etc.) cause the flow pattern to change, however, **E** will also change. The expression for **E** [Eq. (7.12)] may be recognized as being equivalent to the coefficient of velocity, discussed in Chapter 11.

Rearranging Eq. (7.12) to evaluate $1/\mathbf{E}^2$, we get a dimensionless quantity known as the ***pressure coefficient***:

$$C_p = \frac{\Delta p}{\frac{1}{2}\rho V^2} \tag{7.13}$$

This, or half this quantity, is also called the Euler number by some authors; we noted in Sec. 5.8 that some call $\frac{1}{2}\rho V^2$ the dynamic pressure. When Δp is referred to the vapor pressure p_v, the pressure coefficient becomes a dimensionless quantity called the ***cavitation number***:

$$\mathbf{C} = \frac{p - p_v}{\frac{1}{2}\rho V^2} \tag{7.14}$$

Both pressures must be absolute. Some authors omit the $\frac{1}{2}$ from this definition.

SAMPLE PROBLEM 7.2 A certain submerged body is to move horizontally through oil (γ = 52 lb/ft^3, μ = 0.0006 lb·s/ft^2) at a velocity of 45 fps. To study the characteristics of this motion, an enlarged model of the body is tested in 60°F water. The model ratio λ is 8:1. Determine the velocity at which this enlarged model should be pulled through the water to achieve dynamic similarity. If the drag force on the model is 0.80 lb, predict the drag force on the prototype.

[5] Euler is pronounced (oi'lər), to rhyme with boiler.

Solution

The body is submerged; hence there is no wave action. Reynolds' criterion must be satisfied:

$$\left(\frac{DV}{\nu}\right)_p = \left(\frac{DV}{\nu}\right)_m, \quad \text{where} \quad \frac{D_m}{D_p} = \frac{8}{1}$$

$$\nu_m = 1.217 \times 10^{-5} \text{ ft}^2/\text{s (Appendix A, Table A.1)}$$

$$\nu_p = \frac{\mu}{\rho} = \frac{0.0006}{52/32.2} = 0.000372 \text{ ft}^2/\text{s}$$

$$\frac{D_p(45)}{0.000372} = \frac{(8D_p)V_m}{1.217 \times 10^{-5}}$$

$$V_m = 0.1843 \text{ fps} \qquad \textbf{\textit{ANS}}$$

$$F \propto \rho V^2 L^2; \quad \text{hence} \quad \frac{F_p}{F_m} = \frac{\rho_p V_p{}^2 L_p{}^2}{\rho_m V_m{}^2 L_m{}^2}$$

$$\frac{F_p}{F_m} = \frac{(52/32.2)(45)^2 1}{1.94(0.1843)^2(8)^2} = 777$$

$$F_p = 777F_m = 777(0.8) = 621 \text{ lb} \qquad \textbf{\textit{ANS}}$$

EXERCISES

7.4.1 Water at 65°F in a 6-in-diameter pipe flows with a velocity of 6 fps. What is the Reynolds number? Note that $L = D$ and that the physical properties of water are given in Appendix A.

7.4.2 Oil ($s = 0.85$ and $\mu = 0.24$ N·s/m^2) in a 12-cm-diameter pipe flows with a velocity of 4.0 m/s. What is the Reynolds number?

7.4.3 The dimensions of a model airplane are $\frac{1}{25}$ those of its prototype. The model is to be tested in a pressure wind tunnel at the same speed and air temperature as the prototype. What must the pressure in the wind tunnel be relative to the atmospheric pressure if the Reynolds number remains the same?

7.4.4 A wind tunnel test on a 1:40 scale model of a submarine is to be performed to find the drag on the submarine when it is moving at 10 knots through 50°F ocean water. At what velocity should the test be conducted in the wind tunnel if it contains 60°F air at atmospheric pressure? If the drag on the model is 75 lb, what will be the drag on the prototype?

7.4.5 A 500-ft-long ship is to operate at a speed of 20 mph. If a model of this ship is 10 ft long, what should be its speed in fps to give the same Froude number? What is the value of this Froude number?

7.4.6 Water flows over the crest of a 1:30 model spillway, and the velocity measured at a particular point is 0.4 m/s. What velocity does this represent in the prototype? The force exerted on a certain area of the model is measured to be 0.15 N. What would be the force on the corresponding area in the prototype? Develop your own dimensionless ratios.

7.4.7 A vertical jet of water issuing upward from a nozzle at a velocity of 76 fps will rise to a height of approximately 90 ft on the earth. To get a water jet to rise to a height of 120 ft on the moon, where the gravity is one-sixth of that on earth, what must be the jet velocity? Neglect atmospheric resistance.

7.4.8 The flow about a missile that travels at 1500 mph through the atmosphere at an elevation of 15,000 ft is to be modeled in a wind tunnel at standard atmospheric conditions with 70°F air. What is the required air speed in the wind tunnel to achieve dynamic similarity?

7.5 SCALE RATIOS

The Reynolds number, the Froude number, and the Mach number are the dimensionless parameters most commonly encountered in fluid mechanics. In the preceding section the scale ratios for velocity, time, and acceleration for the Reynolds and Froude numbers were developed. Scale ratios for other quantities can be developed in a similar fashion. Such relations are presented in Table 7.1. These enable one to quickly calculate the scale ratio (prototype divided by model) of any desired quantity for the case where the given dimensionless number is the same in both prototype and model. The computed ratio, of course, gives a realistic result only if the flow is predominantly governed by the particular dimensionless number. Thus an important aspect of physical modeling of fluid phenomena is the need to know which dimensionless number is most important.

7.6 COMMENTS ON MODELS

In the use of models it is essential that the fluid velocity should not be so low that laminar flow exists when the flow in the prototype is turbulent. Also, conditions in the model should not be such that surface tension is important if such conditions do not exist in the prototype. For example, the depth of water flowing over the crest of a model spillway should not be too small.

While model studies are very important and valuable, it is necessary to exercise some judgment in transferring results from the model to the prototype. It is not always necessary or desirable that these various dimensionless ratios be adhered to in every case. Thus, in tests of model centrifugal pumps, geometric similarity is essential, but it is desirable to operate at such a rotation speed that the peripheral velocity and all fluid

Table 7.1
Flow characteristics and similitude scale ratios (ratio of prototype quantity to model quantity)

		Scale ratios for laws of		
Characteristic	**Dimension**	**Reynolds**	**Froude**	**Mach**
Geometric				
Length	L	L_r	L_r	L_r
Area	L^2	L_r^2	L_r^2	L_r^2
Volume	L^3	L_r^3	L_r^3	L_r^3
Kinematic				
Time	T	$\left(\frac{L^2\rho}{\mu}\right)_r$	$(L^{1/2}g^{-1/2})_r$	$\left(\frac{L\rho^{1/2}}{E_v^{1/2}}\right)_r$
Velocity	LT^{-1}	$\left(\frac{\mu}{L\rho}\right)_r$	$(L^{1/2}g^{1/2})_r$	$\left(\frac{E_v^{1/2}}{\rho^{1/2}}\right)_r$
Acceleration	LT^{-2}	$\left(\frac{\mu^2}{\rho^2L^3}\right)_r$	g_r	$\left(\frac{E_v}{L\rho}\right)_r$
Discharge	L^3T^{-1}	$\left(\frac{L\mu}{\rho}\right)_r$	$(L^{5/2}g^{1/2})_r$	$\left(\frac{L^2E_v^{1/2}}{\rho^{1/2}}\right)_r$
Dynamic				
Mass	M	$(L^3\rho)_r$	$(L^3\rho)_r$	$(L^3\rho)_r$
Force	MLT^{-2}	$\left(\frac{\mu^2}{\rho}\right)_r$	$(L^3\rho g)_r$	$(L^2E_v)_r$
Pressure	$ML^{-1}T^{-2}$	$\left(\frac{\mu^2}{L^2\rho}\right)_r$	$(L\rho g)_r$	$(E_v)_r$
Impulse and momentum	MLT^{-1}	$(L^2\mu)_r$	$(L^{7/2}\rho g^{1/2})_r$	$(L^3\rho^{1/2}E_v^{1/2})_r$
Energy and work	ML^2T^{-2}	$\left(\frac{L\mu^2}{\rho}\right)_r$	$(L^4\rho g)_r$	$(L^3E_v)_r$
Power	ML^2T^{-3}	$\left(\frac{\mu^3}{L\rho^2}\right)_r$	$(L^{7/2}\rho g^{3/2})_r$	$\left(\frac{L^2E_v^{3/2}}{\rho^{1/2}}\right)_r$

Note: Usually g is the same in model and prototype.

velocities are the same as in the prototype, since only in this way may cavitation be detected.

The roughness of a model should be scaled down in the same ratio as the other linear dimensions, which means that a small model should have surfaces that are much smoother than those in its prototype. But this requirement imposes a limit on the scale that can be used if true geometric similarity is to be achieved. However, in the case of a distorted model with a vertical scale larger than the horizontal scale, it may be necessary to make the model

Figure 7.2
View of the Corps of Engineers Model of San Francisco Bay ($L_r = 1000$, $L_{r'} = 100$), showing the vertical metal tabs that were installed to provide proper frictional flow resistance.

surface rough in order to simulate the flow conditions in the prototype. As any distorted model lacks the proper similitude, no simple rule can be given for this; the roughness should be adjusted by trial until the flow conditions are judged to be typical of those in the prototype. In most distorted models, metal tabs (Fig. 7.2) are used to provide proper frictional boundary effects. The size and spacing of the tabs is determined by trial so as to create flow conditions in the model identical to those observed in the prototype. Vertical distortion disturbs circulation patterns, and the metal tabs create large-scale eddies. Hence mixing (the disposition of pollutants) in distorted models must be interpreted with caution.

In models of systems, such as siphons, involving liquids where large negative pressures are expected and hence the possibility of cavitation exists, the model must be placed in an air-tight chamber in which a partial vacuum is maintained so as to produce an absolute pressure in the model identical to that in the prototype.[6]

---

[6] Hydraulic Models, *Manual of Engineering Practice,* no. 25, American Society of Civil Engineers, 1942.

When modeling a subsonic airplane in a wind tunnel, it is commonly necessary to conduct the test under high pressure in order to satisfy the Reynolds criterion

$$\left(\frac{DV\rho}{\mu}\right)_m = \left(\frac{DV\rho}{\mu}\right)_p$$

without introducing compressibility effects. For example, suppose $L_r = D_p/D_m = 20$. If the viscosity μ and density ρ of the air were the same in the model and prototype then, to satisfy Reynolds' criterion, $V_m = 20 \times V_p$. For an airplane operating at normal speed, this would make the model Mach number much greater than one, and compressibility effects would invalidate the behavior of the model. If, however, the test were conducted under a pressure of 20 atm with identical model and prototype temperatures, $\rho_m = 20 \times \rho_p$ and $\mu_m \approx \mu_p$, since the viscosity of air changes very little with pressure (or density). In this case the model should be operated at a velocity equal to that of the prototype in order for the Reynolds numbers to be the same.

Sample Problem 7.3 A 1:50 model of a boat has a wave resistance of 0.02 N when operating in water at 1.0 m/s. Find the corresponding prototype wave resistance. Find also the horsepower requirement for the prototype. What velocity does this test represent in the prototype?

Solution

Gravity and inertia forces predominate; hence the Froude criterion is applicable.

$$\mathbf{F}_p = \mathbf{F}_m = \left(\frac{V}{\sqrt{gL}}\right)_p = \left(\frac{V}{\sqrt{gL}}\right)_m$$

Since both the model and prototype are acted upon by the earth's gravitational field, the gs can be canceled out. Thus

$$\frac{V_p^2}{L_p} = \frac{V_m^2}{L_m}$$

and

$$\frac{V_p^2}{V_m^2} = \frac{L_p}{L_m} = L_r = 50$$

Since

$$\frac{F_p}{F_m} = \frac{\rho L_p^2 V_p^2}{\rho L_m^2 V_m^2} = L_r^2 L_r = L_r^3$$

$$F_p = L_r^3 F_m = (50)^3(0.02) = 2500 \text{ N} = 562 \text{ lb} \qquad \textbf{\textit{ANS}}$$

$$V_p = \sqrt{L_r} \times V_m = \sqrt{50} \times 1 = 7.07 \text{ m/s} = 23.2 \text{ fps} \qquad \textbf{\textit{ANS}}$$

$$HP_p = \frac{F_p V_p}{550} = \frac{562 \times 23.2}{550} = 23.7 \qquad \textbf{\textit{ANS}}$$

7.7 DIMENSIONAL ANALYSIS

Fluid-mechanics problems may be approached by *dimensional analysis,* a mathematical technique that makes use of the study of dimensions. Dimensional analysis is related to similitude; however, the approach is different. In dimensional analysis, from a general understanding of fluid phenomena, one first predicts the physical parameters that will influence the flow, and then, by grouping these parameters in dimensionless combinations, a better understanding of the flow phenomena is made possible. Dimensional analysis is particularly helpful in experimental work, because it provides a guide to those things that significantly influence the phenomena; thus it indicates the direction in which experimental work should go.

Physical quantities may be expressed in either the force–length–time (FLT) system or the mass–length–time (MLT) system. This is because these two systems are interrelated through Newton's law, which states that force equals mass times acceleration, $F = ma$, or

$$F = M\frac{L}{T^2}$$

Through this relation, conversion can be made from one system to the other. Other than convenience, it makes no difference which system we use, since the results are the same. The dimensions used in either system may be in BG units or SI units. Details on these systems of units and on conversion factors are given in Sec. 1.5 and on the inside covers of the book.

Basic Concepts

All rational equations that relate physical quantities must be dimensionally homogeneous. That is, all the terms in an equation must have the same dimensions. For example, as noted in Sec. 5.9, each of the terms in Bernoulli's equation (5.28) has the dimension of length. This principle is known as the ***principle of dimensional homogeneity*** (PDH), which was first formalized in 1822 by Baron Joseph Fourier (1768–1830), a French mathematician and physicist now best known for his Fourier series. The PDH is a valuable tool for checking equation derivations and engineering calculations, and, as we shall see shortly, it can be of great help in deriving the forms of physical equations.

Nonhomogeneous equations, however, are sometimes used, the best-known example in fluid mechanics being the Manning equation (8.50). Such equations have often resulted from fitting equations to observed data. Generally their use is limited to specialized areas.

Because all the terms in a rational (dimensionally homogeneous) equation have the same dimensions, if we divide them all by a quantity that has the same dimensions then all the terms will become dimensionless. When we do this using various quantities with the proper dimensions, we often find certain dimensionless groups that recur frequently, such as the dimensionless numbers described in Sec. 7.4. Thus many equations can be more simply expressed as relationships between dimensionless groups or numbers.

Dimensional analysis is a powerful scientific procedure that formalizes this process. It readily arranges all the variables involved into an equation containing independent dimensionless groups, so avoiding the experimentation. Another benefit it yields is that the number of independent groups so obtained is less than the number of variables.

To illustrate the basic principles of dimensional analysis, let us explore the equation for the speed V with which a pressure wave travels through a fluid. We must visualize the physical problem to consider what physical factors probably influence the speed. Certainly the compressibility E_v (Sec. 2.5) must be a factor; also the density ρ and the viscosity ν of the fluid (Secs. 2.3 and 2.11) might be factors. The dimensions of these quantities, written in square brackets, are

$$V = \left[\frac{L}{T}\right], \quad E_v = \left[\frac{F}{L^2}\right] = \left[\frac{M}{LT^2}\right], \quad \rho = \left[\frac{M}{L^3}\right], \quad \nu = \left[\frac{L^2}{T}\right]$$

Here the dimensions of E_v were converted into the MLT system using $F = ML/T^2$ as noted earlier. Clearly, adding or subtracting such quantities will not produce dimensionally homogeneous equations. We must therefore multiply them in such a way that their dimensions balance. So let us write

$$V = CE_v^a \rho^b \nu^d$$

where C is a dimensionless constant, and let us solve for the exponents a, b, and d. Substituting the dimensions, we get

$$\frac{L}{T} = \left(\frac{M}{LT^2}\right)^a \left(\frac{M}{L^3}\right)^b \left(\frac{L^2}{T}\right)^d$$

To satisfy dimensional homogeneity, the exponents of each dimension must be identical on both sides of this equation. Thus

For M: $\quad 0 = a + b$

For L: $\quad 1 = -a - 3b + 2d$

For T: $\quad -1 = -2a - d$

Solving these three equations, we get

$$a = \tfrac{1}{2}, \quad b = -\tfrac{1}{2}, \quad d = 0$$

so that

$$V = C\sqrt{\frac{E_v}{\rho}}$$

Thus we have identified the basic form of this relationship, to be treated later as Eqs. (12.9) and (13.17), and we have determined that the wave speed is not affected by the fluid's viscosity.

Dimensional analysis along such lines was developed by Lord Rayleigh, who is mentioned in Sec. 7.4. Although it is a very serviceable method, it has been superseded.

The Pi Theorem

A more generalized method of dimensional analysis developed by E. Buckingham[7] and others is now most widely used. This arranges the variables into a lesser number of dimensionless groups of variables. Because Buckingham used an upper case Π to represent the product of variables in each group, this method is known as the ***pi theorem*** or the Buckingham pi theorem.

Let $X_1, X_2, X_3, \ldots, X_n$ represent n *dimensional* variables, such as velocity, density, and viscosity, which are involved in some physical phenomenon. The dimensionally homogeneous equation relating these variables can be written as

$$f(X_1, X_2, X_3, \ldots, X_n) = 0$$

in which the dimensions of each *term* are the same. It follows that we can rearrange this equation into

$$\phi(\Pi_1, \Pi_2, \ldots, \Pi_{n-k}) = 0$$

where each Π is an independent *dimensionless* product of some of the Xs. The reduction k in the number of terms (from n to $n - k$), is usually equal to and sometimes less than the number of fundamental dimensions m involved in all the variables.

There is a series of seven steps to be followed when applying the pi theorem. As we review these, let us relate them to an example problem, that of the drag force F_D exerted on a submerged sphere as it moves through a viscous fluid.

Step 1 Visualize the physical problem, consider the factors that are of influence, and list and count the n variables.

In our example we must first consider which physical factors influence the drag force. Certainly, the size of the sphere must enter the problem; also, the velocity of the sphere must be important. The fluid properties involved are the density ρ and the viscosity μ. Thus we can write

$$f(F_D, D, V, \rho, \mu) = 0$$

Here D, the sphere diameter, is used to represent sphere size, and f stands for "some function". We see that $n = 5$. Note that the procedure cannot work

[7] E. Buckingham, Model Experiments and the Form of Empirical Equations, *Trans. ASME,* Vol. 37, pp. 263–296, 1915.

if any relevant variables are omitted. Experimentation with the procedure and experience will help determine which variables are relevant.

Step 2 Choose a dimensional system (MLT or FLT) and list the dimensions of each variable. Find m, the number of fundamental dimensions involved in all the variables.

In our example, choosing the MLT system, the dimensions are respectively

$$\frac{ML}{T^2}, \quad L, \quad \frac{L}{T}, \quad \frac{M}{L^3}, \quad \frac{M}{LT}$$

We may get help in identifying these dimensions from Table 7.1 and from the units of quantities on the inside covers and in the appendices, etc. We see that M, L, and T are involved in our example, so $m = 3$.

Step 3 Find the reduction number k. Usually this equals m, which it cannot exceed, but rarely it is less than m. To check this, try to find m dimensional variables that *cannot* be formed into a dimensionless group. If m are found, $k = m$; if not, reduce k by one and retry.

In our example we find that three of the dimensional variables, namely ρ, D, and V, with dimensions M/L^3, L, and L/T, *cannot* be formed into a dimensionless Π group because M and L cannot cancel among them, so $k = 3$. The rare circumstances that reduce k arise when some of the dimensions (usually M and T) occur in the parameters only in fixed combinations.

Step 4 Determine $n - k$, the number of dimensionless Π groups needed. In our example this is $5 - 3 = 2$. So we can write $\phi(\Pi_1, \Pi_2) = 0$.

Step 5 From the list of dimensional variables, select k of them to be so-called primary (repeating) variables. These must contain all of the m fundamental dimensions, and must not form a Π among themselves (see Step 3). It is generally advantageous to choose primary variables that relate to mass, geometry, and kinematics (flow without forces or energy). Form the Π groups by multiplying the product of the primary variables, with unknown exponents, by each of the remaining variables, one at a time.

We choose ρ, D, and V (as in Step 3) as the primary variables for our example. Then the Π terms are

$$\Pi_1 = \rho^{a_1} D^{b_1} V^{c_1} \mu$$
$$\Pi_2 = \rho^{a_2} D^{b_2} V^{c_2} F_D$$

Step 6 To satisfy dimensional homogeneity, equate the exponents of each dimension on both sides of each pi equation, and so solve for the exponents and the forms of the dimensionless groups. Since the Πs are dimensionless, they can be replaced with $M^0L^0T^0$. Experience in fluid mechanics has shown that these dimensionless groups commonly take the form of a Reynolds number, Froude number, or Mach number. Hence one should always be on the lookout for them when using dimensional analysis.

For our example, working with Π_1,

$$M^0L^0T^0 = \left(\frac{M}{L^3}\right)^{a_1} L^{b_1}\left(\frac{L}{T}\right)^{c_1}\left(\frac{M}{LT}\right)$$

M: $\quad 0 = a_1 + 1$

L: $\quad 0 = -3a_1 + b_1 + c_1 - 1$

T: $\quad 0 = -c_1 - 1$

Solving for a_1, b_1, and c_1,

$$a_1 = -1, \quad b_1 = -1, \quad c_1 = -1$$

Thus
$$\Pi_1 = \rho^{-1}D^{-1}V^{-1}\mu = \frac{\mu}{\rho DV} = \left(\frac{\rho DV}{\mu}\right)^{-1}$$

Hence, noting that $(\rho DV/\mu)$ is a Reynolds number,

$$\Pi_1 = \mathbf{R}^{-1}$$

Working in a similar fashion with Π_2, we get

$$\Pi_2 = \frac{F_D}{\rho D^2V^2}$$

Check that all Πs are in fact dimensionless.

Step 7 Rearrange the pi groups as desired. The pi theorem states that the Πs are related, and may be expressed as $f_1(\Pi_1, \Pi_2, \ldots, \Pi_{n-k}) = 0$ or as $\Pi_1 = f_2(\Pi_2, \Pi_3, \ldots, \Pi_{n-k})$, etc. It does not predict the functional form of f_1 or f_2; these relations must be determined experimentally. Further, each Π parameter can be raised to any power, since this will not affect their dimensionless status.

In our example, since we are interested in F_D, we write $\Pi_2 = \phi(\Pi_1^{-1})$ or

$$\frac{F_D}{\rho D^2V^2} = \phi(\mathbf{R})$$

so that
$$F_D = \phi(\mathbf{R})\rho D^2V^2$$

It should be emphasized that dimensional analysis does not provide a complete solution to fluid problems. It provides a partial solution only. The success of dimensional analysis depends entirely on the ability of the individual using it to define the parameters that are applicable. If one omits an important variable, the results are incomplete, and this may lead to incorrect conclusions. For example, with a compressible fluid at high velocities, compressibility effects may be significant, in which case the volume modulus E_v of the fluid must be considered an important physical property. Introducing E_v into the previous example of dimensional analysis of the drag on a sphere will show that for the more general case the drag may depend on the Mach number as well as the Reynolds number. If one includes a variable that is totally unrelated to the problem, an additional insignificant dimensionless group will result. Thus, to use dimensional analysis successfully, one must be familiar with the fluid phenomena involved.

Sample Problem 7.4 Derive an expression for the flow rate q over the spillway shown in the accompanying figure per foot of spillway perpendicular to the sketch. Assume that the sheet of water is relatively thick, so that surface-tension effects may be neglected. Assume also that gravity effects predominate so strongly over viscosity that viscosity may be neglected.

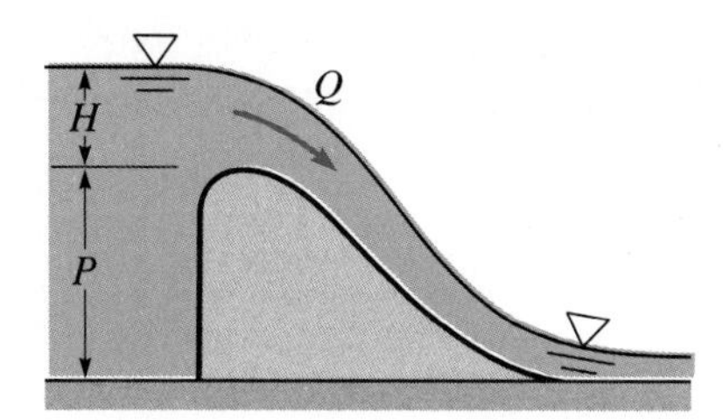

Figure S7.4

Solution

Under the assumed conditions the variables that affect q would be the head H, the acceleration of gravity g, and possibly the spillway height P. Thus

$$q = f(H, g, P)$$

or

$$f_1(q, H, g, P) = 0$$

In this case there are $n = 4$ variables and $m = 2$ dimensions. Two variables can easily be found that cannot be formed into a dimensionless group; therefore $k = m = 2$, and so there are $n - k = 2$ pi groups and

$$\phi(\Pi_1, \Pi_2) = 0$$

Using q and H as the primary variables,

$$\Pi_1 = q^{a_1} H^{b_1} g$$

$$\Pi_2 = q^{a_2} H^{b_2} P$$

Working with Π_1,

$$L^0 T^0 = \left(\frac{L^3}{TL}\right)^{a_1} L^{b_1} \left(\frac{L}{T^2}\right)$$

L: $\quad 0 = 2a_1 + b_1 + 1$

T: $\quad 0 = -a_1 - 2$

Hence

$$a_1 = -2, \quad b_1 = 3$$

$$\Pi_1 = q^{-2} H^3 g = \frac{gH^3}{q^2}$$

Working with Π_2,
$$L^0T^0 = \left(\frac{L^3}{TL}\right)^{a_2} L^{b_2}L$$

L:
$$0 = 2a_2 + b_2 + 1$$

T:
$$0 = -a_2$$

Hence
$$a_2 = 0, \quad b_2 = -1$$

$$\Pi_2 = q^0H^{-1}P = \frac{P}{H}$$

Finally, $\phi(\Pi_1, \Pi_2) = 0$ can be written as

$$\Pi_1^{-1/2} = \phi_1(\Pi_2^{-1})$$

i.e.,
$$\frac{q}{\sqrt{g}H^{3/2}} = \phi_1\left(\frac{H}{P}\right)$$

or
$$q = \phi_1\left(\frac{H}{P}\right)\sqrt{g}H^{3/2}$$

Thus dimensional analysis indicates that the flow rate per unit length of spillway is proportional to $\sqrt{g}$ and to $H^{3/2}$. The flow rate also is affected by the H/P ratio. This relationship is discussed in Sec. 11.13.

If viscosity were included as one of the variables, another dimensionless group would have resulted. This dimensionless group would have had the form of a Reynolds number. With surface tension included as a variable, the resulting dimensionless group would have been a Weber number.

EXERCISES

7.7.1 Use dimensional analysis to arrange the following groups into dimensionless parameters; (*a*) τ, V, ρ; (*b*) V, L, ρ, σ. Use the MLT system.

7.7.2 Use dimensional analysis to arrange the following groups into dimensionless parameters; (*a*) Δp, V, γ, g; (*b*) F, ρ, L, V. Use the MLT system.

7.7.3 Use dimensional analysis to derive an expression for the power developed by an engine in terms of the torque and rotative speed.

7.7.4 Derive an expression for the shear stress at the pipe wall when an incompressible fluid flows through a pipe under pressure. Use dimensional analysis with the following significant parameters: pipe diameter D, flow velocity V, and viscosity μ and density ρ of the fluid.

7.7.5 Refer to the example of the drag on a submerged sphere, used to illustrate the seven-step procedure of Sec. 7.7. Using the same parameters, but also including the acceleration due to gravity g to account for the effect of wave action, derive an expression for the drag on a surface vessel.

PROBLEMS

7.1 What is the Reynolds number for 130°F air at a pressure of 120 psia when it flows at a velocity of 100 fps through a 6-in-diameter pipe? Note that physical properties of air are given in Appendix A.

7.2 What is the Reynolds number for 55°C air at a pressure of 800 kN/m^2 abs when it flows at a velocity of 30 m/s through a 15-cm-diameter pipe? Note that physical properties of air are given in Appendix A.

7.3 Air at 150°C and a pressure of 240 kN/m^2 abs flows at a velocity of 16 m/s through an 18-cm-diameter pipe. What is the Reynolds number?

7.4 Models are to be built of the following prototypes: (*a*) tides; (*b*) oil flowing through a full pipeline; (*c*) a water jet; (*d*) flow over the spillway of a dam; (*e*) a deep submersible vehicle; (*f*) an airplane flying at low speed; (*g*) a supersonic aircraft; (*h*) a supersonic missile. For dynamic similarity, indicate which single dimensionless ratio will govern, and give reasons why.

7.5 The linear dimensions of a model airplane are $\frac{1}{18}$ those of its prototype. If the prototype is to fly at 500 mph, what must be the air velocity in a wind tunnel for the same Reynolds number if the air temperature and pressure are the same?

7.6 Air at 68°F and 60 psia flows in a 1.5-in-diameter pipe. What weight flow rate of this air will give dynamic similarity to 60°F water flowing at 200 gpm through a 3-in-diameter pipe?

7.7 When a submerged sphere moves through 20°C water at 2.0 m/s, the drag force exerted on it is 15 N. Another sphere of three times the diameter is tested in a wind tunnel where the air pressure and temperature are 2.0 MN/m^2 abs and 300 K respectively. What air velocity is required for dynamic similarity, and what then will be the drag force on the larger sphere?

7.8 Air at 80°C and 500 kN/m^2 abs pressure flows in a 6-cm-diameter pipe. What air flow rate (kg/s) will give dynamic similarity to 60 L/s of 60°C water flowing in a 50-cm-diameter pipe?

7.9 Water flows over a spillway at 5000 cfs. For dynamic similarity, what should be the model scale if the flow rate over the model is to be 45 cfs? The force exerted on a certain area of the model is measured to be 1.0 lb. What would be the force on the corresponding area of the prototype?

7.10 A 500 ft long ship is to operate at a speed of 20 mph in ocean water whose viscosity is 1.2 cP and specific weight is 64 lb/ft^3. What should be the kinematic viscosity of the liquid used with a 10-ft-long model of this ship so that both the Reynolds number and the Froude number would be the same? Does such a liquid exist?

7.11 A 1:600 scale model is built to study tides. What length of time in the model corresponds to one day in the prototype? Suppose this model could be tested on the moon where g is one-sixth of that on earth. What then would be the time relationship between the model and prototype?

7.12 A vertical jet of water issuing upward from a nozzle at a velocity of 44 fps will rise to a height of approximately 30 ft on the earth. To get a water jet to rise to a height of 120 ft on the moon, where the gravity is one-sixth of that on earth, what must be the jet velocity? Neglect atmospheric resistance.

7.13 A 3-ft-high sectional model of a spillway is built in a 1-ft-wide laboratory flume. The flow is 0.86 cfs under a head of 0.380 ft. If the model scale is 1:20 and the prototype spillway is 600 ft long, what flow does this represent in the prototype?

7.14 A model spillway has a flow of 0.10 m^3/s per m of width. What is the actual flow for the prototype spillway if the model scale is 1:20?

7.15 An aircraft company is investigating the flow about a model of a supersonic plane in a variable-density wind tunnel at 1500 fps. The air, at 80°F, has a pressure of 20 psia. At what velocity should the model be tested to maintain dynamic similarity if the air temperature is raised to 90°F and the pressure is increased to 30 psia? Solve this two ways: (*a*) using and (*b*) not using specific weights and densities.

7.16 An aircraft company is investigating the flow about a model of a supersonic plane in a variable density wind tunnel at 400 m/s. The air, at 40°C, has a pressure of 150 kN/m^2 abs. At what velocity should the model be tested to maintain dynamic similarity if the air temperature is raised to 75°C and the pressure is increased to 200 kN/m^2 abs?

7.17 A 1:13 model of a ballistic missile is to be tested in a high-speed wind tunnel. The prototype missile will travel at 1250 fps through air at 68°F and 14.5 psia. If the air in the wind tunnel test section has a temperature of 32°F and a pressure of 12.1 psia, what velocity is needed there? Estimate the drag force on the prototype if the drag force on the model is 95 lb?

7.18 A 1:13 model of a ballistic missile is to be tested in a high-speed wind tunnel. The prototype missile will travel at 380 m/s through air at 23°C and 95.0 kN/m^2 abs. If the air in the wind tunnel test section has a temperature of −20°C at a pressure of 89 kN/m^2 abs, what velocity is needed there? Estimate the drag force on the prototype if the drag force on the model is 400 N?

7.19 When traveling at a velocity of 2.7 fps a 1:64 scale model of a ship has a wave resistance of 0.08 lb. This is kinematically similar to the design velocity of the prototype ship. What is the design velocity of the prototype and what is its wave resistance at that velocity?

7.20 Develop the scale ratios given in Table 7.1 for the case where prototype and model Reynolds numbers are the same.

7.21 Develop the scale ratios given in Table 7.1 for the case where prototype and model Froude numbers are identical.

7.22 Develop the scale ratios given in Table 7.1 for the case where prototype and model Mach numbers are the same.

7.23 Gas ($\gamma = 0.32$ lb/ft^3, $\mu = 2.0 \times 10^{-6}$ lb·sec/ft^2) is flowing in a 0.75-in-diameter pipe. When a gas flowmeter measures the flow as being

0.13 lb/sec, it registers a pressure drop of 1.17 psi. An enlarged model that is geometrically similar is to be tested in a 6-in-diameter pipe. What flow rate of 80°F water will achieve dynamic similarity? What would be the pressure drop across the water meter?

7.24 Gas ($\rho = 5.25\ \text{kg/m}^3$, $\nu = 2.0 \times 10^{-5}\ \text{m}^2/\text{s}$) is flowing in a 2-cm-diameter pipe. When a gas flowmeter measures the flow as being 0.064 kg/s, it registers a pressure drop of 8.5 kPa. An enlarged model that is geometrically similar is to be tested in a 18-cm-diameter pipe. What flow rate of 25°C water will achieve dynamic similarity? What would be the pressure drop across the water meter?

7.25 Find the dimensions of force, energy, and momentum in the FLT system. Repeat for the MLT system.

7.26 Find the dimensions of torque, pressure, and power, in the FLT system. Repeat for the MLT system.

7.27 Derive an expression for the velocity of rise of an air bubble in a stationary liquid. Consider the effect of surface tension as well as other variables.

7.28 Derive an expression for the drag on an aircraft flying at supersonic speed.

7.29 Derive an expression for small flow rates over a spillway, in the form of a function including dimensionless quantities. Use dimensional analysis with the following parameters: height of spillway P, head on the spillway H, viscosity of liquid μ, density of liquid ρ, surface tension σ, and acceleration due to gravity g.

7.30 Use dimensional analysis to derive an expression for the height of capillary rise in a glass tube.

CHAPTER 8

Steady Incompressible Flow in Pressure Conduits

In this chapter some aspects of steady flow in pressure conduits are discussed. The discussion is limited to ***incompressible fluids***, that is, to those for which $\rho \approx$ constant. This includes all liquids. Gases flowing with very small pressure changes may be considered incompressible, for then $\rho \approx$ constant. In this chapter isothermal conditions are assumed so as to eliminate thermodynamic effects, some of which are discussed in Chap. 13.

8.1 LAMINAR AND TURBULENT FLOW

If the head loss in a given length of uniform pipe is measured at different velocities, it will be found that, as long as the velocity is low enough to secure laminar flow (Sec. 4.2), the head loss, due to friction, will be directly proportional to the velocity, as shown in Fig. 8.1. But with increasing velocity, at some point B, where visual observation of dye injected in a transparent tube would show that the flow changes from laminar to turbulent (Sec. 4.2), there will be an abrupt increase in the rate at which the head loss varies. If the logarithms of these two variables are plotted on linear scales or, in other words, if the values are plotted directly on log–log paper, it will be found that, after a certain transition region (BCA) has been passed, lines will be obtained with slopes ranging from about 1.75 to 2.00.

It is thus seen that for laminar flow the drop in energy due to friction varies as V, while for turbulent flow the friction varies as V^n, where n ranges from about 1.75 to 2. The lower value of 1.75 for turbulent flow is found for

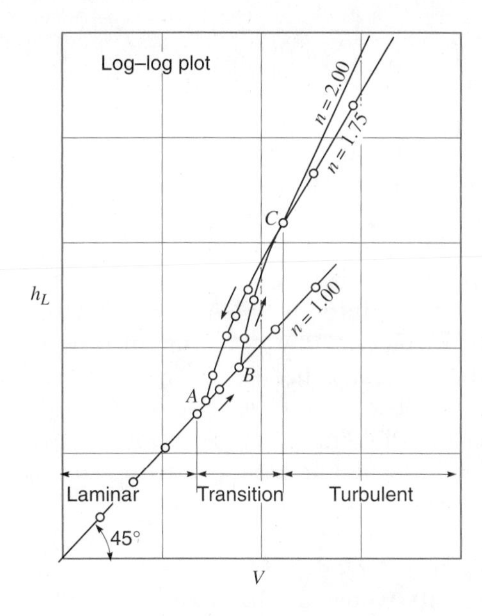

Figure 8.1
Log–log plot for flow in a uniform pipe ($n = 2.00$, rough-wall pipe; $n = 1.75$, smooth-wall pipe).

pipes with very smooth walls; as the wall roughness increases, the value of n increases up to its maximum value of 2.

The points in Fig. 8.1 were plotted directly from measurements made by Osborne Reynolds (Sec. 7.4), and show decided curves in the transition zone where values of n are even greater than 2. If the velocity is gradually reduced from a high value, the line BC will not be retraced. Instead, the points lie along curve CA. Point B is known as the ***higher critical point***, and A as the ***lower critical point***.

However, velocity is not the only factor that determines whether the flow is laminar or turbulent. The criterion is Reynolds number, which has been discussed in Sec. 7.4. For a circular pipe the significant linear dimension L is usually taken as the diameter D, and thus

$$\mathbf{R} = \frac{DV\rho}{\mu} = \frac{DV}{\nu} \tag{8.1}$$

where any consistent system of units may be used, since $\mathbf{R}$ is a dimensionless number.[1]

[1] It is sometimes convenient to use a "hybrid" set of units and compensate with a correction factor. Thus, by substituting $V = Q/A$ and $V = \dot{m}/\rho A$ from Eq. (4.7) into Eq. (8.1), we get

$$\mathbf{R} = 1.273Q/\nu D = 1.273\dot{m}/\mu D,$$

where Q and $\dot{m}$ are defined in the List of Symbols in the front of the book. The last form is especially convenient in the case of gases; it shows that in a pipe of uniform diameter the Reynolds number is constant along the pipe, even for a compressible fluid, where the density and velocity vary, if there is no appreciable variation in temperature to alter the viscosity of the gas.

8.2 CRITICAL REYNOLDS NUMBER

The upper critical Reynolds number, corresponding to point B of Fig. 8.1, is really indeterminate and depends upon the care taken to prevent any initial disturbance from affecting the flow. Its value is normally about 4000, but laminar flow in circular pipes has been maintained up to values of $\mathbf{R}$ as high as 50,000. However, in such cases this type of flow is inherently unstable, and the least disturbance will transform it instantly into turbulent flow. On the other hand, it is practically impossible for turbulent flow in a straight pipe to persist at values of $\mathbf{R}$ much below 2000, because any turbulence that is set up will be damped out by viscous friction. This lower value is thus much more definite than the higher one, and is really the dividing point between the two types of flow. Hence this lower value will be defined as the ***true critical Reynolds number***. However, it is subject to slight variations. Its value will be higher in a converging pipe and lower in a diverging pipe than in a straight pipe. Also, its value will be less for flow in a curved pipe than in a straight one, and even for a straight uniform pipe its value may be as low as 1000, where there is excessive roughness. However, for normal cases of flow in straight pipes of uniform diameter and usual roughness, the critical value may be taken as $\mathbf{R}_{crit} = 2000$.

For water at 75°F, the kinematic viscosity is 1.00×10^{-5} ft²/s, and in this case the critical Reynolds number is obtained when

$$DV_{crit} = \mathbf{R}_{crit}\nu = 2000 \times 10^{-5}\ \text{ft}^2/\text{s} = 0.020\ \text{ft}^2/\text{s}.$$

Thus, for a pipe 1 in (25 mm) in diameter,

$$V_{crit} = 0.02/(1/12) = 0.24\ \text{fps} \quad (0.073\ \text{m/s})$$

Or if the velocity were 2.4 fps (0.73 m/s), the diameter would be only 0.1 in (2.5 mm). Velocities or pipe diameters as small as these are not often encountered with water flowing in practical engineering, though they may be found in certain laboratory instruments. Hence, for such fluids as water and air, practically all cases of engineering importance are in the turbulent-flow region. But if the fluid is a viscous oil, laminar flow is often encountered.

SAMPLE PROBLEM 8.1 An oil ($s = 0.85$, $\nu = 1.8 \times 10^{-5}$ m²/s) in a refinery flows through a 10-cm-diameter pipe at 0.50 L/s. Is the flow laminar or turbulent?

Solution

$$V = \frac{Q}{A} = \frac{0.0005\ \text{m}^3/\text{s}}{\pi(0.1\ \text{m})^2/4} = 0.0637\ \text{m/s}$$

$$\mathbf{R} = \frac{DV}{\nu} = \frac{0.10\ \text{m}(0.0637\ \text{m/s})}{1.8 \times 10^{-5}\ \text{m}^2/\text{s}} = 354$$

Since $\mathbf{R} < \mathbf{R}_{\text{crit}} = 2000$, the flow is laminar. ***ANS***

EXERCISES

8.2.1 Oil with a kinematic viscosity of 0.00018 ft^2/sec is flowing through a 4-in-diameter pipe. Below what velocity will the flow be laminar?

8.2.2 Oil with a kinematic viscosity of 0.142 St is flowing through a 10-cm-diameter pipe. Below what velocity will the flow be laminar?

8.2.3 Oil with a kinematic viscosity of 0.004 ft^2/sec flows through a 4-in-diameter pipe with a velocity of 13.5 fps. Is the flow laminar or turbulent?

8.3 HYDRAULIC RADIUS, HYDRAULIC DIAMETER

For conduits having noncircular cross sections, some value other than the diameter must be used for the linear dimension in the Reynolds number. Such a characteristic is the ***hydraulic radius***, defined as

$$R_h = \frac{A}{P} \tag{8.2}$$

where A is the cross-sectional area of the flowing fluid, and P is the ***wetted perimeter***, that portion of the perimeter of the cross section where there is contact between fluid and solid boundary. For a circular pipe flowing full, $R_h = \pi r^2/2\pi r = \frac{1}{2}r$, or $\frac{1}{4}D$. Thus R_h is not the radius of the pipe, and hence the term "radius" is misleading. If a circular pipe is exactly half full, both the area and the wetted perimeter are half the preceding values; so R_h is $\frac{1}{2}r$, the same as if it were full. But if the depth of flow in a circular pipe is 0.8 times the diameter, for example, $A = 0.674D^2$ and $P = 2.21D$, then $R_h = 0.304D$, or $0.608r$.

The hydraulic radius is a convenient means for expressing the shape as well as the size of a conduit, since, for the same cross-sectional area, the value of R_h will vary with the shape.

In evaluating the Reynolds number for flow in a noncircular conduit (Sec. 8.16), it is customary to substitute $4R_h$ for D in Eq. (8.1).

In some engineering fields $4R_h$ is defined to be the ***hydraulic diameter*** D_h. Thus, for a pipe, $D_h = D$, which is fine, but $D_h = 4R_h$, which seems strange.

EXERCISES

8.3.1 What is the hydraulic radius of a 9 in by 16 in rectangular air duct?

8.3.2 What is the percentage difference between the hydraulic radii of a 30-cm-diameter duct and a 30-cm-square duct?

8.4 HEAD LOSS IN CONDUITS OF CONSTANT CROSS SECTION

The following discussion applies to either laminar or turbulent flow and to any shape of cross section.

Consider steady flow in a conduit of uniform cross section A, not necessarily circular (Fig. 8.2). The pressures at sections 1 and 2 are p_1 and p_2, respectively. The distance between sections is L. For equilibrium in steady flow, the summation of forces acting on any fluid element must be equal to zero (i.e., $\sum F = ma = 0$). Thus, in the direction of flow,

$$p_1A - p_2A - \gamma LA \sin\alpha - \bar{\tau}_0(PL) = 0 \tag{8.3}$$

where $\bar{\tau}_0$, the ***average shear stress*** (average shear force per unit area) at the conduit wall, is defined by

$$\bar{\tau}_0 = \frac{\int_0^P \tau_0 \, dP}{P} \tag{8.4}$$

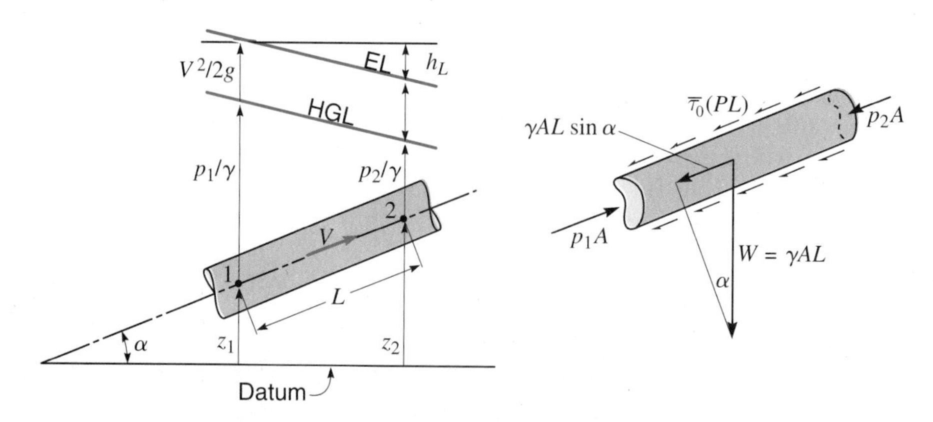

Figure 8.2

in which τ_0 is the local shear stress[2] acting over a small incremental portion dP of the wetted perimeter.

Noting that $\sin \alpha = (z_2 - z_1)/L$ and dividing each term in Eq. (8.3) by γA gives

$$\frac{p_1}{\gamma} - \frac{p_2}{\gamma} - z_2 + z_1 = \bar{\tau}_0 \frac{PL}{\gamma A} \tag{8.5}$$

From the left-hand sketch of Fig. 8.2, it can be seen that

$$h_L = \left(z_1 + \frac{p_1}{\gamma}\right) - \left(z_2 + \frac{p_2}{\gamma}\right)$$

Substituting h_L for the right-hand side of this expression and R_h for A/P in Eq. (8.5), we get

$$h_L = \bar{\tau}_0 \frac{L}{R_h \gamma} \tag{8.6}$$

This equation is applicable to any shape of uniform cross section, regardless of whether the flow is laminar or turbulent.

For a smooth-walled conduit, where wall roughness (discussed in Sec. 8.9) may be neglected, it might be assumed that the average fluid shear stress $\bar{\tau}_0$ at the wall is some function of ρ, μ, V and some characteristic linear dimension, which will here be taken as the hydraulic radius R_h. Thus

$$\bar{\tau}_0 = f(\rho, \mu, V, R_h) \tag{8.7}$$

Using the pi theorem of dimensional analysis (Sec. 7.7) to better determine the form of this relationship, we choose ρ, R_h, and V as primary variables, so that

$$\Pi_1 = \mu \rho^{a_1} R_h^{b_1} V^{c_1}$$
$$\Pi_2 = \bar{\tau}_0 \rho^{a_2} R_h^{b_2} V^{c_2}$$

With the dimensions of the variables being $ML^{-1}T^{-1}$ for μ, $ML^{-1}T^{-2}$ for $\bar{\tau}_0$, ML^{-3} for ρ, L for R_h, and LT^{-1} for V, the dimensions for Π_1 are

For M: $\quad 0 = 1 + a_1$

For L: $\quad 0 = -1 - 3a_1 + b_1 + c_1$

For T: $\quad 0 = -1 - c_1$

The solution of these three simultaneous equations is $a_1 = b_1 = c_1 = -1$, from which

$$\Pi_1 = \frac{\mu}{\rho R_h V} = \mathbf{R}^{-1}$$

[2] The local shear stress varies from point to point around the perimeter of all conduits (irrespective of whether the wall is smooth or rough), except for the case of a circular pipe flowing full where the shear stress at the wall is the same at all points of the perimeter.

where $R_h V\rho/\mu$ is a Reynolds number with R_h as the characteristic length. In a similar manner, we obtain

$$\Pi_2 = \frac{\bar{\tau}_0}{\rho V^2}$$

According to Sec. 7.7, Step 7, we can write $\Pi_2 = \phi(\Pi_1^{-1})$, which results in $\bar{\tau}_0 = \rho V^2 \phi(\mathbf{R})$. Setting the dimensionsionless term $\phi(\mathbf{R}) = \frac{1}{2}C_f$, this yields

$$\bar{\tau}_0 = C_f \rho \frac{V^2}{2} \tag{8.8}$$

Inserting this value of $\bar{\tau}_0$ into Eq. (8.6), and noting that $\gamma = \rho g$,

$$h_L = C_f \frac{L}{R_h} \frac{V^2}{2g} \tag{8.9}$$

which may be applied to any shape of smooth-walled cross section. Later, in Sec. 8.11 it will be shown that this equation also applies to rough-walled conduits.

8.5 FRICTION IN CIRCULAR CONDUITS

In Sec. 8.3 it is shown that for a circular pipe flowing full $R_h = \frac{1}{4}D$. Substituting this value into Eq. (8.9), the result (for both smooth-walled and rough-walled conduits) is

Circular pipe, flowing full:

$$h_L = f \frac{L}{D} \frac{V^2}{2g} \tag{8.10}$$

where

$$f = 4C_f = 8\phi(\mathbf{R}) \tag{8.11}$$

Equation (8.10) is known as the ***pipe-friction equation***, and is also commonly referred to as the ***Darcy–Weisbach equation***.[3] Like the coefficient C_f, the friction factor f is dimensionless and is also some function of Reynolds number. Much research has been directed toward determining the way in which f varies with **R** and also with pipe roughness (to be discussed in Sec. 8.11). The pipe-friction equation expresses the fact that the head lost in friction in a given pipe can be expressed in terms of the velocity head. The equation is dimensionally homogeneous, and may be used with any consistent system of units.

[3] In a slightly different form where D is replaced by the hydraulic radius R_h, Eq. (8.10) is known as the ***Fanning equation***, which is widely used by chemical engineers.

Dimensional analysis gives us the proper form for an equation, but does not yield a numerical result, since it is not concerned with abstract numerical factors. It also shows that Eq. (8.10) is a rational expression for pipe friction. But the exact form of $\phi(\mathbf{R})$ and numerical values for C_f and f must be determined by experiment or other means.

For a circular pipe flowing full, by substituting $R_h = r_0/2$, where r_0 is the radius of the pipe, Eq. (8.6) may be written as

$$h_L = \bar{\tau}_0 \frac{L}{R_h \gamma} = \frac{2\tau_0 L}{r_0 \gamma} \tag{8.12}$$

where the local shear stress at the wall, τ_0, is equal to the average shear stress $\bar{\tau}_0$ because of symmetry.

Following a development similar to that of Eqs. (8.3)–(8.6) and noting that for a circle, $A = \pi r^2$ and $P = 2\pi r$, it can be shown that for a cylindrical body of fluid concentric to the pipe $h_L = 2\tau L/r\gamma$, where τ is the shear stress in the fluid at radius r. Relating this to Eq. (8.12), it follows that the shear stress in the flow in a circular pipe at any radius r is

$$\tau = \tau_0 \frac{r}{r_0} \tag{8.13}$$

or the shear stress is zero at the center of the pipe and increases linearly with the radius to a maximum value τ_0 at the wall as in Fig. 8.3. This is true regardless of whether the flow is laminar or turbulent.

From Eqs. (8.6) and (8.10) and substituting $R_h = D/4$ for a circular pipe, we obtain

$$\tau_0 = \frac{f}{4} \rho \frac{V^2}{2} = \frac{f}{4} \gamma \frac{V^2}{2g} \tag{8.14}$$

With this equation, τ_0 for flow in a circular pipe may be computed for any experimentally determined value of f.

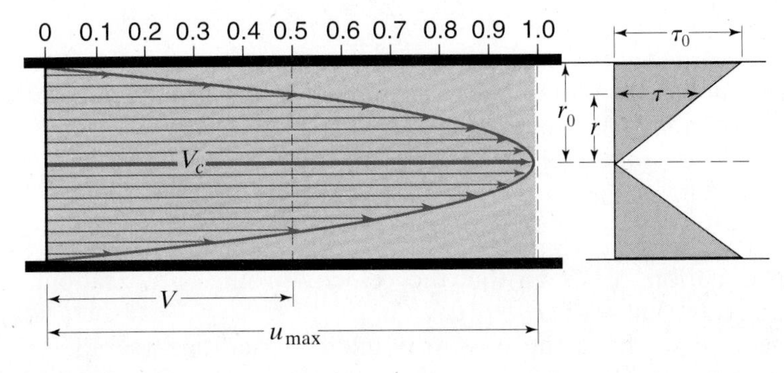

Figure 8.3
Velocity profile in laminar flow and distribution of shear stress.

EXERCISES

8.5.1 Oil ($s = 0.93$) of viscosity 0.004 ft^2/sec flows in a 4-in-diameter pipe at a rate of 6.5 gpm. Find the head loss per unit length.

8.5.2 Oil ($s = 0.93$) of viscosity 0.00034 m^2/s flows in a 10-cm-diameter pipe at a rate of 0.45 L/s. Find the head loss per unit length.

8.5.3 Steam with a specific weight of 0.32 lb/ft^3 is flowing with a velocity of 94 fps through a circular pipe with $f = 0.0171$. What is the shear stress at the pipe wall?

8.5.4 Steam with a specific weight of 38 N/m^3 is flowing with a velocity of 35 m/s through a circular pipe with $f = 0.0154$. What is the shear stress at the pipe wall?

8.6 LAMINAR FLOW IN CIRCULAR PIPES

In Sec. 2.11 it was noted that for laminar flow $\tau = \mu\, du/dy$, where u is the value of the velocity at a distance y from the boundary. As $y = r_0 - r$, it is also seen that $\tau = -\mu\, du/dr$; in other words, the minus sign indicates that u decreases as r increases. The coefficient of viscosity μ is a constant for any particular fluid at a constant temperature, and therefore if the shear varies from zero at the center of the pipe to a maximum at the wall, it follows that the velocity profile must have a zero slope at the center and have a continuously steeper velocity gradient as the wall is approached.

To determine the velocity profile for laminar flow in a circular pipe, the expression $\tau = \mu\, du/dy$ is substituted into the expression $h_L = \tau 2L/r\gamma$. Thus

$$h_L = \frac{\tau 2L}{r\gamma} = \mu \frac{du}{dy}\frac{2L}{r\gamma} = -\mu \frac{du}{dr}\frac{2L}{r\gamma}$$

From this,

$$du = -\frac{h_L \gamma}{2\mu L} r\, dr$$

Integrating and determining the constant of integration from the fact that $u = u_{max}$, when $r = 0$, we obtain

$$u = u_{max} - \frac{h_L \gamma}{4\mu L} r^2 = u_{max} - kr^2 \tag{8.15}$$

From this equation, it is seen that the velocity profile is a parabola, as shown in Fig. 8.3. Note that $k = h_L\gamma/4\mu L$.

At the wall we have the no-slip boundary condition (see Sec. 2.11) that $u = 0$ when $r = r_0$. Substituting this into the second expression of Eq. (8.15) and noting that $u_{max} = V_c$, the centerline velocity, we find $k = V_c/r_0^2$. Thus

Eq. (8.15) can be expressed as

$$u = V_c - \frac{V_c}{r_0^2} r^2 = V_c\left(1 - \frac{r^2}{r_0^2}\right) \tag{8.16}$$

Combining Eqs. (8.15) and (8.16), we get an expression for the centerline velocity as follows

$$V_c = u_{\max} = \frac{h_L \gamma}{4\mu L} r_0^2 = \frac{h_L \gamma}{16\mu L} D^2 \tag{8.17}$$

Equation (8.15) may be multiplied by a differential area $dA = 2\pi r\, dr$ and the product integrated from $r = 0$ to $r = r_0$ to find the rate of discharge. From Eq. (4.3), the rate of discharge is equivalent to the volume of a solid bounded by the velocity profile. In this case the solid is a paraboloid with a maximum height of $u_{\max}$. The mean height of a paraboloid is one-half the maximum height, and hence the mean velocity V is $0.5u_{\max}$. Thus

$$V = \frac{h_L \gamma}{32\mu L} D^2 \tag{8.18}$$

From this last equation, noting that $\gamma = g\rho$ and $\mu/\rho = \nu$, the loss of head in friction is given by

Laminar flow:

$$h_L = 32 \frac{\mu}{\gamma} \frac{L}{D^2} V = 32\nu \frac{L}{gD^2} V \tag{8.19}$$

which is the ***Hagen–Poiseuille law*** for laminar flow in tubes. Hagen, a German engineer, experimented with water flowing through small brass tubes, and published his results in 1839. Poiseuille, a French scientist, experimented with water flowing through capillary tubes in order to determine the laws of flow of blood through the veins of the body, and published his studies in 1840.

From Eq. (8.19) it is seen that in laminar flow the loss of head is proportional to the first power of the velocity. This is verified by experiment, as shown in Fig. 8.1. The striking feature of this equation is that it involves no empirical coefficients or experimental factors of any kind, except for the physical properties of the fluid such as viscosity and density (or specific weight). From this, it would appear that in laminar flow the friction is independent of the roughness of the pipe wall. That this is true is also borne out by experiment.

Dimensional analysis shows that the friction loss may also be expressed by Eq. (8.10). Equating (8.10) and (8.19) and solving for the friction factor f, we obtain for laminar flow under pressure in a circular pipe,

Laminar flow:

$$f = \frac{64\nu}{DV} = \frac{64}{\mathbf{R}} \tag{8.20}$$

Hence, if **R** is less than 2000, we may use Eq. (8.19) to find pipe friction head loss, or we may use the pipe friction equation (8.10) with the value of f as given by Eq. (8.20).

EXERCISES

8.6.1 An oil with kinematic viscosity of 0.004 ft^2/sec weighs 62 lb/ft^3. What will be its flow rate and head loss in a 2750-ft length of a 3-in-diameter pipe when the Reynolds number is 950.

8.6.2 With laminar flow in a circular pipe, at what distance from the centerline (in terms of the pipe radius) does the average velocity occur?

8.6.3 For laminar flow in a two-dimensional passage, find the relation between the average and maximum velocities.

8.7 ENTRANCE CONDITIONS IN LAMINAR FLOW

In the case of a pipe leading from a reservoir, if the entrance is rounded so as to avoid any initial disturbance of the entering stream, all particles will start to flow with the same velocity, except for a very thin film in contact with the wall. Particles in contact with the wall have zero velocity, but the velocity gradient is here extremely steep, and, with this slight exception, the velocity is uniform across the diameter, as shown in Fig. 8.4. As the fluid progresses along the pipe, the streamlines in the vicinity of the wall are slowed down by friction emanating from the wall, but since Q is constant for successive sections, the velocity in the center must be accelerated, until the final velocity

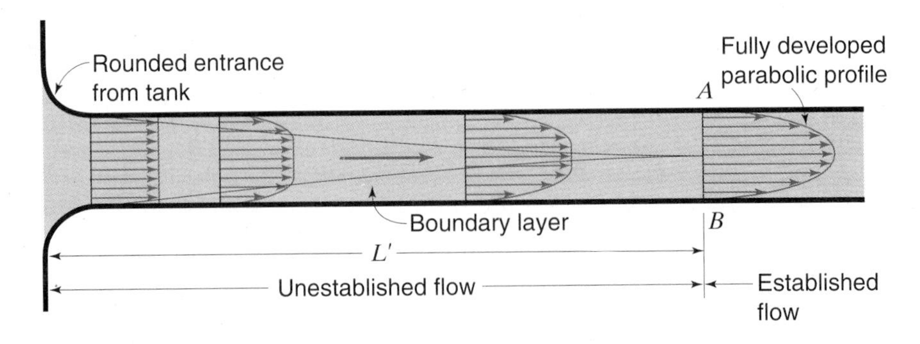

Figure 8.4
Velocity profiles and development of the boundary layer along a pipe in laminar flow.

profile is a parabola, as shown in Fig. 8.3. Theoretically, an infinite distance is required for this, but it has been established both by theory and by observation that the maximum velocity in the center of the pipe will reach 99% of its ultimate value in the distance $L' = 0.058\mathbf{R}D$.[4] Thus, for the critical value $\mathbf{R} = 2000$, the distance L' of Fig. 8.4 equals 116 pipe diameters. In other cases of laminar flow with Reynolds numbers less than 2000, the distance L' will be correspondingly less, in accordance with the expression $L' = 0.058\mathbf{R}D$.

In the entry region of length L' the flow is ***unestablished***; that is, the velocity profile is changing. In this region the flow can be visualized as consisting of a central core in which there are no frictional effects and an outer, annular zone extending from the core to the pipe wall. This outer zone increases in thickness as it moves along the wall, and is known as the ***boundary layer***. Viscosity in the boundary layer acts to transmit the effect of boundary shear inwardly into the flow. At section AB the boundary layer has grown until it occupies the entire section of the pipe. At this point, for laminar flow, the velocity profile is a perfect parabola. Beyond section AB, the velocity profile does not change, and the flow is known as ***established flow***.

The concept of a boundary layer within which viscosity is important, and outside of which friction is unimportant and the fluid may be considered as ideal, was originated in 1904 by Ludwig Prandtl (1875–1953), a German engineering professor. Perhaps the single most significant contribution to fluid mechanics, this concept is particularly important with turbulent flow, and will be discussed further in Secs. 8.9–11 and Chap. 9.

As shown in Sample Prob. 5.1 for a circular pipe, the kinetic energy of a stream with a parabolic velocity profile is $2V^2/2g$ (because $\alpha = 2$), where V is the mean velocity. At the entrance to the pipe the velocity is uniformly V across the diameter, except for an extremely thin layer next to the wall. Thus at the entrance to the pipe the kinetic energy per unit weight is practically $V^2/2g$. Hence in the distance L' there is a continuous increase in kinetic energy accompanied by a corresponding decrease in pressure head. Therefore at a distance L' from the entrance the piezometric head is less than the static value in the reservoir by $2V^2/2g$ plus the friction loss in this distance.

Laminar flow has been dealt with rather fully, not merely because it is of importance in problems involving fluids of very high viscosity, but especially because it permits a simple and accurate rational analysis. The general approach used here is of some assistance in the study of turbulent flow, where conditions are so complex that rigid mathematical treatment is impossible.

---

[4] H. L. Langhaar, Steady Flow in the Transition Length of a Straight Tube, *J. Appl. Mech.*, Vol. 10, p. 55, 1942.

SAMPLE PROBLEM 8.2 For the case of Sample Prob. 8.1, find the centerline velocity, the velocity at $r = 20$ mm, the friction factor, the shear stress at the pipe wall, and the head loss per meter of pipe length.

Solution

Since the flow is laminar,

$$V_c = 2V = 0.1273 \text{ m/s} \qquad \textbf{\textit{ANS}}$$

Eq. (8.15): $$u = u_{\max} - kr^2 \qquad u_{\max} = V_c = 0.1273 \text{ m/s}$$

When $r = r_0 = 50$ mm, $u = 0$; hence $0 = 0.1273 - k(0.05)^2$, from which

$$k = 50.9/(\text{m·s}) \qquad u_{20\text{ mm}} = 0.1273 - 50.9(0.02)^2 = 0.1070 \text{ m/s} \qquad \textbf{\textit{ANS}}$$

Eq. (8.20): $$f = \frac{64}{\mathbf{R}} = \frac{64}{354} = 0.1810 \qquad \textbf{\textit{ANS}}$$

Eq. (8.14): $$\tau_0 = \frac{f}{4}\rho\frac{V^2}{2} = \frac{0.1810}{4}(850 \text{ kg/m}^3)\frac{(0.0637 \text{ m/s})^2}{2}$$

$$= 0.0779 \text{ kg/(m·s}^2)$$

$$\tau_0 = 0.0779 \frac{\text{kg}}{(\text{m·s}^2)} \frac{\text{N·s}^2}{\text{kg·m}} = 0.0779 \text{ N/m}^2 \qquad \textbf{\textit{ANS}}$$

From Eq. (8.10): $$\frac{h_L}{L} = f\frac{1}{D}\frac{V^2}{2g} = 0.1810 \frac{1}{0.10 \text{ m}} \frac{(0.0637 \text{ m/s})^2}{2(9.81 \text{ m/s}^2)}$$

$$= 0.000374 \text{ m/m} \qquad \textbf{\textit{ANS}}$$

EXERCISES

8.7.1 In Exer. 8.2.3 what will be the approximate distance from the pipe entrance to the first point at which the flow is established?

8.7.2 In Exer. 8.5.2 what will be the approximate distance from the pipe entrance to the first point at which the flow is established?

8.8 TURBULENT FLOW

In Sec. 4.2 it was explained that in laminar flow the fluid particles move in straight lines while in turbulent flow they follow random paths. Consider the

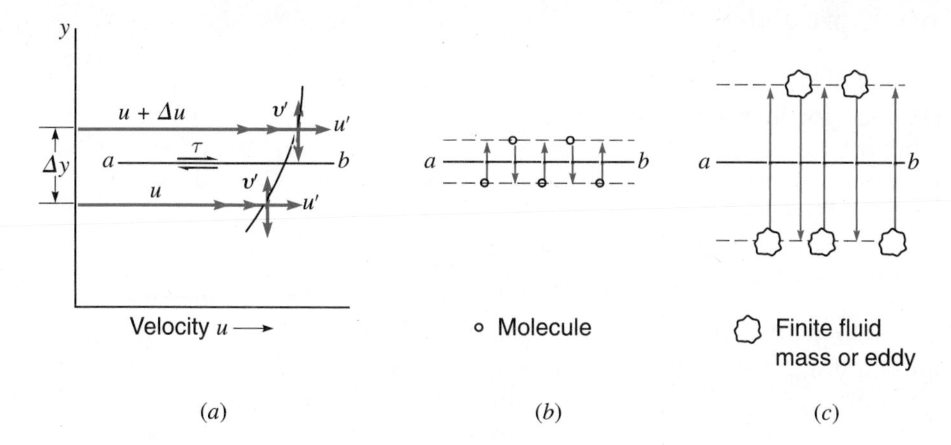

Figure 8.5
(*a*) Velocity profile. (*b*) Laminar flow (transfer of molecules across *ab*). (*c*) Turbulent flow (transfer of finite fluid masses across *ab*).

case of laminar flow as shown in Figs 8.5(*a,b*), where the velocity u increases with y. Even though the fluid particles are moving horizontally to the right, because of molecular motion, molecules will cross line *ab* and will thereby transport momentum. On the average, the velocities of the molecules in the slower-moving fluid below the line will be less than those of the faster-moving fluid above; the result is that the molecules that cross from below tend to slow down the faster-moving fluid. Likewise, the molecules that cross the line *ab* from above tend to speed up the slower-moving fluid below. The result is the production of a shear stress along the surface whose trace is *ab*, the value of which is given in Sec. 2.11 as $\tau = \mu \, du/dy$. This equation is applicable to laminar flow only.

Let us examine some of the characteristics of turbulent flow to see how it differs from laminar flow. In turbulent flow the velocity at a point in the flow field fluctuates in both magnitude and direction.[5] We may observe these fluctuations in accurate velocity measurements (Sec. 11.4), and commonly see their effects on pressure gages and manometers. The fluctuations result from a multitude of small eddies (Sec. 4.2), created by the viscous shear between adjacent particles. These eddies grow in size and then disappear as their particles merge into adjacent eddies. Thus there is a continuous mixing of particles, with a consequent transfer of momentum. Energy is dissipated by the viscosity, generating small amounts of heat.

[5] The velocity at a point in a so-called "steady" turbulent flow can be best visualized as a vector that fluctuates in both direction and magnitude. The mean temporal velocity at that point corresponds to the "average" of those vectors.

First Expression

In the modern conception of turbulent flow, a mechanism similar to that just described for laminar flow is assumed. However, the molecules are replaced by minute but finite masses or eddies (Fig. 8.5*c*). Hence, by analogy, the shear stress along the plane through *ab* in Fig. 8.5 may be defined in the case of turbulent flow as

$$\text{turbulent shear stress} = \eta \frac{du}{dy} \tag{8.21}$$

But unlike μ, the ***eddy viscosity*** η is not a constant for a given fluid at a given temperature, but depends upon the turbulence of the flow. It may be viewed as a coefficient of momentum transfer, expressing the transfer of momentum from points where the velocity is low to points where it is higher, and vice versa. Its magnitude may range from zero to many thousand times the value of μ. However, its numerical value is of less interest than its physical concept. In dealing with turbulent flow it is sometimes convenient to use ***kinematic eddy viscosity*** $\varepsilon = \eta/\rho$, which is a property of the flow alone, analogous to kinematic viscosity.

In general, the total shear stress in turbulent flow is the sum of the laminar shear stress plus the turbulent shear stress, i.e.,

$$\tau = \mu \frac{du}{dy} + \eta \frac{du}{dy} = \rho(\nu + \varepsilon) \frac{du}{dy} \tag{8.22}$$

In turbulent flow the second term of this equation is usually many times larger than the first term.

In turbulent flow the local axial velocity has been shown in Sec. 4.5 (see Fig. 4.6) to have fluctuations of plus and minus u', and there are also fluctuations of plus and minus v' and w' normal to u as shown in Fig. 8.6*b*.

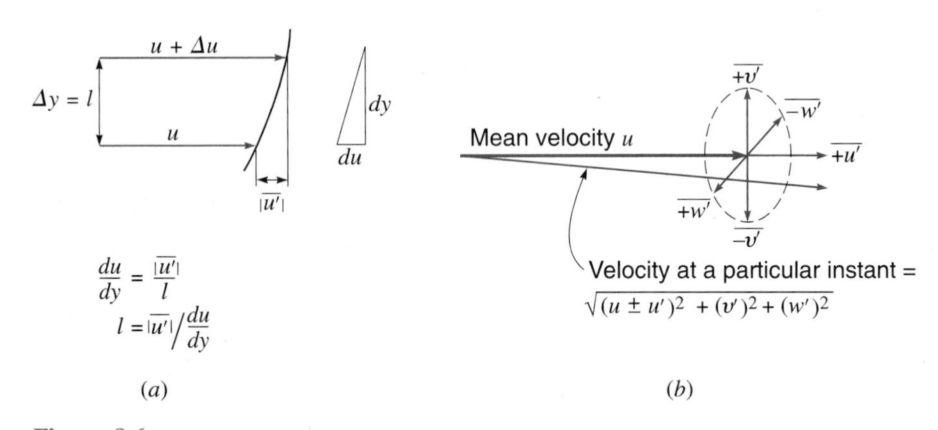

Figure 8.6
(*a*) Mixing length *l*. (b) Instantaneous local velocity in turbulent flow.

As it is obvious that there can be no values of v' next to and perpendicular to a smooth wall, turbulent flow cannot exist there. Hence, near a smooth wall, the shear is due to laminar flow alone, and $\tau = \mu\, du/dy$. It should be noted that the shear stress always acts to cause the velocity distribution to become more uniform.

At some distance from the wall, such as $0.2r$, the value of du/dy becomes small in turbulent flow, and hence the viscous shear becomes negligible in comparison with the turbulent shear. The latter can be large, even though du/dy is small, because of the possibilty of η being very large. This is because of the great turbulence that may exist at an appreciable distance from the wall. But at the center of the pipe, where du/dy is zero, there can be no shear at all. Hence, in turbulent flow as well as in laminar flow, the shear stress is a maximum at the wall and decreases linearly to zero at the axis, as shown in Fig. 8.3 and proved in Sec. 8.5.

Second Expression

Another expression for turbulent shear stress may be obtained that is different from that in Eq. (8.21). Thus in Fig. 8.5*a,* if a mass m of fluid below *ab,* where the temporal mean axial velocity is u, moves upward into a zone where the temporal mean axial velocity is $u + \Delta u$, its initial momentum in the axial direction must be increased by $m\,\Delta u$. Conversely, a mass m that moves from the upper zone to the lower will have its axial momentum decreased by $m\,\Delta u$. Hence this transfer of momentum back and forth across *ab* will produce a shear in the plane through *ab* proportional to Δu. This shear is possible only because of the velocity profile shown. If the latter were vertical, Δu would be zero and there could be no shear.

If the distance Δy in Fig. 8.5*a* is so chosen that the average value of $+u'$ in the upper zone over a time period long enough to include many velocity fluctuations is equal to Δu, i.e., $\Delta u = |u'|$, the two streams will be separated by what is known as the ***mixing length*** l, which will be referred to later. Consider, over a short time interval, a mass moving upward from below *ab* with a velocity v'; it will transport into the upper zone, where the velocity is $u + u'$, a momentum per unit time which is on the average equal to $\rho(v'\,dA)(u)$. The slower-moving mass from below *ab* will tend to retard the flow above *ab*; this creates a shear force along the plane of *ab*. This force can be found by applying the momentum principle [Eq. (6.6)], $F = \tau\,dA = \rho Q(\Delta V) = \rho(v'\,dA)(u + u' - u) = \rho u'v'\,dA$. Thus, over a time period of sufficient length to permit a large number of velocity fluctuations, the shear stress given by

$$\tau = F/dA = -\rho\overline{u'v'} \tag{8.23}$$

where $\overline{u'v'}$ is the temporal average of the product of u' and v'. This is an alternate form for Eq. (8.21), and in modern turbulence theory $-\rho\overline{u'v'}$ is referred to as the ***Reynolds stress.***

The minus sign appears in Eq. (8.23) because the product $\overline{u'v'}$ on the average is negative. By inspecting Fig. 8.5*a*, it can be seen that $+v'$ is associated with $-u'$ values more than with $+u'$ values. The opposite is true for $-v'$. Even though the temporal mean values of u' and v' are individually equal to zero, the temporal mean value of their product is not zero. This is because combinations of $+v'$ and $-u'$ and of $-v'$ and $+u'$ predominate over combinations of $+v'$ and $+u'$ and $-v'$ and $-u'$, respectively.

Prandtl reasoned that in any turbulent flow $\overline{|u'|}$ and $\overline{|v'|}$ must be proportional to each other and of the same order of magnitude. He also introduced the concept of mixing length l, which is defined as the distance one must move transversely to the direction of flow such that $\Delta u = \overline{|u'|}$. From Fig. 8.6*a* it can be seen that $\Delta u = l\,du/dy$ and hence $\overline{|u'|} = l\,du/dy$. If $\overline{|u'|} \propto \overline{|v'|}$ and if one permits l to account for the constant of proportionality, Prandtl[6] showed that $-\overline{u'v'}$ varies as $l^2\,(du/dy)^2$. Thus

$$\tau = -\rho\overline{u'v'} = \rho l^2\left(\frac{du}{dy}\right)^2 \tag{8.24}$$

This equation expresses terms that can be measured. Thus in any experiment where the pipe friction is determined, $\bar{\tau}_0$ can be computed from Eq. (8.6), and τ at any radius is then found from Eq. (8.13). A traverse of the velocity across a pipe diameter will give u at any radius, and the velocity profile will give du/dy at any radius. Thus Eq. (8.24) enables the mixing length l to be found as a function of the pipe radius. The purpose of all of this is to enable us to develop theoretical equations for the velocity profile in turbulent flow, and from this in turn to develop theoretical equations for f, the friction coefficient.

EXERCISE

8.8.1 Tests on 70°F water flowing through a 10-in-diameter pipe showed that when $V = 13$ fps, $f = 0.0162$. (*a*) If, at a distance of 3 in from the center of the pipe, $\tau = 0.398$ psf, and the velocity profile gives a value for du/dy of 6.77/sec, find the viscous shear and the turbulent shear at that radius. (*b*) What is the value of the mixing length l, and what is the value of the ratio l/r_0?

8.9 VISCOUS SUBLAYER IN TURBULENT FLOW

In Fig. 8.4 it is shown that, for laminar flow, if the fluid enters with no initial disturbance, the velocity is uniform across the diameter except for an exceedingly thin film at the wall, inasmuch as the velocity next to any wall is

[6] H. Schlichting: *Boundary Layer Theory*, 7th ed., McGraw-Hill Book Co., 1987, p. 605.

zero. But as flow proceeds down the pipe, the velocity profile changes because of the growth of a ***laminar boundary layer***, which continues until the boundary layers from opposite sides meet at the pipe axis and then there is fully developed laminar flow.

If the Reynolds number is above the critical value (Sec. 8.2), so that the developed flow is turbulent, the initial condition is much like that in Fig. 8.4. But as the laminar boundary layer increases in thickness, a point is soon reached where a transition occurs and the boundary layer becomes turbulent. This ***turbulent boundary layer*** generally increases in thickness much more rapidly, and soon the two layers from opposite sides meet at the pipe axis, and there is then ***fully developed turbulent flow***.

The initial laminar boundary layer may be given a Reynolds number such as $\mathbf{R}_x = Ux/\nu$, where U is the uniform velocity and x is the distance measured from the initial point. When x has such a value that this $\mathbf{R}_x$ is about 500,000, the transition occurs to the turbulent boundary layer. Fully developed turbulent flow will be found at about 50 pipe diameters from the pipe entrance for a smooth pipe with no special disturbance at entrance; otherwise, the turbulent boundary layers from the two sides will meet within a shorter distance. It is this fully developed turbulent flow that we shall consider in all that follows.

As v' must be zero at a smooth wall, turbulence there is inhibited so that a laminar-like sublayer occurs immediately next to the wall. However, the adjacent turbulent flow does repeatedly induce random transient effects that momentarily disrupt this sublayer, even though they are strongly damped out. Because it is therefore not a true laminar layer, and because shear in this layer is predominantly due to viscosity alone, it is called a ***viscous sublayer*** (see Fig. 8.7). This viscous sublayer is extremely thin, usually only a few hundredths of a millimeter, but its effect is great because of the very steep

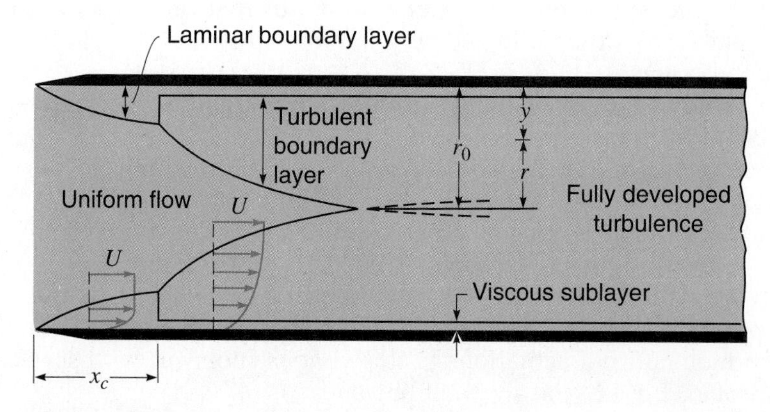

Figure 8.7
Development of boundary layer in a pipe where fully developed flow is turbulent (scales much distorted).

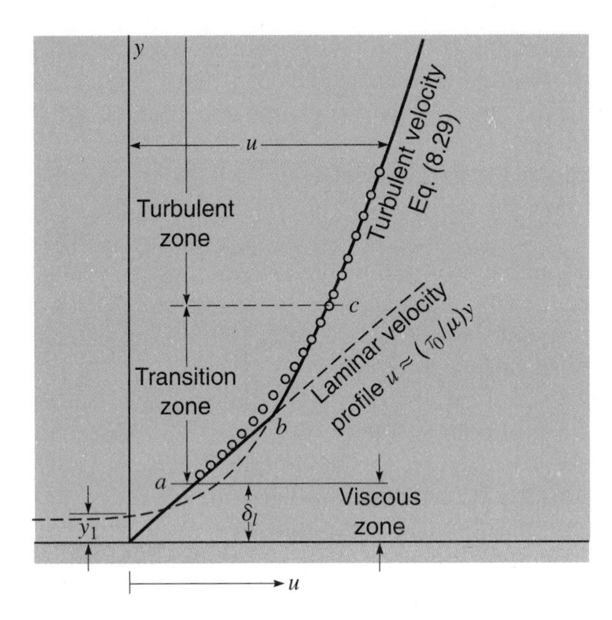

Figure 8.8
Velocity profile near a solid wall (vertical scale greatly exaggerated). Theoretical relations (solid lines) compared with experimental data (small circles).

velocity gradient within it and because $\tau = \mu\, du/dy$ in that region. At a greater distance from the wall the viscous effect becomes negligible, but the turbulent shear is then large. Between the two, there must be a transition zone where both types of shear are significant. It is evident that there can be no sharp lines of demarcation separating these three zones, inasmuch as one must merge gradually into the other.

By plotting a velocity profile from the wall on the assumption that the flow is entirely laminar (Sec. 8.6) and plotting another velocity profile on the assumption that the flow is entirely turbulent (Sec. 8.10), the two will intersect, as shown in Fig. 8.8. It is obvious that there can be no abrupt change in profile at this point of intersection, but that one curve must merge gradually into the other with some kind of transition, as shown by the experimental points.

When studying such velocity profiles, the quantity $\sqrt{\tau_0/\rho}$ frequently occurs. Because it has the dimensions of velocity, it has been named the ***shear-stress velocity*** (or ***friction velocity***) u_*, although it is not a flow velocity.

When the flow in a circular pipe is entirely laminar, the velocity profile has been shown to be a parabola (Fig. 8.3). But when there is only an extremely thin film closest to the wall where viscous shear dominates, the velocity profile in it can scarcely be distinguished from a straight line. If we ignore the momentary fluctuations in this viscous sublayer, with such a linear velocity profile, Eq. (2.9) at the wall becomes

$$\tau_0 = \mu \frac{u}{y}, \qquad \text{or} \qquad \frac{\tau_0}{\rho} = \frac{\nu u}{y}$$

But from the definition of u_*, we have $u_*^2 = \tau_0/\rho$, so, by eliminating τ_0/ρ, we obtain

$$\frac{\nu u}{y} = u_*^2$$

or

$$\frac{u}{u_*} = \frac{yu_*}{\nu} \tag{8.25}$$

This linear relation for $u(y)$ has been found to approximate experimental data well in the range $0 \leqslant yu_*/\nu \leqslant 5$. If we call this imprecise but commonly accepted upper limiting value of y the thickness of the viscous sublayer, δ_l, then $\delta_l = 5\nu/u_*$.

The transition zone may be said to extend from a to c in Fig. 8.8. For the latter point, the value of y has been estimated to be about $70\nu/u_*$ or $14\delta_l$. Beyond this, the flow is so turbulent that viscous shear is negligible.

Noting from Eq. (8.14) that

$$u_* = \sqrt{\frac{\tau_0}{\rho}} = V\sqrt{\frac{f}{8}} \tag{8.26}$$

and that the Reynolds number $\mathbf{R} = DV/\nu$, we see that when $yu_*/\nu = 5$, or $y = \delta_l$,

$$\delta_l = \frac{14.14\nu}{V\sqrt{f}} = \frac{14.14D}{\mathbf{R}\sqrt{f}} \tag{8.27}$$

From this, we see that the higher the velocity or the lower the kinematic viscosity, the thinner is the viscous sublayer. Thus, for a given constant pipe diameter, the thickness of the viscous sublayer decreases as the Reynolds number increases.

Now we can consider what is meant by a smooth wall and a rough wall. There is no such thing in reality as a mathematically smooth surface. But if the irregularities on any actual surface are such that the effects of the projections do not pierce through the viscous sublayer (Fig. 8.8), the surface is ***hydraulically smooth*** from the fluid-mechanics viewpoint. If the effects of the projections extend beyond the sublayer, the laminar layer is broken up and the surface is no longer hydraulically smooth. If the surface roughness projections are large enough to protrude right through the transition layer, it is totally broken up. The resulting flow is completely turbulent, known as ***fully rough***, and friction is independent of Reynolds number. The significance of this is discussed in the next section. If the roughness projections protrude only partially into the transition layer, the flow is said to be ***transitionally rough***, and there is a moderate Reynolds number effect.

To be more specific, if e is the equivalent height of the roughness projections then for $eu_*/\nu < 5$ (or $e < \delta_l$) the surface roughness is completely

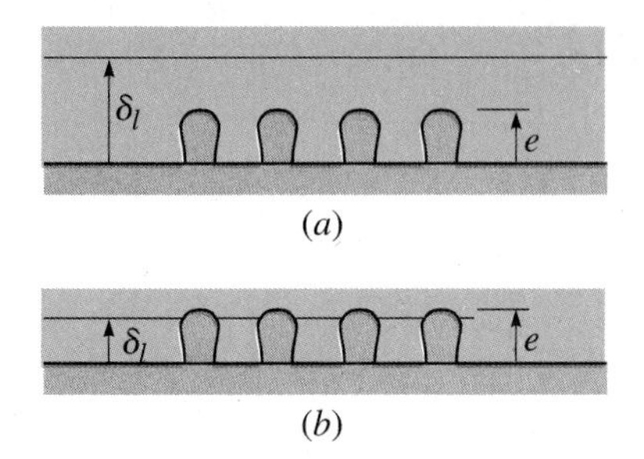

Figure 8.9
Turbulent flow near a boundary. (*a*) Low **R**, $\delta_l > e$; the pipe behaves as a smooth pipe. (*b*) Relatively high **R**, $\delta_l < e$; if $\delta_l < \frac{1}{14}e$, the pipe behaves as a fully rough pipe.

buried in the viscous sublayer, the roughness has no effect on friction, and the pipe is hydraulically smooth. If $eu_*/\nu > 70$ (or $e > 14\delta_l$), the pipe will behave as fully rough. Between these ranges, i.e., when the roughness projections are such that $5 \leqslant eu_*/\nu \leqslant 70$ (or $\delta_l < e < 14\delta_l$), the pipe will behave in a transitional mode, neither hydraulically smooth nor fully rough.

Because the thickness of the viscous sublayer in a given pipe will decrease with an increase in Reynolds number, we see that the same pipe may be hydraulically smooth at low Reynolds numbers and rough at high Reynolds numbers. Thus even a relatively smooth pipe may behave as a rough pipe if the Reynolds number is high enough. It is also apparent that, with increasing Reynolds number, there is a gradual transition from smooth to rough pipe flow. These concepts are depicted schematically in Fig. 8.9.

EXERCISES

8.9.1 Compute δ_l for the data of Sample Prob. 8.3.

8.9.2 Water in a pipe ($f = 0.018$) is at a temperature of 70°F. (*a*) If the mean velocity is 14 fps, what is the nominal thickness δ_l of the viscous sublayer? (*b*) What will δ_l be if the velocity is increased to 24 fps and f does not change?

8.9.3 Water in a pipe ($f = 0.015$) is at a temperature of 15°C. (*a*) If the mean velocity is 3.5 m/s, what is the nominal thickness δ_l of the viscous sublayer? (*b*) What will δ_l be if the velocity is increased to 5.8 m/s and f does not change?

8.9.4 For the data of Exer. 8.9.2(*a*), what is the distance from the wall to the assumed limit of the transition region where true turbulent flow begins?

8.9.5 For the data in Exer. 8.9.3(*a*), what is the distance from the wall to the assumed limit of the transition region where true turbulent flow begins?

8.9.6 Water at 50°C flows in a 15-cm-diameter pipe with $V = 6.5$ m/s and $e = 0.14$ mm. Head loss measurements indicate that $f = 0.020$. (*a*) What is the thickness of the viscous sublayer? (*b*) Is the pipe behaving as a fully rough pipe?

8.10 VELOCITY PROFILE IN TURBULENT FLOW

Prandtl reasoned that turbulent flow in a pipe is strongly influenced by the flow phenomena near the wall. In the vicinity of the wall, $\tau \approx \tau_0$. He assumed that the mixing length l near the wall was proportional to the distance from the wall, that is, $l = Ky$. It has been determined experimentally that K has a value of 0.40. Using these assumptions and applying Eq. (8.24), we get

$$\tau \approx \tau_0 = \rho l^2\left(\frac{du}{dy}\right)^2 = \rho K^2 y^2\left(\frac{du}{dy}\right)^2$$

or

$$du = \frac{1}{K}\sqrt{\frac{\tau_0}{\rho}}\frac{dy}{y} = \frac{u_*}{K}\frac{dy}{y}$$

from which

$$u = 2.5\, u_* \ln y + C$$

The constant C may be evaluated by noting that $u = u_{max}$ = the centerline velocity when $y = r_0$ = the pipe radius. Substituting the expression for C, we get

$$\frac{u_{max} - u}{u_*} = 2.5 \ln \frac{r_0}{y} \tag{8.28}$$

This is known as the velocity defect law, because $u_{max} - u$ is called the velocity defect. Replacing y by $r_0 - r$, and changing the base e logarithm (ln) to a base 10 logarithm (log), the equation becomes

$$u = u_{max} - 2.5u_* \ln \frac{r_0}{r_0 - r} = u_{max} - 5.76u_* \log \frac{r_0}{r_0 - r} \tag{8.29}$$

Although this equation is derived by assuming certain relations very near to the wall, it has been found to hold almost as far out as the pipe axis.

Starting with the derivation of Eq. (8.24), this entire development is open to argument at nearly every step. But the fact remains that Eq. (8.29) agrees very closely with actual measurements of velocity profiles for both smooth and rough pipes. However, there are two zones in which the equation is defective. At the axis of the pipe, du/dy must be zero. But Eq. (8.29) is logarithmic and does not have a zero slope at $r = 0$, and hence the equation gives a velocity profile with a sharp point (or cusp) at the axis, whereas in reality it is rounded at the axis. This discrepancy affects only a very small area and involves very slight error in computing the rate of discharge when using Eq. (8.29).

Equation (8.29) is also not applicable very close to the wall. In fact, it indicates that when $r = r_0$, the value of u is minus infinity. The equation indicates that $u = 0$, not at the wall, but at a small distance from it, shown as y_1 in Fig. 8.8. However, this discrepancy is well within the confines of the viscous sublayer, where the equation is not supposed to apply, and where we have Eq. (8.25). In the intervening transition or overlap zone (Fig. 8.8), where both viscous and turbulent shear are important, experimental velocity profile data are found to follow a logarithmic relation

$$\frac{u}{u_*} = 2.5 \ln\left(\frac{yu_*}{\nu}\right) + 5.0 \tag{8.30}$$

Although Eq. (8.29) is not perfect, it reliably fits the data except for the two small areas mentioned, where it is still close. Thus the rate of discharge may be determined with a high degree of accuracy by using the value of u given by this equation and integrating over the area of the pipe. Thus

$$Q = \int u\, dA = 2\pi \int_0^{r_0} ur\, dr$$

Substituting from the first expression of Eq. (8.29) for u, integrating and dividing by the pipe area πr_0^2, the mean velocity is[7]

$$V = u_{max} - 2.5u_*\left[\ln r_0 - \frac{2}{r_0^2}\int_0^{r_0} r \ln (r_0 - r)\, dr\right]$$

Making use of Eq. (8.26), the above reduces to

$$V = u_{max} - \tfrac{3}{2}\times 2.5u_* = u_{max} - 1.326V\sqrt{f} \tag{8.31}$$

From this last equation, the ***pipe factor***, which is the ratio of the mean to the maximum velocity, may be obtained. It is

$$\frac{V}{u_{max}} = \frac{1}{1 + 1.326\sqrt{f}} \tag{8.32}$$

Using Eq. (8.32) to eliminate u_{max} from Eq. (8.29) and using Eq. (8.26) to eliminate u_*, the result is

$$u = (1 + 1.326\sqrt{f})V - 2.04\sqrt{f}\, V \log \frac{r_0}{r_0 - r} \tag{8.33}$$

[7] The integral results in indeterminate values at $r = r_0$, as we should expect, inasmuch as the equation for u does not really apply close to the wall. However, for all practical purposes, these have been shown to reduce to negligible quantities.

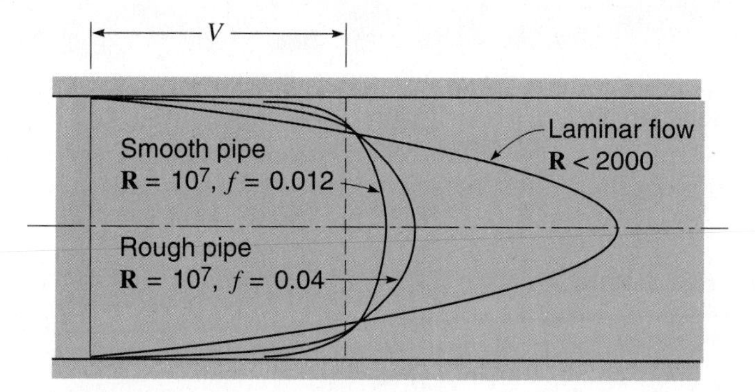

Figure 8.10
Velocity profiles for equal flow rates. The turbulent profiles are plotted from Eq. (8.33).

which enables a velocity profile to be plotted for any mean velocity and any value of f in turbulent flow. In Fig. 8.10 may be seen profiles for both a smooth and a rough pipe plotted from this equation. The only noticeable difference between these and measured profiles is that the latter are more rounded at the axis of the pipe.[8] Of course, the measured turbulent profiles also exhibit turbulent fluctuations everywhere except near the walls.

Comparing the turbulent-flow velocity profiles with the laminar-flow velocity profile (Fig. 8.10) shows the turbulent-flow profiles to be much flatter near the central portion of the pipe and steeper near the wall. It is also noticeable that the turbulent profile for the smooth pipe is flatter near the central section (i.e., blunter) than for the rough pipe. In contrast, the velocity profile in laminar flow is independent of pipe roughness.

As a theoretical equation has now been derived for the velocity profile for turbulent flow in circular pipes, it is also possible to derive equations for the kinetic-energy and momentum-correction factors (Secs. 5.1 and 6.3) when mean velocities are used. Respectively, these equations are[9]

$$\alpha = 1 + 2.7f \tag{8.34a}$$

$$\beta = 1 + 0.98f \tag{8.34b}$$

[8] Although the preceding theory agrees very well with experimental data, it is not absolutely correct throughout the entire range from the axis to the pipe wall, and some slight shifts in the numerical constants could improve agreement with test data. Thus in Eqs. (8.32) and (8.33) the 1.326 may be replaced by 1.44, and in Eq. (8.33) although many writers use 2 instead of 2.04, a better practical value seems to be 2.15.

[9] L. F. Moody, Some Pipe Characteristics of Engineering Interest, *Houille Blanche*, May–June, 1950.

SAMPLE PROBLEM 8.3 The head loss in 200 ft of 6 in-diameter pipe is known to be 25 ft·lb/lb when oil ($s = 0.90$) of viscosity 0.0008 lb·sec/ft^2 flows at 2.0 cfs. Determine the centerline velocity, the shear stress at the wall of the pipe, and the velocity at 2 in from the centerline.

Solution
The first step is to determine whether the flow is laminar or turbulent.

$$V = \frac{Q}{A} = \frac{2}{0.1963} = 10.19 \text{ fps}$$

$$\mathbf{R} = \frac{DV\rho}{\mu} = \frac{0.5(10.19)(0.9 \times 1.94)}{0.0008} = 11{,}120$$

Since $\mathbf{R} > 2000$, the flow is turbulent. Using Eq. (8.10), the friction factor can be found:

$$f = \frac{h_L D(2g)}{LV^2} = \frac{25(0.5)2(32.2)}{200(10.19)^2} = 0.0388$$

From Eq. (8.32), $u_{max} = V(1 + 1.326\sqrt{f}) = 12.85 \text{ fps}$ ***ANS***

Eq. (8.14): $\tau_0 = \dfrac{f\rho V^2}{8} = \dfrac{0.0388(0.9 \times 1.940(10.19)^2}{8} = 0.878 \text{ lb/ft}^2$ ***ANS***

Eq. (8.26): $u_* = V\sqrt{\dfrac{f}{8}} = 10.19\sqrt{\dfrac{0.0388}{8}} = 0.710 \text{ fps}$

Finally, from Eq. (8.29),

$$u_{2\text{ in}} = u_{max} - 5.75u_* \log \tfrac{3}{1} = 12.85 - 1.947 = 10.90 \text{ fps} \qquad \textbf{\textit{ANS}}$$

Note that if the flow had been laminar, the velocity profile would have been parabolic and the centerline velocity would have been twice the average velocity.

EXERCISES

8.10.1 In a 110-cm-diameter pipe velocities are measured as 4.75 m/s on the centerline and 4.36 m/s at $r = 8$ cm. Approximately what is the flow rate?

8.10.2 Oil ($s = 0.91$) with a viscosity of 0.0007 lb·sec/ft^2 flows at a rate of 9 cfs through a 5-in-diameter pipe having $e = 0.047$ in. Find the head loss. Determine the shear stress at the pipe wall and the velocity at 1.5 in from the centerline.

8.10.3 For turbulent flow in a circular pipe, find r/r_0 at the radial distance from the centerline where the mean velocity occurs.

8.11 PIPE ROUGHNESS

Unfortunately, there is as yet no scientific way of measuring or specifying the roughness of commercial pipes. Several experimenters have worked with pipes with artificial roughness produced by various means so that the roughness could be measured and described by geometrical factors, and it has been proved that the friction is dependent not only upon the size and shape of the projections, but also upon their distribution or spacing. Much remains to be done before this problem is completely solved.

The most noteworthy efforts in this direction were made by a German engineer Nikuradse, a student of Prandtl's. He coated several different sizes of pipe with sand grains that had been sorted by sieving so as to obtain different sizes of grain of reasonably uniform diameters. The diameters of the sand grains may be represented by e, which is known as the ***absolute roughness***. In Sec. 8.4 dimensional analysis of pipe flow showed that for a smooth-walled pipe the friction factor f is a function of Reynolds number. A general approach, including e as a parameter, reveals that $f = \phi(\mathbf{R}, e/D)$. The term e/D is known as the ***relative roughness***. In his experimental work Nikuradse had values of e/D ranging from 0.000 985 to 0.0333.

In the case of artificial roughness such as this, the roughness is uniform, whereas in commercial pipes it is irregular both in size and in distribution. However, the roughness of commercial pipe may be described by e, which means that the pipe has the same value of f at a high Reynolds number that would be obtained if a smooth pipe were coated with sand grains of uniform size e.

It has been found for pipes that if the thickness of the viscous sublayer $\delta_l > e$ (i.e., $eu_*/\nu < 5$), the viscous sublayer completely submerges the effect of e (see Sec. 8.9). Prandtl (Secs. 8.7, 8.10), using information from Eq. (8.29) and data from Nikuradse's experiments, developed an equation for the friction factor for such a case:

"Smooth-pipe" flow:

$$\frac{1}{\sqrt{f}} = 2 \log \left(\frac{\mathbf{R}\sqrt{f}}{2.51}\right) \tag{8.35}$$

This equation applies to turbulent flow in any pipe as long as $\delta_1 > e$; when this condition prevails, the flow is known as ***smooth-pipe flow***. The equation has been found to be reliable for smooth pipes for all values of $\mathbf{R}$ over 4000. For such pipes, i.e., drawn tubing, brass, glass, etc., it can be extrapolated with confidence for values of $\mathbf{R}$ far beyond any present experimental values because it is functionally correct, assuming the wall surface to be so smooth that the effects of the projections do not pierce the viscous sublayer, which becomes increasingly thinner with increasing **R**. That this is so is evident from the fact that the formula yields a value of $f = 0$ for $\mathbf{R} = \infty$. This is in accord with the facts, because **R** is infinite for a fluid of zero viscosity, and for such a case f must be zero.

Because of the way that f appears in two places in Eq. (8.35), it is implicit in f and hard to solve; either iteration or a graph of f versus **R** must be used. However, as suggested by Colebrook,[10] it can be approximated by the explicit equation

"Smooth-pipe" flow:

$$\frac{1}{\sqrt{f}} = 1.8 \log \left(\frac{\mathbf{R}}{6.9}\right) \tag{8.36}$$

which differs from Eq. (8.35) by less than ±1.5% for $4000 \leq \mathbf{R} \leq 10^8$.

Blasius[11] has shown that for Reynolds numbers between 3000 and 10^5 the friction factor for a smooth pipe may be expressed approximately as

$$f = \frac{0.316}{\mathbf{R}^{0.25}} \tag{8.37}$$

This can sometimes be used very conveniently to simplify equations. Blasius also found that over the same range of Reynolds numbers, the velocity profile in a smooth pipe is closely approximated by the expression

$$\frac{u}{u_{\max}} = \left(\frac{y}{r_0}\right)^{1/7} \tag{8.38}$$

where $y = r_0 - r$, the distance from the pipe wall. This equation is commonly referred to as the ***seventh-root law*** for turbulent-velocity distribution. Though it is not absolutely accurate, it is useful because it is easy to work with mathematically. At Reynolds numbers above 10^5 an exponent somewhat smaller than $\frac{1}{7}$ must be used to give good results.

At high Reynolds numbers δ_l becomes much smaller, and the roughness elements protrude through the viscous sublayer as in Fig. 8.9*b*. If $\delta_l < \frac{1}{14}e$ (i.e., $eu_*/\nu > 70$), it has been found that the pipe behaves as ***fully rough pipe***, i.e., its friction factor is independent of the Reynolds number. For such

[10] C. F. Colebrook, Turbulent Flow in Pipes, with Particular Reference to the Transition Region between the Smooth and Rough Pipe Laws, *J. Inst. Civil Engrs.* (*London*), Vol. 11, February 1939.

[11] H. Blasius, Das Ähnlichkeitsgesetz bei Reibungsvorgängen in Flüssigkeiten, *Forsch. Gebiete Ingenieurw.*, Vol. 131, 1913.

a case von Kármán found that the friction factor could be expressed as

"Fully-rough-pipe" flow:
$$\frac{1}{\sqrt{f}} = 2\log\left(\frac{3.7}{e/D}\right) \tag{8.39}$$

The values of f from this equation correspond to the right-hand side of the Moody chart (Fig. 8.11), where the curves become horizontal. These values are sometimes referred to as $f_{\min}$.

In the interval where $e > \delta_l > \frac{1}{14}e$ (i.e., $5 < eu_*/\nu < 70$) neither smooth flow Eq. (8.35) nor fully rough flow Eq. (8.39) applies. In 1939 Colebrook[12] combined Eqs. (8.35) and (8.39) to yield

Turbulent flow, all pipes:
$$\frac{1}{\sqrt{f}} = -2\log\left(\frac{e/D}{3.7} + \frac{2.51}{\mathbf{R}\sqrt{f}}\right) \tag{8.40}$$

Besides providing a good approximation to conditions in the intermediate range, for $e = 0$ the Colebrook equation reduces to the smooth-pipe equation (8.35), and for large $\mathbf{R}$ it reduces to the rough-pipe equation (8.39). Values of friction factor f that it predicts are generally accurate to within 10–15% of experimental data. This equation is so useful that it has long been the accepted design formula for turbulent flow; however, it has one major disadvantage. Like Eq. (8.35), it is implicit in f, which makes it inconvenient to use to evaluate f. More recently, in 1983 Haaland[13] combined Eqs (8.36) and (8.39) to provide another approximation,

Turbulent flow, all pipes:
$$\frac{1}{\sqrt{f}} = -1.8\log\left[\left(\frac{e/D}{3.7}\right)^{1.11} + \frac{6.9}{\mathbf{R}}\right] \tag{8.41}$$

which has the great advantage of being explicit in f, it has the same asymptotic behavior, and it differs from Eq. (8.40) by less than ±1.5% for $4000 \leqslant \mathbf{R} \leqslant 10^8$.

EXERCISES

8.11.1 Using the implicit equation (8.35), the approximate equation (8.36), and Blasius' equation (8.37), solve for the smooth-pipe friction factor f using Reynolds numbers of (*a*) 4000, (*b*) 20,000, and (*c*) 10^5. (*d*) For which of these three values do the equations show the most variation in f?

8.11.2 Substitute into Eq. (8.40) the given and computed data of Sample Prob. 8.5*a*. How well does the right-hand side of the equation agree with the left-hand side?

8.11.3 Repeat Exer. 8.11.2 using Eq. (8.41).

[12] See Footnote 10.

[13] S. E. Haaland, Simple and Explicit Formulas for the Friction Factor in Turbulent Pipe Flow, *J. Fluids Eng.*, Vol. 105, March 1983.

■: Programmed computing aids (Appendix B) could help solve problems marked with this icon.

8.12 CHART FOR FRICTION FACTOR

The preceding equations for f have been very inconvenient to use in a number of circumstances, to be discussed further in coming sections, and especially so before Haaland's equation appeared. In the past this inconvenience was partly overcome by reading numerical values from a chart (Fig. 8.11), prepared by Moody[14] in 1944. The chart, often referred to as the ***Moody diagram***,[15] is based on the best information available, and has been plotted with the aid of Eqs. (8.20) and (8.40). All the quantities involved are dimensionless, so the chart may be used for both BG and SI unit systems. For convenience, BG values of DV (diameter times velocity) for water at 60°F and similar SI values for water at 15°C have been placed across the top of the chart to save the need to compute Reynolds number for those common cases.

The Moody chart, and the various flow conditions that it represents, may be divided into four zones: the ***laminar flow zone***; a ***critical zone*** where values are uncertain because the flow might be either laminar or turbulent; a ***transition zone***, where f is a function of both Reynolds number and relative pipe roughness; and a ***zone of complete turbulence*** (fully rough pipe flow), where the value of f is independent of Reynolds number and depends solely upon the relative roughness.

There is no sharp line of demarcation between the transition zone and the zone of complete turbulence. The dashed line of Fig. 8.11 that separates the two zones was suggested by R. J. S. Pigott; the equation of this line is $\mathbf{R} = 3500/(e/D)$. On the right-hand side of the chart the given values of e/D correspond to the curves and not to the grid. Note how their spacing varies. The lowest of the curves in the transition zone is the smooth-pipe curve given by Eqs. (8.36) and (8.37); notice how many of the other curves blend asymptotically into the smooth-pipe curve.

EXERCISES

8.12.1 Water at 20°C flows through a 15-cm-diameter pipe with $e = 0.015$ mm. (*a*) If the mean velocity is 5 m/s, what is the nominal thickness δ_l of the viscous sublayer? (*b*) What will δ_l be if the velocity is increased to 6.2 m/s?

8.12.2 A straight, new 48-in-diameter asphalted cast-iron pipe 700 ft long carries 78°F water at an average velocity of 12 fps. (*a*) Using the value of f as determined from Fig. 8.11, find the shear force on the pipe. (*b*) What will be the shear force if the average velocity is reduced to 4.2 fps?

---

[14] L. F. Moody, Friction Factors for Pipe Flows, *ASME Trans.*, Vol. 66, 1944. Lewis F. Moody (1880–1953), an eminent American engineer and professor, also contributed greatly to our understanding of similitude and cavitation as applied to hydraulic machinery (Secs. 15.4, 15.10, 16.10, and 16.12).

[15] Fig. 8.11 is also referred to as a ***Stanton diagram***, since Stanton first proposed such a plot.

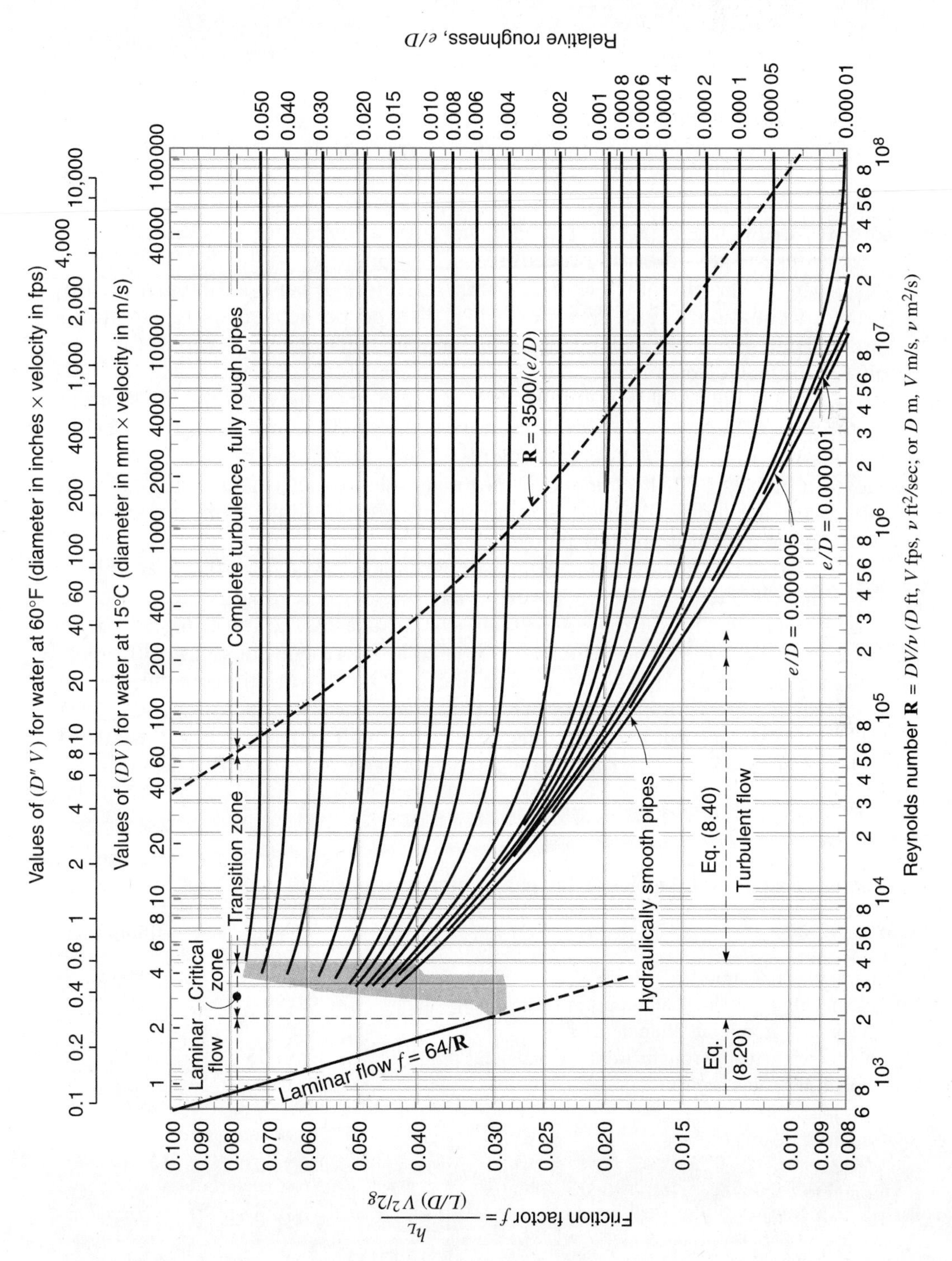

Figure 8.11
Moody chart for pipe friction factor (Stanton diagram).

8.13 SOLUTION OF PIPE FLOW PROBLEMS BY TRIALS

As we have already noted in the preceding two sections, the equations governing pipe flow are often implicit in form, and therefore do not readily permit direct solution. For example, although Haaland's equation (8.41) is explicit in f, it is still implicit in V and D, since these are involved in both **R** and in f through Eq. (8.10). The classical way to overcome this difficulty has been to use trial-and-error procedures in combination with the Moody chart described in the previous section. With the more recent availability of the Haaland equation, these procedures can also be performed using equations alone. Both methods will be described in this section, with applications to different types of problems.

To solve any problem (other than smooth-pipe flow) using the friction factor f, a value of e will be required, and this may be obtained from Table 8.1. As the ratio of e/D is dimensionless, any units may be used, provided they are the same for both e and D. Note also that exact values of the internal pipe diameter should be used. Exact diameters often differ from the nominal sizes, particularly for smaller pipes.

With regard to the values of e, it must be observed that these are given here for new, clean pipes, and even in such cases there may be considerable variation in the values. Consequently, in practical cases, the value of f may be in error by ±5% for smooth pipes and by ±10% for rough ones. For old pipes, values of e may be much higher, but there is much variation in the degree with which pipe roughness increases with age, since so much depends upon the nature of the fluid being transported. In small pipes there is the added factor that deposits materially reduce the internal diameter. In addition, the

Table 8.1
Values of absolute roughness e for new commercial pipes

	Feet	Millimeters
Glass, plastic (smooth)	0.0	0.0
Drawn tubing, brass, lead, copper, centrifugally spun cement, bituminous lining, transite	0.000 005	0.0015
Commercial steel, wrought iron, welded-steel pipe	0.000 15	0.046
Asphalt-dipped cast iron	0.000 4	0.12
Galvanized iron	0.000 5	0.15
Cast iron, average	0.000 85	0.25
Wood stave	0.0006–0.003	0.18–0.9
Concrete	0.001–0.01	0.3–3
Riveted steel	0.003–0.03	0.9–9

Note: $\dfrac{e}{D} = \dfrac{e \text{ in feet}}{D \text{ in feet}} = 12 \times \dfrac{e \text{ in feet}}{D \text{ in inches}} = \dfrac{e \text{ in mm}}{D \text{ in mm}} = 10^{-1} \times \dfrac{e \text{ in mm}}{D \text{ in cm}}.$

effect of the roughness of pipe joints may increase the value of f substantially. Hence judgment must be used in estimating a value of e, and consequently of f.

Less frequently, problems of fully rough pipe flow (complete turbulence) occur, where the friction is directly proportional to V^2 and is independent of the Reynolds number, i.e., f = constant. Then values of f may be determined from the relative roughness alone [Eq. (8.39)]. But most practical problems fall within the transition zone, where conditions also depend on the Reynolds number, and so there it is necessary to also have a definite value of **R**. But when V or D is unknown, so is **R**. The solution procedure varies with the type of problem, and most single-pipe flow problems can be categorized into one of the following three types:

Type	Find	Given	
1. Head-loss problem	h_L	D, Q or V,	and g, L, e, ν
2. Discharge problem	Q or V	D, h_L,	and g, L, e, ν
3. Sizing problem	D	Q, h_L,	and g, L, e, ν

Here g is the gravitational acceleration, L is the pipe length, and ν is the kinematic viscosity. Of course, μ and ρ may be given instead of ν, in which case ν may be obtained from Eq. (2.11).

Type 1 problems are the simplest. If Q is given, V may easily be obtained from the continuity equation (4.7), where $A = \frac{1}{4}\pi D^2$. The Reynolds number and e/D can then easily be obtained from the given data. Figure 8.11 is next entered vertically with **R** and along a curve (possibly interpolated) for e/D from the right, to identify an operating point for which the friction factor f can be read off horizontally to the left. Alternatively, f can be computed from Eq. (8.41). With this value of f, the friction head loss h_L can be directly computed from Eq. (8.10).

Because either V or D is unknown in Type 2 and 3 problems, the Reynolds number **R** is not known at the outset and so a direct solution is not possible. However, we notice in Fig. 8.11 that the value of f changes very slowly with large changes in **R**. So the problem may be solved most effectively by assuming an initial value of f, and then obtaining the final solution by successive trials (trial and error). If D is known (Type 2), the rough-pipe f value ($f_{\min}$) given on the right-hand side of Fig. 8.11 or by Eq. (8.39) provides a good starting point. If D is not known (Type 3), a value near the middle of the f range on Fig. 8.11, such as 0.03, makes a good start. Note that if values of some other variable besides f are assumed, convergence is usually much slower.

For Type 2 problems, from Eq. (8.10) we write $V = K/\sqrt{f}$, where $K = \sqrt{2gDh_L/L}$ is known. Assuming an f then yields a V, which enables the diagram to be entered or Eq. (8.41) to be used as before to obtain an improved value of f. If this is different from the assumed f, the procedure must be repeated assuming the just-obtained value, and successively repeated until the two values converge. This usually only requires two or three trials, by which time all the values are correct, including the required value of V.

For Type 3 problems, since D is unknown, neither e/D nor $\mathbf{R}$ are known initially. We substitute $V = 4Q/(\pi D^2)$ into Eq. (8.10) and rearrange it to obtain $D = (fK)^{1/5}$, where $K = 8LQ^2/(\pi^2 g h_L)$ is known. Then assuming an f yields a D, which allows us to proceed with repetition in a manner similar to that used for Type 2.

Experience has shown that new users of the Moody diagram frequently misread it. This most probably occurs because none of the scales are linear, and because the intervals between grid lines and chart curves keep changing. So special care is needed in reading the chart, and it is best to confirm an interpolated value by comparing it with nearby grid values or curves in both directions.[16]

The following sample problems illustrate the use of trial and error.

SAMPLE PROBLEM 8.4 A 20-in-diameter galvanized iron pipe 2 miles long carries 4 cfs of water at 60°F. Find the friction head loss: (*a*) using Fig. 8.11 and the Reynolds number; (*b*) using Fig. 8.11 and its $D''V$ scale; (*c*) using only a basic scientific calculator,[17] without Fig. 8.11.

Solution

This is a Type 1 problem, to find h_L. From Table 8.1 for galvanized iron,

$$e = 0.0005 \text{ ft}; \quad e/D = 0.0005(12)/20 = 0.0003$$

$$L = 2 \text{ mi}(5280 \text{ ft/mi}) = 10{,}560 \text{ ft.} \quad A = \pi D^2/4 = (\pi/4)(20/12)^2 = 2.18 \text{ ft}^2$$

$$V = Q/A = 4/2.18 = 1.833 \text{ fps}$$

Table A.1 for water at 60°F gives

$$\nu = 1.217 \times 10^{-5} \text{ ft}^2/\text{sec}$$

$$\mathbf{R} = \frac{DV}{\nu} = \frac{(20/12)1.833}{1.217 \times 10^{-5}} = 2.51 \times 10^5 \quad (>\mathbf{R}_{\text{crit}}, \text{ i.e., flow is turbulent})$$

(*a*) Enter Fig. 8.11 at the right-hand side with $e/D = 0.0003$, by interpolating between 0.0002 and 0.0004; note that the e/D spacing varies. Follow this (unplotted) e/D curve to the left until it crosses a vertical line at $\mathbf{R} = 2.51 \times 10^5$ (*caution*: this is between 10^5 and 10^6). For this operating point, reading horizontally to the left, $f = 0.0172$.

$$\text{Eq. (8.10):} \quad h_L = \frac{fL}{D}\frac{V^2}{2g} = \frac{0.0172(10{,}560)(1.833)^2}{(20/12)2(32.2)} = 5.69 \text{ ft} \qquad \textit{ANS}$$

[16] Charts involving the same functional relations may be plotted with different coordinates from those in Fig. 8.11 and may be more convenient for certain specific purposes, but it is believed that the form shown is best both for instruction purposes and for general use.

[17] A basic scientific calculator is here defined to be one that is not programmable and does not have automatic equation solving capabilities.

(*b*) $D''V = 20(1.833) = 36.7$. Find this value on the scale across the top of Fig. 8.11; note that this scale is varying. Find where the (interpolated) curve for $e/D = 0.0003$ crosses the vertical line at $D''V = 36.7$. From this point, read horizontally to the left, to find $f = 0.0172$. Compute h_L as for part (*a*).

Note: From the operating point on Fig. 8.11, we see that flow conditions are in the transition zone of turbulent flow, which is typical.

(*c*) Eq. (8.41): $$\frac{1}{\sqrt{f}} = -1.8 \log\left[\left(\frac{0.0003}{3.7}\right)^{1.11} + \frac{6.9}{2.51\times 10^5}\right] = 7.65$$

from which $$f = 0.01709$$

Eq. (8.10): $$h_L = 0.01709\frac{10{,}560(1.883)^2}{(20/12)2(32.2)} = 5.65 \text{ ft} \qquad \textit{ANS}$$

Sample Problem 8.5 Water at 20°C flows in a 500-mm-diameter welded steel pipe. If the friction loss gradient is 0.006, determine the flow rate: (*a*) using Fig. 8.11; (*b*) using only a basic scientific calculator,[17] without Fig. 8.11.

Solution

This is a Type 2 problem, to find Q.

Table 8.1 for welded steel: $e = 0.046$ mm; $e/D = 0.046/500 = 0.000\,092$

Table A.1 at 20°C: $\nu = 1.003\times 10^{-6}$ m²/s, $h_L/L = 0.006$ is given

From Eq. (8.10): $$\frac{h_L}{L} = \frac{f}{D}\frac{V^2}{2g}, \quad \text{i.e.} \quad 0.006 = \frac{fV^2}{0.5(2)9.81}$$

from which $V = 0.243/f^{1/2}$.

(*a*) Fig. 8.11 for $e/D = 0.000\,092$: $f_{\min} \approx 0.0117$.

Try $f = 0.0117$. Then $V = 0.243/(0.0117)^{1/2} = 2.25$ m/s.

Eq. (8.1): $$\mathbf{R} = \frac{DV}{\nu} = \frac{0.5(2.25)}{1.003\times 10^{-6}} = 1.120\times 10^6 \quad \text{(turbulent flow)}$$

Figure 8.11 with $e/D = 0.000\,092$ and $\mathbf{R} = 1.120\times 10^6$: $f = 0.0131$. Assumed and obtained f values are different, so we must try again. Tabulating this and subsequent trials:

Try f	V, m/s	**R**	Obtained f	
0.0117	2.25	1.120×10^6	0.0131	Try again
0.0131	2.12	1.059×10^6	0.0131	Converged!

f values now agree, so we have the true operating point.

$$Q = AV = (\pi/4)D^2V = (\pi/4)(0.5)^2 2.12 = 0.416 \text{ m}^3\text{/s} \qquad \textit{ANS}$$

Note: Care is needed to read Fig. 8.11 accurately.

(*b*) Eq. (8.39) for $e/D = 0.000\,092$: $f_{min} = 0.01179$. Calculate V and **R** as above, then obtain improved f from Eq. (8.41). Tabulating this and subsequent trials:

Try f	V, m/s	**R**	Obtained f	
0.01179	2.23	1.114×10^6	0.013 09	Try again
0.01309	2.12	1.057×10^6	0.013 15	Converged!

$$Q = AV = (\pi/4)(0.5)^2(2.12) = 0.416 \text{ m}^3/\text{s} \qquad \textbf{\textit{ANS}}$$

SAMPLE PROBLEM 8.6 A galvanized iron pipe 18,000 ft long is to convey ethyl alcohol ($\nu = 2.3 \times 10^{-5}$ ft²/sec) at a rate of 135 gpm. If the friction head loss is to be 215 ft, determine the pipe size theoretically required: (*a*) using Fig. 8.11; (*b*) using only a basic scientific calculator, without Fig. 8.11.

Solution

This is a Type 3 problem, to find D.

$$Q = 135 \text{ gpm}(2.23 \text{ cfs}/1000 \text{ gpm}) = 0.301 \text{ cfs}$$

Table 8.1 for galvanized iron: $e = 0.0005$ ft.

$$\frac{e}{D} = \frac{0.0005}{D}, \qquad V = \frac{Q}{A} = \frac{0.301}{\pi D^2/4} = \frac{0.383}{D^2}$$

Eq. (8.1): $$\mathbf{R} = \frac{DV}{\nu} = \frac{D}{2.3 \times 10^{-5}}\left(\frac{0.383}{D^2}\right) = \frac{16{,}666}{D}$$

Eq. (8.10): $$215 = h_L = f\frac{18{,}000}{2D(32.2)}\left(\frac{0.383}{D^2}\right)^2, \text{ from which } D^5 = 0.1910f$$

(*a*) Start by assuming a mid-range value of f

Try f	*D*, ft	*e*/*D*	**R**	Chart f	
0.0300	0.356	0.001404	4.68×10^4	0.0253	Try again
0.0253	0.344	0.001453	4.84×10^4	0.0253	Converged!

Values of f now agree, so we have the true operating point. Convergence is rapid!

$$D = 0.344 \text{ ft} = 4.13 \text{ in} \qquad \textbf{\textit{ANS}}$$

(*b*) Start by assuming a mid-range value of f. Calculate V and **R** as above, then obtain improved f from Eq. (8.41). Tabulating this and subsequent trials:

Try f	*D*, ft	*e*/*D*	**R**	Eq. (8.41) f	
0.0300	0.356	0.001404	46,800	0.02501	Try again
0.0250	0.343	0.001456	48,500	0.02504	Converged!

$$D = 0.343 \text{ ft} = 4.12 \text{ in} \qquad \textbf{\textit{ANS}}$$

As noted earlier, in practice there can be a considerable amount of uncertainty in the size of the absolute roughness *e*. So it is important to have some idea what effects changes in *e* will have on h_L, Q, and D. From Eq. (8.10) and continuity, it follows that $h_L \propto f$, $Q \propto f^{-1/2}$, and $D \propto f^{1/5}$. The variation of f with e may be seen in Fig. 8.11 or Eq. (8.40). As an example, in Sample Prob. 8.4, if e had been 20% larger then the h_L would have been 2.3% larger. This change in the head loss would be larger for larger f (or larger e/D), and vice versa. Changes in Q and D will be smaller than those in h_L.

EXERCISES

8.13.1 Compute the friction head per 100 ft of 3-in-diameter pipe for a Reynolds number of 50,000 if (*a*) the flow is laminar (achievable with great care); (*b*) the flow is turbulent in a smooth pipe; (*c*) the flow is turbulent in a rough pipe with $e/D = 0.05$. Consider two situations: one where the fluid is 70°F water, the other where the fluid is SAE 10 (Western) oil at 150°F.

8.13.2 California crude oil, warmed until its kinematic viscosity is 0.0004 ft^2/sec and its specific weight is 53.45 lb/ft^3, is pumped through a 3-in pipe ($e = 0.001$ in). (*a*) For laminar flow with $\mathbf{R} = \mathbf{R}_{\text{crit}} = 2000$, what would be the loss in energy head in pounds per square inch per 1000 ft of pipe? (*b*) What would be the loss in head per 1000 ft if the velocity were three times the value in (*a*)?

8.13.3 Water at 50°F flowing through 80 ft of 4-in-diameter average cast-iron pipe causes a head loss of 0.27 ft. Find the flow rate.

8.13.4 When gasoline with a kinematic viscosity of 5×10^{-7} m^2/s flows in a 20-cm-diameter smooth pipe, the head loss is 0.43 m per 100 m. Find the flow rate.

8.14 RIGOROUS SOLUTION OF PIPE FLOW PROBLEMS

With progress in technology and solution techniques, rigorous methods have become available for solving pipe-flow problems that involve the complicated equations for roughness described in Sec. 8.11. These methods provide accurate alternatives to the somewhat tedious manual trial-and-error methods described in the previous section. They are particularly attractive in comparison to the chart-based methods, which are less precise and in which the chart may easily be misread.

One important new solution method that has recently become available makes use of preprogrammed equation solvers available on certain more advanced programmable scientific calculators and in some mathematics software packages (see Appendix B). The equation, explicit or implicit and in any form, must first be entered into the calculator's memory or into a computer file. Such calculators now often have "equation writers," which present the equation on the screen in the same form as we write it on paper. Some of the mathematics software packages also present the equation in the same form. In either case the equations can be saved, and they can be edited if necessary. Values are then assigned to each of the known variables, and estimated values are assigned to the unknown variables.

When the calculator or software solves for the unknown variable(s), it uses an internally programmed trial-and-error procedure to find them, to high precision. Of great practical convenience is the fact that, once everything is stored in memory, it is very easy to change which of the variables are to be treated as the unknowns. An important difference between the advanced programmable calculators and the mathematics software is that at present (1997) the calculators can solve only one nonlinear (and implicit) equation for one unknown, whereas some software like Mathcad can automatically solve a system of many nonlinear equations in an equal number of unknowns.

For people who do not have access to advanced programmable calculators or computers with mathematics software, ways have been found to solve each of the three types of pipe-flow problem mentioned in Sec. 8.13 without trial and error, so that they may be evaluated directly on a basic scientific calculator.[18] This of course requires an explicit equation in the sought quantity. Such ways, and methods using automated equation solvers, are described further below.

The head-loss problem (Type 1). The Haaland equation (8.41) now makes possible a direct computation of f, from which h_L may be found using Eq. (8.10). This is described more fully in Sec. 8.13 and by Sample Prob. 8.4. For a faster solution f may be eliminated between Eqs. (8.10) and (8.41).

Equation solvers are not necessary for Type 1 problems, but they may be used as described below for Types 2 and 3. They are necessary if the implicit Colebrook equation (8.40) is to be used.

The discharge problem (Type 2). A single equation for the velocity (or discharge) may be obtained as follows. First we rearrange Eq. (8.10) to obtain

$$\frac{1}{\sqrt{f}} = V\sqrt{\frac{L}{2gDh_L}}$$

Then, by substituting this and $\mathbf{R} = DV/\nu$ into the Colebrook equation (8.40) and rearranging, we get

Turbulent flow, Type 2:
$$V = -2\sqrt{\frac{2gDh_L}{L}}\log\left(\frac{e/D}{3.7} + \frac{2.51\nu}{D}\sqrt{\frac{L}{2gDh_L}}\right) \qquad (8.42a)$$

---

[18] A basic scientific calculator is here defined to be one that is not programmable and does not have automatic equation solving capabilities.

This equation is explicit in V, which fortunately results because the Vs from $\mathbf{R}$ and $\sqrt{f}$ cancel in the last term of Eq. (8.40) [the same substitution into the Haaland equation (8.41) does not produce such a desirable result]. If Q is required rather than V, we know that from the continuity equation (4.7) that $Q = \pi D^2 V/4$, so, by eliminating V, Eq. (8.42a) becomes

Turbulent flow, Type 2:
$$\frac{4Q}{\pi D^2} = -2\sqrt{\frac{2gDh_L}{L}}\log\left(\frac{e/D}{3.7} + \frac{2.51\nu}{D}\sqrt{\frac{L}{2gDh_L}}\right) \quad (8.42b)$$

Using Eq. (8.42), Q or V may be calculated directly with a basic scientific calculator. Remember to use Eq. (8.1) to confirm that $\mathbf{R}$ is in the turbulent range. As a convenience for this purpose, note that in some programmable calculators equations can be "linked", so that the same numerical values will be automatically used for the same variables in each linked equation. If $\mathbf{R}$ is in the laminar range, i.e., less than 2000, V or Q must instead be found from Eq. (8.19) rearranged.

When using mathematics software such as Mathcad to solve the discharge problem, the four governing equations are entered, in their familiar forms, into a "solve block." These equations are the Colebrook equation (8.40), the head loss equation (8.10), Eq. (8.1) for the Reynolds number, and, from Eq. (4.7), the continuity equation $Q = \pi D^2 V/4$. Values must be assigned to the known variables g, ν, h_L, L, e, and D, and initial estimates must be provided for the unknown variables f (0.03 suggested), V, $\mathbf{R}$ (> 2000), and Q. The solver then simply solves the four equations (some nonlinear, some implicit) for the four unknowns.

***The sizing problem* (*Type* 3).** When using mathematics software such as Mathcad, the same procedure is followed as for the discharge problem just described, with the one difference that Q and D are interchanged on the lists of known and unknown variables.

With an equation solver on a programmable calculator, Eq. (8.42b) may be solved for D directly, even though it is implicit in D. The value of $\mathbf{R}$ must then be checked.

If only a basic scientific calculator is available, iteration may be eliminated by reformulation as follows. The dimensionless quantity

$$\mathbf{N}_1 = f\mathbf{R}^5$$

is seen to be independent of the unknown D; substituting from Eqs. (8.10) and (8.1) while also writing the unknown V as $4Q/\pi D^2$ gives

$$\mathbf{N}_1 = \frac{h_L D 2g}{L}\left(\frac{\pi D^2}{4Q}\right)^2\left(\frac{D}{\nu}\frac{4Q}{\pi D^2}\right)^5 = \frac{128g}{\pi^3}\frac{h_L}{L}\frac{Q^3}{\nu^5} \quad (8.43)$$

$\mathbf{N}_1$ is a known quantity. The relative roughness can also be transformed into a known dimensionless quantity by dividing it by $\mathbf{R}$, i.e.

$$\mathbf{N}_2 = \frac{e/D}{\mathbf{R}} = \frac{e}{D}\left(\frac{\nu}{DV}\right) = \frac{e}{D}\left(\frac{\nu}{D}\right)\frac{\pi D^2}{4Q} = \frac{\pi e \nu}{4Q} \quad (8.44)$$

We may use $\mathbf{N}_1$ to eliminate f from Eq. (8.20) to obtain

Laminar flow,
Type 3:

$$\mathbf{R} = \left(\frac{\mathbf{N}_1}{64}\right)^{0.25} \tag{8.45}$$

and we may use $\mathbf{N}_1$ and $\mathbf{N}_2$ to eliminate f and e/D from the Colebrook equation (8.40) to obtain

$$\mathbf{R}^{2.5} = -2\mathbf{N}_1^{0.5} \log\left(\frac{\mathbf{N}_2\mathbf{R}}{3.7} + \frac{2.51}{\mathbf{N}_1^{0.5}}\mathbf{R}^{1.5}\right) \tag{8.46}$$

Because $\mathbf{R}$ occurs in three places, Eq. (8.46) appears to be not a very useful result. However, a plot of $\mathbf{N}_1$ versus $\mathbf{R}$ is found to collapse the Moody chart into a very narrow band that is closely approximated by the formula $\mathbf{R} \approx 1.43\mathbf{N}_1^{0.208}$. This can be substituted into the right-hand side of Eq. (8.46) to result in the more accurate equation:

Turbulent flow,
Type 3:

$$\mathbf{R}^{2.5} = -2\mathbf{N}_1^{0.5} \log\left(\frac{\mathbf{N}_2\mathbf{N}_1^{0.208}}{2.59} + \frac{4.29}{\mathbf{N}_1^{0.188}}\right) \tag{8.47}$$

We see that this reformulation has converted the Colebrook equation into an explicit form.

Comparison of the two Reynolds numbers from Eqs. (8.45) and (8.47) with $\mathbf{R}_{crit} = 2000$ will indicate which is applicable for a given Type 3 problem.

Finally, from Eq. (8.1) combined with $V = Q/A = 4Q/(\pi D^2)$, we have

$$\mathbf{R} = \frac{DV}{\nu} = \frac{D}{\nu}\left(\frac{4Q}{\pi D^2}\right) = \frac{4Q}{\pi \nu D}$$

and so, by rearrangement, the required diameter D can be obtained from

Type 3:

$$D = \frac{4Q}{\pi \nu \mathbf{R}} \tag{8.48}$$

SAMPLE PROBLEM 8.7 Solve Sample Prob. 8.5 using only a basic scientific calculator, without trial and error.

Solution

This is a Type 2 problem, to find Q. As in Sample Prob. 8.5, $D = 0.5$ m, $h_L/L = 0.006$, $e/D = 0.000\,092$, and $\nu = 1.003 \times 10^{-6}$ m²/s. The quantity

$$\sqrt{\frac{2gDh_L}{L}} = \sqrt{2(9.81)0.5(0.006)} = 0.243 \text{ m/s},$$

so in Eq. (8.42*a*):

$$V = -2(0.243)\log\left[\frac{0.000\,092}{3.7} + \frac{2.51(1.003\times10^{-6})}{0.5}\frac{1}{0.243}\right] = 2.11 \text{ m/s}$$

$$\text{Check:}\quad \mathbf{R} = \frac{DV}{\nu} = \frac{0.5(2.11)}{1.003\times10^{-6}} = 1.050\times10^{6} \qquad (> \mathbf{R}_{crit} = 2000)$$

so flow is turbulent, confirming the use of Eq. (8.42*a*). Finally,

$$Q = AV = (\pi/4)D^2V = 0.25\pi(0.5)^2 2.11 = 0.414 \text{ m}^3/\text{s} \qquad \textbf{\textit{ANS}}$$

Sample Problem 8.8 Solve Sample Prob. 8.6 using (*a*) an equation solver on a programmable scientific calculator; (*b*) Mathcad; (*c*) only a basic scientific calculator, without trial and error.

Solution

This is a Type 3 problem, to find D. As in Sample Prob 8.6, $h_L = 215$ ft, $L = 18{,}000$ ft, $Q = 0.301$ cfs, $e = 0.0005$ ft, $g = 32.2$ ft/sec^2, and $\nu = 2.3\times10^{-5}$ ft^2/sec.

(*a*) Using an "equation writer" on the programmable calculator, key in Eq. (8.42*b*) and store it permanently, assigning it a name such as EQ842B. Then activate the equation solving feature, select the desired equation from the catalog of equations, and make it the "current" equation.

Display the menu of variables in the equation, and enter numeric values for them all (g, ν, h_L, L, e, Q, D) as above, plus an estimate of D (1 ft, say).

Instruct the solver to solve for D.

Result: $D = 0.344$ ft ***ANS***

(*b*) Using Mathcad, open a file (or "worksheet") and establish a "solve block" beginning with "Given" and ending with a statement including "Find(f, V, R, D)." Within the solve block, type the four governing equations: (8.40), (8.10), (8.1), and $Q = \pi D^2V/4$. These four equations involve ten variables, four of which are those to be found.

Above the solve block, assign numerical values to the six known variables (g, ν, h_L, L, e, Q), as above. Then assign estimated values to the four unknown variables, for example, $f = 0.03$, $V = 5$, $R = 100{,}000$, $D = 0.5$. It is helpful to keep these two sets of assignments on separate lines; they may be labeled with text. The success of the procedure may be sensitive to the estimated values, in which case different ones should be tried.

If the "Find" statement ends in an equals sign, the program will write the four results immediately after it, in the form of a vertical array or vector.

Results:

$$f = 0.0253, \quad V = 3.235 \text{ ft/s}, \quad \mathbf{R} = 48{,}411, \quad D = 0.344 \text{ ft} \qquad \textbf{\textit{ANS}}$$

■: Programmed computing aids (Appendix B) could help solve problems marked with this icon.

(*c*) Using only a basic scientific calculator, we have

$$\text{Eq. (8.43):}\quad \mathbf{N}_1 = \frac{128(32.2)}{\pi^3}\left(\frac{215}{18{,}000}\right)\frac{(0.301)^3}{(2.3\times10^{-5})^5} = 6.73\times10^{21}$$

$$\text{Eq. (8.44):}\quad \mathbf{N}_2 = \frac{\pi e \nu}{4Q} = \frac{\pi(0.0005)2.3\times10^{-5}}{4(0.301)} = 3.00\times10^{-8}$$

$$\text{Laminar flow Eq. (8.45):}\quad \mathbf{R} = \left(\frac{\mathbf{N}_1}{64}\right)^{0.25} = 101{,}300 > \mathbf{R}_{\text{crit}} = 2000$$

Turbulent flow Eq. (8.47):

$$\mathbf{R}^{2.5} = -2(6.73\times10^{21})^{0.5}\log\left[\frac{3.00\times10^{-8}(6.73\times10^{21})^{0.208}}{2.59} + \frac{4.29}{(6.73\times10^{21})^{0.188}}\right]$$

$$= 5.14\times10^{11}$$

from which, $\mathbf{R} = 48{,}300$. Clearly the laminar flow equation is invalid, flow is turbulent, and $\mathbf{R} = 48{,}300$. Thus, from Eq. (8.48),

$$D = \frac{4Q}{\pi\nu\mathbf{R}} = \frac{4(0.301)}{\pi(2.3\times10^{-5})48{,}300} = 0.345\text{ ft} = 4.14\text{ in}\qquad \textit{ANS}$$

EXERCISES

8.14.1 Solve Sample Prob. 8.5 using (*a*) an equation solver on a programmable scientific calculator; (*b*) Mathcad or similar software.

8.14.2 Solve Exer. 8.13.3 using (*a*) a programmable scientific calculator with an equation solver; (*b*) Mathcad or similar mathematics software; (*c*) only a basic scientific calculator, without trial and error.

8.14.3 Solve Exer. 8.13.4 using (*a*) a programmable scientific calculator with an equation solver; (*b*) Mathcad or similar mathematics software; (*c*) only a basic scientific calculator, without trial and error.

8.15 EMPIRICAL EQUATIONS FOR PIPE FLOW

The presentation of friction loss in pipes given in Secs. 8.1–8.14 incorporates the best knowledge available on this subject, as far as application to Newtonian fluids (Sec. 2.11) is concerned. Admittedly, however, the trial-and-error type of solution, especially when encumbered with computations for

---

■: Programmed computing aids (Appendix B) could help solve problems marked with this icon.

relative roughness and Reynolds number, becomes tedious when repeated often for similar conditions, as with a single fluid such as water. It is natural, therefore, that simple and convenient-to-use design formulas were developed, based on experiments and observations but limited to specific fluids and conditions. Such equations, based on observed data rather than theory, are known as ***empirical equations***. Perhaps the best example of such a formula is that of Hazen and Williams, applicable only to the flow of water in pipes larger than 2 in (5 cm) and at velocities less than 10 fps (3 m/s), but widely used in the waterworks industry. This formula is given in the form

BG units: $$V = 1.318 C_{HW} R_h^{0.63} S^{0.54} \tag{8.49a}$$

SI units: $$V = 0.849 C_{HW} R_h^{0.63} S^{0.54} \tag{8.49b}$$

where R_h (ft or m) is the hydraulic radius (Sec. 8.3), and $S = h_L/L$, the energy gradient. The advantage of this formula over the standard pipe-friction formula is that the roughness coefficient C_{HW} is not a function of the Reynolds number, and trial solutions are therefore eliminated. Values of C_{HW} range from 140 for very smooth, straight pipe down to 110 for new riveted-steel and vitrified pipe and to 90 or 80 for old and tuberculated pipe.

Another empirical formula, which is discussed in detail in Sec. 10.2, is the Manning formula, which is

BG units: $$V = \frac{1.486}{n} R_h^{2/3} S^{1/2} \tag{8.50a}$$

SI units: $$V = \frac{1}{n} R_h^{2/3} S^{1/2} \tag{8.50b}$$

where n is a roughness coefficient, varying from 0.008 for the smoothest brass or plastic pipe, to 0.014 for average drainage tile or vitrified sewer pipe, to 0.021–0.030 for corrugated metal, and up to 0.035 for tuberculated cast-iron pipe (Table 10.1). The Manning formula applies to about the same flow range as does the Hazen–Williams formula.

For some problems, it is more convenient to work with the above equations in the form of expressions for head loss, like Eq. (8.10). Because the equations can also be expressed in terms of V or Q, depending on which is given or sought, and expressed in BG or SI units, the number of alternative forms is quite large. Because each variable occurs only once in the above empirical equations, they are always explicit regardless of how they are rearranged or which variable is unknown. This gives them their distinct advantage, that they can always be solved directly.

Nomographic charts and diagrams have been developed for the application of Eqs. (8.49) and (8.50). The attendant lack of accuracy in using these formulas is not important in the design of water distribution systems, since it is seldom possible to predict the capacity requirements with high precision, and flows vary considerably throughout the day.

EXERCISES

8.15.1 When water flows at 2.5 cfs through a 20-in-diameter pipeline, the head loss is 0.0004 ft/ft. Find the value of the Hazen–Williams coefficient.

8.15.2 Water flows at 0.20 m^3/s through a 50-cm-diameter pipeline with a head loss of 0.0032 m/m. Find the value of the Hazen–Williams coefficient.

8.16 FLUID FRICTION IN NONCIRCULAR CONDUITS

Most closed conduits used in engineering practice are of circular cross section; however, rectangular ducts and cross sections of other geometry are occasionally used. Some of the foregoing equations may be modified for application to noncircular sections by use of the concept of hydraulic radius.

The hydraulic radius was defined (Sec. 8.3) as $R_h = A/P$, where A is the cross-sectional area and P is the wetted perimeter. To apply the empirical equations (Sec. 8.15), the appropriate value of R_h is simply determined for the conduit. For the Darcy–Weisbach equations involving $f = f(\mathbf{R}, e/D)$ (Secs. 8.5, 8.6 and 8.11–8.14), we notice that all the equations are expressed in terms of a pipe diameter, D. But for a circular pipe flowing full,

$$R_h = \frac{A}{P} = \frac{\pi D^2/4}{\pi D} = \frac{D}{4} \tag{8.51}$$

or

$$D = 4R_h \tag{8.52}$$

This may be substituted into Eq. (8.10) to yield

$$h_L = \frac{f}{4}\frac{L}{R_h}\frac{V^2}{2g} \tag{8.53}$$

and into the expression for Reynolds number to give

$$\mathbf{R} = \frac{(4R_h)V\rho}{\mu} \tag{8.54}$$

and into any other of the equations associated with the friction factor f [Eqs. (8.19)–(8.20) and Eqs. (8.35)–(8.48)]. Similarly replacing e/D by $e/4R_h$, problems may then be solved in the same way as with D, using either equations or the Moody diagram (Fig. 8.11).

This approach gives reasonably accurate results for turbulent flow, but the results are poor for laminar flow, because in such flow frictional phenomena are caused by viscous action throughout the body of the fluid, while in turbulent flow the frictional effect is accounted for largely by the region close to the wall; i.e., it depends on the wetted perimeter.

EXERCISES

8.16.1 When fluid of specific weight 40 lb/ft^3 flows in a 9-in-diameter pipe, the frictional stress between the fluid and the pipe wall is 0.6 lb/ft^2. Calculate the head loss per foot of pipe. If the flow rate is 2.4 cfs, how much power is lost per foot of pipe?

8.16.2 When fluid of specific weight 8.2 kN/m^3 flows in a 15-cm-diameter pipe, the frictional stress between the fluid and the pipe wall is 25 N/m^2. Calculate the head loss per meter of pipe. If the flow rate is 42 L/s, how much power is lost per meter of pipe?

8.17 MINOR LOSSES IN TURBULENT FLOW

Losses due to the *local* disturbances of the flow in conduits such as changes in cross section, projecting gaskets, elbows, valves, and similar items are called ***minor losses***. In the case of a very long pipe or channel, these losses are usually insignificant in comparison with the fluid friction in the length considered. But if the length of pipe or channel is very short, these so-called minor losses may actually be major losses. Thus, in the case of the suction pipe of a pump, the loss of head at entrance, especially if a strainer and a foot valve are installed, may be very much greater than the friction loss in the short inlet pipe.

Whenever the velocity of a flowing stream is altered either in direction or in magnitude in turbulent flow, ***eddy currents*** are set up and a loss of energy in excess of the pipe friction in that same length is created.[19] Head loss in decelerating (i.e., diverging) flow is much larger than that in accelerating (i.e., converging) flow (Sec. 8.21). In addition, head loss generally increases with an increase in the geometric distortion of the flow. Though the causes of minor losses are usually confined to a very short length of the flow path, the effects may not disappear for a considerable distance downstream. Thus an elbow in a pipe may occupy only a small length, but the disturbance in the flow will extend for a long distance downstream.

The most common sources of minor loss are described in the remainder of this chapter. Such losses may be represented in one of two ways. They may be expressed as $kV^2/2g$, where the ***loss coefficient*** k must be determined for each case. Or they may be represented as being equivalent to a certain length of straight pipe, usually expressed in terms of the number of pipe diameters, N. Since

$$k\frac{V^2}{2g} = \frac{f(ND)}{D}\frac{V^2}{2g}$$

[19] In laminar flow these losses are insignificant, because irregularities in the flow boundary create a minimal disturbance to the flow and separation is essentially nonexistent.

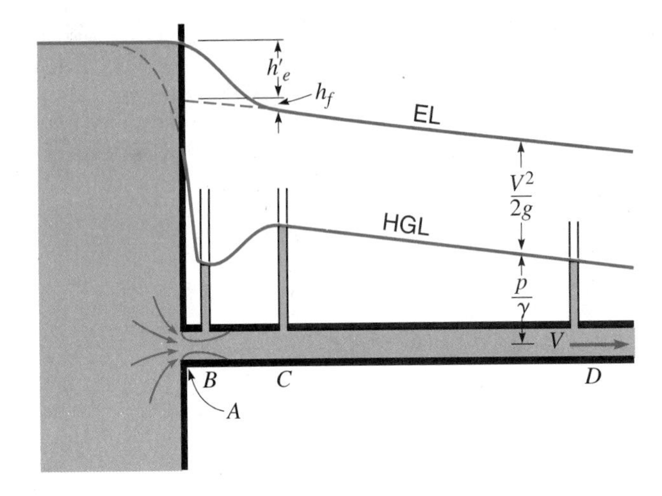

Figure 8.12
Conditions at entrance.

it follows that $k = Nf$.

Because minor losses are losses of energy, we will frequently be relating them to the energy line (EL) and the hydraulic grade line (HGL). These were originally defined and explained in Secs. 5.12 and 5.14.

8.18 LOSS OF HEAD AT ENTRANCE

Referring to Fig. 8.12, it may be seen that, as fluid from the reservoir enters the pipe, the streamlines continue to converge for a while, much as though this were a jet issuing from a sharped-edged orifice (Sec. 11.6). As a result, we find a cross section with maximum velocity and minimum pressure at B. This minimum flow area is known as the ***vena contracta***. At B, the contracted flowing stream is surrounded by fluid that is in a state of turbulence but has very little forward motion. Between B and C the fluid is in a very disturbed condition because the stream expands and the velocity decreases while the pressure rises. From C to D the flow is normal.

It is seen that the loss of energy at entrance is distributed along the length AC, a distance of several diameters. The increased turbulence and vortex motion in this portion of the pipe cause the friction loss to be much greater than in a corresponding length where the flow is normal, as is shown by the drop of the total-energy line. Of this total loss, a small portion h_f would be due to the normal pipe friction (see Fig. 8.12). Hence the difference between this and the total, or h_e', is the true value of the extra loss caused at entrance.

The loss of head at entrance may be expressed as

$$h_e' = k_e \frac{V^2}{2g} \tag{8.55}$$

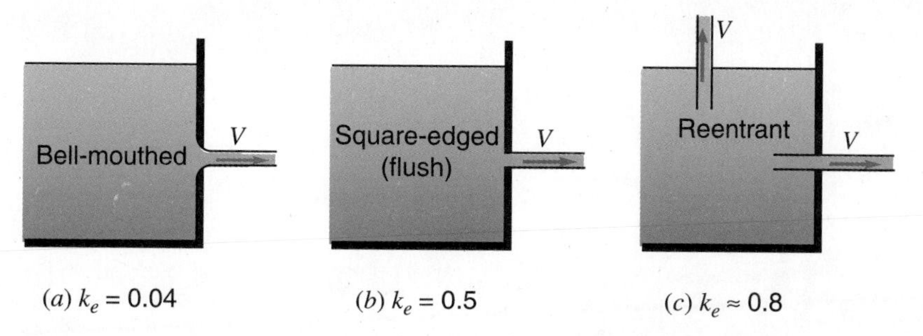

Figure 8.13
Entrance loss coefficients.

where V is the mean velocity in the pipe, and k_e is the loss coefficient, whose general values are shown in Fig. 8.13.

The entrance loss is caused primarily by the turbulence created by the enlargement of the stream after it passes section B, and this enlargement in turn depends upon how much the stream contracts as it enters the pipe. Thus it is very much affected by the conditions at the entrance to the pipe. Values of the entrance-loss coefficients have been determined experimentally. If the entrance to the pipe is well rounded or ***bell-mouthed*** (Fig. 8.13*a*), there is no contraction of the stream entering and the coefficient of loss is correspondingly small. For a ***flush*** or ***square-edged entrance***, such as shown in Fig. 8.13*b*, k_e has a value of about 0.5. A ***reentrant tube***, such as shown in Fig. 8.13*c*, produces a maximum contraction of the entering stream, because the streamlines come from around the outside wall of the pipe, as well as more directly from the fluid in front of the entrance. The degree of the contraction depends upon how far the pipe may project within the reservoir and also upon how thick the pipe walls are, compared with its diameter. With very thick walls, the conditions approach that of a square-edged entrance. For these reasons, the loss coefficients for reentrant tubes vary; for very thin tubes, $k_e \approx 0.8$.

8.19 LOSS OF HEAD AT SUBMERGED DISCHARGE[20]

When a fluid with a velocity V is discharged from the end of a pipe into a closed tank or reservoir which is so large that the velocity within it is negligible, the entire kinetic energy of the flow is dissipated. Hence the

[20] This topic was first discussed in Sec. 5.13.

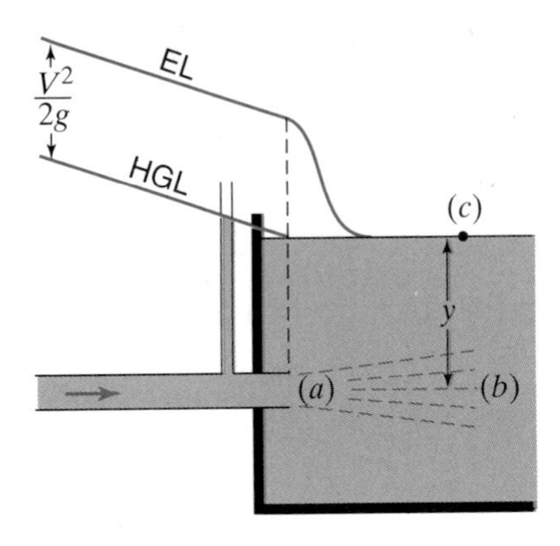

Figure 8.14
Submerged discharge loss.

discharge loss is

$$h'_d = \frac{V^2}{2g} \tag{8.56}$$

That this is true may be shown by writing an energy equation between (a) and (c) in Fig. 8.14. Taking the datum plane through (a) and recognizing that the pressure head of the fluid at (a) is y, its depth below the surface, $H_a = y + 0 + V^2/2g$ and $H_c = 0 + y + 0$. Therefore we obtain

$$h'_d = H_a - H_c = \frac{V^2}{2g}$$

Thus the discharge loss coefficient $k_d = 1.0$ under all conditions; hence the only way to reduce the discharge loss is to reduce the magnitude of V by means of a diverging tube. This is the reason for a diverging draft tube that discharges the flow from a reaction turbine (Sec. 16.8).

As contrasted with entrance loss, it must here be emphasized that discharge loss occurs *after* the fluid *leaves* the pipe,[21] while entrance loss occurs *after* the fluid *enters* the pipe.

EXERCISES

8.19.1 A 12-in-diameter pipe ($f = 0.02$) carries fluid between two tanks at 8 fps. The entrance and exit conditions to and from the pipe are square-edged and flush with the wall of the tank. Find the ratio of the minor losses divided by the pipe friction loss if the length of the pipe is (a) 4 ft; (b) 80 ft; (c) 1600 ft.

8.19.2 A 30-cm-diameter pipe ($f = 0.015$) carries fluid between two tanks at 2.4 m/s. The entrance and exit conditions to and from the pipe are

[21] In a short pipe, where the discharge loss may be a major factor, greater accuracy is obtained by using the correction factor α, as explained in Sec. 5.1 [see also Eq. (8.34a)].

reentrant. Find the ratio of the minor losses divided by the pipe friction loss if the length of the pipe is (*a*) 2 m; (*b*) 50 m; (*c*) 1000 m.

8.19.3 A smooth 15-in-diameter pipe is 400 ft long and has a flush entrance and a submerged discharge. It carries 70°F water at a velocity of 8 fps. What is the total head loss?

8.19.4 A smooth 30-cm-diameter pipe is 90 m long and has a flush entrance and a submerged discharge. It carries 15°C water at a velocity of 3 m/s. What is the total head loss?

8.19.5 Suppose the fluid in Exer. 8.19.3 is oil with a kinematic viscosity of 0.001 ft^2/sec and a specific gravity of 0.92. What would be the head loss in feet of oil and in pounds per square inch?

8.19.6 Suppose the fluid in Exer. 8.19.4 is oil with a kinematic viscosity of 9.7×10^{-5} m^2/s and a specific gravity of 0.94. What would be the head loss in meters of oil and in kN/m^2?

8.20 LOSS DUE TO CONTRACTION

Sudden Contraction

The phenomena attending the sudden contraction of a flow are shown in Fig. 8.15. There is a marked drop in pressure due to the increase in velocity and to

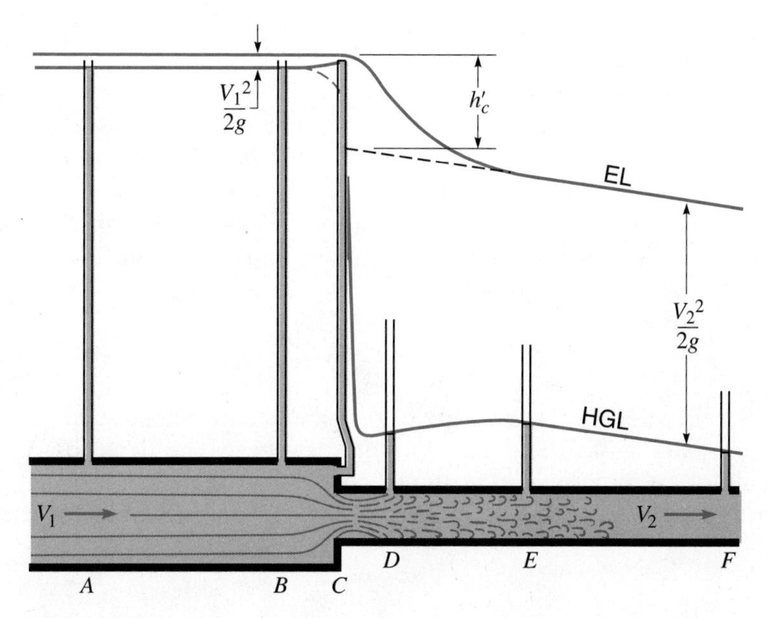

Figure 8.15
Loss due to sudden contraction. (*Plotted to scale.*)

Table 8.2
Loss coefficients for sudden contraction

D_2/D_1	0.0	0.1	0.2	0.3	0.4	0.5	0.6	0.7	0.8	0.9	1.0
k_c	0.50	0.45	0.42	0.39	0.36	0.33	0.28	0.22	0.15	0.06	0.00

the loss of energy in turbulence. It is noted that in the corner upstream at section C there is a rise in pressure because the streamlines here are curving, so that the centrifugal action causes the pressure at the pipe wall to be greater than in the center of the stream. The dashed line indicates the pressure variation along the centerline streamline from sections B to C.

From C to E, the conditions are similar to those described for entrance. The loss of head for a sudden contraction may be represented by

$$h'_c = k_c \frac{V_2^2}{2g} \tag{8.57}$$

where k_c has the values given in Table 8.2.

The entrance loss of Sec. 8.18 is a special case where $D_2/D_1 = 0$.

Gradual Contraction

In order to reduce the foregoing losses, abrupt changes of cross section should be avoided. This may be accomplished by changing from one diameter to the other by means of a smoothly curved transition or by employing the frustrum of a cone. With a smoothly curved transition, a loss coefficient k_c as small as 0.05 is possible. For conical reducers, a minimum k_c of about 0.10 is obtained, with a total cone angle of 20–40°. Smaller or larger total cone angles result in higher values of k_c.

The nozzle at the end of a pipeline (Fig. 8.22*b*) is a special case of gradual contraction. The head loss through a nozzle at the end of a pipeline is given by Eq. (8.57), where k_c is the nozzle loss coefficient whose value commonly ranges from 0.04 to 0.20 and V_j is the jet velocity.[22] The head loss through a nozzle cannot be regarded as a minor loss because the jet velocity head is usually quite large. More details on the flow through nozzles is presented in Sec. 11.6.

[22] See also Eq. (11.14).

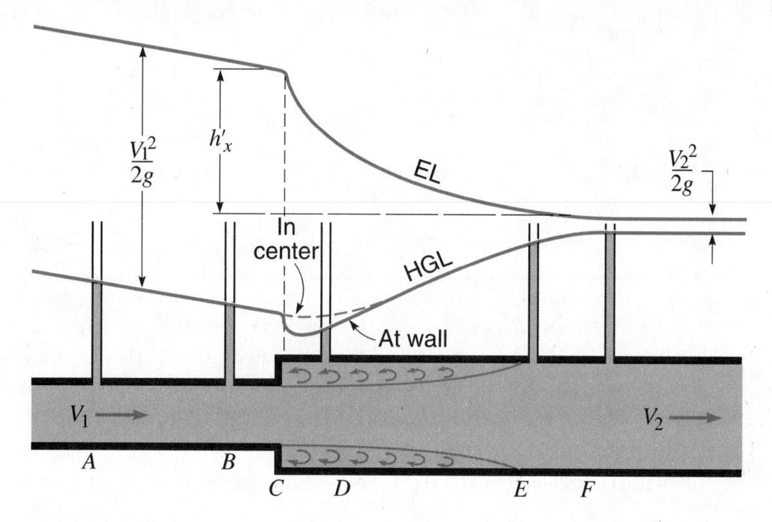

Figure 8.16
Loss due to sudden enlargement. (*Plotted to scale. Velocity the same as in Fig. 8.15.*)

8.21 LOSS DUE TO EXPANSION

Sudden Expansion

The conditions at a sudden expansion are shown in Fig. 8.16. There is a rise in pressure because of the decrease in velocity, but this rise is not so great as it would be if it were not for the loss in energy. There is a state of excessive turbulence from *C* to *F,* beyond which the flow is normal. The drop in pressure just beyond section *C,* which was measured by a piezometer not shown in the illustration, is due to the fact that the pressures at the wall of the pipe are in this case less than those in the center of the pipe because of centrifugal effects.

Figures 8.15 and 8.16 are both drawn to scale from test measurements for the same diameter ratios and the same velocities, and show that the loss due to sudden expansion is greater than the loss due to a corresponding contraction. This is so because of the inherent instability of flow in an expansion where the diverging paths of the flow tend to encourage the formation of eddies within the flow. Moreover, separation of the flow from the wall of the conduit induces pockets of eddying turbulence outside the flow region. In converging flow there is a dampening effect on eddy formation, and the conversion from pressure energy to kinetic energy is quite efficient.

An expression for the loss of head in a sudden enlargement can be

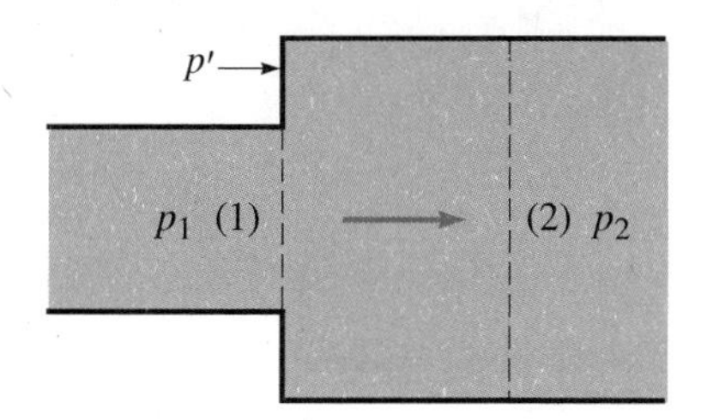

Figure 8.17

derived as follows. In Fig. 8.17, section 2 corresponds to section F in Fig. 8.16, which is a section where the velocity profile has become normal again, and marks the end of the region of excess energy loss due to the turbulence created by the sudden enlargement. In Fig. 8.17 assume that the pressure at section 2 in the ideal case without friction is p_0. Then in this ideal case

$$\frac{p_0}{\gamma} = \frac{p_1}{\gamma} + \frac{V_1^2}{2g} - \frac{V_2^2}{2g}$$

If in the actual case the pressure at section 2 is p_2 while the average pressure on the annular ring is p' then, equating the resultant force on the body of fluid between sections 1 and 2 to the time rate of change of momentum between sections 1 and 2, we obtain

$$p_1A_1 + p'(A_2 - A_1) - p_2A_2 = \frac{\gamma}{g}(A_2V_2^2 - A_1V_1^2)$$

From this,

$$\frac{p_2}{\gamma} = \frac{A_1}{A_2}\frac{p_1}{\gamma} + \frac{A_2 - A_1}{A_2}\frac{p'}{\gamma} + \frac{A_1}{A_2}\frac{V_1^2}{g} - \frac{V_2^2}{g}$$

The loss of head is given by the difference between the ideal and actual pressure heads at section 2. Thus $h_x' = (p_0 - p_2)/\gamma$, and noting that

$$A_1V_1 = A_2V_2$$

and that $A_1V_1^2 = A_1V_1V_1 = A_2V_2V_1$, we obtain, from substituting the above expressions for p_0/γ and p_2/γ into $(p_0 - p_2)/\gamma$,

$$h_x' = \frac{(V_1 - V_2)^2}{2g} + \left(1 - \frac{A_1}{A_2}\right)\left(\frac{p_1}{\gamma} - \frac{p'}{\gamma}\right)$$

It is usually assumed that $p' = p_1$, in which case the loss of head due to sudden enlargement is

$$h_x' = \frac{(V_1 - V_2)^2}{2g} \tag{8.58}$$

Although it is possible that, under some conditions, p' will equal p_1, it is also possible for it to be either more or less than that value, in which case the loss of head will be either less or more than that given by Eq. (8.58). The exact value of p' will depend upon the manner in which the fluid eddies around in the corner adjacent to this annular ring. However, the deviation from Eq. (8.58) is quite small and of negligible importance.

The discharge loss of Sec. 8.19 is seen to be a special case where A_2 is infinite compared with A_1, or $V_2 = 0$, so that Eq. (8.58) will reduce to Eq. (8.56).

It may become necessary to express this minor loss all in terms of one velocity (see, e.g., Sec. 8.27). If so, we can use continuity, $(\pi/4)D_1^2V_1 = (\pi/4)D_2^2V_2$, to obtain

$$(V_1 - V_2)^2 = \left(1 - \frac{D_1^2}{D_2^2}\right)^2 V_1^2 = \left(\frac{D_2^2}{D_1^2} - 1\right)^2 V_2^2$$

Gradual Expansion

To minimize the loss accompanying a reduction in velocity, a ***diffuser*** such as that shown in Fig. 8.18 may be used. The diffuser may be given a curved outline, or it may be a frustum of a cone. In Fig. 8.18 the loss of head will be some function of the angle of divergence and also of the ratio of the two areas, the length of the diffuser being determined by these two variables.

In flow through a diffuser the total loss may be considered as made up of two factors. One is the ordinary pipe-friction loss, which may be represented by

$$h_L = \int \frac{f}{D} \frac{V^2}{2g} \, dL$$

In order to integrate the foregoing, it is necessary to express the variables f, D, and V as functions of L. For our present purpose, it is sufficient, however, merely to note that the friction loss increases with the length of the cone. Hence, for given values of D_1 and D_2, the larger the angle of the cone, the

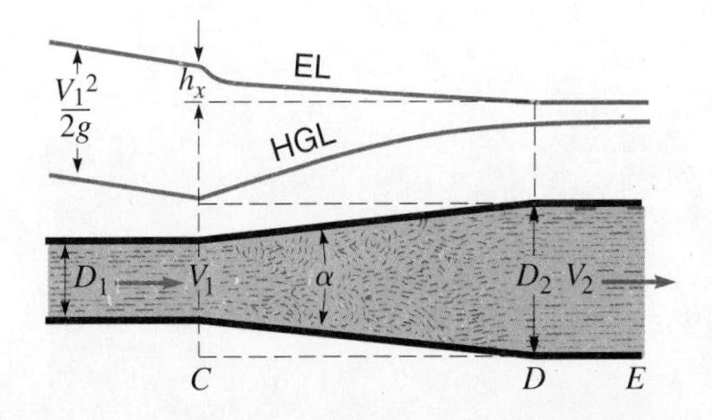

Figure 8.18
Loss due to gradual enlargement.

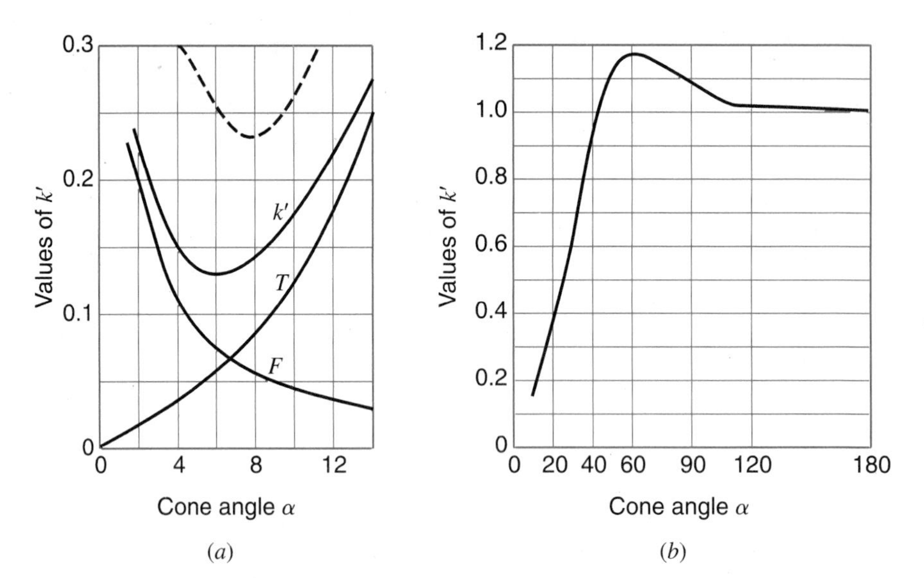

Figure 8.19
Loss coefficient for conical diffusers.

less its length and the less the pipe friction, which is indicated by the curve marked F in Fig. 8.19*a*. However, in flow through a diffuser there is an additional turbulence loss set up by induced currents that produce a vortex motion over and above that which normally exists. This additional turbulence loss will naturally increase with the degree of divergence, as indicated by the curve marked T in Fig. 8.19*a*, and if the rate of divergence is great enough then there may be a separation at the walls and eddies flowing backward along the walls. The total loss in the diverging cone is then represented by the sum of these two losses, marked k'. This is seen to have a minimum value at 6° for the particular case chosen, which is for a very smooth surface. If the surface were rougher, the value of the friction F would be increased. This increases the value of k', which is indicated by the dotted curve, and also shifts the angle for minimum loss to 8°. Thus the best angle of divergence increases with the roughness of the surface.

It has been seen that the loss due to a sudden enlargement is very nearly represented by $(V_1 - V_2)^2/2g$. The loss due to a gradual enlargement is expressed as

$$h' = k' \frac{(V_1 - V_2)^2}{2g} \tag{8.59}$$

Values of k' as a function of the cone angle α are shown in Fig. 8.19*b*, for a wider range than appears in Fig. 8.19*a*. It is of interest to note that at an angle slightly above 40° the loss is the same as that for a sudden enlargement, which is 180°, and that between these two the loss is greater than for a sudden enlargement, being a maximum at about 60°.

EXERCISES

8.21.1 Two pipes with a diameter ratio of 1:2 are connected in series. With a velocity of 7.5 m/s in the smaller pipe, find the loss of head due to (*a*) sudden contraction; (*b*) sudden enlargement; (*c*) expansion in a conical diffuser with a total angle of 30°, and of 10°.

8.21.2 A 5-in-diameter pipe ($f = 0.033$) 110 ft long connects two reservoirs whose water-surface elevations differ by 12 ft. The pipe entrance is flush, and the discharge is submerged. (*a*) Compute the flow rate. (*b*) If the last 10 ft of pipe were replaced with a conical diffuser with a cone angle of 10°, compute the flow rate.

8.22 LOSS IN PIPE FITTINGS

The loss of head in pipe fittings may be expressed as $kV^2/2g$, where V is the velocity in a pipe of the nominal size of the fitting. Typical values of k are given in Table 8.3. As an alternative, the head loss due to a fitting may be found by increasing the pipe length by using values of L/D given in the table. However, it must be recognized that these fittings create so much turbulence that the loss caused by them is proportional to V^2, and hence this latter

Table 8.3
Values of loss factors for pipe fittings

Fitting	k	L/D
Globe valve, wide open	10	350
Angle valve, wide open	5	175
Close-return bend	2.2	75
T, through side outlet	1.8	67
Short-radius elbow	0.9	32
Medium-radius elbow	0.75	27
Long-radius elbow	0.60	20
45° elbow	0.42	15
Gate valve, wide open	0.19	7
half open	2.06	72

method should be restricted to the case where the pipe friction itself is in the zone of complete turbulence. For very smooth pipes, it is better to use the k values when determining the loss through fittings.

8.23 LOSS IN BENDS AND ELBOWS

In flow around a bend or elbow, because of centrifugal effects [Eq. (5.40)], there is an increase in pressure along the outer wall and a decrease in pressure along the inner wall. The centrifugal force on a number of fluid particles, each of mass m, along the diameter CE of the pipe that is normal to the plane of curvature of the pipe is shown in Fig. 8.20. The centrifugal force on the particles near the center of the pipe, where the velocities are high, is larger than the centrifugal force on the particles near the walls of the pipe, where the velocities are low. Because of this unbalanced condition, a secondary flow[23] develops, as shown in Fig. 8.20. This combines with the axial velocity to form a double spiral flow, which persists for some distance. Thus not only is there some loss of energy within the bend itself, but this distorted

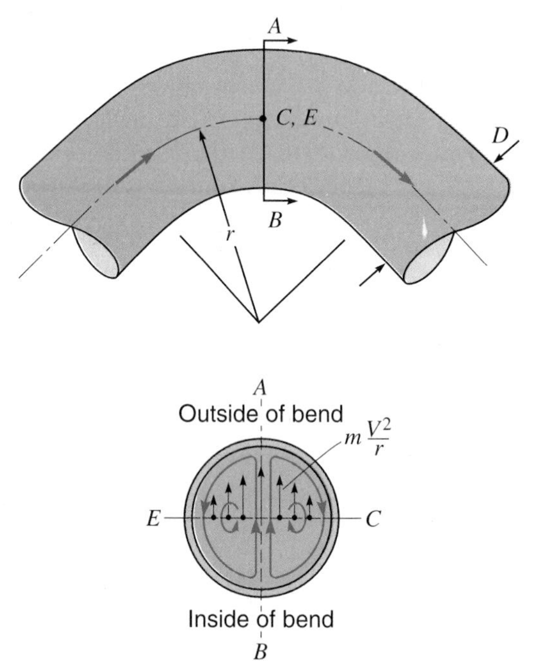

Figure 8.20
Secondary flow in bend.

[23] Secondary flow in the bends of open channels is discussed in Sec. 10.21.

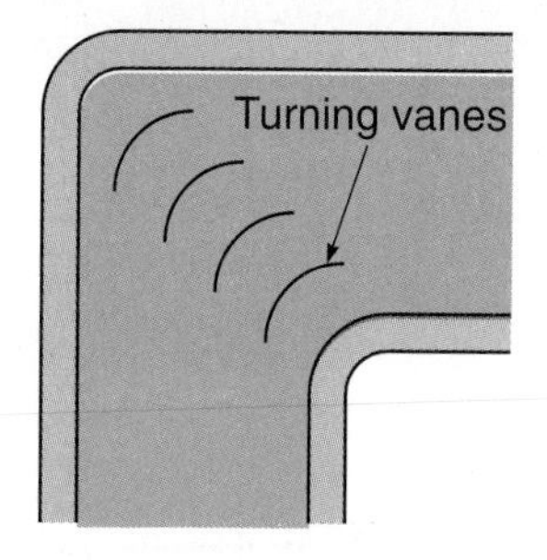

Figure 8.21
Vaned elbow.

flow condition persists for some distance downstream until dissipated by viscous friction. The velocity in the pipe may not become normal again within as much as 100 pipe diameters downstream from the bend. In fact, more than half the friction loss produced by a bend or elbow takes place in the straight pipe following it.

Most of the loss of head in a sharp bend may be eliminated by the use of a vaned elbow, such as is shown in Fig. 8.21. The vanes tend to impede the formation of the secondary flows that would otherwise occur.

The head loss produced by a bend or elbow ($h_b = k_b V^2/2g$) *in excess of the loss for an equal length of straight pipe* is greatly dependent upon the ratio of the radius of curvature r to the diameter of the pipe D. Also, combinations of different pipe bends placed close together cannot be treated by adding up the losses of each one considered separately. The total loss depends not only upon the spacing between the bends, but also upon the relations of the directions of the bends and the planes in which they are located. Bend loss is not proportional to the angle of the bend; for 22.5° and 45° bends the losses are respectively about 40% and 80% of the loss in a 90° bend. Values of k_b for 90° bends, varying with bend radius and pipe roughness, are given in Fig. 8.22.

EXERCISES

8.23.1 Water at 72°F flows through a 110-ft-long, 5-in-diameter wrought-iron pipe that contains the following fittings: one open globe valve, one medium-radius elbow, and one 90° pipe bend ($k_b = 0.13$) with a radius of curvature of 45 in. The length of the bend is not included in the 110 ft. There are no entrance or discharge losses. Find the total head loss if the flow velocity is 4.8 fps.

8.23.2 Water at 18°C flows through a 25-m-long, 7.5-cm-diameter commercial steel pipe that contains the following fittings: one open angle valve, one short-radius elbow, and one 90° pipe bend ($k_b = 0.12$) with a radius of curvature of 60 cm. The length of the bend is not included in the 25 m. There are no entrance or discharge losses. Find the total head loss if the flow velocity is 2 m/s.

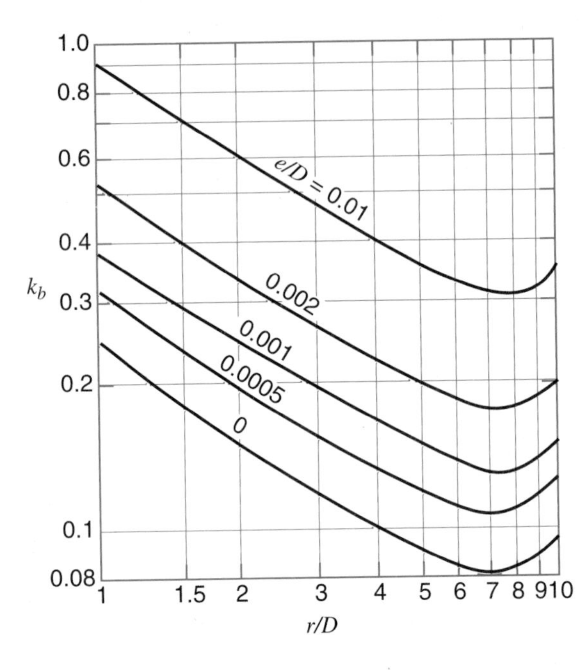

Figure 8.22
Resistance coefficients for 90° bends (resistance due to length of pipe $\pi r/2$ in the bend must be added, where r is the radius to the pipe centerline). (*From Ref. 30*).

8.24 SOLUTION OF SINGLE-PIPE FLOW PROBLEMS WITH MINOR LOSSES

We have examined the fundamental fluid mechanics associated with the frictional loss of energy in single-pipe flow, caused by both the wall roughness of the pipes and by pipe fittings that disturb the flow (minor losses). While the interest of the scientist extends very little beyond this, it is the task of the engineer to apply these fundamentals to various types of practical problems.

It is generally conceded that for pipes of length greater than 1000 diameters, the error incurred by neglecting minor losses is less than that inherent in selecting a value for the friction factor (f, n, or C_{HW}). In applying this rule, one must of course use common sense and recall that a valve, for example, is a minor loss only when it is wide open. Partially closed, it may be the most important loss in the system.

When minor losses are negligible, as they often are, pipe flow problems may be solved by the methods of Secs. 8.12–8.15. Available are the Hazen–Williams equation (8.49), the Manning equation (8.50) or the Darcy–Weisbach equation (8.10). The latter is to be preferred, since it will provide greater accuracy because its application utilizes the basic parameters that influence pipe friction, namely, Reynolds number **R** and relative roughness e/D. To get good results with the Hazen–Williams and Manning equations, the user must select proper values for C_{HW} and n, respectively. This is more difficult than estimating the e/D ratio for a pipe as required by the Darcy–Weisbach equation.

When minor losses are *included*, the total head loss between two points is the sum of the pipe friction loss plus the minor losses, or

$$h_L = h_{L_f} + \sum h' \tag{8.60}$$

The pipe friction loss can be represented by a number of different equations, and may be dependent on a number of different factors, as noted in Secs. 8.1–8.15. And, as we noted in Sec. 8.17, the minor losses may be represented as a coefficient multiplied by the velocity head, $kV^2/2g$, or as an equivalent length of pipe, expressed as a number of diameters, ND. As a result, the right-hand side of Eq. (8.60) can take on many forms. Because of the extra term added by the minor losses, such forms of this equation are often harder to solve than those that omit minor losses.

In problems where f is given, Eq. (8.60) still has only one unknown, namely, h_L or V or Q or D. In most cases this equation is explicit in the unknown, and so is easy to solve. But for sizing problems, the resulting equation in D is of the fifth degree, requiring trial and error or an equation solver as discussed in Sec. 8.14.

In practice, we may need to take into account the variation of f, which can vary by as much as a factor of five for smoother pipes (Fig. 8.11). With minor losses included, methods involving the Darcy–Weisbach equation (8.10) and an unknown friction factor f remain explicit or direct for Type 1 problems (find h_L), but for the other two types of problems (find Q or V or D) they are always implicit. These problem categories were first discussed in Sec. 8.13.

The empirical equations (Sec. 8.15) do not necessarily continue to yield direct solutions when minor losses are included. By adding $kV^2/2g$ or ND to Eqs. (8.49) and (8.50) rearranged for h_L, we find that they are always explicit for Type 1, usually explicit for Type 2 (find Q or V), and never explicit for Type 3 (find D). The exception with Type 2 occurs when using the Hazen–Williams equation and representing minor losses by $kV^2/2g$; this is because, upon rearranging Eq. (8.49), we find that the pipe friction loss is proportional to $V^{1.852}$, so the two terms contain V to different inconvenient powers. Type 3 problems are more complex in part owing to the fact that Q is given rather than V, since this causes V to also be a function of D.

To solve the many implicit equations that arise when minor losses are included, it is clear we must use iterative procedures, either manually or programmed. Manual iterative procedures (trial and error) are generally like those described in Sec. 8.13.

Preprogrammed iterative procedures are available in mathematics software packages, and in programmable calculators with an equation solving function, as discussed in Sec. 8.14 and Appendix B. These unquestionably are the most convenient, and they can be expected to become more accessible in time.

When using a mathematics software package like Mathcad, we just need to specify n equations in n unknowns, as described in Sec. 8.14. The terms added to account for minor losses merely have to be typed in with the energy equation, and values must be assigned to any new loss coefficients. Otherwise, the procedure is unchanged.

To use a program solver in a programmable calculator, we recall from Sec. 8.14 that for the implicit, nonlinear equations we must deal with there can be only one equation in one unknown. Similarly to the development of Eq. (8.42), we can develop a "universal" turbulent flow equation for use in an equation solver, including minor losses, by eliminating h_{L_f} and Eq. (8.60) with the help of Eqs. (8.10) and (8.40), and by replacing V by $4Q/(\pi D^2)$. Expressing minor losses $\Sigma h'$ in terms of $\Sigma kV^2/2g$, we obtain

Turbulent flow:

$$\sqrt{\frac{L/D}{\dfrac{\pi^2 g D^4 h_L}{8Q^2} - \sum k}} = -2\log\left(\frac{e/D}{3.7} + \frac{2.51\pi\nu D^2}{4Q}\sqrt{\frac{L/D}{\dfrac{\pi^2 g D^4 h_L}{8Q^2} - \sum k}}\right) \tag{8.61}$$

If we express minor losses in terms of an equivalent length of pipe, *ND*, then we obtain

Turbulent flow:

$$\frac{4Q}{\pi D^2}\sqrt{\frac{(L/D)+N}{2gh_L}} = -2\log\left(\frac{e/D}{3.7} + \frac{2.51\nu}{D}\sqrt{\frac{(L/D)+N}{2gh_L}}\right) \tag{8.62}$$

Each of these equations involves eight variables: h_L, Q, D, L, e, ν, g, and Σk or N. We note that the equations are implicit in h_L, Q, and D, but, with an equation solver we may easily solve for any one of these three quantities if the rest of the variables are known. Similar and less complex "universal" equations can also be developed using the Haaland equation (8.41); however, they sometimes yield false results when large diameters cause the argument of the logarithm to be very close to unity. An important reminder when using these equations is to use Eqs. (4.7) and (8.1) to check the Reynolds number and confirm that the flow is turbulent. If $\mathbf{R} < 2000$, the flow is laminar (Sec. 8.2), and the problem must instead be solved with Eq. (8.19) and Eq. (4.3) if needed.

The following sample problem illustrates the method of solution for flow through a pipeline of uniform diameter with minor losses.

SAMPLE PROBLEM 8.9 Water at 60°F flows through the new 10-in-diameter cast-iron pipe sketched in Fig. 8.23*a*. The pipe is 5000 ft long, its entrance is sharp-cornered but nonprojecting, and $\Delta z = 260$ ft. Find the flow rate using (*a*)

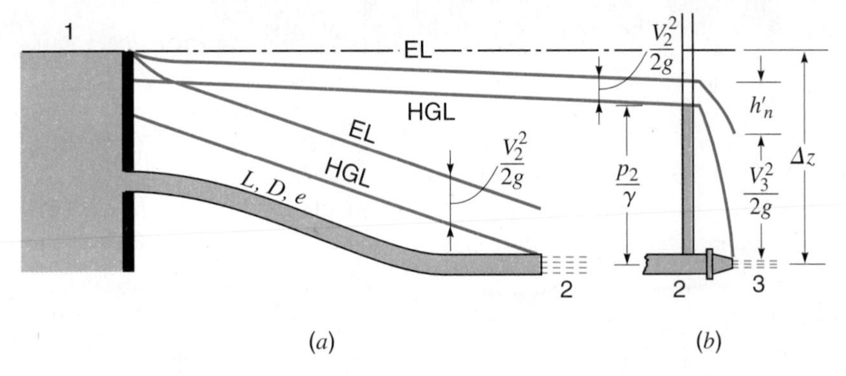

Figure 8.23
Discharge from a reservoir. (*a*) Free discharge. (*b*) With nozzle. As L/D gets larger the EL and HGL approach one another.

only Fig. 8.11 and a basic scientific calculator; (*b*) an equation solver on a programmable calculator; (*c*) Mathcad.

Solution
This is a Type 2 problem with minor losses.

Table 8.1 for cast iron: $e = 0.00085$ ft; $e/D = 0.00085/(10/12) = 0.00102$

Sec. 8.18 for square-edged entrance: $k_e = 0.5$

Energy Eq. (5.15) from water surface to free jet:

$$260 + 0 + 0 = 0 + 0 + \frac{V_2^2}{2g} + \left(0.5 + f\frac{5000}{10/12}\right)\frac{V_2^2}{2g} \tag{1}$$

(*a*) Rearranging: $\dfrac{V_2^2}{2g} = \dfrac{260}{1.5 + 6000f}; \quad V_2 = \sqrt{\dfrac{2(32.2)260}{1.5 + 6000f}}$

Fig. 8.11, right-hand side, for $e/D = 0.00102$: $f_{\min} \approx 0.0196$

$$\text{Try } f = 0.0196 \rightarrow V_2 = \sqrt{16{,}744/[1.5 + 6\,000(0.0196)]} = 11.86 \text{ fps}$$

$$D''V = 10(11.86) = 118.6$$

Fig. 8.11 for $e/D = 0.00102$ and $D''V = 118.6$: $f = 0.020$. Obtained and assumed f values are different, so we must try again. Tabulating this and subsequent trials:

Try f	V, ft/s	$D''V$	Obtained f	
0.0196	11.86	118.6	0.020	Try again
0.020	11.74	117.4	0.020	Converged!

$$Q = \frac{\pi}{4}\left(\frac{10}{12}\right)^2(11.74) = 6.40 \text{ cfs} \qquad \textbf{\textit{ANS}}$$

Alternatively, replace the chart by equations as follows:

Table A.1 for water at 60°F: $\nu = 1.217 \times 10^{-5}$ ft²/sec

Eq. (8.39): $f_{\min} = 0.01973$

Energy equation with assumed $f \rightarrow V$; $\mathbf{R} = DV/\nu$

e/D and $\mathbf{R}$ into Haaland equation (8.41) $\rightarrow$ calculated f; compare with assumed f. If the difference is less than about 2%, we have convergence; otherwise repeat.

(*b*) Using an equation solver on a programmable calculator, follow the procedure of Sample Prob. 8.8*a*, but with an equation that includes minor losses. Equation (8.61) appears to be inappropriate, because it does not include for the change in velocity head from zero at point 1 to $V_2^2/(2g)$ at point 2. However, from Eq. (1) above we see that we can include for this by increasing $\sum k$ by one. The seven known variables are assigned the values (without units) $g = 32.2$ ft/sec², $\nu = 1.217 \times 10^{-5}$ ft²/sec, $h_L = \Delta z = 260$ ft, $\sum k = k_e + 1 = 1.5$, $L = 5000$ ft, $e = 0.00085$ ft, and $D = 10/12$ ft. An estimated value of 10 cfs is assigned to the unknown Q.

The solver is instructed to solve for Q. The result is $Q = 6.39$ cfs. But we must check the $\mathbf{R}$: $V = 4Q/(\pi D^2) = 11.71$ fps, so $\mathbf{R} = DV/\nu = 8.02 \times 10^5$, and the turbulent flow assumption is correct. Therefore

$$Q = 6.39 \text{ cfs} \qquad \textbf{\textit{ANS}}$$

(*c*) Using Mathcad, we establish a "solve block" containing four equations as in Sample Prob. 8.8*b*, except that in this case Eq. (8.10) is replaced by the energy equation (1) above. Also, the "Find" statement must be changed to Find(f, V, R, Q). The seven known variables are assigned the same values as in solution (*b*) above, and estimated values (without units) are assigned to the four unknown variables, say $f = 0.030$, $V = 5$ fps, $\mathbf{R} = 100{,}000$, and $Q = 10$ cfs.

The program delivers the result

$$f = 0.0201, \quad V = 11.71 \text{ fps}, \quad \mathbf{R} = 801{,}731, \quad Q = 6.39 \text{ cfs} \qquad \textbf{\textit{ANS}}$$

Because this $\mathbf{R}$ is for turbulent flow, the equation (8.40) used for f is valid, and so is the answer obtained for Q.

Note the simplicity and clarity of the Mathcad solution in comparison with the other two solutions.

If the pipe in this example discharged into a fluid that was at a pressure other than atmospheric, the proper value of p_2/γ would have to be used in the energy equation.

Another example of flow from a reservoir is that of a pipeline leading to an impulse turbine. In this case the pipe does not discharge freely, but ends in a nozzle (Fig. 8.23b), which has a known or assumed loss coefficient. The head loss in the nozzle, h'_n, is associated with the high issuing velocity head, and is therefore not a minor loss. The procedure is to employ the equation of continuity to place all losses in terms of the velocity head in the pipe. This is the logical choice for the "common unknown," because the trial-and-error solution will again be built around the pipe friction loss rather than the nozzle loss.

Sample Problem 8.10 As in Fig. 8.23*b*, suppose that the pipeline of Sample Prob. 8.9 is now fitted with a nozzle, at the end of which discharges a 2.5-in-diameter jet and that has a loss coefficient of 0.11. Find the flow rate.

Solution

Let point 2 now refer to the pipe at the base of the nozzle and point 3 be in the jet. The head loss in the nozzle is $0.11V_3^2/2g$. Writing the energy equation between 1 and 3, neglecting entrance loss,

$$260 + 0 + 0 = 0 + 0 + \frac{V_3^2}{2g} + 6000f\frac{V_2^2}{2g} + 0.11\frac{V_3^2}{2g}$$

By the continuity equation, $V_3^2/2g = (10/2.5)^4 V_2^2/2g = 256V_2^2/2g$. Thus

$$260 = (1.11 \times 256 + 6000f)\frac{V_2^2}{2g}$$

A trial value of f is selected. Let $f = 0.02$ for the first assumption. Then $260 = (284 + 120)V_2^2/2g$, from which

$$\frac{V_2^2}{2g} = \frac{260}{404} = 0.644 \text{ ft}$$

and $V_2 = \sqrt{(2(32.2)0.644} = 6.44$ fps. With $D''V = 10 \times 6.44 = 64.4$ and $e/D = 0.001$, Fig. 8.11 shows $f = 0.02$. In this case the first solution may be considered sufficiently accurate, but in general the value of f determined from the chart may be materially different from that assumed, and a second trial may be necessary.

The rate of discharge is $Q = A_2V_2 = 0.545 \times 6.44 = 3.51$ cfs, and

$$V_3 = 16V_2 = 16 \times 6.44 = 103.0 \text{ fps}$$

As additional information, $H_2 = p_2/\gamma + V_2^2/2g = 260 - 0.02 \times 6000 \times 0.644 = 182.72$ ft, and the pressure head $p_2/\gamma = 182.72 - 0.644 = 182.08$ ft.

This example shows that the addition of the nozzle has reduced the discharge from 6.40 to 3.51 cfs, but has increased the jet velocity from 11.74 to 103.0 fps. The head loss due to pipe friction is 77.3 ft and the head loss through the nozzle is 18.1 ft. (The head loss at entrance, which was neglected in the calculations, is approximately 0.3 ft.)

▭: Programmed computing aids (Appendix B) could help solve problems marked with this icon.

When solving Type 3 sizing problems, in general, the diameter so obtained will not be a standard pipe size, and the size selected will usually be the next largest commercially available size. In planning for the future, it must be recalled that scale deposits will increase the roughness and reduce the cross-sectional area. For pipes in water service, the absolute roughness e of old pipes (20 years and more) may increase over that of new pipes by three-fold for concrete or cement-lined steel, up to 20-fold for cast iron, and even to 40-fold for tuberculated wrought-iron and steel pipe. Substituting $V = 4Q/(\pi D^2)$ into Eq. (8.10) shows that for a constant value of f, Q varies as $D^{5/2}$. Hence for the case where minor losses are negligible and f is constant, to achieve a 100% increase in flow, the diameter need be increased by only 32%. This amounts to a 74% increase in cross-sectional area.

EXERCISES

8.24.1 An 8-in-diameter pipeline ($f = 0.028$) 500 ft long discharges a 3-in-diameter jet of water into the atmosphere at a point that is 250 ft below the water surface at intake. The entrance to the pipe is reentrant, with $k_e = 0.9$, and the nozzle loss coefficient is 0.045. Find the flow rate and the pressure head at the base of the nozzle.

8.24.2 An 18-cm-diameter pipeline ($f = 0.032$) 150 m long discharges a 6-cm-diameter jet of water into the atmosphere at a point that is 80 m below the water surface at intake. The entrance to the pipe is reentrant, with $k_e = 0.9$, and the nozzle loss coefficient is 0.055. Find the flow rate and the pressure head at the base of the nozzle.

8.24.3 A horizontal 4-in-diameter pipe ($f = 0.028$) projects into a body of water 2.5 ft below the surface. Considering all losses, find the pressure at a point 15 ft from the end of the pipe if the velocity is 12 fps and the flow is (*a*) into the body of water; (*b*) out of the body of water.

8.24.4 A horizontal 10-cm-diameter pipe ($f = 0.027$) projects into a body of water 1 m below the surface. Considering all losses, find the pressure at a point 5 m from the end of the pipe if the velocity is 4 m/s and the flow is (*a*) into the body of water; (*b*) out of the body of water.

8.24.5 A 450-ft-long pipeline runs between two reservoirs, both ends being under water, and the intake end is square-edged and nonprojecting. The difference between the water levels in the two reservoirs is 150 ft. (*a*) What is the discharge if the pipe diameter is 12 in and $f = 0.028$? (*b*) When this same pipe is old, assume that the growth of tubercles has reduced the diameter to 11.25 in and that $f = 0.06$. What will the rate of discharge be then?

8.24.6 Solve Exer. 8.24.5*a* when f is unknown but given that $e = 0.005$ ft and that the water temperature is 60°F. What, then, is the value of f?

: Programmed computing aids (Appendix B) could help solve problems marked with this icon.

8.25 PIPELINE WITH PUMP OR TURBINE

If a pump lifts a fluid from one reservoir to another, as in Fig. 8.24, not only does it do work in lifting the fluid the height Δz, but also it has to overcome the friction loss in the suction and discharge piping. This friction head is equivalent to some added lift, so that the effect is the same as if the pump lifted the fluid a height $\Delta z + \sum h_L$. Hence the power delivered to the liquid by the pump is $\gamma Q(\Delta z + \sum h_L)$. The power required to run the pump is greater than this, depending on the efficiency of the pump. The total pumping head h_p for this case is

$$h_p = \Delta z + \sum h_L \tag{8.63}$$

If the pump discharges a stream through a nozzle, as shown in Fig. 8.25, not only has the liquid been lifted a height Δz, but also it has received a kinetic energy head of $V_2^2/2g$, where V_2 is the velocity of the jet. Thus the total pumping head is now

$$h_p = \Delta z + \frac{V_2^2}{2g} + \sum h_L \tag{8.64}$$

In any case the total pumping head may be determined by writing the energy equation between any point upstream from the pump and any other point downstream, as in Eq. (5.30). For example, if the upstream reservoir were at a higher elevation than the downstream one, then the Δzs in the two foregoing equations would have negative signs.

The machine that is employed for converting the energy of flow into mechanical work is called a ***turbine***. In flowing from the upper tank in Fig. 8.26 to the lower, the fluid loses potential energy head equivalent to Δz. This

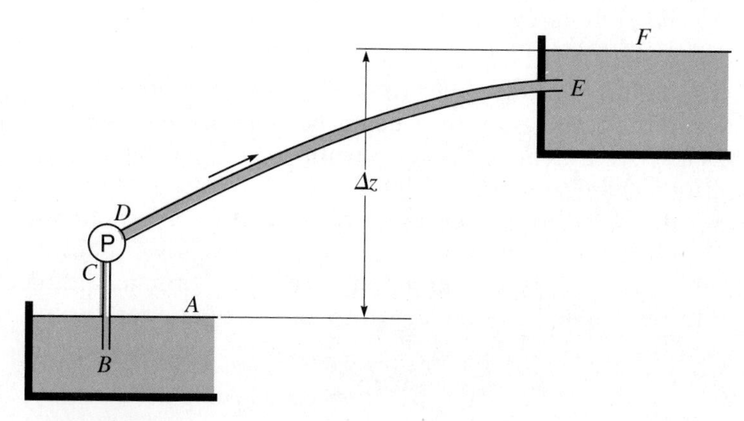

Figure 8.24
Pipeline with pump between two reservoirs.

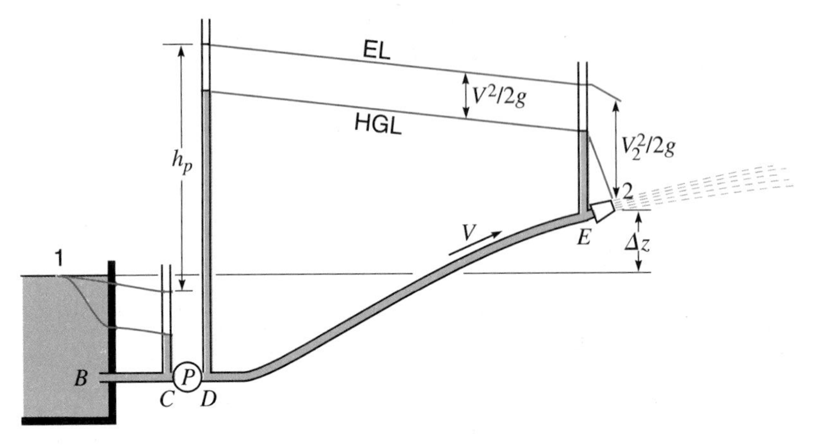

Figure 8.25
Pipeline with pump and nozzle.

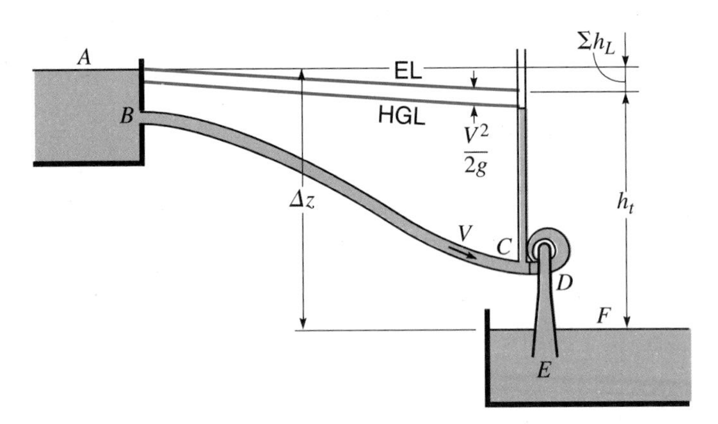

Figure 8.26
Pipeline with turbine.

energy is expended in two ways, part of it in hydraulic friction in the pipe and the remainder in the turbine. Of that energy which is delivered to the turbine, a portion is lost in hydraulic friction within the turbine and the rest is converted into mechanical work.

The power delivered to the turbine is decreased by the friction loss in the pipeline, and its value is given by $\gamma Q(\Delta z - \sum h_L)$. The power delivered by the machine is less than this, depending upon both the hydraulic and mechanical losses of the turbine. The head under which the turbine operates is

$$h_t = \Delta z - \sum h_L \tag{8.65}$$

where $\sum h_L$ is the loss of head in the supply line plus the submerged discharge loss, but does *not* include the head loss in the ***draft tube*** (DE in

Fig. 8.26), since the draft tube is considered an integral part of the turbine. A draft tube has a gradually increasing cross-sectional area, which reduces the velocity at discharge. This enhances the efficiency of the turbine because of the reduced head loss at discharge (Secs. 8.19 and 16.8). It should be noted that the h_t of Eq. (8.65) represents the energy head removed from the fluid by the turbine; this, of course, is identical to the energy head transferred to the turbine from the fluid.

Sample Problem 8.11 In this problem we shall assume that the Reynolds number is high enough to assure turbulent flow. A pump is located 15 ft above the surface of a liquid ($\gamma = 52$ lb/ft^3) in a closed tank. The pressure in the space above the liquid surface is 5 psi. The suction line to the pump is 50 ft of 6-in-diameter pipe ($f = 0.025$). The discharge from the pump is 200 ft of 8-in-diameter pipe ($f = 0.030$). This pipe discharges in a submerged fashion to an open tank whose free liquid surface is 10 ft lower than the liquid surface in the pressure tank. If the pump puts 2.0 hp into the liquid, determine the flow rate and find the pressure in the pipe on the suction side of the pump.

Solution

From Eq. (5.32):
$$\text{HP} = \frac{\gamma Q h_p}{550} = 2 = \frac{52 Q h_p}{550}$$

Thus
$$h_p = \frac{21.2}{Q}$$

Writing the energy equation from one liquid surface to the other,

$$\frac{5(144)}{52} - 0.5\frac{V_6^2}{2g} - 0.025\left(\frac{50}{6/12}\right)\frac{V_6^2}{2g} + h_p - 0.030\left(\frac{200}{8/12}\right)\frac{V_8^2}{2g} = -10 + \frac{V_8^2}{2g}$$

Substituting $V_6 = Q/0.1963$ and $V_8 = Q/0.349$, this reduces to

$$23.9 + h_p - 2.49Q^2 = 0$$

or

$$23.9 + \frac{21.2}{Q} - 2.49Q^2 = 0$$

By trial, $Q = -2.48, -0.987$, or 3.47 cfs (3 roots); so

$$Q = 3.47 \text{ cfs} \qquad \textit{ANS}$$

To obtain the pressure at the suction side of the pump,

$$\frac{5(144)}{52} - 0.5\frac{V_6^2}{2g} - 0.025\left(\frac{50}{6/12}\right)\frac{V_6^2}{2g} = 15 + \frac{p}{\gamma} + \frac{V_6^2}{2g}$$

where
$$V_6 = \frac{3.47}{0.1963} = 17.68 \text{ fps}$$

from which
$$\frac{p}{\gamma} = -20.6 \text{ ft} \qquad \textit{ANS}$$

▣: Programmed computing aids (Appendix B) could help solve problems marked with this icon.

or $p_2 = -20.6(52/144) = -7.43$ psi, which is equivalent to $(7.43/14.70)29.9 =$ 15.12 inHg vacuum.

In this type of problem one should check the absolute pressure against the vapor pressure of the liquid to see that vaporization does not occur.

EXERCISES

8.25.1 An 80-in-diameter pipe ($f = 0.025$) 7252 ft long delivers water to a powerhouse at a point 1500 ft lower in elevation than the water surface at intake. When the pipe delivers 450 cfs, what is the horsepower delivered to the plant?

8.25.2 A 10-in pipeline ($f = 0.020$) is 3 miles long. If 4 cfs of water is to be pumped through it, with a total actual lift of 25 ft, what will be the power required? The pump efficiency is 72%.

8.25.3 A 25-cm pipeline ($f = 0.025$) is 4.7 km long. If 0.1 m^3/s of water is to be pumped through it, with a total actual lift of 10.5 m, what will be the power required? The pump efficiency is 75%.

8.25.4 A 12-in pipe 9200 ft long ($f = 0.024$) discharges freely into the air at an elevation 18 ft below the surface of the water at intake. It is necessary that the flow be doubled by inserting a pump. If the efficiency of the pump is 73%, what will be the power required?

8.25.5 A 30-cm pipe 3400 m long ($f = 0.022$) discharges freely into the air at an elevation 5.6 m below the surface of the water at intake. It is necessary that the flow be doubled by inserting a pump. If the efficiency of the pump is 76%, what will be the power required?

8.25.6 Refer to Fig. 8.25. When the pump is delivering 1.2 cfs of water, a pressure gage at D reads 25 psi, while a vacuum gage at C reads 10 inHg. The pressure gage at D is 2 ft higher than the vacuum gage at C. The pipe diameters are 4 in for the suction pipe and 3 in for the discharge pipe. Find the power delivered to the water.

8.25.7 Refer to Fig. 8.25. When the pump is delivering 35 L/s of water, a pressure gage at D reads 175 kN/m^2, while a vacuum gage at C reads 25 cmHg. The pressure gage at D is 60 cm higher than the vacuum gage at C. The pipe diameters are 10 cm for the suction pipe and 7.5 cm for the discharge pipe. Find the power delivered to the water.

8.25.8 If in Sample Problem 8.11 the vapor pressure of the liquid is 1.9 psia and the atmospheric pressure is 14.5 psia, what is the maximum theoretical flow rate?

8.25.9 In Fig. 8.26 assume the pipe diameter is 12 in, $f = 0.021$, $BC = 200$ ft, and $\Delta z = 120$ ft. The entrance to the pipe at the intake is flush with the wall, and discharge losses are negligible. (a) If $Q = 8$ cfs of water, what is the head supplied to the turbine? (*b*) What is the power delivered by the turbine if its efficiency is 75%.

8.25.10 In Fig. 8.26 assume the pipe diameter is 30 cm, $f = 0.021$, $BC = 60$ m, and $\Delta z = 36.5$ m. The entrance to the pipe at the intake is flush with the wall, and discharge losses are negligible. (*a*) If $Q = 225$ L/s of water, what is the head supplied to the turbine? (*b*) What is the power delivered by the turbine if its efficiency is 75%.

8.26 BRANCHING PIPES

For convenience, let us consider three pipes connected to three reservoirs as in Fig. 8.27 and connected together or branching at the common junction point *J*. Actually, any of the pipes may be considered to be connected to some other destination than a reservoir by simply replacing the reservoir with a piezometer tube in which the water level is the same as the reservoir surface. We shall suppose that all the pipes are sufficiently long that minor losses and velocity heads may be neglected.

We name the pipes and flows and corresponding friction losses as shown in the diagram. The continuity and energy equations require that the flow entering the junction equal the flow leaving it and that the pressure head at *J* (which may be represented schematically by the open piezometer tube shown, with water at elevation *P*) be common to all pipes.

There being no pumps, the elevation of *P* must lie between the surfaces of reservoirs *A* and *C*. If *P* is level with the surface of reservoir B then h_2 and Q_2 are both zero. If *P* is above the surface of reservoir *B* then water must flow into *B* and $Q_1 = Q_2 + Q_3$. If *P* is below the surface of reservoir *B* then the flow must be out of *B* and $Q_1 + Q_2 = Q_3$. So for the situation shown in Fig. 8.27 we have the following governing conditions:

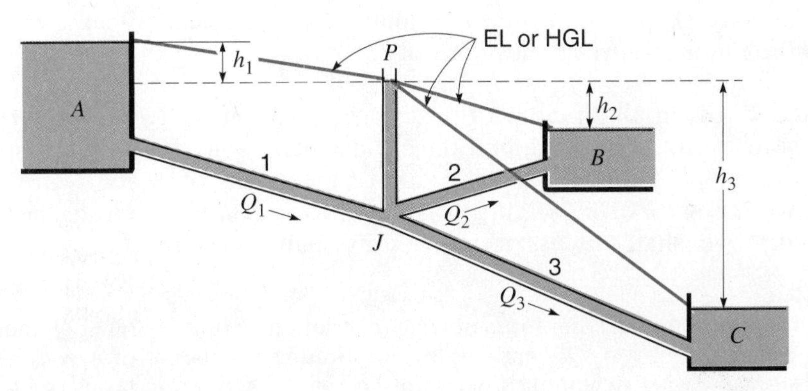

Figure 8.27
Branching pipes.

1. $Q_1 = Q_2 + Q_3$.
2. Elevation of P is common to all.

The diagram suggests several different problems or cases, three of which will be discussed below using different methods of solution.

Rigorous Solutions

When we know the pipe wall material, we can estimate its e value (Table 8.1) and we know that the friction factor f varies with the e/D of the pipe and the Reynolds number of the flow. Because we are not considering minor losses, we can use the equations and methods of Secs. 8.11–8.14. In particular, using only a basic scientific calculator, we can solve pipelines for h_L (Type 1 problems) using the Haaland equation (8.41) with Eq. (8.10); we can solve for V or Q (Type 2 problems) using Eq. (8.42); and more rarely, we can solve for D (Type 3 problems) using Eqs. (8.43)–(8.48). These equations are preferred because they avoid trial and error, which can become quite confusing when combined with other trial-and-error techniques needed to solve for branching flow. These, and the variety of approaches used to "manually" solve the different types of problems that can occur, are illustrated in the following cases.

Case **1.** Given all pipe data (lengths, diameters, and materials for e values), the surface elevations of two reservoirs, and the flow to or from one of these two, find the surface elevation of the third reservoir.

This problem can be solved directly. Suppose that Q_1 and the elevations of A and B are given. Then the head loss h_1 may be determined directly (Type 1), using Fig. 8.11 or Eq. (8.41) to find the proper value of f. Knowing h_1 fixes P, so h_2 is now easily obtained. Knowing h_2 enables the flow in pipe 2 to be determined directly using the Type 2 equation (8.42). Condition 1 (continuity at junction J) then determines Q_3, which in turn enables h_3 to be found directly (Type 1), in the same manner as for line 1. Finally, P and h_3 define the required surface elevation of C.

Case **2.** Given all pipe data, the surface elevations of two reservoirs, and the flow to or from the third, find the surface elevation of the third reservoir.

Let us suppose that Q_2 and the surface elevations of A and C are given. Then we know $h_1 + h_3 = \Delta h_{13}$, say. Various solution approaches may be used;[24] we shall discuss a more convenient one. In this, we assume

[24] Other approaches include (a) assuming distributions of the flows Q_1 and Q_3, knowing that $Q_1 - Q_3 = Q_2$, and (b) by substituting for the hs in $h_1 + h_3 = \Delta h_{13}$ using Eq. (8.10) with V_3 written in terms of Q_3, and V_1 written in terms of $Q_2 + Q_3$, and successively solving the resulting quadratic equation for Q_3 while converging on f values.

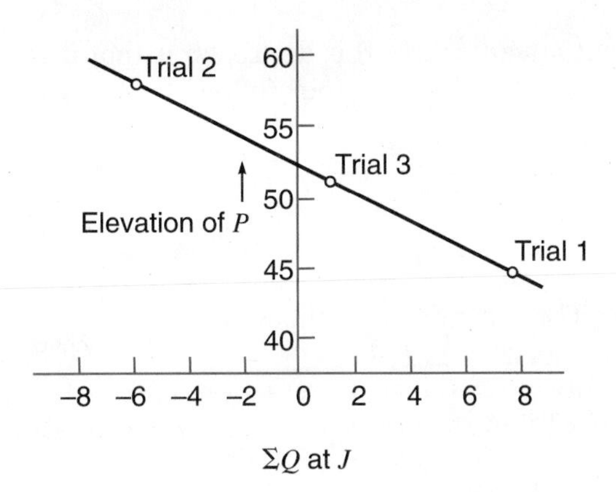

Figure 8.28

the elevation of P, which yields values for h_1 and h_3, and so Q_1 and Q_3 via Eq. (8.42). If these do not satisfy the discharge relation at J then P must be adjusted until they do. To help us converge on the correct elevation of P, we can plot the results of each assumption on a graph like Fig. 8.28. For $\sum Q$ at J, inflows to J are taken as positive and outflows as negative. Two or three points, with one fairly close to the vertical axis, determine a curve that intersects that axis at the equilibrium level of P, where $\sum Q = 0$, as required. Last, h_2 can be determined from Q_2 and Eqs. (8.41) and (8.10), and the required surface elevation of B found.

Case* 3.** Given all pipe lengths and diameters and the elevations of all three reservoirs, find the flow in each pipe. This is the classic ***three-reservoir problem, and it differs from the foregoing cases in that it is not immediately evident whether the flow is *into* or *out of* reservoir B. This direction is readily determined by first assuming no flow in pipe 2; that is, the piezometer level P is assumed at the elevation of the surface of B. The head losses h_1 and h_3 then determine the flows Q_1 and Q_3 via Eq. (8.42). If $Q_1 > Q_3$ then P must be raised to satisfy continuity at J, causing water to flow *into* reservoir B, and we shall have $Q_1 = Q_2 + Q_3$; if $Q_1 < Q_3$ then P must be lowered to satisfy continuity at J, causing water to flow *out* of reservoir B, and we shall have $Q_1 + Q_2 = Q_3$. From here on the solution proceeds by adjusting P as for Case 2 above.

***Note*.** For any of the above three cases, manual trial and error can be avoided by setting up the governing equations and solving them simultaneously using mathematics software like Mathcad. There will be the usual four governing equations for each line, a flow continuity equation for the junction, and, depending on the case addressed, up to two equations relating the various head losses. With so many unknowns to solve for, the success of the procedure becomes more sensitive to the guessed values, and different guesses may have to be tried. The great advantage of this approach is that it is so straightforward. It is illustrated in part (*b*) of the following sample problem.

SAMPLE PROBLEM 8.12 Given that, in Fig. 8.27, pipe 1 is 6000 ft of 15 in diameter, pipe 2 is 1500 ft of 10 in diameter, and pipe 3 is 4500 ft of 8 in diameter, all asphalt-dipped cast iron. The elevations of the water surfaces in reservoirs *A* and *C* are 250 ft and 160 ft respectively, and the discharge Q_2 of 60°F water into reservoir *B* is 3.3 cfs. Find the surface elevation of reservoir *B*: (*a*) using only a basic scientific calculator; (*b*) using Mathcad.

Solution

This is a Case 2 problem.
Table A.1 for water at 60°F: $\nu = 1.217 \times 10^{-5}$ ft²/sec.

Line:	1	2	3
L, ft	6000	1500	4500
D, ft	1.25	10/12	8/12
e, ft (Table 8.1)	0.0004	0.0004	0.0004

(*a*) Find the elevation of *P* by trial and error.

L/D	4800	1800	6750
A, ft²	1.227	0.545	0.349
e/D	0.000 32	0.000 48	0.0006

Elevation of P lies between 160 and 250 ft. Calculate *V* from Eq. (8.42). Trials:

El. *P*	h_1	h_3	V_1	V_3	Q_1	Q_3	ΣQ	Move *P*?
200	50	40	6.444	4.481	7.907	1.564	+3.04	Up
230	20	70	4.013	5.984	4.925	2.088	−0.463	Down

Interpolation (Fig. 8.28): (230 − El. *P*)/(230 − 200) = 0.463/(0.463 + 3.04); El. *P* = 226.03.

226	24	66	4.412	5.805	5.414	2.026	+0.088	Up

Interpolation (Fig. 8.28): (230 − El. *P*)/(230 − 226) = 0.463/(0.463 + 0.088); El. *P* = 226.64.

: Programmed computing aids (Appendix B) could help solve problems marked with this icon.

Close enough! *Note:* these adjustments are very suitable for making on a spreadsheet.

$$V_2 = \frac{Q_2}{A_2} = \frac{3.3}{0.545} = 6.055 \text{ fps}; \quad \mathbf{R}_2 = \frac{D_2 V_2}{\nu} = 416{,}500$$

Eq. (8.41): $f_2 = 0.01761$; Eq. (8.10): $h_2 = 18.05$ ft

$$\text{El. } B = \text{El. } P - h_2 = 226.64 - 18.05 = 208.59 \text{ ft} \qquad \textit{ANS}$$

(*b*) Using Mathcad, we note that there are 14 governing equations. Two of these are

$$h_1 + h_3 = 250 - 160 = 90, \qquad Q_1 = Q_2 + Q_3$$

The remaining 12 equations are, for each of the three pipes: the head loss Eq. (8.10), the Colebrook Eq. (8.40), Eq. (8.1) for Reynolds number, and the continuity equation $Q = \pi D^2 V/4$. Open a Mathcad file and place these 14 equations in a "solve block," which begins with "Given" and ends with "Find(*f1, V1, R1, Q1, h1, f2, V2, R2, h2, f3, V3, R3, Q3, h3*) =" (note that the known *Q2* is omitted).

The known values are $g = 32.2$, $\nu = 1.217 \times 10^{-5}$, $Q_2 = 3.3$, and the nine values tabulated above for *L*, *D*, and *e* in each of the three lines. Assign these numerical values above the solve block.

Guessed values of the 14 unknowns are (say) 0.03, 10, 800,000, and 10 for each of the three values of *f*, *V*, *R*, and *h* respectively, and 10 for *Q1* and *Q3*. Assign these above the solve block.

Mathcad solves the 14 equations for the 14 unknowns, with the following results:

	f	V	$\mathbf{R}$	Q	h_L
Line 1:	0.0166	4.35	446,670	5.34	23.34
Line 2:	0.0177	6.05	414,300	(given)	18.13
Line 3:	0.0187	5.83	319,632	2.04	66.66

So El. B = El. A $- h_1 - h_2 = 250 - 23.34 - 18.13 = 208.53$ ft *ANS*

Note: observe how simple and clear the Mathcad solution is.

Sample Problem 8.13 With the sizes, lengths, and material of pipes given in Sample Prob. 8.12, suppose that the surface elevations of reservoirs *A*, *B*, and *C* are 525 ft, 500 ft, and 430 ft respectively. (*a*) Does water enter or leave reservoir *B*? (*b*) Find the flow rates of 60°F water in each pipe.

Solution

This is a Case 3 problem. Find the elevation of *P* by trial and error.

Table A.1 for water at 60°F: $\nu = 1.217 \times 10^{-5}$ ft²/sec.

The tabulated line data are the same as for Sample Prob. 8.12.

: Programmed computing aids (Appendix B) could help solve problems marked with this icon.

(*a*) *Trial 1.* Try P at elevation of reservoir surface $B = 500$ ft:

Line:	1	2	3
h, ft	25	0	70
$\sqrt{2gDh_L/L}$, fps	0.579	0	0.817
V, fps [Eq. (8.42)]	4.51	0	5.98
$Q = AV$, cfs	5.53	0	2.09

At J, $\sum Q$ = inflow − outflow = 5.53 − 2.09 = 3.44 cfs. This must be zero, so P must be raised (to reduce Q_1 and increase Q_3); then water will flow *into* reservoir B. ***ANS***

(*b*) *Trial 2.* Try P at elevation 510 ft:

	1	2	3
h, ft	15	10	80
$\sqrt{2gDh_L/L}$, fps	0.449	0.598	0.874
V, fps [Eq. (8.42)]	3.46	4.49	6.41
$Q = AV$, cfs	4.24	2.42	2.24

At J, $\sum Q = 4.23 - 2.42 - 2.24 = -0.42$ cfs. By interpolation (using Fig. 8.28),

$$\frac{510 - \text{EL. } P}{510 - 500} = \frac{0.42}{0.42 + 3.44}; \qquad \text{El. } P = 508.91 \text{ ft}$$

Trial 3. Try P at elevation 508.9 ft:

	1	2	3	
h, ft	16.1	8.9	78.9	
$\sqrt{2gDh_L/L}$, fps	0.465	0.564	0.868	
V, fps [Eq. (8.42)]	3.59	4.19	6.36	
$Q = AV$, cfs	4.40	2.28	2.22	***ANS***

At J, $\sum Q = 4.40 - 2.28 - 2.22 = -0.10$ cfs. Close enough!
Note: these adjustments are very suitable for making on a spreadsheet.

Approximate Solutions

If the value of the friction factor (f, C_{HW}, or Manning's n) is given for each pipeline, this will lead to a less accurate solution because we know that pipe friction in fact varies with Reynolds number **R**. Although we have called such solutions approximate, we see that they are more equivalent to the empirical methods of Sec. 8.15.

When the friction factor does not vary with **R**, it is most convenient to represent the head loss equations in the form

$$h_L = KQ^n \tag{8.66}$$

where we note that n is an exponent and not Manning's n, which we shall here represent by n_m. By rearranging Eq. (8.10) when f is constant and replacing V by $4Q/(\pi D^2)$, we obtain

$$\text{Darcy–Weisbach:} \qquad K = \frac{8fL}{\pi^2 gD^5}, \qquad n = 2$$

By rearranging the empirical equation (8.49*a*), we obtain

Hazen–Williams, BG units:

$$K = \frac{4.727L}{C_{HW}^{1.852}D^{4.87}}, \quad n = 1.852$$

For SI units we must replace the 4.727 by 10.675 in the Hazen–Williams K. By rearranging the empirical equation (8.50*a*), we obtain

Manning, BG units:

$$K = \frac{4.66\ Ln_m^{\ 2}}{D^{16/3}}, \quad n = 2$$

For SI units, we must replace the 4.66 by 10.29 in the Manning K.

We see that $h_L = KQ^2$ except when working with the Hazen–Williams coefficient, and we notice that K is a property of the pipe alone. When we need to solve for discharge, we can of course rearrange Eq. (8.66) into

$$Q = \left(\frac{h_L}{K}\right)^{1/n} \tag{8.67}$$

With these approximate solutions, the approaches used to solve the above three cases follow exactly the same steps as for the rigorous solutions, but instead of using the equations of Secs. 8.11–8.14, we simply use Eqs. (8.66) and (8.67). Of course, the K and n value for each pipeline must be determined first. It is interesting that Case 2 can easily be solved *directly* if $n = 2$ (Darcy– Weisbach or Manning) by writing the known elevation difference $\Delta h_{13} = h_1 + h_3 = K_1Q_1^{\ 2} + K_3Q_3^{\ 2} = K_1(Q_2 + Q_3)^2 + K_3Q_3^{\ 2}$, which is a quadratic equation in Q_3, the only unknown.

Sample Problem 8.14 In Fig. 8.27 pipe 1 is 30 cm diameter and 900 m long, pipe 2 is 20 cm diameter and 250 m long, and pipe 3 is 15 cm diameter and 700 m long. The Hazen–Williams coefficient for all pipes is 120. The surface elevations of reservoirs *A, B,* and *C* are 160 m, 150 m, and 120 m, respectively. (*a*) Does water enter or leave reservoir *B*? (*b*) Find the flow rate in each pipe.

Solution

This is a Case 3 problem. Find the elevation of *P* by trial and error. Use the Hazen–Williams form of Eq. (8.66) in SI units:

$$K = \frac{10.675L}{C_{HW}^{\ \ 1.852}D^{4.87}}, \quad n = 1.852$$

Line:	1	2	3
L, m	900	250	700
D, m	0.3	0.2	0.15
L/D	3 000	1 250	4 667
C_{HW}	120	120	120
n	1.852	1.852	1.852
K	480	961	10 924

: Programmed computing aids (Appendix B) could help solve problems marked with this icon.

(*a*) *Trial 1.* Try P at elevation of reservoir surface $B = 150$ m:

h, m	10	0	30
Q, m³/s [Eq. (8.67)]	0.1236	0	0.0414

At J, ΣQ = inflow − outflow = 0.1236 − 0.0414 = 0.0822 m³/s. This must be zero, so P must be raised (to reduce Q_1 and increase Q_3), then water will flow *into* reservoir B ***ANS***

(*b*) *Trial 2.* Try P at elevation 155 m:

h, m	5	5	35
Q, m³/s [Eq. (8.67)]	0.0850	0.0584	0.0450

At J, $\Sigma Q = 0.0850 - 0.0584 - 0.1450 = -0.0184$ m³/s. This must be zero, so P must be lowered. By interpolation (Fig. 8.28),

$$\frac{155 - \text{El. } P}{155 - 150} = \frac{0.0184}{0.0184 + 0.0822}; \qquad \text{El. } P = 154.09 \text{ m}$$

Trial 3. Try P at elevation 154 m:

h, m	6	4	34	
Q, m³/s [Eq. (8.67)]	0.0938	0.0518	0.0443	***ANS***

At J, $\Sigma Q = -0.0023$ m³/s. This is close enough.
Note: these adjustments are very suitable for making on a spreadsheet.

8.27 PIPES IN SERIES

The discussion in Sec. 8.24 was restricted to the case of a single pipe. If the pipe is made up of sections of different diameters, as shown in Fig. 8.29, the

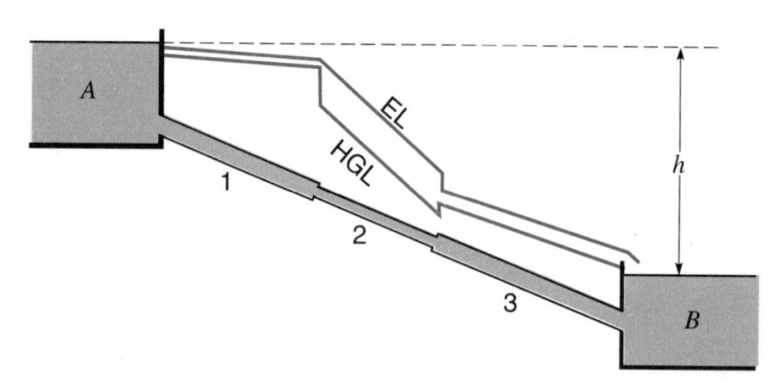

Figure 8.29
Pipes in series.

continuity and energy equations establish the following two simple relations that must be satisfied:

$$Q = Q_1 = Q_2 = Q_3 = \dots \tag{8.68}$$

$$h_L = h_{L1} + h_{L2} + h_{L3} + \dots \tag{8.69}$$

If the rate of discharge Q is given, the problem is straightforward. The head loss may be found directly by adding the contributions from the various sections, as in Eq. (8.69). If empirical coefficients or constant f values are given, we can do this using Eq. (8.66) and the appropriate values of K and n. If, however, the pipe material or e is given, this is preferred, because the Darcy–Weisbach approach is more accurate. Then we use Eq. (8.10) to find the individual head loss contributions after finding e/D, V, **R**, and f for each pipe.

If the total head loss h_L is given and the flow is required, the problem is a little more involved. Using the empirical equations, we again substitute Eq. (8.66) into Eq. (8.69), to get

$$h_L = K_1 Q_1^n + K_2 Q_2^n + K_3 Q_3^n + \dots$$

But since all the Qs are equal from Eq. (8.68), this becomes

$$h_L = (K_1 + K_2 + K_3 + \dots)Q^n = \left(\sum K\right)Q^n \tag{8.70}$$

So, knowing the pipe information and the empirical equation to be used, we can solve for Q.

If we wish to use the more accurate Darcy–Weisbach approach to find Q, we must note that in Eq. (8.70) each K has now become a function of a different f. The preferred manual method of solution is similar to the above, and is known as the ***equivalent-velocity-head method***. Substituting from Eq. (8.10) into Eq. (8.69) and including minor losses if needed (usually if $L/D < 1000$),

$$h_L = \left(f_1 \frac{L_1}{D_1} + \sum k_1\right)\frac{V_1^2}{2g} + \left(f_2 \frac{L_2}{D_2} + \sum k_2\right)\frac{V_2^2}{2g} + \dots$$

Using the equation of continuity, we know $D_1^2 V_1 = D_2^2 V_2 = D_3^2 V_3$, etc., from which all the velocities may be expressed in terms of one chosen velocity. So, by assuming reasonable values for each f (e.g., from Eq. (8.39) or Fig. 8.11), for any pipeline, however complex, the total head loss may be written as

$$h_L = K \frac{V^2}{2g} \tag{8.71}$$

where V is the chosen velocity. This equation may be solved for the chosen V, and so the V and **R** and f values obtained for each pipe. For better accuracy, the assumed values of f should be replaced by the values just obtained, and an improved solution obtained. When the f values have converged V is correct, and Q may be calculated.

SAMPLE PROBLEM 8.15 Suppose in Fig. 8.29 the pipes 1, 2, and 3 are 300 m of 30 cm diameter, 150 m of 20 cm diameter, and 250 m of 25 cm diameter, respectively, of new cast iron and are conveying 15°C water. If $h = 10$ m, find the rate of flow from A to B.

Solution

Table 8.1 for cast iron pipe: $e = 0.25$ mm $= 0.000\ 25$ m.

Pipe:	1	2	3
L, m	300	150	250
D, m	0.3	0.2	0.25
e/D	0.000 833	0.001 25	0.001 00
f_{min} (Fig. 8.11)	0.019	0.021	0.020

Assuming these friction factor values,

$$h = 10 = 0.019\left(\frac{300}{0.3}\right)\frac{V_1^2}{2g} + 0.021\left(\frac{150}{0.2}\right)\frac{V_2^2}{2g} + 0.020\left(\frac{250}{0.25}\right)\frac{V_3^2}{2g}$$

From continuity, $$\frac{V_2^2}{2g} = \frac{V_1^2}{2g}\left(\frac{D_1}{D_2}\right)^4 = \frac{V_1^2}{2g}\left(\frac{30}{20}\right)^4 = 5.06\frac{V_1^2}{2g}$$

Similarly, $$\frac{V_3^2}{2g} = 2.07\frac{V_1^2}{2g}$$

and thus $$10 = \frac{V_1^2}{2g}\left(0.019\frac{1000}{1} + 0.021\frac{750}{1}5.06 + 0.020\frac{1000}{1}2.07\right)$$

from which $$\frac{V_1^2}{2g} = 0.071 \text{ m}$$

Hence $$V_1 = \sqrt{2(9.81 \text{ m/s}^2)(0.072 \text{ m})} = 1.18 \text{ m/s}$$

The corresponding values of **R** are 0.31×10^6, 0.47×10^6, and 0.37×10^6; the corresponding friction factors are only slightly different from those originally

: Programmed computing aids (Appendix B) could help solve problems marked with this icon.

assumed, since the flow is at Reynolds numbers very close to those at which the pipes behave as rough pipes. Hence

$$Q = A_1V_1 = \tfrac{1}{4}\pi(0.30)^2 1.18 = 0.083 \text{ m}^3/\text{s} \qquad \textbf{\textit{ANS}}$$

Greater accuracy would have been obtained if the friction factors had been adjusted to match the pipe-friction chart more closely or were calculated by Eq. (8.41), and if minor losses had been included. In that case $Q = 0.081 \text{ m}^3/\text{s}$.

Manual iteration for f may be avoided by solving simultaneous equations using mathematics software like Mathcad. There are the usual four equations for each pipe, plus Eq. (8.69); if necessary, minor losses may easily be accounted for by using head loss equations with the form of Eq. (8.60). For the three pipes of Fig. 8.29, for example, there are therefore 13 simultaneous equations, which may be solved in the usual manner (see Sample Prob. 8.12*b*) for 13 unknowns. These unknowns are either the flow rate or the total head loss, and three values each of h_L, V, $\mathbf{R}$, and f.

8.28 PIPES IN PARALLEL

In the case of flow through two or more parallel pipes, as in Fig. 8.30, the continuity and energy equations establish the following relations that must be satisfied:

$$Q = Q_1 + Q_2 + Q_3 \tag{8.72}$$

$$h_L = h_{L1} = h_{L2} = h_{L3} \tag{8.73}$$

as the pressures at A and B are common to all pipes. Problems may be posed in various ways.

If the head loss h_L is given, the problem is straightforward. The head loss may be found directly by adding the contributions from the various pipes, as in Eq. (8.72). If empirical coefficients or constant f values are given, we can

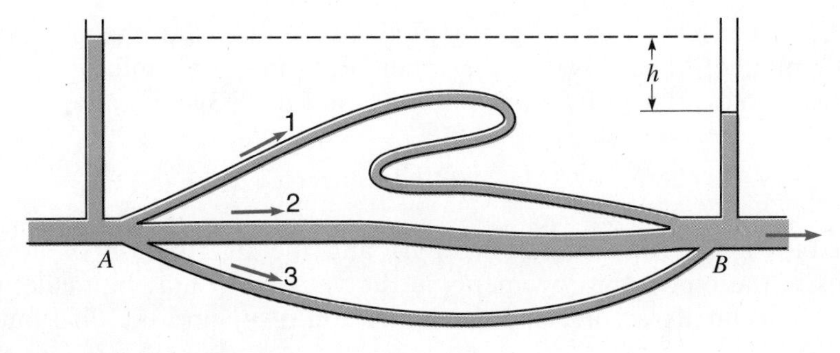

Figure 8.30
Pipes in parallel.

do this using Eq. (8.67) and the appropriate values of K and n. If, however, the pipe material or e is given, this is preferred because the Darcy–Weisbach approach is more accurate. Then we have an independent Type 2 problem for each pipe (see Secs. 8.13–8.14), which can be solved directly by Eq. (8.42) for example.

If the total flow Q is given and the head loss and individual flows are required, the problem is a little more involved. Using the empirical equations, we again substitute Eq. (8.67) into Eq. (8.72), to get

$$Q = \left(\frac{h_{L1}}{K_1}\right)^{1/n} + \left(\frac{h_{L2}}{K_2}\right)^{1/n} + \left(\frac{h_{L3}}{K_3}\right)^{1/n} + \ldots$$

But since all the h_Ls are equal from Eq. (8.73), this becomes

$$Q = (h_L)^{1/n}\left[\left(\frac{1}{K_1}\right)^{1/n} + \left(\frac{1}{K_2}\right)^{1/n} + \left(\frac{1}{K_3}\right)^{1/n} + \ldots\right] = (h_L)^{1/n} \sum \left(\frac{1}{K}\right)^{1/n} \tag{8.74}$$

So, knowing the pipe information and the empirical equation to be used, we can solve for h_L. We can then find the individual flows using Eq. (8.67).

If we wish to use the more accurate Darcy–Weisbach approach to find h_L and the individual Qs, we must note that in Eq. (8.74) now each K has become a function of a different f. The preferred manual method of solution is similar to the above. Writing Eq. (8.10) for each line, including minor losses if needed,

$$h_L = \left(f\frac{L}{D} + \sum k\right)\frac{V^2}{2g}$$

where $\sum k$ is the sum of the minor-loss coefficients, which may usually be neglected if the pipe is longer than 1000 diameters. Solving for V and then Q, the following is obtained for pipe 1:

$$Q_1 = A_1 V_1 = A_1\sqrt{\frac{2gh_L}{f_1(L_1/D_1) + \sum k_1}} = C_1\sqrt{h_L} \tag{8.75}$$

where C_1 is constant for the given pipe, except for the change in f with Reynolds number. The flows in the other pipes may be similarly expressed, using reasonable values of f from Fig. 8.11 or Eq. (8.39). Finally, Eq. (8.72) becomes

$$Q = C_1\sqrt{h_L} + C_2\sqrt{h_L} + C_3\sqrt{h_L} = (C_1 + C_2 + C_3)\sqrt{h_L} \tag{8.76}$$

This enables a first determination of h_L and the distribution of flows and velocities in the pipes. Improvements in the values of f may be made next, if indicated, and finally a corrected determination of h_L and the distribution of flows.

If the turbulent flow equation (8.40) or (8.41) is used to obtain f, it is important to remember to confirm that the Reynolds number is in the turbulent range. We can pre-check the likelihood of laminar flow occurring in any of the pipes by calculating an "average" flow velocity from the total flow divided by the total area of all the pipes, and using this to obtain an indicator **R** for each pipe.

Sample Problem 8.16 Three pipes *A*, *B*, and *C* are interconnected as shown. The pipe characteristics are given as follows:

Pipe	D (in)	L (ft)	f
A	6	2000	0.020
B	4	1600	0.032
C	8	4000	0.024

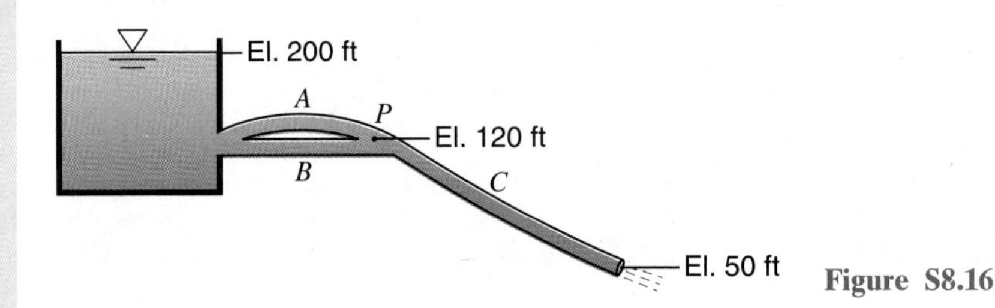

Figure S8.16

Find the rate at which water will flow in each pipe. Find also the pressure at point *P*. All pipe lengths are much greater than 1000 diameters, therefore minor losses may be neglected.

Solution

Energy Eq.:
$$200 - 0.020\frac{2000}{6/12}\frac{V_A^2}{2g} - 0.024\frac{4000}{8/12}\frac{V_C^2}{2g} = 50 + \frac{V_C^2}{2g}$$

i.e.,
$$150 = 80\frac{V_A^2}{2g} + 145\frac{V_C^2}{2g}$$

Continuity:
$$6^2V_A + 4^2V_B = 8^2V_C$$

i.e.
$$36V_A + 16V_B = 64V_C$$

Also,
$$h_{L_A} = h_{L_B} = 0.020\frac{2000}{6/12}\frac{V_A^2}{2g} = 0.032\frac{1600}{4/12}\frac{V_B^2}{2g}$$

i.e.
$$80V_A^2 = 153.6V_B^2, \quad V_B = 0.722V_A$$

[icon]: Programmed computing aids (Appendix B) could help solve problems marked with this icon.

Substituting into the continuity equation,

$$36V_A + 16(0.722V_A) = 64V_C$$
$$47.5V_A = 64V_C, \quad V_A = 1.346V_C$$

Substituting into the energy equation,

$$150 = 80\frac{(1.346V_C)^2}{2g} + 145\frac{V_C^2}{2g} = 289.9\frac{V_C^2}{2g}$$
$$V_C^2 = 2(32.2)150/289.9 = 33.3$$
$$V_C = 5.77 \text{ fps}, \quad Q_C = A_CV_C = (0.349)5.77 = 2.01 \text{ cfs} \qquad \textbf{\textit{ANS}}$$
$$V_A = 1.346V_C = 7.77 \text{ fps} \quad Q_A = (0.1963)7.77 = 1.526 \text{ cfs} \qquad \textbf{\textit{ANS}}$$

Continuity:

$$36(7.77) + 16V_B = 64(5.77)$$
$$279.7 + 16V_B = 369.4$$
$$V_B = 89.7/16 = 5.61 \text{ fps}$$
$$Q_B = A_BV_B = (0.0873)5.61 = 0.489 \text{ cfs} \qquad \textbf{\textit{ANS}}$$

As a check, note that $Q_A + Q_B = Q_C$.
To find the pressure at P

$$200 - 80\frac{V_A^2}{2g} = 120 + p_P/\gamma$$
$$p_P/\gamma = 80 - 80\frac{(7.77)^2}{2g} = 5.01 \text{ ft}$$

Check:

$$120 + p_P/\gamma - 144\frac{V_C^2}{2g} = 50 + \frac{V_C^2}{2g}$$
$$p_P/\gamma = 145\frac{(5.77)^2}{2g} - 70 = 5.01 \text{ ft}$$

So $p_P/\gamma = 5.01$ ft and $p_P = (62.4/144)5.01 = 2.17$ psi. ***ANS***

In this example it was assumed that the values of f for each pipe were known. Actually f depends on **R** [Fig. 8.11 or Eq. (8.41)]. Usually the absolute roughness e of each pipe is known or assumed, and an accurate solution is achieved through trial and error until the fs and **R**s for each pipe have converged.

Manual iteration for f may be avoided by solving simultaneous equations using mathematics software like Mathcad. There are the usual four equations for each pipe, plus Eq. (8.72); if necessary, minor losses may easily be accounted for by using head loss equations with the form of Eq. (8.60). For the three pipes of Fig. 8.30, for example, there are therefore 13 simultaneous equations, which may be solved in the usual manner (see Sample Prob.

8.12*b*) for 13 unknowns. These unknowns are either the head loss or the total flow rate, and three values each of Q, V, $\mathbf{R}$, and f.

It is instructive to compare the solution methods for pipes in parallel with those for pipes in series. The role of the head loss in one case becomes that of the discharge rate in the other, and vice versa. The student should be already familiar with this situation from the elementary theory of dc circuits. The flow corresponds to the electrical current, the head loss to the voltage drop, and the frictional resistance to the ohmic resistance. The outstanding deficiency in this analogy occurs in the variation of potential drop with flow, which is with the first power in the electrical case ($E = IR$) and with the second power in the hydraulic case ($h_L \propto V^2 \propto Q^2$) for fully developed turbulent flow.

8.29 PIPE NETWORKS

An extension of pipes in parallel is a case frequently encountered in municipal distribution systems, in which the pipes are interconnected so that the flow to a given outlet may come by several different paths, as shown in Fig. 8.31. Indeed, it is frequently impossible to tell by inspection which way the flow travels, as in pipe *BE*. Nevertheless, the flow in any network, however complicated, must satisfy the basic relations of continuity and energy as follows:

1. The flow into any junction must equal the flow out of it.
2. The flow in each pipe must satisfy the pipe-friction laws for flow in a single pipe.
3. The algebraic sum of the head losses around any closed loop must be zero.

Pipe networks are generally too complicated to solve analytically, as was possible in the simpler cases of parallel pipes (Sec. 8.28). Furthermore, it is seldom possible to predict the capacity requirements of water distribution

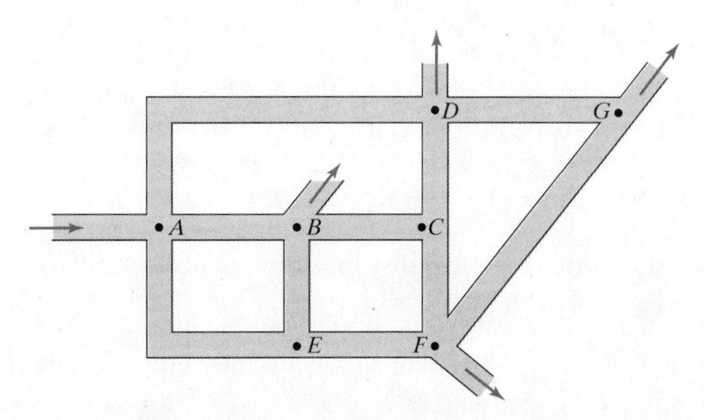

Figure 8.31
Pipe network.

systems with high precision, and flows in them vary considerably throughout the day, so high accuracy in calculating their flows is not important. As a result, the use of empirical equations (Sec. 8.15) and constant values of f are very acceptable for this purpose. A practical procedure is the method of successive approximations, introduced by Cross.[25] It consists of the following elements, in order:

1. By careful inspection assume the most reasonable distribution of flows that satisfies condition 1.

2. Write condition 2 for each pipe in the form

$$h_L = KQ^n \tag{8.77}$$

   where K and n are constants for each pipe as described in Sec. 8.26 under Approximate Solutions [Eq. (8.66), etc.]. If minor losses are important they may be included as in Eq. (8.75), which yields $K = 1/C^2$ and $n = 2$ for constant f. Minor losses may be included within any pipe or loop, but they are neglected at the junction points.

3. To investigate condition 3, compute the algebraic sum of the head losses around each elementary loop, $\sum h_L = \sum KQ^n$. Consider losses from clockwise flows as positive, counterclockwise negative. Only by good luck will these add to zero on the first trial.

4. Adjust the flow in each loop by a correction ΔQ to balance the head in that loop and give $\sum KQ^n = 0$. The heart of this method lies in the determination of ΔQ. For any pipe, we may write

$$Q = Q_0 + \Delta Q$$

   where Q is the correct discharge and Q_0 is the assumed discharge. Then, for each pipe,

$$h_L = KQ^n = K(Q_0 + \Delta Q)^n = K(Q_0{}^n + nQ_0^{n-1}\Delta Q + \ldots)$$

   If ΔQ is small compared with Q_0, the terms of the binomial series after the second one may be neglected. Now, for a circuit, with ΔQ the same for all pipes,

$$\sum h_L = \sum KQ^n = \sum KQ_0{}^n + \Delta Q \sum KnQ_0^{n-1} = 0$$

   As the corrections of head loss in all pipes must be summed *arithmetically*,

[25] H. Cross, Analysis of Flow in Networks of Conduits or Conductors, *Univ. Ill. Eng. Expt. Sta. Bull.* 286, 1936.

we may solve this equation for ΔQ,

$$\Delta Q = \frac{-\Sigma\, KQ_0^{\,n}}{\Sigma\, |KnQ_0^{\,n-1}|} = \frac{-\Sigma\, h_L}{n\, \Sigma\, |h_L/Q_0|} \tag{8.78}$$

since, from Eq. (8.77), $h_L/Q = KQ^{n-1}$. It must be emphasized again that the numerator of Eq. (8.78) is to be summed algebraically, with due account of sign, while the denominator is summed arithmetically. The negative sign in Eq. (8.78) indicates that when there is an excess of head loss around a loop in the clockwise direction, the ΔQ must be subtracted from clockwise Q_0s and added to counterclockwise ones. The reverse is true if there is a deficiency of head loss around a loop in the clockwise direction.

5. After each circuit is given a first correction, the losses will still not balance, because of the interaction of one circuit upon another (pipes which are common to two circuits receive two independent corrections, one for each circuit). The procedure is repeated, arriving at a second correction, and so on, until the corrections become negligible.

Either form of Eq. (8.78) may be used to find ΔQ. As values of K appear in both numerator and denominator of the first form, values proportional to the actual K may be used to find the distribution. The second form will be found most convenient for use with pipe-friction diagrams for water pipes.

An attractive feature of the approximation method is that errors in computation have the same effect as errors in judgment and will eventually be corrected by the process.

Sample Problem 8.17 If the flow into and out of a two-loop pipe system are as shown in Fig. S8.17, determine the flow in each pipe. The K values for each pipe were calculated from the pipe and minor loss characteristics and from an assumed value of f.

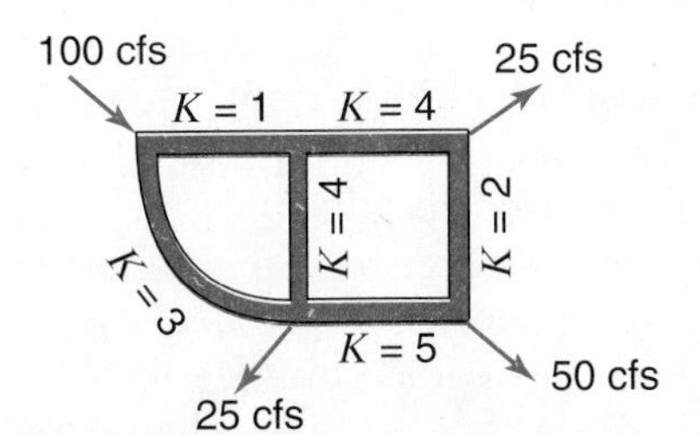

Figure S8.17

: Programmed computing aids (Appendix B) could help solve problems marked with this icon.

Solution

As a first step, assume flow in each pipe such that continuity is satisfied at all junctions. Calculate ΔQ for each loop, make corrections to the assumed Qs, and repeat several times until the ΔQs are quite small.

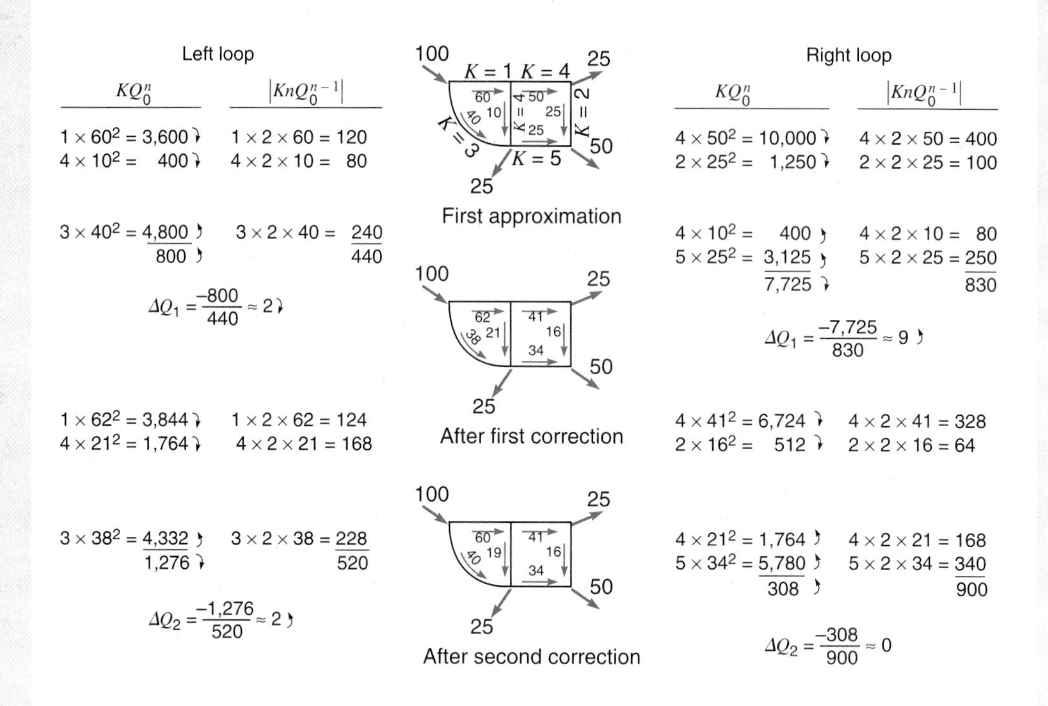

Further corrections can be made if greater accuracy is desired.

As noted earlier, varying demand rates usually make high solution accuracy unnecessary with pipe networks. However, if high accuracy is required for some reason, the problem can be first solved in a similar manner to the preceding example using the Darcy–Weisbach K in Eq. (8.66) and constant f values. Then the resulting flows may be used to adjust the f and K values, and the process repeated (more than once if necessary) to refine the answers. The value of such refinement is questionable, not only because of uncertainties in the demand flows, but also because of uncertainties in the e values (pipe roughness) to be used (see Sec. 8.13). Usually f values change by only a few percent when they are adjusted, but we can see in Fig. 8.11 that for smoother pipe it is possible for them to change by as much as a factor of five.

Simple networks can be solved without approximation and manual iteration by solving simultaneous equations using mathematics software like

Mathcad. For networks containing i pipes, $5i$ equations are required if using the Darcy–Weisbach equation with variable f, and $2i$ equations are required if using the simplified equation (8.77) with constant friction factors. These required equations include (*a*) the usual (condition 2) flow equations for each pipe (four or one per pipe, depending on the equations used); (*b*) flow continuity equations (condition 1) at all but one of the j nodes (as these imply continuity at the last node); (*c*) equations for the sum of the head losses around $i - j + 1$ loops (condition 3). The unknowns to be determined for each pipe are h_L, Q, V, $\mathbf{R}$, and f using the Darcy–Weisbach equation, or only h_L and Q using Eq. (8.77).

The pipe-network problem lends itself well to solution by use of a digital computer. Programming takes time and care, but once set up, there is great flexibility and many hours of repetitive labor can be saved. Many software packages are now available to simulate water distribution networks; see Appendix B.

8.30 FURTHER TOPICS IN PIPE FLOW

The basics of steady incompressible flow in pressure conduits discussed in this chapter are just an introduction to many more advanced subjects involving flow in conduits. Flow through submerged culverts is a special case discussed in Sec. 10.23. Where sewers must dip to flow under obstacles like streams, they flow full and are known as ***inverted siphons*** or ***sag pipes***; turbulence, and therefore velocities, must be maintained when flow rates vary in order to prevent suspended solids from accumulating at such low points. Later in this text, we discuss ***unsteady flow*** in pipes (Chap. 12), involving both moderate rates of change and the very rapid changes associated with ***water hammer***. Such flows are often investigated or simulated using ***numerical methods***, usually on a computer. The analysis of the flow of ***compressible fluids*** (gases) is introduced in Chap. 13, and methods of ***flow measurement*** in pipes are described in Chap. 11.

When a different fluid is injected into a pipeline, besides being transported by the main flow (***advection***), it mixes and spreads through the main fluid by the processes of ionic and molecular ***diffusion***, by which dissolved species move relatively slowly from areas of higher concentration to areas of lower concentration, and by ***dispersion***, also known as ***eddy diffusion*** or ***turbulent mixing***. As these latter names suggest, dispersion is caused by turbulence, and as a result it is usually a far more rapid process than diffusion. Injection of easily detectable fluids into pipelines, such as salts (measured by conductivity) or dyes, have been used to measure flow rates and velocities. If dye dissolved in water, for example, is injected continuously at a known flow rate and concentration into a pipeline and given sufficient opportunity to mix completely, the dye concentration downstream indicates the relative magnitude of the pipe flow rate to the flow rate of the injected dye solution. If a "slug" of concentrated salt solution is rapidly injected into a pipeline, because of diffusion and dispersion the salt concentration at downstream points will be observed to vary with time, and this variation will be somewhat like a skewed bell-shaped curve. By taking the centroid of the area under the curve as the true arrival time, the flow velocity can be calculated (see Sec. 11.16).

Pumping with pipelines, discussed in Sec. 8.25, has many engineering applications. Pump stations must be provided at low areas such as at underpasses which would otherwise flood during storms. Often, multiple pumps are provided to enable a more efficient response to varying inflows. In sanitary sewer systems pump stations become necessary if the slopes needed for gravity drainage require excessive excavation; sometimes, local topography requires sewage to be pumped over a hill, through a ***force main***. In such cases precautions must be taken to keep solids in suspension, and to prevent accumulation of dangerous gases. Water distribution systems, discussed in the previous section, must often supply areas of uneven topography, requiring them to be divided into subareas or ***pressure zones*** served by separate pump stations which boost the pressures to the appropriate levels. Advanced software has been developed to analyze such conditions; see Appendix B.

To aid the dilution of sewage effluent, submarine and river ***outfalls*** are often provided with a ***diffuser section***. (Submarine discharge as a single jet was discussed in Sec. 5.16.) Diffusers have a number of outlets or ***ports*** along the length of the outfall, so that the flow in the main line or ***manifold*** decreases with distance. Manifolds are also used in navigation locks, to facilitate filling lock chambers as rapidly as possible while minimizing disturbance to moored vessels. The hydraulics of manifold flow is an advanced subject, particularly when applied to outfalls where port sizes must be adjusted to maintain a uniform discharge along the length of the diffuser.

PROBLEMS

8.1 Hydrogen at 60°F and atmospheric pressure has a kinematic viscosity of 0.0012 ft^2/s. Determine the minimum turbulent flow rate in pounds per second through a 3-in-diameter pipe. At this flow rate, what is the average velocity?

8.2 Air at 80°C and a pressure of approximately 1350 kN/m^2 abs flows in a 2-cm-diameter tube. What is the minimum turbulent flow rate? Express the answer in liters per second, newtons per second, and kilograms per second. At this flow rate what is the average velocity?

8.3 Two pipes, one circular and one square, have the same cross-sectional area. Which has the larger hydraulic radius, and by what percentage

8.4 Prove that for a constant rate of discharge and a constant value of f the friction head loss in a pipe varies inversely as the fifth power of the diameter.

8.5 Two long pipes convey water between two reservoirs whose water surfaces are at different elevations. One pipe has a diameter twice that of the other; both pipes have the same length and the same value of f. If minor losses are neglected, what is the ratio of the flow rates through the two pipes?

8.6 Tests were made with 70°F water flowing in a 10-in-diameter pipe. They showed that, when $V = 13$ fps, $f = 0.0162$. Find the unit shear at the pipe wall and at radii of 0, 0.25, 0.4, 0.6, 0.85 times the pipe radius.

8.7 For laminar flow in a circular pipe, find the velocities at $0.1r$, $0.3r$, $0.5r$, $0.7r$, and $0.9r$. Plot the velocity profile.

8.8 Prove that the centerline velocity is twice the average velocity when laminar flow occurs in a circular pipe.

8.9 For laminar flow between two parallel, flat plates a small distance d apart, at what distance from the centerline (in terms of d) will the velocity be equal to the mean velocity?

8.10 Oil with a viscosity of 0.18 N·s/m^2 and a density of 915 kg/m^3 is flowing in a 25-cm-diameter pipe at 0.75 L/s. How much power is lost per meter of pipe length?

8.11 Water at 50°F enters a pipe with a uniform velocity of 12 fps. (*a*) What is the distance at which the transition occurs from a laminar to a turbulent boundary layer? (*b*) If the thickness of this initial laminar boundary layer is given by $4.91\sqrt{\nu x/U}$, what is the thickness reached by it at the point of transition?

8.12 Water at 130°F flows in an 0.5-in-diameter copper tube at 1.2 gpm. Find the head loss per 100 ft. What is the centerline velocity, and what is the value of δ_l?

8.13 Water at 60°C flows in a 1.5-cm-diameter copper tube at 0.06 L/s. Find the head loss per 10 m. What is the centerline velocity, and what is the value of δ_l?

8.14 Repeat Prob. 8.12 for flow rates of 0.08 and 18 gpm.

8.15 Repeat Prob. 8.13 for flow rates of 0.004 and 0.9 L/s.

8.16 Oil ($s = 0.85$) with a viscosity of 0.0056 N·s/m^2 flows at a rate of 80 L/s in a 15-cm-diameter pipe having $e = 0.90$ mm. Find the head loss. Determine the shear stress at the pipe wall. Find the velocity 2.5 cm from the centerline. Under these conditions is the pipe behaving as a fully rough, transitional, or smooth pipe?

8.17 In a 42-in-diameter pipe velocities are measured as 17.0 fps at $r = 0$ and 16.5 fps at $r = 3.5$ in. Approximately what is the flow rate?

8.18 The flow rate in a 10-in-diameter pipe is 7 cfs. The flow is known to be turbulent, and the centerline velocity is 15.2 fps. Plot the velocity profile, and determine the head loss per foot of pipe.

8.19 The flow rate in a 25-cm-diameter pipe is 0.20 m^3/s. The flow is known to be turbulent, and the centerline velocity is 4.75 m/s. Plot the velocity profile, and determine the head loss per meter of pipe.

8.20 If the diameter of a pipe is doubled, what effect does this have on the flow rate for a given head loss? Consider (*a*) laminar flow; (*b*) turbulent flow.

8.22 Refer to the data of Prob. 8.16. Above what flow rate will this pipe behave as a fully rough pipe? Below what flow rate will it behave as a smooth pipe?

8.23 Kerosene ($s = 0.813$) at a temperature of 70°F flows in a 3-in-diameter smooth brass pipeline at a rate of 15 gpm. (*a*) Find the head loss per foot. (*b*) For the same head loss, what would be the flow rate if the temperature of the kerosene were raised to 100°F?

8.24 Kerosene ($s = 0.813$) at a temperature of 20°C flows in a 7.5-cm-diameter smooth brass pipeline at a rate of 0.80 L/s. (*a*) Find the head loss per meter. (*b*) For the same head loss, what would be the flow rate if the temperature of the kerosene were raised to 40°C?

8.25 Water at 50°C flows in a 15-cm-diameter pipe with $V = 7.5$ m/s. Head loss measurements indicate that $f = 0.020$. Determine the value of e. Find the shear stress at the pipe wall and at $r = 3$ cm. What is the value of du/dy at $r = 3$ cm?

8.26 Water at 40°F flows in a 42-in-diameter concrete pipe ($e = 0.022$ in). Determine $\mathbf{R}$, τ_0, u_{max}/V, δ_l, δ_l/e, and the flow regime (hydraulic smoothness) for flow rates of 250, 25, 2.5, 0.25, and 0.025 cfs.

8.27 For $\mathbf{R}$ ranging from 10^2 to 10^7, make a plot of the values of α and β versus $\mathbf{R}$ for smooth brass pipes. On the same plot also show values of u_{max}/V.

8.28 When water at 50°F flows at 2.5 cfs in a 20-in pipeline, the head loss is 0.0004 ft/ft. What will be the head loss when glycerin at 68°F flows through this same pipe at the same rate?

8.29 Oil ($s = 0.92$) with viscosity 1.8×10^{-4} lb·sec/ft^2 flows in a 3-in-diameter welded-steel pipe at 0.15 cfs. What is the head loss per foot of pipe?

8.30 Oil ($s = 0.91$) with viscosity 0.0102 N·s/m^2 flows in a 7.5-cm-diameter welded-steel pipe at 4.9 L/s. What is the head loss per meter of pipe?

8.31 Crude oil ($s = 0.855$) at 50°C flows at 0.30 m^3/s through a 45-cm-diameter pipe ($e = 0.054$ mm) 1500 m long. Find the kilowatt loss.

8.32 Air at 90 psia and 70°F flows through a 4.5-in-diameter welded-steel pipe

---

: Programmed computing aids (Appendix B) could help solve problems marked with this icon.

at 60 lb/min. Find the head loss and pressure drop in 200 ft of this pipe. Assume the air to be of constant density.

8.33 Air flows at an average velocity of 0.7 m/s through a long 3.8-m-diameter circular tunnel (e = 1.5 mm). Find the head-loss gradient at a point where the air temperature and pressure are 20°C and 102 kN/m^2 abs, respectively. Find also the shear stress at the pipe wall and the thickness δ_l of the viscous sublayer.

8.34 Repeat Prob. 8.33 for the case where the average velocity is 7.0 m/s.

8.35 Air at 70°C and atmospheric pressure flows with a velocity of 22 fps through a 3-in-diameter pipe (e = 0.000 15 in). Find the head loss in 75 ft of pipe.

8.36 Air at 30°C and atmospheric pressure flows with a velocity of 6.5 m/s through a 75-mm-diameter pipe (e = 0.002 mm). Find the head loss in 30 m of pipe.

8.37 Water at 15°C flowing through 25 m of 10-cm-diameter galvanized iron pipe causes a head loss of 7.5 cm. Find the flow rate.

8.38 When gasoline with a kinematic viscosity of 6×10^{-6} ft^2/sec flows in a 10-in-diameter smooth pipe, the head loss is 0.36 ft per 100 ft. Find the flow rate.

8.39 Oil with a kinematic viscosity of 0.000 22 ft^2/sec is to flow at 7.4 cfs with a head loss of 0.37 ft·lb/lb per 100 ft of pipe length. What size pipe (e = 0.00013 ft) is theoretically required?

8.40 Oil with a kinematic viscosity of 2.0×10^{-5} m^2/s is to flow at 0.21 m^3/s, with a head loss of 0.42 N·m/N per 100 m of pipe length. What size pipe (e = 0.038 mm) is theoretically required?

8.41 Water at 150°F flows in a 350-ft-long 0.863-in-diameter iron pipe (e = 0.000 15 ft) between points A and B. At A the elevation of the pipe is 112.0 ft and the pressure is 8.35 psi. At B the elevation of the pipe is 105 ft and the pressure is 8.80 psi. Compute the flow rate as accurately as you can.

8.42 Water at 60°C flows in a 100-m-long 2-cm-diameter pipe (e = 0.060 mm) between points A and B. At A the elevation of the pipe is 54.1 m and the pressure is 88.7 kPa. At B the elevation of the pipe is 52.0 m and the pressure is 91.8 kPa. Compute the flow rate as accurately as you can.

8.43 A steel pipe (e = 0.0002 ft) 13,450 ft long is to convey oil (ν = 0.000 54 ft^2/sec) at 13 cfs from a reservoir with surface elevation 705 ft to one with surface elevation 390 ft. Theoretically, what pipe size is required?

8.44 A steel pipe (e = 0.065 mm) 4200 m long is to convey oil (ν = 5.2×10^{-5} m^2/s) at 0.3 m^3/s from a reservoir with surface elevation 247 m to one with surface elevation 156 m. Theoretically, what pipe size is required?

8.45 Using only a basic scientific calculator, solve Prob. 8.37 without trial and error.

8.46 Using only a basic scientific calculator, solve Prob. 8.40 without trial and error.

8.47 Using only a basic scientific calculator, solve Prob. 8.42 without trial and error.

8.48 Using only a basic scientific calculator, solve Prob. 8.44 without trial and error.

8.49 Using an equation solver on a programmable scientific calculator, or in mathematics software such as Mathcad, solve the following without manual trial and error: (*a*) Prob. 8.37; (*b*) Prob. 8.38; (*c*) Prob. 8.41; (*d*) Prob. 8.42.

8.50 Using an equation solver on a programmable scientific calculator, or in mathematics software such as Mathcad, solve the following without manual trial and error: (*a*) Prob. 8.39; (*b*) Prob. 8.40; (*c*) Prob. 8.43; (*d*) Prob. 8.44.

8.51 In a field test of the 16-ft-diameter Colorado River aqueduct flowing full, Manning's n was found to have a value of 0.0132 when 50°F water was flowing at a Reynolds number of 10.5×10^6. Determine the average value of e for this conduit.

8.52 Measurements were taken on pressure flow through a 2.5-m-diameter aqueduct built of concrete. When the water temperature was 10°C and the Reynolds number was 2.0×10^6, Manning's n was determined to be 0.0140. Find the average value of e for this aqueduct.

8.53 Air at 160°F and standard sea-level atmospheric pressure flows in a 15 in by 21 in rectangular air duct (e = 0.0007 in) at the rate of 1.2 lb/min. Find the head loss per 100 ft of duct. Express the answer in feet of air flowing and in pounds per square inch.

8.54 Water at 80°F flows in a conduit with a cross section shaped in the form of an equilateral triangle. The cross-sectional area of the duct is 160 in^2 and e = 0.0018 in. If the head loss is 3 ft in 150 ft, find the approximate flow rate.

8.55 Water at 20°C flows in a conduit with a cross section shaped in the form of an equilateral triangle. The cross-sectional area of the duct is 1000 cm^2 and e = 0.045 mm. If the head loss is 1 m in 50 m, find the approximate flow rate.

8.56 A smooth pipe consists of 52 ft of 9-in pipe followed by 310 ft of 18-in pipe, with an abrupt change of cross section at the junction. The entrance is flush and the discharge is submerged. If it carries water at 60°F, with a velocity of 19 fps in the smaller pipe, what is the total head loss?

8.57 A smooth pipe consists of 12 m of 18-cm pipe followed by 75 m of 55-cm pipe, with an abrupt change of cross section at the junction. The entrance

---

■: Programmed computing aids (Appendix B) could help solve problems marked with this icon.

is flush and the discharge is submerged. If it carries water at 15°C, with a velocity of 5.7 m/s in the smaller pipe, what is the total head loss?

8.58 Water flows at 12 fps through a vertical 4-in-diameter pipe standing in a body of water with its lower end 5 ft below the surface. Considering all losses and with $f = 0.026$, find the pressure in the pipe at a point 15 ft above the surface of the water when the flow is (*a*) downward; (*b*) upward.

8.59 Water flows at 2.5 m/s through a vertical 10-cm-diameter pipe standing in a body of water with its lower end 1 m below the surface. Considering all losses and with $f = 0.024$, find the pressure in the pipe at a point 4 m above the surface of the water when the flow is (*a*) downward; (*b*) upward.

8.60 A 12-in-diameter pipe ($f = 0.028$) 450 ft long runs from one reservoir to another, both ends of the pipe being under water. The intake is square-edged. The difference between the water levels in the two reservoirs is 150 ft. Find (*a*) the flow rate; (*b*) the pressure in the pipe at a point 320 ft from the intake, when the elevation is 135 ft lower than the surface of the water in the upper reservoir.

8.61 A 32-cm-diameter pipe ($f = 0.025$) 140 m long runs from one reservoir to another, both ends of the pipe being under water. The intake is square-edged. The difference between the water levels in the two reservoirs is 36 m. Find (*a*) the flow rate; (*b*) the pressure in the pipe at a point 95 m from the intake, where the elevation is 39 m lower than the surface of the water in the upper reservoir.

8.62 A pump delivers water through 300 ft of 4-in fire hose ($f = 0.025$) to a nozzle that throws a 1-in-diameter jet. The loss coefficient of the nozzle is 0.04. The nozzle is 20 ft higher than the pump, and a jet velocity of 70 fps is required. What must be the pressure in the hose at the pump?

8.63 A pump delivers water through 100 m of 10-cm fire hose ($f = 0.025$) to a nozzle that throws a 2.5-cm-diameter jet. The loss coefficient of the nozzle is 0.04. The nozzle is 6 m higher than the pump, and a jet velocity of 20 m/s is required. What must be the pressure in the hose at the pump?

8.64 A 6-in-diameter horizontal jet of water is discharged into air through a nozzle (loss coefficient of 0.15) at a point 150 ft lower than the water level above the intake. The 12-in-diameter pipeline ($f = 0.015$) is 600 ft long, with a square-edged nonprojecting entrance. What is (*a*) the jet velocity; (*b*) the pressure at the base of the nozzle?

▣ **8.65** Refer to Fig. 8.23*b*. Suppose $\Delta z = 50$ ft, the line is 600 ft of 8-in-diameter pipe ($f = 0.025$), and the nozzle loss coefficient is 0.05. Find the jet diameter that will result in the greatest jet horsepower.

▣ **8.66** Refer to Fig. 8.23*b*. Suppose $\Delta z = 15$ m, the line is 180 m of 20-cm-diameter pipe ($f = 0.025$), and the nozzle loss coefficient is 0.05. Find the jet diameter that will result in the greatest jet power.

▣: Programmed computing aids (Appendix B) could help solve problems marked with this icon.

8.67 Water at 60°F flows through 800 ft of 12-in-diameter pipe between two reservoirs whose water-surface elevation difference is 16 ft. The pipe entrance is flush and square-edged, and there is a half-open gate valve in the line. Using only a basic scientific calculator, find the flow rate (*a*) if $e = 0.0018$ in; (*b*) if e is 20 times larger.

8.68 Solve Prob. 8.67 without manual trial and error, by using an equation solver (*i*) on a programmable calculator; (*ii*) in mathematics software like Mathcad.

8.69 A 150 m long commercial steel pipe is to convey 30 L/s of oil ($s = 0.9$, $\mu = 0.038$ N·s/m^2) from one tank to another where the difference in elevation of the free liquid surfaces is 2 m. The pipe entrance is flush and square-edged, and there is a fully open gate valve in the line. Using only a basic scientific calculator, find the diameter theoretically required.

8.70 Solve Prob. 8.69 without manual trial and error, by using an equation solver (*a*) on a programmable calculator; (*b*) in mathematics software like Mathcad.

8.71 In Fig. 8.24 assume the pipe diameter is 9 in, $f = 0.025$, $BC = 20$ ft, $DE = 2800$ ft, and $\Delta z = 125$ ft. (*a*) If $Q = 5.5$ cfs of water and the pump efficiency is 78%, what horsepower is required? (*b*) If the elevation of C above the lower water surface is 12 ft, that of D is 15 ft, and that of E is 100 ft, compute the pressure heads at C, D, and E. (c) Sketch the energy line and the hydraulic grade line.

8.72 In Fig. 8.24 assume a pipe diameter of 35 cm, $f = 0.016$, $BC = 12$ m, $DE = 920$ m, and $\Delta z = 48$ m. Find the maximum theoretical flow rate if 15°C water is being pumped at an altitude of 1000 m above sea level. Point C is 6.0 m above the lower water surface.

8.73 In Fig. 8.24 assume that the pipe diameter is 4 in, $f = 0.035$, $BC = 15$ ft, $DE = 180$ ft, and $\Delta z = 60$ ft. The elevation of C is 10 ft above the lower water surface. (*a*) If the pressure head at C is to be no less than -20 ft, what is the maximum rate at which the water can be pumped? (*b*) If the efficiency of the pump is 65%, what is the horsepower required?

8.74 Refer to Fig. 8.25. Suppose that the water-surface elevation, elevation of the pump, and elevation of the nozzle tip are 100, 90, and 120 ft, respectively. Pipe BC is 40 ft long, has a diameter of 8 in, with $f = 0.025$; pipe DE is 200 ft long, has a diameter of 10 in, with $f = 0.030$; the jet diameter is 6 in, and the nozzle loss coefficient is 0.04. Assume the pump is 80% efficient under all conditions of operation. Make a plot of flow rate and p_C/γ versus pump horsepower input. At what flow rate will cavitation begin in the pipe at C if the water temperature is 50°F and the atmospheric pressure is 13.9 psia?

8.75 In a testing laboratory, a certain turbine has been found to discharge 10 cfs under a head h_t of 45 ft. In the field, it is to be installed near the end of a pipe 360 ft long. The supply line (flush entrance) and discharge line

---

☐: Programmed computing aids (Appendix B) could help solve problems marked with this icon.

(submerged exit) will both be 10 in in diameter, with $f = 0.025$. The total fall from the surface of the water at intake to the surface of the tailwater will be 52 ft. What will be the head on the turbine, the rate of discharge, and the power delivered to the flow? Note that for turbines, $Q \propto \sqrt{h_t}$ [Eq. (16.17)].

8.76 In a testing laboratory, a certain turbine has been found to discharge 285 L/s under a head h_t of 13.5 m. In the field, it is to be installed near the end of a pipe 110 m long. The supply line (flush entrance) and discharge line (submerged exit) will both be 25 cm in diameter, with $f = 0.024$. The total fall from the surface of the water at intake to the surface of the tailwater will be 15.5 m. What will be the head on the turbine, the rate of discharge, and the power delivered to the flow? Note that for turbines, $Q \propto \sqrt{h_t}$ [Eq. (16.17)].

8.77 Assume the total fall from one body of water to another is 130 ft. The water is conveyed by 250 ft of 15-in pipe ($f = 0.020$) that has its entrance flush with the wall. At the end of the pipe is a turbine and draft tube that discharged 6 cfs of water when tested under a head of 63.1 ft in another location. Discharge losses are negligible. What is the rate of discharge through the turbine and the head on it under the present conditions? Note that for turbines, $Q \propto \sqrt{h_t}$ [Eq. (16.17)].

8.78 A pump is installed to deliver water from a reservoir of surface elevation zero to another of elevation 200 ft. The 12-in-diameter suction pipe ($f = 0.020$) is 40 ft long, and the 10-in-diameter discharge pipe ($f = 0.032$) is 4500 ft long. The pump head may be defined as $h_p = 300 - 20Q^2$, where the pump head h_p is in feet and Q is in cubic feet per second. Compute the rate at which this pump will deliver the water. Also, what is the horsepower input to the water?

8.79 In Fig. 8.27, suppose that pipe 1 is 36-in smooth concrete, 5000 ft long; pipe 2 is 24-in cast iron, 3000 ft long; and pipe 3 is 20-in cast iron, 1300 ft long. The elevations of the the water surfaces in reservoirs A and B are 225 and 200 ft, respectively, and the discharge through pipe 1 is 42 cfs. The water temperature is 60°F. Using a basic scientific calculator only, find the elevation of the surface of reservoir C. Neglect minor losses, and assume the energy line and hydraulic grade line are coincident.

8.80 In Fig. 8.27, suppose that pipe 1 is 90-cm smooth concrete, 1500 m long, pipe 2 is 60-cm cast iron, 900 m long; and pipe 3 is 50-cm cast iron, 400 m long. The elevations of the water surfaces in reservoirs A and B are 75 and 67 m, respectively, and the discharge through pipe 1 is 1.2 m^3/s. The water temperature is 15°C. Using a basic scientific calculator only, find the elevation of the surface of reservoir C. Neglect minor losses and assume the energy line and hydraulic grade line are coincident.

8.81 Solve Prob. 8.80 without manual trial and error, by using an equation solver in mathematics software like Mathcad.

: Programmed computing aids (Appendix B) could help solve problems marked with this icon.

8.82 With the sizes and lengths and materials of pipes given in Prob. 8.79, suppose that the surface elevations of reservoirs *A* and *C* are 250 and 180 ft, respectively, and the discharge through pipe 2 is 10 cfs of water at 60°F into reservoir *B*. Using a basic scientific calculator only, find the surface elevation of reservoir *B*.

8.83 Solve Prob. 8.82 without manual trial and error, by using an equation solver in mathematics software like Mathcad.

8.84 With the sizes and lengths and materials of pipes given in Prob. 8.80, suppose that the surface elevations of reservoirs *A* and *C* are 60 and 38 m, respectively, and the discharge through pipe 2 is 0.3 m^3/s of water at 15°C into reservoir *B*. Using a basic scientific calculator only, find the surface elevation of reservoir *B*.

8.85 Repeat Prob. 8.82, except that the 10 cfs discharge through pipe 2 is now from (not into) reservoir *B*.

8.86 Repeat Prob. 8.84, except that the 0.3 m^3/s discharge through pipe 2 is now from (not into) reservoir *B*.

8.87 Suppose in Fig. 8.27 that pipes 1, 2, and 3 are 900 m of 60 cm, 300 m of 45 cm, and 1200 m of 40 cm, respectively, of new welded-steel pipe. The surface elevations of reservoirs *A*, *B*, and *C* are 36, 22 and 0 m, respectively. The water temperature is 15°C. Using a basic scientific calculator only, find the approximate water flow in all pipes.

8.88 Solve Prob. 8.87 accurately without manual trial and error, by using an equation solver in mathematics software like Mathcad.

8.89 Suppose, in Fig. 8.27, that pipes 1, 2, and 3 are 3000 ft of 24 in, 1000 ft of 18 in and 4000 ft of 16 in, respectively, of new welded-steel pipe. The surface elevations of reservoirs *A*, *B*, and *C* are 120, 75 and 0 ft, respectively. The water temperature is 60°F. Using a basic scientific calculator only, find the approximate water flow in all pipes.

8.90 Solve Prob. 8.89 accurately without manual trial and error, by using an equation solver in mathematics software like Mathcad.

8.91 Suppose that, in Fig. 8.27, pipe 1 is 1500 ft of 12-in new cast-iron pipe, pipe 2 is 800 ft of 6-in wrought-iron pipe, and pipe 3 is 1200 ft of 8-in wrought-iron pipe. The water surface of reservoir *B* is 20 ft below that of *A*, while the junction *J* is 35 ft below the surface of *A*. In place of reservoir *C*, pipe 3 leads away to some other destination but its elevation at *C* is 60 ft below *A*. Find the flow of 60°F water in each pipe and the pressure head at *C* when the pressure head at *J* is 25 ft. Neglect velocity heads.

8.92 Suppose that in Fig. 8.29 pipes 1, 2, and 3 are 750 ft of 4-in, 250 ft of 2-in, and 300 ft of 3-in, asphalt-dipped cast-iron pipe. With a total head loss of 25 ft between *A* and *B*, find the flow of 60°F water.

: Programmed computing aids (Appendix B) could help solve problems marked with this icon.

8.93 Suppose that in Fig. 8.29 pipes 1, 2, and 3 are 150 m of 80-mm, 60 m of 50-mm, and 120 m of 60-mm wrought-iron pipe. With a total head loss of 6 m between A and B, find the flow of 15°C water.

8.94 Two pipes connected in series are respectively 150 ft of 2-in (e = 0.000 006 ft) and 450 ft of 8-in (e = 0.0009 ft). With a total head loss of 30 ft, find the flow of 60°F water.

8.95 Repeat Prob. 8.92 for the case where the fluid is an oil with $s = 0.92$, $\mu = 0.00096$ lb·sec/ft^2.

8.96 Repeat Prob. 8.93 for the case where the fluid is an oil with $s = 0.94$, $\mu = 0.04$ N·s/m^2.

8.97 A pipeline 900 ft long discharges freely at a point 200 ft below the water level at intake. The pipe intake projects into the reservoir. The first 600 ft is of 10-in diameter, and the remaining 300 ft is of 6-in diameter. Find the rate of discharge, assuming $f = 0.06$. If the junction point C of the two sizes of pipe is 150 ft below the intake water surface level, find the pressure head just upstream of C and just downstream of C. Assume a sudden contraction at C.

8.98 A pipeline 300 m long discharges freely at a point 50 m below the water level at intake. The pipe intake projects into the reservoir. The first 200 m is of 35 cm diameter, and the remaining 100 m is of 25 cm diameter. Find the rate of discharge, assuming $f = 0.06$. If the junction point C of the two sizes of pipe is 40 m below the intake water surface level, find the pressure head just upstream of C and just downstream of C. Assume a sudden contraction at C.

8.99 Repeat Prob. 8.97, neglecting minor losses.

8.100 Repeat Prob. 8.98, neglecting minor losses.

8.101 Three new cast-iron pipes, having diameters of 24, 21, and 18 in, respectively, each 450 ft long, are connected in series. The 24-in pipe leads from a reservoir (flush entrance), and the 18-in pipe discharges into the air at a point 15 ft below the reservoir water surface level. Assuming all changes in section to be abrupt, find the rate of discharge of water at 60°F.

8.102 Suppose that in Fig. 8.30 pipes 1, 2, and 3 are of smooth brass as follows: 500 ft of 2-in, 350 ft of 3-in, and 750 ft of 4-in, respectively. When the total flow of 70°F crude oil ($s = 0.855$) is 0.7 cfs, find the head loss from A to B and the flow in each pipe. *Note*: This problem can be solved without trial and error, using a basic scientific calculator only.

■: Programmed computing aids (Appendix B) could help solve problems marked with this icon.

8.103 Repeat Prob. 8.102 for the case where the total flow rate is 0.07 cfs.

8.104 Repeat Prob. 8.102 for the case where all the pipes are of galvanized iron. Does the "Note" still apply?

8.105 Suppose that in Fig. 8.30 pipes 1, 2, and 3 are of drawn copper tubing as follows: 90 m of 2 cm, 150 m of 4 cm and 80 m of 6 cm, respectively. When the total flow of 50°C crude oil (s = 0.855) is 7 L/s, find the head loss from A to B and the flow in each pipe. *Note*: This problem can be solved without trial and error, using a basic scientific calculator only.

8.106 Repeat Prob. 8.105 for the case where the total flow rate is 0.35 L/s.

8.107 Repeat Prob. 8.105 for the case where all the pipes are of galvanized iron. Does the "Note" still apply?

8.108 The pipes in the system shown in the figure are all galvanized iron. (*a*) With a flow of 15 cfs, find the head loss from A to D. (*b*) What should be the diameter of a single pipe from B to C such that it replaces pipes 2, 3, and 4 without altering the capacity for the same head loss from A to D?

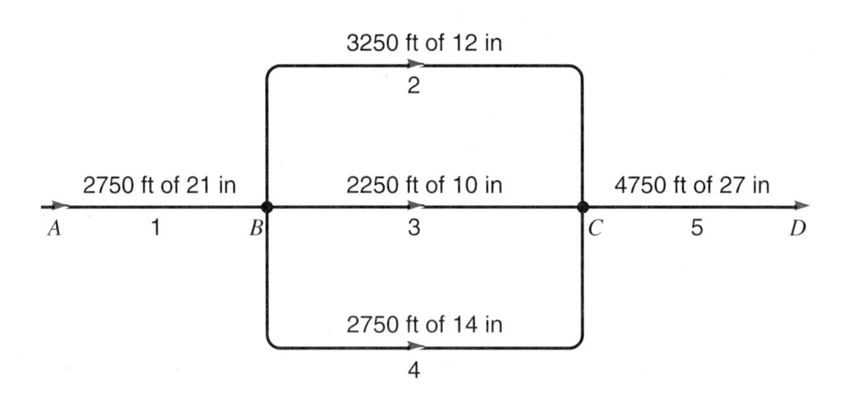

Figure P8.108

8.109 (*a*) With the same pipe lengths, sizes, and connections as in Prob. 8.108, find the flow in each pipe if the head loss from A to D is 150 ft and if all pipes have f = 0.020. Also find the head losses from A to B, B to C, and C to D. (*b*) Find the new head loss distributions and the percentage increase in the capacity of the system achieved by adding another 12-in pipe 3250 ft long between B and C.

8.110 In the figure pipe AB is 1200 ft long, of 8 in diameter, with f = 0.035; pipe BC (upper) is 800 ft long, of 6 in diameter, with f = 0.025; pipe BC (lower) is 900 ft long, of 4 in diameter, with f = 0.045; and pipe CD is 550 ft long, of 6 in diameter, with f = 0.025. The elevations are reservoir water surface = 150 ft, A = 120 ft, B = 70 ft, C = 60 ft, and D =

: Programmed computing aids (Appendix B) could help solve problems marked with this icon.

30 ft. There is free discharge to the atmosphere at D. Neglecting velocity heads, (a) compute the flow in each pipe; (b) determine the pressures at B and C.

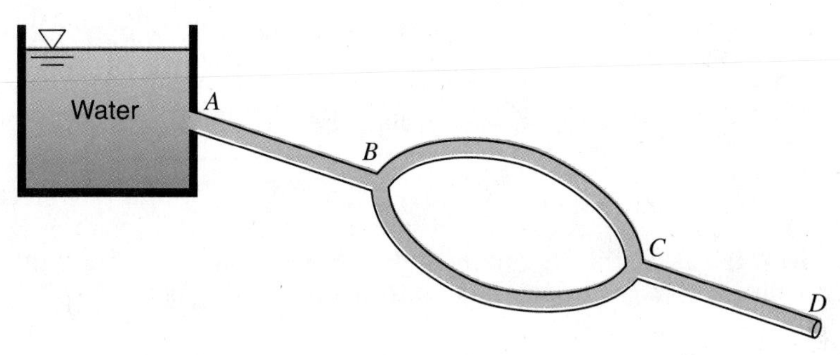

Figure P8.110

8.111 In Fig. P8.110, pipe AB is 600 m long, of 18 cm diameter, with $f = 0.035$; pipe BC (upper) is 500 m long, of 12 cm diameter, with $f = 0.025$; pipe BC (lower) is 400 m long, of 16 cm diameter, with $f = 0.030$; and pipe CD is 900 m long, of 32 cm diameter, with $f = 0.020$. The elevations are reservoir water surface = 150 m, A = 100 m, B = 60 m, C = 50 m, D = 20 m. There is free discharge to the atmosphere at D. Neglecting velocity heads, (a) compute the flow in each pipe; (b) determine the pressures at B and C. Comment on the practicality of this system.

8.112 Referring to the figure, the pipe characteristics are as follows: pipe A is 5000 ft long, of 3 ft diameter, with $f = 0.035$; pipe B is 4500 ft long, of 2 ft diameter, with $f = 0.035$; and pipe C is 5000 ft long, of 3 ft diameter, with $f = 0.025$. When the pump develops 30 ft of head, the velocity of flow in pipe C is 5 fps. Neglecting minor losses, find (a) the flow rates in cubic feet per second in pipes A and B under these conditions; (b) the elevation of the discharge end of pipe B.

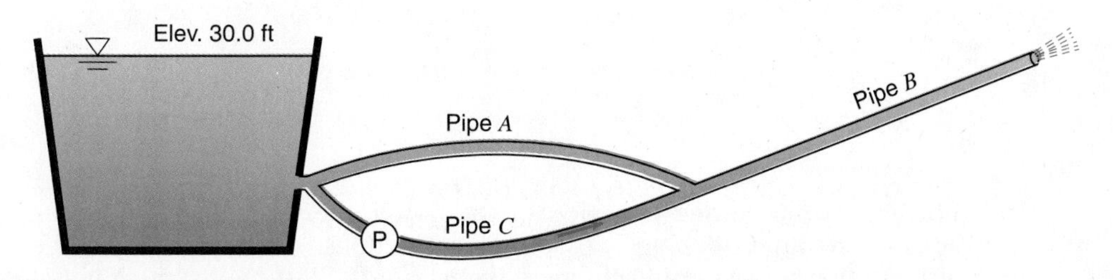

Figure P8.112

: Programmed computing aids (Appendix B) could help solve problems marked with this icon.

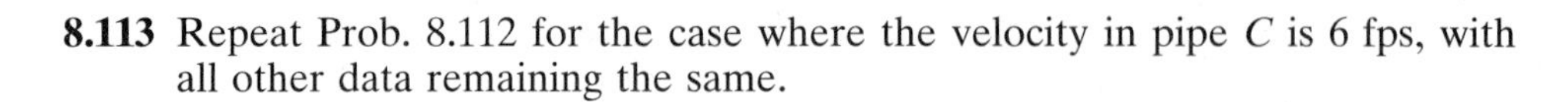

8.113 Repeat Prob. 8.112 for the case where the velocity in pipe C is 6 fps, with all other data remaining the same.

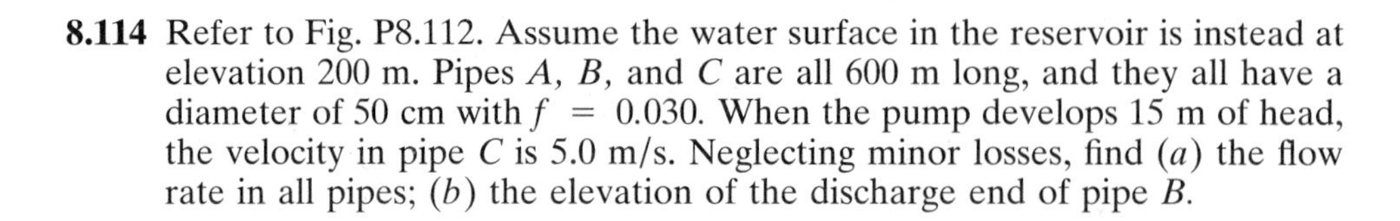

8.114 Refer to Fig. P8.112. Assume the water surface in the reservoir is instead at elevation 200 m. Pipes A, B, and C are all 600 m long, and they all have a diameter of 50 cm with $f = 0.030$. When the pump develops 15 m of head, the velocity in pipe C is 5.0 m/s. Neglecting minor losses, find (*a*) the flow rate in all pipes; (*b*) the elevation of the discharge end of pipe B.

8.115 A 10-in cast-iron pipe 1500 ft long forms one link in a pipe network. If the velocities to be encountered are assumed to fall within the range of 4–10 fps, derive an equation for the flow of water at 60°F in this pipe in the form $h_L = KQ^n$. *Hint:* Using information from Fig. 8.11 and Table 8.1, set up two simultaneous equations corresponding to the ends of the desired velocity range; then solve for the unknowns K and n.

8.116 A 25-cm cast-iron pipe 400 m long forms one link in a pipe network. If the velocities to be encountered are assumed to fall within the range of 0.75–2.3 m/s, derive an equation for the flow of water at 15°C in this pipe in the form $h_L = KQ^n$. *Hint:* See Prob. 8.115.

8.117 Find the magnitude and direction of the flow in network lines *ab* and *bc* (Fig. P8.117) after making two sets of corrections. The numbers on the figure are the K values of each line; take $n = 2.0$. Start by assuming initial flows as follows: 9 cfs in lines *ab* and *cd*, 6 cfs in lines *ac* and *bd*, and 3 cfs in line *bc*.

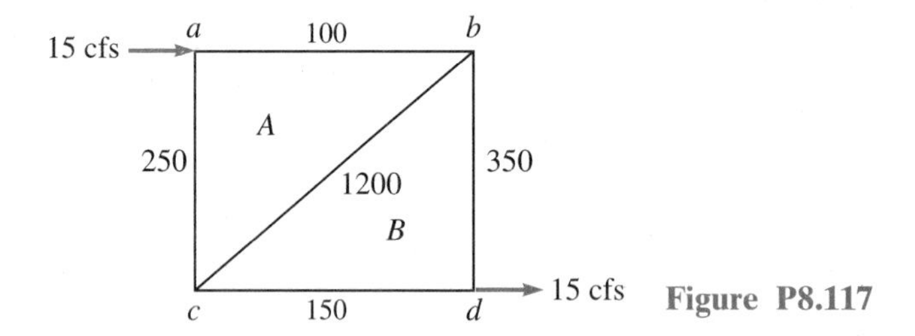

Figure P8.117

8.118 Find the magnitude and direction of the flow in network lines *ab* and *bc* (Fig. P8.118) after making two sets of corrections. The numbers on the figure are the K values of each line; take $n = 2.0$. Start by assuming initial

---

▢: Programmed computing aids (Appendix B) could help solve problems marked with this icon.

flows as follows: 0.3 m^3/s in lines *ab* and *cd*, 0.2 m^3/s in lines *ac* and *bd*, and 0.1 m^3/s in line *bc*.

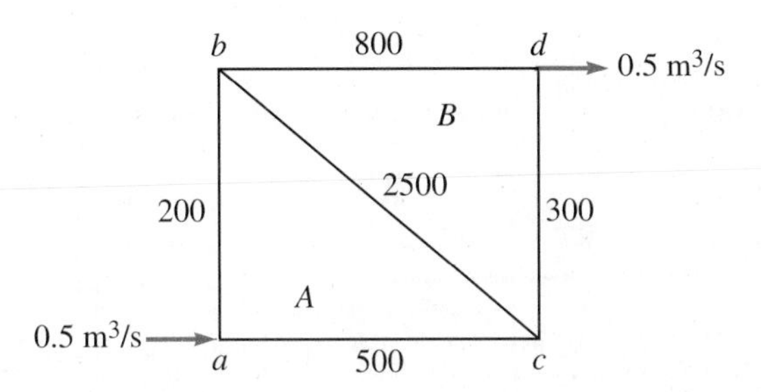

Figure P8.118

8.119 Solve the pipe network shown in Fig. P8.119 using four approximations, to find the flow in each pipe. For simplicity, take $n = 2.0$ and use the value of f for complete turbulence, as given in Fig. 8.11. All pipe is cast iron. For initial flows, assume only values of 30, 15, and 0 L/s (the latter in *dg* and *fh*). If the pressure head at *a* is 40 m, find the pressure head at *d* (which might represent a fire demand, for example) neglecting velocity heads.

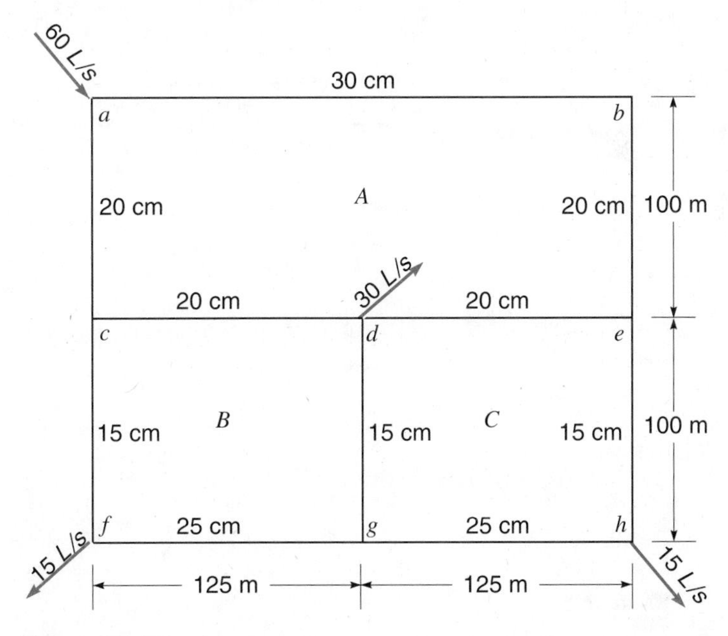

Figure P8.119

8.120 Solve the pipe network shown in Fig. P8.120 using five trials. The 12-in and 16-in pipes are of average cast iron, while the 18-in and 24-in sizes are

: Programmed computing aids (Appendix B) could help solve problems marked with this icon.

of average concrete ($e = 0.003$ ft). Assume $n = 2.0$, and use the values of f from Fig. 8.11 for complete turbulence. If the pressure at h is 80 psi, find the pressure at f.

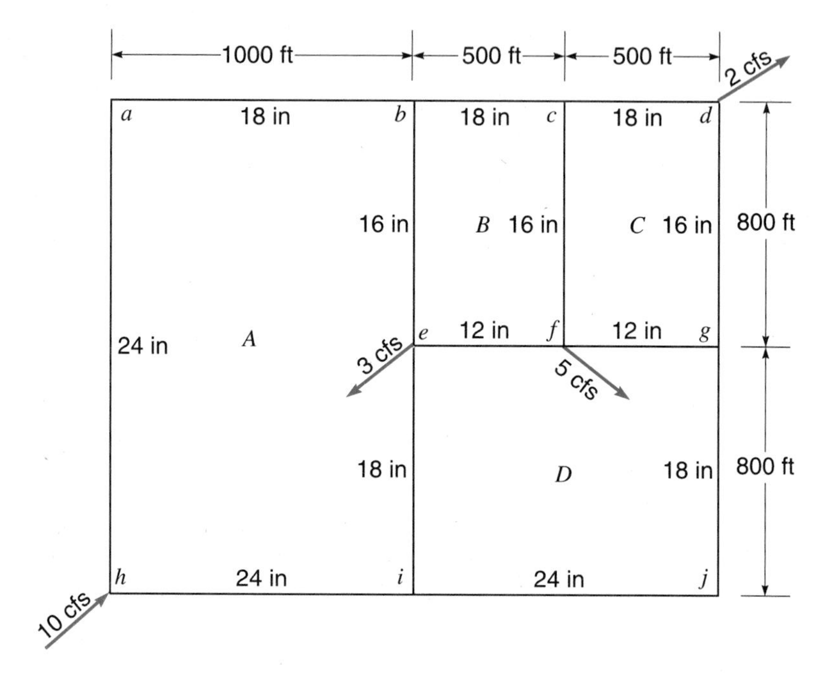

Figure P8.120

CHAPTER 9

Forces on Immersed Bodies

In this chapter the discussion relates primarily to fluid phenomena encountered in incompressible flow or in low-velocity compressible flow where the effects of compressibility are negligible. Near the end of the chapter in Sec. 9.14 there is a brief discussion of the effects of compressibility on drag and lift. These become important at Mach numbers above 0.7.

9.1 INTRODUCTION

When bodies are completely surrounded by a flowing fluid, as for example are airplanes, birds, automobiles, raindrops, submarines, and fish, the flows are known as ***external flows***. Such a body, wholly immersed in a homogeneous fluid, may be subject to two kinds of forces arising from relative motion between the body and the fluid. These forces are termed the ***drag*** and the ***lift***, depending on whether the force is parallel to the motion or perpendicular to it, respectively. Fluid mechanics draws no distinction between two cases of relative motion, namely, when a body moves rectilinearly at constant speed through a stationary fluid or when a fluid travels at constant velocity past a stationary body. Thus it is possible to test airplane models in wind tunnels and torpedo models in water tunnels and predict with confidence the behavior of their prototypes when moving through still fluid. For instructional purposes, it is somewhat simpler to fix our ideas on the stationary body in the moving fluid, while the practical result desired is more frequently associated with a body moving through still fluid.

In this chapter we shall first consider the drag, or resistance, forces. As we shall not be concerned with wave action at a free surface, gravity does not enter the problem, and the forces involved are those due to inertia and viscosity.[1] The drag forces on a submerged body can be viewed as having two components: a ***pressure drag*** F_p and a ***friction drag*** (or ***surface drag***) F_f. The pressure drag, often referred to as the ***form drag*** because it depends largely on the form or shape of the body, is equal to the integration of the components in the direction of motion of all the pressure forces exerted on the surface of the body. It may be expressed[2] as the dynamic component of the stagnation pressure [from Eq. (5.26)] acting on the projected area A of the body *normal to the flow* times a coefficient C_p that is dependent on the geometric form of the body and generally determined by experiment. Thus,

$$F_p = C_p \rho \frac{V^2}{2} A \tag{9.1}$$

The friction drag along a body surface is equal to the integral of the components of the shear stress along the surface in the direction of motion. For convenience, the friction drag is commonly expressed in the same general form as Eq. (9.1). Thus,

$$F_f = C_f \rho \frac{V^2}{2} BL \tag{9.2}$$

where C_f = friction-drag coefficient, dependent on viscosity among other factors

L = length of surface parallel to flow

B = transverse width, conveniently approximated for irregular shapes by dividing total surface area by L

It is important to note that for a body such as a plate with both sides immersed in the fluid, Eq. (9.2) gives the drag for *one side* only.

From our experience with pipe flow, we should expect that the friction drag would be more amenable to a theoretical approach than pressure drag. It turns out that this is not necessarily the case. In previous chapters the ***boundary layer*** was described as a very thin layer of fluid adjacent to a surface, in which viscosity is important, while outside this layer the fluid can be considered as frictionless or ideal. This concept, originated by Ludwig

[1] Actually, without viscosity there could be no drag force at all. The flow of a frictionless fluid about any body, as constructed mathematically or by the flow-net techniques of Chap. 4, produces opposing stagnation points at the nose and tail of the body. The pressure distribution, as computed from the Bernoulli theorem and integrated over the entire body, always adds up to zero in the direction of the flow. This situation is known as ***d'Alembert's paradox***.

[2] Note that Eq. (9.1) is of the same general form as the expression for the drag on a sphere that was developed by dimensional analysis in Sec. 7.7. The comparison indicates that $C_p = \phi(\mathbf{R})$. If the effects of compressibility had been considered in Sec. 7.7, the comparison would have shown that $C_p = \phi(\mathbf{R}, \mathbf{M})$. Similarly, Eq. (9.1) may be compared with Eq. (7.13).

Prandtl in 1904 (see Sec. 8.7), is one of the important advances in modern fluid mechanics. It means that the mathematical theory of ideal fluid flow, including the flow-net method discussed in Chap. 4, can actually be used to determine the streamlines in the real fluid at a short distance from a solid boundary. The Bernoulli theorem may then be used to determine the normal pressures on the surface, for such pressures are practically the same as those outside this thin layer.

The boundary layer may be entirely laminar, or it may be primarily turbulent with a viscous sublayer, as in Fig. 8.7. The thickness δ of the boundary layer is usually defined as the distance from the boundary to the point where the velocity is 99% of the undisturbed velocity, i.e., to where $u = 0.99U$. This thickness increases with distance from the leading edge of a surface, as shown in both Figs. 8.4 and 8.7. At the leading tip, as in Figs. 9.1 and 8.7 for a plate and a reentrant pipe, there is a stagnation point (Sec. 4.10), and because the flow approaching this point must decelerate, the boundary layer begins a short distance upstream. At higher approach velocities the boundary layer becomes thinner in accordance with Eq. (8.27), and this upstream distance becomes very small. Commonly it is neglected, and the boundary layer is assumed to originate from the leading edge, as we shall assume in the following sections.

There is an important difference between flow around immersed bodies and pipe flow. In pipe flow the boundary layers from the opposite walls of the pipe merge together after a certain distance and the flow becomes "all boundary layer," while with airplanes, submarines, trains, etc., the boundary layer, even though it may reach a thickness of several inches, is still small compared with the dimensions of the "ideal fluid" outside the boundary layer.

9.2 FRICTION DRAG OF BOUNDARY LAYER—INCOMPRESSIBLE FLOW

Figure 9.1 illustrates the growth of a boundary layer along one side of a smooth plate in steady flow of an incompressible fluid. Let us consider the

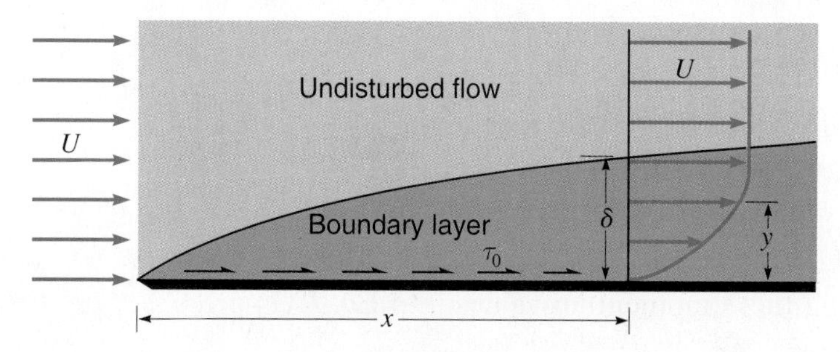

Figure 9.1
Growth of a boundary layer along a smooth plate (vertical scale exaggerated).

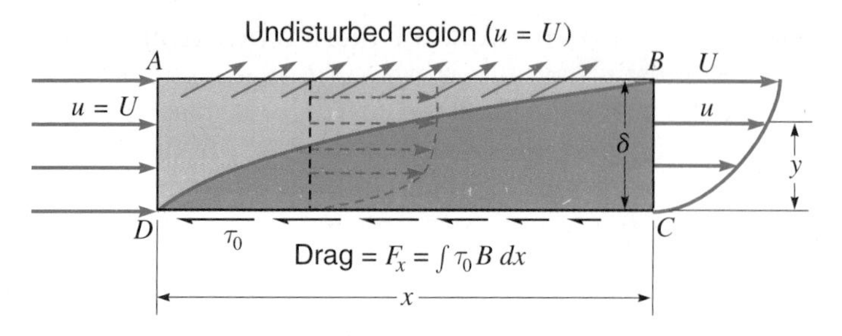

Figure 9.2
Control volume *ABCD* for flow over one side of a flat plate.

control volume *ABCD* shown in Fig. 9.2, which extends a distance δ from the plate, where δ is the thickness of the boundary layer at a distance x along the plate. For convenience, in this analysis let us assume that the velocity $u = U$ at the edge of the boundary layer, instead of the usual $0.99U$. Along control surface *AB*, the undisturbed velocity U exists. The pressure forces around the periphery of the control volume will cancel one another out, since the undisturbed flow field pressure must exist along *AB* and *DA*, and the distance *BC* $(=\delta)$ is so small it will have a negligible effect on pressure variations.

Applying Eq. (6.7*a*), we get

$$
\begin{aligned}
-F_x = -\text{drag} &= \text{rate of momentum in } x \text{ direction leaving through } BC \\
&\quad + \text{rate of momentum in } x \text{ direction leaving through } AB \\
&\quad - \text{rate of momentum in } x \text{ direction entering through } DA
\end{aligned} \tag{9.3}
$$

Since the flow $Q_{BC} < Q_{DA}$, there is flow out of the control volume across control surface *AB* and $Q_{AB} = Q_{DA} - Q_{BC}$.

If the width of the plate is B, neglecting edge effects, the flows and momentums across the control surfaces can be expressed as follows:

Control surface	Flow rate	Rate of momentum in x direction
DA	$UB\delta$	$\rho(UB\delta)U$
BC	$B\int_0^\delta u\, dy$	$\rho B\int_0^\delta u^2\, dy$
AB	$UB\delta - B\int_0^\delta u\, dy$	$\rho\left(UB\delta - B\int_0^\delta u\, dy\right)U$

Substituting these momentum values into Eq. (9.3) gives

$$
F_x = \rho B \int_0^\delta u(U - u)\, dy \tag{9.4}
$$

where F_x is the total friction drag of the plate on the fluid *from the leading edge up to x* directed to the left, as shown on Fig. 9.2. Equal and opposite to this is the drag of the fluid on the plate.

It will now be assumed that the velocity profiles within the boundary layer at various distances along the plate are similar to one another, i.e.,

$$\frac{u}{U} = f\left(\frac{y}{\delta}\right) = f(\eta), \qquad \eta = \frac{y}{\delta}$$

There is experimental evidence that this assumption is valid if there is no pressure gradient along the surface and if the boundary layer does not change from laminar to turbulent within the region considered. Then, substituting $u = Uf(\eta)$, $y = \delta\eta$ and $dy = \delta\, d\eta$ into Eq. (9.4) and noting that the limits of integration become 0 to 1, we get

$$F_x = \rho B U^2 \delta \int_0^1 f(\eta)[1 - f(\eta)]\, d\eta$$

which, for convenience, may be written as

$$F_x = \rho B U^2 \delta \alpha \tag{9.5}$$

where α is a function of the boundary-layer velocity distribution only and is given by the indicated integral (not to be confused with the kinetic-energy factor α discussed in Chap. 5).

We next investigate the local wall shear stress τ_0 at distance x from the leading edge. From the definition of surface resistance, $dF_x = \tau_0 B\, dx$, or

$$\tau_0 = \frac{1}{B}\frac{dF_x}{dx} = \frac{1}{B}\frac{d}{dx}(\rho B U^2 \delta \alpha)$$

and as all terms in the expression for F_x are constant except δ,

$$\tau_0 = \rho U^2 \alpha \frac{d\delta}{dx} \tag{9.6}$$

This expression for the shear stress is valid for either laminar or turbulent flow in the boundary layer, but in this form it is not useful until the quantities α and $d\delta/dx$ have been evaluated.

9.3 LAMINAR BOUNDARY LAYER FOR INCOMPRESSIBLE FLOW ALONG A SMOOTH FLAT PLATE

As in the case of laminar flow in pipes, we may examine the shear stress at the plate wall with the aid of the velocity gradient and the definition of viscosity.

$$\tau_0 = \mu\left(\frac{du}{dy}\right)_{y=0} = \frac{\mu}{\delta}\left(\frac{du}{d\eta}\right)_{\eta=0} = \frac{\mu U}{\delta}\left[\frac{df(\eta)}{d\eta}\right]_{\eta=0}$$

which may be abbreviated to

$$\tau_0 = \frac{\mu U \beta}{\delta} \tag{9.7}$$

where β, like α, is a dimensionless function of the velocity-distribution curve and is given by the expression in brackets.

Equations (9.6) and (9.7) are two independent expressions for τ_0. Equating them to one another results in a simple differential equation,

$$\delta \, d\delta = \frac{\mu \beta}{\rho U \alpha} dx$$

with solution

$$\frac{\delta^2}{2} = \frac{\mu \beta}{\rho U \alpha} x + C$$

where $C = 0$, since $\delta \approx 0$ at $x = 0$ for large U. Therefore

$$\delta = \sqrt{\frac{2\mu\beta x}{\rho U \alpha}} = \sqrt{\frac{2\beta}{\alpha}} \frac{x}{\sqrt{\mathbf{R}_x}} \tag{9.8}$$

where $\mathbf{R}_x = xU\rho/\mu$ may be called the ***local Reynolds number.*** It should be noted that $\mathbf{R}_x$ increases linearly in the downstream direction. Examination of the first expression of Eq. (9.8) shows that the thickness of the laminar boundary layer increases with distance from the leading edge; thus the shear stress [Eq. (9.7)] decreases as the layer grows along the plate.

To evaluate Eq. (9.8), we must know or assume the velocity profile in the laminar boundary layer. The velocity distribution may be closely represented by a parabola, as shown in Fig. 9.3. In dimensionless terms, this curve

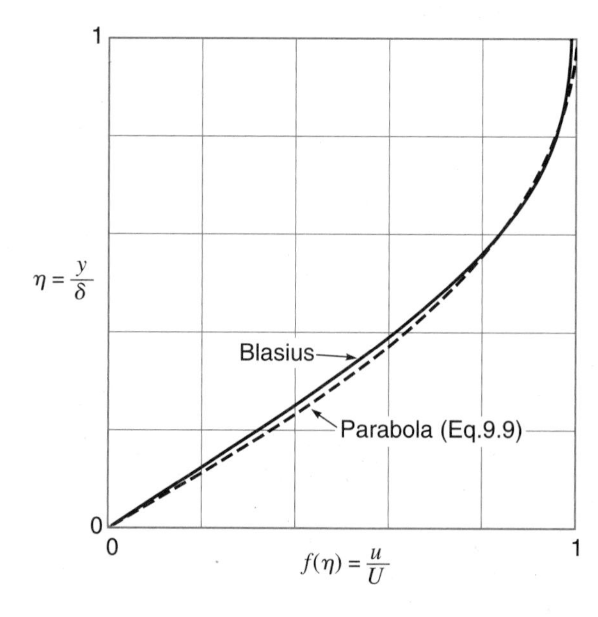

Figure 9.3
Velocity distributions in the laminar boundary layer on a flat plate.

becomes

$$\frac{u}{U} = f(\eta) = 2\eta - \eta^2 \tag{9.9}$$

The other velocity profile in Fig. 9.3 was derived by Blasius from the fundamental equations of viscous flow, with all factors considered, and has been closely checked by experiment. This curve is based on the thickness δ being defined as that for which $u = 0.99U$.

As can be demonstrated by Prob. 9.2, the parabolic distribution will give numerical values for α and β of 0.133 and 2.0, respectively. The Blasius curve yields $\alpha = 0.135$ and $\beta = 1.63$, the principal difference lying in the milder slope of the velocity gradient at the wall. With the Blasius values substituted in Eq. (9.8) we obtain

$$\frac{\delta}{x} = \sqrt{\frac{2 \times 1.63}{0.135}}\frac{1}{\sqrt{\mathbf{R}_x}} = \frac{4.91}{\sqrt{\mathbf{R}_x}} \tag{9.10}$$

If the value of δ from Eq. (9.10) is substituted into Eq. (9.7) with $\beta = 1.63$, we get for the shear stress

$$\tau_0 = 0.332 \frac{\mu U}{x} \sqrt{\mathbf{R}_x} \tag{9.11}$$

But we have another expression for shear stress, given in Eq. (8.8), $\tau_0 = c_f \rho U^2/2$.[3] Setting this equal to Eq. (9.11) allows a determination of the *local* friction coefficient

$$c_f = \frac{0.332\mu U \sqrt{\mathbf{R}_x}}{\rho x U^2/2} = \frac{0.664}{\sqrt{\mathbf{R}_x}} \tag{9.12}$$

If the boundary layer remains laminar over a length L of the plate, the total friction drag on one side of the plate is given by integrating Eq. (9.11):

$$F_f = B \int_0^L \tau_0 \, dx = 0.332B\sqrt{\rho\mu U^3} \int_0^L x^{-1/2} \, dx = 0.664B\sqrt{\rho\mu L U^3} \tag{9.13}$$

Comparing Eq. (9.13) with a standard friction-drag equation (9.2), and substituting U for the more general velocity V, it may be seen that for a laminar boundary layer,

Laminar layer:

$$C_f = 1.328 \sqrt{\frac{\mu}{\rho L U}} = \frac{1.328}{\sqrt{\mathbf{R}}} \tag{9.14}$$

[3] The reader will observe an apparent inconsistency between the notation used here and that used in Chap. 8. Thus, in pipe flow, the significant reference velocity is the mean velocity V in the pipe, while in flow over a plate, it is the *uniform* velocity U of the undisturbed fluid. Likewise, C_f has been employed in Chap. 8 to denote a friction coefficient for the fully developed boundary layer in a pipe, while c_f is used here to denote the *local* friction coefficient of the growing layer.

where it is noted that **R** is based on the characteristic length of the whole plate. The laminar boundary layer will remain laminar if undisturbed, up to a value of $\mathbf{R}_x$ of about 500,000. In this region the layer becomes turbulent, increasing noticeably in thickness and displaying a marked change in velocity distribution.

SAMPLE PROBLEM 9.1 (*a*) Find the friction drag on one side of a smooth flat plate 6 in (15 cm) wide and 18 in (50 cm) long, placed longitudinally in a stream of crude oil ($s = 0.925$) at 60°F (20°C) flowing with undisturbed velocity of 2 fps (60 cm/s). (*b*) Find the thickness of the boundary layer and the shear stress at the trailing edge of the plate.

Solution (BG units)

(*a*) From Fig. 2.4: $\nu = 0.0010 \text{ ft}^2/\text{sec}$

Then, at $x = L$:

$$\mathbf{R} = \frac{LU}{\nu} = \frac{1.5 \times 2}{0.0010} = 3000$$

which is well within the laminar range; that is, $\mathbf{R} < 500{,}000$.

From Eq. (9.14):

$$C_f = \frac{1.328}{\sqrt{\mathbf{R}}} = \frac{1.328}{\sqrt{3000}} = 0.0242$$

From Eq. (9.2): $$F_f = 0.0242 \times 0.925 \times 1.94 \frac{2^2}{2} \frac{6 \times 18}{144} = 0.0652 \text{ lb} \qquad \textit{ANS}$$

(*b*) From Eq. (9.10):

$$\frac{\delta}{x} = \frac{4.91}{\sqrt{3000}} = 0.0896$$

$$\delta = 0.0896 \times 1.5 = 0.1345 \text{ ft} = 1.614 \text{ in} \qquad \textit{ANS}$$

From Eq. (9.11), at $x = L$:

$$\tau_0 = 0.332 \frac{\nu\rho U}{L} \sqrt{\mathbf{R}} = 0.332 \times \frac{(0.0010 \times 0.925 \times 1.94)2}{1.5} \sqrt{3000}$$

$$= 0.0435 \text{ lb/ft}^2 \qquad \textit{ANS}$$

Solution (SI units)

(*a*) From Fig. 2.4: $\nu = 0.93 \times 10^{-4} \text{ m}^2/\text{s}$

Then, at $x = L$: $$\mathbf{R} = \frac{LU}{\nu} = \frac{(0.50\ \text{m})(0.60\ \text{m/s})}{0.93\times 10^{-4}} = 3220$$

which is well within the laminar range; that is, $\mathbf{R} < 500\,000$.

From Eq. (9.14): $$C_f = \frac{1.328}{\sqrt{\mathbf{R}}} = 0.0234$$

From Eq. (9.2):

$$F_f = 0.0234\times 0.925\,\frac{10^3\ \text{kg}}{\text{m}^3}\times\frac{1}{2}\,(0.6\ \text{m/s})^2(0.15\times 0.50)\ \text{m}^2$$

$$F_f = 0.292\,\frac{\text{kg}\cdot\text{m}}{\text{s}^2}\times\frac{\text{N}}{\text{kg}\cdot\text{m/s}^2} = 0.292\ \text{N} \qquad \textbf{\textit{ANS}}$$

(*b*) From Eq. (9.10):

$$\frac{\delta}{x} = \frac{4.91}{\sqrt{3220}} = 0.0864, \qquad \delta = 0.0864\times 50\ \text{cm} = 4.32\ \text{cm} \qquad \textbf{\textit{ANS}}$$

From Eq. (9.11), at $x = L$:

$$\tau_0 = 0.332\,\frac{\mu U}{L}\sqrt{\mathbf{R}}\times 0.332\,\frac{\rho\nu U}{L}\sqrt{\mathbf{R}}$$

$$\tau_0 = 0.332\,\frac{(0.925\times 10^3)(0.93\times 10^{-4})0.6}{0.50}\sqrt{3220} = 1.947\ \text{N/m}^2 \qquad \textbf{\textit{ANS}}$$

EXERCISES

9.3.1 Determine the shear stress 9 in and 18 in back from the leading edge of the plate in Sample Prob. 9.1.

9.3.2 Find the shear stress and the thickness of the boundary layer (*a*) at the center and (*b*) at the trailing edge of a smooth, flat plate 3.0 m wide and 0.6 m long parallel to the flow, immersed in 15°C water flowing at an undisturbed velocity of 0.9 m/s. Assume a laminar boundary layer over the whole plate. Also, (*c*) find the total friction drag on one side of the plate.

9.3.3 For the critical Reynolds number of 500,000 for transition from laminar to turbulent flow in the boundary layer, find the corresponding critical Reynolds number for flow in a circular pipe. How does this compare with the value given in Chap. 8? *Hint*: Let the boundary-layer thickness correspond to the radius of the pipe in laminar flow, and let the undisturbed velocity U of the boundary layer theory represent the centerline velocity u_{max} of the pipe flow.

9.4 TURBULENT BOUNDARY LAYER FOR INCOMPRESSIBLE FLOW ALONG A SMOOTH FLAT PLATE

Comparing the laminar and turbulent boundary layers in Fig. 9.4, the velocity distribution in the turbulent boundary layer shows a much steeper gradient near the wall and a flatter gradient throughout the remainder of the layer. As would be expected, then, the wall shear stress is greater in the turbulent boundary layer than in the laminar layer at the same Reynolds number. In this case, however, it is not practical to proceed along the lines of Eqs. (9.6) and (9.7), determining the shear stress from the velocity gradient at the wall. As an approximation, we turn instead to turbulent flow in a circular pipe because of the wealth of experimental information on that subject compared with that of flow in a turbulent boundary layer along a smooth flat plate. We learned in Eq. (8.14) that the shear stress at the wall of a pipe is given by

$$\tau_0 = f\rho \frac{V^2}{8} \tag{9.15}$$

where V again denotes the average velocity in the pipe. Now we shall assume that the turbulent boundary layer occupies all the region between the wall and the center line of the pipe, as in Fig. 9.5. The radius of the pipe then

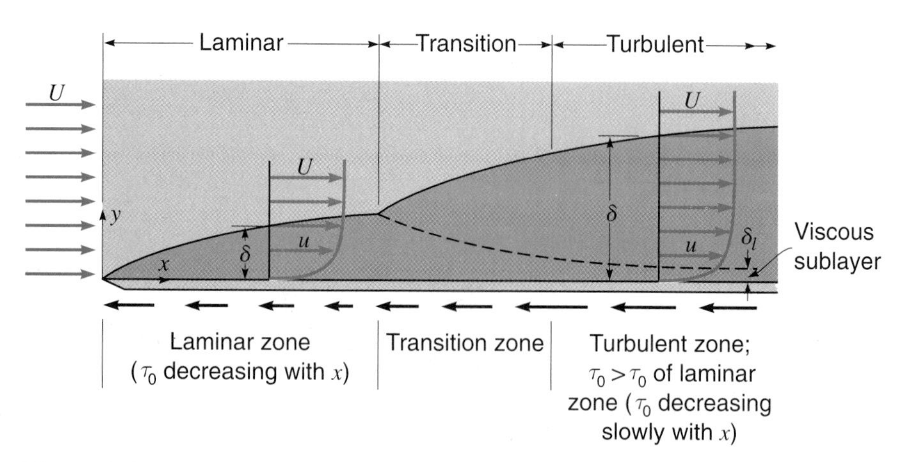

Figure 9.4
Laminar and turbulent boundary layers along a smooth flat plate (vertical scale greatly exaggerated).

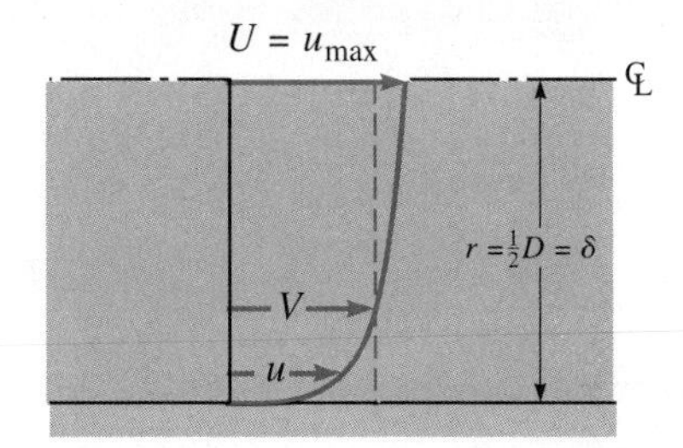

Figure 9.5
Flow in a pipe as a turbulent boundary layer.

becomes the thickness of the boundary layer, and, by analogy, the velocity at the center of the pipe, here denoted by U, corresponds to the undisturbed velocity at the outer edge of the boundary layer. We may obtain an approximate relation between V and U by use of the pipe-factor equation (8.32). Taking a middle value of $f = 0.028$ and allowing for the 1% difference in velocity between the edge of the boundary layer and the free stream, we have

$$U = 1.235V \tag{9.16}$$

To proceed further, we need a simple relation between f and $\mathbf{R}$ for turbulent pipe flow in a smooth pipe. The Blasius equation (8.37) provides a useful relationship. Substituting Eqs. (8.37) and (9.16) into (9.15) gives

$$\tau_0 = \frac{0.316}{(DV/\nu)^{1/4}}\frac{\rho V^2}{8} = \frac{0.316}{[(2\delta/\nu)(U/1.235)]^{1/4}}\frac{\rho}{8}\left(\frac{U}{1.235}\right)^2 = \frac{0.0230\rho U^2}{(\delta U/\nu)^{1/4}} \tag{9.17}$$

If we now equate the two expressions for τ_0 [Eqs. (9.6) and (9.17)], we obtain

$$\rho U^2 \alpha \frac{d\delta}{dx} = \frac{0.023\rho U^2}{(\delta U/\nu)^{1/4}} \tag{9.18}$$

Separating variables and integrating this expression (with the condition $\delta = 0$ at $x = 0$), yields

$$\delta = \left(\frac{0.0287}{\alpha}\right)^{4/5}\left(\frac{\nu}{Ux}\right)^{1/5} x \tag{9.19}$$

In using Eq. (9.6) we assume that we know the velocity distribution in the turbulent boundary layer. Of the many formulas for this distribution that have been proposed, the most convenient for our purpose is Eq. (8.38), the seventh-root law, which can be expressed as

$$u = Uf(\eta) = U(\eta)^{1/7} = U(y/\delta)^{1/7} \tag{9.20}$$

For the case of such a velocity distribution, $\alpha = 0.0972$. This is found by evaluating the integral definition of α (Sec. 9.2):

$$\alpha = \int_0^1 f(\eta)[1 - f(\eta)]\, d\eta$$

where $\eta = y/\delta$

$d\eta = dy/\delta$

$f(\eta) = (y/\delta)^{1/7}$

Substituting this value for α into Eq. (9.19) gives

$$\frac{\delta}{x} = 0.377\left(\frac{\nu}{Ux}\right)^{1/5} = \frac{0.377}{\mathbf{R}_x^{\ 1/5}} \tag{9.21}$$

and substituting this value of δ into Eq. (9.17) yields

$$\tau_0 = 0.0587\rho\,\frac{U^2/2}{\mathbf{R}_x^{\ 1/5}} \tag{9.22}$$

And since (Sec. 9.3) $\tau_0 = c_f\rho U^2/2$,

$$c_f = \frac{0.0587}{\mathbf{R}_x^{\ 1/5}} \tag{9.23}$$

With this value for c_f (the ***local friction coefficient***), we can express the total friction drag on one side of the plate as

$$F_f = B\int_0^L \tau_0\,dx = 0.0735\rho\,\frac{U^2}{2}\left(\frac{\nu}{UL}\right)^{1/5} BL \tag{9.24}$$

or in Eq. (9.2) the friction-drag coefficient for a turbulent boundary layer is

Turbulent layer:

$$C_f = \frac{0.0735}{\mathbf{R}^{1/5}} \quad \text{for } 500{,}000 < \mathbf{R} < 10^7 \tag{9.25}$$

where it is again noted that the characteristic length in **R** is L, the total length of the plate parallel to the flow. For Reynolds numbers above 10^7, Schlichting has proposed a modification of Eq. (9.25) that agrees better with experimental results:[4]

Turbulent layer:

$$C_f = \frac{0.455}{(\log \mathbf{R})^{2.58}} \quad \text{for } \mathbf{R} > 10^7 \tag{9.26}$$

For Reynolds numbers less than 10^7, this equation gives values for C_f that are very close to those given by Eq. (9.25), and consequently Eq. (9.26) is commonly employed over the entire range of Reynolds numbers above 500,000.

---

[4] H. Schlichting, Boundary Layer Theory, Part II, *NACA Tech. Memo.* 1218, p. 39, 1949.

Sample Problem 9.2 (*a*) Find the frictional drag on the top and sides of a box-shaped moving van 8 ft wide, 10 ft high, and 35 ft long, traveling at 60 mph through air at 50°F. Assume that the vehicle has a rounded nose so that the flow does not separate from the top and sides (see Fig. 9.12*b*). Assume also that even though the top and sides of the van are relatively smooth, there is enough roughness that, for all practical purposes, a turbulent boundary layer starts immediately at the leading edge. (*b*) Find the thickness of the boundary layer and the shear stress at the trailing edge.

Solution

Table A.2 for air at 50°F: $\nu = 0.000152\ \text{ft}^2/\text{sec}$, $\rho = 0.00242\ \text{slug/ft}^3$. 60 mph = 88 fps.

(*a*)
$$\mathbf{R} = \frac{LU}{\nu} = \frac{35 \times 88}{0.000\,152} = 20{,}260{,}000, \qquad \log \mathbf{R} = 7.31$$

As $\mathbf{R} > 10^7$, use Eq. (9.26),

$$C_f = \frac{0.455}{(7.31)^{2.58}} = 0.00269$$

Then, by Eq. (9.2),

$$F_f = 0.00269(0.00242)\frac{(88)^2}{2}(10 + 8 + 10)35 = 24.7\ \text{lb} \qquad \textbf{\textit{ANS}}$$

(*b*) Applying Eq. (9.21), at $x = 35$ ft,

$$\delta_{35} = \frac{36 \times 0.377}{(20{,}260{,}000)^{1/5}} = 0.456\ \text{ft} \qquad \textbf{\textit{ANS}}$$

From Eq. (9.22), with $x = 35$ ft,

$$(\tau_0)_{35} = 0.0587(0.00242)\frac{88^2}{2}\left(\frac{0.000152}{88 \times 35}\right)^{1/5} = 0.01901\ \text{lb/ft}^2 \qquad \textbf{\textit{ANS}}$$

EXERCISES

9.4.1 Find the shear stress on the sides of the van in Sample Prob. 9.2 at (*a*) 2 ft, (*b*) 12 ft, and (*c*) 22 ft back from the leading edge of the sides.

9.4.2 Assume that the boundary layer of Exer. 9.3.2 is disturbed near the leading edge. Compute the corresponding quantities for the turbulent boundary layer covering the entire plate, and compare the results.

9.4.3 A 280-ft-long streamlined train has 8.5-ft-high sides and an 8-ft-wide top. Compute the power required to overcome the skin-friction drag when the train is traveling at 90 mph through the ICAO standard atmosphere at sea level, assuming the drag on the sides and top to be equal to that on one side of a flat plate 25 ft wide and 280 ft long.

9.4.4 A lifeguard determines the wind velocity 6 ft above the beach to be 25 fps. If one wishes to get out of the wind by lying down, what would be the velocity at (*a*) 0.5 ft, and (*b*) at 1.0 ft above beach level?

9.4.5 Compute C_f for $\mathbf{R} = 10^7$ using Eqs. (9.25) and (9.26), and compare the two values.

9.5 FRICTION DRAG FOR INCOMPRESSIBLE FLOW ALONG A SMOOTH FLAT PLATE WITH A TRANSITION REGIME

In the two preceding sections we have treated separately the resistance due to laminar and turbulent boundary layers on a smooth flat plate. We now wish to determine how to compute the total friction drag when there is a transition from the laminar to the turbulent boundary layer part way along the plate surface.

Let x_c in Fig. 9.6 be the distance from the leading edge to the point where the boundary layer becomes turbulent, which will normally occur at a value of $\mathbf{R}_x$ of about 500,000. The drag of the turbulent portion of the boundary layer may be approximated as the drag that would occur if a turbulent boundary layer extended along the whole plate, minus the drag of a fictitious turbulent layer from the leading edge to x_c. Thus

$$F_{\text{turb}} \approx F_{\text{total turb}} - F_{\text{turb to } x_c}$$

When this is added to the drag from the laminar boundary layer up to x_c, we have, from Eqs. (9.2), (9.14), (9.26), and (9.25), for the total drag, assuming the plate is long enough[5] that $\mathbf{R} > 10^7$,

$$F_f = \rho \frac{U^2}{2} B \left[\frac{1.328 x_c}{\sqrt{\mathbf{R}_c}} + \frac{0.455 L}{(\log \mathbf{R})^{2.58}} - \frac{0.0735 x_c}{\mathbf{R}_c^{1/5}} \right]$$

where $\mathbf{R}_c$ is based on the length x_c to the point of transition, while $\mathbf{R}$ is based on the total length L of the plate, as before. Next we observe that

$$\frac{\mathbf{R}_c}{\mathbf{R}} = \frac{x_c}{L}, \qquad \text{or} \qquad x_c = \frac{\mathbf{R}_c}{\mathbf{R}} L$$

and thus
$$F_f = \rho \frac{U^2}{2} BL \left[1.328 \frac{\sqrt{\mathbf{R}_c}}{\mathbf{R}} + \frac{0.455}{(\log \mathbf{R})^{2.58}} - \frac{0.0735 \mathbf{R}_c^{4/5}}{\mathbf{R}} \right]$$

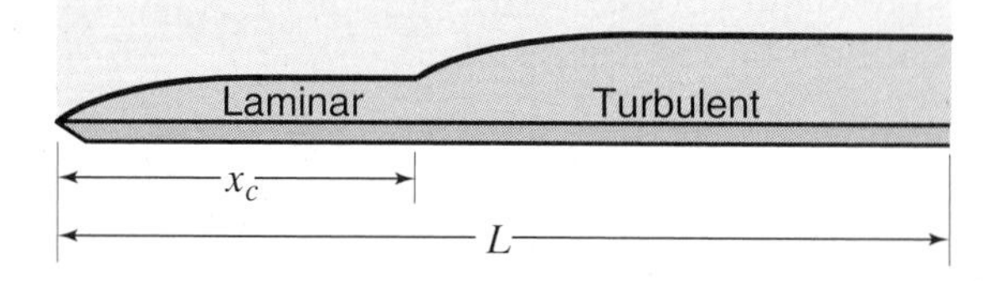

Figure 9.6
Boundary layers along a smooth flat plate of finite length.

[5] This expression for F_f is also quite accurate in the undisturbed transition regime for a shorter plate with $\mathbf{R} < 10^7$ because Eqs. (9.26) and (9.25) give almost identical values for C_f when $\mathbf{R} < 10^7$.

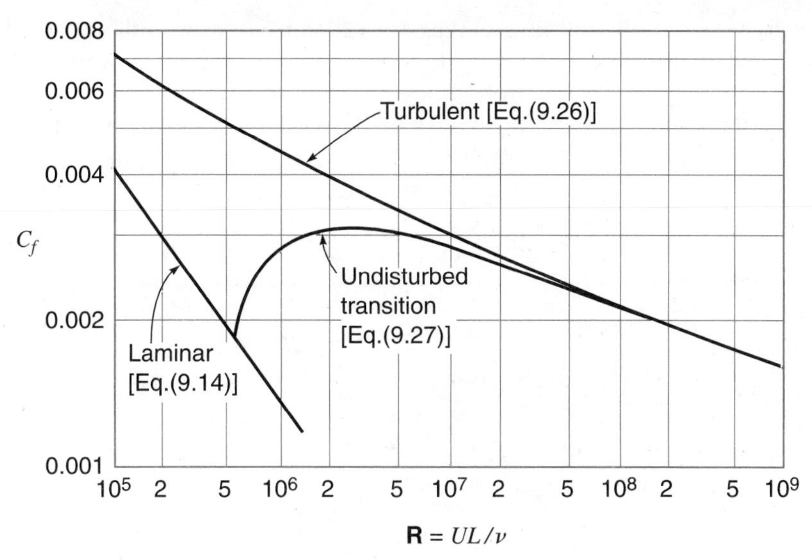

Figure 9.7
Drag coefficients for a smooth flat plate.

and, since the quantity in brackets is the friction-drag coefficient C_f in Eq. (9.2), we get for $\mathbf{R}_c = 500{,}000$,

With transition:

$$C_f = \frac{0.455}{(\log \mathbf{R})^{2.58}} - \frac{1700}{\mathbf{R}} \tag{9.27}$$

Equations (9.14), (9.26), and (9.27) are plotted in Fig. 9.7, for comparison.

All the treatment of laminar and turbulent boundary layers has so far been based upon the surface of the immersed body being smooth. However, a local region of excess roughness in the laminar zone can "trip" the laminar layer into becoming a turbulent layer at Reynolds numbers less than 500,000. The height of the roughness that will cause this tripping is known as the ***critical roughness***. This is given approximately by[6]

$$e_c = \frac{15\nu}{\sqrt{\tau_0/\rho}} = 26.0\frac{\nu}{U}\mathbf{R}_x^{1/4} \tag{9.28}$$

where τ_0 is determined by the *laminar* equation (9.11), not by the turbulent equation (9.22).

We see that the height of the critical roughness depends on its distance from the leading edge. As the laminar boundary layer grows along the plate, the roughness must be greater in order to upset the stability of the layer. Recall that when $\mathbf{R}_x$ reaches a value in the neighborhood of 500,000, the laminar layer of itself becomes unstable, however smooth the surface, and

[6] I. Tani, J. Hama, and S. Mituisi, On the Permissible Roughness in the Laminar Boundary Layer, *Aeronaut. Res. Inst., Tokyo Imp. Univ., Rept.* 15, p. 419, 1940.

changes to a turbulent boundary layer with a thin viscous sublayer. As in the case of flow in pipes, the surface is considered hydraulically smooth if the effect of the roughness projections does not extend through this sublayer.

In the turbulent zone the thickness of the viscous sublayer is not a clearly determinable quantity, but it appears to be agreed that the thickness of this predominantly laminar film is given approximately (see Sec. 8.9 for pipe flow) by

$$\delta_l = \frac{5\nu}{u_*} \tag{9.29}$$

while the transition layer extends out to

$$\delta_t = \frac{70\nu}{u_*} \tag{9.30}$$

where the shear-stress velocity u_* was first defined in Sec. 8.9, and τ_0 is given by the *turbulent* equation (9.22), so that

$$u_* = \sqrt{\frac{\tau_0}{\rho}} = \frac{0.1713U}{\mathbf{R}_x^{\ 0.1}} \tag{9.31}$$

If the roughness height is only of the order of δ_l, the surface may still be considered smooth, but if the roughness height is greater than δ_t, the surface is truly rough and the drag is materially increased.

It may be remarked finally that a plate or wing that is to incur minimum drag must be very smooth near the leading edge, where the laminar layer or sublayer is thinnest, while greater roughness may be tolerated near the trailing edge. Since the wall shear is so much greater in a turbulent boundary layer than in a laminar one, anything that can be done to delay the breakdown of the laminar boundary layer will greatly reduce the frictional drag force on a body. The ***laminar flow wing*** for aircraft is one for which suction slots along the leading edge of the wing together with a smooth leading edge and a properly shaped wing profile help to maintain a favorable pressure gradient (Sec. 9.6) along the upper surface of the wing. This delays the breakdown of the laminar boundary layer, and thus such wings have much less drag than conventional ones.

SAMPLE PROBLEM 9.3 A small submarine, which may be supposed to approximate a cylinder 10 ft in diameter and 50 ft long, travels submerged at 3 knots (5.06 fps) in sea water at 40°F. (*a*) Find the friction drag assuming no separation from the sides. (*b*) Find the value of the critical roughness for a point 1 ft from the nose of the submarine. (*c*) Find the height of roughness at the mid-section of the submarine that would class the surface as truly rough.

Solution

Viscosity of sea water ≈ viscosity of fresh water; specific weight ≈ 64 lb/ft³. Table A.1 at 40°F: $\nu = 0.000\,016\,64$ ft²/sec.

(*a*)
$$\mathbf{R} = \frac{50 \times 5.06}{0.000\,016\,64} = 1.525 \times 10^7$$

Eq. (9.27) or Fig. 9.7: $C_f = 0.00270$

$$F_f = 0.002\,70\left(\frac{64}{32.2}\right)\frac{(5.06)^2}{2}(\pi\times 10)\,50 = 108.0 \text{ lb} \qquad \textit{ANS}$$

(*b*) At $x = 1$ ft: $$\mathbf{R}_x = \frac{5.06\times 1}{0.000\,016\,64} = 305{,}000$$

Eq. (9.28): $$e_c = \frac{26\times 0.000\,016\,64}{5.06}(305{,}000)^{1/4} = 0.002\,00 \text{ ft} \qquad \textit{ANS}$$

(*c*) At $x = 25$ft: $$\mathbf{R}_x = \frac{5.06\times 25}{0.000\,016\,64} = 7.63\times 10^6$$

Eq. (9.31): $$u_* = \frac{0.1713\times 5.06}{(7.63\times 10^6)^{0.1}} = 0.1778 \text{ fps}$$

Eq. (9.30): $$\delta_t = \frac{70\times 0.000\,016\,64}{0.1778} = 0.006\,53 \text{ ft}; \quad e = 0.006\,53 \text{ ft} \qquad \textit{ANS}$$

EXERCISES

9.5.1 A 7.5-ft by 1.5-ft smooth, thin, flat plate with sharpened edges is submerged in 60°F water moving with a velocity of 1.4 fps in the direction of the 7.5 ft length. What is the total drag?

9.5.2 An 1-in-diameter harpoon 6 ft long, with a sharp tip, is launched at 20 fps into 60°F water. Find (*a*) the friction drag; (*b*) the maximum thickness of the boundary layer.

9.5.3 An airplane wing with a chord length of 3 m parallel to the flow moves through standard atmospheric air at an altitude of 6 km with a speed of 350 km/h. Find (*a*) the critical roughness for a point one-tenth the chord length back from the leading edge; (*b*) the surface drag on an 8-m-span section of this wing.

9.6 BOUNDARY-LAYER SEPARATION AND PRESSURE DRAG

The motion of a thin stratum of fluid lying wholly inside the boundary layer is determined by three forces:

1. The forward pull of the outer free-moving fluid, transmitted through the laminar boundary layer by viscous shear and through the turbulent boundary layer by momentum transfer (Sec. 8.8);
2. The viscous retarding effect of the solid boundary, which must, by definition, hold the fluid stratum immediately adjacent to it at rest;

3. The pressure gradient along the boundary: the stratum is accelerated by a pressure gradient whose pressure decreases in the direction of flow and is retarded by an adverse gradient.

The treatment of fluid resistance in the foregoing sections has been restricted to the drag of the boundary layer along a smooth flat plate located in an unconfined fluid, that is to say, in the *absence of a pressure gradient.* In the presence of a favorable pressure gradient, the boundary layer is "held" in place. This is what occurs in the accelerated flow around the ***forebody,*** or upstream portion, of a cylinder, sphere, or other object, such as that of Fig. 4.12. If a particle enters the boundary layer near the forward stagnation point with a low velocity and high pressure, its velocity will increase as it flows into the lower-pressure region along the side of the body. But there will be some retardation from wall friction (force 2 above), so that its total useful energy will be reduced by a corresponding conversion into thermal energy.

What happens next may best be explained by reference to Fig. 9.8. Let A represent a point in the region of accelerated flow with a normal velocity distribution in the boundary layer (either laminar or turbulent), while B is the point where the velocity outside the boundary layer reaches a maximum. Then C, D, and E are points downstream where the velocity outside the boundary layer decreases, resulting in an increase in pressure in accordance with ideal-flow theory. Thus the velocity of the layer close to the wall is reduced at C and finally brought to a stop at D. Now the increasing pressure calls for further retardation; but this is impossible, and so the boundary layer actually ***separates*** from the wall. At E there is a ***backflow*** next to the wall, driven in the direction of decreasing pressure—upstream in this case—and feeding fluid into the boundary layer that has left the wall at D.

Downstream from the point of separation, the flow is characterized by irregular turbulent eddies, formed as the separated boundary layer becomes rolled up in the reversed flow. This condition generally extends for some

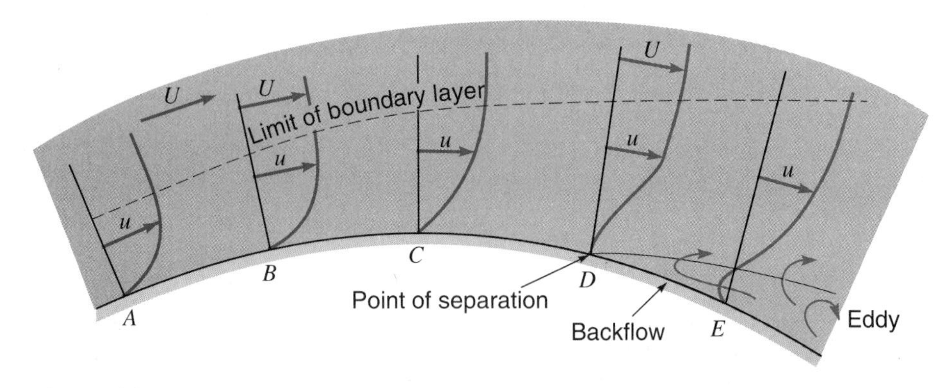

Figure 9.8
Growth and separation of boundary layer owing to increasing pressure gradient. Note that U has its maximum value at B and then gets smaller.

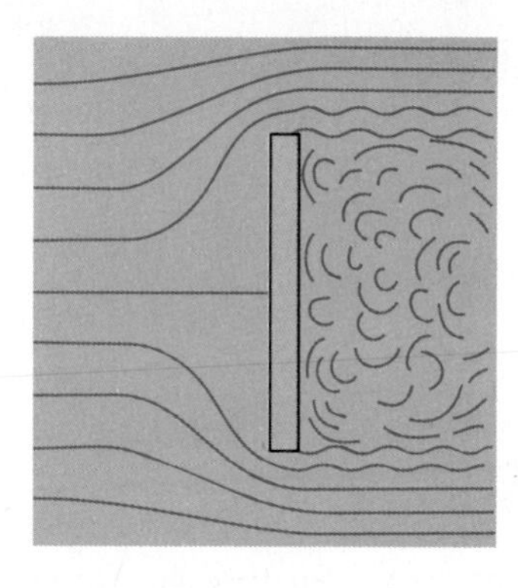

Figure 9.9
Turbulent wake behind a flat plate held normal to the flow.

distance downstream until the eddies are worn away by viscous attrition. The whole disturbed region is called the ***turbulent wake*** of the body (Fig. 9.9).

Because the eddies cannot convert their kinetic energy of rotation into an increased pressure, as the ideal-fluid theory would dictate, the pressure within the wake remains close to that at the separation point. Since this is always less than the pressure at the forward stagnation point, there results a net pressure difference tending to move the body with the flow, and this force is the pressure drag.

Although the laminar and turbulent boundary layers behave in essentially the same manner at a point of separation, the *location* of the separation point on a given curved surface will be very different for the two cases. In the laminar layer the transfer of momentum from the rapidly moving outer strata through the viscous-shear process to the inner strata is slow and ineffective. Consequently, the laminar boundary layer is "weak" and cannot long stick to the wall against an adverse pressure gradient. The transition to a turbulent boundary layer, on the other hand, brings a violent mixing of the faster-moving outer strata into the slower-moving inner strata, and vice versa. The mean velocity close to the boundary is greatly increased, as shown in Fig. 9.4. This added energy enables the boundary layer to better withstand the adverse pressure gradient, with the result that *with a turbulent boundary layer the point of separation is moved downstream* to a region of higher pressure. An example of this is shown in Fig. 9.11.

9.7 DRAG ON THREE-DIMENSIONAL BODIES (INCOMPRESSIBLE FLOW)

The total drag on a body is the sum of the friction drag and the pressure drag:

$$F_D = F_f + F_p$$

In the case of a well-streamlined body, such as an airplane wing or the hull of a submarine, the friction drag is the major part of the total drag, and may be estimated by the methods of the preceding sections on the boundary layer. Only rarely is it desired to compute the pressure drag separately from the friction drag. Usually, when the wake resistance becomes significant, one is interested in the total drag only. Indeed, it is customary to employ a single equation that gives the total drag[7]

$$F_D = C_D \rho \frac{V^2}{2} A \tag{9.32}$$

in terms of an overall drag coefficient C_D, with the other quantities the same as in Eq. (9.1), except that in the case of the lifting vane (like an airplane wing) the area A is defined as the product of the span and the mean chord (Figs. 9.15 and 9.21). In such a case the area is neither strictly parallel to nor normal to the flow.

In the case of a body with sharp corners, such as the plate of Fig. 9.9, set normal to the flow, separation always occurs at the same point, and the wake extends across the full projected width of the body. This results in a relatively constant value of C_D, as may be seen from the plot for the flat disk in Fig. 9.10. If the body has curved sides, however, the location of the separation point will be determined by whether the boundary layer is laminar or turbulent. This location in turn determines the size of the wake and the amount of the pressure drag.

The foregoing principles are vividly illustrated in the case of the flow around a sphere. For very low Reynolds numbers ($DV/\nu < 1$, in which D is the diameter of the sphere), the flow about the sphere is completely viscous, and the friction drag is given by Stokes' law,

$$F_D = 3\pi\mu VD \tag{9.33}$$

Equating this equation to Eq. (9.32), where A is defined as $\pi D^2/4$, the frontal area of the projected sphere, gives the result that $C_D = 24/\mathbf{R}$. The similarity between this case and the value of the friction factor for laminar flow in pipes is at once apparent. This regime of the flow about a sphere is shown as the straight line at the left of the log–log plot of C_D versus $\mathbf{R}$ in Fig. 9.10.

As $\mathbf{R}$ is increased beyond 1, the laminar boundary layer separates from the surface of the sphere, beginning first at the rear stagnation point, where the adverse pressure gradient is the strongest. The curve of C_D in Fig. 9.10 begins to level off as the pressure drag becomes of increasing importance, and the drag becomes more proportional to V^2. With further increase in $\mathbf{R}$, the point of separation moves forward on the sphere, until at $\mathbf{R} \approx 1000$ the point of separation becomes fairly stable at about 80° from the forward stagnation point.

[7] In the equations for total drag on submerged bodies, we revert to the use of V to designate a general reference velocity.

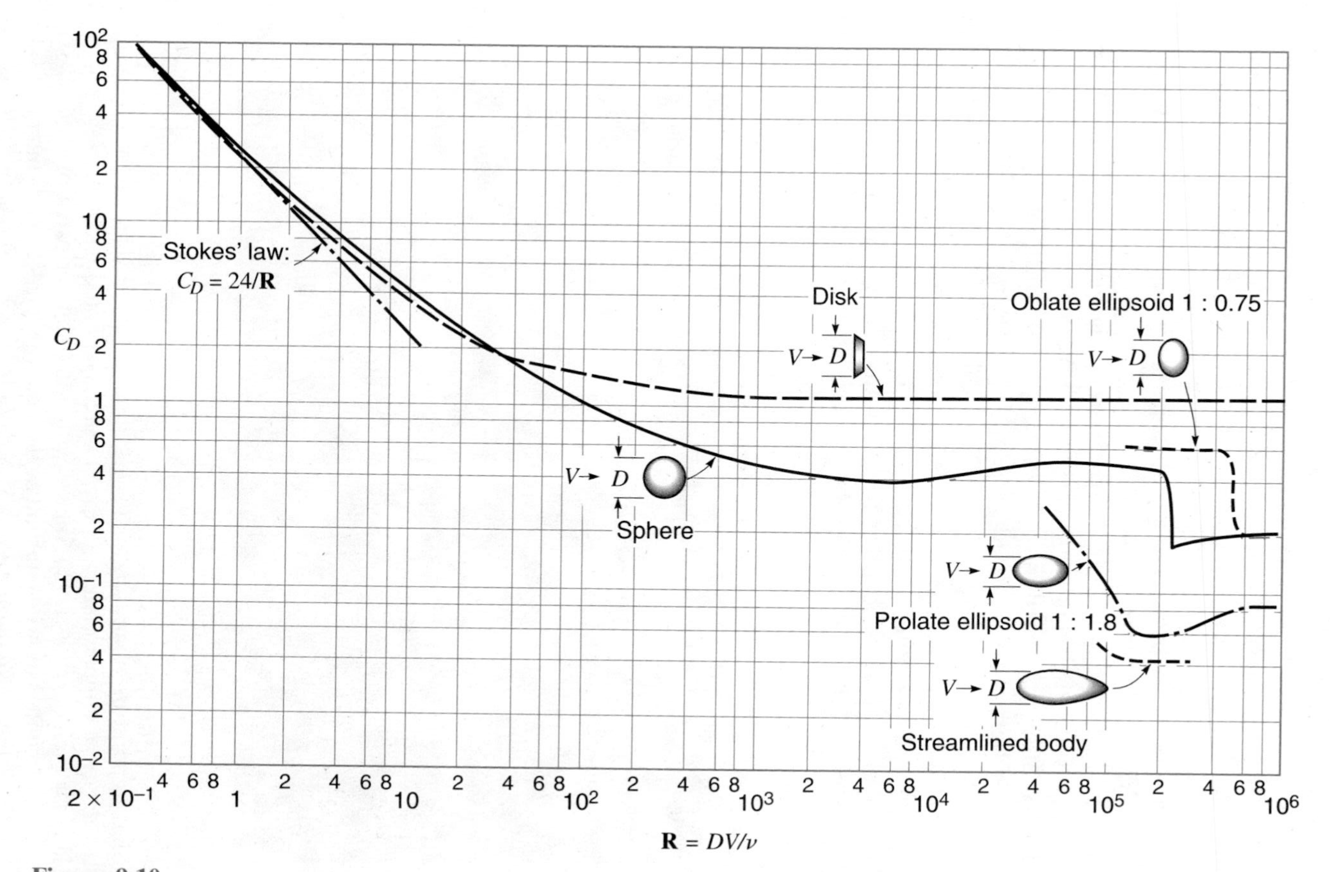

Figure 9.10
Drag coefficients for axisymmetric bodies.

For a considerable range of Reynolds numbers, conditions remain fairly stable, the laminar boundary layer separating from the forward half of the sphere and C_D remaining fairly constant at about 0.45. At a value of **R** of about 250,000 for the smooth sphere, however, the drag coefficient is suddenly reduced by about 50%, as may be seen in Fig. 9.10. The reason for this lies in a change from a laminar to a turbulent boundary layer on the sphere. The point of separation is moved back to something like 115°, from the stagnation point, with a consequent decrease in the size of the wake and the pressure drag.

If the level of turbulence in the free stream is high, the transition from laminar to turbulent boundary layer will take place at lower Reynolds numbers. Because this phenomenon of shift in separation point is so well defined, the sphere is often used as a turbulence indicator. The Reynolds number producing a value of C_D of 0.3—which lies in the middle of the rapid-drop range—becomes an accurate measure of the turbulence.

As we noted previously, the transition from a laminar to a turbulent boundary layer may also be prematurely induced by artificially roughening the surface over a local region. The two pictures of Fig. 9.11 clearly show the

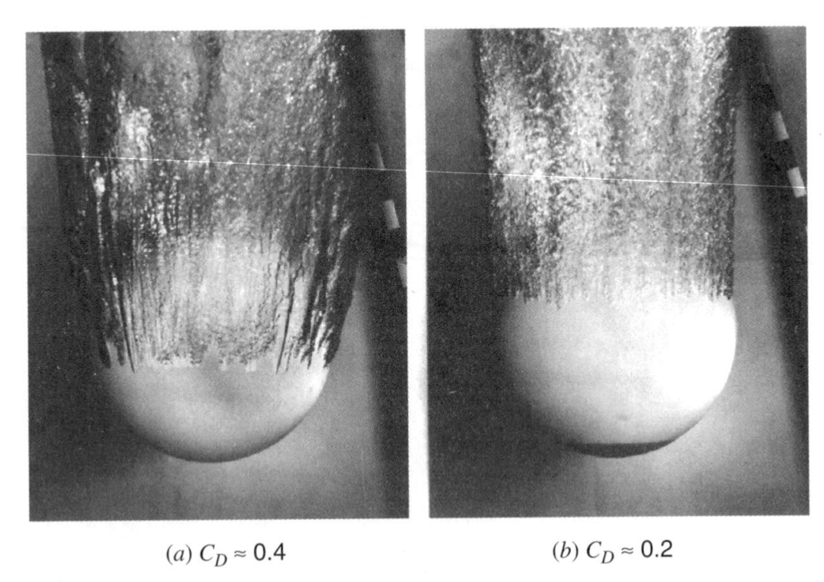

(*a*) $C_D \approx 0.4$ (*b*) $C_D \approx 0.2$

Figure 9.11
Shift in point of separation on 8.5-in-diameter sphere (bowling ball) at a velocity of approximately 25 fps in water. (*a*) Smooth sphere—laminar boundary layer. (*b*) Sphere with 4-in-diameter patch of sand grains cemented to nose—turbulent boundary layer. Reynolds numbers are the same. (*Photograph by U.S. Naval Ordnance Test Station, Pasadena Annex.*)

effectiveness of this procedure. By roughening the nose of the sphere, the boundary layer is made turbulent and the separation point moved back. The added roughness and turbulent boundary layer cause an increase in friction drag, to be sure, but this is of secondary importance compared with the marked decrease in the size and effect of the wake. This is the main reason why the surface of a golf ball is dimpled. A smooth-surfaced ball would have greater overall drag and would travel only about 60% as far when driven.

Plots of C_D versus **R** for various other three-dimensional shapes are also shown in Fig. 9.10. It may be pointed out here that the object of ***streamlining*** a body is to move the point of separation as far back as possible and thus to produce the minimum size of turbulent wake. This decreases the pressure drag, but by making the body longer so as to promote a gradual increase in pressure, the friction drag is increased. The optimum amount of streamlining, then, is that for which the sum of the friction and pressure drag is a minimum. Quite evidently, from what we have learned, attention in streamlining must be given to the rear end, or downstream part, of a body as well as to the front. The shape of the forebody is important principally to the extent that it governs the location of the separation point(s) on the afterbody. A rounded nose produces the least disturbance in the streamlines, and is therefore the best for incompressible or compressible flow at subsonic velocities.[8] This is illustrated in Fig. 9.12, where flow about a blunt-nosed motor vehicle is compared with that about a round-nosed vehicle.

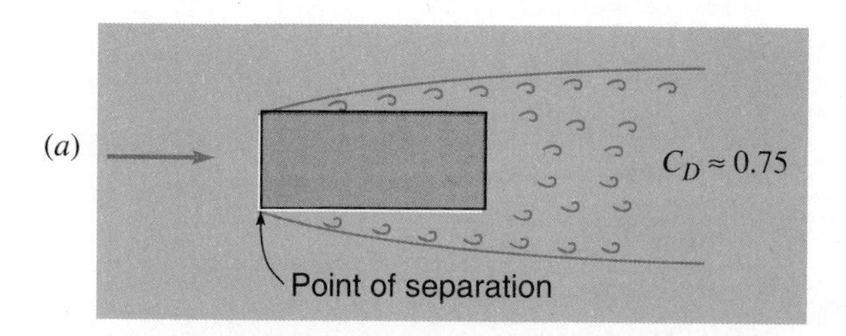

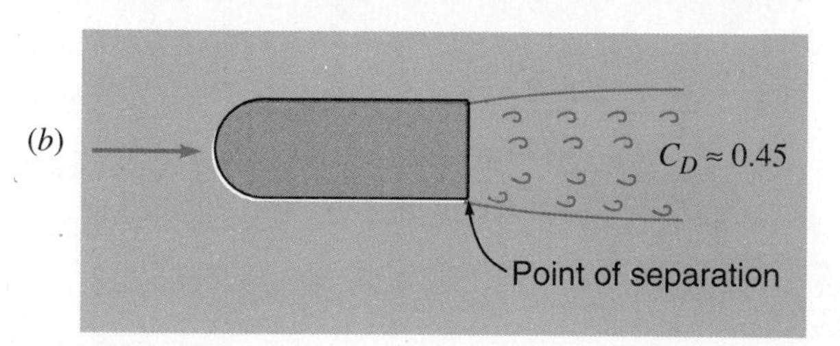

Figure 9.12
Plan view of flow about a motor vehicle (delivery van). (*a*) Blunt nose with separated flow along the entire side wall and a large drag coefficient $C_D = 0.75$. (*b*) Round nose with separation at the rear of the vehicle and smaller drag coefficient $C_D = 0.45$. (Adapted from H. Schlichting, *Boundary Layer Theory,* 4th ed., p. 34, McGraw-Hill, New York, 1960.)

[8] For supersonic flow, a sharp-nosed body has less drag than a round-nosed body (Fig. 9.27).

Efforts to reduce the drag force on automobiles and trucks, and thereby improve their fuel efficiency, have strongly influenced body design. Examples of historical body shapes and the resulting drag coefficients are given in Table 9.1.

Table 9.1
The effects of automobile body shapes on aerodynamic drag

	Body shape	C_D based on frontal area
1920		0.80
1940–45		0.54–0.58
1968–69		0.36–0.38
1990–92		0.29–0.30
Tractor–trailer		0.75–0.95
With rounded cab, fairing		0.55–0.75

Sample Problem 9.4 Using the data of Sample Prob. 9.2 determine the total drag exerted by the air on the van. Assume that $C_D \approx 0.45$ (see Fig. 9.12).

Solution

$$F_D = C_D \rho \frac{V^2}{2} A = 0.45(0.002\,42) \frac{(88)^2}{2} (8 \times 10) = 337 \text{ lb} \qquad \textbf{\textit{ANS}}$$

Thus the pressure drag $= 337 - 25 = 312$ lb; in this case the pressure drag is responsible for about 93% of the total drag, while the friction drag comprises only 7% of the total.

Sample Problem 9.5 Find the "free-fall" velocity of a 8.5-in-diameter sphere (bowling ball) weighing 16 lb falling through the following fluids under the action of gravity: (*a*) the standard atmosphere at sea level; (*b*) the standard atmosphere at 10,000 ft elevation; (*c*) water at 60°F; (*d*) crude oil ($s = 0.925$) at 60°F.

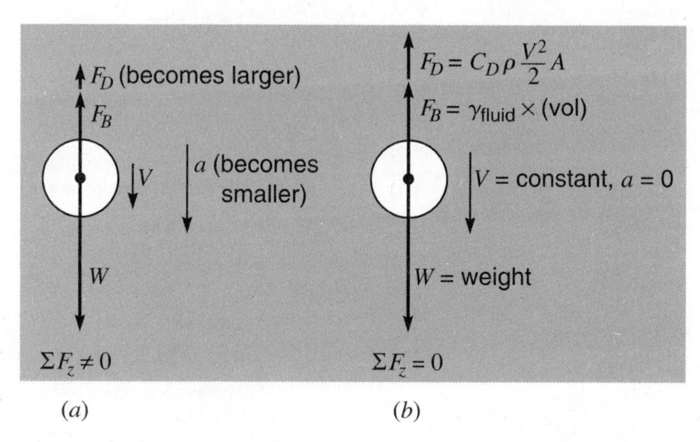

Figure S9.5

Solution

When first released, the sphere will accelerate (left-hand side of Fig. S9.5), because the forces acting on it are out of balance. This acceleration results in an increase in velocity, which causes an increase in the drag force. After a while, the drag force will increase to the point where the forces acting on the sphere are in balance, as indicated in the right-hand side of Fig. S9.5. When that point is reached, the sphere will attain a constant or terminal (free-fall) velocity. Thus for free-fall conditions,

$$\sum F_x = W - F_B - F_D = \text{mass} \times \text{acceleration} = 0$$

where W is the weight, F_B the buoyant force, and F_D the drag force. The buoyant force is equal to the unit weight of the fluid multiplied by the volume ($\pi D^3/6 = 0.186$ ft³) of the sphere. The basic data are approximately as follows:

Fluid	γ, lb/ft^3	ρ, slug/ft^3	$\nu = \mu/\rho$, ft^2/sec	F_B, lb
Air (sea level)	0.0765	0.002 38	1.57×10^{-4}	0.0142
Air (10,000 ft)	0.0564	0.001 76	2.01×10^{-4}	0.0105
Water, 60°F	62.4	1.94	1.22×10^{-5}	11.6
Oil, 60°F	57.7	1.79	0.001	10.7

The detailed analysis for the sphere falling through the standard sea-level atmosphere is as follows:

$$16 - 0.0142 - C_D \rho \frac{V^2}{2} A = 0$$

where $\rho = 0.002\ 38$ slugs/ft^3, $A = \frac{1}{4}\pi(8.5/12)^2 = 0.394$ ft^2

or $$15.986 = C_D(0.002\ 38)\frac{V^2}{2}(0.394) = 0.000\ 47 C_D V^2$$

A trial-and-error solution is required. Let $C_D = 0.2$; then $V = 412$ fps, and

$$\mathbf{R} = \frac{DV}{\nu} = \frac{(8.5/12)412}{1.57 \times 10^{-4}} = 1.86 \times 10^6$$

The values of C_D and $\mathbf{R}$ check (Fig. 9.10); hence

$$C_D = 0.2 \quad \text{and} \quad V = 412 \text{ fps} \qquad \textit{ANS}$$

Following a similar procedure for the other three cases gives the following free-fall velocities:

Fluid	C_D	$\mathbf{R}$	V_{fall}	
Standard atmosphere at 10,000 ft	0.20	1.69×10^6	480 fps	
Water at 60°F	0.19	453,000	7.4 fps*	*ANS*
Crude oil ($s = 0.925$) at 60°F	0.39	4,390	6.2 fps	

* In this instance the Reynolds number is 453,000, which, for the case of a sphere, generally indicates a turbulent boundary layer (Fig. 9.10). This is very close to the point where the boundary layer changes from laminar to turbulent. If the water had been at a somewhat lower temperature, and hence more viscous, a laminar boundary layer might have developed, in which case the free-fall velocity would have been only about 5.2 fps.

EXERCISES

9.7.1 An 0.3-in-diameter steel sphere ($s = 8.0$) is released in a large tank of oil ($s = 0.80$). The terminal velocity of the sphere is determined to be 2.5 fpm. Calculate the viscosity of the oil?

9.7.2 A 5-mm-diameter steel sphere ($s = 8.0$) is released in a large tank of oil

(s = 0.80). The terminal velocity of the sphere is determined to be 0.7 m/min. Calculate the viscosity of the oil?

9.7.3 For a 38-cm-diameter sphere, compute the drag from wind under sea level conditions in a standard ICAO atmosphere. Plot drag versus velocity for 0, 10, 20, and 30 m/s.

9.7.4 A poorly streamlined automobile has a body form corresponding roughly to the 1:0.75 oblate ellipsoid of Fig. 9.10, while a well-streamlined car has a body approximating the streamlined body in the same figure, each with a diameter of 5 ft. If the velocity is 50 mph through standard air at sea level (Appendix A, Table A.3), find the horsepower required to overcome air resistance in each case.

9.7.5 For the streamlined train in Exer. 9.4.3, calculate the pressure drag. As a rough approximation, assume that the nose and tail of the train are of the shape of the two halves of the prolate ellipsoid of Fig. 9.10, of 8.5 ft diameter. Find the drag on the ellipsoid (pressure drag on the train), and compare this with the skin-friction drag on the train determined earlier.

9.7.6 A hemispherical shell with its concave side upstream has a drag coefficient of approximately 1.33 if $\mathbf{R} > 10^3$. If the total load is 250 lb, find the diameter of a hemispherical parachute needed to provide a fall velocity no greater than that caused by jumping from a height of 10 ft. Assume standard air at sea level.

9.7.7 A hemispherical shell with its concave side upstream has a drag coefficient of approximately 1.33 if $\mathbf{R} > 10^3$. If the total load is 1000 N, find the diameter of a hemispherical parachute needed to provide fall velocity no greater than that caused by jumping from a height of 3 m. Assume standard air at sea level.

9.7.8 Assuming C_D = 1.25 and a speed of 30 m/s, find the drag force exerted by a 3.5-m-diameter braking parachute at sea level? At what speed will the same braking force be exerted by this parachute at an elevation of 2000 m. Assume C_D remains constant.

9.8 DRAG ON TWO-DIMENSIONAL BODIES (INCOMPRESSIBLE FLOW)

Two-dimensional bodies are also subject to friction and pressure drag. However, the flow about a two-dimensional body exhibits some peculiar properties that are not ordinarily found in the three-dimensional case of flow around a sphere. For example, with Reynolds numbers less than 1, the flow around a cylinder is completely viscous, and the drag coefficient is given by the straight-line part of the curve at the left of Fig. 9.13. As the Reynolds

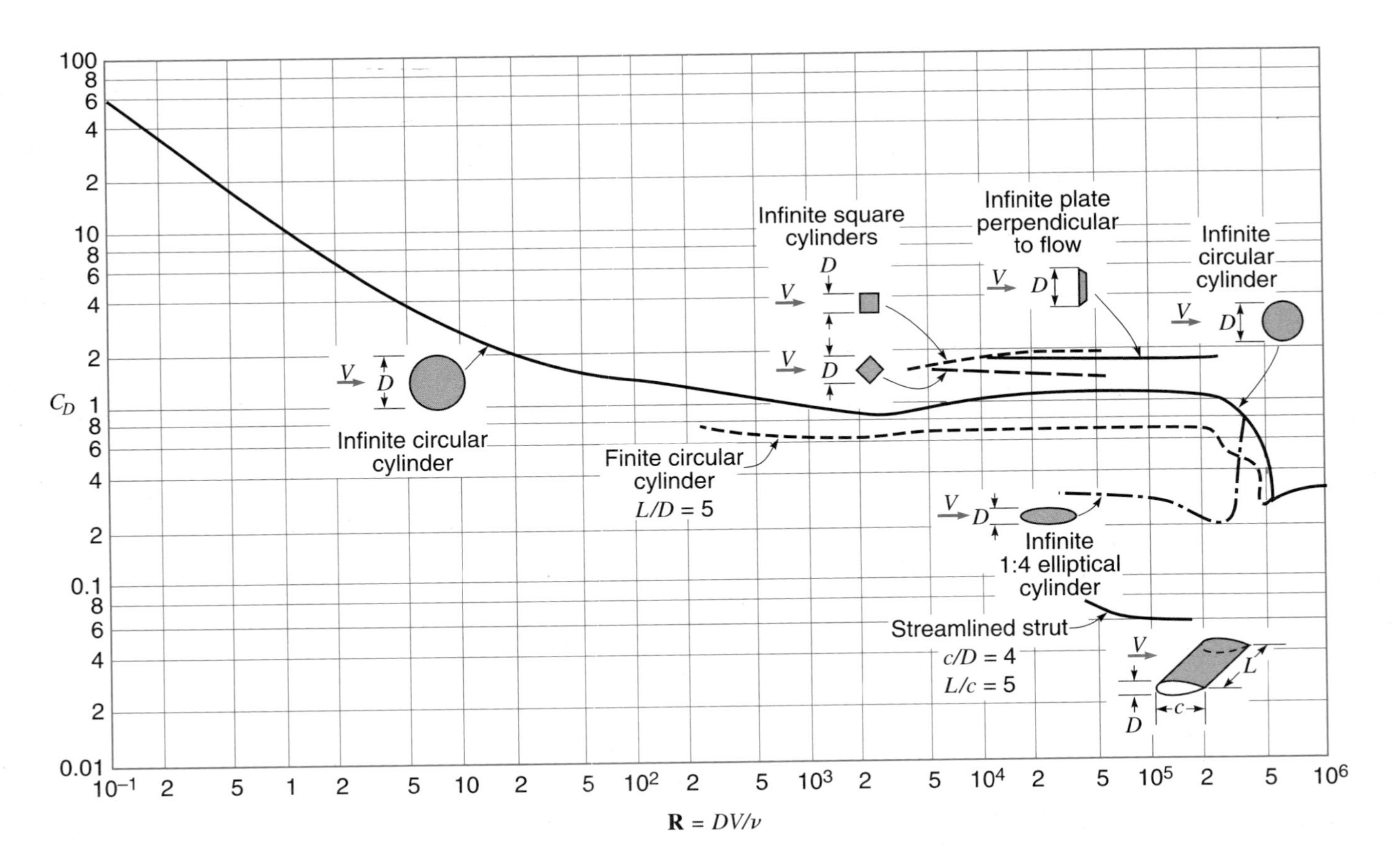

Figure 9.13
Drag coefficients for two-dimensional bodies.

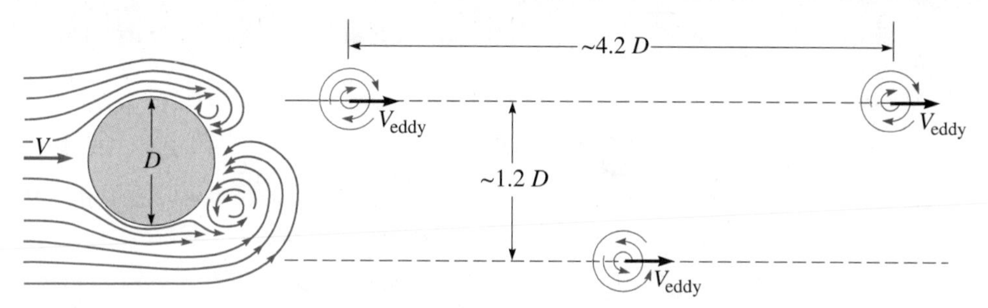

Figure 9.14
The Kármán vortex street following a cylinder. Velocity of eddies $V_{eddy} < V$, velocity of the fluid.

number increases from 2 to about 30, the boundary layer separates symmetrically from the two sides of the cylinder and two weak, symmetrical, standing large eddies are formed. The equilibrium of the standing eddies is maintained by the flow from the separated boundary layer, and if the cylinder is of finite length then the eddies increase in length with increase in velocity in order to dissipate their rotational energy to the free-streaming fluid.

At some limiting Reynolds number, usually about 60, depending on the shape of the cylinder (not necessarily circular), the width of the confining channel, and the turbulence in the stream, the eddies break off, having become too long to hang on, and wash downstream. This gives rise to the beginning of the so-called ***Kármán vortex street***. Above this critical **R**, and visibly up to a value of about 120, the vortices are shed first from one side of the cylinder and then from the other. The result is a staggered double row of vortices in the wake of the object spaced approximately as shown in Fig. 9.14. This alternating shedding of vortices and the accompanying forces gives rise to the phenomenon of aerodynamic instability, of much importance in the design of tall smoke stacks and suspension bridges. It is also understood to account for the "singing" of wind blowing across wires. The frequency at which the vortices are shed was given by G. I. Taylor and substantiated by Lord Rayleigh to be about

$$\text{frequency} \approx 0.20 \frac{V}{D}\left(1 - \frac{20}{\mathbf{R}}\right) \tag{9.34}$$

A serious condition may result if the natural frequency of vibration of the structure is equal or close to the frequency of the shedding of the vortices.

For Reynolds numbers above 120 or so it is difficult to perceive the vortex street, but the eddies continue to be shed alternately, from each side up to a value of **R** of about 10,000. Beyond this, the viscous forces become negligible, and it is not possible to say how the eddies form and leave the cylinder. As in the case of the sphere, the boundary layer for a circular cylinder becomes turbulent at a value of **R** of about 350,000. The corresponding sharp drop in C_D may be seen in Fig. 9.13.

Values of C_D for various other two-dimensional shapes are given in Fig. 9.13. As may be noted from the curve for the finite cylinder, the resistance is decreased if three-dimensional flow can take place around the ends. This decrease in C_D occurs because the vortices can extend laterally into the flow field and permit dissipation of energy over a larger region.

SAMPLE PROBLEM 9.6 What frequency of oscillation is produced by a 15 m/s wind at −20°C blowing across a 2-mm-diameter telephone wire at sea level?

Solution

Table A.2 at −20°C: $\nu_{air} = 1.15\times 10^{-5}\ \text{m}^2/\text{s}$

$$\mathbf{R} = \frac{DV}{\nu} = \frac{2\times 10^{-3}(15)}{1.15\times 10^{-5}} = 2610$$

Eq. (9.34): $$\text{frequency} = 0.20\frac{15}{2\times 10^{-3}}\left(1-\frac{20}{2610}\right) = 1489\ \text{Hz} \qquad \textit{ANS}$$

EXERCISES

9.8.1 What is the bending moment at the base of a vertical 0.3-in-diameter cylindrical ratio antenna 6 ft tall on an automobile traveling at 80 mph through standard air at sea level?

9.8.2 A 70-mph wind blows across an 0.1-in-diameter wire. Find the frequency of oscillation that results in standard atmosphere: (*a*) at sea level; (*b*) at 10,000 ft elevation.

9.9 LIFT AND CIRCULATION

At the start of this chapter, we briefly mentioned the lift as a force that acts on an immersed body normal to the relative motion between the fluid and the body. The most commonly observed example of lift is that of the airplane wing supported in the air by this force. The elementary explanation for such a lift force is that the air velocity over the top of the wing is faster than the mean velocity, while that along the underside is slower than the mean

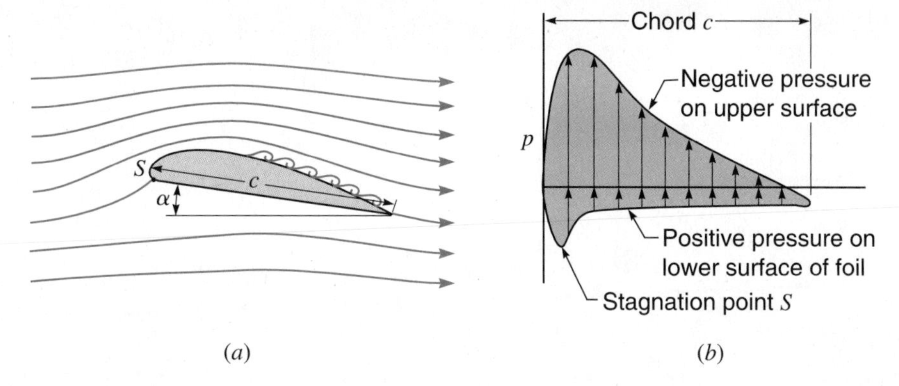

Figure 9.15
(*a*) Streamlines and (*b*) pressure distribution about a cambered airfoil, at an angle of attack of $\alpha = 8.6°$.

(Fig. 9.15*a*). The Bernoulli theorem then shows a lower pressure on the top and a higher pressure on the bottom (Fig. 9.15*b*), resulting in a net upward lift.

The increased velocity over the top of the wing of Fig. 9.15*a* and the decreased velocity around the bottom of the wing can be explained by noting that a ***circulation*** (Sec. 14.3) is induced as the wing moves relative to the flow field. The strength of the circulation depends, in the real case, on the shape of the wing and its velocity and orientation with respect to the flow field. A schematic diagram of the situation is presented in Fig. 9.16.

The relationship between lift and circulation is one that has been studied exhaustively for years by many investigators. An understanding of this relationship is essential to the analysis of various aerodynamic and hydrodynamic problems. To illustrate the theory of lift, we shall consider the flow of an ideal fluid past a cylinder and assume that a circulation about the cylinder is

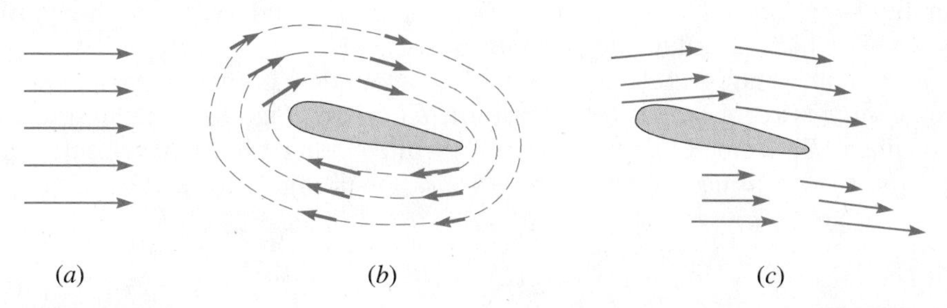

Figure 9.16
Schematic superposition of circulation on a uniform rectilinear flow field. (*a*) Uniform rectilinear flow field. (*b*) Circulation. (*c*) Net effect (sums of velocity vectors).

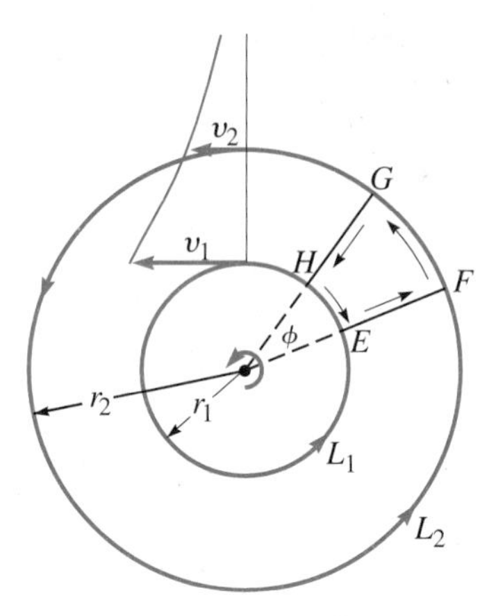

Figure 9.17
Circulation about a free vortex center.

imposed on the flow. First, though, let us consider the velocity field surrounding a free vortex (Fig. 9.17). The equation for this field was given in Sec. 5.19 as $vr = C$, a constant. The circulation Γ can be readily computed by application of Eq. (14.9) if we choose the closed path as the circular streamline L_1 concentric with the center of the vortex. The velocity v_1 is constant around this path and tangent to it ($\cos \beta = 1$), and the line integral of dL is simply the circumference of the circle. Applying the same treatment to another concentric circle L_2, we get

$$\Gamma = v \oint_L dL = v_1(2\pi r_1) = v_2(2\pi r_2)$$

But, from the free-vortex velocity field, $v_1 r_1 = v_2 r_2 = C$, and hence

$$\Gamma = 2\pi C = 2\pi v r \tag{9.35}$$

which demonstrates that the circulation around two different curves, each completely enclosing the vortex center, is the same. It may be proved more rigorously that the circulation around *any path enclosing the vortex center* is given by $\Gamma = 2\pi C$. The circulation is seen to depend only on the vortex constant C, which is called the ***strength*** of the vortex.

A corollary states that the circulation around a path *not enclosing* the vortex center is zero. Let us take the path *EFGH* of Fig. 9.17. Along the two radial lines *EF* and *GH*, $\cos \beta = 0$, while along the two circular arc segments, $\Gamma_{FG} = v_2 \phi r_2$ and $\Gamma_{HE} = -v_1 \phi r_1$, resulting in a net circulation of $\phi(v_2 r_2 - v_1 r_1) = 0$.

EXERCISE

9.9.1 Estimate the lift per unit length of span if the mean velocity along the top of a wing having a 10 ft chord is 100 mph and that along the bottom of the wing is 80 mph when the wing moves at 85 mph through still air ($\gamma = 0.072$ lb/ft^3).

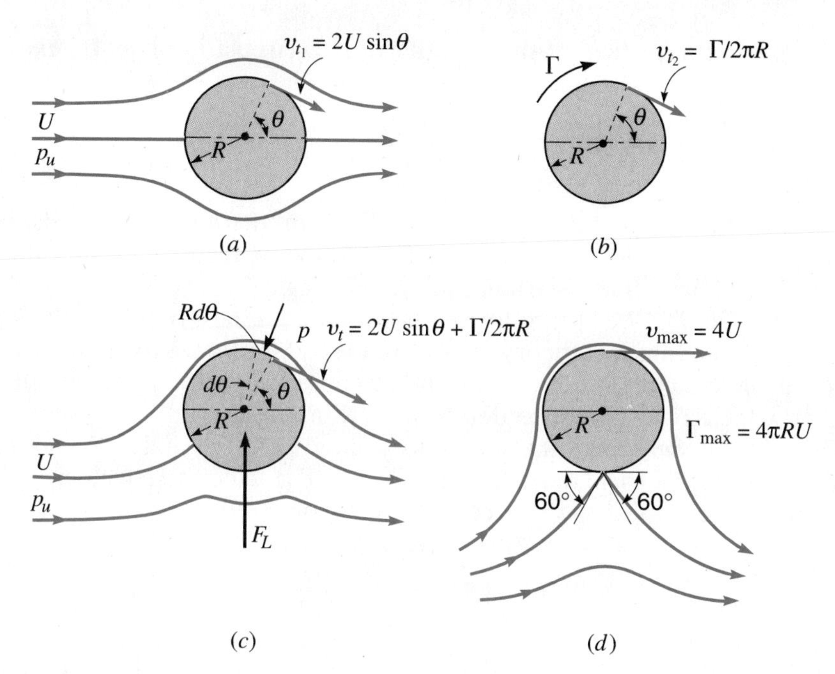

Figure 9.18
Circulation and lift from unsymmetrical flow about a cylinder.

9.10 IDEAL FLOW ABOUT A CYLINDER

Let us first consider uniform flow of an ideal fluid about a cylinder that is infinitely long. From classical hydrodynamics,[9] it has been shown that with steady flow of uniform velocity U (Fig. 9.18*a*), the clockwise velocity v_{t_1} at the periphery ($r = R$) of the cylinder is given by

$$v_{t_1} = 2U \sin \theta \tag{9.36}$$

The pressure distribution on the cylinder may be computed by writing the Bernoulli theorem between a point at infinity in the free-streaming fluid and a point on the cylinder wall. Since the pressure distribution is completely symmetrical about the cylinder, there is no net lift or drag for this ideal case.

Putting this uniform flow aside for the moment, we next suppose a circulatory flow about the cylinder (Fig. 9.18*b*). Adopting the positive clockwise direction of circulation Γ, from Eq. (9.35), the peripheral clockwise

[9] Equation (9.36) can be obtained by noting that $v_t = -\partial\psi/\partial r$ (by definition, Sec. 14.4) with v_t positive counterclockwise, where ψ is given by Eq. (14.19), in which $y = r \sin \theta$ and $m = 2\pi UR^2$ (Sec. 14.5).

velocity v_{t_2} on the surface of the cylinder due to circulation may be expressed as

$$v_{t_2} = \frac{\Gamma}{2\pi R} \tag{9.37}$$

where R is the radius of the cylinder. Thus in the flow field outside the cylinder, where $r > R$, $v_{t_2} = \Gamma/2\pi r$. This velocity distribution produces a pressure variation that is radially symmetrical, in accordance with the free-vortex theory. We see that the solid cylinder has replaced the vortex center in the circulation theory. If the reader wishes an explanation for the existence of the circulation, it may be supposed to arise from rotating the cylinder, which indeed may be demonstrated in a real fluid.

Next let us ***superpose*** the circulatory flow onto the uniform motion, to form the unsymmetrical flow of Fig. 9.18*c*. The clockwise velocity at the periphery is the sum of the two contributions, or

$$v_t = 2U \sin\theta + \frac{\Gamma}{2\pi R} \tag{9.38}$$

The general equation for the pressure p at any point on the circumference of the cylinder is obtained from the energy equation (5.15) as follows:

$$\frac{p_u}{\gamma} + \frac{U^2}{2g} = \frac{p}{\gamma} + \frac{v_t^2}{2g}$$

where p_u is the pressure at some distance away where the velocity is uniform. From these two equations,

$$p - p_u = \frac{\rho}{2}\left[U^2 - \left(2U \sin\theta + \frac{\Gamma}{2\pi R}\right)^2\right] \tag{9.39}$$

Since the elementary area per unit of length of the cylinder is $R\,d\theta$ and the lift dF_L on this elementary area is normal to the direction of U and equal to $R\,d\theta \sin\theta(p_u - p)$, the resulting total lift is

$$F_L = -B\int_0^{2\pi} (p - p_u)R \sin\theta\, d\theta$$

Substituting the expression for $p - p_u$ from Eq. (9.39) and integrating, this reduces to

$$F_L = \rho B U \Gamma \tag{9.40}$$

where F_L is the lift force and B is the length of the cylinder.

The existence of this transverse force on a rotating cylinder is known as the ***Magnus effect***, after Heinrich G. Magnus (1802–1870), the German scientist who first observed it in 1852. Equation (9.40) is known as the ***Kutta–Joukowski theorem***, after W. M. Kutta and N. E. Joukowski, who independently pioneered the quantitative investigation of the lift force in 1902 and 1906, respectively. The great importance of this theorem is that it applies not only to the circular cylinder but to a cylinder of any shape, including in particular the lifting vane, or ***airfoil***, as shown in Fig. 9.15.

It is clear from Fig. 9.18*c* that the stagnation points have shifted downward from the horizontal axis, but they are still symmetrical about the vertical axis. At the point of stagnation on the cylinder, v_t in Eq. (9.38) will be zero. Thus we have at the stagnation point

$$-2U \sin \theta = \frac{\Gamma}{2\pi R}$$

This shows that if we can measure the angle to the stagnation point and know the free-stream velocity, we may obtain the circulation from

$$\Gamma = -4\pi RU \sin \theta_0 \qquad (9.41)$$

where θ_0 is the angle between the horizontal diameter and the stagnation point in Fig. 9.18*c*. Figure 9.18*c* illustrates a case where $\Gamma < 4\pi RU$, that is, where $|\sin \theta| < 1$. For the case of $\Gamma = 4\pi RU$, $\sin \theta = -1$, and the two stagnation points meet together at the bottom of the cylinder as shown in Fig. 9.18*d*. The two streamlines make angles of 60° with the tangent to the cylinder, and the maximum velocity in the flow for this case occurs at the top of the cylinder and is equal to

$$v_{\max} = 2U + \frac{4\pi RU}{2\pi R} = 2U + 2U = 4U$$

Thus, according to ideal-flow theory,[10] if the cylinder is rotated so that $v_t = 2U$ (that is, with $\omega = v_t/R = 2U/R$), the circulation thus produced will cause the stagnation point to occur at the bottom of the cylinder, as in Fig. 9.18*d*. If the cylinder is rotated at still greater speed, the stagnation point is removed entirely from the cylinder surface, and a ring of fluid is dragged around with the cylinder.

Sample Problem 9.7 A cylinder 4 ft in diameter and 25 ft long rotates at 90 rpm with its axis perpendicular to an airstream with a wind velocity of 120 fps. The specific weight of the air is 0.0765 lb/ft³. Assuming no slip between the cylinder and the circulatory flow, find (*a*) the value of the circulation: (*b*) the transverse or lift force; (*c*) the position of the stagnation points.

[10] In the case of a real fluid, because of viscosity, the required velocity to bring the stagnation point to the bottom of the cylinder is about twice that indicated by ideal-flow theory.

Solution

(*a*) Peripheral velocity: $v_t = \dfrac{2\pi Rn}{60} = 2\pi(2)\dfrac{90}{60} = 18.85$ fps

Eq. (9.35): $\Gamma = 2\pi R v_t = 2\pi(2)18.85 = 237\ \text{ft}^2/\text{sec}$ ***ANS***

(*b*) Eq. (9.40): $F_L = \rho BU\Gamma = \dfrac{0.0765}{32.2}(25)120(237) = 1688\ \text{lb}$ ***ANS***

(*c*) Eq. (9.41): $\sin\theta_0 = -\dfrac{\Gamma}{4\pi RU} = -\dfrac{237}{4\pi 2\times 120} = -0.0785$

Therefore $\theta_0 = 184.5°, 355.5°$ ***ANS***

Note: The real circulation actually produced by surface drag of the rotating cylinder would be only about one-half of that obtained above for the no-slip assumption.

EXERCISES

9.10.1 Imagine that a circulation of 25 ft^2/sec is superimposed around a 1-ft-diameter cylinder immersed in 60°F water flowing at 10 fps perpendicular to the cylinder axis. Find the location of the stagnation points, and the lift on a 20-ft length of cylinder.

9.10.2 Imagine that a circulation of 2.4 m^2/s is superimposed around a 30-cm-diameter cylinder immersed in 15°C water flowing at 3.5 m/s perpendicular to the cylinder axis. Find the location of the stagnation points, and the lift on a 10-m length of cylinder.

9.11 LIFT OF AN AIRFOIL

The reader may well ask why so much attention is given to the flow about a cylinder when it is obvious that there are few practical applications of the lift on a cylinder.[11] The answer is that one of the most remarkable applications of mathematics to engineering is ***conformal transformation***,[12] by which the flow

---

[11] In the early 1920s A. Flettner in Germany developed the "rotorship," which substituted motor-driven cylindrical rotors for sails. The ship was then driven by the Magnus effect but still required wind. A few trans-Atlantic crossings were made, but the rotorship was ultimately found to be uneconomical. See A. Flettner, *The Story of the Rotor,* Willhoft, New York, 1926.

[12] For an excellent discussion of conformal transformation, see H. R. Vallentine, *Applied Hydrodynamics,* 2d ed., Butterworth, London, and Plenum, New York, 1967.

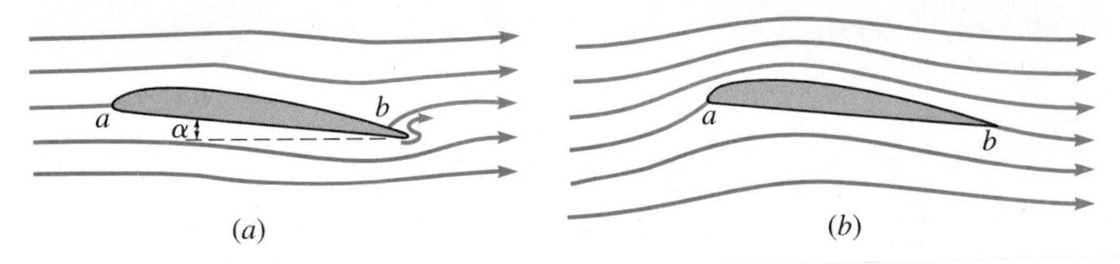

Figure 9.19
Adjustment of stagnation points to avoid infinite velocity at trailing edge.

about one body may be mapped into the flow about a body of different (though mathematically related) shape. Certain quantities, notably the circulation and relative position of the stagnation points, remain unchanged under the mapping. The importance of the circular cylinder, then, is that it can be mapped into a perfectly workable airfoil by the so-called ***Joukowski transformation***. The position of the stagnation points is determined from the physical requirements of the flow about the airfoil, and these stagnation points, mapped back onto the cylinder, determine the circulation, by Eq. (9.41), and the lift, by Eq. (9.40).[13]

Let us examine the airfoil of Fig. 9.19. As fluid flows past the foil, there will be a tendency for stagnation points to form at the points of the foil, corresponding to the 0 and 180° points of the corresponding cylinder (Fig. 9.18*a*). Just where these points occur on the foil depends on the ***angle of attack*** α, or the attitude of the foil with respect to the oncoming flow, as shown in the figure. We shall assume a positive angle of attack in Fig. 9.19*a*, with corresponding initial stagnation points *a* and *b*. While the location of these stagnation points involves no difficulty in the case of the ideal fluid, we see at once that the condition at the trailing edge—with the air from the underside trying to flow around the sharp cusp of the foil—becomes a point of violent separation in real fluid flow.

This condition lasts no more than an instant, however, for stagnation point *b* is soon swept back to the trailing edge of the foil (Fig. 9.19*b*), where it stays. This stable position of the point *b* is necessary, according to the so-called ***stagnation hypothesis*** of N. Joukowski, in order to avoid an infinite velocity around the sharp cusp of the foil. Now this shift in the rear stagnation point of the foil corresponds to a shift in the rear stagnation point of the related cylinder to a negative angle, somewhat as shown in Fig. 9.18*c*. Vertical symmetry of the flow about the cylinder requires that the forward stagnation point move downward by the same angle. This in turn maps a new location of the forward stagnation point on the airfoil, and such a shift also takes place

[13] It must be understood that while the Joukowski profile is a workable lifting vane, the modern airfoil has undergone many modifications to improve its performance for various special purposes. The mapping theory described here applies exactly to the Joukowski foil and in principle to any lifting vane.

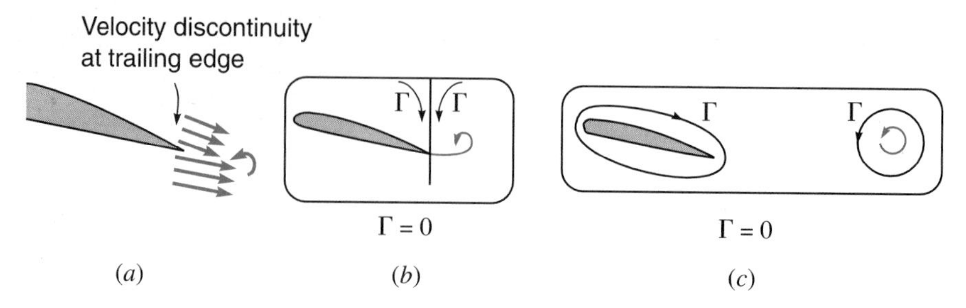

Figure 9.20
Life history of the starting vortex.

in the real flow. We see, then, that a circulation has become established about the airfoil, the magnitude of which is determined by the location of the stagnation points on the corresponding cylinder. The lift may then be determined analytically by Eq. (9.40).

Although the Joukowski hypothesis appears perfectly reasonable, we must investigate whether or not nature will actually perform this adjustment of the stagnation point to the cusp of the airfoil profile. Our acceptance of this hypothesis is complicated by the perfectly valid theorem of the Scottish physicist and mathematician W. Thomson (Baron Kelvin) 1824–1907, which states that "the circulation around a closed curve in the fluid does not change with time if one moves with the fluid." How is a circulation created around the airfoil where none existed before? The answer was first suggested by Prandtl, and has been well substantiated with photographs. He showed that the initial separation point at the cusp caused a ***starting vortex*** to form, as shown in Fig. 9.20*a*. In order to satisfy Thomson's theorem, an equal and opposite circulation must automatically be generated around the foil (Fig. 9.20*b*). After this circulation has been established, the starting vortex breaks off and is left behind as the airplane moves forward, but just to satisfy Thomson's theorem, the starting vortex keeps whirling around (Fig. 9.20*c*) until it dies out from viscous effects. The net circulation around a curve including the profile and this vortex is still zero. When the airfoil comes to a stop or changes its angle of attack, new vortices are formed to effect the necessary change in circulation.

9.12
INDUCED DRAG ON AIRFOIL OF FINITE LENGTH

The discussion of lift has so far been limited to strictly two-dimensional flow. When the foil or lifting vane is of finite length in a free fluid, however, there are end conditions that affect both the lift and the drag. Since the pressure on the underside of the vane is greater than that on the upper side, fluid will

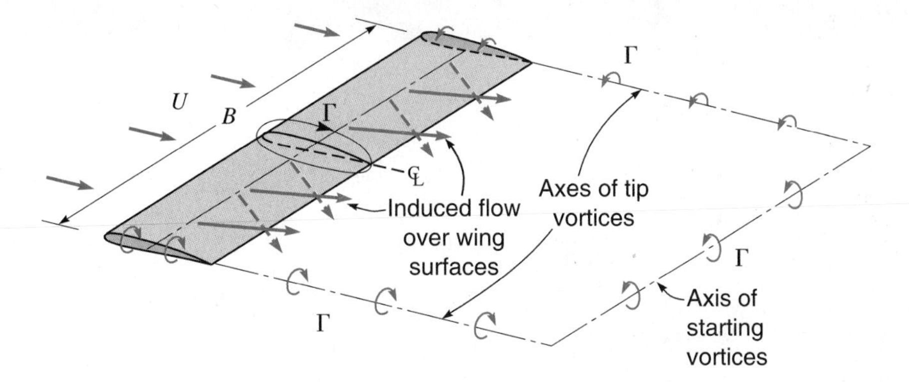

Figure 9.21
Wing of finite span.

escape around the ends of the vane, and there will be a general flow outward from the center to the ends along the bottom of the vane and inward from the ends to the center along the top. The movement of the fluid upward around the ends of the vane results in small ***tip vortices***, which are cast off from the wing tips. In theory, Thomson's theorem still holds, for the tip vortices are of equal and opposite magnitude. If the circulation is computed about a closed path passing along the axis of the airfoil and along the axes of the tip and starting vortices, as shown in Fig. 9.21, it will still add up to zero. Practically, of course, the circulation about the foil continues to exist, but the tip and starting vortices soon die out from viscous resistance.

The closed path consisting of the finite wing, the tip vortices, and the starting vortices of Fig. 9.21 constitutes a large vortex ring, inside of which there is a downward velocity induced by the vortices. Prandtl showed this induced, or ***downwash***, velocity U_i to be a constant if the wing is so constructed as to produce an elliptical distribution of lift over a given span. The downwash changes the direction of the flow in the vicinity of the foil from U to U_0, thus *decreasing the effective angle of attack* from α to α_0. The decrease in the effective angle of attack $\alpha_i = \alpha - \alpha_0 = \arctan(U_i/U)$, as shown in Fig. 9.22.

The wing may be analyzed on the basis of a foil of infinite length set in a stream of uniform velocity U_0, at angle of attack α_0. The lift F_{L0} generated from the circulation about the infinite foil must be normal to U_0. This force is seen to be resolved into two components: the true lift F_L normal to U and a component parallel to U called the ***induced drag*** F_{Di}. In conformity with our other drag terms, we represent the induced drag in the standard form

$$F_{Di} = C_{Di}\rho \frac{V^2}{2} A \tag{9.42}$$

We now need to distinguish between the two- and three-dimensional

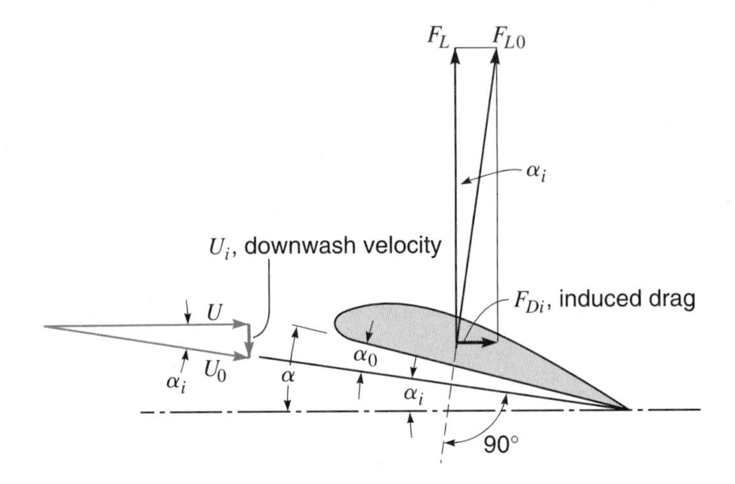

Figure 9.22
Definition sketch for induced drag.

cases of drag. The skin friction and pressure drag discussed earlier in this chapter will be lumped into the ***profile drag*** F_{D0}, which includes all drag forces acting on the profile of infinite length. The total drag on the finite span is then the sum of the profile and induced drags:

$$F_D = F_{D0} + F_{Di} \tag{9.43}$$

As the angle α_i is small,

$$U_0 \approx U, \qquad F_{L0} \approx F_L, \qquad F_{Di} \approx \alpha_i F_L$$

It should be noted at this point that, in addition to expressing the lift force by Eq. (9.40), it is convenient to express it as

$$F_L = C_L \rho \frac{V^2}{2} A \tag{9.44}$$

where C_L is the lift coefficient, whose value depends primarily on the angle of attack and the shape of the airfoil, and A is the projected area of the airfoil or body normal to the lift vector.

The computations for the elliptical distribution of lift are too complex to include here, but they result in the simple relation

$$\frac{U_i}{U} = \alpha_i \text{ (radians)} = \frac{C_L}{\pi B^2/A} \tag{9.45}$$

where B is the span of the airfoil and A is its plane area. The quantity

$B^2/A = B/c$, where c is the mean chord length, is referred to as the ***aspect ratio***.

From Eqs. (9.42)–(9.44), together with the above expression for F_{Di}, we have

$$C_{Di} = \frac{C_L^2}{\pi B^2/A} \tag{9.46}$$

Dividing Eq. (9.43) by $\rho V^2 A/2$ and substituting equations similar to Eq. (9.42) and making use of Eq. (9.46) gives for the coefficient of total drag on a foil of finite length,

$$C_D = C_{D0} + C_{Di} = C_{D0} + \frac{C_L^2}{\pi B^2/A} \tag{9.47}$$

As would be expected, C_{Di} is seen to depend on the lift coefficient, i.e., the angle of attack α_0 and the aspect ratio B^2/A. For zero lift or infinite aspect ratio, the induced drag would be zero. These equations are important in comparing data for an airfoil tested at one aspect ratio with data for another foil at a different aspect ratio.

The explanation of how the induced drag, occurring as it does in the ideal-fluid theory, fits into d'Alembert's paradox—which states that there is no drag on a body in ideal flow—is that the work done against the flow by the induced drag is conserved in the kinetic energy of the tip vortices cast from the ends of the foil. In a real fluid, evidence of the tip vortices may frequently be seen in the form of ***vapor trails*** extending for miles across the sky. The decreased temperature caused by the decreased pressure at the center of the vortex causes condensation of the moisture in the air.

EXERCISE

9.12.1 A wing with a 15-m span and a 45-m^2 "planform" area moves horizontally through the standard atmosphere at 6000 m with a velocity of 750 km/hr. If the wing supports 180 kN, find (*a*) the required value of the lift coefficient; (*b*) the downwash velocity, assuming an elliptical distribution of lift over the span; (*c*) the induced drag.

9.13 LIFT AND DRAG DIAGRAMS

A wealth of data on the lift and drag of various airfoils has been obtained from wind-tunnel tests. The results of such tests may be presented graphically as plots of the lift and drag coefficients versus the angle of attack. Since the efficiency of the airfoil is measured by the ratio of lift to drag, the value of C_L/C_D is generally plotted also. These three curves can be combined

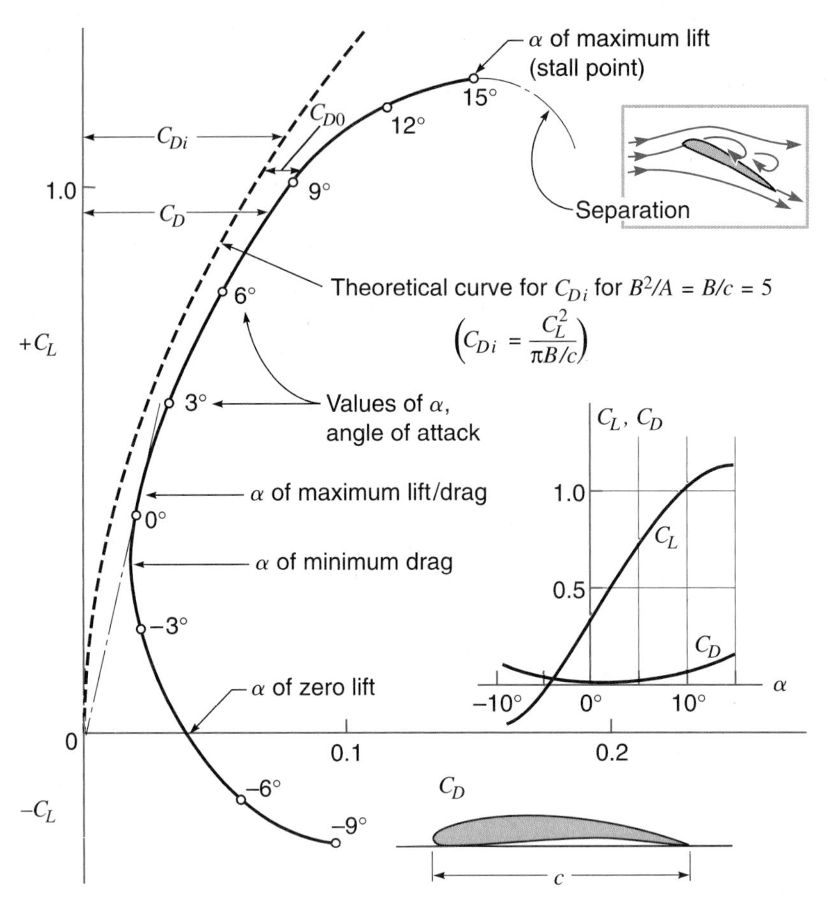

Figure 9.23
Polar diagram for wing of aspect ratio 5. (Curve from L. Prandtl and O. G. Tietjens, *Applied Hydro- and Aeromechanics,* p. 152, McGraw-Hill, New York, 1934.)

neatly into a single curve, suggested by Prandtl, known as a ***polar diagram*** (Fig. 9.23).

The coordinates of the polar diagram are the lift and drag coefficients, while the angles of attack are represented by different points along the curve. The ratio of lift to drag is the slope of the line from the origin to the curve at any point. Clearly, the maximum value of the ratio occurs when this line is tangent to the curve. The lift increases with the angle of attack up to the ***point of stall***. Beyond this point the boundary layer along the upper surfaces of the foil separates and creates a deep turbulent wake.

The polar diagram is notably instructive with regard to the drag coefficient, consisting of the coefficients of profile and induced drag as shown in Eq. (9.47). The dashed line in Fig. 9.23 is the parabola of Eq. (9.46). For

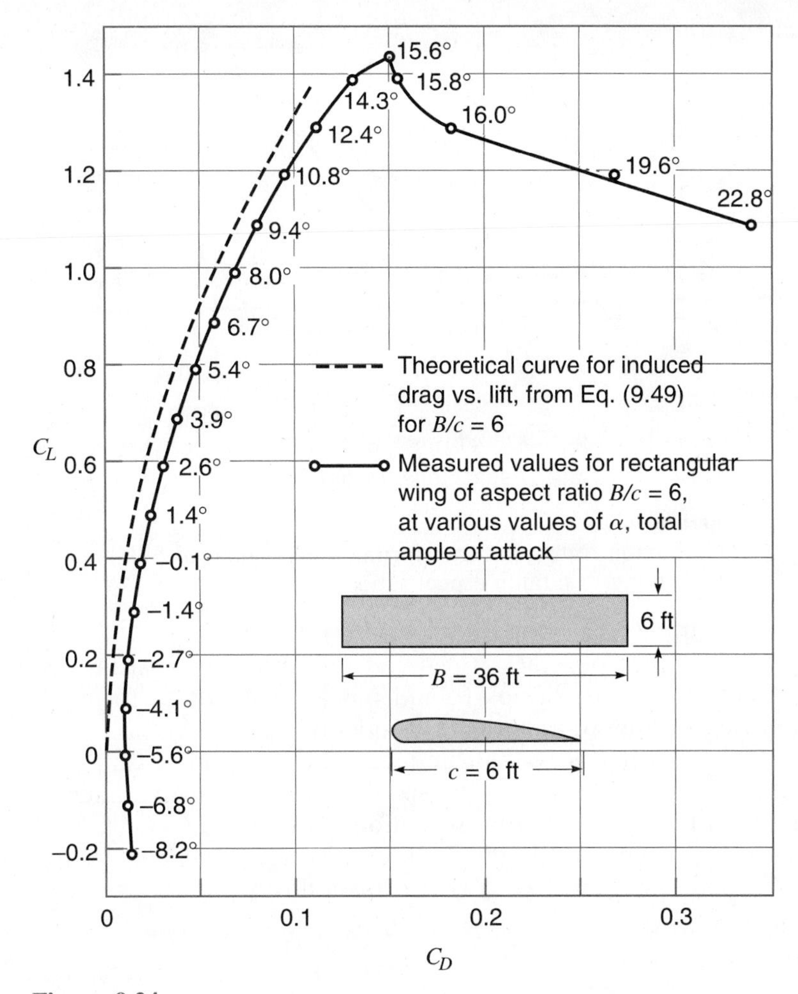

Figure 9.24
Polar diagram for rectangular Clark Y airfoil of 6-ft chord by 36-ft span.

an aspect ratio of 5, as shown, the induced drag is a major part of the total drag. For larger aspect ratios the parabola remains closer to the vertical axis, and the total drag is correspondingly decreased.

The polar diagram of a Clark Y airfoil, rectangular in plan, 6-ft chord by 36-ft span, is shown in Fig. 9.24. Notice that the angle of attack is read from a geometric reference which has little meaning by itself. The important reference angle is the angle of attack for zero lift, in this case −5.6°. In general, this is also the angle for minimum drag. The lift coefficient can be shown theoretically to be given by

$$C_L = 2\pi\eta\alpha_0' \tag{9.48}$$

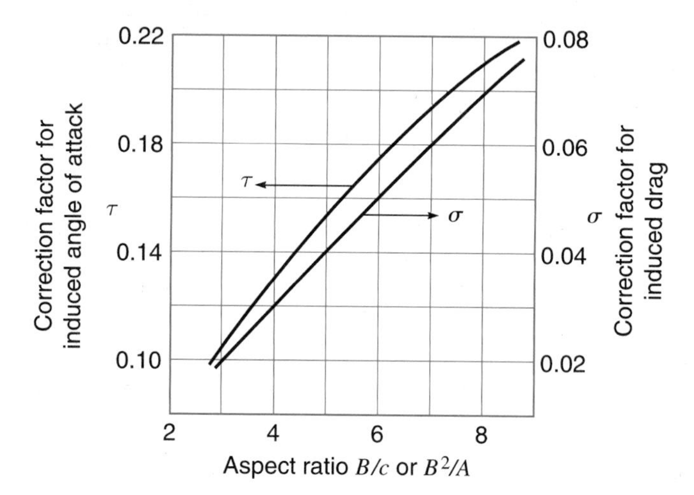

Figure 9.25
Correction factors for transforming rectangular airfoils from finite to infinite aspect ratio.

where α_0' is the angle of attack (for the airfoil of infinite span) measured in radians from the attitude of no lift, and η is a correction factor for frictional effects, having a value of about 0.9 for modern airfoil sections.

We recall from Sec. 9.12 that the induced-drag theory assumed an elliptical distribution of lift over the span of the finite airfoil. Such a distribution of lift is only an approximation, and for the rectangular airfoil the expressions for induced angle of attack and induced-drag coefficient given in Eqs. (9.45) and (9.46) must be corrected as follows:

$$\alpha_i \text{ (radians)} = \frac{C_L}{\pi(B/c)}(1+\tau) \tag{9.49}$$

and

$$C_{Di} = \frac{C_L^2}{\pi(B/c)}(1+\sigma) \tag{9.50}$$

where τ and σ are correction factors given in Fig. 9.25. Information on airfoils of shapes other than rectangular and lift and drag on the fuselage, stabilizers, and control surfaces of aircraft[14] may be found in the literature.

SAMPLE PROBLEM 9.8 For a rectangular Clark Y airfoil of 6-ft chord by 36-ft span, find (*a*) the value of the friction correction factor η if the angle of attack $\alpha = 5.4°$ when the wing is moving at 300 fps through standard atmosphere at altitude 10,000 ft. Find (*b*) the weight which the wing will carry and (*c*) the horsepower required to drive it.

[14] H. Schlichting and E. Truckenbrodt, *Aerodynamics of the Airplane,* translated by Heinrich J. Ramm, McGraw-Hill International Book Co., New York, 1979.

Solution
(*a*) Fig. 9.24, with $\alpha = 5.4°$: $C_L = 0.8$ and $C_D = 0.047$. From Fig. 9.25, for $B/c = 36/6 = 6$: $\tau = 0.175$. From Eq. (9.49):

$$\alpha_i = \frac{0.8}{\pi(36/6)}(1 + 0.175) = 0.0499 \text{ rad} = 2.86°$$

From Fig. 9.22: $\alpha_0 = \alpha - \alpha_i = 5.40 - 2.86 = 2.54°$. Since the angle of zero lift is $-5.6°$,

$$\alpha_0' = 2.54 + 5.6 = 8.14° = 0.1421 \text{ rad}$$

From Eq. (9.48): $$\eta = \frac{C_L}{2\pi\alpha_0'} = \frac{0.8}{2\pi \times 0.1421} = 0.896 \qquad \textit{ANS}$$

(*b*) Table A.3 at 10,000 ft: $\rho = 0.001\,756 \text{ slug/ft}^3$.

The wing will support a weight equal to the lift force:

Eq. (9.44): $$W = F_L = 0.8(0.001\,756)\frac{(300)^2}{2}(36)6 = 13{,}650 \text{ lb} \qquad \textit{ANS}$$

(*c*) From Eqs. (9.32) and (9.44): $$F_D = \left(\frac{C_D}{C_L}\right)F_L = \frac{0.047}{0.8} \times 13{,}650 = 802 \text{ lb}$$

$$\text{Horsepower required} = \frac{FV}{550} = \frac{802 \times 300}{550} = 438 \text{ hp} \qquad \textit{ANS}$$

EXERCISES

9.13.1 A 1400-lb airplane has a Clark Y airfoil wing of 6-ft chord by 36-ft span, with polar diagram given in Fig. 9.24. Find (*a*) the speed required to get the plane off the ground; (*b*) the horsepower required; (*c*) the circulation about the wing; (*d*) the strength of the starting vortex. Assume standard air at sea level and the angle of attack for maximum ratio of lift to drag. Neglect aerodynamic forces on the fuselage and tail.

9.13.2 A sailplane including its load weighs 300 lb. It has a Clark Y section wing with a 3.5-ft chord by 21-ft span. Given that it has the same characteristics as the larger wing of the same aspect ratio shown in Fig. 9.24, find the angle of glide through standard air at 2000 ft that will produce the greatest horizontal distance range. Neglect air forces on the fuselage and the tail. (*Note*: The aspect ratio of 6 is here chosen to facilitate working the problem with the available data. Actually, the sailplane may be constructed with an aspect ratio of about twice this, in order to reduce the drag to a minimum.)

9.13.3 Solve Exer. 9.13.2 for a sailplane with a 1-m chord by 6-m span wing, flying at 600 m altitude with a total weight of 1800 N.

9.14
EFFECTS OF COMPRESSIBILITY ON DRAG AND LIFT

In Sec. 13.11 we shall see that when a body moves through a fluid at supersonic velocity a shock wave is formed. The wave pattern (Fig. 13.9) is the same whether the body is moving through the fluid or whether the fluid is moving past the body. Unlike subsonic flow (Fig. 4.12), in supersonic flow the streamlines in front of the body are unaffected, because the body is moving faster than the disturbance can be transmitted ahead. This is demonstrated in Fig. 13.10.

With most bodies, the drag coefficient tends to increase drastically at a Mach number of about 0.70. This happens because the body is encountering ***transonic flow*** phenomena, which means that at some place in the flow field supersonic flow is occurring. With a streamlined body, the highest velocity in the flow field occurs at some point such as b in Fig. 9.26 near the body and away from its nose. The local Mach number at b will reach unity when the free-stream Mach number at a has a value of perhaps only 0.7 or 0.8. Thus a shock wave will form at b. Through the shock wave, there is a sudden jump in pressure, which causes an adverse pressure gradient in the boundary layer, resulting in separation and an increase in drag. Drag coefficients for several bodies as a function of the free-stream Mach number are given in Fig. 9.27. The increased drag is caused not only by the separation effects; a substantial amount of energy is dissipated in the shock wave. Skin friction also contributes to the drag, and at Mach numbers above 2 or 3 heating in the boundary layer from skin friction may be an important phenomenon. As the value of the Mach number increases beyond about 2, for most bodies there is a drop in the value of the drag coefficient because of a shift in the point of separation.

Earlier we noted that, to streamline for subsonic flow, a rounded nose and a long, tapered afterbody generally result in minimum drag. *In supersonic flow the best nose form is a sharp point.* This tends to minimize the effect of the shock wave.

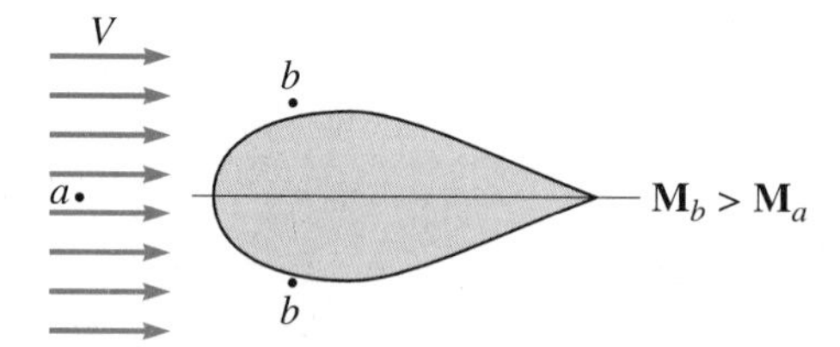

Figure 9.26
Local Mach number greater than freestream Mach number. When $\mathbf{M}_b = 1.0$, $\mathbf{M}_a \approx 0.7$.

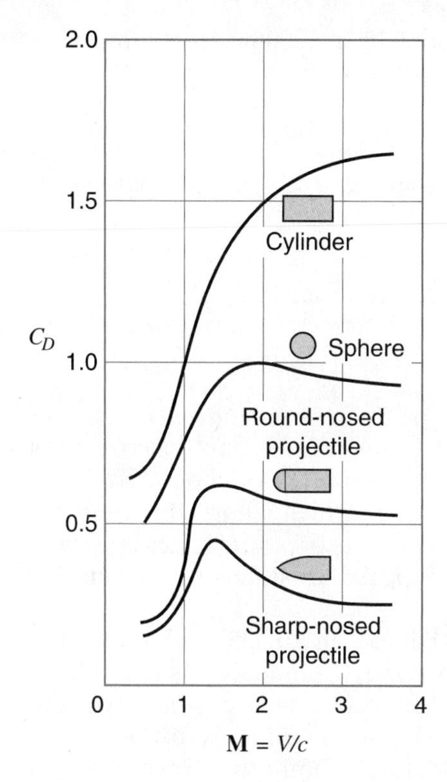

Figure 9.27
Drag coefficients as a function of Mach number.

With respect to lift, it has been found that for Mach numbers less than about 0.7, the lift coefficient for compressible fluids may be estimated by dividing the lift coefficient for incompressible flow by $\sqrt{1-\mathbf{M}^2}$, where $\mathbf{M}$ is the Mach number of the free stream.[15] The reduced pressures on the top of the airfoil are responsible for this trend.

At $\mathbf{M} \approx 0.8$, there is a rather abrupt drop in the lift coefficient, because the shock wave induced by the local supersonic flow creates high pressures on the top side of the airfoil which may result in ***shock stall***. At somewhat higher values of the Mach number, a shock wave forms on the bottom of the airfoil, which tends to compensate for the preceding action, and there is an increase in the value of the lift coefficient. Although there have been tremendous advances in theory, the best way to predict the drag and lift of a particular airfoil is by conducting model tests in a wind tunnel.

EXERCISES

9.14.1 The blunt-nosed projectile (cylinder) of Fig. 9.27 has a diameter of 16 in and weighs 500 lb. Find its rate of deceleration when moving through

[15] This is known as the ***Prandtl–Glauert rule***.

standard sea level atmosphere (*a*) horizontally at 1200 mph; (*b*) upward at an angle of 40° with the horizontal at a velocity of 1200 mph.

9.14.2 The blunt-nosed projectile (cylinder) of Fig. 9.27 has a diameter of 0.45 m and weighs 2500 N. Find its rate of deceleration when moving through standard sea level atmosphere (*a*) horizontally at 450 m/s; (*b*) upward at an angle of 45° with the horizontal at a velocity of 450 m/s.

9.15 CONCLUDING REMARKS

There are some other aspects of drag that ought to be mentioned. One of these is drag under conditions of ***supercavitation***. This occurs when bodies move at high speeds through liquids. It is particularly common with blunt bodies. A large cavity forms behind the body, which alters the pressure distribution because the limiting minimum absolute pressure in the cavity is the vapor pressure. Such problems require special treatment; experimental data provide the best information.

Another interesting drag problem is that of an object moving at the interface of two fluids of different density. A good example of this is a ship or boat moving through water. In such cases energy is also expended in the generation of waves. The drag of the ship is caused primarily by both skin friction and wave action; hence in a model test both the Reynolds and Froude criteria ought to be satisfied, but this is usually not practical (see Sec. 7.4). The modeling procedure usually employed involves determining the total drag of the model by testing it at the prototype Froude number. The frictional drag of the model is then estimated by using boundary-layer theory, as presented in Secs. 9.3–9.5, and subtracted from the total drag to get an estimate of the drag on the model caused by wave action. This is then translated by model laws to an estimate of the wave-action drag on the prototype, to which is added the prototype friction drag as estimated by boundary-layer theory to give the total prototype drag.

SAMPLE PROBLEM 9.9 A 20-cm-diameter round-nosed projectile whose drag coefficient is shown in Fig. 9.27 travels at 600 m/s through the standard atmosphere at an altitude of 6000 m. Find the drag.

Solution

Table A.3 at 6000 m:

$$\rho = 0.660 \text{ kg/m}^3, \text{ and temperature } = -24°\text{C or } 273 - 24 = 249 \text{ K}.$$

Table A.5 for air: $k = 1.40$ and $R \approx 287 \text{ m}^2/(\text{s}^2{\cdot}\text{K})$.

Eq. (13.15) per Sec. 7.4: $c = \sqrt{kRT} = \sqrt{1.40(287)249} = 316 \text{ m/s}$

Eq. (7.10):
$$\mathbf{M} = \frac{600}{316} = 1.897$$

Fig. 9.27 with $\mathbf{M} = 1.897$:
$$C_D = 0.62$$

Eq. (9.32):
$$F_D = 0.62(0.660)\frac{(600)^2}{2}\frac{\pi(0.20)^2}{4} = 2310\text{ N} \qquad \textit{ANS}$$

PROBLEMS

9.1 Given the general equation for a parabola, $u = ay^2 + by + c$, derive the dimensionless velocity distribution of Eq. (9.9).

9.2 Given the parabolic velocity distribution of Eq. (9.9), prove that $\alpha = 0.1333$ [in Eq. (9.5)] and $\beta = 2.0$ [in Eq. (9.7)].

9.3 Determine the shear stress at 15 cm and 30 cm back from the leading edge of the plate in Sample Prob. 9.1.

9.4 Find the shear stress and the thickness of the boundary layer (*a*) at the center and (*b*) at the trailing edge of a smooth, flat plate 10 ft wide and 2 ft long parallel to the flow, immersed in 60°F water flowing at an undisturbed velocity of 3 fps. Assume a laminar boundary layer over the whole plate. Also, (*c*) find the total friction drag on one side of the plate.

9.5 Derive Eq. (9.19) along the lines suggested in the text.

9.6 For the turbulent boundary layer, derive the value of $\alpha = 0.0972$ from the seventh-root law given in Eq. (9.20).

9.7 From the information given, derive Eq. (9.22).

9.8 Assume that the boundary layer of Prob. 9.4 is disturbed near the leading edge. Compute the corresponding quantities for the turbulent boundary layer covering the entire plate, and compare the results.

9.9 A 85-m-long streamlined train has 2.5-m-high sides and a 2.5-m-wide top. Compute the power required to overcome the skin-friction drag when the train is traveling at 40 m/s through the ICAO standard atmosphere at sea level, assuming the drag on the sides and top to be equal to that on one side of a flat plate 7.5 m wide and 85 m long.

9.10 A lifeguard determines the wind velocity 2 m above the beach to be 8 m/s. If one wishes to get out of the wind by lying down, what would be the velocity at (*a*) 0.15 m; (*b*) 0.3 m above beach level?

9.11 Verify that the two expressions of Eq. (9.28) are equal.

9.12 A 20-mm-diameter harpoon 1.8 m long, with a sharp tip, is launched at 6 m/s into 15°C water. Find (*a*) the friction drag; (*b*) the maximum thickness of the boundary layer.

9.13 An airplane wing with a chord length of 7 ft parallel to the flow moves through standard atmospheric air at an altitude of 15,000 ft with a speed of 300 mph. Find (*a*) the critical roughness for a point one-tenth the chord length back from the leading edge; (*b*) the surface drag on a 25-ft-span section of this wing.

9.14 A flat plate 25 ft long and 1.5 ft wide is towed at 8 fps through a liquid ($\gamma = 50$ lb/ft^3, $\mu = 0.00026$ lb·sec/ft^2). Determine the total drag on the plate. Plot the thickness of the boundary layer and the local shear stress τ_0 as functions of x along the plate. Determine the area under the stress curve, and compare it with the previously computed value of the drag. Assume that the boundary layer changes from laminar to turbulent at a Reynolds number of 300,000.

9.15 For velocities of 0, 10, 20, 30, 40, 50, and 60 fps, make the necessary calculations to plot drag vs. velocity for the plate and data of Prob. 9.14.

9.16 Determine the drag on the harpoon of Exer. 9.5.2 for velocities of 0, 10, 30, and 50 fps, through (*a*) 60°F water; (*b*) 60°F air at standard sea-level pressure. Calculate the length x_c of the laminar zone, and plot curves of drag versus velocity for each case.

9.17 Determine the drag on the harpoon of Prob. 9.12 for velocities of 0, 3, 9, and 15 m/s, through (*a*) 15°C water; (*b*) 15°C air at standard sea-level pressure. Calculate the length x_c of the laminar zone, and plot curves of drag versus velocity for each case.

9.18 What will be the terminal velocity of the sphere of Exer. 9.7.1 in 120°F water?

9.19 For a 18-in-diameter sphere, compute the drag from wind under sea level conditions in a standard ICAO atmosphere. Plot drag versus velocity for 0, 25, 50, and 100 fps.

9.20 Repeat Prob. 9.19 for wind at a 10,000-ft elevation in a standard ICAO atmosphere.

9.21 A 1.5 ft diameter metal ball weighing 150 lb is dropped into the ocean from a boat. Determine the maximum velocity of the ball as it falls through sea water with $\rho = 2.0$ slugs/ft^3 and $\mu = 3.3 \times 10^{-5}$ lb·sec/ft^2.

9.22 A 40 cm diameter metal ball weighing 500 N is dropped into the ocean from a boat. Determine the maximum velocity of the ball as it falls through sea water with $\rho = 1030$ kg/m^3 and $\mu = 0.0016$ N·s/m^2.

9.23 A poorly-streamlined automobile has a body form corresponding roughly to the 1:0.75 oblate ellipsoid of Fig. 9.10, while a well-streamlined car has a body approximating the streamlined body in the same figure, each with a diameter of 1.5 m. If the velocity is 30 m/s through standard air at sea level (Appendix A, Table A.3), find the power in kW required to overcome air resistance in each case.

9.24 For the streamlined train in Prob. 9.9, calculate the pressure drag. As a rough approximation, assume that the nose and tail of the train are of the shape of the two halves of the prolate ellipsoid of Fig. 9.10, of 2.5-m diameter. Find the drag on the ellipsoid (pressure drag on the train), and compare this with the skin-friction drag on the train determined earlier.

9.25 A submerged 7.5-ft by 1.5-ft flat plate is dragged through 60°F water at 1.4 fps with the flat side normal to the direction of motion. (*a*) What is the approximate drag force? (*b*) How does this compare with the drag force of 0.1124 lb when pulled in the direction of the 7.5 ft length? *Hint*: Assume that the drag coefficient for the plate of finite length is in the same ratio to the coefficient for the infinite plate as is the ratio of coefficients for the finite cylinder ($L/D = 5$) and the infinite cylinder of Fig. 9.13, for the same Reynolds number.

9.26 A submerged 2.5-m by 0.5-m flat plate is dragged through 15°C water at 0.5 m/s with the flat side normal to the direction of motion. (*a*) What is the approximate drag force? (*b*) How does this compare with the drag force of 0.888 N when pulled in the direction of the 2.5 m length? *Hint*: See Prob. 9.25.

9.27 Compare the velocity of an 0.15-in-diameter drop of water falling through standard air at sea level, with that of a spherical bubble of air of the same size rising through water at the same temperature.

9.28 Compare the velocity of an 3-mm-diameter drop of water falling through standard air at sea level, with that of a spherical bubble of air of the same size rising through water at the same temperature.

9.29 Find the rate of fall of a spherical particle of sand ($s = 2.65$) in 15°C water if the diameter is (*a*) 0.1 mm; (*b*) 1.0 mm; (*c*) 10 mm. Express answers in mm/s.

9.30 A regulation football has a shape very similar to the prolate ellipsoid of Fig. 9.10, with a diameter of 6.78 in and a weight of 14.5 oz. (*a*) Find the resistance when the ball is passed through still air (14.5 psia and 80°F) at a velocity of 45 fps. Neglect the effect of spin about the longitudinal axis. What is the deceleration at the beginning of the trajectory? (*b*) Find the percentage change in resistance if the air temperature is 30°F rather than 80°F, and the drag coefficient is unchanged.

9.31 An eight-oar racing shell is traveling through 60°F water at a mean velocity of 12 mph. Each oar is 9 ft long, with a length of 6 ft from the oarlock to the center of the "spoon", which has a projected area of 120 in^2. Assume that all drag is due to the spoon, and that this drag is equal to that of a disk of equal area. (*a*) If the "stroke" is 32 per min, and if each oarsman sweeps a right-angle in one-fourth of his rowing cycle, what is the maximum thrust of the oars? It must be assumed that the shell moves at something less (say, 20%) than its mean velocity when the oars are in the water. (*b*) The maximum velocity occurs during the backstroke when the oarsmen shift their weight toward the stern. Why? (*c*) The oarsman on his backstroke moves at half the angular velocity of his forward stroke, while the shell moves at perhaps 10% above its mean velocity. Find the drag in 60°F air resulting from a "feathered" oar (turbulent boundary layer, Fig. 9.7) and an unfeathered one, in percentage of the forward thrust from part (*a*). What is the ratio of these two drag forces?

9.32 (*a*) Find the bending moment at the base of a 30-cm-diameter cylindrical light post 12 m high when it is subject to a uniform wind velocity of 30 m/s at standard sea level. Neglect end effects. (*b*) Discuss the consequences of considering the atmospheric boundary layer above the surface of the earth.

9.33 (*a*) Repeat Prob. 9.32(*a*) for the case where the pole has a uniform taper from 35 cm diameter at the base to 25 cm at its top. (*b*) Determine if the drag force would be larger or smaller if the pole had instead been tapered from 35 cm down to 15 cm. Why?

9.34 Beginning with the expression above Eq. (9.40), fill in the steps leading to Eq. (9.40). Take care to account for all changes in sign.

9.35 For the rotating cylinder in Sample Prob. 9.7, calculate the value of the lift coefficient. Assume the effective circulation is half the theoretical. Assuming the drag coefficient to be unchanged by the rotation of the cylinder, find the magnitude and direction of the total wind force on the rotor.

9.36 Assume the rotor in Sample Problem 9.7 to be installed upright on a ship traveling due north at 20 knots. The wind has an *absolute* velocity of 35 mph due east. If the drag coefficient of the cylinder is 1.0 and the "lines" of stagnation are separated by 120° on the rotor, find approximately the component of the total air force on the rotor in the direction of the ship's motion. Assume standard air.

9.37 Consider a cylinder of radius a in a stream of ideal fluid in which the undisturbed velocity and pressure are U and p_u and the density is ρ. (*a*) Using Eq. (9.36) and the Bernoulli theorem, evaluate the dimensionless pressure coefficient $(p - p_u)/(\rho U^2/2)$ for every 10° over the surface of one quadrant of the cylinder, measuring the angle from the forward stagnation point. Plot the pressure coefficient to scale, plotting it radially from the cylinder surface. (*b*) What is the actual pressure in pounds per square inch on the surface of a 1-ft-diameter cylinder, 70° from the forward stagnation point, if the cylinder is 15 ft below the free surface of a stream of water at 60°F, flowing at 12 fps?

9.38 Consider a cylinder of radius a in a stream of ideal fluid in which the undisturbed velocity and pressure are U and p_u and the density is ρ. (*a*) Using Eq. (9.36) and the Bernoulli theorem, evaluate the dimensionless pressure coefficient $(p - p_u)/\rho U^2/2$ for every 10° over the surface of one quadrant of the cylinder, measuring the angle from the forward stagnation point. Plot the pressure coefficient to scale, plotting it radially from the cylinder surface. (*b*) What is the actual pressure in newtons per square meter on the surface of a 30-cm-diameter cylinder, 70° from the forward stagnation point, if the cylinder is 5 m below the free surface of a stream of water at 15°C, flowing at 4 m/s?

9.39 A double stagnation point is observed to occur on a 4-ft-diameter cylinder rotating in an air stream moving at 60 fps. Given standard atmospheric conditions at sea level, find (*a*) the lift force per foot length of the cylinder; (*b*) the lift coefficient.

9.40 A double stagnation point is observed to occur on a 1.2-m-diameter cylinder rotating in an air stream moving at 18 m/s. Given standard atmospheric conditions at sea level, find (*a*) the lift force per meter length of the cylinder; (*b*) the lift coefficient.

9.41 There have been many arguments over the validity and extent of the curve of a pitched baseball. According to tests (*Life,* July 27, 1953), a pitched baseball was found to rotate at 1400 rpm while traveling at 43 mph. The horizontal projection of the trajectory revealed a smooth curve of about 800 ft radius. If the ball had a circumference of 9 in and weighed 5 oz, find the transverse force required to produce the observed curvature. Assume the shape of the ball to be roughly that of a cylinder having a diameter equal to the ball's diameter and a length of two-thirds its diameter. Find the circulation that would be required to produce the transverse force. Compare this with that obtained by assuming no slip at the equator of the ball. Assume standard air at sea level.

9.42 Solve Prob. 9.41 for a speed of 69 km/h and a trajectory radius of 245 m, with a ball weighing 1.4 N and having a circumference of 23 cm.

9.43 A wing of 44-ft span and a 350-ft^2 "planform" area moves horizontally through standard atmosphere at 20,000 ft with a velocity of 250 mph. If the wing supports 9000 lb, find (*a*) the required value of the lift coefficient; (*b*) the downwash velocity, assuming an elliptical distribution of lift over the span; (*c*) the induced drag.

9.44 Determine the induced angle of attack and the induced drag if the plan form of the wing in Exer. 9.12.1 were rectangular.

9.45 Determine the induced angle of attack and the induced drag if the plan form of the wing in Prob. 9.43 were rectangular.

9.46 For the Clark Y airfoil of Fig. 9.24, evaluate the friction coefficient η of Eq. (9.48) for the values of C_L of 0.6, 1.0, and 1.4.

9.47 A kite has the shape of a rectangular airfoil with a chord length of 2.5 ft and a span of 4.5 ft. When rigged and oriented as shown in Fig. P9.47, the guideline exerts a tension T of 14 lb when the wind velocity V is 28 mph in standard air at 1500 ft altitude. Find C_L, C_{D0}, and the friction coefficient η.

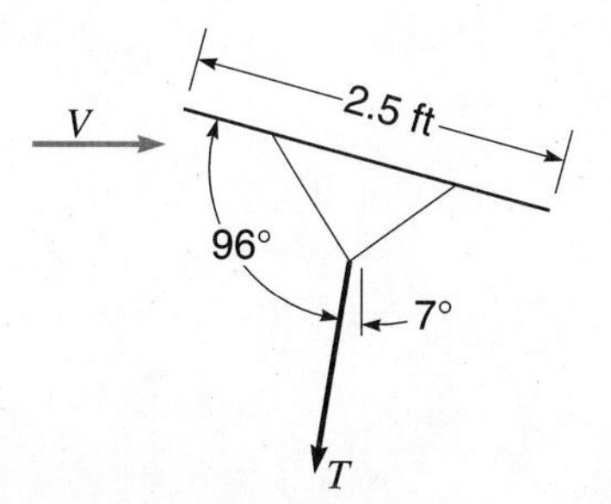

Figure P9.47

9.48 A rectangular airfoil with a 2.5-m chord and 15-m span has a drag coefficient of 0.062 and a lift coefficient of 0.95 at an angle of attack of 6.7°. What would be the corresponding (*a*) lift coefficient, (*b*) drag coefficient, and (*c*) angle of attack for a wing having the same profile but with an aspect ratio of 7.6?

9.49 Determine the angle of glide that will allow the sailplane in Exer. 9.13.2 to remain in the air for the longest time? (*Note*: A trial-and-error solution will be required here.)

9.50 A supersonic jet plane is flying horizontally at 1500 mph. How soon after it passes overhead at an elevation of 5000 ft will the shock wave be felt at sea level?

9.51 A supersonic jet plane is flying horizontally at 2800 km/h. How soon after it passes overhead at an elevation of 2 km will the shock wave be felt at sea level?

9.52 If the round-nosed and sharp-nosed projectiles of Fig. 9.27 each represent a 850-lb bomb with a diameter of 18 in, and assuming they travel vertically downward nose first, find their terminal velocities in standard air at sea level.

9.53 If the round-nosed and sharp-nosed projectiles of Fig. 9.27 each represent a 3500-N bomb with a diameter of 45 cm, and assuming they travel vertically downward nose first, find their terminal velocities in standard air at sea level.

CHAPTER 10

Steady Flow in Open Channels

10.1 OPEN CHANNELS

An open channel is one in which the stream is not completely enclosed by solid boundaries and therefore has a free surface subjected only to atmospheric pressure. The flow in such a channel is caused not by some external head, but rather only by the gravity component along the slope of the channel. Thus open-channel flow is often referred to as ***free-surface flow*** or ***gravity flow***. This chapter will deal only with *steady* flow in open channels.

The principal types of open channel are natural streams and rivers; artificial canals; and sewers, tunnels, and pipelines flowing not completely full. Artificial canals may be built to convey water for purposes of water-power development, irrigation or city water supply, and drainage or flood control and for numerous other purposes. While there are examples of open channels carrying liquids other than water, there are few experimental data for such, and the numerical coefficients given here apply only to water at natural temperatures.

For convenience in dealing with large channel systems, they are often divided into reaches. A ***reach*** is a continuous stretch of a waterway, often chosen to have reasonably uniform properties like cross section, slope, and discharge.

The accurate solution of problems of flow in open channels is much more difficult than in the case of pressure pipes. Not only are reliable experimental data more difficult to secure, but we meet a wider range of conditions than we do with pipes. Practically all pipes are round, but the cross sections of open channels may be of any shape, from circular to the irregular forms of natural streams. In pipes the degree of roughness ordinarily ranges from that of new smooth metal, on the one hand, to that of old corroded iron or steel pipes, on the other. But with open channels the surfaces vary from that of smooth timber or concrete to that of the rough or irregular beds of some rivers. Hence the friction coefficients to be used are more uncertain with open channels than with pipes.

Uniform flow was described in Sec. 4.3 as it applies to hydraulic phenomena in general. In the case of open channels uniform flow means that the water cross section and depth remain constant over a certain reach of the channel as well as over time. This requires that the drop in potential energy due to the fall in elevation along the channel be exactly that consumed by the energy dissipation through boundary friction and turbulence.

Uniform flow is an equilibrium condition that flow tends to if the channel is sufficiently long with constant slope, cross section, and roughness. This may be stated in another way, as follows. For any channel of given roughness, cross section, and slope, there exists for a given flow rate one and only one water depth, y_0, at which the flow will be uniform. Thus, in Fig. 10.1, the flow is accelerating in the reach from A to C, becomes established as uniform flow from C to D, suffers a violent deceleration due to the change of slope between D and E, and finally achieves a new depth of uniform flow somewhere beyond E. There is acceleration in the reach from B to C because the gravity component along the slope is greater than the boundary shear resistance. As the flow accelerates, the boundary shear increases because of the increase in velocity, until at C the boundary shear resistance becomes equal to the gravity component along the slope. Beyond C there is no acceleration, the velocity is constant, and the flow is uniform. The depth in uniform flow is commonly referred to as the ***normal depth***, y_0.

Open-channel flow is usually fully rough; that is, it occurs at high Reynolds numbers. For open channels the Reynolds number is defined by $\mathbf{R} = R_h V/\nu$, where R_h is the hydraulic radius. Since $R_h = D/4$, the critical value of Reynolds number at which the changeover occurs from laminar flow

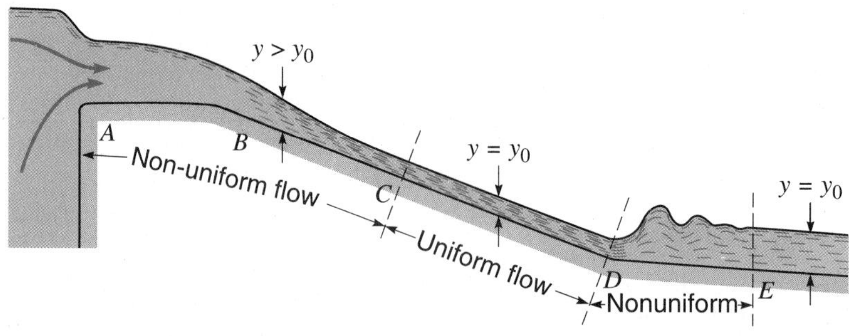

Figure 10.1
Steady flow down a chute or spillway.

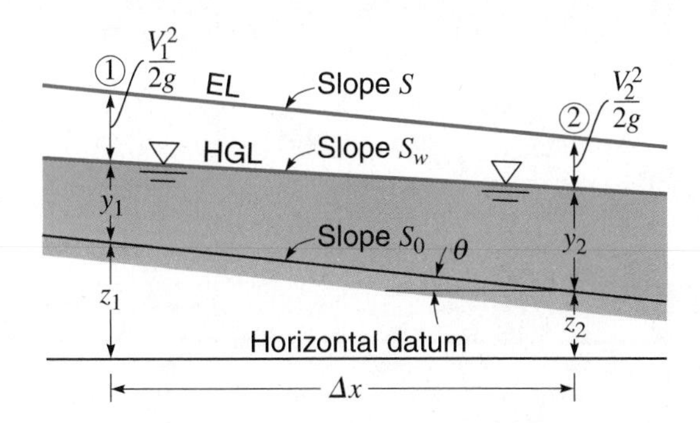

Figure 10.2
Open channel flow—definition sketch (L = distance along the channel bed between sections 1 and 2).

to turbulent flow in open channels is 500, whereas in pressure conduits the critical value is 2000.

In open-channel flow (Fig. 10.2) we refer to the slope of the *channel bed* S_0, the slope of the *water surface* S_w, and the slope of the *energy line* S. It is quite evident that in the case of flow in an open channel the hydraulic grade line coincides with the water surface, provided there is no unusual curvature in the streamlines or stream tubes, for if a piezometer tube is attached to the side of the channel, the water will rise in it until its surface is level with that of the water in the channel at that point. Water depth y is always measured vertically and the distance between sections is commonly defined by the horizontal distance Δx between them. The various slopes mentioned above are therefore defined as follows:

$$S_0 = \frac{z_1 - z_2}{\Delta x} = -\frac{\Delta z}{\Delta x} \tag{10.1}$$

$$S_w = \frac{(z_1 + y_1) - (z_2 + y_2)}{\Delta x} = -\frac{\Delta(z + y)}{\Delta x} \tag{10.2}$$

$$S = \frac{h_L}{L} = \frac{(z_1 + y_1 + V_1^2/2g) - (z_2 + y_2 + V_2^2/2g)}{L} \tag{10.3}$$

where x, y, Δx, and L are defined in Fig. 10.2. Note that the energy gradient S is defined as the head loss per length of flow path and that it is usually assumed that $\alpha = 1.0$. This assumption is reasonable when the flow depth is less than the channel width. Also, for all practical purposes, the angle θ between the channel bed and the horizontal is small; hence L, the distance along the channel bed between the two sections, is almost equal to Δx, the horizontal distance between the two sections.

10.2 UNIFORM FLOW

In uniform flow (Fig. 10.3) the cross section through which flow occurs is constant along the channel, and so also is the velocity. Thus, $y_1 = y_2$ and

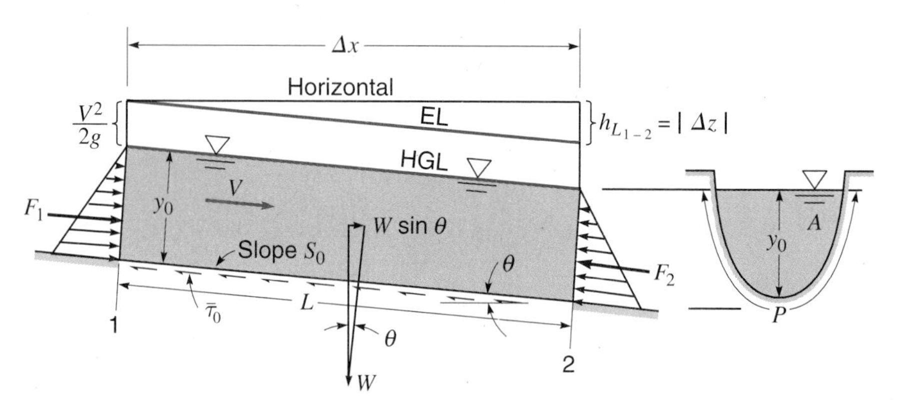

Figure 10.3
Uniform flow in open channel.

$V_1 = V_2$ and the channel bed, water surface, and energy line are parallel to one another. Also, $S_w = S_0 = -\Delta z/\Delta x = \tan\theta$, while $S = h_L/L = \sin\theta$, where θ is the angle the channel bed makes with the horizontal.

In most open channels (rivers, canals, and ditches) the bed slope S_0 $(=\Delta z/\Delta x)$ falls in the range of 0.01–0.0001. For some chutes, spillways, and sewers it may be larger, and for some large rivers it may be smaller. Because, therefore, usually $S_0 < 0.1$ so that $\theta < 5.7°$, it follows that in most cases $\sin\theta \approx \tan\theta$ (within 0.5%) and $S_0 = S_w \approx S$. In this chapter it will be assumed that $\theta < 5.7°$ and thus that $S \approx S_0$. In channels where $\theta > 5.7°$ a distinction must be made in the difference between S and S_0. Moreover, on such slopes air entrainment and pulsating flow (successions of traveling waves) are common.

In Sec. 8.4 a general equation for frictional resistance in a pressure conduit was developed. The same reasoning may now be applied to uniform flow with a free surface. Consider the short reach of length L along the channel between stations 1 and 2 in uniform flow with water cross section of area A (Fig. 10.3). As the flow is neither accelerating nor decelerating, we may consider the body of water contained in the reach in static equilibrium. Summing forces along the channel, the hydrostatic-pressure forces F_1 and F_2 balance each other, since there is no change in the depth y between the stations. The only force in the direction of motion is the gravity component, and this must be resisted by the average boundary shear stress $\bar{\tau}_0$, acting over the area PL, where P is the wetted perimeter of the section. Thus

$$\gamma AL \sin\theta = \bar{\tau}_0 PL$$

But $\sin\theta = h_L/L = S$. Solving for $\bar{\tau}_0$, we have

$$\bar{\tau}_0 = \gamma \frac{A}{P} S = \gamma R_h S \tag{10.4}$$

where R_h is the hydraulic radius,[1] discussed in Sec. 8.3, and for most slopes (with $\theta < 5.7°$) S_0 may be taken as equal to S. Substituting the value of $\bar{\tau}_0$ from Eq. (8.8) and replacing S with S_0,

$$\bar{\tau}_0 = C_f \rho \frac{V^2}{2} = \gamma R_h S_0$$

This may be solved for V in terms of either the friction coefficient C_f or the conventional friction factor f [Eq. (8.11)] to give

$$V = \sqrt{\frac{2g}{C_f} R_h S_0} = \sqrt{\frac{8g}{f} R_h S_0} \tag{10.5}$$

The Chézy Formula

Antoine de Chézy (1718–1798), a French bridge engineer and hydraulics expert, proposed in 1775 that the velocity in an open channel varied as $\sqrt{R_h S_0}$. This led to the formula

$$V = C\sqrt{R_h S_0} \tag{10.6}$$

which is known by his name. It has been widely used both for open channels and for pipes under pressure. Comparing Eqs. (10.6) and (10.5), it is seen that $C = \sqrt{8g/f}$. Despite the simplicity of Eq. (10.6), it has the distinct drawback that C is not a pure number but has the dimensions $L^{1/2}T^{-1}$, requiring that values of C in SI units be converted before being used with BG units in the rest of the formula.

As C and f are related, the same considerations that have been presented in Chap. 8 regarding the determination of a value for f apply also to C. For a small open channel with smooth sides, the problem of determining f or C is the same as that in the case of a pipe. But most channels are relatively large compared with pipes, thus giving Reynolds numbers that are higher than those commonly encountered in pipes. Also, open channels are frequently rougher than pipes, especially in the case of natural streams. A study of Fig. 8.11 reveals that, as the Reynolds number and the relative roughness both increase, the value of f becomes practically independent of **R** and depends only on the relative roughness.

---

[1] Strictly speaking, A and P should be measured in a plane at right angles to L. However, since depth is measured vertically, values of A and P are calculated as those in the vertical plane. The resulting values of R_h are identical regardless of which way A and P are determined.

The Manning Formula

One of the best as well as one of the most widely used formulas for uniform flow in open channels is that published by the Irish engineer Robert Manning[2] (1816–1897). Manning had found from many tests that the value of C in the Chézy formula varied approximately as $R_h^{1/6}$, and others observed that the proportionality factor was very close to the reciprocal of n, the coefficient of roughness in the previously used, but complicated and inaccurate, Kutter formula. This led to the formula that has since spread to all parts of the world. In SI units, the Manning formula is

In SI units:
$$V(m/s) = \frac{1}{n} R_h^{2/3} S_0^{1/2} \tag{10.7a}$$

The dimensions of the friction factor n are seen to be $TL^{-1/3}$. To avoid converting the numerical value of n for use with BG units, the formula itself is changed so as to leave the value of n unaffected. Thus, in BG units, the Manning formula is

In BG units:
$$V(\text{fps}) = \frac{1.486}{n} R_h^{2/3} S_0^{1/2} \tag{10.7b}$$

where 1.486 is the cube root of 3.28, the number of feet in a meter. Despite the dimensional difficulties of the Manning formula, which have long plagued those attempting to put all fluid mechanics on a rational dimensionless basis, it continues to be popular because it is simple to use and reasonably accurate. Representative values of n for various surfaces are given in Table 10.1.

In terms of flow rate, Eqs (10.7) may be expressed as

In BG units:
$$Q(\text{cfs}) = \frac{1.486}{n} A R_h^{2/3} S_0^{1/2} \tag{10.8a}$$

In SI units:
$$Q(\text{m}^3/\text{s}) = \frac{1}{n} A R_h^{2/3} S_0^{1/2} \tag{10.8b}$$

Variation of *n*

We mentioned in Sec. 8.11 that e is a measure of the absolute roughness of the inside of a pipe. The question naturally arises as to whether e and n may be functionally related to each other. Such a relation has been proposed by Powell (Fig. 10.4) on the basis of experimental data. In terms of the hydraulic radius Powell's relation is

$$\frac{1}{\sqrt{f}} = 2 \log \left(14.8 \frac{R_h}{e}\right) \tag{10.9}$$

[2] Robert Manning, Flow of Water in Open Channels and Pipes, *Trans. Inst. Civil Engrs.* (*Ireland*), vol. 20, 1890.

Table 10.1
Values of n in Manning's formula

Nature of surface	n Min	n Max
Lucite	0.008	0.010
Glass	0.009	0.013
Neat cement surface	0.010	0.013
Wood-stave pipe	0.010	0.013
Plank flumes, planed	0.010	0.014
Vitrified sewer pipe	0.010	0.017
Concrete, precast	0.011	0.013
Metal flumes, smooth	0.011	0.015
Cement mortar surfaces	0.011	0.015
Plank flumes, unplaned	0.011	0.015
Common-clay drainage tile	0.011	0.017
Concrete, monolithic	0.012	0.016
Brick with cement mortar	0.012	0.017
Cast iron, new	0.013	0.017
Riveted steel	0.017	0.020
Cement rubble surfaces	0.017	0.030
Canals and ditches, smooth earth	0.017	0.025
Corrugated metal pipe	0.021	0.030
Metal flumes, corrugated	0.022	0.030
Canals:		
Dredged in earth, smooth	0.025	0.033
In rock cuts, smooth	0.025	0.035
Rough beds and weeds on sides	0.025	0.040
Rock cuts, jagged and irregular	0.035	0.045
Natural streams:		
Smoothest	0.025	0.033
Roughest	0.045	0.060
Very weedy	0.075	0.150

Combining Eq. (10.5) with Eq. (10.7), we get

In BG units: $$n = 1.486 R_h^{1/6} \sqrt{\frac{f}{8g}}$$

In SI units: $$n = R_h^{1/6} \sqrt{\frac{f}{8g}}$$

Substituting numerical values for g gives

In BG units: $$n = 0.0928 f^{1/2} R_h^{1/6} \qquad (10.10a)$$

In SI units: $$n = 0.113 f^{1/2} R_h^{1/6} \qquad (10.10b)$$

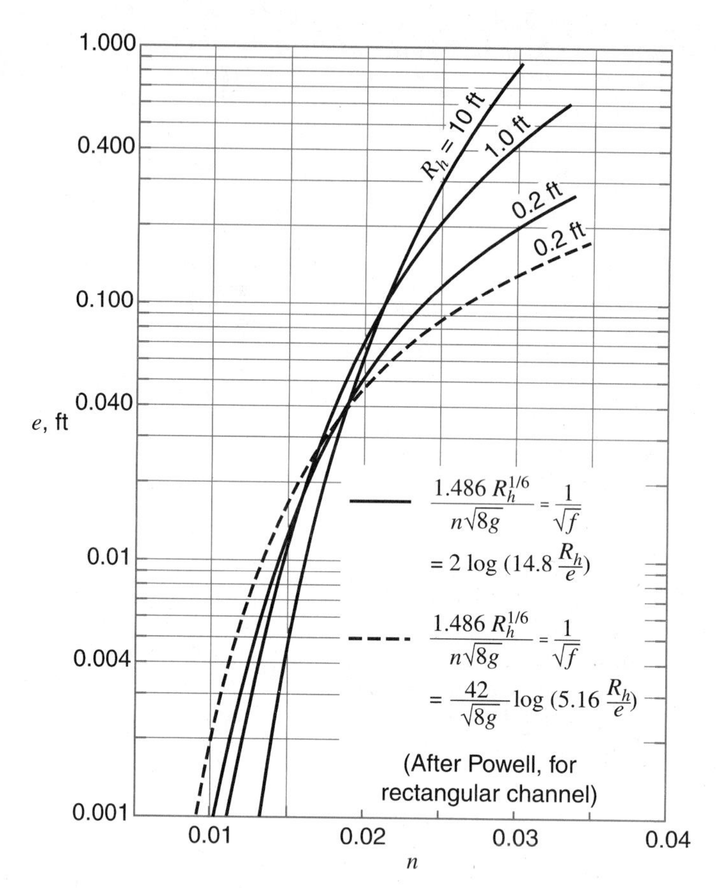

Figure 10.4
Correlation of n with absolute roughness e.

Thus we see that n is related to the friction factor, which depends on the relative roughness and Reynolds number and on the hydraulic radius, which is indicative of the size of the channel.

Equating Eq. (10.10*a*) with Eq. (10.9) provides a correlation between e and n, which is plotted as the solid lines in Fig. 10.4 for three representative values of the hydraulic radius. The dashed line is the plot of another correlation proposed by Powell that gave a better fit to experimental data for small values of hydraulic radius.[3] One notable feature of these curves is that a large relative error in e results in only a small error in n. Another important observation concerns the variation of n with channel size for the same absolute roughness. For channels that are quite rough, like natural channels

[3] R. W. Powell, Resistance to Flow in Rough Channels, *Trans. Am. Geophys. Union*, vol. 31, no. 4, pp. 575–582, 1950.

with $e > 0.10$ ft, we see in Fig. 10.4 that n gets smaller with increasing hydraulic radius. This, being similar to the behavior of relative roughness e/D in pipes, has long been understood. However, at lower e values the curves cross, with the result that for smoother, artificial surfaces with $e < 0.02$ ft we find that n *increases* with increasing hydraulic radius. This contrary behavior was formerly not understood, with the result that some very large concrete-lined canal systems were designed using inappropriately low n values; when put into service, it was found that the higher true n values reduced the capacity of the canals by as much as 15%.

10.3 SOLUTION OF UNIFORM FLOW PROBLEMS

Uniform flow problems usually involve the application of Manning's equation [Eqs. (10.8)]. The selection of an appropriate value for the Manning roughness factor n is critical to the accuracy of the results of a problem. When the channel surface is concrete or some other structural material, it is possible to select a reasonably accurate value for n, but for the case of a natural channel one must rely on judgment and experience, and in many instances the selected value may be quite inaccurate.

There are a number of different types of problems that are encountered when using Manning's equation. For example, to find the normal depth of flow for a particular flow rate in a given channel, a trial-and-error solution is required, because y_0 is involved in A and R_h in complicated ways (see, for example, Sample Prob. 10.1). On the other hand, the expected flow in a particular channel under given conditions can be solved for directly. Various types of sliderules, nomographs, tables,[4] and computer programs,[5] are available to serve as an aid to the solution of open-channel problems. Some of these provide helpful visual representations of the interdependence of the various factors, but their applicability is usually limited.

Sample Problem 10.1 This channel has a bed slope of 0.0006 and a Manning's n of 0.016. Using only a basic scientific calculator, (a) find the uniform flow depth when the flow rate is 225 cfs and (b) compute the corresponding value of the absolute roughness e. Also, find the uniform flow depth using (c) Mathcad and (d) an equation solver on a programmable scientific calculator.

[4] E. F. Brater, H. W. King, J. E. Lindell, and C. Y. Wei, *Handbook of Applied Hydraulics,* 7th ed., McGraw-Hill Book Company, New York, 1996.
[5] See citations in Appxs. B and C.
: Programmed computing aids (Appx. B) could help solve problems marked with this icon.

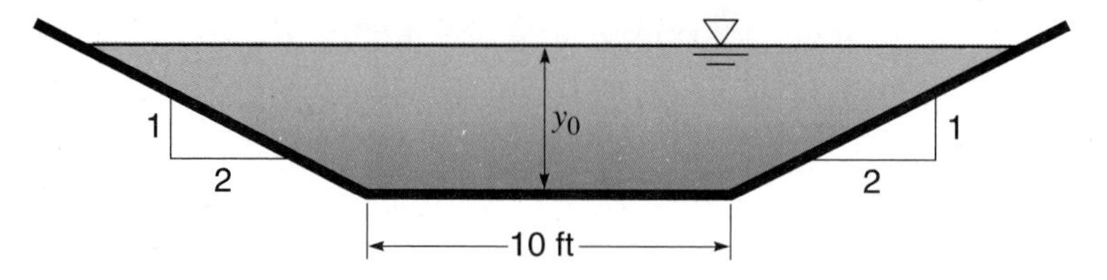

Figure S10.1

Solution

(*a*)

$$A = (10 + 2y)y$$

and

$$R_h = \frac{A}{P} = \frac{(10 + 2y)y}{10 + 2\sqrt{5}y}$$

Eq. (10.8*a*):

$$Q = 225 = \frac{1.486}{0.016}(10 + 2y)y\left[\frac{(10 + 2y)y}{10 + 2\sqrt{5}y}\right]^{2/3}(0.0006)^{1/2}$$

By trial (see Sample Probs. 3.5 and 5.7), $y_0 = 3.41$ ft, uniform flow depth ***ANS***

Note: The accuracy with which flow depth can be measured can differ considerably, depending on many factors. In a laboratory, depths can sometimes be measured to 0.001 ft (0.3 mm), whereas in the field, with turbulent flow and wind, surface fluctuations (which can be averaged) may make it difficult to achieve an accuracy of 0.1 ft (30 mm).

(*b*)

$$A = [10 + 2(3.41)]3.41 = 57.3 \text{ ft}^2$$

$$P = 10 + 2\sqrt{5}(3.41) = 25.2 \text{ ft}$$

$$R_h = \frac{A}{P} = \frac{57.3}{25.2} = 2.27 \text{ ft}$$

Eq. (10.10*a*):

$$0.016 = 0.0928 f^{1/2}(2.27)^{1/6}$$

from which

$$f^{1/2} = 0.1503 \quad (f = 0.0226)$$

Eq. (10.9):

$$\frac{1}{0.1503} = 2 \log\left(\frac{14.8 \times 2.27}{e}\right)$$

from which

$$e = 0.015\,85 \text{ ft} \qquad \textbf{\textit{ANS}}$$

(*c*) Using Mathcad, open a file (worksheet) and establish a "solve block" beginning with "Given" and ending with a statement including "Find(*y, A, P, Rh*)." Within the solve block, type the four governing equations,

Eq. (10.8):

$$Q = \frac{C}{n} A R_h^{2/3} S^{1/2}$$

and

$$A = (b + my)y, \qquad P = b + 2y\sqrt{1 + m^2}, \qquad R_h = \frac{A}{P}$$

These four equations involve ten variables, four of which are those to be found. The quantity c must be 1.486 for problems in BG units and 1.0 for those in SI units. The side slope m is defined as the horizontal run over the vertical rise; for rectangular channels it is zero.

Above the solve block, assign numerical values to the six known variables (Q, c, b, m, n, S) as above. Then assign estimated values to the four unknown variables, for example $y = 1.0$, $A = 10$, $P = 10$, $Rh = 1.0$. These assignments may be labeled with text.

If the "Find" statement ends in an equals sign, the program will write the four results immediately after it, in the form of a vertical array or vector. The results are

$$y = 3.406 \text{ ft}, \quad A = 57.27 \text{ ft}^2, \quad P = 25.23 \text{ ft}, \quad R_h = 2.270 \text{ ft} \qquad \textbf{\textit{ANS}}$$

Note: Other mathematics software packages that may be used are discussed in Appendix B.

(*d*) To use an equation solver on a programmable scientific calculator, we recall from Sec. 8.14 that such solvers can solve only one nonlinear (and implicit) equation for one unknown. We can create such a single equation by substituting the last three equations above (for Mathcad) into Eq. (10.8) to eliminate A, P, and Rh, which yields

$$Q = \frac{c}{n} \frac{(by + my^2)^{5/3}}{(b + 2y\sqrt{1 + m^2})^{2/3}} S^{1/2}$$

Using an "equation writer," key in this equation, and store it permanently, assigning it a name. Then activate the equation solving feature, select the desired equation from the catalog of equations, and makes it the "current" equation.

Display the menu of variables in the equation, and enter numeric values for them all (Q, c, b, m, n, S) as above plus an estimate of y (1 ft, say). Instruct the solver to solve for y. The result is

$$y = 3.406 \text{ ft} \qquad \textbf{\textit{ANS}}$$

In applying Manning's equation to channel shapes such as that in Fig. 10.5, which simulates a river with overbank flow conditions, the usual procedure is to break the section into several parts, as indicated in the figure.

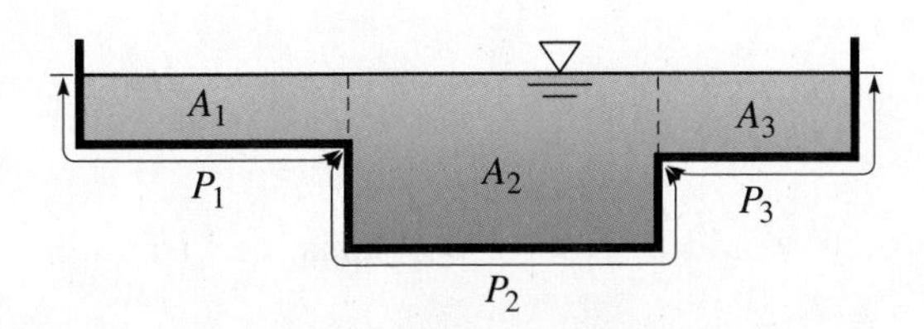

Figure 10.5

It is assumed that there is no resistance along the dashed vertical line. Actually the flow in area A_2 tends to speed up the flow in area A_1, while the flow in A_1 tends to slow down the flow in area A_2. These two effects come very close to balancing out one another. If A/P for the total cross section had been computed by the usual method, that is, $R_h = (A_1 + A_2 + A_3)/(P_1 + P_2 + P_3)$, it would imply that the effect of boundary resistance is uniformly distributed over the flow cross section, which, of course, is not the case.

Another advantage of breaking the total section into parts is that possible variations in Manning's n can be taken into consideration. Thus, for the channel shown in Fig. 10.5, in English units

$$Q = \frac{1.486}{n_1} A_1 R_{h_1}^{2/3} S_0^{1/2} + \frac{1.486}{n_2} A_2 R_{h_2}^{2/3} S_0^{1/2} + \cdots \tag{10.11}$$

where $R_{h_1} = A_1/P_1$, $R_{h_2} = A_2/P_2$, etc. The A and P are defined in Fig. 10.5. An equation of similar form can be written for SI units in which case the 1.486 is replaced by unity.

EXERCISES

10.3.1 For the channel of Sample Prob. 10.1, compute the flow rate for depths of 2, 4, 6, and 8 ft. Plot a curve of Q versus y.

10.3.2 Water flows uniformly in a trapezoidal channel with a 15-ft-wide bed and 1.8:1 (H:V) side slopes. If the bed slope is 1.25 ft/mile and $n = 0.016$, find the flow rate when the depth is 8 ft.

10.3.3 Water flows uniformly in a 3-m-wide rectangular channel at a depth of 40 cm. The channel slope is 0.003 and $n = 0.014$. Find the flow rate in m^3/s.

10.3.4 At what depth will water flow in a 4-m-wide rectangular channel if $n = 0.018$, $S = 0.0009$, and $Q = 7\ m^3/s$?

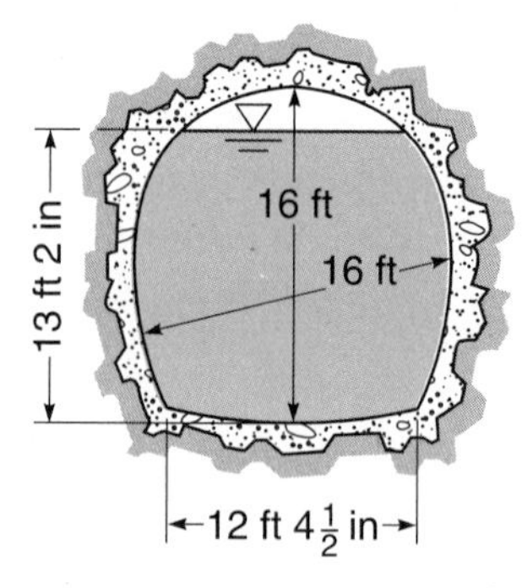

Figure X10.3.5

: Programmed computing aids (Appx. B) could help solve problems marked with this icon.

10.3.5 Figure X.10.3.5 shows a tunnel section on the Colorado River Aqueduct. The area of the water cross section is 191 ft^2, and the wetted perimeter is 39.1 ft. The flow is 1600 cfs. If $n = 0.013$ for its concrete lining, find the slope.

10.3.6 If $a = 1$ m, $b = 4$ m, $d = 2$ m, $w = 20$ m, and $n = 0.016$, what slope is required so that the flow will be 25 m^3/s when the depth of flow is 1.50 m?

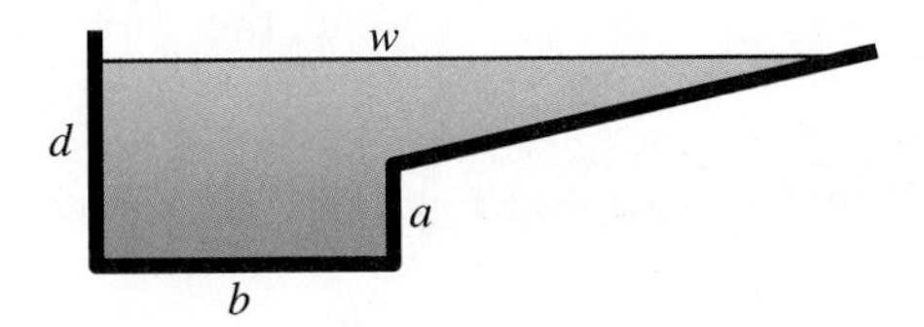

Figure X10.3.6

10.3.7 Water flows in a rectangular flume 6 ft wide made of unplaned timber ($n = 0.013$). Find the necessary channel slope if the water flows uniformly at a depth of 3 ft and at 20 fps.

10.3.8 Water flows in a rectangular flume 1.5 m wide made of unplaned timber ($n = 0.013$). Find the necessary channel slope if the water flows uniformly at a depth of 0.6 m and at 4.5 m/s.

10.3.9 Refer to Fig. 10.5. Suppose the widths of A_1, A_2, and A_3 are 120, 50, and 240 ft and the total depths are 3, 12, and 4 ft. Compute the flow rate if $S = 0.0018$, $n_1 = n_3 = 0.06$, and $n_2 = 0.030$.

10.3.10 Refer to Fig. 10.5. Suppose the widths of A_1, A_2, and A_3 are 30, 10, and 60 m and the total depths are 0.5, 3.0, and 1.0 m. Compute the flow rate if $S = 0.0016$, $n_1 = n_3 = 0.04$, and $n_2 = 0.025$.

10.4 VELOCITY DISTRIBUTION IN OPEN CHANNELS

Due to friction around the wetted perimeter, we expect strong velocity variations across flow cross sections in open channels. A small frictional effect must also occur at the water surface, due to air drag.

Vanoni[6] has demonstrated that the universal logarithmic velocity-distribution law for pipes [Eq. (8.29)] also applies to a two-dimensional open

[6] V. A. Vanoni, Velocity Distribution in Open Channels, *Civil Eng.*, vol. 11, pp. 356–357, 1941.

channel, i.e., one of uniform depth that is infinitely wide. This equation may be written as

$$\frac{u - u_{max}}{\sqrt{gy_0S}} = \frac{2.3}{K} \log \frac{y}{y_0}$$

where y_0 = depth of water in channel
u = velocity at a distance y from channel bed
K = von Kármán constant, having a value of about 0.40
S = channel bed slope

This expression can be integrated over the depth to yield the more useful relation

$$u = V + \frac{1}{K}\sqrt{gy_0S}\left(1 + 2.3 \log \frac{y}{y_0}\right) \tag{10.12}$$

which expresses the distribution law in terms of the mean velocity V. This equation is plotted in Fig. 10.6, together with velocity measurements that were made on the center line of a rectangular flume 2.77 ft (0.844 m) wide with a water depth of 0.59 ft (0.180 m). The filament whose velocity u is equal to V is seen to lie at a distance of $0.632y_0$ beneath the surface.

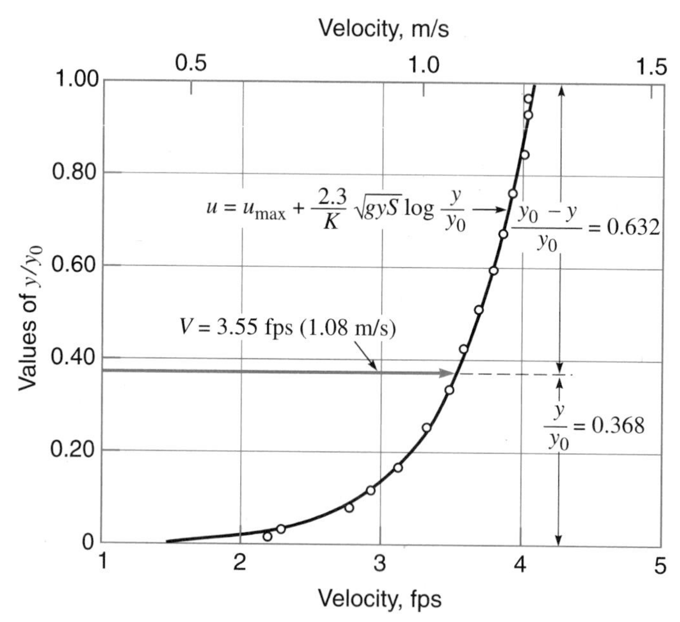

Figure 10.6
Velocity profile at the center of a flume 2.77 ft (0.844 m) wide for a flow 0.59 ft (0.180 m) deep (after Vanoni). The circles represent measurements.

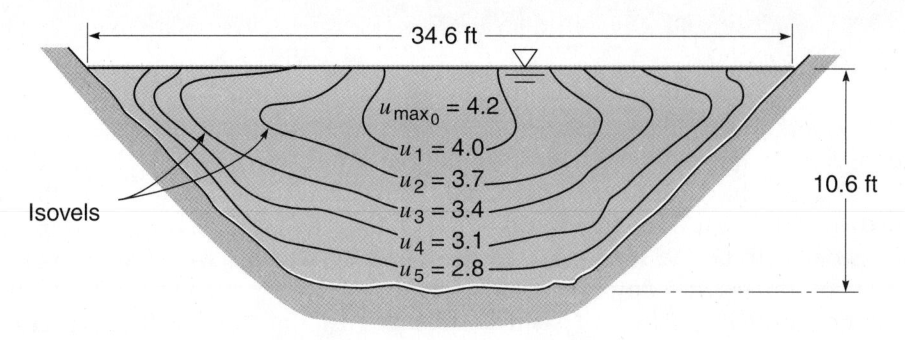

Figure 10.7
Velocity distribution in a trapezoidal canal. $V = 3.32$ fps, $A = 230.5$ ft², $S_0 = 0.000\,057$, $\alpha = 1.105$, $\beta = 1.048$.

Velocity measurements made in a trapezoidal channel resulted in the isovels (contours of equal velocity) shown in Fig. 10.7. The point of maximum velocity is seen to lie beneath the surface, the result of air drag. From this two-dimensional velocity distribution the values of the correction factors for kinetic energy and momentum were computed, and are given with the figure. Although they are greater than the corresponding values for pipe flow [Eqs (8.34)], the treatment in this chapter follows the earlier procedure of assuming the values of α and β to be unity, unless stated otherwise. Any thorough analysis would of course have to take account of their true values.

EXERCISES

10.4.1 In Fig. X10.4.1 with uniform flow in the wide open channel, $a = 3.20$ ft. Find V and b if $n = 0.015$.

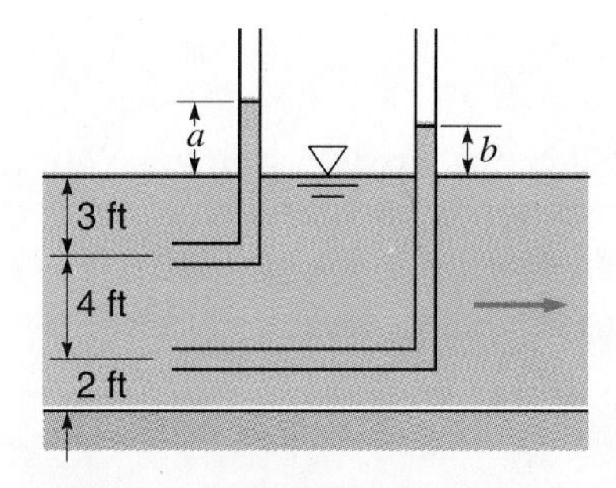

Figure X10.4.1

10.4.2 Water flows uniformly in a very wide rectangular channel ($R_h = y$) at a depth y_0 of 2.0 m. $S = 0.005$ and $n = 0.018$. Calculate the velocities at y-values of 0.2, 0.4, 0.8, 1.2, and 2.0 m above the bed and plot the velocity profile. Note the value of the maximum velocity at the water surface.

10.5 "WIDE AND SHALLOW" OPEN CHANNELS

If a channel is sufficiently wide compared with its depth, it is reasonable to expect that the sides will have negligible effect on the flow in the central portion of the channel. Experiments have confirmed this to be the case provided the width-to-depth ratio $b/y > 10$. Then the central flow may be considered to be the same as in a channel of infinite width—in other words, to be two-dimensional flow.

For a small portion of such a two-dimensional flow with width Δb, the flow cross-sectional area is $y\ \Delta b$ and the wetted perimeter is Δb, so that the hydraulic radius is

Wide and shallow:

$$R_h = \frac{A}{P} = \frac{y\ \Delta b}{\Delta b} = y \qquad (10.13)$$

This convenient result is commonly assumed for ***wide and shallow open channels.*** The same result is obtained by considering the hydraulic radius of a rectangular channel,

$$R_h = \frac{A}{P} = \frac{by}{b + 2y} = \frac{y}{1 + 2y/b}$$

As the ratio $b/y \to \infty$, we see that $R_h \to y$.

Analytically, the percentage error in R_h when using y as the hydraulic radius for a rectangular channel is

$$\frac{100(y - R_h)}{R_h} = 100\left(y - \frac{by}{b + 2y}\right)\frac{b + 2y}{by} = 200\,\frac{y}{b}$$

This results in a 20% error for $b/y = 10$, a 5% error for $b/y = 40$, and a 2% error for $b/y = 100$. If we make the assumption that a channel is "wide and shallow" and replace R_h by y in our analysis, it is important to check that the b/y ratio is sufficiently large to justify this.

SAMPLE PROBLEM 10.2 A 12-m-wide stream has a reasonably rectangular cross section, and the average depth is computed to be 11.2 cm. What would be the percentage error in the hydraulic radius if this stream were assumed to be "wide and shallow"? Would it be reasonable to make this assumption?

Solution

$$\frac{b}{y} = \frac{12}{0.112} = 107.1$$

$$\text{Error percent} = \frac{200}{107.1} = 1.867\% \qquad \textbf{\textit{ANS}}$$

This error is small enough that the assumption is reasonable. ***ANS***

EXERCISES

10.5.1 Water flows $\frac{1}{8}$ in deep in a 9-in-wide rectangular flume. What would be the percentage error in the hydraulic radius if this were assumed to be wide and shallow open channel flow?

10.5.2 Water flows 6.9 mm deep in a 0.4-m-wide rectangular flume. What would be the percentage error in the hydraulic radius if this were assumed to be wide and shallow open channel flow?

10.5.3 At a location where the cross section of a stream is reasonably rectangular, the width is 13 ft and the average depth is computed to be 18.4 in. What would be the percentage error in the hydraulic radius if this stream were assumed to be "wide and shallow"? Would it be reasonable to make this assumption?

10.6 MOST EFFICIENT CROSS SECTION

Any of the open-channel formulas given above show that, for a given slope and roughness, the velocity increases with the hydraulic radius. Therefore, for a given area of water cross section, the rate of discharge will be a maximum when R_h is a maximum, which is to say, when the wetted perimeter P and so the frictional resistance is a minimum. Such a section is called the ***most efficient cross section*** for the given area, and we emphasize that here we are addressing only the *hydraulic efficiency*. Or for a given rate of discharge, the cross-sectional area will be a minimum when the design is such as to make R_h a maximum (and thus P a minimum). This section would be the most efficient cross section for the given rate of discharge. These issues are important, because reducing P tends to reduce channel construction costs. However, other factors influence the overall cost of an open channel. For example, the cost of deep excavation is higher than shallow excavation. Consequently most channels are designed to flow with depths less than the hydraulic optimum, resulting in R_h values a few percent smaller than $R_{h_{max}}$. In situations where right-of-way costs are very high, or where right-of-way width is limited, open channels are sometimes deeper than the hydraulic optimum.

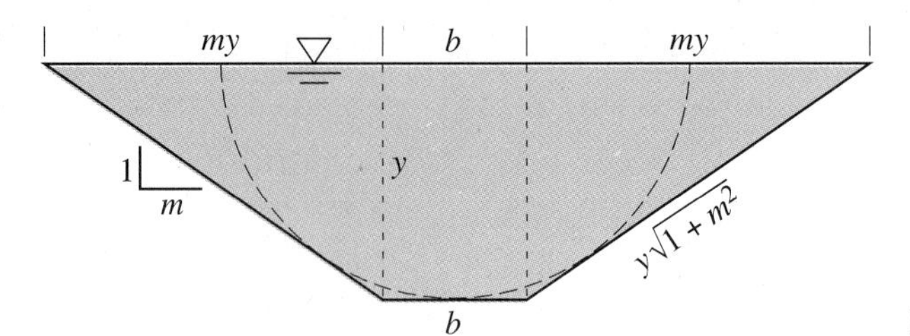

Figure 10.8

Of all geometric figures, the circle has the least perimeter for a given area. The hydraulic radius of a semicircle is the same as that of a circle. Hence a semicircular open channel will discharge more water than one of any other shape, assuming that the area, slope, and surface roughness are the same. Semicircular open channels are often built of pressed steel and other forms of metal, but for other types of construction such a shape is impractical. For wooden flumes the rectangular shape is usually employed. Canals excavated in earth usually have a trapezoidal cross section, with side slopes less than the ***angle of repose*** of the saturated bank material. Besides hydraulic efficiency, such factors as those mentioned help determine the best cross section.

The shape of the most efficient trapezoidal cross section can be determined by expressing the wetted perimeter P as a function of the section area A, the depth y, and the side slope. Consider the trapezoid of Fig. 10.8, with bed width b and side slope m:1. It follows that $A = (b + my)y$ and $P = b + 2y\sqrt{1+m^2}$. Eliminating b, we obtain

$$P = \frac{A}{y} + y(2\sqrt{1+m^2} - m)$$

and

$$R_h = \frac{A}{P} = \frac{A}{\dfrac{A}{y} + y(2\sqrt{1+m^2} - m)}$$

For a given A and m, we find $R_{h_{max}}$ by differentiating R_h with respect to y while holding A and m constant, and setting the result equal to zero. We find that then

$$\frac{A}{y} = y(2\sqrt{1+m^2} - m)$$

so that

$$P = 2y(2\sqrt{1+m^2} - m), \qquad b = 2y(\sqrt{1+m^2} - m)$$

$$y = \sqrt{\frac{A}{2\sqrt{1+m^2} - m}}$$

$$R_{h_{max}} = \frac{y}{2}$$

Of these results for any trapezoid, the last is easy to remember. It is also interesting (see Prob. 10.12) that the optimum trapezoid always envelopes a semicircle whose center is in the water surface (shown dashed in Fig. 10.8).

For the common rectangular cross section, this is a trapezoid with $m = 0$, so the above results then reduce to $A = 2y^2$, $P = 4y$, $b = 2y$, and $R_h = y/2$. Thus the most efficient rectangular section has its depth equal to half its width, and this of course still envelopes a semicircle.

The above results for trapezoidal cross sections are for given values of A and side slope m. The hydraulically best trapezoid can be found by expressing R_h as a function of only A and m, i.e.

$$R_h = \frac{y}{2} = \frac{1}{2}\sqrt{\frac{A}{2\sqrt{1+m^2}-m}}$$

and then differentiating R_h with respect to m while holding A constant. The result is that dR_h/dm is zero when $m = \infty$ or

$$\left(\frac{2m}{\sqrt{1+m^2}} - 1\right) = 0$$

Thus R_h is maximum when $m = 1/\sqrt{3}$, i.e., the side slopes are at 60° to the horizontal and the trapezoid is a half-hexagon. We note that this is the trapezoid that most closely envelopes the semicircle.

In a similar manner we can show that the most efficient triangular cross section is the one that has a total vertex angle of 90°.

Sample Problem 10.3 The first 95 miles of the Delta–Mendota Canal in California, designed to carry 4600 cfs, has the following dimensions (referring to Fig. 10.8): $b = 48$ ft, $m = 1.5$, $y = 16.7$ ft. Using the same cross sectional area and side slopes, how much would the most efficient cross section increase the present hydraulic radius and flow capacity? What would be the corresponding depth and bed width?

Solution

Existing:
$$A = (48 + 1.5\times 16.7)16.7 = 1220 \text{ ft}^2$$

$$P = 48 + 2\left(\frac{\sqrt{13}}{2}\right)16.7 = 108.2 \text{ ft}$$

so
$$R_h = \frac{A}{P} = \frac{1220}{108.2} = 11.27 \text{ ft}$$

which we note is considerably different from $y/2 = 8.35$ ft. For an optimum cross section:

$$A = 1220 = by + 1.5y^2 \tag{1}$$

$$(R_h)_{opt} = \frac{y}{2} = \frac{1220}{b + 2\left(\frac{\sqrt{13}}{2}\right)y} \tag{2}$$

Eliminating b between (1) and (2) yields $y^2 = 1220/(\sqrt{13} - 1.5)$

from which $$y_{opt} = 24.07 \text{ ft} \qquad \textbf{\textit{ANS}}$$

We note that this is 44% greater than the existing depth.

Also $$(R_h)_{opt} = \frac{y}{2} = 12.04 \text{ ft}$$

This is only 6.8% greater than the existing hydraulic radius. ***ANS***

From Manning's Eq. (10.8) for the same cross-sectional area:

$$\frac{Q_{opt}}{Q} = \left[\frac{(R_h)_{opt}}{R_h}\right]^{2/3} = \left(\frac{12.04}{11.27}\right)^{2/3} = 1.0450$$

Therefore Q_{opt} is 4.5% greater than the existing capacity. ***ANS***

Substituting y into (1): $$b_{opt} = 14.58 \text{ ft} \qquad \textbf{\textit{ANS}}$$

EXERCISES

10.6.1 What diameter of semicircular channel will have the same capacity as a rectangular channel of width 10 ft and depth 4 ft. Assume S and n are the same for both channels. Compare the lengths of the wetted perimeters.

10.6.2 A rectangular flume of planed timber ($n = 0.012$) slopes 0.5 ft per 1000 ft. (a) Compute the rate of discharge if the width is 7 ft and the depth of water is 3.5 ft. (b) What would be the rate of discharge if the width were 3.5 ft and the depth of water 7 ft? (c) Which of the two forms would have the greater capacity and which would require less lumber?

10.6.3 A rectangular flume of planed timber ($n = 0.012$) slopes 1 m per km. (a) Compute the rate of discharge if the width is 2 m and the depth of water is 1 m. (b) What would be the rate of discharge if the width were 1 m and the depth of water 2 m? (*c*) Which of the two forms would have the greater capacity and which would require less lumber?

10.7 CIRCULAR SECTIONS NOT FLOWING FULL

In circular pipes flow frequently occurs at partial depth, in which case they are behaving as open channels. As we can visualize from Fig. 10.9, at depths just slightly less than full depth the wetted perimeter (frictional resistance) is reduced in greater proportion than the flow area, so the hydraulic radius A/P is greater than when full. In accordance with the Manning formula, therefore, the maximum rate of discharge in such a section occurs at slightly less than full depth.

Such maximum condition can be explored analytically. Referring to Fig. 10.9,

$$y = 0.5D(1 - \cos\theta) = D\sin^2(\theta/2)$$

$$A = \frac{D^2}{4}(\theta - \sin\theta\cos\theta) = \frac{D^2}{4}(\theta - \tfrac{1}{2}\sin 2\theta)$$

$$P = D\theta$$

where θ is expressed in radians. This gives

$$R_h = \frac{A}{P} = \frac{D}{4}\left(1 - \frac{\sin\theta\cos\theta}{\theta}\right) = \frac{D}{4}\left(1 - \frac{\sin 2\theta}{2\theta}\right)$$

For the maximum rate of discharge, the Manning formula indicates that $AR_h^{2/3}$ must be a maximum. Substituting the preceding expressions for A and R_h into $AR_h^{2/3}$ and differentiating with respect to θ, setting equal to zero and solving for θ gives $\theta = 151.2°$, which corresponds to $y = 0.938D$ for the condition of maximum discharge. By differentiating $R_h^{2/3}$ the maximum velocity is found to occur at $0.813D$.

The above analyses of optimum conditions *are theoretical only*, because

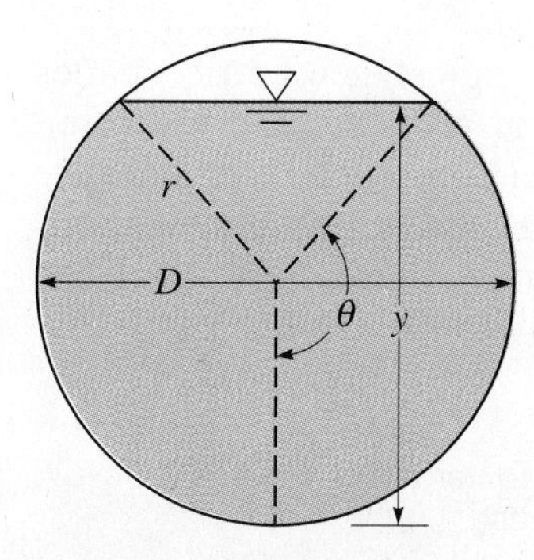

Figure 10.9
Circular section not full.

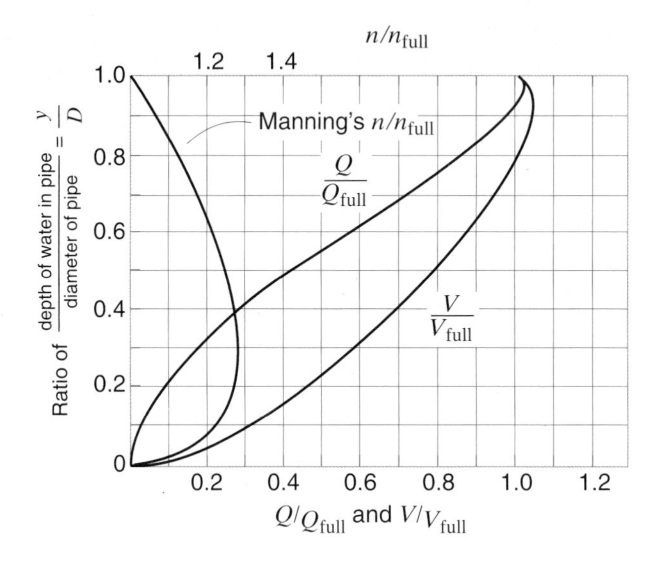

Figure 10.10
Hydraulic characteristics of circular pipe flowing partly full (n variable with depth).

they assume a constant roughness coefficient in Manning's equation, whereas in Sec. 10.2 we saw how, for constant e, the friction factor n varied with R_h. Actually in circular pipes the value of n has been shown to increase by as much as 28% from the full to about one-quarter full depth, where it appears to be a maximum. This effect causes the actual maximum discharge and velocity to occur at water depths of about 0.97 and 0.94 full depth, respectively as indicated in Fig. 10.10.[7] The interesting result is that in the small topmost depth range where the discharge is equal to or greater than Q_{full}, for a given discharge two different depths may occur.

The simplest way to handle the problem of a partially full circular section is to compute the velocity or flow rate for the pipe-full condition and then adjust to partly full conditions by using a chart such as Fig. 10.10. On this chart the effect of the variation of Manning's n with depth is already taken into consideration.

Despite the analysis just made of optimum conditions, circular sections are often designed to carry the design capacity when flowing full, since the conditions producing maximum flow frequently include sufficient backwater to place the conduit under slight pressure.

The rectangle, trapezoid, and circle are the simplest geometric shapes from the standpoint of hydraulics, but other forms of cross section are often used, either because they have certain advantages in construction or because they are desirable from other standpoints. Thus oval- or egg-shaped sections are common for sewers and similar channels where there may be large fluctuations in the rate of discharge. They are used because it is desirable that

[7] See Design and Construction of Urban Stormwater Management Systems, *ASCE Manuals and Reports on Engineering Practice No.* 77, p. 146, 1992.

the velocity, when a small quantity is flowing, be kept high enough to prevent the deposit of sediment, and when the conduit is full, the velocity should not be so high as to cause excessive wear of the pipe wall. For example, in sanitary sewers it is desirable to maintain a velocity of at least 2 fps to prevent deposition of sediment, while in concrete channels conveying storm water, velocities in excess of 10 or 12 fps may result in excessive abrasion of the channel sides and bottom.

Sample Problem 10.4 A 36-in-diameter pipe flows just full when it carries 30 cfs. What will be the discharge and depth when the velocity is 2 fps?

Solution

$$V_{\text{full}} = \frac{Q_{\text{full}}}{A} = \frac{30}{\pi 3^2/4} = 4.24 \text{ fps}$$

At 2 fps: $V/V_{\text{full}} = 2/4.24 = 0.471$

Fig. 10.10 for $V/V_{\text{full}} = 0.471$: $y/D \approx 0.20$

So $y \approx 0.20(36) = 7.2$ in ***ANS***

Fig. 10.10 for $y/D = 0.20$: $Q/Q_{\text{full}} \approx 0.077$

So $Q \approx 0.077(30) = 2.31$ cfs ***ANS***

EXERCISES

10.7.1 Water flows uniformly in a circular concrete pipe ($n = 0.014$) of diameter 14 ft at a depth of 6 ft. Using Fig. 10.10, determine the flow rate and the average velocity of flow. $S = 0.0004$.

10.7.2 At what depth will water flow at 0.30 m^3/s in a 120-cm-diameter concrete pipe on a slope of 0.003? (*a*) Assume $n = 0.013$. (*b*) Repeat with $n = 0.015$ and compare results.

10.8 LAMINAR FLOW IN OPEN CHANNELS

Laminar flow in open channels is sometimes encountered in industrial processes where a very viscous liquid is flowing in a trough or similar conveyance structure. More commonly, however, laminar flow occurs as ***sheet flow***, a thin sheet of flowing liquid, such as that in drainage from sidewalks, streets, and airport runways.

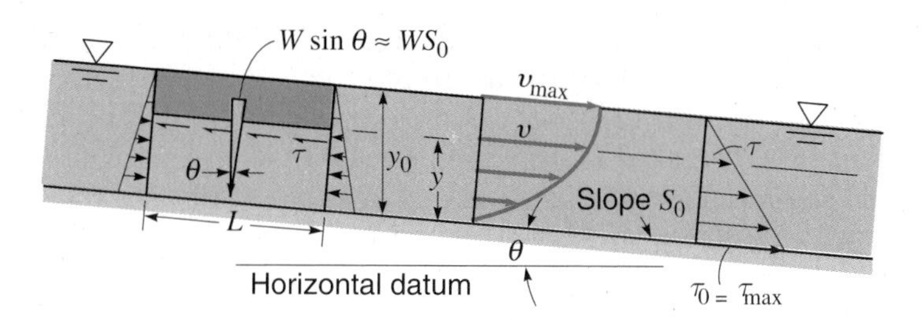

Figure 10.11
Laminar flow in open channel of infinite width and uniform depth, showing the velocity profile and the shear stress distribution.

Sheet flow is essentially two-dimensional, and can be analyzed in that fashion. Consider two-dimensional uniform laminar flow at depth y_0 as shown in Fig. 10.11. At a distance y above the channel bed the shear stress in the flow is τ and the velocity is v. The forces acting on a unit width of the liquid represented by the shaded area include the hydrostatic forces which cancel out, the gravity force[8] component parallel to the slope $\gamma(y_0 - y)LS$ and the shear force τL exerted along the lower boundary of the shaded liquid by the liquid below it. Since the flow is uniform, there is no acceleration and thus these two forces must balance one another. Hence

$$\gamma(y_0 - y)LS = \tau L$$

From this expression, it can be seen that the shear stress must vary linearly from zero at the liquid surface to a maximum value at the channel bed.

Since the flow is laminar, we can replace τ with $\mu\, dv/dy$ (Sec. 2.11). Making this substitution, we get

$$\gamma(y_0 - y)S = \mu \frac{dv}{dy}$$

Separating variables and integrating, noting that $v = 0$ when $y = 0$, gives

$$dv = \frac{\gamma}{\mu}(y_0 - y)S\, dy$$

and

$$v = \frac{\gamma}{\mu}\left(y_0 y - \frac{y^2}{2}\right)S = \frac{g}{\nu}\left(y_0 y - \frac{y^2}{2}\right)S \qquad (10.14)$$

This is the equation of a parabola; thus the velocity profile is parabolic, as it was for laminar flow in a pipe (Sec. 8.6).

[8] Depth y is always measured vertically. Thus $\gamma(y_0 - y)L$ is not precisely the weight of the shaded volume of liquid. However, for the small slopes usually encountered it is an excellent approximation.

We can now integrate Eq. (10.14) over the depth from $y = 0$ to $y = y_0$ to obtain an expression for q, the flow rate per unit width:

$$dq = \int_0^{y_0} v\, dy$$

$$q = \frac{g}{\nu}\frac{y_0^3}{3} S \tag{10.15}$$

From Eqs. (10.14) and (10.15), it can be shown that the average velocity V for this case is equal to $\frac{2}{3}v_{\max}$. In contrast, for laminar flow in a pipe flowing full it was shown (Sec. 8.6) that $V = \frac{1}{2}v_{\max}$. The total flow Q through any width B will be qB.

Equation (10.15) shows that if the flow is laminar, the flow rate is independent of the roughness. However, in situations where there are significant irregularities in the surface over which the liquid flows, the flow may be disturbed such that it is not everywhere laminar. Consequently Eqs. (10.14) and (10.15) must be applied with caution.

Sample Problem 10.5 Water is observed to flow uniformly across a glass roof at a depth of 1 mm. (*a*) If the temperature is 15°C and the slope of the roof is 0.08, what is the discharge rate from the edge of the roof per meter length? (*b*) How fast should we expect an oil mark on the water to move?

Solution

Table A.1 for 15°C: $\nu = 1.139 \times 10^{-6}\ \text{m}^2/\text{s}$

(*a*) Assuming laminar flow:

Eq. (10.15): $$q = \frac{9.81}{1.139 \times 10^{-6}}\frac{(0.001)^3}{3} 0.08 = 0.000\ 230\ \text{m}^3/\text{s per meter}$$

Since $b/y > 1/0.001 = 1000$, per Eq. (10.13): $R_h = y_0 = 0.001$ m

$$V = \frac{Q}{A} = \frac{q}{y_0} = \frac{0.000\ 230}{0.001} = 0.230\ \text{m/s}$$

Sec. 10.1: $$\mathbf{R} = \frac{R_h V}{\nu} = \frac{(0.001)0.230}{1.139 \times 10^{-6}} = 201$$

As $\mathbf{R} < \mathbf{R}_c = 500$, flow is laminar and Eqs. (10.14) and (10.15) apply. Therefore

$q = 0.000\ 230\ \text{m}^3/\text{s}$ per meter ***ANS***

(*b*) Laminar flow Eq. (10.14) at the surface, where $y = y_0$:

$$v = \frac{g}{\nu}\frac{y_0^2}{2} S = \frac{9.81}{1.139 \times 10^{-6}}\frac{(0.001)^2}{2} 0.08 = 0.345\ \text{m/s} \qquad \textbf{\textit{ANS}}$$

EXERCISES

10.8.1 Eastern lubricating oil (SAE 30) at 100°F flows down a flat plate 12 ft wide. (*a*) What is the maximum rate of discharge at which laminar flow may be ensured, assuming that the critical Reynolds number is 500? (*b*) What should be the slope of the plate to secure a depth of 8 in at this flow rate?

10.8.2 Eastern lubricating oil (SAE 30) at 30°C flows down a flat plate 3 m wide. (*a*) What is the maximum rate of discharge at which laminar flow may be ensured, assuming that the critical Reynolds number is 500? (*b*) What should be the slope of the plate to secure a depth of 15 cm at this flow rate?

10.8.3 At what rate (cfs/ft of width) will 70°F water flow in a wide, smooth, rectangular channel on a slope of 0.000 10 if the depth is 0.02 ft? Assume laminar flow and justify this assumption by computing Reynolds number.

10.8.4 At what rate (L/s per meter of width) will water at 20°C flow in a wide, smooth, rectangular channel on a slope of 0.0004, if the depth is 6.0 mm? Assume laminar flow and justify the assumption by computing Reynolds number.

10.9 SPECIFIC ENERGY AND ALTERNATE DEPTHS OF FLOW IN RECTANGULAR CHANNELS

For *any* cross-section shape, the ***specific energy*** E at a particular section is defined as the energy head referred to the channel bed as datum. Thus

$$E = y + \alpha \frac{V^2}{2g} \qquad (10.16)$$

where α is the kinetic energy correction factor (Sec. 5.1), which accounts for velocity variations across the section. Friction at the channel walls reduces velocities near the wetted perimeter, as indicated in Fig. 10.7 for which $\alpha = 1.105$. As noted in Sec. 10.4, the value of α is usually assumed to be unity; for typical velocities this results in only a small error in E.

For *rectangular* channels, provided they are not unusually narrow so that α is large, a representative average value of the flow q per unit width can be expressed as $q = Q/b$. The average velocity $V = Q/A = qb/by = q/y$ and so Eq. (10.16) with $\alpha = 1$ can be expressed as

$$E = y + \frac{1}{2g}\frac{q^2}{y^2} \qquad (10.17)$$

Let us consider how E will vary with y if q remains constant. Physically, this situation would occur if the slope of a rectangular channel could be changed with the flow rate remaining constant. Manning's equation [Eq. (10.8)] indicates that on a steep slope, with a given flow rate, the normal depth of flow y_0 will be relatively small in contrast to a larger depth on a flatter slope. That this is so can be seen more easily for a wide and shallow channel by writing the Manning equation in terms of the flow q per unit width, noting that $A = by$ and $R_h \approx y$ for $b \gg y$. Thus, for uniform flow $(y = y_0)$, in BG units

Wide and shallow, uniform flow:

$$q = \frac{Q}{b} = \frac{1.486}{nb} A R_h^{2/3} S_0^{1/2} = \frac{1.486}{n} \frac{by_0}{b} y_0^{2/3} S_0^{1/2} = \frac{1.486}{n} y_0^{5/3} S_0^{1/2} \quad (10.18)$$

For the case of constant q, Eq. (10.17) can be restated as follows:

$$(E - y)y^2 = \frac{q^2}{2g} = \text{constant} \quad (10.19)$$

A plot of E vs. y is hyperbola-like with asymptotes $(E - y) = 0$ (that is, $E = y$) and $y = 0$. Such a curve, shown in Fig. 10.12, is known as the ***specific energy diagram***. Actually each different value of q will give a different curve, as shown in Fig. 10.12. For a particular q, we see there are two possible values of y for a given value of E. These are known as ***alternate depths***. Equation (10.19) is a cubic equation with three roots, the third root being negative has no physical meaning. The two alternate depths represent two

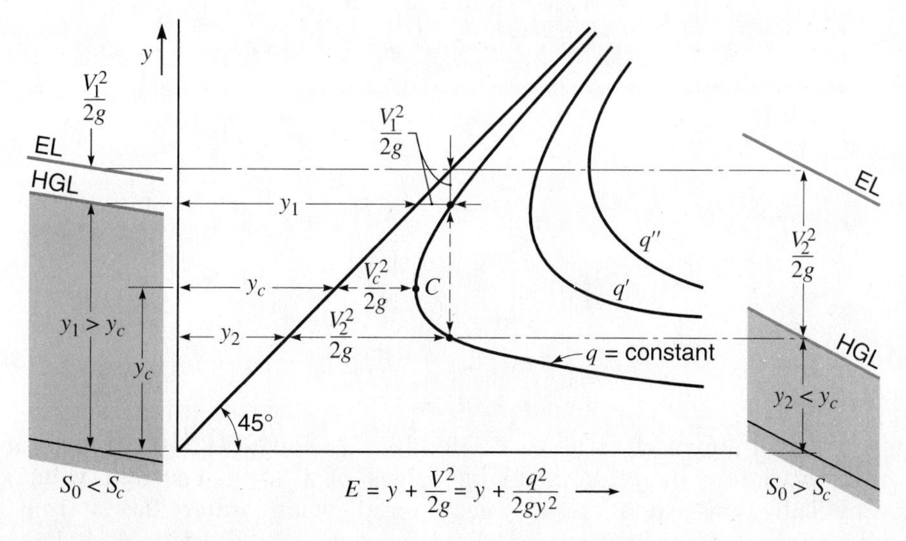

Figure 10.12
Specific-energy diagram for three constant rates of discharge in a rectangular channel. (Bed slopes are greatly exaggerated.)

totally different flow regimes—slow and deep on the upper limb of the curve and fast and shallow on the lower limb of the curve. Point C represents the dividing point between the two regimes of flow. At C, for a given q, the value of E *is a minimum* and the flow at this point is referred to as ***critical flow***. The depth of flow at that point is the ***critical depth*** y_c and the velocity is the ***critical velocity*** V_c. A relation for critical depth in a wide rectangular channel can be found by differentiating E of Eq. (10.17) with respect to y to find the value of y for which E is a minimum. Thus

$$\frac{dE}{dy} = 1 - \frac{q^2}{gy^3} \tag{10.20}$$

and when E is a minimum, $y = y_c$ and $dE/dy = 0$, so that

$$0 = 1 - \frac{q^2}{gy_c^{\,3}}, \qquad \text{or} \qquad q^2 = gy_c^{\,3} \tag{10.21}$$

Substituting $q = Vy = V_c y_c$ gives

$$V_c^2 = gy_c \qquad \text{and} \qquad V_c = \sqrt{gy_c} = \frac{q}{y_c} \tag{10.22}$$

where the subscript c indicates critical flow conditions (minimum specific energy for a given q).

Equation (10.22) may also be expressed as

$$y_c = \frac{V_c^2}{g} = \left(\frac{q^2}{g}\right)^{1/3} \tag{10.23}$$

From Eq. (10.22),

$$\frac{V_c^2}{2g} = \tfrac{1}{2}y_c \tag{10.24}$$

Hence

$$E_c = E_{\min} = y_c + \frac{V_c^2}{2g} = \tfrac{3}{2}y_c \tag{10.25}$$

and

$$y_c = \tfrac{2}{3}E_c = \tfrac{2}{3}E_{\min} \tag{10.26}$$

A different approach to alternate depths is to solve Eq. (10.19) for q and note the variation in q for changing values of y for a constant value of E. Physically this situation is encountered when water flows from a larger reservoir of constant surface elevation over a high, frictionless, broad-crested weir provided with a movable sluice gate near its downstream end

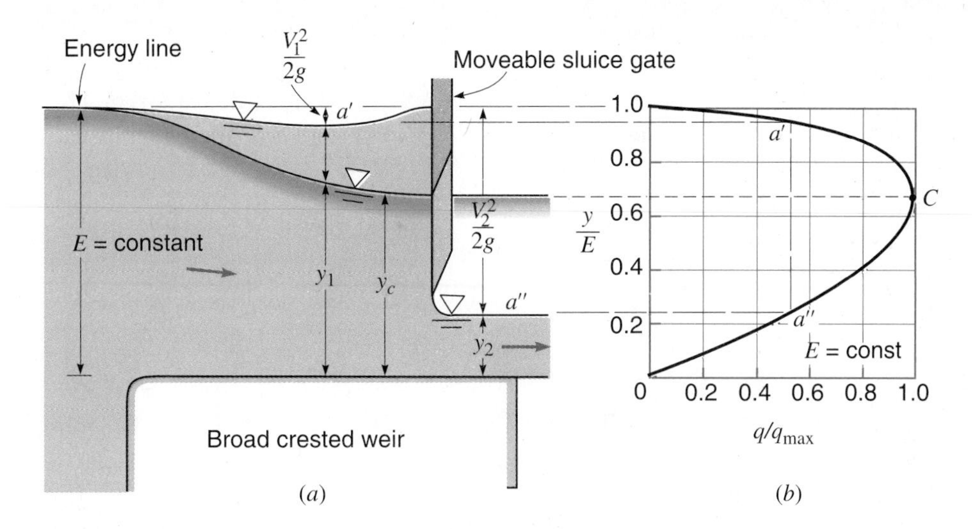

Figure 10.13
Variation of depth y and discharge per unit width q for constant specific energy E. (a) Side view of flow over a broad-crested weir. (b) The dimensionless discharge curve.

(Fig. 10.13a). As the gate is opened the flow rate increases until the opening becomes just large enough for critical depth to occur. With further opening of the gate, there is no increase in flow rate. As long as the water impinges on the gate (Fig. 10.13a), the flow is subcritical (a') upstream of the gate and supercritical (a'') downstream of the gate.

Rewriting Eq. (10.19) gives

$$q = y\sqrt{2g(E - y)} \tag{10.27}$$

This is the equation of the curve that is shown in dimensionless form in Fig. 10.13b. It is seen that maximum discharge for a given specific energy occurs when the depth is between $0.6E$ and $0.7E$. This may be established more exactly by differentiating Eq. (10.27) with respect to y and equating to zero. Thus

$$\frac{dq}{dy} = \sqrt{2g}\left(\sqrt{E - y} - \frac{1}{2}\frac{y}{\sqrt{E - y}}\right) = 0$$

from which

$$y_c = \tfrac{2}{3}E \tag{10.28}$$

where y_c is called the ***critical depth*** *for the given specific energy.* This equation is identical to Eq. (10.26). Thus there is a maximum value of q for a given E as indicated by point C of Fig. 10.13b. This curve is often referred to as the ***discharge curve***. The flow depicted by the upper limb of the curve has characteristics similar to those of the upper limb of that in Fig. 10.12. Similarly, the flow depicted by the lower limb of the curve has characteristics similar to the lower limb of that of Fig. 10.12. The significance of these two regimes of flow is discussed in Sec. 10.10. Point C of Fig. 10.13b represents critical flow conditions.

The discharge is a maximum when $y = y_c$, as indicated on the discharge curve of Fig. 10.13b. An expression for q_{max} may be obtained by substituting $E = 1.5y_c$ from Eq. (10.28) into Eq. (10.27), to obtain

$$q_{max} = \sqrt{g y_c^3} \tag{10.29}$$

In Fig. (10.12) we can see that when the flow is near critical, a small change in specific energy results in a large change in depth. Hence flow at or near critical depth is unstable and there will be an undulating water surface. Because of this it is undesirable to design channels to flow near critical depth.

We summarize the foregoing discussion as axioms of open-channel flow related to conditions at a given section in a *rectangular channel*:

1. A flow condition, i.e., a certain rate of discharge flowing at a certain depth, is completely specified by any two of the variables y, q, V, and E, except the combination q and E, which yields in general two alternate stages of flow.
2. For any value of E there exists a critical depth, given by Eq. (10.26), for which the flow is a maximum.
3. For any value of q there exists a critical depth, given by Eq. (10.23), for which the specific energy is a minimum.
4. When flow occurs at critical depth, both Eq. (10.22) and (10.26) are satisfied and the velocity head is one-half the depth.
5. For any flow condition other than critical, there exists an alternate stage at which the same rate of discharge is carried by the same specific energy. The alternate depth may be found from either the specific-energy diagram (Fig. 10.12) or the discharge curve (Fig. 10.13b), by extending a vertical line to the alternate limb of the curve. Analytically, the alternate depth is found by solving Eq. (10.17).

SAMPLE PROBLEM 10.6 Water in a rectangular channel flows 0.5 m deep at a velocity of 6 m/s. Find (a) the alternate depth, (b) the critical depth for this discharge, and (c) the percentage by which the specific energy exceeds the minimum.

: Programmed computing aids (Appx. B) could help solve problems marked with this icon.

Solution

(*a*) Eq. (10.16):
$$E = 0.5 + \frac{6^2}{2g} = 2.33 \text{ m}$$

$$q = Vy = 6(0.5) = 3 \text{ m}^3\text{/s per m}$$

Eq. (10.17):
$$2.33 = y + \frac{3^2}{2(9.81)y^2} \qquad (1)$$

i.e.,
$$y^3 - 2.33y^2 + 0.459 = 0$$

Knowing that 0.5 m is one root of this cubic equation, we may divide through by $(y - 0.5)$ to obtain the quadratic

$$y^2 - 1.835y - 0.917 = 0$$

which yields $\quad y = -0.409$ m (meaningless) or 2.24 m $\qquad$ ***ANS***

(*b*) Eq. (10.23):
$$y_c = \left(\frac{3^2}{9.81}\right)^{1/3} = 0.972 \text{ m} \qquad \textit{ANS}$$

(*c*) Eq. (10.25): $\quad E_c = E_{\min} = 1.5y_c = 1.5(0.972) = 1.458$ m

$$\frac{E}{E_{\min}} = \frac{2.33}{1.458} = 1.602, \qquad \text{so exceedance} = 60.2\% \qquad \textit{ANS}$$

Note: A popular alternative manual approach is to solve (1) by trial and error (see Sample Prob. 3.5); values of y are tried until the right side equals the left. As there are two meaningful roots, helpful guidance will be provided by first finding y_c and $E_{\min}$ and making a sketch of the specific energy diagram. Some programmable scientific calculators have equation solvers, which in effect automate the trial and error. Other calculators have cubic equation solvers that deliver the three roots. Mathematics software packages can also conveniently solve such equations.

EXERCISES

10.9.1 A long straight rectangular channel 10 ft wide is observed to have a wavy water surface at a depth of about 6 ft. Estimate the rate of discharge.

10.9.2 A long straight rectangular channel 5 m wide is observed to have a wavy water surface at a depth of about 3 m. Estimate the rate of discharge.

10.9.3 Water flows down a wide and shallow rectangular channel of concrete ($n = 0.015$) laid on a slope of 0.004. Find the depth and rate of flow in SI units for critical conditions in this channel.

10.10 SUBCRITICAL AND SUPERCRITICAL FLOW

In Sec. 10.9 we referred to the upper and lower portions of the specific-energy diagram (Fig. 10.12) and the discharge curve (Fig. 10.13). The upper limb of those curves, where velocities are less than critical, represent ***subcritical*** (also known as tranquil or upper-stage) flow, while the lower limb of the curves where velocities are greater than critical represent ***supercritical*** (also known as rapid or lower-stage) flow. We discussed how to identify the critical point separating these two portions in Sec. 10.9.

The slope required to give uniform flow at critical depth ($y_0 = y_c$) for a given discharge is known as the ***critical slope*** S_c. Note that S_c varies with discharge. We obtain an expression for the critical slope of a *wide and shallow* rectangular channel ($R_h = y$) when we combine Eq. (10.23) for critical flow with Eq. (10.18) for uniform flow, eliminating q as follows:

Wide and shallow, BG units:
$$S_c = \left(\frac{n}{1.486}\right)^2 \frac{g}{y_c^{1/3}} \qquad (10.30)$$

Substituting for y_c from Eq. (10.23), this can also be written as

Wide and shallow, BG units:
$$S_c = \left(\frac{n}{1.486}\right)^2 \frac{g^{10/9}}{q^{2/9}} \qquad (10.31)$$

When using SI units, we simply replace the 1.486 by 1 in these equations.

For channels with cross sections other than wide and shallow, an expression for the critical slope is more complex than the above. However, we seldom need to calculate S_c because, as we shall see, we can more easily determine how S_0 compares with S_c by comparing y_0 with y_c.

If the bed slope $S_0 > S_c$, the slope is known as a ***steep slope*** for the given discharge. Normal depth for uniform flow on such a slope will be less than critical depth and hence normal flow will be supercritical. In contrast, if $S_0 < S_c$, the normal depth will be greater than critical and normal flow is subcritical. Such a slope is referred to as a ***mild slope.*** By referring to Eq. (10.30), we see that the hydraulic steepness of a channel slope is determined by more than its elevation gradient. A steep slope for a channel with a smooth lining could be a mild slope for the same flow with a rough lining. Even for a given channel with a given boundary roughness, the slope may be mild for a low rate of discharge and steep for a higher one.

It may be recalled that the Froude number (Sec. 7.4) is defined as $V/\sqrt{gL}$. If for rectangular channels the depth of flow is used to represent the significant length parameter in the Froude number (that is, $\mathbf{F} = V/\sqrt{gy}$), we find by comparing this with Eq. (10.22) that the flow is critical if $\mathbf{F} = 1.0$, the flow is subcritical if $\mathbf{F} < 1.0$, and the flow is supercritcal if $\mathbf{F} > 1.0$.

Another convenient way to determine the type of flow in rectangular channels is to compare the velocity head with half the depth. Figure 10.12 clearly indicates that if $V^2/2g > y/2$ the flow is supercritical, and if $V^2/2g < y/2$

Table 10.2
Characteristics of subcritical, critical, and supercritical flow in rectangular channels

Characteristic	Subcritical	Critical	Supercritical
Depth of flow, y	$y > y_c$	$y = y_c = \left(\frac{q^2}{g}\right)^{1/3}$	$y < y_c$
Velocity of flow, V	$V < V_c$	$V = V_c = \sqrt{gy}$	$V > V_c$
Slope for uniform flow, S_0	Mild slope $S_0 < S_c$	Critical slope $S_0 = S_c$ [Eq. (10.30) if wide and shallow]	Steep slope $S_0 > S_c$
Froude number, $\mathbf{F} = \frac{V}{\sqrt{gy}} = \frac{q}{\sqrt{gy^3}}$	$\mathbf{F} < 1.0$	$\mathbf{F} = 1.0$	$\mathbf{F} > 1.0$
Disturbance waves (Sec. 10.20)	Will propagate in all directions	Will hold fast, not propagate upstream	Will form standing wave with $\sin \beta = c/V$ downstream only
Velocity head compared with half-depth	$\frac{V^2}{2g} < \frac{y}{2}$	$\frac{V^2}{2g} = \frac{y}{2}$	$\frac{V^2}{2g} > \frac{y}{2}$
Can be followed by a hydraulic jump? (Sec. 10.16)	No	No	Yes

the flow is subcritical. A brief summary of some of the characteristics of subcritical, critical, and supercritical flow in rectangular channels is given in Table 10.2.

EXERCISE

10.10.1 A rectangular channel 50 ft wide ($n = 0.015$) carries a flow of 375 cfs. For this flow, find the critical depth, the critical velocity, and the critical bed slope.

10.11 CRITICAL DEPTH IN NONRECTANGULAR CHANNELS

For nonrectangular channels the Froude number is defined by $\mathbf{F} = V/\sqrt{gy_h}$ where y_h *is the* ***hydraulic depth*** (or ***hydraulic mean depth***) defined by

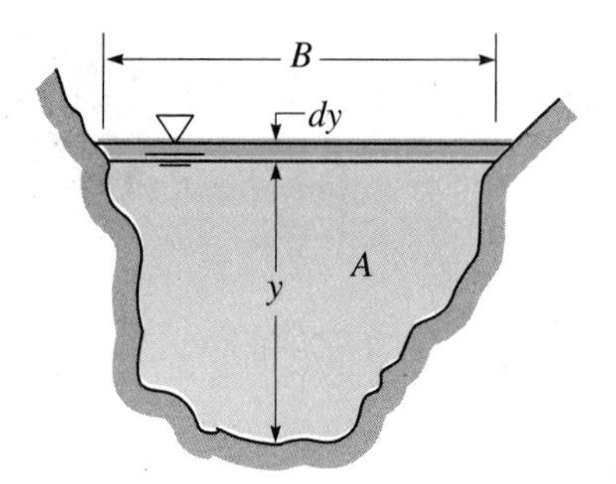

Figure 10.14

$y_h = A/B$ where A is the cross-sectional flow area and B is the width of the flow area at the water surface. The expression just given for Froude number is the same as $\mathbf{F} = Q\sqrt{B/(gA^3)}$.

Let us now consider an irregularly shaped flow section (Fig. 10.14) of area A carrying a flow Q. Thus Eq. (10.17) becomes

$$E = y + \frac{Q^2}{2gA^2} \tag{10.32}$$

Differentiating with respect to y and setting to zero to find y_c, that is, the value of y for which E is a minimum,

$$\frac{dE}{dy} = 1 - \frac{Q^2}{2g}\left(\frac{2}{A^3}\frac{dA}{dy}\right) = 0$$

As A may or may not be a reasonable function of y, it is helpful to observe that $dA = B\,dy$, and thus $dA/dy = B$, the ***width of the water surface***. Substituting this in the above expression results in

$$\frac{Q^2}{g} = \left(\frac{A^3}{B}\right)_{y = y_c} \tag{10.33}$$

as the equation that must be satisfied for critical flow. For a given cross section, the right-hand side is a function of y only. A trial-and-error solution is generally required to find y_c, which is the value of y that causes A and B to satisfy Eq. (10.33). When found, A and B may be named A_c and B_c. It is easy to confirm that Eq. (10.33) causes $\mathbf{F} = 1$.

We may next solve for V_c, the critical velocity in the irregular channel, by observing that $Q = A_c V_e$. Substituting this in Eq. (10.33) yields

$$\frac{V_c^2}{g} = \frac{A_c}{B_c}, \quad \text{or} \quad V_c = \sqrt{\frac{gA_c}{B_c}} = \sqrt{g(y_h)_c} \tag{10.34}$$

If the channel is rectangular, $A_c = B_c y_c$, and the above is seen to reduce to Eq. (10.22); also, then $Q = B_c q$, and Eq. (10.33) reduces to Eq. (10.23).

The cross sections most commonly encountered in open-channel hydraulics are not rectangular but are ***trapezoidal*** or part-full circular sections. The latter have been discussed in Sec. 10.17. For the very common trapezoidal cross section, practicing hydraulic engineers in the past made use of numerous tables, curves, and computer programs to avoid tedious repeated trial-and-error solutions of Eq. (10.33) for y_c. Now it is more convenient to replace A and B by the appropriate functions of y_c, b, and m (defined as in Fig. 10.8), and to solve the resulting equation

$$\frac{Q^2}{g} = \frac{(by_c + my_c^2)^3}{b + 2my_c}$$

using an equation solver in a programmable scientific calculator or in a mathematics software package.

SAMPLE PROBLEM 10.7 In Fig. S10.7 water flows uniformly at a steady rate of 14.0 cfs in a very long triangular flume that has side slopes 1:1. The flume is on a slope of 0.006, and $n = 0.012$. (*a*) Is the flow subcritical or supercritical? (*b*) Find the relation between $V_c^2/2g$ and y_c for this channel.

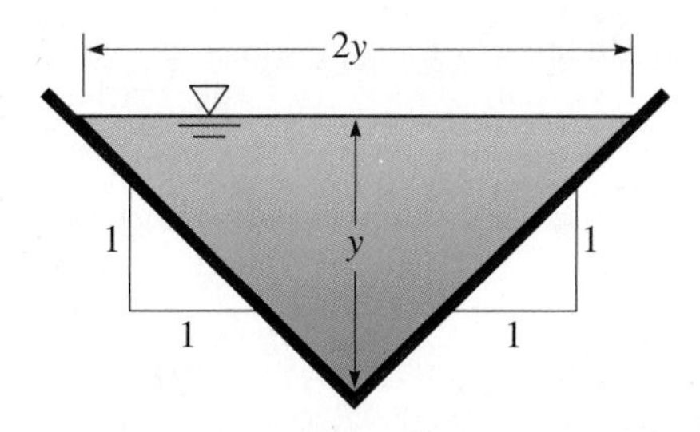

Figure S10.7

Solution

(*a*)

$$A = \tfrac{1}{2}(2y)y = y^2$$

$$P = 2\sqrt{2}y = 2.83y$$

$$R_h = \frac{A}{P} = 0.354y$$

Eq. (10.8*a*) for uniform flow: $$14 = \frac{1.486}{0.012} y_0^2(0.354y_0)^{2/3}(0.006)^{1/2}$$

from which $y_0 = 1.494$ ft. When $y = y_c$, from Eq. (10.33):

$$\frac{(14)^2}{32.2} = \frac{(y_c^2)^3}{2y_c}$$

and $$y_c = 1.648 \text{ ft}$$

Since y_c is greater than uniform flow depth, the flow is supercritical. ***ANS***

Note: If the data in this problem had been given in SI units rather than in BG units, the procedure for solution would have been the same except that Eq. (10.8*b*) would have been used instead of Eq. (10.8*a*).

(*b*) Eq. (10.34):
$$\frac{V_c^2}{g} = \frac{A_c}{B_c} = \frac{y^2}{2y} = \frac{y}{2}$$

so
$$\frac{V_c^2}{2g} = \frac{y_c}{4} \qquad \textbf{\textit{ANS}}$$

Consequently we see that the relation between $V^2/2g$ and y for critical flow conditions depends on the geometry of the flow section. If the vertex angle of the triangle had been different from 90°, the relation would have been different.

EXERCISES

10.11.1 A flow of 12 cfs of water is carried in a 90° triangular flume built of planed timber ($n = 0.011$). Find the critical depth and the critical slope.

10.11.2 A flow of 0.28 m^3/s of water is carried in a 90° triangular flume built of planed timber ($n = 0.011$). Find the critical depth and the critical slope.

10.11.3 A circular conduit of well-laid (smooth) brickwork when flowing half full is to carry 320 cfs at a velocity of 8 fps. (*a*) What will be the necessary fall per mile? (*b*) Will the flow be subcritical or supercritical?

10.11.4 A trapezoidal channel has a 4-m-wide bed and $m = 2$ (i.e., $2H{:}1V$). What is the critical flow depth when it is carrying 85 m^3/s?

10.12 OCCURRENCE OF CRITICAL DEPTH

When flow changes from subcritical to supercritical or vice versa, the depth must pass through critical depth. In the former, the phenonemon gives rise to what is known as a ***control section***. In the latter a hydraulc jump (Sec. 10.18) usually occurs. In Fig. 10.15 is depicted a situation where the flow changes from subcritical to supercritical. Upstream of the break in slope there is a ***mild*** slope, the flow is subcritical, and $y_{0_1} > y_c$. Downstream of the break there is a ***steep*** slope, the flow is supercritical, and $y_{0_2} < y_c$. At the break in slope the depth passes through critical depth. This point in the stream is referred to as a control section since the depth at the break controls the depth upstream. A similar situation occurs when water from a reservoir enters a canal in which the uniform depth is less than critical. In such an instance (Fig. 10.16), the depth passes through critical depth in the vicinity of the entrance to the

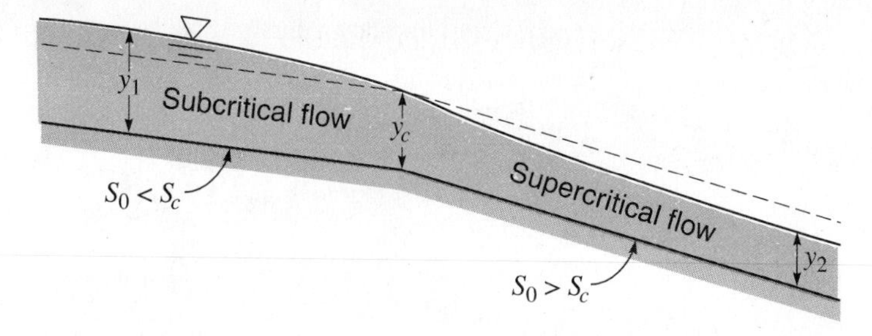

Figure 10.15
Change in flow from subcritical to supercritical at a break in slope.

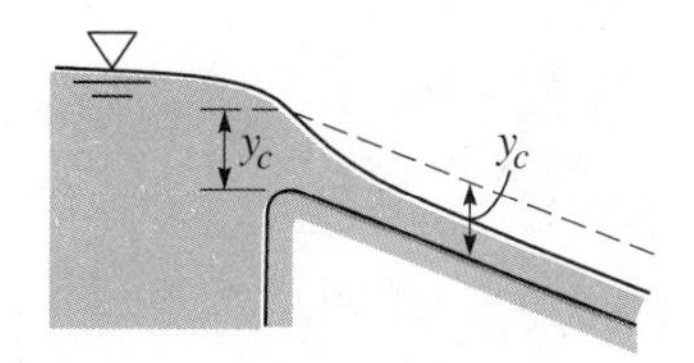

Figure 10.16
Hydraulic drop entering a steep slope.

canal. Once again, this section is known as a control section. By measuring the depth at a control section, one can compute a reasonably accurate value of Q by application of Eq. (10.23) for rectangular channels or Eq. (10.33) for nonrectangular channels.

At a control section there is a specific relationship between the stage and the discharge, regardless of the channel roughness and slope. This indicates that such locations are desirable sites for gaging stations. It also means that when the upstream or downstream flow conditions change, the transmission of such changes through a control section are restricted. The student should note that not only critical depth creates a control section; control structures, such as sluice gates and orifices, may also be used to create control sections.

Another instance where critical depth occurs is that of a free outfall (Fig. 10.17*a*) with subcritical flow in the channel prior to the outfall. Since friction produces a constant diminution in energy in the direction of flow, it is obvious that at the point of outfall the total energy must be less than at any point upstream. As critical depth is the value for which the specific energy is a minimum, one would expect critical depth to occur at the outfall. However, the value for the critical depth is derived on the assumption that the water is flowing in straight lines. In the free outfall, gravity creates a curvature of the streamlines, with the result that the depth at the brink is less than critical depth. It has been found by experiment that the depth at the brink $y_b \approx 0.72y_c$. Also, critical depth generally occurs upstream of the brink a distance of somewhere between $3y_c$ and $10y_c$. If the flow is supercritical, there is no drop-down curve (Fig. 10.17*b*).

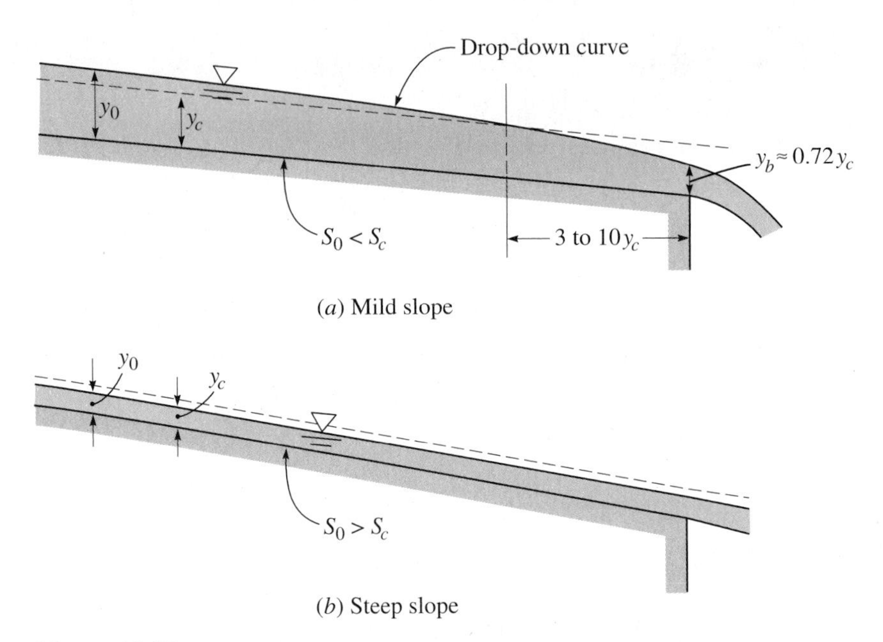

Figure 10.17
Free outfall.

Critical depth *may* occur in a channel if the bottom is humped or if the sidewalls are moved in to form a contracted section. However, in such cases critical depth will not always occur (Sample Prob. 10.8).

10.13 HUMPS AND CONTRACTIONS

Let us examine the case of a hump in a rectangular channel (Fig. 10.18). We shall assume the hump is streamlined and neglect head loss. In Fig. 10.18*a* a

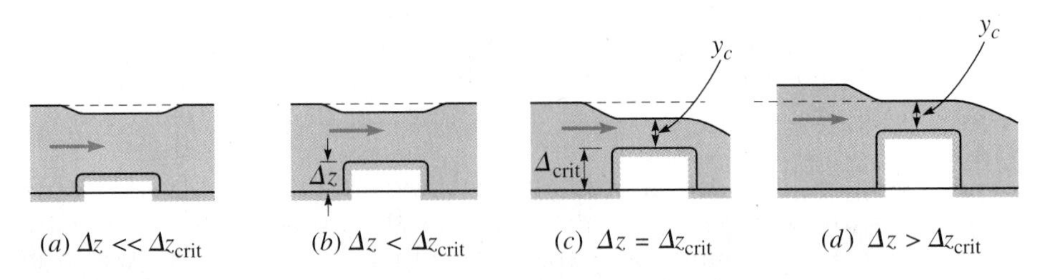

Figure 10.18
Subcritical flow over a hump in a rectangular channel. Flow rate is the same in all four cases. Δz = hump height; Δz_{crit} = critical hump height.

small hump is shown. If the flow upstream of the hump is *subcritical,* there will be a slight depression in the water surface over the hump. Inspection of the subcritical (upper) portion of the specific-energy diagram (Fig. 10.12) confirms this statement. The hump (locally raised bed) causes a drop in specific energy E, since the energy line is unchanged. A decrease in E with unchanged q is accompanied by a decrease in y that exceeds the decrease in E, so there must be a depression in the water surface over the hump. If the height of the hump Δz is increased with no change in q, there will be a further drop in the water surface over the hump (Fig. 10.18*b*) for the same reason. Further increases in hump height create further depression of the water surface over the hump until finally the depth on the hump becomes critical (Fig. 10.18*c*). The minimum height of the hump that causes critical depth is here referred to as the ***critical hump height***, Δz_{crit}. If the hump is made still higher (Fig. 10.18*d*), critical depth is maintained on the hump and the depth upstream of the hump increases until it gains sufficient energy to be able to flow over at the same rate. This phenonemon is referred to as ***damming action***. With damming action, immediately downstream of y_c the flow must be supercritical (to comply with Fig. 10.20). In such a situation, the hump, if it has a flat top, becomes a ***broad-crested weir*** (Sec. 11.12).

Let us next consider contractions in rectangular channels. We shall again assume them to be streamlined so that we can neglect any head loss they may cause. Contractions change q, the flow per unit width, and we note that curves for increasing q move to the right on Fig. 10.12. So it follows from Fig. 10.12 that a contraction, which causes q to increase, has a similar effect on y as does a hump, which causes E to decrease. If the approaching flow is subcritical, a small contraction will cause a slight depression in the water surface through the contraction. Further contraction creates further depression in the water surface until critical depth occurs in the contraction. This happens when point C on Fig. 10.12 has energy E. Upon further contraction, the depth remains critical in the contracted section. However, unlike for humps, the depth in the contracted section continues to increase. This is because the critical depth there depends upon q via Eq. (10.23), and q increases with contraction. Figure 10.12 indicates that further contraction beyond the onset of critical depth requires the flow to have extra energy, and this is again provided by damming action, causing the upstream water surface to rise. The maximum width (i.e., minimum contraction) at which critical depth occurs in the contracted section is referred to here as the ***critical contracted width***.

With *supercritical* flow, humps and contractions behave differently than they do with subcritical flow according to the trends indicated by the supercritical (lower) limbs of Figs. 10.12 and 10.13. Thus, if the flow is supercritical, the water depth at the hump or contraction will *increase* as the hump height is made larger or as the amount of contraction is increased until critical depth is reached. Beyond that, further increases in hump height or increases in the amount of contraction cause damming action.

A hump may be combined with a contraction to give a section not unlike a venturi meter (Sec. 11.7). The same principles discussed above for the hump and the contraction may be applied for such a situation. It is important to note that the hump changes E and the contraction changes q in the *single* equation (10.17) to be solved; using a separate equation for each effect would be incorrect. Accompanying such calculations by a labeled sketch of the specific energy diagram is extremely helpful.

SAMPLE PROBLEM 10.8 In Fig. S10.8 uniform flow of water occurs at 135 cfs in a 20-ft-wide rectangular channel at a depth of 2.00 ft. (*a*) Is the flow subcritical or supercritical? (*b*) If a hump of height $\Delta z = 0.30$ ft is placed in the bottom of the channel, calculate the water depth on the hump, and the change in the water surface level at the hump. (*c*) If the hump height is raised to $\Delta z = 0.60$ ft, what then are the water depths upstream and downstream of the hump? (*d*) If the 0.30-ft hump is accompanied by a local contraction to 18 ft, find the water depth on the hump. In all cases neglect head losses over the hump and through the contraction.

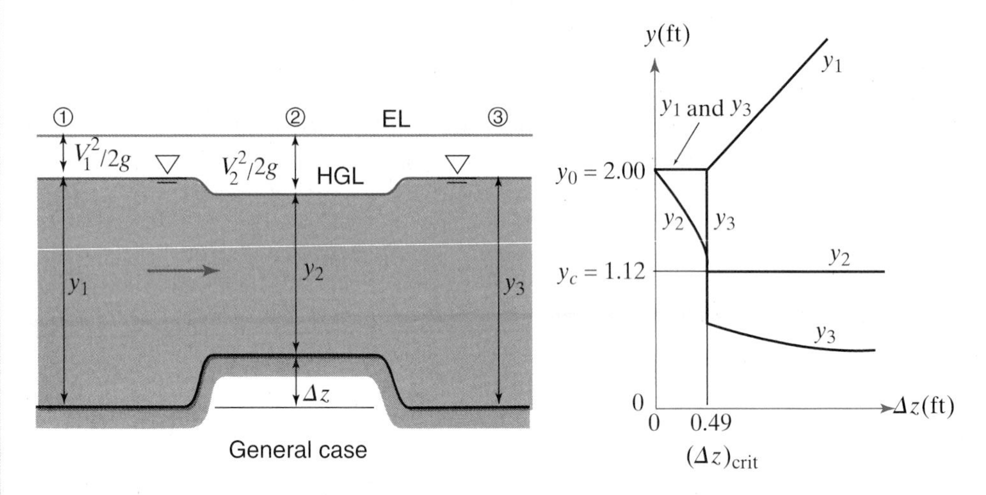

Figure S10.8

Solution

(*a*)

$$q = 135/20 = 6.75 \text{ cfs/ft}$$

Eq. (10.23):

$$y_c = \left(\frac{q^2}{g}\right)^{1/3} = \left[\frac{(6.75)^2}{32.2}\right]^{1/3} = 1.123 \text{ ft}$$

Since the normal depth (2.00 ft) is greater than the critical depth, the flow is subcritical and the channel slope is mild. ***ANS***

▭ : Programmed computing aids (Appx. B) could help solve problems marked with this icon.

(b) First find the critical hump height. Write the energy equation between sections 1 and 2, assuming critical flow on the hump.

Eq. (10.17): $$E_1 = 2.00 + \frac{1}{2(32.2)}\left(\frac{6.75}{2.00}\right)^2 = 2.18 \text{ ft}$$

Energy: $$E_1 = (\Delta z)_{crit} + (E_{min})_2$$

Using Eq. (10.25): $$2.18 = (\Delta z)_{crit} + 1.5(1.123)$$

$$(\Delta z)_{crit} = 0.493 \text{ ft}$$

Thus the minimum-height hump that will produce critical depth on the hump is 0.493 ft.

Since the actual hump height, $\Delta z = 0.30$ ft, is less than the critical hump height, 0.493 ft, critical flow does not occur on the hump and there is no damming action.

Let us now find the depth y_2 on the hump.

Energy from section 1 to section 2:

$$2.18 = 0.30 + y_2 + \frac{1}{2(32.2)}\left(\frac{6.75}{y_2}\right)^2$$

We may solve this cubic equation by trial and error (see Sample Prob. 3.5) or by using an equation solver (see Sec. 8.14 and/or Sample Prob. 10.1). We obtain three roots, 1.601 ft, 0.817 ft, and a negative answer (−0.541 ft) that has no physical meaning. Since $\Delta z < (\Delta z)_{crit}$, the flow on the hump must be subcritical; that is, $y_2 > y_c$. Hence $y_2 = 1.601$ ft. ***ANS***
The change in the water surface over the hump = 2.00 − (0.30 + 1.60) = 0.10 ft drop. ***ANS***

(c) In this case the hump height $\Delta z = 0.60$ ft, which is greater than the critical hump height. Hence there will be damming action and critical depth (y_c = 1.123 ft) will occur on the hump. Writing the energy equation for this case, from section 1 to section 2:

$$y_1 + \frac{1}{2(32.2)}\left(\frac{6.75}{y_1}\right)^2 = 0.60 + 1.5(1.123) = 2.28$$

Solving this cubic for y_1 gives three roots, 2.13 ft, 0.660 ft, and a negative answer that has no physical meaning. In this case damming action occurs and the depth y_1 upstream of the hump is 2.13 ft. On the hump the depth passes through critical depth of 1.123 ft and just downstream of the hump the depth will be 0.66 ft. ***ANS***

The depth will then increase in the downstream direction following an M_3 water surface profile (Fig. 10.20) until a hydraulic jump (Sec. 10.18) occurs to return the depth to the normal uniform flow depth of 2.00 ft. The depth just downstream of the hump cannot be 2.13 ft, because then there would be no way for the flow to dissipate its extra energy and return to normal depth; this is explained in Sec. 10.16 and Fig. 10.20.

This higher type of hump is commonly built of concrete and is used for flow measurement purposes. Such structures are discussed under broad-crested weirs, in Sec. 11.12. The right-hand sketch of this example shows the relation between the hump height Δz, and the water depths y_1, y_2, and y_3 for the condition where $Q = 135$ cfs.

(*d*) At section 2:

$$q = 135/18 = 7.50 \text{ cfs/ft}$$

$$y_{c2} = \left(\frac{7.50^2}{32.2}\right)^{1/3} = 1.204 \text{ ft}$$

$$(E_{\min})_2 = 1.5(1.204) = 1.807 \text{ ft}$$

$$(\Delta z)_{\text{crit}} = E_1 - (E_{\min})_2 = 2.18 - 1.807 = 0.37 \text{ ft}$$

So Δz is less than $(\Delta z)_{\text{crit}}$; therefore the combination of hump and contraction will not cause critical depth or damming action.
Energy from section 1 to section 2:

$$E_2 = E_1 - \Delta z = 2.18 - 0.30 = 1.88 \text{ ft}$$

Eq. (10.17):

$$1.88 = y_2 + \frac{1}{2(32.2)}\left(\frac{7.50}{y_2}\right)^2$$

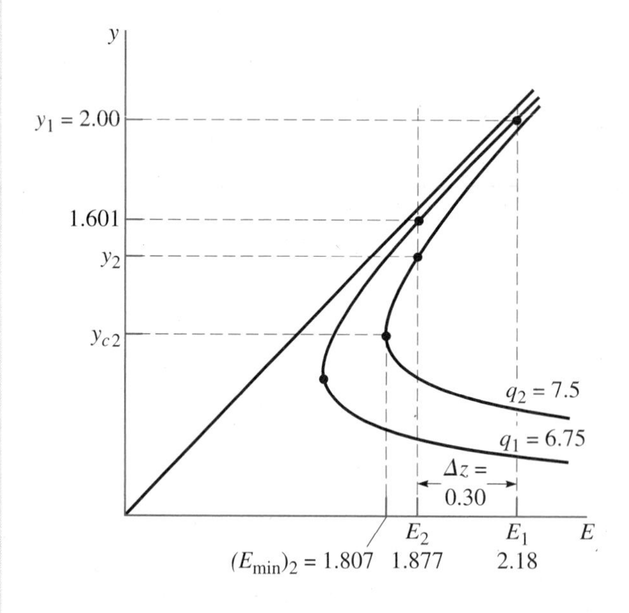

From the sketch of the specific energy diagram, we expect $1.204 < y_2 < 1.601$ ft. The sketch helps direct manual trials.
By trials or by equation solver, $y_2 = 1.48$ ft. ***ANS***

An unnecessary difficulty sometimes baffles students trying to solve problems like part (*c*) of the preceding example. It happens if they forget to first calculate $(\Delta z)_{\text{crit}}$, and forget that damming action may have occurred. When they try to solve for the water depth on the hump, they cannot get a solution, or get only negative and imaginary roots. This is because without damming action E on the hump is less than $E_{\min}$ (see Fig. 10.12), and they are trying to solve an impossible situation.

If the flow had been supercritical rather than subcritical in the preceding example, a reverse situation would have occurred. With supercritical flow, the depth on a low hump will be slightly greater than the depth upstream of the hump. As the hump height is increased, the depth on the hump will increase until critical depth is reached. With further increase in the hump height, critical depth will prevail on the hump and damming action will occur upstream. That this is so can again can be reasoned out by referring to Fig. 10.12.

Similar logic can be applied to solve problems involving depressions in the channel bed (negative humps) and channel expansions (negative contractions). Of course opposite effects are produced to those that occur with positive humps and contractions.

EXERCISES

10.13.1 A rectangular channel 3 m wide carries 4 m^3/s of water in subcritical uniform flow at a depth of 1.2 m. A frictionless hump is to be installed across the bed. Find the critical hump height (i.e., the minimum hump which causes y_c on it).

10.13.2 Suppose that the depth of uniform flow in a 4-ft-wide ractangular channel is 1.10 ft. Find the change in water-surface elevation caused by a 1-ft-wide bridge pier placed in the middle of the channel. The flow rate is 50 cfs.

10.13.3 Suppose that the depth of uniform flow in a 1.2-m-wide rectangular channel is 27 cm. Find the change in water-surface elevation caused by a 30-cm-wide bridge pier placed in the middle of the channel. The flow rate is 1.1 m^3/s.

10.13.4 A rectangular channel 12 ft wide carries 24 cfs in uniform flow at a depth of 0.80 ft. Find the local change in water-surface elevation caused by a hump 0.12 ft high across the floor of the channel.

: Programmed computing aids (Appx. B) could help solve problems marked with this icon.

10.14 NONUNIFORM, OR VARIED, FLOW

As a rule, uniform flow is found only in artificial channels of constant shape and slope, although even under these conditions the flow for some distance may be nonuniform, as shown in Fig. 10.1. But with a natural stream, the slope of the bed and the shape and size of the cross section usually vary to such an extent that true uniform flow is rare. Hence the application of the equations given in Sec. 10.2 to natural streams can be expected to yield results that are only approximations to the truth. In order to apply these equations at all, the stream must be divided into lengths (reaches) within which the conditions are approximately the same.

In the case of artificial channels that are free from the irregularities found in natural streams, it is possible to apply analytical methods to the various problems of nonuniform flow. In many instances, however, particularly when dealing with natural channels, we must resort to trial solutions and graphical methods.[9]

In the case of pressure conduits, we have dealt with uniform and nonuniform flow without drawing much distinction between them. This can be done because in a closed pipe the area of the water section, and hence the mean velocity, is fixed at every point. But in an open channel these conditions are not fixed, and the stream adjusts itself to the size of cross section that the energy gradient (i.e., slope of the energy line) requires.

In an open stream on a falling grade, without friction, the effect of gravity is to tend to produce a flow with a continually increasing velocity along the path, as in the case of a freely falling body. However, the gravity force is opposed by frictional resistance. The frictional force increases with velocity, while gravity is constant; so eventually the two will be in balance, and uniform flow will occur. When the two forces are not in balance, the flow is nonuniform.

There are two types of nonuniform flow. In one the changing conditions extend over a long distance, and this may be called ***gradually varied flow***. In the other the change may take place very abruptly, and the transition is thus confined to a short distance. This may be designated as a ***local nonuniform phenomenon*** or ***rapidly varied flow***. Gradually varied flow can occur with either subcritical or supercritical flow, but the transition from one condition to the other is ordinarily abrupt, as between D and E in Fig. 10.1. Other cases of local nonuniform flow occur at the entrance and exit of a channel, at changes in cross section, at bends, and at obstructions such as dams, weirs, or bridge piers.

[9] For details on various methods of solving gradually varied open-channel flow, see Richard H. French, *Open-Channel Hydraulics*, McGraw-Hill Book Company, New York, pp. 201–247, 1985.

10.15 ENERGY EQUATION FOR GRADUALLY VARIED FLOW

The principal forces involved in flow in an open channel are inertia, gravity, net hydrostatic force due to change in depth, and friction. The first three represent the useful kinetic and potential energies of the liquid, while the fourth dissipates useful energy into the useless kinetic energy of turbulence and eventually into heat because of the action of viscosity. Referring to Fig. 10.19, the total energy of the elementary volume of liquid shown is proportional to

$$H = z + y + \alpha \frac{V^2}{2g} \tag{10.35}$$

where $z + y$ is the potential energy head above the arbitrary datum, and $\alpha V^2/2g$ is the kinetic energy head, V being the mean velocity in the section. Each term of the equation represents energy in foot-pounds per pound of fluid (or newton-meters per newton of fluid in SI units). Once again, as in Secs. 10.1 and 10.2, we define L as the distance along the channel bed between any two sections and Δx as the horizontal distance between the sections. For all practical purposes, these two distances can be considered as being equal to one another.

The value of α will generally be found to be higher in open channels than in pipes, as was explained in Sec. 10.4. It may range from 1.05 to 1.40, and in the case of a channel with an obstruction the value of α just upstream may be as high as 2.00 or even more. As the value of α is not known unless the velocity distribution is determined, it is often omitted from the equations, but an effort should be made to employ it if a high degree of accuracy is

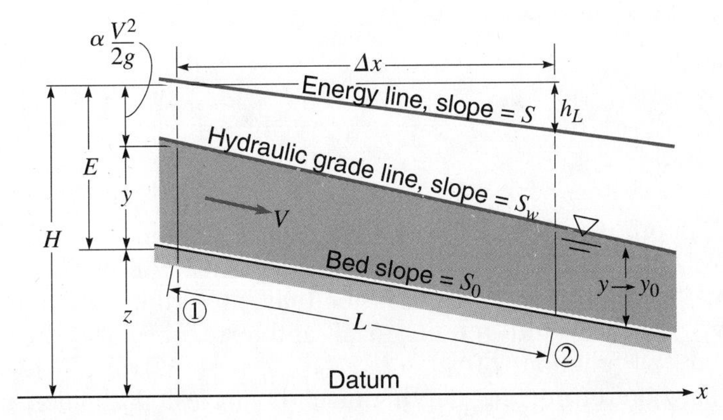

Figure 10.19
Energy relations for nonuniform flow.

necessary. In the numerical problems in this chapter, α is assumed to be unity.

The energy equation for steady flow between two sections (1) and (2) of Fig. 10.19 a distance Δx apart is

$$z_1 + y_1 + \alpha_1 \frac{V_1^2}{2g} = z_2 + y_2 + \alpha_2 \frac{V_2^2}{2g} + h_L \tag{10.36}$$

As $z_1 - z_2 = S_0 \Delta x$ and $h_L \approx S\Delta x$, the energy equation may also be written (with $\alpha_1 = \alpha_2 = 1$) in the form

$$y_1 + \frac{V_1^2}{2g} = y_2 + \frac{V_2^2}{2g} + (S - S_0)\,\Delta x \tag{10.37}$$

An approximate analysis of gradually varied, nonuniform flow can be achieved by considering a length of stream as consisting of a number of successive reaches, in each of which uniform flow occurs. Greater accuracy results from smaller depth variations in each reach. The Manning formula [Eq. (10.7)] is rearranged and applied to average conditions in each reach to provide an estimate of the value of the energy gradient S for that reach as follows:

In BG units:
$$S = \left(\frac{nV_m}{1.486R_{hm}^{2/3}}\right)^2 \tag{10.38a}$$

In SI units:
$$S = \left(\frac{nV_m}{R_{hm}^{2/3}}\right)^2 \tag{10.38b}$$

where V_m and R_{hm} are the means of the respective values at the two ends of the reach. With this value of S, with S_0 and n known, and with the depth and velocity at both ends of the reach known, the length Δx can be computed from Eq. (10.37), rearranged as follows:

$$\Delta x = \frac{(y_1 + V_1^2/2g) - (y_2 + V_2^2/2g)}{S - S_0} = \frac{E_1 - E_2}{S - S_0} \tag{10.39}$$

Only the depth y must be prescribed at each end; then the values of V and R_h follow from the discharge and cross section shape. As the calculation for Δx is direct, this approach is much easier than prescribing Δx and trying to solve for y_2, which would require trial and error because V_2, R_{h2}, and therefore S are all functions of y_2.

In practice, therefore, the depth range of interest is divided into small increments, usually equal, which define reaches whose lengths can be found by using Eq. (10.39). The discretization process must be started from a section where y is known, i.e., from calculations for a control section (Sec. 10.12) or from a field measurement at any section. If the resulting value of Δx is negative, this means that section (2) is upstream of section (1), not downstream as assumed.

Sample Problem 10.9 At a certain section in a very smooth rectangular channel of 6 ft width the depth is 3.00 ft when the flow rate is 160 cfs. Compute the distance to the section where the depth is 3.20 ft if $S_0 = 0.0020$ and $n = 0.012$.

Solution

The calculations are shown in the following table, where $V_m = (V_1 + V_2)/2$ and $R_{hm} = (R_{h1} + R_{h2})/2$. The total distance is calculated to be 71.3 ft. The accuracy could be improved by taking more steps.

y, ft	$A = 6y$, ft^2	$P = 6+2y$, ft	$R_h = A/P$, ft	$V = Q/A$, fps	$V^2/2g$, ft	$E = y + \frac{V^2}{2g}$, ft	Numerator ΔE, ft	V_m, fps	R_{hm}, ft	S Eq. (10.38)	Denominator $S - S_0$	$\Delta x = \frac{\Delta E}{S - S_0}$, ft
3.00	18.00	12.00	1.500	8.889	1.2269	4.2269						
							0.0221	8.746	1.512	0.002 873	0.000 873	25.3
3.10	18.60	12.20	1.525	8.602	1.1490	4.2490						
							0.0293	8.468	1.536	0.002 637	0.000 637	46.0
3.20	19.20	12.40	1.548	8.333	1.0783	4.2783						
											$\Sigma(\Delta x) =$	71.3

ANS

An important issue in calculations of this type is working with sufficient precision, i.e., carrying enough significant figures. In two instances we must deal with the small difference between two numbers whose values are close to one another. In the example just presented, we see that we lost two significant figures in calculating ΔE, and we lost one in calculating $S - S_0$. Therefore the calculations for these quantities, and for V and R_h upon which they are based, need to be carried to more significant figures so that the final results are always reliable to three significant figures.

If the data for the preceding example had been given in SI rather than BG units, the procedure for solution would have been the same except that Eqs. (10.38*b*) and (10.7*a*) would have been used rather than Eqs. (10.38*a*) and (10.7*b*). Note that V_m is *not* calculated from Q/A_m, nor from $y_m = (y_1 + y_2)/2$; these would give different results. Because of the extensive calculations required for many reaches, this type of problem is commonly

: Programmed computing aids (Appx. B) could help solve problems marked with this icon.

solved by computer (using applications software, or writing a program, or using a spreadsheet).

Section 10.16 will explain that depth ranges for this procedure must not cross the normal or critical depths. Therefore, to check this, first y_0 should be computed from Eq. (10.8) and y_c from Eq. (10.23) for rectangular channels or from Eq. (10.33) for other channels.

EXERCISES

10.15.1 A rectangular flume of planed timber ($n = 0.012$) is 5 ft wide and carries 60 cfs of water. The bed slope is 0.0006, and at a certain section the depth is 3 ft. Find the distance (in one reach) to the section where the depth is 2.5 ft. Is the distance upstream or downstream?

10.15.2 A rectangular flume of timber ($n = 0.013$) is 1.6 m wide and carries 1.7 m^3/s of water. The bed slope is 0.0005, and at a certain section the depth is 1.0 m. Find the distance (in one reach) to the section where the depth is 0.85 m. Is the distance upstream or downstream?

10.15.3 A test on a rectangular glass flume 9 in wide yielded the following data on a reach of 40-ft length: with still water, $z_1 - z_2 = 0.009$ ft; with a measured flow of 0.1446 cfs, $y_1 = 0.381$ ft, $y_2 = 0.386$ ft. Find the value of the roughness coefficient n using only one reach.

10.15.4 A test on a rectangular glass flume 25 cm wide yielded the following data on a reach of 9-m length: with still water, $z_1 - z_2 = 2.7$ mm; with a measured flow of 4.3 L/s, $y_1 = 11.02$ cm, $y_2 = 11.17$ cm. Find the value of the roughness coefficient n using only one reach.

10.16 WATER-SURFACE PROFILES IN GRADUALLY VARIED FLOW

As there are some 12 different circumstances giving rise to as many different fundamental types of gradually varied flow, it is helpful to have a logical scheme of type classification. In general, any problem of varied flow, no matter how complex it may appear, with the stream passing over dams, under sluice gates, down steep chutes, on the level, or even on an upgrade, can be broken down into reaches such that the flow within any reach is either uniform or falls within one of the given nonuniform classifications. The stream is then analyzed one reach at a time, proceeding from one reach to the next until the desired result is obtained.

The following treatment is based, for simplicity, on channels of rectangular section. The section will be considered sufficiently wide and shallow so that we may confine our attention to a section 1 ft (or 1 m) wide through which the velocity is essentially uniform. It is important to bear in mind that the following development is based on a constant value of q, the discharge per unit width, and upon one value of n, the roughness coefficient.

Differentiating Eq. (10.35) with respect to x, the horizontal distance along the channel, the rate of energy dissipation is found (with $\alpha = 1$) to be

$$\frac{dH}{dx} = \frac{dz}{dx} + \frac{dy}{dx} + \frac{1}{2g}\frac{d(V^2)}{dx} \tag{10.40}$$

The energy gradient $S = -dH/dL \approx -dH/dx$, while the slope of the channel bed is $S_0 = -dz/dx$, and the slope of the hydraulic grade line or water surface is given by $S_w = -dz/dx - dy/dx$.

Commencing with the last term of Eq. (10.40), we may observe that since $V = q/y$,

$$\frac{1}{2g}\frac{d(V^2)}{dx} = \frac{1}{2g}\frac{d}{dx}\left(\frac{q^2}{y^2}\right) = -\frac{q^2}{g}\frac{1}{y^3}\frac{dy}{dx}$$

Substituting this, plus the S and S_0 terms, into Eq. (10.40), yields

$$-S = -S_0 + \frac{dy}{dx}\left(1 - \frac{q^2}{gy^3}\right)$$

or

$$\frac{dy}{dx} = \frac{S_0 - S}{1 - q^2/gy^3} = \frac{S_0 - S}{1 - V^2/gy} = \frac{S_0 - S}{1 - \mathbf{F}^2} \tag{10.41}$$

Evidently, if the value of dy/dx as determined by Eq. (10.41) is positive, the water depth will be increasing along the channel; if negative, it will be decreasing. Let us examing the numerator and denominator of Eq. (10.41).

Looking first at the numerator, S may be considered as the energy gradient [such as would be obtained from Eq. (10.38)] that would carry the given discharge at depth y with uniform flow. For a wide and shallow rectangular channel, substituting $V = q/y$ and $R_h = y$ into Eq. (10.38*a*), we obtain

Wide and shallow: $$S = \left(\frac{nq}{1.486y^{5/3}}\right)^2$$

Similarly substituting these into Eq. (10.7*b*), we obtain

Wide and shallow: $$S_0 = \left(\frac{nq}{1.486y_0^{5/3}}\right)^2$$

Comparing these two expressions, we see that $S/S_0 = (y_0/y)^{10/3}$. Consequently, for constant q and n, when $y > y_0$, $S < S_0$, and the numerator

$(S_0 - S)$ is positive. Conversely, when $y < y_0$, $S > S_0$, and the numerator is negative.

To investigate the denominator of Eq. (10.41), we observe that if $\mathbf{F} = 1$ (critical flow, Sec. 10.10, $y = y_c$), the denominator is zero and $dy/dx = \infty$; if $\mathbf{F} > 1$ (supercritical flow, $y < y_c$), the denominator is negative; and if $\mathbf{F} < 1$ (subcritical flow, $y > y_c$), the denominator is positive.

Thus the signs of the numerator and the denominator of Eq. (10.41) can be found for any depth by comparing it with y_0 and y_c. These two signs together give the sign of dy/dx, which in turn defines the slope of the water surface.

The foregoing analyses have been combined graphically into a series of water-surface profiles [Fig. (10.20)]. The surface profiles are classified according to slope and depth as follows. If S_0 is positive, the bed slope is termed ***mild*** (M) when $y_0 > y_c$, ***critical*** (C) when $y_0 = y_c$, and ***steep*** (S) when $y_0 < y_c$; if $S_0 = 0$, the channel is ***horizontal*** (H); and if S_0 is negative, the bed slope is called ***adverse*** (A). If the stream surface lies above both the normal (uniform flow) and critical depth lines, it is of type 1; if between these lines, it is of type 2; and it below both lines, it is of type 3. The 12 forms of surface curvature are labeled accordingly in Fig. 10.20. Each is discussed in more detail in the next section.

It must be pointed out that the scale of the drawings in Fig. 10.20 is greatly reduced in the horizontal direction. The problems at the end of this chapter demonstrate that gradually varied flow generally extends over many hundreds of feet, and if plotted to an undistorted scale, the rate of change in depth would be scarcely discernible. It may be noted further that, since even a hydraulically steep slope varies but a few degrees from the horizontal, it makes little difference whether the depth y is measured vertically (as shown) or perpendicular to the bed.

It will be observed that some of the curves of Fig. 10.20 are concave upward while others are concave downward. Although the mathematical proof for this is not given, the physical explanation is not hard to find. In the case of the type 1 curves, the surface must approach a horizontal asymptote as the velocity is progressively slowed down because of the increasing depth. Likewise all curves that approach the normal or uniform depth line must approach it asymptotically, because uniform flow will only prevail at sections remote from disturbances, as pointed out in Sec. 10.1. Theoretically the curves that cross the critical-depth line must do so vertically, as the denominator of Eq. (10.41) becomes zero in this case. So these end conditions govern the surface concavity. The critical-slope curves, for which $y_0 = y_c$, constitute exceptions, since it is not possible for a curve to be both tangent and perpendicular to the same critical-uniform depth line.

To the right of each water-surface profile is shown a representative example of how this particular curve can occur. Many of the examples show a rapid change from a depth below the critical to a depth above the critical. This is a local phenomenon, known as the hydraulic jump, which is discussed in detail in Sec. 10.18.

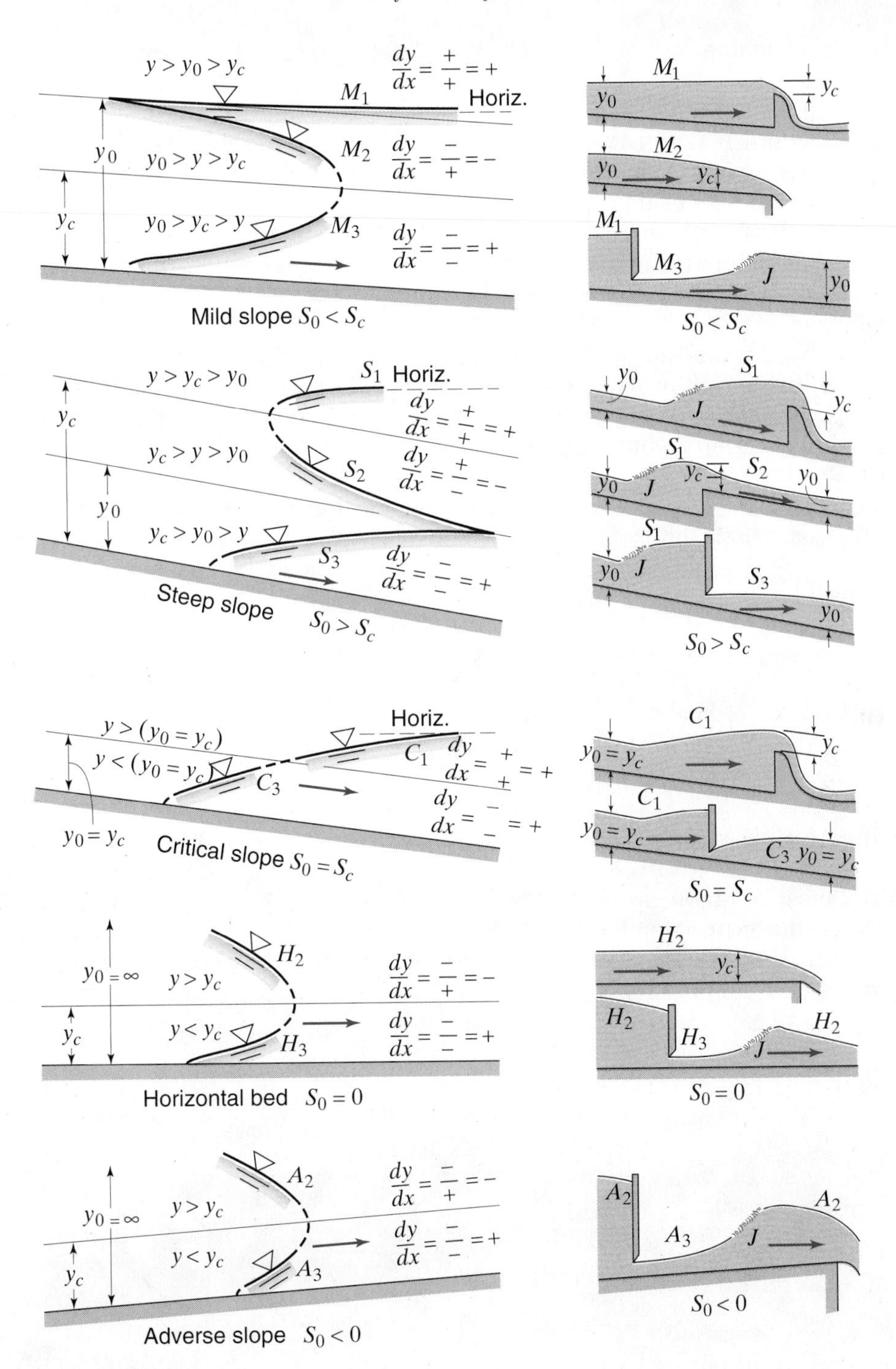

Figure 10.20
Various types of nonuniform flow, with flow from left to right.

The qualitative analysis of water-surface profiles has been restricted to rectangular sections. The curve forms of Fig. 10.20 are, however, applicable to any channel of uniform cross section if y_0 is the depth for uniform flow and y_c is the depth that satisfies Eq. (10.33). The surface profiles can even be used qualitatively in the analysis of natural stream surfaces as well, provided that local variations in slope, shape, and roughness of cross section, etc., are taken into account. The step-by-step method presented in Sec. 10.15 for the solution of steady nonuniform-flow problems is not restricted to uniform channels, and is therefore suited to water-surface-profile computations for any stream whatever.

A very important precaution, when using the stepwise procedure to solve for water-surface profiles, is to *always first determine the critical and normal depths,* y_c and y_0. Then it will be easy to confirm that the profile sought does not cross either of these depths. For if the ends of a reach are on different sides of either of these depths, the results are always invalid, even though sometimes they may look numerically reasonable. Along similar lines, if part of a reach is at uniform flow, this can also lead to impossible results when solving for depth. Such a situation can be avoided by first calculating the gradually varied profile distance to y_0.

10.17 EXAMPLES OF WATER-SURFACE PROFILES

The M_1 Curve

The most common case of gradually varied flow is where the depth is already above the critical and is increased still further by a dam, as indicated in Fig. 10.21. Referring to the specific energy diagram of Fig. 10.12, this case is found on the upper limb of the diagram, for here also, as the depth increases,

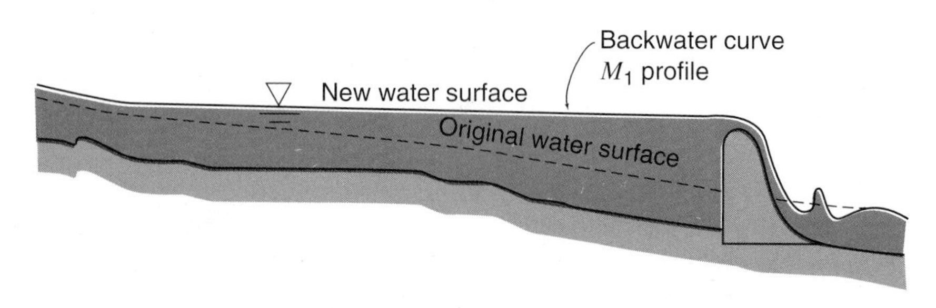

Figure 10.21
Backwater curve in a natural stream.

the velocity diminishes without any abrupt transitions, so that a smooth surface curve is obtained. In the case of flow in an artificial channel with a constant bed slope, the water-surface curve would be asymptotic at infinity to the surface for uniform flow, as we noted before. But the problems that are usually of more important interest are those concerned with the effect of a dam on a natural stream and the extent to which it raises the water surface at various points upstream. The resulting water-surface profile in such a case is commonly known as a ***backwater curve***.

For an artificial channel where the conditions are uniform, save for the variation in water depth, the problem may be solved by use of Eqs. (10.38) and (10.39). Usually, the solution commences at the dam, where conditions are assumed to be known, and the lengths of successive reaches upstream, corresponding to assumed increments of depth, are computed. A tabular type of solution (Sample Prob. 10.9) is the most helpful, with column headings corresponding to the various elements of Eqs. (10.38) and (10.39), the last column being $\sum \Delta x$, which sums up the length from the dam to the point in question. It is important, if accuracy is desired, to keep the depth increment small within any reach; a depth change of 10% or less is fairly satisfactory. The smaller the depth increment used in this step-by-step procedure, the greater the accuracy of the final result. This type of problem where successive calculations are required can advantageously be solved through use of a computer.

For a natural stream, such as that shown in Fig. 11.11, the solution is not so direct, because the form and dimensions of a cross section cannot be assumed and then the distance to its location computed. As there are various slopes and cross sections at different distances upstream, the value of Δx in Eq. (10.37) must be assumed, and then the depth of stream at this section can be computed by trials, as in Prob. 10.45. The solution is then pursued in similar fashion on a reach-by-reach basis. The accuracy of the results depends on the selection of a proper value for Manning's n, which is difficult when dealing with natural streams. For this reason and because of irregularities in the flow cross sections, the refinements of Eq. (10.37) are not always justified, and is often satisfactory to assume uniform flow by applying Eq. (10.8) to each successive reach.

The M_2 Curve

This curve, representing accelerated subcritical flow on a slope that is flatter than critical, exists, like the M_1 curve, because of a control condition *downstream.* In this case, however, the control is not an obstruction but the removal of the hydrostatic resistance of the water downstream, as in the case of the free overfall shown in Fig. 10.17*a*. As in the M_1 curve, the surface will approach the depth for uniform flow at an infinite distance upstream. Practically, because of slight wave action and other irregularities, the distinction between the M_2, or ***drop-down curve***, and the curve for uniform flow disappears within a finite distance.

The M_3 Curve

This occurs because of an upstream control, like the sluice gate shown in Fig. 10.20. The bed slope is not sufficient to sustain lower-stage flow, and at a certain point determined by energy and momentum relations, the surface will pass through a hydraulic jump unless this is made unnecessary by the existence of a free overfall before the M_3 curve reaches critical depth.

The S Curves

These may be analyzed in much the same fashion as the M curves, having due regard for downstream control in the case of subcritical flow and for upstream control for supercritical flow. Thus a dam or an obstruction on a steep slope produces an S_1 curve upstream (Fig. 10.20), which approaches the horizontal asymptotically but cannot so approach the uniform depth line, which lies below the critical depth. Therefore this curve must be preceded by a hydraulic jump. The S_2 curve shows accelerated lower-stage flow, smoothly approaching uniform depth. Such a curve will occur whenever a steep channel receives flow at critical depth, as from an obstruction (as shown) or reservoir. The sluice gate on a steep channel will produce the S_2 curve, which also approaches smoothly the uniform depth line.

The C Curves

These curves, with the anomalous condition $(dy/dx = \infty)$ at $y_0 = y_c$, do not occur frequently. As noted in Sec. 10.9, conditions that produce them are to be avoided because of the inherent instability of such flows.

The H and the A Curves

These curves have in common the fact that there is no condition of uniform flow possible. The H_2 and A_2 drop-down curves are similar to the M_2 curve, but even more noticeable. The value of $y_b = 0.72y_c$ given in Sec. 10.12 applies strictly only to the H_2 curve, but it is approximately true for the M_2 curve also. The sluice gate on the horizontal and adverse slopes produces H_3 and A_3 curves that are like the M_3 curve, but they do not exist for as great a distance as the M_3 curve before a hydraulic jump occurs. Of course, it is not possible to have a channel of any appreciable length carry water on a horizontal grade, and even less so on an adverse grade.

Other Examples

Some other interesting water-surface profiles occur when the slope of a channel of uniform section changes abruptly from a mild to a milder slope or to a less mild slope. In this case the flow is everywhere subcritical. Similar water-surface profiles occur when a channel on a constant slope that is mild throughout its entire length has an abrupt change in width to an either narrower or wider channel. These possibilities are depicted in Fig. 10.22.

Other water-surface profiles include those that occur when the slope of a channel changes abruptly from steep to either steeper or less steep. In this case the flow is supercritical. Similar profiles occur when a channel on a

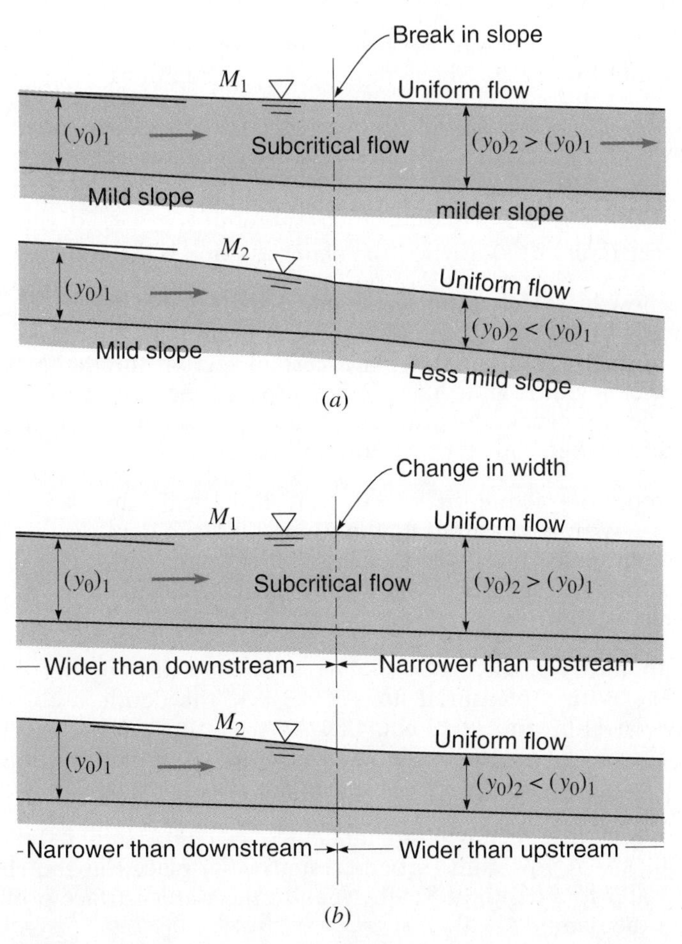

Figure 10.22
Subcritical flow water-surface profiles. (*a*) Constant section with change in slope. (*b*) Constant mild slope with change in width.

constant slope that is steep throughout its entire length has an abrupt change in width to an either wider or narrower channel. As an exercise it is suggested that the reader sketch profiles similar to Fig. 10.22 for the steep-slope situations. In these cases it will be found that, with steep slopes (supercritical flow), uniform flow occurs upstream of the change in either slope or width, while with mild slopes (subcritical flow), uniform flow prevails downstream of the change.

EXERCISES

10.17.1 Under what different conditions can supercritical flow leaving a hump ($\Delta z > \Delta z_{crit}$) return to normal depth without passing through a hydraulic jump?

10.17.2 Classify the water-surface profile of Exer. 10.15.1 as one of the forms shown in Fig. 10.20. Show all necessary calculations.

10.17.3 Classify the water-surface profile of Exer. 10.15.2 as one of the forms shown in Fig. 10.20. Show all necessary calculations.

10.17.4 Repeat Exer. 10.17.2 when the channel slope is −0.000 40.

10.17.5 Repeat Exer. 10.17.3 when the channel slope is −0.000 40.

10.17.6 The flow in a 15-ft-wide rectangular channel that has a constant bottom slope is 1400 cfs. A computation using Manning's equation indicates that the normal depth is 6.0 ft. At a certain section the depth of flow in the channel is 2.8 ft. Does the depth increase, decrease, or remain the same as one proceeds downstream from this section? Sketch a physical situation where this type of flow will occur.

10.17.7 A laboratory flume ($n = 0.012$) is 10 in wide and set on a slope of 0.0003. With a measured flow of 0.1516 cfs, the depth is observed to vary between 0.361 and 0.366 ft. Classify the water-surface profile as one of the forms of Fig. 10.20. Show all necessary calculations, and sketch the profile.

10.17.8 A laboratory flume ($n = 0.012$) is 25 cm wide and set on a slope of 0.0003. With a measured flow of 4.3 L/s, the depth is observed to vary between 11.02 and 11.17 cm. Classify the water-surface profile as one of the forms of Fig. 10.20. Show all necessary calculations, and sketch the profile.

10.17.9 In an 8-ft-wide rectangular channel ($S_0 = 0.003$, $n = 0.015$) water flows at 300 cfs. A low dam (broad-crested weir) placed in the channel raises the water to a depth of 8.0 ft. Analyze the water-surface profile upstream from the dam.

: Programmed computing aids (Appx. B) could help solve problems marked with this icon.

10.18 THE HYDRAULIC JUMP

By far the most important of the local nonuniform-flow phenomena is that which occurs when supercritical flow has its velocity reduced to subcritical. We have seen in the surface profiles of Fig. 10.20 that there is no ordinary means of changing from supercritical to subcritical flow with a smooth transition, because theory calls for a vertical slope of the water surface. The result, then, is a marked discontinuity in the surface, characterized by a steep upward slope of the profile, broken throughout with violent turbulence, and known universally as the ***hydraulic jump***.

The specific reason for the occurrence of the hydraulic jump can perhaps best be explained by reference to the M_3 curve of Fig. 10.20. Downstream of the sluice gate the flow decelerates because the slope is not great enough to maintain supercritical flow. The specific energy decreases as the depth increases (proceeding to the left along the lower limb of the specific-energy diagram, Fig. 10.12). Were this condition to progress until the flow reached critical depth, an increase in specific energy would be required as the depth increased from the critical to the uniform flow depth downstream. But this is a physical impossibility. Therefore the jump forms before the necessary energy is lost.

The hydraulic jump can also occur from an upstream condition of uniform supercritical flow to a nonuniform S_1 curve downstream when there is an obstruction on a steep slope, as illustrated in Fig. 10.20, or again from a nonuniform upstream condition to a nonuniform downstream condition, as illustrated by the H_3–H_2 or the A_3–A_2 combinations. An M_3–M_2 combination is also possible. In addition to the foregoing cases, where the channel bed continues at a uniform slope, a jump will form when the slope changes from steep to mild, as on the apron at the base of the spillway, illustrated in Fig. 10.23. This is an excellent example of the jump serving a useful purpose, for

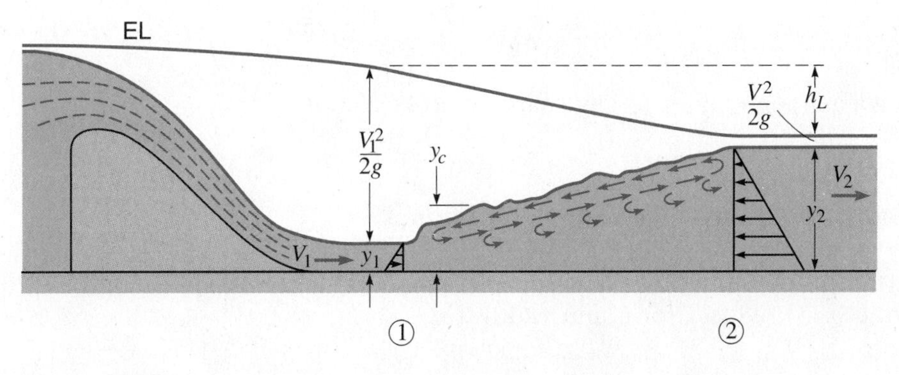

Figure 10.23
Hydraulic jump on horizontal bed following a spillway; horizontal scale foreshortened between sections 1 and 2 approximately $2\frac{1}{2}$:1.

it dissipates much of the destructive energy of the high-velocity water, thereby reducing downstream erosion. The turbulence within hydraulic jumps has also been found to be very useful and effective for mixing fluids, and jumps have been used for this purpose in water treatment plants and sewage treatment plants.

Depth Relations—General

The equation relating the depths before and after the hydraulic jump will be derived for the case of a horizontal channel bottom (the H_3–H_2 combination of Fig. 10.20). For channels on a gradual slope (i.e., less than about 3° or $S_0 = 0.05$) the gravity component of the weight is relatively small and may be neglected without introducing significant error. The friction forces acting are negligible because of the short length of channel involved and therefore the only significant forces are hydrostatic forces. Applying the impulse–momentum principle [Eq. (6.7*a*)] to the volume of fluid between sections 1 and 2 of Fig. 10.23, and using Eq. (3.16) for the end forces, we obtain for any shape of channel cross section

$$\sum F_x = \gamma h_{c_1} A_1 - \gamma h_{c_2} A_2 = \frac{\gamma}{g} Q(V_2 - V_1)$$

where h_c is the depth to the centroid of the end area. This equation can be rearranged to give

$$\frac{\gamma}{g} QV_1 + \gamma h_{c_1} A_1 = \frac{\gamma}{g} QV_2 + \gamma h_{c_2} A_2 \qquad (10.42)$$

This states that the momentum plus the pressure force on the cross-sectional area is constant, or dividing by γ and observing that $V = Q/A$,

$$F_m = \frac{Q^2}{Ag} + Ah_c = \text{constant} \qquad (10.43)$$

This equation applies to any shape of cross section.

Depth Relations—Rectangular Channel

In the case of a rectangular channel, using $h_c = y/2$, $A = by$, and $Q = bq$, Eq. (10.43) reduces for a unit width to

$$f_m = \frac{q^2}{y_1 g} + \frac{y_1^2}{2} = \frac{q^2}{y_2 g} + \frac{y_2^2}{2} \qquad (10.44)$$

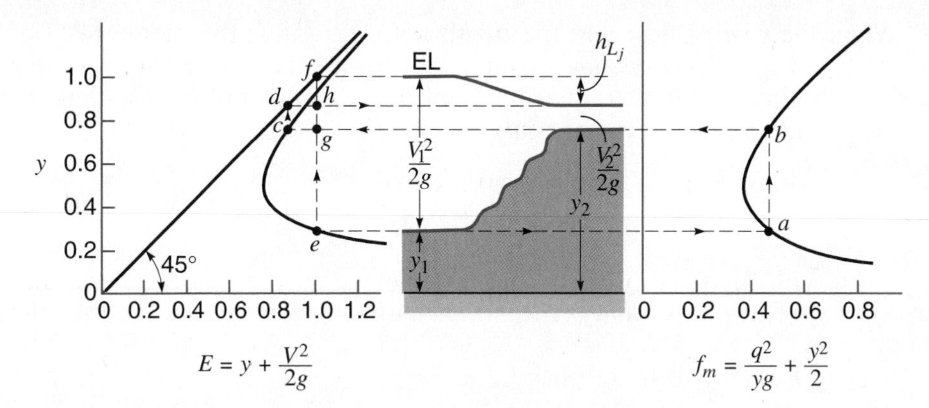

Figure 10.24
Energy and momentum relations in hydraulic jump.

A curve of values of f_m for different values of y is plotted to the right of the specific-energy diagram shown in Fig. 10.24. Both curves are plotted for the condition of 2 cfs/ft of width. As the loss of energy in the jump does not affect the "force" quantity f_m, the latter is the same after the jump as before, and therefore any vertical line intersecting the f_m curve serves to locate two ***conjugate depths***, y_1 and y_2. These depths represent possible combinations of depth that could occur before and after the jump.

Thus, in Fig. 10.24, the line for the initial water level y_1 intersects the f_m curve at a as shown, giving the value of f_m, which must be the same after the jump. The vertical line ab then fixes the value of y_2. This depth is then transposed to the specific-energy diagram to determine the value cd of $V_2^2/2g$. The value of $V_1^2/2g$ is the vertical distance ef.

It can be shown that for a given flow rate in a rectangular channel the minimum value of f_m [Eq. (10.44)] occurs at the same depth as the minimum value of E. Differentiating f_m with respect to y and equating to zero gives

$$f_m = \frac{q^2}{yg} + \frac{y^2}{2}$$

$$\frac{df_m}{dy} = -\frac{q^2}{gy^2} + \frac{2y}{2} = 0$$

and

$$y = \left(\frac{q^2}{g}\right)^{1/3}$$

This expression is identical to Eq. (10.23). Thus we have shown that for a given q the minimum value of f_m occurs at the same depth as does the minimum value of E. This is indicated on Fig. 10.24.

When the rate of flow and the depth before or after the jump are given, we see that Eq. (10.44) becomes a cubic equation when solving for the other depth. This may readily be reduced to a quadratic, however, by observing that $y_2^2 - y_1^2 = (y_2 + y_1)(y_2 - y_1)$, so that

$$\frac{q^2}{g} = y_1 y_2 \frac{y_1 + y_2}{2} \tag{10.45}$$

From Eq. (10.23) we also know that $q^2/g = y_c^3$, so if $y_1 < y_c$ then we must have $y_2 > y_c$. This proves that the hydraulic jump must always cross the critical depth.

Solving Eq. (10.45) by the quadratic formula gives

$$y_2 = \frac{y_1}{2}\left(-1 + \sqrt{1 + \frac{8q^2}{gy_1^3}}\right) \tag{10.46a}$$

or

$$y_1 = \frac{y_2}{2}\left(-1 + \sqrt{1 + \frac{8q^2}{gy_2^3}}\right) \tag{10.46b}$$

These equations relate the depths before and after hydraulic jump (i.e., the conjugate depths) in a rectangular channel. They give good results if the channel slope is less than about 0.05. For steeper channel slopes the effect of the gravity component of the weight of liquid between sections 1 and 2 of Fig. 10.23 must be considered.

Energy Loss

The head loss h_{L_j} caused by the jump is the drop in energy from 1 to 2. Or

$$h_{L_j} = E_1 - E_2 = \left(y_1 + \frac{V_1^2}{2g}\right) - \left(y_2 + \frac{V_2^2}{2g}\right) \tag{10.47}$$

On Fig. 10.24 points e and c on the specific energy diagram represent the conjugate depths, and the horizontal distance between them ($= cg = fh$) is the head loss. Replacing V by q/y and using Eq. (10.45), it can be shown that also

Rectangular channel:

$$h_{L_j} = \frac{(y_2 - y_1)^3}{4y_1 y_2} \tag{10.48}$$

Examples of depth and energy loss calculations are included in Sample Prob. 10.10.

Figure 10.25
Hydraulic jump. (*Photograph by Hydrodynamics Laboratory, California Institute of Technology.*)

Jump Length

Although the length of a jump is difficult to predict, a good approximation for jump length is about $5y_2$. This relation may be seen to be approximately true by examination of Fig. 10.25, a photograph of a hydraulic jump in a horizontal channel, caused by a sluice gate upstream. In most cases $4 < L_j / y_2 < 6$.

Types of Jump

Since the flow must be supercritical in order for a jump to occur, $\mathbf{F}_1$, the Froude number of the flow just upstream of the jump, must be greater than 1.0. As $\mathbf{F}_1$ increases, the jump becomes more turbulent and more energy is dissipated. Ranges of different jump behaviors have been identified, some more desirable than others. These are summarized in Table 10.3. In particular, the oscillating jump is to be avoided; the main stream of water

Table 10.3
Types of hydraulic jump[a]

Name	$\mathbf{F}_1$	Energy dissipation	Characteristics
Undular jump	1.0–1.7	<5%	Standing waves
Weak jump	1.7–2.5	5–15%	Smooth rise
Oscillating jump	2.5–4.5	15–45%	Unstable; avoid
Steady jump	4.5–9.0	45–70%	Best design range
Strong jump	>9.0	70–85%	Choppy, intermittent

[a] U.S. Bureau of Reclamation, Research Studies on Stilling Basins, Energy Dissipators, and Associated Appurtenances, *Hydraul. Lab. Rept.* Hyd-399, 1955.

flowing through the jump oscillates between the bed and the surface, generating waves that can be damaging for great distances downstream.

In some areas of the world with high tides, when they enter estuaries and the like, a small hydraulic jump forms, which travels upstream and can cause minor damage. This phenomenon is known as a ***surge*** or a ***tidal bore***. It may be analyzed by similar methods to those used for gravity waves (Sec. 10.20).

Stilling Basins

When water from elevated water bodies like reservoirs flows over spillways, through outlet works, or down chutes, it can have tremendous energy. This energy can cause great damage to the structure itself, to the ground supporting the structure (and therefore ultimately to the structure), and to the downstream channels. The hydraulic jump is one of the most effective methods of dissipating such flow energy, as we noted earlier. The objective of a ***stilling basin*** is to initiate a jump for this purpose and to contain it within a structure that will minimize damage.

The basin floor may be recessed or have an adverse slope to provide sufficient tailwater depth; a ***sill*** across the downstream end is also commonly used for this purpose. To increase the turbulence and to encourage the jump to form immediately as the water enters the basin, rows of ***chute blocks*** and/or staggered ***baffle blocks*** are commonly arranged across the basin. Sometimes those blocks with an end sill can reduce the length of the jump, and so the basin.

Because high entry velocities can cause cavitation damage and erosion by sediment, the basin must be finished with quality concrete brought to a smooth surface.

EXERCISES

10.18.1 A hydraulic jump occurs in a 15-ft wide rectangular channel carrying 200 cfs on a slope of 0.005. The depth after jump is 4.5 ft. (*a*) What must be the depth before jump? (*b*) What are the losses of energy and power in the jump?

10.18.2 A hydraulic jump occurs in a 6-m-wide rectangular channel carrying 8 m^3/s on a slope of 0.005. The depth after jump is 1.5 m. (*a*) What must be the depth before jump? (*b*) What are the losses of energy and power in the jump?

10.18.3 The hydraulic jump may be used as a crude flowmeter. Suppose that in a horizontal rectangular channel 6 ft wide the observed depths before and after a hydraulic jump are 0.80 and 3.60 ft, respectively. Find the rate of flow and the head loss.

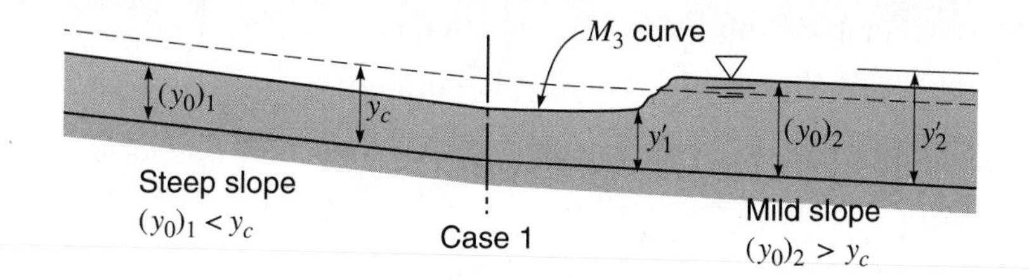

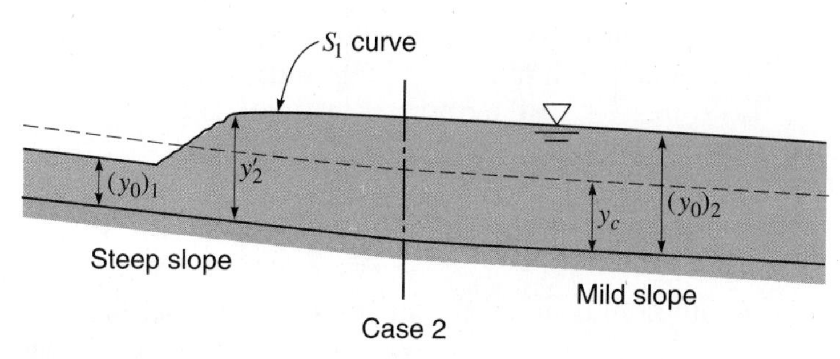

Figure 10.26
Examples of the location of a hydraulic jump where one conjugate depth is normal. [*Note*: y_2' is the conjugate depth of $(y_0)_1$ and y_1' is the conjugate depth of $(y_0)_2$.]

10.19 LOCATION OF HYDRAULIC JUMP

The problem of determining the location of a hydraulic jump involves a combined application of the principles discussed in Secs. 10.17 and 10.18. Examples of the location of a hydraulic jump are shown in Fig. 10.26. In both these cases one of the conjugate depths is a normal depth. In case (1) the jump occurs downstream of the break. The reasons for this are illustrated by the following example.

Sample Problem 10.10 Analyze the water-surface profile in a long rectangular channel with concrete lining ($n = 0.013$). The channel is 10 ft wide, the flow rate is 400 cfs, and there is an abrupt change in channel slope from 0.0150 to 0.0016. Find also the horsepower loss in the resulting jump.

Solution

Eq. (10.8*a*):
$$400 = \frac{1.486}{0.013}(10y_{0_1})\left(\frac{10y_{0_1}}{10+2y_{0_1}}\right)^{2/3}(0.015)^{1/2}$$

: Programmed computing aids (Appx. B) could help solve problems marked with this icon.

By trial (see Sample Prob. 3.5) or by equation solver,

$$y_{0_1} = 2.16 \text{ ft} \qquad \text{(normal depth on the upper slope)}$$

Using a similar procedure, the normal depth y_{0_2} on the lower slope is found to be 4.81 ft.

Eq. (10.23): $$y_c = \left(\frac{q^2}{g}\right)^{1/3} = \left[\frac{\left(\frac{400}{10}\right)^2}{32.2}\right]^{1/3} = 3.68 \text{ ft}$$

Thus flow is supercritical ($y_{0_1} < y_c$) before the break in slope and subcritical ($y_{0_2} > y_c$) after the break, so a hydraulic jump must occur.

Applying Eq. (10.46*a*) to determine the depth conjugate to the 2.16-ft (upper-slope) normal depth, we get

$$y_2' = \frac{2.16}{2}\left\{-1 + \left[1 + \frac{8(40)^2}{32.2(2.16)^3}\right]^{1/2}\right\} = 5.78 \text{ ft}$$

Therefore the depth conjugate to the upper-slope normal depth of 2.16 ft is 5.78 ft. This is $> y_{02}$, so it is in zone 1. A jump to this depth cannot occur, because the normal depth y_{0_2} on the lower slope is less than 5.78 ft, and an M_1 curve cannot bring the surface down to it.

Applying Eq. (10.46*b*) to determine the depth conjugate to the 4.80-ft (lower-slope) normal depth, we get

$$y_1' = \frac{4.80}{2}\left\{-1 + \left[1 + \frac{8(40)^2}{32.2(4.80)^3}\right]^{1/2}\right\} = 2.74 \text{ ft}$$

The lower conjugate depth of 2.74 ft will occur downstream of the break in slope. Thus the condition here is similar to that depicted in Fig. 10.26 (case 1). The location of the jump (i.e., its distance below the break in slope) may be found by applying Eq. (10.39) to the M_3 curve;

$$\Delta x = \frac{E_1 - E_2}{S - S_0}$$

$$E_1 = 2.16 + \frac{(400/21.6)^2}{2(32.2)} = 7.47 \text{ ft}$$

$$E_2 = 2.74 + \frac{(400/27.4)^2}{2(32.2)} = 6.04 \text{ ft}$$

$$V_m = \frac{1}{2}\left(\frac{400}{21.6} + \frac{400}{27.4}\right) = 16.53 \text{ fps}$$

$$R_m = \frac{1}{2}\left(\frac{21.6}{14.33} + \frac{27.4}{15.49}\right) = 1.641 \text{ ft}$$

From Eq. (10.38*a*) $$S = \left[\frac{(0.013)(16.53)}{1.486(1.641)^{2/3}}\right]^2 = 0.010\,75$$

Eq. (10.39): $$\Delta x = \frac{7.47 - 6.04}{0.010\,75 - 0.001\,60} = 155.6 \text{ ft}$$

Note that this distance could be computed more accurately by dividing it into more reaches, i.e., by using more, smaller depth increments.

Thus the depth on the upper slope slope is 2.16 ft; downstream of the break the depth increases gradually (M_3 curve) to 2.74 ft over a distance of approximately 155 ft; then a hydraulic jump occurs from a depth of 2.74 ft to 4.80 ft; downstream of the jump the depth remains constant (i.e., normal) at 4.80 ft. ***ANS***

Eq. (10.48): $$h_{L_j} = \frac{(4.80 - 2.74)^3}{4(4.80)2.74} = 0.1655 \text{ ft}$$

Hence $$\text{HP loss} = \frac{\gamma Q h_{L_j}}{550} = \frac{62.4(400)0.1655}{550} = 7.51 \qquad \textbf{\textit{ANS}}$$

In the two examples shown in Fig. 10.26 the flow is at normal depth either before or after the hydraulic jump. These situations are straightforward, and are easy to handle as shown in the preceding illustrative example. There are instances, however, where the flow is not normal either before or after the jump. Such a situation will occur, for example, when water flows out from under a sluice gate and there is an overflow weir downstream of the gate. If the channel is on a steep slope, a hydraulic jump will occur between an S_3 and an S_1 water surface profile, while if the channel is on a mild slope, the jump will occur between an M_3 and an M_1 profile (Fig. 10.20). Another instance would be between M_3 and M_2 profiles, as shown in Fig. 10.27. This

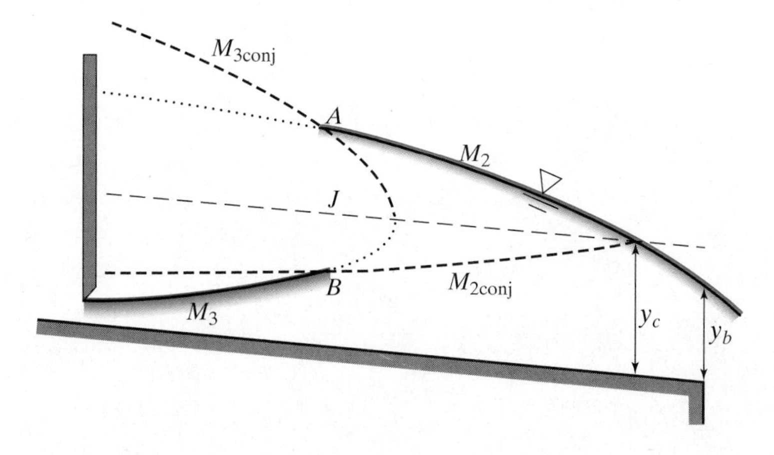

Figure 10.27
Example of location of hydraulic jump when neither conjugate depth is normal.

combination is similar to the H_3–H_2 and A_3–A_2 combinations shown on the right-hand side of Fig. 10.20. To find the location of such a jump, varied flow calculations must be made in the downstream direction from the sluice gate and in the upstream direction from the free overfall (or from the overflow weir in the other cases), at the same time computing the conjugate depths for one of these curves using Eq. (10.46). Where the conjugate of one profile crosses the other profile is the only place where the energies are correct for a hydraulic jump to occur. This is at point A or B (the same location) in Fig. 10.27. Graphically plotting these various curves greatly aids the solution process. Also, it will be helpful to reduce the size of the varied flow increments in the neighborhood of the intersection point, which will probably have to be found by interpolation.

10.20 VELOCITY OF GRAVITY WAVES

Consider a small wave of height Δy traveling at a velocity (or ***celerity***) c to the left across the surface of stationary water whose depth is y (Fig. 10.28*a*). Let us now replace this with an equivalent steady flow to the right having a velocity $V_1 = -c$ (Fig. 10.28*b*). Because the wave *velocity or celerity is relative to the water,* the wave is now standing still with respect to the observer. Applying the principle of continuity to Fig. 10.28*b* we have $V_1 y = V_2(y + \Delta y)$, where V_2 is the average velocity of flow past section 2 in Fig. 10.28*b*.

Applying the impulse–momentum principle, $F = \rho Q\, \Delta V$, to the control volume between sections 1 and 2 for a unit width of channel gives

$$\gamma \frac{y^2}{2} - \gamma \frac{(y + \Delta y)^2}{2} = \frac{\gamma}{g} y V_1 (V_2 - V_1)$$

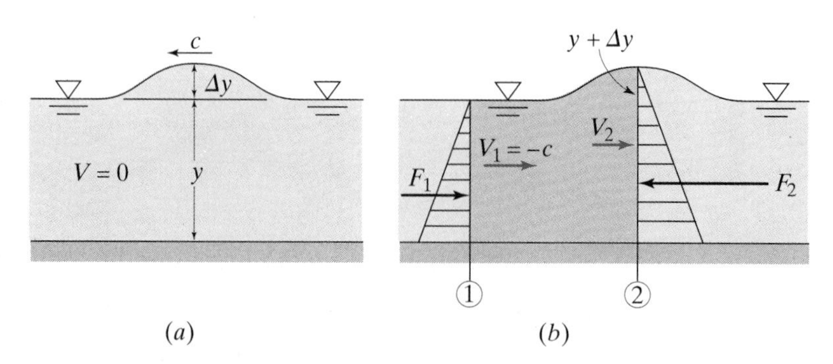

Figure 10.28
Gravity wave of small amplitude. (*a*) As seen by a stationary observer. (*b*) As seen by an observer moving with the wave.

Substituting the expression for V_2 from continuity and letting $c = -V_1$ in the above gives

$$c = \sqrt{g(y + \Delta y)\left(\frac{y + \frac{1}{2}\Delta y}{y}\right)} \approx \sqrt{g(y + \Delta y)} \approx \sqrt{gy} \qquad (10.49)$$

The latter expressions apply if surface disturbance is relatively small, that is, $\Delta y \ll y$. We see that the velocity (celerity) of a wave will increase as the depth of the water increases. But this is the velocity relative to the water. If the water is flowing, the absolute speed of travel of the wave will be the resultant of the two velocities. Thus

$$c_{\text{abs}} = c \pm V \qquad (10.50)$$

Let us now consider only "small" waves, for which $c = \sqrt{gy}$. We are all familiar with the circular waves that spread on a still water surface when disturbed, as when we throw a stone into a pond. When the water is also moving, these spreading patterns are carried with the water, in accordance with Eq. (10.50). In Fig. 10.29 we see how the wave pattern produced by a repeated, regular disturbance varies with water velocity. The familiar pattern of Fig. 10.29*a* is produced when the water is static. In Fig. 10.29*b* the water

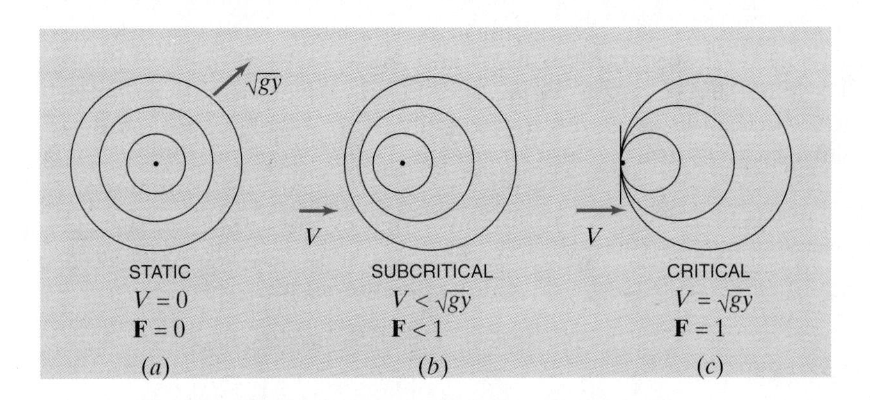

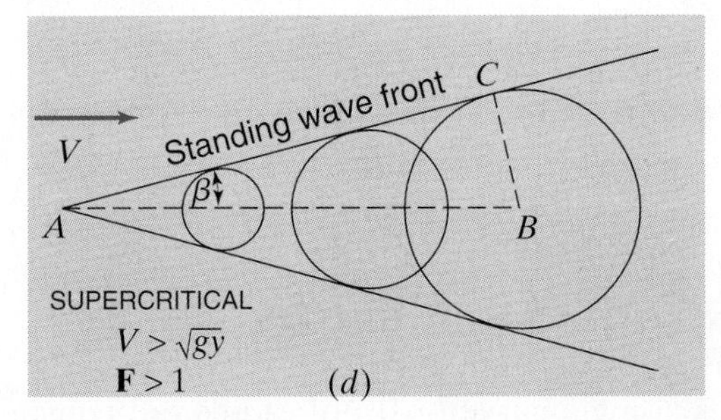

Figure 10.29
Variation of gravity wave patterns with flow conditions.

moving with a low, subcritical velocity carries the circular waves to the right. For the somewhat higher velocity of $V = \sqrt{gy}$, conditions are critical (Sec. 10.10), and $V = c$, so that waves trying to move upstream (to the left in Fig. 10.29*c*) are brought to a standstill as shown; they "hold fast." Thus we see that small disturbance waves can only travel upstream if $V < c$, i.e., if the flow is subcritical. For still higher, supercritical velocities, the circles are carried downstream as shown in Fig. 10.29*d*; they *cannot travel upstream.* As the center of the disturbance is carried in time t from A to B for example, the radius grows from B to C. So we see that

Small waves:
$$\sin\beta = \frac{BC}{AB} = \frac{ct}{Vt} = \frac{\sqrt{gy}}{V} = \frac{1}{\mathbf{F}} \qquad (10.51)$$

If the disturbance is not intermittent but is continuous, as is usual, we shall not see the individual circles, but we shall see where they reinforce one another. This occurs perpendicular to the flow at the disturbance site for critical flow, see Fig. 10.29*c*, and along line AC in Fig. 10.29*d* for supercritical flow. The perpendicular wave front with critical flow is a special case of Fig. 10.29*d*, because when $V = c$ we have $\mathbf{F} = 1$ and $\beta = 90°$. AC is known as a ***standing wave front*** or a ***disturbance line***, and is comparable to the oblique shock wave discussed in Sec. 13.11. The angle β is called the ***wave angle***, and this obviously decreases as V increases. When the channel shape must change in any way, such as at bridge piers, bends, and transitions, these standing waves form with supercritical flow, and because they reflect off side walls, complex patterns of ***cross waves*** can result. Every small irregularity will similarly cause oblique standing waves in supercritical flow. There are many irregularities in natural streams of course, and in order to run rapids canoeists learn to "read" the water surface to interpret subsurface conditions.

If the waves are larger, in other words not small compared to the water depth, we see from Eq. (10.49) that the wave velocity (celerity) is greater than $\sqrt{gy}$, and from Eq. (10.51) that β must also be larger.

EXERCISE

10.20.1 A thin rod is placed vertically in a stream that is 4 ft deep, and the resulting small disturbance wave is observed to make an angle of about 60° with the axis of the stream. Find the approximate velocity of the stream.

10.21 FLOW AROUND CHANNEL BENDS

When a body moves along a curved path of radius r at constant speed, it has a normal acceleration $a_n = V^2/r$ toward the center of the curve, and hence the body must be acted on by a force in that direction. In Fig. 10.30 it may be

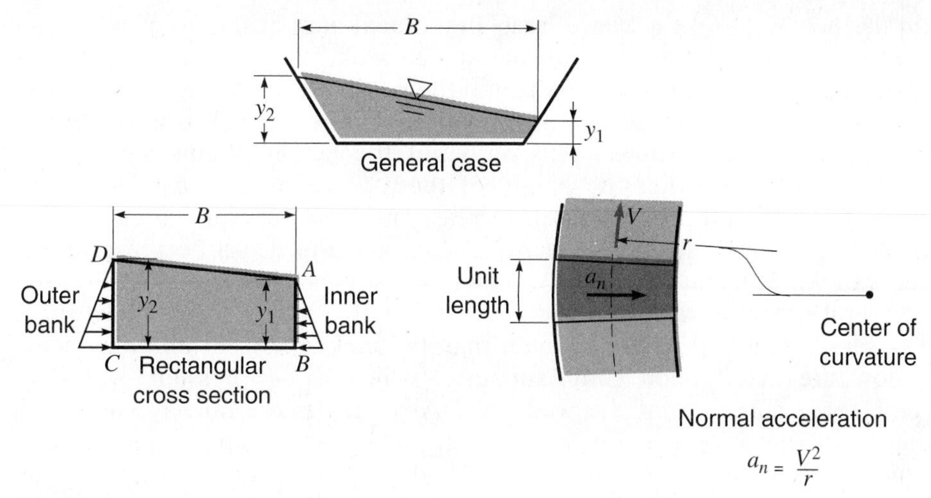

Figure 10.30
Flow in an open-channel bend.

seen that this force comes from the unbalanced pressure forces due to the difference in liquid levels between the outer and inner banks of the channel. Assuming that the velocity V across the rectangular section is uniform and that $r \gg B$, we can write, for a unit length of channel, $\sum F_n = ma_n$,

$$\frac{\gamma y_2^2}{2} - \frac{\gamma y_1^2}{2} = \frac{\gamma}{g} B \frac{y_1 + y_2}{2} \frac{V^2}{r}$$

which by algebraic transformation can be written as

$$\Delta y = y_2 - y_1 = \frac{V^2 B}{gr} \tag{10.52}$$

where B is the top width of the water surface as shown in Fig. 10.30. It can be shown that Eq. (10.52) applies to any shape of cross section. If the effects of velocity distribution and variations in curvature across the stream are considered, the difference in water depths between the outer and inner banks may be as much as 20% more than that given by Eq. (10.52). If the actual velocity distribution across the stream is known, the width may be divided into sections and the difference in elevation computed for each section. The total difference in surface elevation across the stream is the sum of the differences for the individual sections.

With supercritical flow the complicating factor is the effect of disturbance waves, generated by the very start of the curve. These waves, one from the outside wall and one from the inside, traverse the channel, making an angle β with the original direction of flow, as discussed in Sec. 10.20. The result is a criss-cross wave pattern. The water surface along the outside wall will rise from the beginning of the curve, reaching a maximum at the point where the wave from the inside wall reaches the outside wall. The wave is then reflected back to the inside wall and so on around the bend. Hence in supercritical flow around a bend the increase in water depth on the outer wall is equal to that created by centrifugal effects plus or minus the height of the waves. Field observations indicate that the height of these waves is approximately $\Delta y/2$ [Eq. (10.52)]. Thus the water-depth increase on the outer wall varies between zero and Δy, i.e., double the subcritical rise, and the depth decrease on the inner wall varies similarly.

Several schemes to lessen the surface rise from wave effects have been investigated. The bed of the channel may be banked so that all elements of the flow are acted upon simultaneously, which is not possible when the turning force comes from the wall only. As in a banked-railway curve, this requires a transition section with a gradually increasing superelevation preceding the main curve. Another method is to introduce a counterdisturbance to offset the disturbance wave caused by the curve. Such a counterdisturbance can be provided by a section of curved channel of twice the radius of the main section, by a spiral transition curve, or by diagonal sills on the channel bed, all preceding the main curve.

Flow around a bend in an open channel is complicated by the development of a secondary flow as the liquid travels around the curve. The development of secondary flow in the bends of pressure conduits was discussed in Sec. 8.23.

The flow around a channel bend is called ***spiral flow*** because superposition of the secondary flow on the forward motion of the liquid causes the liquid to follow a path like a corkscrew (or spiral). The existence of spiral flow is readily explained by reference to Fig. 10.31. The water surface is superelevated at the outside wall. The element EF is subjected to a centrifugal force mV^2/r, which is balanced by an increased hydrostatic force on the left side, due to the superelevation of the water surface at C above that at D. The element GH is subjected to the same net hydrostatic force inward, but the centrifugal force outward is much less because the velocity is

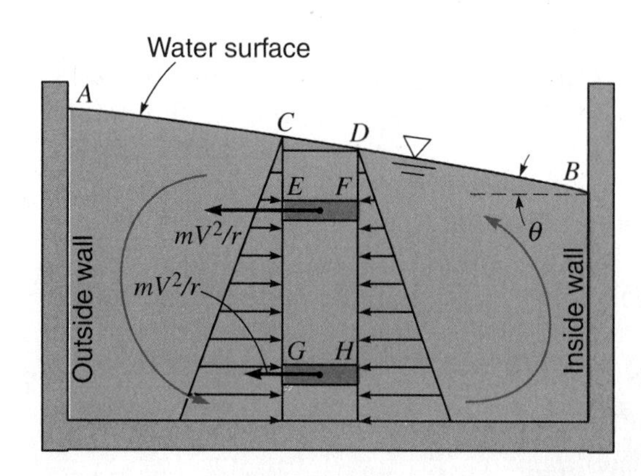

Figure 10.31
Schematic sketch of flow around a bend in a rectangular channel curving to the right, looking downstream, with spiral flow counterclockwise.

decreased by friction near the bottom. This results in a cross flow inward along the bottom of the channel, which is balanced by an outward flow near the water surface; hence the spiral. This spiral flow is largely responsible for the commonly observed erosion of the outside bank of a river bend, with consequent deposition and building up of a sand bar near the inside bank.

Sample Problem 10.11 A rectangular channel 2 m wide carries 4.2 m³/s in uniform flow at a depth of 0.5 m. What will be the maximum and minimum water depths at the inside and outside walls of a bend in the channel of radius 10 m to the centerline?

Solution

$$V = Q/A = 4.2/(2\times 0.5) = 4.2 \text{ m/s}$$

Eq. (10.51):
$$\mathbf{F} = \frac{V}{\sqrt{gy}} = \frac{4.2}{\sqrt{9.81\times 0.5}} = 1.896$$

$\mathbf{F} > 1$, so the flow is supercritical.

Eq. (10.52):
$$\Delta y = \frac{(4.2)^2 2}{9.81(10)} = 0.360 \text{ m}$$

Because of wave action with supercritical flow, Δy varies from zero to double the value computed. Thus the extreme water depths are as follows:

	Inside wall	Outside wall	
y_{max}	0.500 m	0.860 m	
y_{min}	0.140 m	0.500 m	***ANS***

EXERCISE

10.21.1 A rectangular channel 5 m wide carries 8 m³/s in uniform flow at a depth of 2 m. What will be the maximum difference in water-surface elevations between the inside and outside of a circular bend in this channel if the radius of the bend at the centerline is 20 m?

10.22 TRANSITIONS

Special transition sections are often used to join channels of different size and shape in order to avoid undesirable flow conditions and to minimize head loss. If the flow is subcritical, a straightline transition (Fig. 10.32) with an angle θ of about 12.5° is fairly satisfactory. This is also known as a ***wedge transition***. Without the transition, i.e., with an abrupt change in section with square corners, the corresponding head losses are about two to four times larger. Intermediate between these two types is to replace the abrupt corners

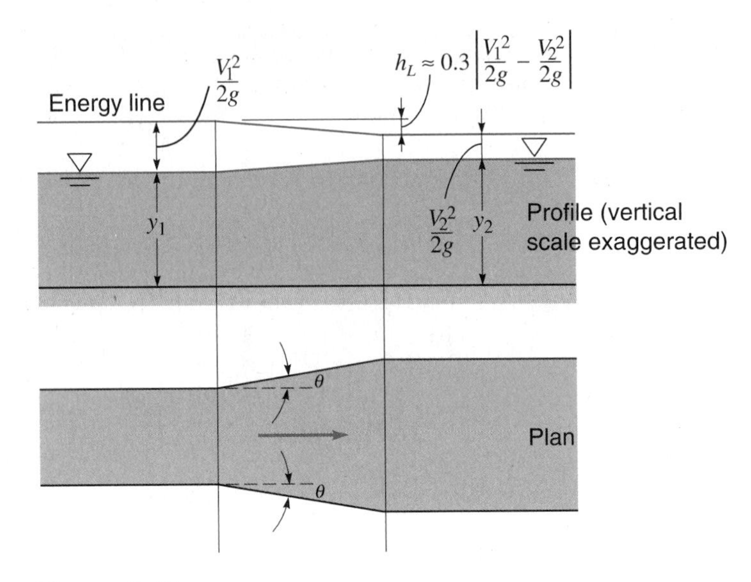

Figure 10.32
Simple open-channel wedge transition for decelerating flow.

with vertical quarter-cylinder segments; this is known as a ***cylinder-quadrant transition***. At Froude numbers between about 0.5 and 1.0, complex ***warped transitions*** that avoid any sharp angles are advisable.

The head loss in transitions is proportional to the change in the velocity head, and is traditionally expressed as

$$h_{L_t} = k_t \left| \frac{V_2^2}{2g} - \frac{V_1^2}{2g} \right| \tag{10.53}$$

Approximate values of the loss coefficient k_t for the transitions discussed are given in Table 10.4. Notice that the more efficient transitions are the ones

Table 10.4
Loss coefficients for channel transitions with subcritical flow

Transition type	Loss coefficient k_t	
	Contracting	**Expanding**
Abrupt	0.4–0.5	0.75–1.00
Cylinder-quadrant	0.2	0.5
Wedge	0.1–0.2	0.3–0.5
Warped	0.1	0.3

that are more expensive to build, and that losses are greater with expanding flows, as usual. Experimentally it has been found that transition lengths need to be at least 2.25 times the change in the water surface width.

With supercritical flow ($\mathbf{F} > 1$), surface waves are formed as described in Sec. 10.20, and special procedures are required for transition design.

At a channel entrance from a reservoir or from a larger channel the head loss for a square-edged entrance is about 0.5 times the velocity head. By rounding the entrance, the head loss can be reduced to slightly less than 0.2 times the velocity head.

SAMPLE PROBLEM 10.12 Refer to Fig. 10.32. A rectangular channel changes in width from 4 ft to 6 ft. Measurements indicate that $y_1 = 2.50$ ft and $Q = 50$ cfs. Determine the depth y_2 by (*a*) neglecting head loss, and (*b*) considering the head loss to be given as shown on the figure.

Solution

$$V_1 = 50/(2.5 \times 4) = 5 \text{ fps}; \; V_2 = 50/6y_2 = 8.33/y_2 \text{ fps}$$

(*a*) $h_L = 0$, so $E_1 = E_2$, i.e., $2.5 + \dfrac{5^2}{2(32.2)} = y_2 + \dfrac{(8.33/y_2)^2}{2(32.2)}$

By trial (see Sample Probs. 3.5 and 5.7), $y_2 = 2.75$ ft ***ANS***

(*b*)
$$E_1 = E_2 + h_L = E_2 + 0.3\left(\frac{V_1^2}{2g} - \frac{V_2^2}{2g}\right)$$

$$y_1 + \frac{V_1^2}{2g} = y_2 + \frac{V_2^2}{2g} + 0.3\left(\frac{V_1^2}{2g} - \frac{V_2^2}{2g}\right), \quad \text{so} \quad y_1 + 0.7\frac{V_1^2}{2g} = y_2 + 0.7\frac{V_2^2}{2g}$$

$$2.5 + 0.7\frac{5^2}{2(32.2)} = 2.77 = y_2 + \frac{0.755}{y_2^2}; \quad \text{by trial, } y_2 = 2.67 \text{ ft} \quad \textbf{\textit{ANS}}$$

Note: Because trial and error is needed to solve the resulting cubic equations, an equation solver in mathematics software or in a programmable scientific calculator can be helpful in such problems.

EXERCISES

10.22.1 Water flows at 180 cfs from a trapezoidal channel ($b = 10$ ft, side slopes $= 2.5\,H{:}1V$) into a rectangular channel with the same bed width. If the upstream depth is 3.5 ft and the transition is abrupt ($k_t = 0.4$), find the downstream depth.

: Programmed computing aids (Appx. B) could help solve problems marked with this icon.

10.22.2 A rectangular, 3-m-wide channel is connected to a trapezoidal channel ($b = 1$ m, side slopes = $1.5H:1V$) by a warped transition. When 7.5 m^3/s of water is flowing, the depth in the upstream channel is 1.5 m. (*a*) Find y_2. (*b*) Is the flow through the transition expanding or contracting?

10.23 HYDRAULICS OF CULVERTS

A culvert is a conduit passing under a road or highway. In section, culverts may be circular, rectangular, or oval. Culverts may operate with either a submerged entrance (Fig. 10.33) or a free entrance (Fig. 10.34).

Submerged Entrance

In the case of a submerged entrance there are three possible regimes of flow as indicated in Fig. 10.33. Under conditions (*a*) and (*b*) of the figure the culvert is said to be flowing under ***outlet control***, while condition (*c*) represents ***entrance control.*** In (*a*) the outlet is submerged, possibly because

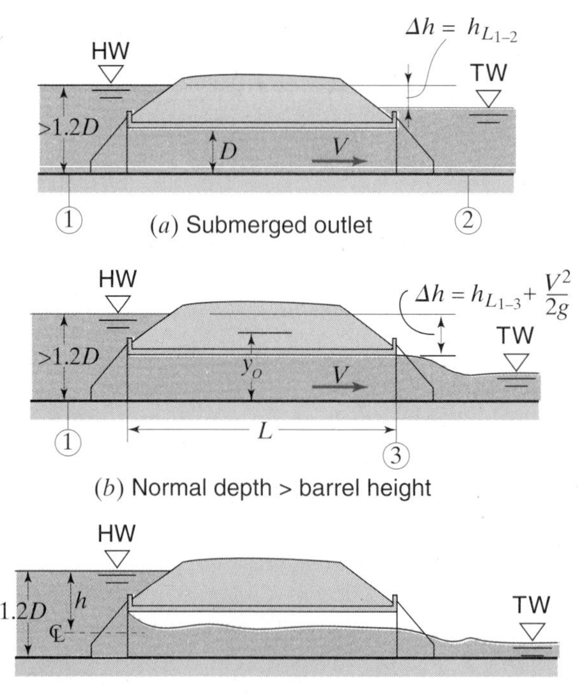

Figure 10.33
Flow conditions in culverts with submerged entrance.

▢ : Programmed computing aids (Appx. B) could help solve problems marked with this icon.

of inadequate channel capacity downstream or due to backwater from a connecting stream. In (*b*) the normal depth y_0 of the flow is greater than the culvert height D, causing the culvert to flow full.[10] The same equation is applicable to both cases (*a*) and (*b*), namely

$$\Delta h = h_e + h_f + h_v \tag{10.54}$$

where Δh is defined[11] in Figs. 10.33*a* and 10.33*b*, h_e is the entrance head loss, h_f is the friction head loss in the culvert barrel, and h_v is the velocity head loss at submerged discharge in case (*a*) or the residual velocity head at discharge in case (*b*).

Entrance loss is a function of the velocity head in the culvert, while friction loss may be computed using Manning's equation [Eq. (10.8)]. Thus, in BG units,[12]

Outlet control, case (*a*) or (*b*):
$$\Delta h = k_e \frac{V^2}{2g} + \frac{n^2 V^2 L}{2.21 R_h^{4/3}} + \frac{V^2}{2g} \tag{10.55}$$

This expression in BG units can be reduced to

Outlet control, case (*a*) or (*b*):
$$\Delta h = \left(k_e + \frac{29.2 n^2 L}{R_h^{4/3}} + 1\right)\frac{V^2}{2g} \tag{10.56}$$

The entrance coefficient k_e is about 0.5 for a square-edged entrance and about 0.05 if the entrance is well rounded. If the outlet is submerged, the head loss may be reduced somewhat by flaring the culvert outlet so that the outlet velocity is reduced and some of the velocity head recovered. Tests show that the flare angle should not exceed about 6° for maximum effectiveness.

To determine which of cases (*b*) or (*c*) occurs when the outlet is free (not submerged), we need to find if normal flow in the barrel will fill it. Usually the discharge is known or assumed. For rectangular culverts the normal depth can be solved in the usual way from Manning's Eq. (10.8) by trial and error. For circular cross sections it is easier to use Manning's equation to find the diameter which would just flow full ($R_h = D/4$), and to compare that with the actual or proposed diameter. If alternative slopes are being considered with a given barrel diameter, the algebra can be rearranged to solve for the slope that just causes the barrel to flow full.

If normal depth in the culvert is less than the barrel height, with the inlet submerged and the outlet free, the condition (*c*) illustrated in Fig. 10.33*c* will normally result. This culvert is said to be flowing under ***entrance control***, i.e.,

[10] If there is a contraction of flow at the culvert entrance, reexpansion will require about six diameters. Hence a very short culvert may not flow full even though $y_0 \geq D$.

[11] The energy grade line at the entrance and exit is above the actual water surface by the velocity head of the approaching or leaving water. With water ponded at the entrance or exit, this velocity head is usually negligible but should be included in computations if it is of significant magnitude.

[12] In SI units, the 2.21 disappears from the denominator of the second term on the right-hand side of Eq. (10.55) and the constant 29.2 in the numerator of the second term in parentheses of Eq. (10.56) becomes 19.62.

the entrance will not admit water fast enough to fill the barrel, and the discharge is determined by the entrance conditions. The inlet functions like an orifice for which

Entrance control, case (c):
$$Q = C_d A\sqrt{2gh} \tag{10.57}$$

where h is the head on the center of the orifice[13] and C_d is the orifice coefficient of discharge. The head required for a given flow Q is therefore

Entrance control, case (c):
$$h = \frac{1}{C_d^2}\frac{Q^2}{2gA^2} \tag{10.58}$$

It is impractical to cite appropriate values of C_d, because of the wide variety of entrance conditions which may be encountered; for a specific design this must be determined from model tests or tests of similar entrances. For a sharp-edged entrance without suppression of the contraction $C_d = 0.62$, while for a well-rounded entrance C_d approaches unity. If the culvert is set with its invert at stream-bed level, the contraction is suppressed at the bottom. Flared wingwalls may also cause partial suppression of the side contractions.

Free Entrance

Some box culverts may be designed so that the top of the box forms the roadway. In this case the headwater should not submerge the inlet, and one of the flow conditions of Fig. 10.34 (free entrance) will exist. In cases (a) and

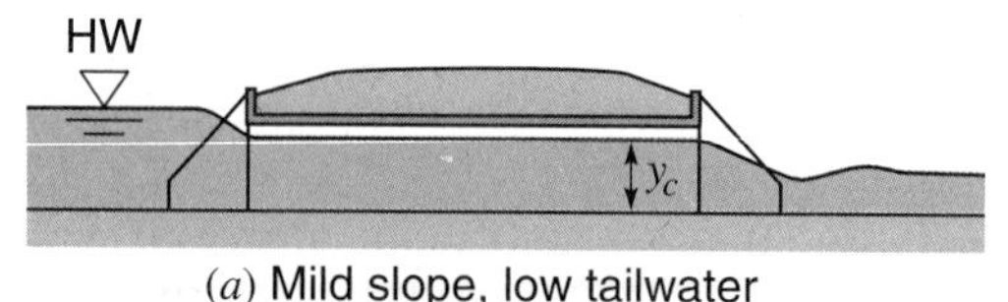

(a) Mild slope, low tailwater

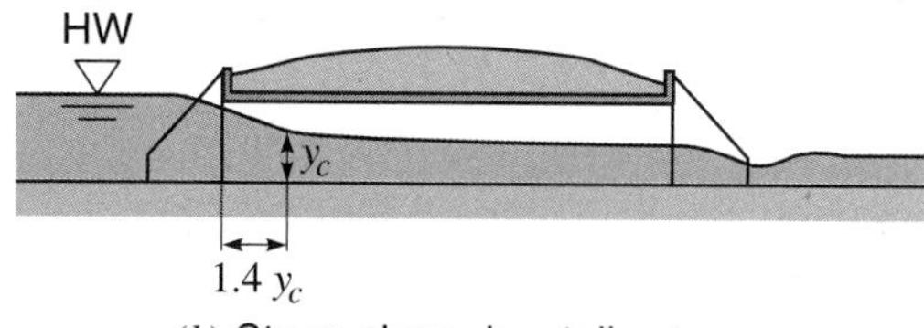

(b) Steep slope, low tailwater

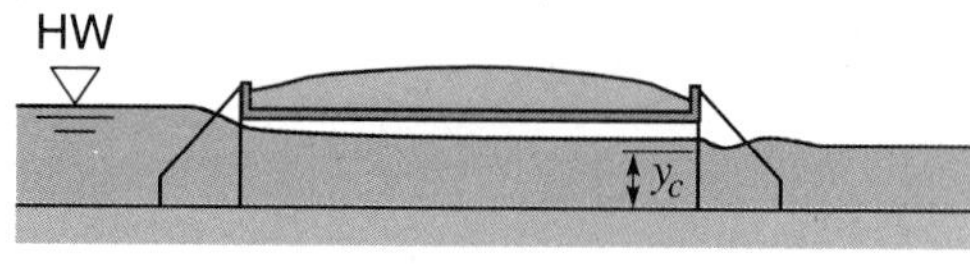

(c) Mild slope, tailwater submerges y_c

Figure 10.34
Flow conditions in culverts with free entrance.

[13] Equation (10.57) applies for an orifice on which the head h to the center of the orifice is large compared with the orifice height D. When the headwater depth is $1.20D$ (that is, $h = 0.7D$), an error of 2% results. In the light of other uncertainties in the design this error can be ignored.

(*b*) critical depth in the barrel controls the headwater elevation, while in case (*c*) the tailwater elevation is the control. In all cases the headwater elevation may be computed using the principles of open-channel flow discussed in this chapter with an allowance for entrance and exit losses.

When the culvert is on a steep slope [case (*b*)], critical depth will occur at about $1.4y_c$ downstream from the entrance. The water surface will impinge on the headwall when the headwater depth is about $1.2D$ if y_c is $0.8D$ or more. Since it would be inefficient to design a culvert with y_c much less than $0.8D$, a headwater depth of $1.2D$ is approximately the boundary between free-entrance conditions (Fig. 10.34) and submerged-entrance conditions (Fig. 10.33).

Entrance and Outlet Structures

The geometry of the entrance structure is an important aspect of culvert design. Entrance structures (Fig. 10.35) serve to protect the embankment from erosion and, if properly designed, may improve the hydraulic characteristics of the culvert. The straight endwall (*a*) is used for small culverts on flat slopes when the axis of the stream coincides with the culvert axis. If an abrupt change in flow direction is necessary, the L endwall (*b*) is used. The U-shaped endwall (*c*) is the least efficient form from the hydraulic viewpoint

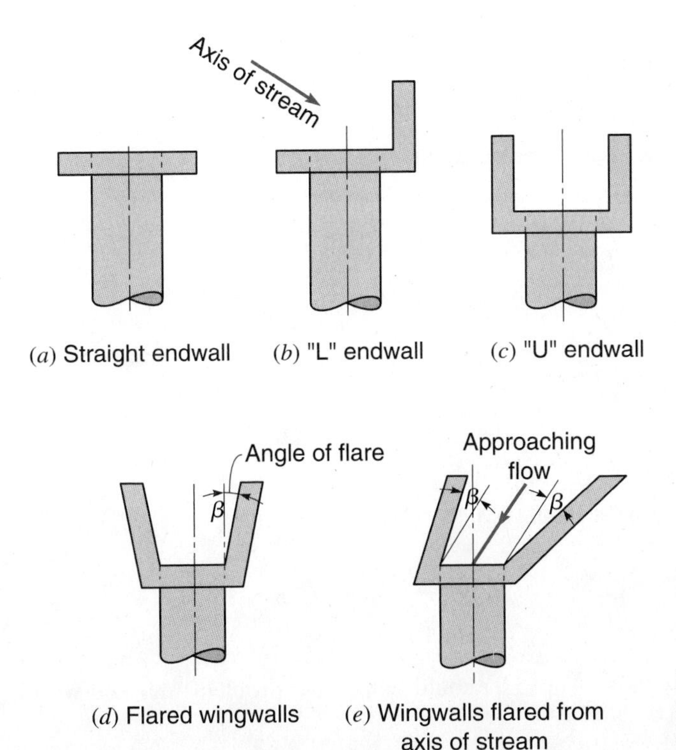

Figure 10.35 Culvert endwalls and wingwalls.

and has the sole advantage of economy of construction. Where flows are large, the flared wingwall (*d*) is preferable. The flare should, however, be made from the axis of the approaching stream (*e*) rather than from the culvert axis. Some hydraulic advantage is gained by warping wingwalls into a smooth transition, but the gain is not usually sufficient to offset the cost of the complex forming required for such warped surfaces.

The purpose of the culvert outlet is to protect the downstream slope of the fill from erosion and prevent undercutting of the culvert barrel. Where the discharge velocity is low or the channel below the outlet is not subject to erosion, a straight endwall or a U-shaped endwall may be quite sufficient. At higher velocities, lateral scour of the embankment or channel banks may result from eddies at the end of the walls, especially when the culvert is much narrower than the outlet channel. With moderate velocities, flaring of outlet wingwalls is helpful, but the flare angle β must be small enough so that the stream from the culvert will adhere to the walls of the transition.

SAMPLE PROBLEM 10.13 A culvert under a road must be 30 m long, have a slope of 0.003, and carry 4.3 m^3/s. If the maximum permissible headwater level is 3.6 m above the culvert invert, what size of corrugated-pipe culvert ($n = 0.025$) would you select? Neglect velocity of approach. Assume a square-edged inlet with $k_e = 0.5$ and $C_d = 0.65$. The outlet will discharge freely.

Solution

Sec. 10.23: Assume $D < 3.0$ m, i.e., that headwater depth/$D > 3.6/3.0 = 1.2$, i.e., assume that the entrance is submerged.

Given the discharge is free, so Fig. 10.33*a* cannot apply.

Conditions are those of either Fig. 10.33*b* or 10.33*c*.

Assume case (*b*), Fig. 10.33, i.e., that the barrel flows full.

$$V = \frac{Q}{A} = \frac{4.3}{\pi D^2/4} = \frac{5.47}{D^2}; \qquad R = D/4$$

Fig. 10.33(*b*):

$$\Delta h = h_{L_{1-3}} + \frac{V^2}{2g} = (y_1 - y_3) + (z_1 - z_3)$$

$$= y_1 - y_3 + S_0 L$$

$$= 3.6 - D + 0.003(30)$$

$$= 3.69 - D$$

: Programmed computing aids (Appx. B) could help solve problems marked with this icon.

Eq. (10.56): $$\Delta h = \left[0.5 + \frac{19.62(0.025)^2 30}{(D/4)^{4/3}} + 1\right]\frac{5.47^2}{2(9.81)D^4}$$

$$= \left(1.5 + \frac{2.34}{D^{4/3}}\right)\frac{1.528}{D^4}$$

Equating the two Δh expressions and simplifying;

$$3.69 = D + \left(1.5 + \frac{2.34}{D^{4/3}}\right)\frac{1.528}{D^4}$$

By trial and error (see Sample Prob. 3.5) or by equation solver, $D = 1.196$ m. Thus the first assumption ($D < 3$ m) is OK.

Now to determine if we have case (b) or case (c); find the maximum diameter d_0 that will just flow full with normal (uniform) flow:

Eq. (10.8b): $$4.3 = \frac{1}{0.025}\frac{\pi d_0^2}{4}\left(\frac{d_0}{4}\right)^{2/3}(0.003)^{1/2}; \quad d_0 = 1.994 \text{ m}$$

As $D < d_0$, the culvert runs full, we do have case (b), as assumed. The above assumptions and analysis are valid. $D = 1.196$ m. Use standard $D =$ 1.2 m ***ANS***

EXERCISE

10.23.1 Repeat Sample Prob. 10.13 for the situation where the culvert must be 100 m long.

10.24 FURTHER TOPICS IN OPEN-CHANNEL FLOW

The basics of open-channel flow discussed in this chapter are just an introduction to many more advanced subjects.

With steady gradually varied flow, a variety of specialized techniques have been devised to solve for water-surface profiles in irregular, natural channels. Ways have been developed to treat the dividing of flow of around islands, and the joining of flows at ***confluences***. When a canal connects two reservoirs

: Programmed computing aids (Appx. B) could help solve problems marked with this icon.

having varying levels, the discharge in the canal, which depends on these levels, is called its ***delivery***, and ***delivery curves*** may be developed to relate the discharge to the depth at each end. When flow from a reservoir or lake enters a channel with a mild slope, the discharge may depend on the length of the channel if there is a downstream control such as a contraction or a free overfall; this is known as ***the discharge problem***. Steady flow along a channel may vary with distance, such as in gutters and with side-channel spillways. This is known as ***spatially varied flow***, for which the water surface profiles are usually solved by trial and error procedures.

Near channel ***appurtenances*** like weirs (Secs. 11.11 and 11.12), spillways (Sec. 11.13), bridge piers, flumes, transition structures (Sec. 10.22), culverts (Sec. 10.23), and stilling basins (Sec. 10.18), an analysis of the resulting steady rapidly varied flow conditions is often quite involved. These are specialized subjects, which depend strongly on details of the appurtenances. ***Standing waves*** and ***cross waves***, introduced in Sec. 10.20, are rapidly varying flow phenomena that often occur with supercritical flow. They can be very complex, and are strongly influenced by the channel shape.

With ***unsteady flow*** in open channels the depth and/or the velocity varies with time at a point. This is an advanced subject not addressed in this text. The flow may be gradually varied or rapidly varied. Gradually varied unsteady flow includes flood waves, tidal flows, and waves resulting from the gradual change of channel control structures such as gates. It is governed by differential equations that usually require numerical methods and computers to solve them. Rapidly varied unsteady flow includes such flows as ***pulsating flow (roll waves)***, which occurs in supercritical flow on very steep slopes, and ***surges*** or ***surge waves*** (moving hydraulic jumps) caused by the rapid operation of a control structure or the sudden failure of a dam. In some locations surges are caused by tidal action, when the surge is known as a ***tidal bore***. Although some very simplified problems of these types can be solved analytically, in most cases more realistic problems must be solved by numerical methods using computers.

Phenomena governed by density include ***salt wedges*** at estuaries where fresh river water overlies saline ocean water, and similar ***cooling water wedges*** in rivers caused by waste heat from power plants. Density effects in lakes and reservoirs in many cases result in an annual overturning of the contents due to seasonal temperature changes. Density currents may also be caused by sediment loads, which in turn relate to the subjects of ***scour*** and ***sediment transport***.

Other more advanced phenomena of interest arise when analyzing the fate of pollutants, and include ***advection*** and mixing processes known as ***turbulent diffusion*** and ***dispersion***. Again, solution of these processes usually requires numerical methods using computers.

Excellent reference books specializing in open channel flow include *Open-Channel Hydraulics* by Richard H. French, McGraw-Hill, Inc., New York, 1985, *Open-Channel Hydraulics* by Ven Te Chow, McGraw-Hill Book Co., Inc., New York, 1959, and *Open Channel Flow* by F. M. Henderson, Macmillan Co., New York, 1966.

PROBLEMS

10.1 For the channel of Sample Prob. 10.1 compute the "open-channel Reynolds number" assuming that water at 50°F is flowing. Refer to Fig. 8.11 to verify whether or not the flow is fully rough. Determine e from Fig. 8.11 and compare it with the value computed in the example.

10.2 Assuming the values of f versus **R** and e/D given for pipes in Fig. 8.11 also apply to open channels, find the rate of discharge of water at 60°F in a 100-in-diameter smooth concrete pipe flowing half full, if the pipe is laid on a grade of 2 ft/mi. Why must D be replaced by $4R_h$?

10.3 Figure P10.3 shows a cross section of a canal that is to carry 2000 cfs. The canal is lined with concrete, for which n is 0.014. (*a*) What is the slope of this canal, and what is the drop in elevation per mile? (*b*) If the flow in the canal were to decrease to 1000 cfs, all other data, including the slope and n, being the same, what would be the depth of the water?

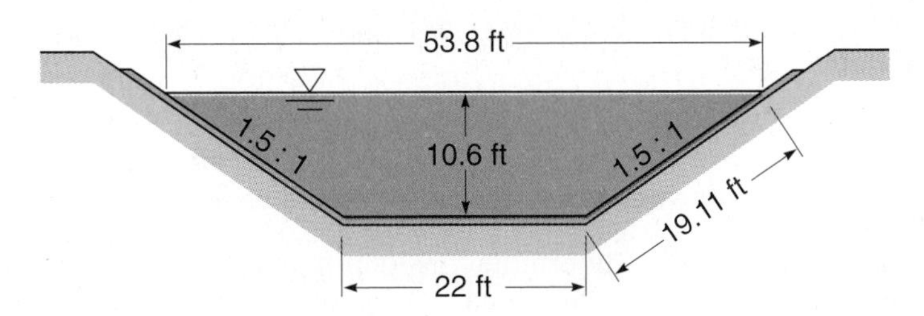

Figure P10.3

10.4 In Prob. 10.3 find the corresponding value of e and compare it with values previously given for concrete pipe. Does it fall in the range given?

10.5 Figure P10.5 shows a cross section of a canal forming a portion of the Colorado River Aqueduct, which is to carry 44 m^3/s. The canal is lined with concrete, for which n is 0.014. (*a*) What is the slope of this canal, and what is the drop in elevation per kilometer? (*b*) If the flow in the canal were to decrease to 22 m^3/s, all other data, including the slope and n, being the same, what would be the depth of the water?

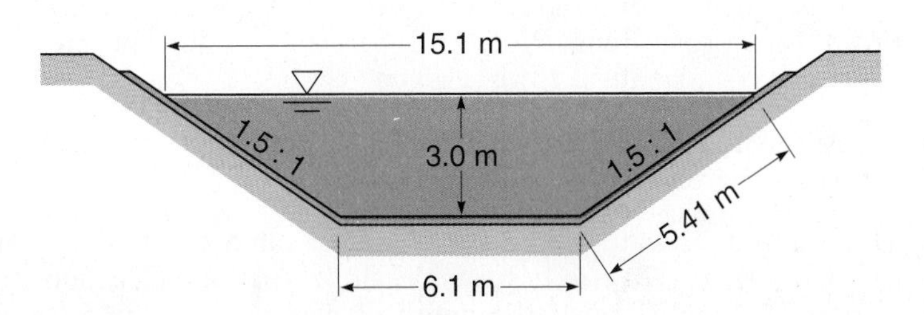

Figure P10.5

■: Programmed computing aids (Appx. B) could help solve problems marked with this icon.

10.6 In Prob. 10.5 find the corresponding value of e and compare it with values previously given for concrete pipe. Does it fall in the range given?

10.7 A monolithic concrete inverted siphon is circular in cross section and is 20 ft in diameter. Obviously, it is completely filled with water. (a) If $n = 0.013$, find the slope of the hydraulic grade line for a flow of 2100 cfs. (b) Solve the same problem using the methods of Chap. 8 and assuming a mean value of e from Table 8.1 for concrete pipe. Compare the result with that of part (a).

10.8 A monolithic concrete inverted siphon on the Colorado River Aqueduct is circular in cross section and is 4.88 m in diameter. Obviously, it is completely filled with water. (a) If $n = 0.013$, find the slope of the hydraulic grade line for a flow of 45 m^3/s. (b) Solve the same problem using the methods of Chap. 8 and assuming a mean value of e from Table 8.1 for concrete pipe. Compare the result with that of part (a).

10.9 A 36-in-diameter pipe is known to have a Manning's n of 0.024. What is Manning's n for a 108-in-diameter pipe having exactly the same e as the smaller pipe?

10.10 Refer to Fig. X10.3.6. Find the flow rate at water depths of 1, 2.5, 4, 5.5, and 7 ft if $n = 0.025$ and $S = 0.0020$. The dimensions are as follows: $a = 4$ ft, $b = 5$ ft, $d = 7$ ft, and $w = 35$ ft.

10.11 Using Eq. (10.12), determine the depth below the surface at which the velocity is equal to the mean velocity. Also find the average of the velocities at 0.2 and 0.8 depths. Let $y = 6$ ft, $S = 0.0005$, and $n = 0.024$.

10.12 Prove that the most efficient hydraulic trapezoidal cross section discussed in Sec. 10.6 envelopes a semicircle with its center in the water surface.

10.13 Consider a variety of rectangular sections, all of which have a cross-sectional area of 20 ft^2. Plot the hydraulic radii versus channel widths for a range of channel widths from 2 to 20 ft and note the depth:width ratio when R_h is maximum.

10.14 Set up a general expression for the wetted perimeter P of a trapezoidal channel in terms of the cross-sectional area A, depth y, and angle of side slope ϕ. Then differentiate P with respect to y with A and ϕ held constant. From this prove that $R_h = y/2$ for the section of greatest hydraulic efficiency (i.e., smallest P for a given A).

10.15 Using the results of Prob. 10.14, prove that the most efficient triangular section is the one with a 90° vertex angle.

10.16 The amount of water to be carried by a canal excavated in smooth earth ($n = 0.030$) is 370 cfs. It has side slopes of 2 :1 (see figure), and a bed slope of 2.5 ft/mi. (a) If the depth of water y is to be 5 ft, what must be

---

▣: Programmed computing aids (Appx. B) could help solve problems marked with this icon.

the bottom width b? (b) How does this compare with the most efficient trapezoidal section for these side slopes?

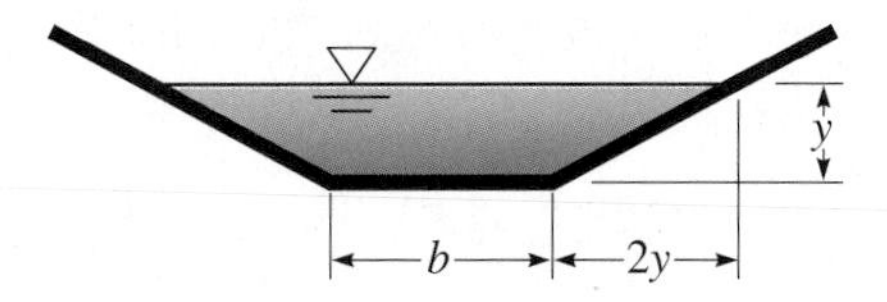

Figure P10.16

10.17 The amount of water to be carried by a canal excavated in smooth earth ($n = 0.030$) is 12 m^3/s. It has side slopes of 2:1 (see Fig. P10.16), and a bed slope of 60 cm/km. (a) If the depth of water y is to be 2.0 m, what must be the bottom width b? (b) How does this compare with the most efficient trapezoidal section for these side slopes?

10.18 Refer to Fig. P10.16. If the discharge in the canal ($n = 0.030$) is to be 200 cfs while the depth y is 5 ft and the velocity is not to exceed 150 ft/min, what must be the bottom width b and the drop in elevation per mile? Compare this with the bottom width for maximum efficiency (same slope).

10.19 Refer to Fig. P10.16. If the discharge in the canal ($n = 0.030$) is to be 6 m^3/s while the depth y is 1.8 m and the velocity is not to exceed 50 m/min, what must be the bottom width b and the drop in elevation per km? Compare this with the bottom width for maximum efficiency (same S).

10.20 Prove that the value of θ given in Sec. 10.7 for the point of maximum discharge (assuming constant n) is correct. After differentiating, a trial-and-error type of solution will be found most practical here.

10.21 Evaluate the friction factor f for laminar flow in a wide and shallow open channel in terms of the Reynolds number, and compare it with f for pipe flow using Eq. (8.20). (*Note*: Recall that for a wide and shallow channel the hydraulic radius is approximately equal to the depth.)

10.22 In Sec. 10.8 the velocity distribution in laminar sheet flow was found to be given by $u = (gS/2\nu)y_0^2[1 - (y/y_0)^2]$, where y in this case is the variable distance downward from the surface. Evaluate α, and compare it with the result of Sample Prob. 5.1.

10.23 Consider a wide rectangular channel on a given slope. With what power of the discharge does the depth vary? With what power of the discharge does the critical depth vary? As the flow increases, does the Froude number increase or decrease? Assume Manning's equation, with a constant value of n.

10.24 Differentiate Eq. (10.17) to obtain the expression for y_c given in Eq. (10.23).

: Programmed computing aids (Appx. B) could help solve problems marked with this icon.

10.25 A flow of 140 cfs is carried in a rectangular channel 12 ft wide at a depth of 1.8 ft. If the channel is made of smooth concrete ($n = 0.012$), find the slope necessary to sustain uniform flow at this depth. What roughness coefficient would be required to produce uniform critical flow for the given rate of discharge on this slope?

10.26 A rectangular channel 12 ft wide carries a flow of 240 cfs. Find the critical depth and the critical velocity for this flow. Find also the critical slope if $n = 0.015$.

10.27 Water flows with a velocity of 6 fps and at a depth of 3 ft in a wide rectangular channel. Is the flow subcritical or supercritical? Find the alternate depth for the same discharge and specific energy by two methods: (*a*) by direct solution of Eq. (10.17); (*b*) by use of Fig. 10.13.

10.28 Water is released from a sluice gate in a rectangular channel 6 ft wide such that the depth is 3 ft and the velocity is 18 fps. Find (*a*) the critical depth for this specific energy; (*b*) the critical depth for this rate of discharge; (*c*) the type of flow and the alternate depth by either direct solution or the discharge curve.

10.29 Water is released from a sluice gate in a rectangular channel 1.5 m wide such that the depth is 0.6 m and the velocity is 4.5 m/s. Find (*a*) the critical depth for this specific energy; (*b*) the critical depth for this rate of discharge; (*c*) the type of flow and the alternate depth by either direct solution or the discharge curve.

■ **10.30** A trapezoidal canal with side slopes of 2:1 ($H:V$) has a bottom width of 10 ft and carries a flow of 600 cfs. (*a*) Find the critical depth and critical velocity. (*b*) If the canal is lined with brick ($n = 0.015$), find the critical slope for the same rate of discharge.

■ **10.31** A trapezoidal canal with side slopes of 1:1 has a bottom width of 4 m and carries a flow of 25 m^3/s. (*a*) Find the critical depth and critical velocity. (*b*) If the canal is lined with brick ($n = 0.015$), find the critical slope for the same rate of discharge.

■ **10.32** For a circular conduit with a diameter of 12 ft, compute the specific energy for a flow of 120 cfs at depths of 2, 4, 6, and 10 ft assuming $\alpha = 1.0$. At what depth is E the least? Check to see if Eq. (10.33) is satisfied at this depth.

■ **10.33** Figure P10.33 describes the cross section of an open channel for which $S_0 = 0.03$ and $n = 0.020$. The sketch is drawn to the scale shown. When the flow rate is 120 cfs, find (*a*) the depth for uniform flow and (*b*) the critical depth. *Hint*: The cross-sectional area may be found by planimetry or counting squares and the wetted perimeter by use of dividers.

■: Programmed computing aids (Appx. B) could help solve problems marked with this icon.

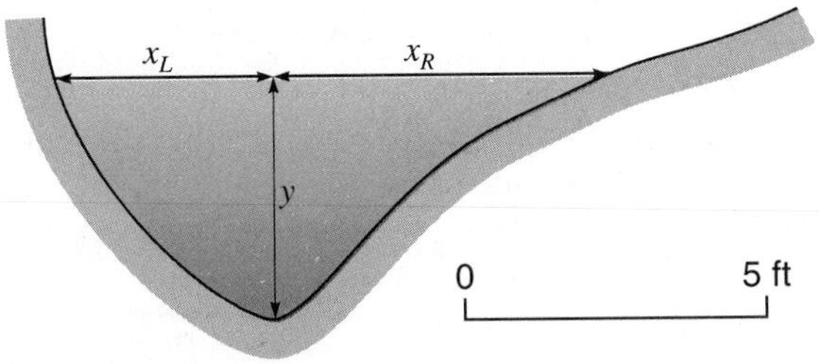

y, ft	x_L, ft	x_R, ft
0	0	0
1	1.5	1.1
2	2.6	2.0
3	3.25	3.3
4	3.6	5.5
5	3.85	8.2

x_L x_R y 0 5 ft

Figure P10.33

10.34 Refer to the figure, table, and hint for Prob. 10.33, replacing feet dimensions with meters. Let the slope be 0.009 with $n = 0.020$. When the flow rate is 60 m^3/s, find (*a*) the depth for uniform flow; (*b*) the critical depth.

10.35 Work Sample Prob. 10.8 ($y_0 = 2.0$ ft) for the case where the flow rate is 275 cfs.

10.36 A flow of 2.0 m^3/s is carried in a rectangular channel 1.8 m wide at a depth of 1.0 m. Will critical depth occur at a section where (*a*) a frictionless hump 18 cm high is installed across the bed, (*b*) a frictionless sidewall constriction (with no hump) reduces the channel width to 1.4 m, and (*c*) the hump and the sidewall constriction are installed together? Show calculations.

10.37 Work Sample Prob. 10.8 ($y_0 = 2.0$ ft) for the case where the flow rate is 90 cfs.

10.38 A rectangular channel 5 ft wide carries 50 cfs of water in uniform flow at a depth of 3.20 ft. If a bridge pier 1.5 ft wide is placed in the middle of this channel, find the local change in the water-surface elevation. What is the minimum width of constricted channel which will not cause a rise in water surface upstream?

10.39 A rectangular channel 1.2 m wide carries 1.1 m^3/s of water in uniform flow at a depth of 0.85 m. If a bridge pier 0.3 m wide is placed in the middle of this channel, find the local change in the water-surface elevation. What is the minimum width of constricted channel which will not cause a rise in water surface upstream?

10.40 A rectangular channel 12 ft wide carries 24 cfs in uniform flow at a depth of 0.32 ft. Find the local change in water-surface elevation caused by a 0.15-ft-high obstruction across the channel bed.

10.41 Fifty cubic feet per second of water flows uniformly in a 6-ft-wide channel at a depth of 2.5 ft. What is the change in water-surface elevation at a section contracted to a 4-ft width with a 0.2-ft depression in the bottom?

: Programmed computing aids (Appx. B) could help solve problems marked with this icon.

10.42 Water flows uniformly in a 2.0-m-wide channel at a rate of 1.6 m^3/s and a depth of 0.75 m. What is the change in water-surface elevation at a section contracted to a 1.4-m width with an 8-cm depression in the bottom?

10.43 A 5-ft-wide rectangular flume ($n = 0.012$) carries 60 cfs of water. At one point the water depth is found to be 4 ft; 1000 ft downstream it is measured at 3 ft. Calculate the bed slope of the flume, using one reach. Sketch the flume bed, water surface, and energy line, to check that the answer is reasonable.

10.44 A 1.6-m-wide rectangular flume ($n = 0.013$) carries 1.9 m^3/s of water. At one point the water depth is found to be 1.3 m; 320 m downstream it is measured at 1.0 m. Calculate the bed slope of the flume, using one reach. Sketch the flume bed, water surface, and energy line, to check that the answer is reasonable.

10.45 Suppose that the slope of the flume in Prob. 10.43 is now 0.01. With the same flow as before, find the depth 50 ft downstream from a section where the flow is 1.5 ft deep. Use only one reach. Is the flow subcritical or supercritical? [*Note*: In this case it will not be possible to make a direct solution from Eq. (10.37). A trial-and-error solution may best be set in the form of a table with the headings y_2, V_2, V_m, A_2, P_2, R_2, R_m, $R_m^{2/3}$, S, etc.]

10.46 Suppose that the slope of the flume in Prob. 10.44 is now 0.008. With the same flow as before, find the depth 17 m downstream from a section where the flow is 50 cm deep. Use only one reach. Is the flow subcritical or supercritical? [*Note*: In this case it will not be possible to make a direct solution from Eq. (10.37). A trial-and-error solution may best be set in the form of a table with the headings y_2, V_2, V_m, A_2, P_2, R_2, R_m, $R_m^{2/3}$, S, etc.]

10.47 Repeat Prob. 10.45, but increase the 50-ft distance to 500 ft.

10.48 Repeat Prob. 10.46, but increase the 17-m distance to 170 m.

10.49 A rectangular flume 9 ft wide is built of planed timber ($n = 0.012$) on a bed slope of 0.3 ft per 1000 ft, ending in a free overfall. If the measured depth at the fall is 1.67 ft, find (*a*) the rate of flow; (*b*) the distance upstream from the fall to where the depth is 3.6 ft. [*Note*: Assume that critical depth occurs at a distance of $4y_c$ upstream from the fall, and employ reaches with end depths of 2.5, 2.8, 3.2, and 3.6 ft.]

10.50 A rectangular flume 3 m wide is built of planed timber ($n = 0.012$) on a bed slope of 20 cm per km, ending in a free overfall. If the measured depth at the fall is 0.55 m, find (*a*) the rate of flow; (*b*) the distance upstream from the fall to where the depth is 1.2 m. [*Note*: Assume that critical depth occurs at a distance of $4y_c$ upstream from the fall, and employ reaches with end depths of 0.82, 0.9, 1.0, and 1.2 m.]

: Programmed computing aids (Appx. B) could help solve problems marked with this icon.

10.51 A wide and shallow rectangular channel dredged in earth ($n = 0.035$) is laid on a slope of 9 ft/mi and carries a flow of 90 cfs/ft of width. (*a*) Find the water depth 2 mi upstream of a section where the depth is 26.1 ft, using a single reach. (*b*) Compare the result with that obtained using three reaches of equal length.

10.52 The slope of a stream of rectangular cross section is $S_0 = 0.0003$, the width is 170 ft, and the value of the Chézy C is 78.3 $ft^{1/2}/s$. (*a*) Find the depth for a uniform flow of 101.54 cfs per foot of width of the stream. (*b*) If a dam raises the water level so that at a certain distance upstream the increase is 5 ft, how far from this latter section will the increase by only 1 ft? Use reaches with 1-ft depth increments.

10.53 A portion of an outfall sewer is approximately a circular conduit 5 ft in diameter and with a slope of 1 ft in 1100 ft. It is of brick, for which $n = 0.013$. (*a*) What would be its maximum capacity for uniform flow. (*b*) If it discharges 120 cfs with a depth at the end of 2.90 ft, how far back from the end must it become a pressure conduit? Proceeding from the mouth upstream, find by tabular solution the length of sewer that is not flowing full. Use three reaches with equal depth increments. (*c*) Find the pressure at the top and bottom of the pipe at the section that is 600 ft upstream from the point at which the sewer no longer flows full. Sketch the energy line and hydraulic grade line from this point to the end of the sewer.

10.54 For the channel of Prob. 10.33 find the distance between a section where the depth is 3.5 ft to another where the depth is 3.0 ft. Which section is upstream? Use only one reach. See the hint of Prob. 10.33.

10.55 When the flow in a certain natural stream is 7600 cfs, it is required to find the elevation of the water surface at different sections upstream from a certain initial point. A survey of the channel shows that conditions are fairly similar for a length of 1500 ft upstream from the initial point, and then beyond that there is another stretch of 2200 ft, and so on. Assuming a rise in the water surface in the distance of 1500 ft to be 0.20 ft, a study of the stream bed shows the average values of the area and wetted perimeter to be as given in the table below. Using Eq. (10.6), the computed head loss, based on average velocity and hydraulic radius, is seen to be 0.283 ft, which is greater than that assumed. Hence assume a larger value, and repeat. Note that average areas and wetted perimeters have also been determined for alternative water-surface slopes. Complete the following table, and find the probable rise in elevation in the first 1500 ft. [In a similar manner, the rises in other lengths may be computed, and the sum of all of them up to the desired point will give the elevation at that point above the initial.] Assume $n = 0.036$.

■: Programmed computing aids (Appx. B) could help solve problems marked with this icon.

Assumed rise SL, ft	A average, ft^2	P average, ft	R average, ft	V average, fps	$SL = LV^2/C^2R$, ft
0.20	3100	350	8.86	2.45	0.283
0.25	3180	359	8.86		
0.26	3190	360	8.86		
0.27	3220	363	8.86		
0.28	3230	364	8.86		

10.56 Classify the water-surface profile of Prob. 10.45 as one of the forms shown in Fig. 10.20. Show all necessary calculations. Sketch the profile.

10.57 Classify the water-surface profile of Prob. 10.46 as one of the forms shown in Fig. 10.20. Show all necessary calculations. Sketch the profile.

10.58 A trapezoidal canal dreged in smooth earth ($n = 0.030$) has a bottom width of 15 ft, side slopes of 1 : 1, and a bed slope $S_0 = 0.0003$. With a flow of 800 cfs, $y_c = 4.05$ ft, and $y_0 = 10.8$ ft. Find the length of M_2 curve extending from a free overfall back to where the depth is 10 ft. Use reaches with end depths of 6, 8, and 10 ft.

10.59 A rectangular channel 10 ft wide carries 120 cfs in uniform flow at a depth of 2.00 ft. Suppose that an obstruction such as a submerged weir is placed across the channel, rising to a height of 8 in above the bottom. (*a*) Does this obstruction cause a hydraulic jump to form upstream? Why or why not? (*b*) Find the water depth over the obstruction, and just upstream of it. Classify the surface profile, if possible, to be found upstream from the weir. Sketch the resulting water surface profile and energy line, showing y_c and y_0.

10.60 Suppose that the slope and roughness of the channel in Prob. 10.59 are such that uniform flow of 120 cfs occurs at a depth of 1.20 ft. Consider an obstruction rising 3 in above the bottom of the channel. Will a hydraulic jump form upstream? Classify the surface profile found just upstream from the obstruction.

10.61 Analyze the water-surface profile in a long rectangular channel ($n = 0.014$). The channel is 12 ft wide, the flow rate is 480 cfs, and there is an abrupt change in slope from 0.0019 to 0.0180. Make a sketch showing normal depths, critical depths, and water surface profile types.

10.62 Repeat Prob. 10.61 for the case where the flow rate is 180 cfs.

10.63 Repeat Prob. 10.61 for the case where the slope change is from 0.0019 to 0.0007, and the flow rate is 180 cfs. Compute the approximate distance upstream from the break to the point where normal depth occurs, using one reach.

: Programmed computing aids (Appx. B) could help solve problems marked with this icon.

10.64 In an 8-ft-wide rectangular channel ($n = 0.015$) water flows at 300 cfs. A low dam (broad-crested weir) placed in the channel raises the water to a depth of 8.0 ft. Analyze the water-surface profile upstream from the dam if the channel slope is (*a*) 0.0006, (*b*) 0.0009, and (*c*) 0.006.

10.65 Derive Eq. (10.48) in the manner suggested in Sec. 10.18.

10.66 In a rectangular channel 10 ft wide with a flow of 200 cfs the depth is 1 ft. If a hydraulic jump is produced, (*a*) what will be the depth after it? (*b*) What will be the loss of energy?

10.67 In a rectangular channel 4 m wide with a flow of 7.65 m^3/s the depth is 0.4 m. If a hydraulic jump is produced, (*a*) what will be the depth after it? (*b*) What will be the loss of energy?

10.68 Repeat Prob. 10.66 for the case where the channel bed slopes at 10°. For this slope, jump length $\approx 4y_2$. Assume friction force = 400 lb/ft of width. Also find the horsepower loss.

10.69 Repeat Prob. 10.67 for the case where the channel bed slopes at 10°. For this slope, jump length $\approx 4y_2$. Assume friction force = 6 kN/m of width. Also find the power (kW) loss.

10.70 The tidal bore, which carries the tide into the estuary of a large river, is an example of an abrupt translatory wave, or moving hydraulic jump. Suppose such a bore is observed to rise to a height of 13 ft above the normal low-tide river depth of 10 ft. The speed of travel of the bore upstream is observed to be 13 mph. Find the approximate velocity of the undisturbed river. Does this represent subcritical or supercritical flow? (*Note*: The theory developed in Sec. 10.18 is based on the hydraulic jump in a fixed position. In the case of a moving jump, all kinematic terms must be considered relative to the moving wave as a frame of reference.)

10.71 A hydraulic jump occurs in a triangular flume having side slopes at 1:1. The flow rate is 18 cfs and the depth before the jump is 1.2 ft. Find the depth after the jump and the power loss in the jump.

10.72 A hydraulic jump occurs in a triangular flume having side slopes at 1:1. The flow rate is 0.45 m^3/s and the depth before the jump is 0.3 m. Find the depth after the jump and the power loss in the jump.

10.73 A wide and shallow rectangular channel with bed slope $S_0 = 0.0004$ and roughness $n = 0.022$ carries a steady flow of 65 cfs/ft of width. If a sluice gate (Fig. 11.34) is so adjusted as to produce a minimum depth of 1.6 ft in the channel, determine whether a hydraulic jump will form downstream, and if so, find (using one reach) the distance from the gate to the jump.

10.74 At a point in a shallow lake, the wave from a passing boat is observed to rise 0.5 ft above the undisturbed water surface. The observed speed of the wave is 8 mph. Find the approximate depth of the lake at this point. Compute it three ways and comment on them.

■: Programmed computing aids (Appx. B) could help solve problems marked with this icon.

10.75 At a point in a shallow lake, the wave from a passing boat is observed to rise 30 cm above the undisturbed water surface. The observed speed of the wave is 16 km/h. Find the approximate depth of the lake at this point. Compute it three ways and comment on them.

10.76 A rectangular channel 12 ft wide carries 360 cfs under uniform flow conditions. By how much should the outside wall be elevated above the inside wall for a bend of 45-ft radius to the center line of the channel if (*a*) the normal depth is 4.4 ft and (*b*) the normal depth is 2.2 ft.

10.77 Refer to Fig. 10.32. A rectangular channel changes in width from 5 to 7.5 ft. Measurements indicate that $y_1 = 3.00$ ft and $Q = 62$ cfs. Determine the depth y_2 by (*a*) neglecting head loss; (*b*) considering the head loss to be given as shown on the figure.

10.78 A rectangular channel changes in width from 1.35 m to 2.0 m with the wedge-shaped transition shown in Fig. 10.32. Given that $y_1 = 0.75$ m and $Q = 1.5\ \mathrm{m^3/s}$, find the depth y_2 by (*a*) neglecting head loss and (*b*) using a loss coefficient k_t of 0.4. (*c*) Does the head loss increase or decrease the flow expansion?

10.79 Flow from a trapezoidal channel (20-ft bed width, 1:1 side slopes) enters a 20-ft-wide rectangular channel via a cylinder–quadrant transition. (*a*) If the upstream depth is 4 ft when the flow rate is 530 cfs, find the resulting downstream depth. (*b*) If the flow increases to 550 cfs, causing the upstream depth to increase to 4.1 ft, does the downstream depth increase or decrease? By how much?

10.80 What is the capacity of a 5 ft by 5 ft concrete box culvert ($n = 0.013$) with a rounded entrance ($k_e = 0.05$, $C_d = 0.95$) if the culvert slope is 0.004, the length is 180 ft, and the headwater level is 7 ft above the culvert invert? Assume (*a*) free outlet conditions, (*b*) tailwater elevation 1 ft above top of box at outlet, and (*c*) tailwater elevation 2 ft above top of box at outlet. Neglect headwater and tailwater velocity heads.

10.81 Repeat Prob. 10.80(*a*) for the cases where the culvert slope is (*a*) 0.03 and (*b*) 0.07.

10.82 A culvert under a road must carry $4.5\ \mathrm{m^3/s}$. (*a*) If the culvert length is 32 m, the slope is 0.004, and the maximum permissible headwater level above the culvert invert is 3.8 m, what size of corrugated-pipe culvert ($n = 0.025$) would you select? The outlet will discharge freely. Neglect velocity of approach. Assume square-edged entrance with $k_e = 0.5$, $C_d = 0.65$. (*b*) Repeat for a culvert length of 110 m.

10.83 A 120-ft-long corrugated metal pipe ($n = 0.022$) of 30-in diameter is tested in a laboratory. The headwater is maintained at a level which is 5 ft above the pipe invert at entrance, and the outlet cannot be submerged. Assume a square-edged entrance with $ke = 0.5$, $C_d = 0.68$, and neglect headwater and tailwater velocity heads. Compute values of Q for S_0 values of 0.0, 0.01, 0.03, and 0.08. As the slope is increased, at what slope does the flow change to condition (*c*) of Fig. 10.33?

: Programmed computing aids (Appx. B) could help solve problems marked with this icon.

CHAPTER 11

Fluid Measurements

Both the engineer and the scientific investigator are often faced with the problem of measuring various fluid properties such as density, viscosity, and surface tension. Also, measurements are often required of various fluid phenomena, as pressure, velocity, and flow rate. In this chapter only the principles and theory of such measurements will be discussed. Detailed information on the various measuring devices may be found elsewhere in the literature.[1] This chapter draws on many concepts presented in previous chapters; hence it also serves as a good review.

11.1 MEASUREMENT OF FLUID PROPERTIES

The measurement of ***liquid density***, is most commonly determined by weighing a known volume of the liquid to find its specific weight (Sec. 2.3), and then dividing by the acceleration due to gravity. Another technique makes use of hydrostatic weighing, where a nonporous solid object of known volume is weighed (*a*) in air and then (*b*) in the liquid whose density is to be determined. The ***hydrometer*** is a variation of this technique. The densities are calculated from fluid statics. Another, though not very accurate, technique of determining liquid density is achieved by placing two immiscible liquids in a U tube, one of known density, the other of unknown density. From fluid statics the unknown density may be found. These various techniques of

[1] See, for example, *Fluid Meters: Their Theory and Application*, 6th ed., American Society of Mechanical Engineers, New York, 1971.

measuring liquid density are illustrated in Exers. 11.1.1–11.1.4 and Probs. 11.1 and 11.2.

The measurement of ***viscosity*** is generally made with a device known as a ***viscometer.*** Various types of viscometers are available. They all depend on the creation of laminar-flow conditions. We shall confine this discussion to the measurement of the viscosity of liquids. Since viscosity varies considerably with temperature, it is essential that the fluid be at a constant temperature when a measurement is being made. This is generally accomplished by immersing the device in a constant-temperature bath.

Several types of ***rotational viscometers*** are available. These generally consist of two concentric cylinders that are rotated with respect to one another. The narrow space between them is filled with liquid whose viscosity is to be measured. The rate of rotation under the influence of a given torque is indicative of the viscosity of the liquid. One difficulty with this type of viscometer is that mechanical friction must be accounted for, and this is difficult to deal with accurately.

The ***tube-type viscometer*** is perhaps the most reliable. Figure 11.1 shows the Saybolt viscometer. In this device the liquid is originally at M, with the bottom of the tube plugged. The plug is removed, and the time required for a certain volume of liquid to pass through the tube is a measure of the kinematic viscosity of the liquid. In this device the flow is unsteady and the tube is of such small diameter that the flow may be assumed to be laminar. Substituting the expression for mean velocity V for laminar flow in tubes [Eq. (8.19)] into the continuity expression [Eq. (4.3)] gives

$$Q = \frac{\pi D^4 \gamma h_L}{128 \mu L} \tag{11.1}$$

As an approximation, let h_L be the average imposed head during the flow

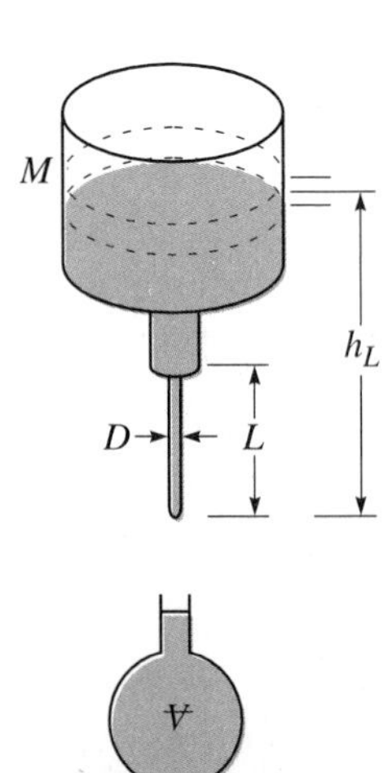

Figure 11.1
Falling-head tube-type viscometer.

period and let $Q \approx \forall/t$, where $\forall$ is the volume of liquid that flows out of the tube in time t. Substituting $Q = \forall/t$ into Eq. (11.1), we get

$$\nu = \frac{\pi D^4 h_L}{128 \forall L} gt \tag{11.2}$$

Since D, L, $\forall$, and h_L are constants of the device, $\nu = Kgt$, and the kinematic viscosity is seen to be proportional to the measured time. Equation (11.2) gives good results if the tube is relatively long. However, for a short tube, as with the Saybolt viscometer, a correction factor must be applied if the tube is too short for the establishment of laminar flow (Sec. 8.7).

There are several other types of tube viscometers, but they are all based on the same principle. Some come with a set of tubes of various diameters so that measurements can be made on liquids with a wide range of viscosities in a convenient time period. Because the dimensions of such fine tubes cannot be perfectly duplicated, each tube is individually calibrated by measuring the time for a liquid of known viscosity at a given temperature to discharge the standard volume.

The ***falling-sphere viscometer*** is a third type. In such a device the liquid is placed in a tall transparent cylinder and a sphere of known weight and diameter is dropped in it. If the sphere is small enough, Stokes' law (Sec. 9.7) will prevail and the fall velocity of the sphere will be approximately inversely proportional to the absolute viscosity of the liquid. That this is so may be seen by examining the free-body diagram of such a falling sphere (Fig. 11.2). The forces acting include gravity, buoyancy, and drag. Stokes' law states that if $DV/\nu < 1$, the drag force on a sphere is given by $F_D = 3\pi\mu VD$, where V is the velocity of the sphere and D is its diameter. When the sphere is dropped in a liquid, it will quickly accelerate to terminal velocity, at which $\sum F_z = 0$. Then

$$W - F_B - F_D = \gamma_s \frac{\pi D^3}{6} - \gamma \frac{\pi D^3}{6} - 3\pi\mu VD = 0$$

where γ_s and γ represent the specific weight of the sphere and liquid,

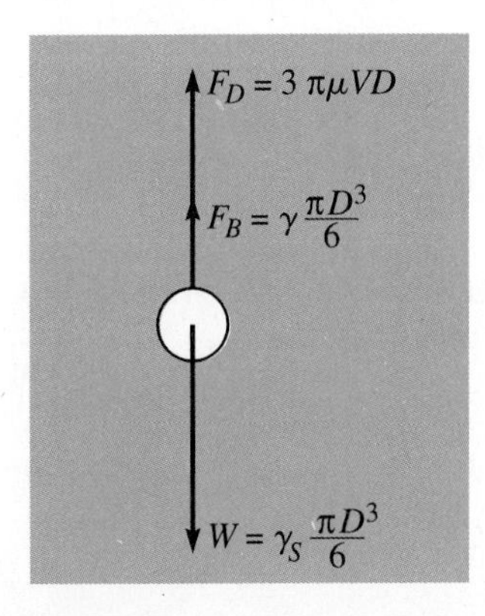

Figure 11.2
Free-body diagram of sphere falling at terminal velocity.

respectively. Solving the above equation, we get

$$\mu = \frac{D^2(\gamma_s - \gamma)}{18V} \tag{11.3}$$

In the preceding development it was assumed that the sphere was dropped into a liquid of infinite extent. In actuality, the liquid will be contained in a tube and a ***wall effect*** will influence the drag force and hence the fall velocity. It has been found that the wall effect[2] can be expressed approximately as

$$\frac{V}{V_t} \approx 1 + \frac{9D}{4D_t} + \left(\frac{9D}{4D_t}\right)^2 \tag{11.4}$$

where D_t is the tube diameter, and V_t represents the fall velocity in the tube. Equation (11.4) is reliable only if $D/D_t < \frac{1}{3}$.

Other fluid properties such as surface tension, elasticity, vapor pressure, specific heats at constant pressure and constant temperature, and gas constant are commonly determined by physicists, and the techniques for the their measurement will not be discussed here.

EXERCISES

11.1.1 A small object has a volume of 0.0070 ft^3 and weighs 1.50 lb in air and 1.18 lb in a liquid. What is the density of the liquid?

11.1.2 A small object has a volume of 300 cm^3 and weighs 15.50 N in air and 10.50 N in a liquid. What is the density of the liquid?

11.1.3 Carbon tetrachloride ($s = 1.59$) is placed in an open U tube. A liquid is poured into one of the legs of the tube. A column of this liquid 19.4 in high balances a carbon tetrachloride column 15.0 in high. What is the specific weight of the liquid? In Sec. 11.1 it is mentioned that this method will give only approximate values. Why is this so?

11.1.4 Carbon tetrachloride ($s = 1.59$) is placed in an open U tube. A liquid is poured into one of the legs of the tube. A column of this liquid 39.2 cm high balances a carbon tetrachloride column 30.0 cm high. What is the specific weight of the liquid? In Sec. 11.1 it is mentioned that this method will give only approximate values. Why is this so?

11.1.5 A rotational viscometer is constructed of two concentric cylinders of height 15.0 in. The OD of the inner cylinder is 4.00 in, and the ID of the outer cylinder is 4.10 in. When a torque of 6.0 ft·lb is applied to the outer cylinder, it was found to rotate at 1 revolution per 3.8 sec. Find the (absolute) viscosity of the fluid. Neglect mechanical friction.

---

[2] J. S. McNown, H. M. Lee, M. B. McPherson, and S. M. Engez, Influence of Boundary Proximity on the Drag of Spheres, *Proc. 7th Intern. Congr. Appl. Mech.*, 1948.

11.1.6 A rotational viscometer is constructed of two concentric cylinders of height 30.0 cm. The OD of the inner cylinder is 10.00 cm, and the ID of the outer cylinder is 10.20 cm. When a torque of 8.0 N·m is applied to the outer cylinder, it was found to rotate at 1 revolution per 4.0 s. Find the (absolute) viscosity of the fluid. Neglect mechanical friction.

11.1.7 A tube viscometer similar to the one of Fig. 11.1 has a diameter of 0.0420 in and a length of 3.10 in. A volume of 60 cm^3 flowed through the tube in 128.7 s, causing the vertical distance from the liquid surface in the reservoir to the tube outlet to change from 10.00 to 9.50 in. Find the kinematic viscosity of the liquid.

11.1.8 If 60°F water flows through a tube-type viscometer in 120 sec, how long will it take the same volume of 100°F water to pass through the same viscometer?

11.1.9 If 15°C water flows through a tube-type viscometer in 85 s, how long will it take the same volume of 40°C water to pass though the same viscometer?

11.1.10 A lead sphere ($s = 11.4$) with a diameter of 0.25 in falls through oil ($s = 0.86$) in a 2.25-in-diameter cylinder. If the sphere moves with a constant velocity of 0.150 fps, what is the absolute viscosity of the oil? Check **R** to see if it is less than 1.0.

11.2 MEASUREMENT OF STATIC PRESSURE

The static pressue at a point in a flowing fluid was first defined in Secs. 5.8 and 5.12; it is the pressure in the fluid unchanged by the measuring instrument. To get an accurate measurement of static pressure in a flowing fluid, it is important that the measuring device fit the streamlines perfectly so as to create no disturbance to the flow. In a straight reach of conduit the static pressure is ordinarily measured by attaching to the piezometer a pressure gage or a U-tube manometer, as described in Sec. 3.6. The piezometer opening in the side of the conduit should be normal to and flush with the surface. Any projection, such as (c) in Fig. 11.3, will result in error. It has been found, for example, that a projection of 0.10 in (2.5 mm) will cause a 16% change in the local velocity head. In this case the recorded pressure is depressed below the pressure in the undisturbed fluid because the disturbance of the streamline pattern increases the velocity, hence decreasing the pressure according to the Bernoulli equation.

When measuring the static pressure in a pipe, it is desirable to have two or more openings around the periphery of the section to account for possible imperfections of the wall. For this purpose a ***piezometer ring*** (Fig. 11.4) is used.

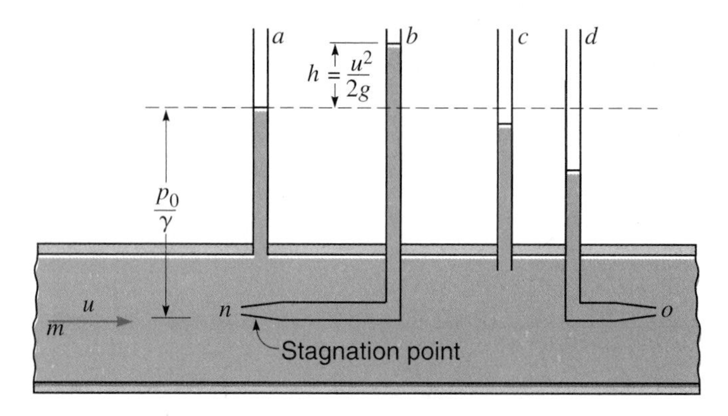

Figure 11.3

To measure the static pressure in a flow field, the ***static tube*** (Fig. 11.5) is used. In this device the pressure is transmitted to a gage or manometer through piezometric holes that are evenly spaced around the circumference of the tube. This device will give good results if it is perfectly aligned with the flow. Actually, the mean velocity past the piezometer holes will be slightly larger than that of the undisturbed flow field; hence the pressure at the holes will generally be somewhat below the pressure of the undisturbed fluid. This error can be minimized by making the diameter of the tube as small as possible. If the direction of the flow is unknown for two-dimensional flows, a ***direction-finding tube*** (Fig. 11.6) may be used. This device is a cylindrical tube having two piezometer holes located as shown. Each piezometer is connected to its own measuring device. The tube may be rotated until each tube shows the same reading. Then, from symmetry, one can determine the direction of flow. It has been found that if the piezometer openings are located as shown, the recorded pressures will correspond very closely to those of the undisturbed flow.

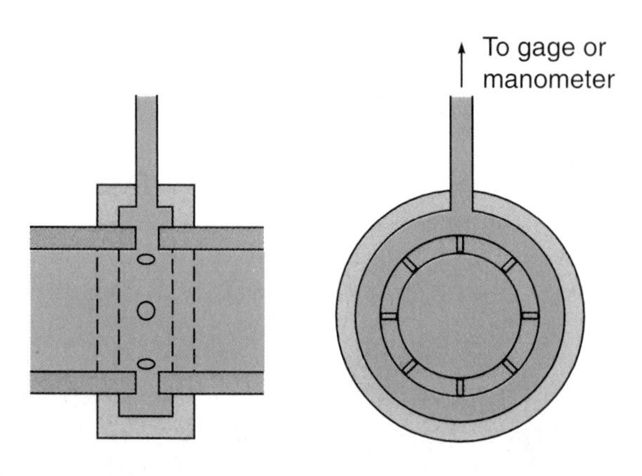

Figure 11.4
Piezometer ring.

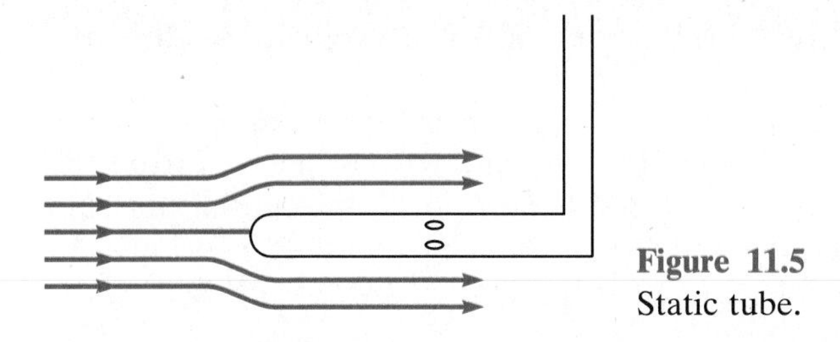

Figure 11.5
Static tube.

To obtain a reading of the fluid pressure, a piezometer tube may be connected to a bourdon gage (Sec. 3.6) or to a pressure transducer (Fig. 3.9). The latter is sometimes connected to a strip-chart recorder or the pressure reading may be displayed on a panel in digital form.

11.3 MEASUREMENT OF VELOCITY WITH PITOT TUBES

One means of measuring the local velocity u in a flowing fluid is the pitot tube, named after its inventor Henri Pitot (1695–1771), a French physicist who used a bent tube in 1732 to measure velocities in the River Seine. In Sec. 5.8 we saw that the pressure at the forward stagnation point of a stationary body in a flowing fluid is $p_0 = p + \rho u^2/2$, where p and u are the pressure and velocity, respectively, in the undisturbed flow upstream from the body. If $p_0 - p$ can be measured, the velocity at a point is determined by this relation. The stagnation pressure can be measured by a tube facing upstream, such as (b) in Fig. 11.3. For a liquid jet or open stream with parallel streamlines, only this single tube is necessary, since the height h to which the liquid rises in the

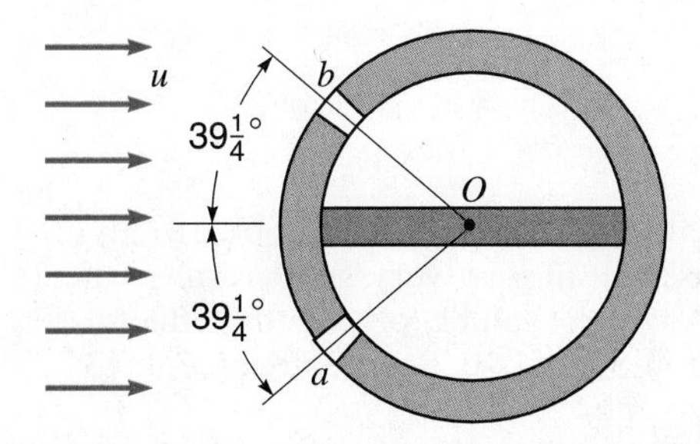

Figure 11.6
Direction-finding tube.

tube above the surrounding free surface is equal to the velocity head in the stream approaching the tip of the tube.

For a closed conduit under pressure it is necessary to measure the static pressure also, as shown by tube (*a*) in Fig. 11.3, and to subtract this from the total pitot reading to secure the differential head h. The differential pressure may be measured with any suitable manometer arrangement. The formula for the pitot tube for incompressible flow may be derived by writing the energy equation between points m and n of Fig. 11.3,

$$\frac{p}{\gamma} + \frac{u^2}{2g} = \frac{p_0}{\gamma} \tag{11.5}$$

from which

$$u = \sqrt{2g\left(\frac{p_0}{\gamma} - \frac{p}{\gamma}\right)} \tag{11.6}$$

This equation gives the ideal velocity of flow[3] at the point in the stream where the pitot tube is located. In actuality, directional velocity fluctuations due to turbulence increase pitot tube readings, so that the right-hand side of Eq. (11.6) must be multiplied by a factor varying from 0.98 to 0.995 to give the true velocity.

Where conditions are such that it is impractical to measure static pressure at the wall, a combined ***pitot-static tube***, as in Fig. 11.7, may be used. The static pressure is measured through two or more holes drilled through an outer tube into an annular space. Rarely are the piezometer holes located in precisely the correct position to indicate the true value of p/γ. Hence Eq.

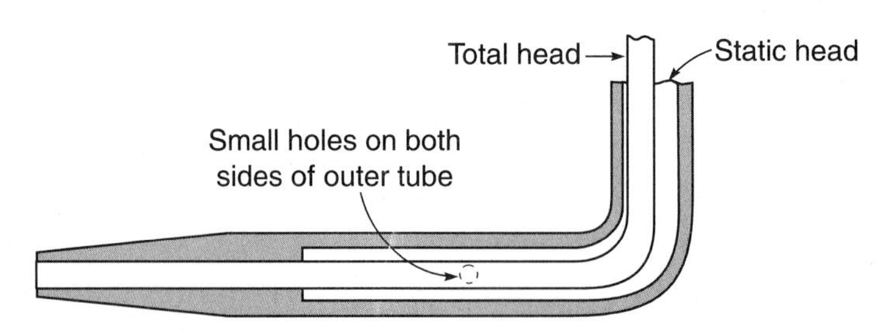

Figure 11.7
Pitot-static tube.

[3] Equations (11.5)–(11.7) as well as those presented in Secs. 11.6–11.9 apply strictly to incompressible fluids. However these equations will all give very good results when applied to compressible fluids if $\mathbf{M} < 0.1$. At high values of $\mathbf{M}$ the effects of compressibility must be considered as discussed in Sec. 11.10. See also Secs. 7.4, 13.2, and 13.3.

(11.6) is modified as follows:

$$u = C_I \sqrt{2g\left(\frac{p_0}{\gamma} - \frac{p}{\gamma}\right)} \tag{11.7}$$

where C_I, a coefficient of instrument, is introduced to account for this discrepancy. Either BG or SI units may be used with this equation, since C_I is dimensionless. However, when a coefficient possesses dimensions [see Eq. (11.27), for example], an equation developed for BG units must be modified for application to SI units, and vice versa. A particular type of pitot-static tube with a blunt nose, the ***Prandtl tube***, is designed so that $C_I = 1$. For other pitot-static tubes, the coefficient C_I must be determined by calibration in the laboratory.

Another instrument, the ***pitometer***, consists of two tubes, one pointing upstream and the other downstream, such as tubes (b) and (d) of Fig. 11.3. The reading for tube (d) will be considerably below the level of the static head. The equation applicable to a pitometer is identical to Eq. (11.7), except that p/γ is replaced by the pressure head sensed by the downstream tube.

Most of these devices will give reasonably accurate results even if the tube is as much as ±15° out of alignment with the direction of flow.

Still greater insensitivity to angularity may be obtained by guiding the flow past the pitot tube by means of a shroud, as shown in Fig. 11.8. Such an arrangement, called a ***Kiel probe***, is used extensively in aeronautics. The stagnation-pressure measurement with this device is accurate to within 1% of the dynamic pressure for yaw angles up to ±54°. A disadvantage is that the static pressure must be measured independently.

The direction-finding tube (Fig. 11.6) may be used to determine velocity.

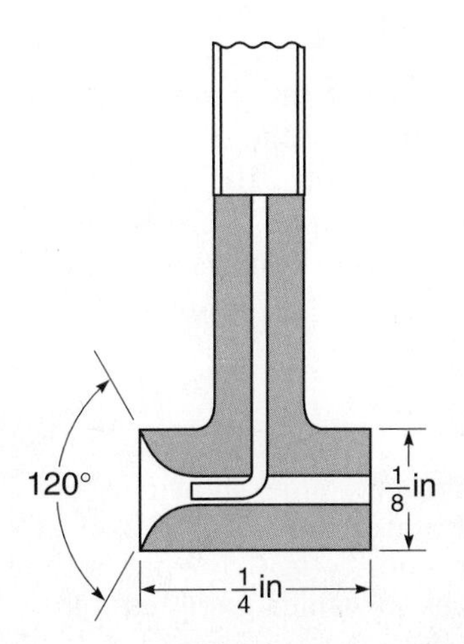

Figure 11.8
Kiel probe.

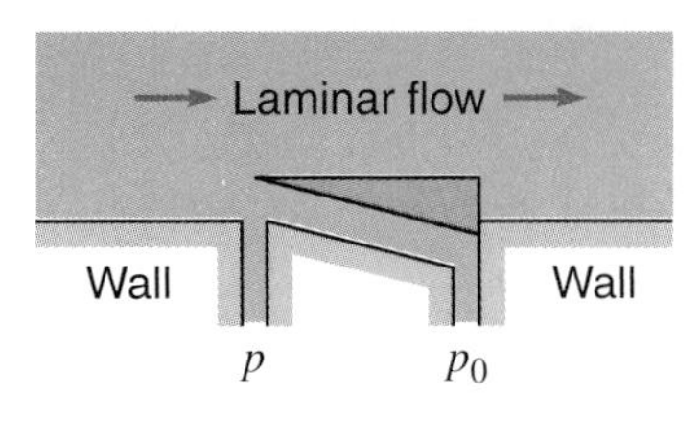

(*a*) Stanton tube

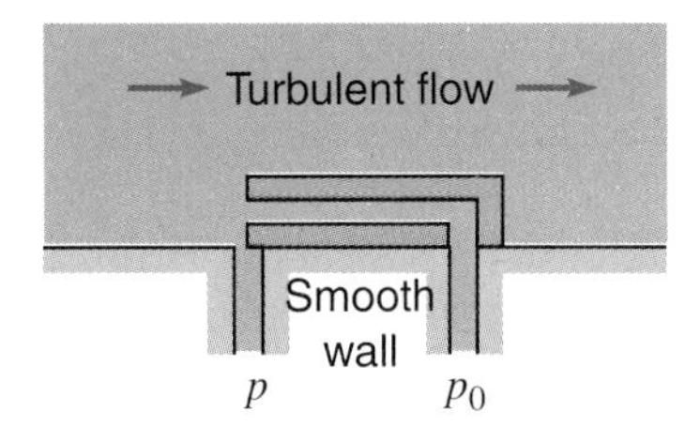

(*b*) Preston tube

Figure 11.9
Wall pitot tubes.

The procedure is to orient it properly so that both piezometers give the same reading. This reading is the static head. Then turn the tube through $39\frac{1}{4}°$ to obtain the stagnation pressure head. The difference in the two readings is the velocity head. This device has been used extensively in wind tunnels and in the investigations of hydraulic machinery.

Specialized, very small pitot tubes are used in experimental work to measure velocities very close to a boundary. The ***Stanton tube*** (Fig. 11.9*a*) uses the wall to form one side of the tube, and it may only be used within the viscous sublayer of turbulent flows (Fig. 8.8) or with laminar flow. The ***Preston tube*** (Fig. 11.9*b*) is designed for use in the transition zone (see Fig. 8.8), not submerged in the viscous sublayer, and may be used for turbulent flow over smooth surfaces. Both are used to collect data for the determination of the boundary shear stress τ_0, which has been found to depend on $p_0 - p$. The boundary shear stress is calculated by substituting the measured velocities into the theory of velocity profiles in circular pipes. It is assumed that this theory holds for flat boundaries as well as curved boundaries in the region very close to the boundary.

SAMPLE PROBLEM 11.1 Air at 20°C is flowing through the pipe shown in Fig. S11.1, resulting in pressure gage readings of 70.2 kPa at *A* and 71.1 kPa at *B*. Atmospheric pressure is 684 mmHg. (*a*) Find the air velocity *u*. (*b*) What is the largest pressure difference (kPa) between the two gages for which compressibility effects can be safely neglected?

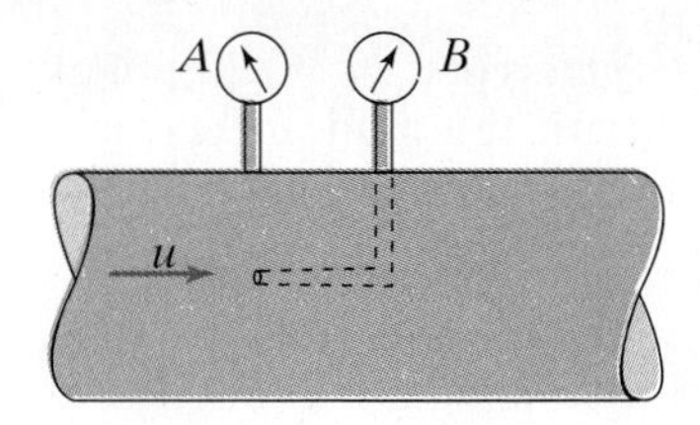

Figure S11.1

Solution
Table A.5 for air: $R = 287\ \text{m}^2/(\text{s}^2\cdot\text{K})$

$$(a)\qquad p_{abs} = p_{at} + p_{gage} = 101.32(684/760) + 70.2 = 161.4\ \text{kPa}$$

$$\text{Eq. (2.5):}\qquad \gamma = \frac{gp}{RT} = \frac{9.81(161.4)}{287(273+20)} = 0.018\,83\ \text{kN/m}^3$$

$$\text{Eq. (11.6):}\qquad u = \sqrt{2(9.81)\left(\frac{71.1-70.2}{0.01883}\right)} = 30.6\ \text{m/s}\qquad \textbf{\textit{ANS}}$$

(b) Sec. 11.3, footnote 3: We need $\mathbf{M} < 0.1$
Sec. 5.8: $\mathbf{M} = V/c$; at 20°C, $c = 345$ m/s [or use Eq. (13.15)].
So we need $V < 0.1c = u_{max} = 34.5$ m/s

$$\text{From Eq. (11.5):}\qquad \Delta p = p_0 - p = \gamma\frac{u^2}{2g}$$

$$\text{So}\qquad \Delta p_{max} = \frac{\gamma(u_{max})^2}{2g} = \frac{0.01883(34.5)^2}{2(9.81)} = 1.142\ \text{kPa}\qquad \textbf{\textit{ANS}}$$

This compares with the observed Δp of $71.1 - 70.2 = 0.9$ kPa. Therefore it was safe to neglect compressibility effects in part (a).

EXERCISES

11.3.1 In Fig. S11.1 kerosene ($s = 0.81$) is flowing. The pressure gages at A and B read 70 and 150 N/m^2. Find the velocity u, assuming $C_I = 1.0$.

11.3.2 A pitot tube and a wall piezometer tube are installed in a pipe carrying 60°F water. The tubes are connected to a water-mercury manometer which registers a differential of 3.4 in. Given that $C_I = 0.98$, find the flow velocity approaching the tube.

11.3.3 A pitot tube and a wall piezometer tube are installed in a pipe carrying 15°C water. The tubes are connected to a water–mercury manometer, which registers a differential of 78 mm. Given that $C_I = 0.97$, find the flow velocity approaching the tube.

11.3.4 The fluids of Exer. 11.3.2 are reversed so that mercury ($s = 13.56$) is flowing in the pipe and water is the gage fluid (with the manometer now inverted). With the same gage differential, find the flow velocity of the mercury.

11.3.5 The fluids of Exer. 11.3.3 are reversed so that mercury ($s = 13.56$) is flowing in the pipe and water is the gage fluid (with the manometer now inverted). With the same gage differential, find the flow velocity of the mercury.

11.3.6 A pitot-static tube ($C_I = 0.985$) is connected to an inverted U tube containing oil ($s = 0.875$). Find the velocity of flowing water if the manometer reading is 3.6 in.

11.3.7 A pitot-static tube ($C_I = 0.992$) is connected to an inverted U tube containing oil ($s = 0.91$). Find the velocity of flowing water if the manometer reading is 127 mm.

11.4 MEASUREMENT OF VELOCITY BY OTHER METHODS

Other methods for measuring local velocity will be discussed in this section.

Current Meter and Rotating Anemometer

These two instruments, which are the same in principle, determine the velocity as a function of the speed at which a series of cups or vanes rotate about an axis either parallel or normal to the flow. The instrument used in water is called a ***current meter***, and when designed for use in air, it is called an ***anemometer***. As the force exerted depends upon the density of the fluid as well as upon its velocity, the anemometer must be so made as to operate with less friction than the current meter.

If the meter is made with cups that move in a circular path about an axis perpendicular to the flow, it always rotates in the same direction and at the same rate regardless of the direction of the velocity, whether positive or negative, and it even rotates when the velocity is at right angles to its plane of rotation. Thus this type is not suitable where there are eddies or other irregularities in the flow. If the meter is constructed of vanes rotating about an axis parallel to the flow, like a propeller, it will register the component of velocity along its axis, especially if it is surrounded by a shielding cylinder. It will rotate in an opposite direction for negative flow, and is therefore a more dependable type of meter.

Hot-Wire and Hot-Film Anemometer

The ***hot-wire anemometer*** measures the instantaneous velocity at a point. It consists of a small sensing element that is placed in the flow field at the point where the velocity is to be measured. The sensing element is a short thin wire, which is generally of plantinum or tungsten, connected to a suitable electronic circuit. The operation depends on the fact that the electrical resistance of a wire is a function of its temperature; that the temperature, in turn, depends upon the heat transfer to the surrounding fluid; and that the rate of heat transfer increases with increasing velocity of flow past the wire.

In one type of hot-wire anemometer the wire is maintained at a constant temperature by a variable voltage, which changes the current through the wire. Thus, when an increase in velocity tends to cool the wire, a balancing device creates an increase in voltage to increase the current through the wire. This tends to heat up the wire to counteract the cooling and thus maintain it at constant temperature. The voltage provides a measure of the velocity of the fluid. The hot-wire anemometer is a very sensitive instrument particularly adapted to the measurement of turbulent velocity fluctuations as in Fig. 4.6. A ***hot-film anemometer***, though similar to the hot-wire, is more rugged in that its sensing element consists of a metal film laid over a glass rod and provided with a protecting coating.

Float Measurements

A crude technique for estimating the average velocity of flow in a river or stream is to observe the velocity at which a float will travel down a stream. To get good results the reach of stream should be straight and uniform with a minimum of surface disturbances. The average velocity of flow V will generally be about (0.85 ± 0.05) times the float velocity.

Photographic and Optical Methods

The camera is one of the most valuable tools in a fluid-mechanics research laboratory. In studying the motion of water, for example, a series of small spheres consisting of a mixture of benzene and carbon tetrachloride adjusted to the same specific gravity as the water can be introduced into the flow through suitable nozzles. When illuminated from the direction of the camera, these spheres will stand out in a picture. If successive exposures are taken on the same film, the velocities and accelerations of the particles can be determined. A similar technique involves the use of hydrogen bubbles generated through use of a fine wire which serves as the negative electrode of a dc electric circuit. By pulsing the voltage across the wire, the water is electrolyzed, thus releasing hydrogen bubbles. Short uninsulated sections of wire will permit the bubbles to be emitted at fixed points along the wire. This, when combined with intermittent pulsing, will aid in flow visualization.

In the study of compressible fluids many techniques have been devised to measure optically the variations in density, as given by the ***interferometer***, or the rate at which density changes in space, as determined in the ***shadowgraph*** and ***schlieren*** methods.[4] From such measurements of density and density gradient, it is possible to locate shock waves. Although of great importance, these photographic methods are too complex to warrant further description here.

Other Methods

Other devices for measuring velocity of flow include magnetic flowmeters, acoustic flowmeters and laser–Doppler anemometers. ***Magnetic flowmeters*** are used to measure velocity of flow in liquids. The liquid serves as a conductor and develops a voltage as it travels through a magnetic field. With proper calibration, this device can be used to measure the average velocity in pipes. Small magnetic flowmeters can be used to measure local velocities in a flowing liquid. However, their accuracy drops off in the vicinity of boundaries.

Acoustic flow meters and ***laser–Doppler anemometers*** depend on the effect of the moving fluid on waves: the former on sound waves and the latter on light waves. These devices are expensive, and are used primarily for research. One of their advantages is that they can be employed so as not to disturb the flow.

11.5 MEASUREMENT OF DISCHARGE

There are various ways of measuring discharge. In a pipe, for example, the velocity may be determined at various radii using a pitot-static tube or a pitot tube in combination with a wall piezometer. The cross section of a pipe may then be considered as a series of concentric rings, each with a known velocity. The flow through these rings is summed up, as in Fig. 11.10, to determine the total flow rate.

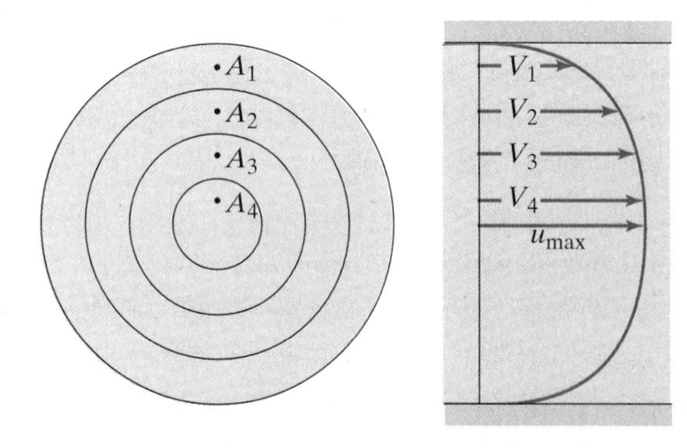

Figure 11.10
Determination of pipe discharge.

$$Q = \sum A_i V_i$$
$$= A_1 V_1 + A_2 V_2 + \cdots$$

[4] For an excellent discussion of optical methods used in the study of fluid flow, see Michel A. Saad, *Compressible Flow,* 2d ed., Sec. 11.6, Prentice-Hall, Englewood Cliffs, New Jersey, 1993.

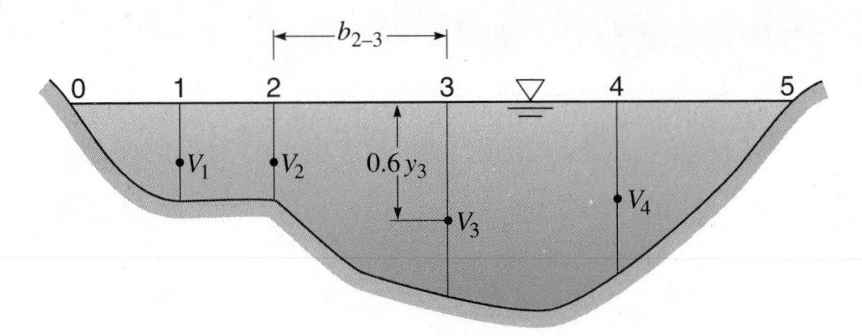

Figure 11.11
Determination of discharge in a stream.

To determine the flow in a river or stream, a similar technique is used. The stream is divided into a number of convenient sections, as in Fig. 11.11, and the average velocity in each section is measured. A pitot tube could be used for such measurements, but a current meter is more commonly used. It has been found that the average velocity occurs at about $0.6 \times$ depth (Sec. 10.4), so the velocity is generally measured at that level. As a result, in Fig. 11.11

$$Q_{2-3} = \left(\frac{y_2 + y_3}{2}\right) b_{2-3} \frac{V_2 + V_3}{2}$$

and $Q_{\text{total}} = \Sigma\, Q$. Another widely used method is to replace the velocity at $0.6 \times$ depth with the average of the velocities at $0.2 \times$ depth and $0.8 \times$ depth. A crude estimate of the flow in a river or stream can be made by multiplying ($0.85 \times$ float velocity) times the area of the average cross section in the reach of stream over which the float measurement was made.

Devices for the direct measurement of discharge can be divided into two categories, those that measure by weight or positive displacement a certain quantity of fluid and those that employ some aspect of fluid mechanics. An example of the first type of device is the household water meter in which a nutating disk oscillates in a chamber. On each oscillation a known quantity of water passes through the meter. The second type of flow-measuring device, which depends on basic principles of fluid mechanics combined with empirical data, will be discussed in the following sections.

SAMPLE PROBLEM 11.2 Measurements made on the cross section of Fig. 11.11 are as follows:

			Velocity V, fps, at			
Position	y, ft	b, ft	Surf	$0.2y$	$0.6y$	$0.8y$
0	—		—	—	—	—
		4.09				
1	3.60		—	2.88	2.80	1.50
		3.53				
2	3.44		—	3.15	3.02	2.50
		6.55				
3	7.20		4.05	4.18	3.48	3.10
		6.45				
4	6.72		—	3.40	3.15	2.85
		6.58				
5	—		—	—	—	—

Calculate the total discharge in the stream by the three different methods described.

Solution

Using $Q = AV$:

Method:		(1)	(1)	(2)	(2)	(2)
	$A = b\bar{y}$	$V_{0.6}$	$Q_{0.6}$	$\frac{V_{0.2} + V_{0.8}}{2}$	$\bar{V}_{0.2,0.8}$	$Q_{0.2,0.8}$
Position	ft²	fps	cfs	fps	fps	cfs
0				—		
	7.36	1.40	10.31		1.10	8.06
1				2.19		
	12.43	2.91	36.16		2.51	31.16
2				2.83		
	34.85	3.25	113.25		3.23	112.64
3				3.64		
	43.28	3.32	143.47		3.38	146.39
4				3.13		
	20.46	1.58	32.23		1.56	31.97
5				—		
Total	118.38		335.4	***ANS***		330.2 ***ANS***

(3) Assume float velocity = surface velocity at position 3:

$$Q = 0.85AV_s = 0.85(118.38)4.05 = 407.5 \text{ cfs} \qquad \textbf{\textit{ANS}}$$

11.6 ORIFICES, NOZZLES, AND TUBES

Among the devices used for the measurement of discharge are orifices and nozzles. Tubes are rarely used in this way but we include them here because their theory is the same and experiments on tubes provide information about entrance losses from reservoirs into pipelines. An ***orifice*** is an opening

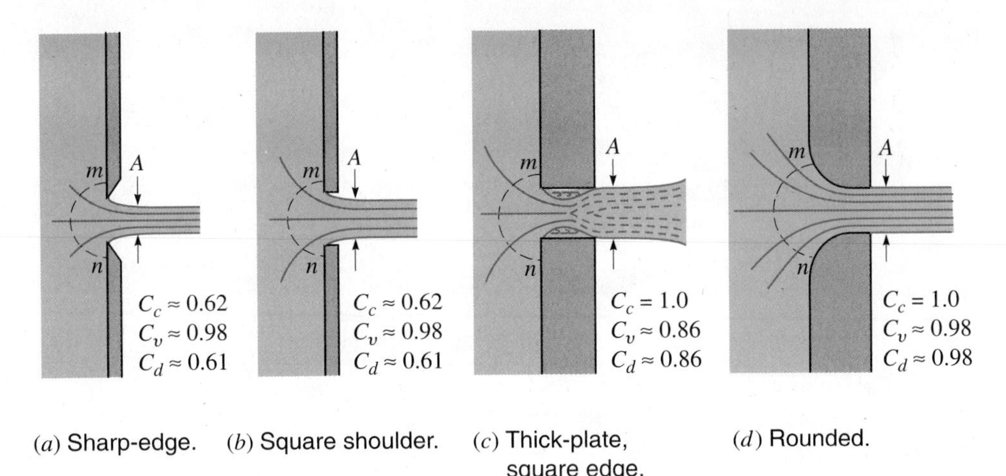

Figure 11.12
Orifices.

(usually circular) in the wall of a tank or in a plate normal to the axis of a pipe, the plate being either at the end of the pipe or in some intermediate location. An orifice is characterized by the fact that the thickness of the wall or plate is very small relative to the size of the opening. A ***standard orifice*** is one with a sharp edge as in Fig. 11.12*a* or an absolutely square shoulder as in Fig. 11.12*b*, so that there is only a line contact with the fluid. Those shown in Fig. 11.12*c* and *d* are not standard, because the flow through them is affected by the thickness of the plate, the roughness of the surface, and, for *d*, the radius of curvature. Hence such orifices should be calibrated if high accuracy is desired.

A ***nozzle*** is a converging tube, as in Fig. 11.13, if it is used for liquids; but for a gas or a vapor a nozzle may first converge and then diverge (Sec. 13.9) to produce supersonic flow. In addition to possible use as a flow measuring device a nozzle has other important uses, such as providing a high-velocity stream for fire fighting or for power in a steam turbine or an impulse turbine (Pelton wheel, see Sec. 16.2).

A ***tube*** is a short pipe whose length is not more than two or three diameters. There is no sharp distinction between a tube and the thick-walled orifices of Fig. 11.12*c* and *d*. A tube may be of uniform diameter, or it may diverge.

A ***jet*** is a stream issuing from an orifice, nozzle, or tube. It is not enclosed by solid boundary walls but is surrounded by a fluid whose velocity is less than its own. The two fluids may be different or they may be of the same kind. A ***free jet*** is a stream of *liquid* surrounded by a *gas* and is therefore directly under the influence of gravity. A ***submerged jet*** is a stream of any fluid surrounded by a fluid of the same type, that is, a gas jet discharging into

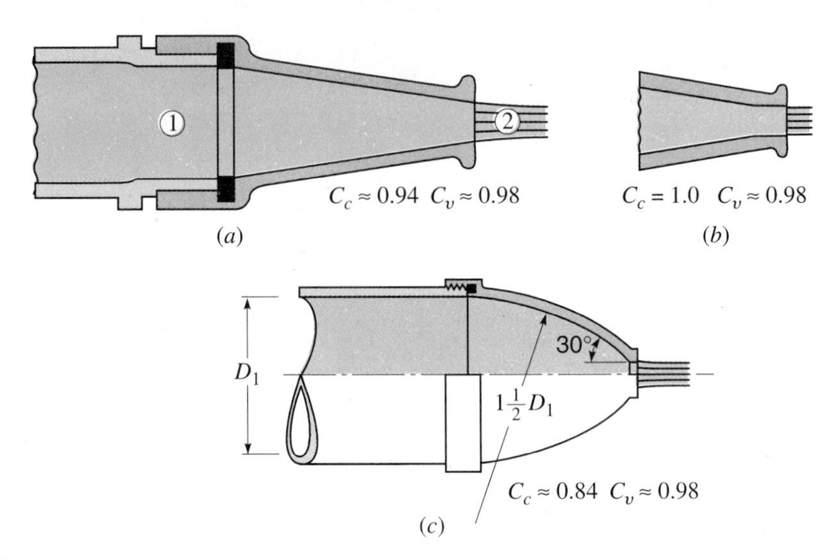

Figure 11.13
Nozzles. (*a*) Conical. (*b*) Straight-tip. (*c*) Fire.

a gas or a liquid jet discharging into a liquid. A submerged jet is buoyed up by the surrounding fluid and is not directly under the action of gravity.

Jet Contraction

Where the streamlines converge in approaching an orifice, as shown in Fig. 11.14, they continue to converge beyond the upstream section of the orifice until they reach the section *xy*, where they become parallel. Commonly this section is about $0.5D_o$ from the upstream edge of the opening, where D_o is

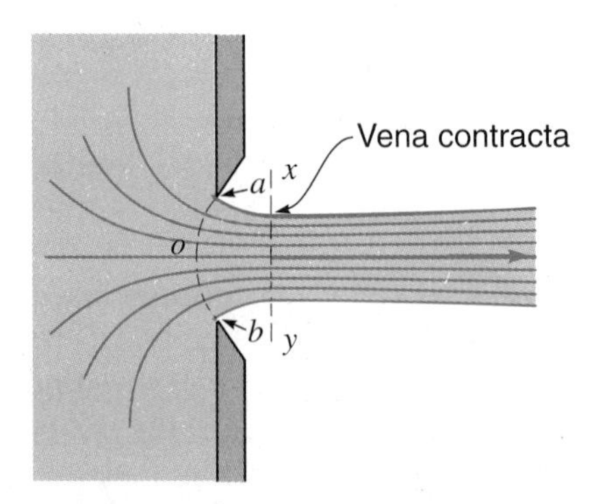

Figure 11.14
Jet contraction.

the diameter of the orifice. The section xy is then a section of minimum area, and is called the ***vena contracta***. Beyond the vena contracta the streamlines commonly diverge because of frictional effects.[5] In Fig. 11.12*c* the minimum section is referred to as a ***submerged vena contracta***, since it is surrounded by its own fluid. In Fig. 11.12*d* there is no vena contracta as the rounded entry to the opening permits the streamlines to gradually converge to the cross-sectional area of the orifice.

Jet Velocity and Pressure

Jet velocity is defined as the average velocity at the vena contracta in Fig. 11.12*a* and *b*, and at the downstream edge of the orifices in Fig. 11.12*c* and *d*. The velocity at these sections is practically constant across the section except for a small annular region around the outside (Fig. 11.15*b*). In all four of the jets of Fig. 11.12 the pressure is practically constant across the diameter of the jet wherever the streamlines are parallel, and this pressure must be equal to that in the medium surrounding the jet at that section. At sections mn in Fig. 11.12 where the streamlines are curved, the effective cross-sectional area of the flow (at right angles to streamlines) is greater than at the minimum section, and hence the average velocities at sections mn are considerably less than the jet velocities. The same is true of section aob of Fig. 11.14. In Fig. 11.15*a* the velocity and pressure distributions at section aob of Fig. 11.14 are shown. These variations are the result of the curvature of the streamlines and centrifugal effects (Sec. 5.17).

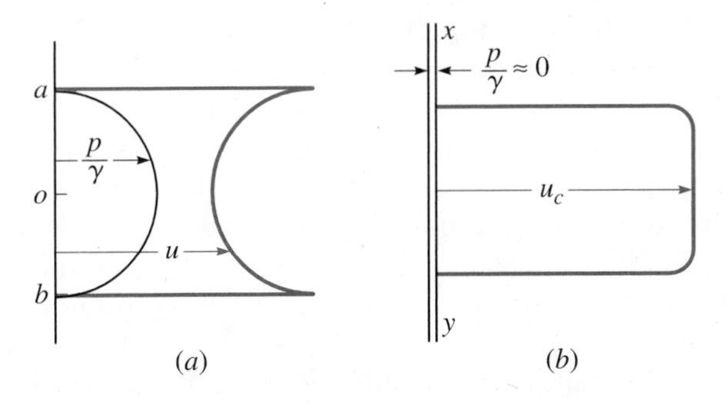

Figure 11.15
Pressure and velocity variations in a jet. (*a*) At section *aob* of Fig. 11.14. (*b*) At vena contracta (section *xy*) in Fig. 11.14.

[5] Of course, if a free jet is discharged vertically downward, the acceleration due to gravity will cause its velocity to increase and the area to decrease continuously, so that there may be no apparent section of minimum area. In such special cases, the vena contracta should be taken as the place where marked contraction ceases and before the place where gravity has increased the velocity to any appreciable extent above the true jet velocity.

Coefficient of Contraction C_c

The ratio of the area A of a jet (Fig. 11.12), to the area A_o of the orifice or other opening, is called the ***coefficient of contraction*.** Thus $A = C_c A_o$.

Coefficient of Velocity C_v

The velocity that would be attained in the jet if friction did not exist may be termed the ideal velocity V_i.[6] It is practically the value of u_c in Fig. 11.15. Because of friction, the actual average velocity V is less than the ideal velocity, and the ratio V/V_i is called the ***coefficient of velocity***. Thus $V = C_v V_i$.

Coefficient of Discharge C_d

The ratio of the actual rate of discharge Q to the ideal rate of discharge Q_i (the flow that would occur if there were no friction and no contraction) is defined as the coefficient of discharge. Thus $Q = C_d Q_i$. By observing that $Q = AV$ and $Q_i = A_o V_i$, it is seen that $C_d = C_c C_v$.

Determining the Coefficients

The coefficient of contraction can be determined by using outside calipers to measure the jet diameter at the vena contracta and then comparing the jet area with the orifice area. The contraction coefficient is very sensitive to small variations in the edge of the orifice or in the upstream face of the plate. Thus slightly rounding the edge of the orifice in Fig. 11.12*b* or roughening the orifice plate will increase the contraction coefficient materially.

The average velocity V of a free jet may be determined by a velocity traverse of the jet with a fine pitot tube or it may be obtained by measuring the flow rate and dividing by the cross-sectional area of the jet. The velocity may also be computed approximately from the coordinates of the trajectory of the jet, as discussed in Sec. 5.16. The ideal velocity V_i is computed by the Bernoulli theorem. Thus C_v for an orifice, nozzle, or tube may be computed by dividing V by V_i.

The coefficient of discharge is the one that can most readily be obtained and with a high degree of accuracy. It is also the one that is of the most practical value. For a liquid the actual Q can be determined by some

---

[6] This is frequently called the ***theoretical velocity***, but the authors feel this is a misuse of the word "theoretical." Any correct theory should allow for the fact that friction exists and affects the result. Otherwise it is not correct theory but merely an incorrect hypothesis.

standard method such as a volume or a weight measurement over a known time. For a gas one can note the change in pressure and temperature in a container of known volume from which the gas may flow. Obviously, if any two of the coefficients are measured, the third can be computed from them. Thus, in equation form,

Ideal flow rate: $$Q_i = A_i V_i = A_o\sqrt{2g(\Delta H)} \quad (11.8)$$

Actual flow rate: $$Q = AV = C_c A_o (C_v \sqrt{2g(\Delta H)}) \quad (11.9)$$

or $$Q = C_d Q_i = C_d A_o \sqrt{2g(\Delta H)} \quad (11.10)$$

and $$C_d = \frac{Q}{Q_i} = C_c C_v \quad (11.11)$$

where ΔH is the total difference in energy head between the upstream section and the minimum section of the jet (section A of Fig. 11.12). It should be recalled that the total energy head $H = z + p/\gamma + V^2/2g$. If the flow is from a tank, the velocity of approach is negligible and may be neglected. If the discharge is to the atmosphere (free jet), the downstream pressure head is zero, whereas if the jet is submerged, the downstream pressure head is equal to the depth of submergence[7] (Fig. 11.18) in the case of a liquid or to the pressure head surrounding the jet in the case of a gas.

Typical values of the coefficients for orifices, nozzles, and tubes are as indicated in Figs. 11.12,[8] 11.13, and 11.16, respectively. It is apparent from Fig. 11.16 that rounding the entrance to a tube increases the coefficient of velocity. Any device that provides a uniform diameter for a long enough

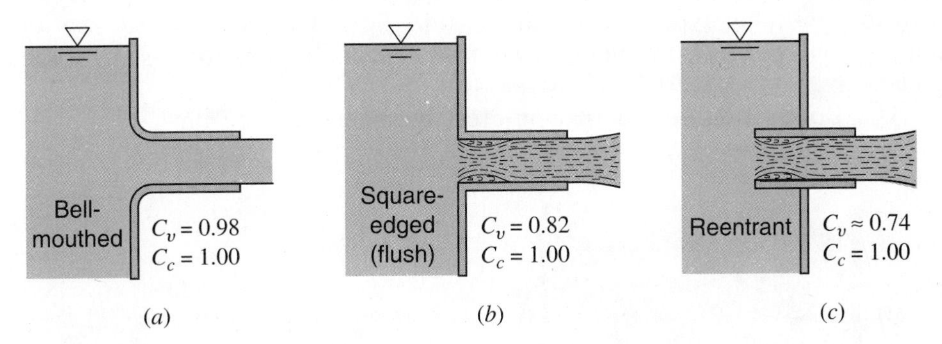

Figure 11.16
Coefficients for tubes.

[7] If the jet discharges into a different liquid, the depth of submergence must be converted to equivalent depth of the *flowing liquid.*

[8] Surface tension may become important when orifices operate under low heads. The coefficient of contraction of small, sharp-edged, and square-edged orifices such as those of Fig. 11.12*a* and *b* have values of C_c as high as 0.72 rather than the usual 0.62 when operating under heads less than about 0.5 ft.

distance before exit, such as the tubes of Fig. 11.16 or the nozzle tipe of Fig. 11.13*b*, will usually create a $C_c = 1.0$. Although this increases the size of the jet from the given area, it also tends to produce more friction.

If the geometry of the orifice, nozzle or tube is standard such as those of Figs. 11.12, 11.13, and 11.16, the coefficients should be very close to the values indicated on the figures. However, the best way to determine the coefficients of a device, particularly one of unusual shape, is by experiment in the laboratory. Also, one can make a fair estimate of the contraction by sketching the flow net. If one wishes to estimate the coefficient of discharge of an orifice, nozzle, or tube it is usually best to estimate velocity and contraction coefficients separately and calculate the discharge coefficient from them.

Borda Tube

Tubes (*b*) and (*c*) in Fig. 11.16 are shown as flowing full, and because of the turbulence, the jets issuing from them will have a "broomy" appearance. Because of the contraction of the jet at entrance to these tubes the local velocity in the central portion of the stream will be higher than that at exit from the tubes, and hence the pressure will be lower. If the pressure is lowered to that of the vapor pressure of the liquid, the streamlines will then no longer follow the walls. In such a case tube (*b*) of Fig. 11.16 becomes equivalent to orifice (*b*) in Fig. 11.12, while tube (*c*) of Fig. 11.16 behaves as shown in Fig. 11.17. If its length is less than its diameter, the reentrant tube is called a ***Borda mouthpiece***. Because of the greater curvature of the streamlines for a reentrant tube, the velocity coefficient is lower (Fig. 11.16*c*) than for any other type of entry, if the tube flows full. But if the jet springs clear as in Fig. 11.17, the velocity coefficient is as high as for a sharp-edged orifice.

The Borda mouthpiece is of interest because it is one device for which

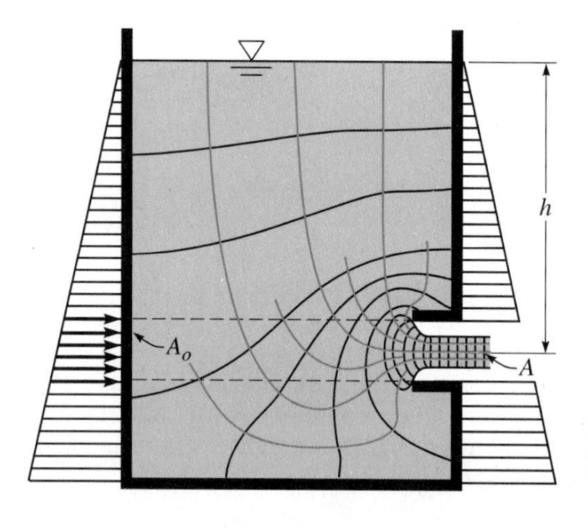

Figure 11.17
Borda tube.

the contraction coefficient can be very simply calculated. For all other orifices and tubes there is a reduction of the pressure on the walls adjacent to the opening, but the exact pressure values are unknown. But for the reentrant tube the velocity along the wall of the tank is almost zero at all points, and hence the pressure is essentially hydrostatic. In the case of a Borda tube the only unbalanced pressure is that on an equal area A_o opposite to the tube (Fig. 11.17), and the force due to this pressure is $\gamma h A_o$. The time rate of change of momentum due to the flow out of the tube is $\rho Q V = \gamma A V^2/g$, where A is the area of the jet. Equating force to time rate of change of momentum, $\gamma h A_o = \gamma A V^2/g$, and thus $V^2 = g h A_o/A$. Ideally, $V^2 = 2gh$, and thus, ideally, $C_c = A/A_o = 0.5$. The actual values of the coefficients for a Borda tube are $C_c = 0.52$, $C_v = 0.98$, and $C_d = 0.51$.

Head Loss

The relationship between the head loss and the coefficient of velocity of an orifice, nozzle, or tube may be found by comparing the ideal energy equation with the actual (or real) energy equation between points 1 and 2 in Fig. 11.13*a*. The ideal energy equation is

$$z_1 + \frac{p_1}{\gamma} + \frac{V_1^2}{2g} = z_2 + \frac{p_2}{\gamma} + \frac{V_2^2}{2g}$$

In the case of a free jet $p_2 = 0$, while for the most general case of a submerged jet $p_2 \neq 0$. From continuity, $A_1 V_1 = A_2 V_2$; hence we can write

$$z_1 + \frac{p_1}{\gamma} + \left(\frac{A_2}{A_1}\right)^2 \frac{V_2^2}{2g} = z_2 + \frac{p_2}{\gamma} + \frac{V_2^2}{2g}$$

which leads to

$$(V_2)_{\text{ideal}} = \frac{1}{\sqrt{1-(A_2/A_1)^2}} \sqrt{2g\left[\left(z_1 + \frac{p_1}{\gamma}\right) + \left(z_2 + \frac{p_2}{\gamma}\right)\right]} \qquad (11.12)$$

The real energy equation accounts for head loss and is expressed as

$$z_1 + \frac{p_1}{\gamma} + \frac{V_1^2}{2g} - h_{L_{1-2}} = z_2 + \frac{p_2}{\gamma} + \frac{V_2^2}{2g}$$

which leads to the actual velocity

$$V_2 = \frac{1}{\sqrt{1-(A_2/A_1)^2}} \sqrt{2g\left[\left(z_1 + \frac{p_1}{\gamma}\right) - \left(z_2 + \frac{p_2}{\gamma}\right) - h_{L_{1-2}}\right]} \qquad (11.13)$$

Remembering that the actual $V = C_v V_{\text{ideal}}$, and combining this with the above expressions for V_{ideal} and actual V gives

$$h_{L_{1-2}} = \left(\frac{1}{C_v^2} - 1\right)\left[1 - \left(\frac{A_2}{A_1}\right)^2\right]\frac{V_2^2}{2g} \qquad (11.14)$$

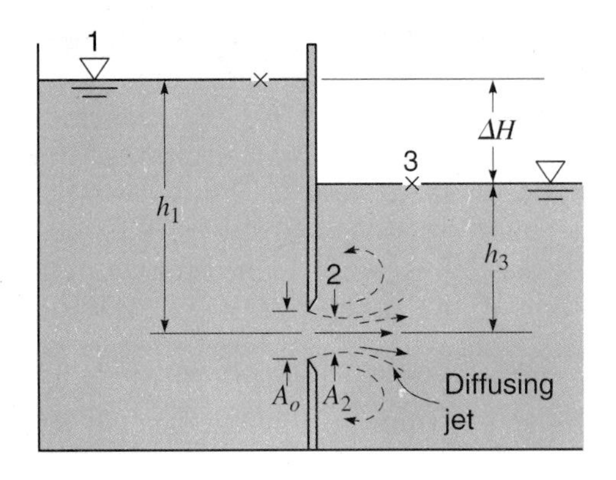

Figure 11.18
Submerged jet.

where V_2 is an actual velocity. This equation is perfectly general; it expresses the head loss between a section upstream of an orifice and the jet (section *A* in Fig. 11.12) or between sections 1 and 2 in Fig. 11.13*a*, etc. If the orifice or nozzle takes off directly from a tank where $A_1 \gg A_2$, the velocity of approach is negligible and Eq. (11.14) reduces to

$$h_{L_{1-2}} = \left(\frac{1}{C_v^2} - 1\right)\frac{V_2^2}{2g} \tag{11.15}$$

Note that for the tubes of Fig. 11.16 with $C_v = 0.98$, 0.82, and 0.74, Eq. (11.15) yields $h_L = 0.04V_2^2/2g$, $0.5V_2^2/2g$, and $0.8V_2^2/2g$, respectively. These correspond to the values for minor loss at entrance shown in Fig. 8.13.

Submerged Jet

For the case of a submerged jet, as shown in Fig. 11.18, the ideal energy equation is written between 1 and 2, realizing that the pressure head on the jet at 2 is equal to h_3. Thus

$$h_1 = h_3 + \frac{V_i^2}{2g}$$

or

$$V_i = \sqrt{2g(h_1 - h_3)} = \sqrt{2g(\Delta H)}$$

where V_i is the ideal velocity at the vena contracta of the submerged jet and ΔH is the net head differential expressed in terms of the flowing liquid. Hence $Q = C_c C_v A_o \sqrt{2g(\Delta H)}$ as in Eq. (11.9).

For a submerged orifice, nozzle, or tube the coefficients are practically the same as for a free jet, except that, for heads less than 10 ft (3 m) and for very small openings, the discharge coefficient may be slightly less. It is of interest to observe that, if the energy equation is written between 1 and 3, the result is $h_{L_{1-3}} = h_1 - h_3 = \Delta H$. Actually, the head loss in this case is that of Eq.

(11.15) plus that of a submerged discharge, as described in Sec. 8.19. Hence, for submerged orifices, nozzles, and tubes,

$$h_{L_{1-3}} = \left(\frac{1}{C_v^2} - 1\right)\frac{V_2^2}{2g} + \frac{V_2^2}{2g} = \frac{1}{C_v^2}\frac{V_2^2}{2g} = \frac{V_i^2}{2g} = \Delta H$$

where the actual $V_2 = C_v V_i$, the velocity at the vena contracta.

Sample Problem 11.3 A 2-in circular orifice (not standard) at the end of the 3-in-diameter pipe shown in Fig S11.3 discharges into the atmosphere a measured flow of 0.60 cfs of water when the pressure in the pipe is 10.0 psi. The jet velocity is determined by a pitot tube to be 39.2 fps. Find the values of the coefficients C_v, C_c, and C_d. Find also the head loss from inlet to vena contracta.

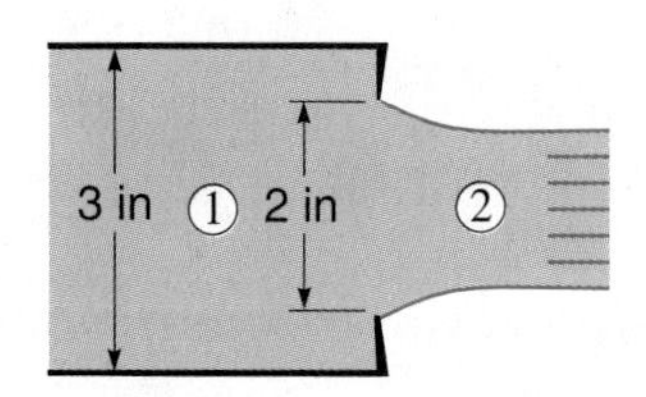

Figure S11.3

Solution

Define the inlet as section 1 and the throat as section 2.

$$\frac{p_1}{\gamma} = 10\left(\frac{144}{62.4}\right) = 23.1 \text{ ft}$$

$$V_1 = \frac{Q}{A_1} = \frac{0.60}{\pi(1.5/12)^2} = 12.22 \text{ fps}, \qquad \frac{V_1^2}{2g} = 2.32 \text{ ft}$$

Express the ideal energy equation from 1 to 2 to determine the ideal velocity V_{2i} at 2

$$\frac{p_1}{\gamma} + \frac{V_1^2}{2g} = \frac{V_{2i}^2}{2g}$$

$$\frac{V_{2i}^2}{2g} = 23.1 + 2.32 = 25.4; \qquad V_{2i} = 40.4 \text{ fps}$$

$$C_v = \frac{V_2}{V_{2i}} = \frac{39.2}{40.4} = 0.969 \qquad \textit{ANS}$$

Area of jet $$A_2 = \frac{Q}{V_2} = \frac{0.60}{39.2} = 0.01531 \text{ ft}^2$$

$$C_c = \frac{A_2}{A_o} = \frac{0.01531}{\pi(1/12)^2} = 0.702 \qquad \textit{ANS}$$

Hence $$C_d = C_c C_v = 0.680 \qquad \textit{ANS}$$

From Eq. (11.14):

$$h_{L_{1-2}} = \left(\frac{1}{(0.969)^2} - 1\right)\left[1 - \left(\frac{2}{3}\right)^4\right]\frac{V_2^2}{2g} = 0.0517\frac{V_2^2}{2g}$$

$$= 0.0517\,\frac{(39.2)^2}{2(32.2)} = 1.233 \text{ ft} \qquad \textit{ANS}$$

Check: calculate the actual velocity V_2 at 2 by expressing the real energy equation from 1 to 2:

$$\frac{p_1}{\gamma} + \frac{V_1^2}{2g} - h_{L_{1-2}} = \frac{V_2^2}{2g}$$

$$23.1 + 2.32 - 1.233 = \frac{V_2^2}{2g}; \qquad \text{actual } V_2 = 39.4 \text{ fps}$$

This checks well with the measured velocity of 39.2 fps.

EXERCISES

11.6.1 Water issues from a circular 5-in-diameter orifice under a head of 45 ft. If 534 ft^3 are discharged in 2 min, what is the coefficient of discharge? If the diameter at the vena contracta is measured to be 3.93 in, what is the coefficient of contraction and what is the coefficient of velocity?

11.6.2 A jet discharges 0.131 cfs from a 1.25-in-diameter orifice in a vertical plane under a head of 10 ft. The jet center line passes through the point 11.47 ft horizontally from the vena contracta and 3.5 ft below the center of the orifice. Find the coefficients of (*a*) discharge, (*b*) velocity, and (*c*) contraction.

11.6.3 A jet discharges 5.19 L/s from a 35 mm-diameter orifice in a vertical plane under a head of 4 m. The jet center line passes through the point 4.28 m horizontally from the vena contracta and 1.2 m below the center of the orifice. Find the coefficients of (*a*) discharge, (*b*) velocity, and (*c*) contraction.

11.6.4 A 4-in-diameter water jet discharges from a nozzle whose velocity coefficient is 0.96. The pressure in the 9-in-diameter pipe is 11 psi. Assuming the jet does not contract, what is the velocity at the tip of the nozzle? What is the flow rate?

11.6.5 In Fig. X11.6.5 the area A is twice the area A_o. If the diverging tube discharges water when $h = 6$ ft, find (*a*) the velocity at the throat; (*b*) the pressure head at the throat. Neglect all friction losses.

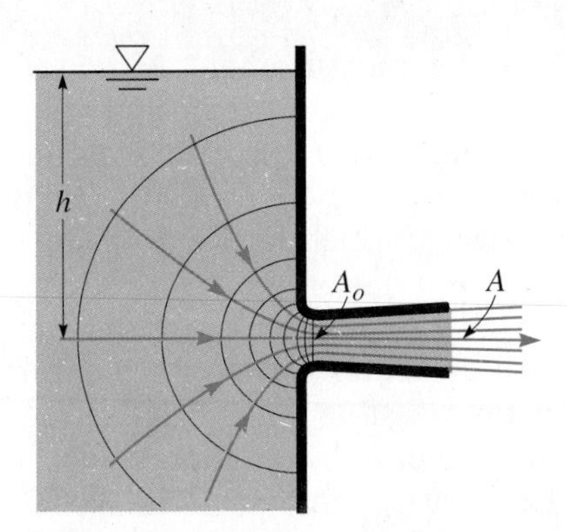

Figure X11.6.5

11.6.6 For the same data as Exer. 11.6.5, with a barometric pressure of 14.8 psia and a water temperature of 90°F, what is the maximum value of h at which the tube will flow full? What will happen if h is made greater than this?

11.6.7 For the rounded entrance and tube flowing full as in Fig. X11.6.5, $C_c = 1.0$ for both the throat and the exit, and thus $C_v = C_d$ for both sections. For the throat, assume the value of C_v as given for Fig. 11.16*a*, and assume that for the tube as a whole the discharge coefficient applied to the exit end is 0.72. Find the velocity at the throat and the pressure head at the throat if $h = 6$ ft. Compare your answer with Exer. 11.6.5, which neglected friction.

11.6.8 Given that the diverging tube shown in Fig. X11.6.5 has $A = 1.75A_0$. Find the throat velocity and pressure head when $h = 3$ m of water. Neglect all friction losses.

11.6.9 Given that the diverging tube shown in Fig. X11.6.5 has $A = 1.75A_0$. If it is operating at 1000 m elevation under standard atmospheric conditions, what will be the maximum value of h for which the tube will flow full?

11.6.10 Water flows through a 60-mm-diameter sharp-edged orifice which connects two adjacent tanks. The head on one side of the orifice is 2.5 m and 0.5 m on the other. Given $C_c = 0.62$ and $C_v = 0.95$, calculate the flow rate.

11.7 VENTURI METER

The converging tube is an efficient device for converting pressure head to velocity head, while the diverging tube converts velocity head to pressure head. The two may be combined to form a ***venturi tube***, named after Giovanni B. Venturi (1746–1822), an Italian physicist who investigated its principle about 1791. It was applied to the measurement of water by an American engineer, Clemens Herschel, in 1886. As shown in Fig. 11.19, it consists of a tube with a constricted ***throat***, which produces an increased velocity accompanied by a reduction in pressure, followed by a gradually diverging portion in which the velocity is transformed back into pressure with slight friction loss. As there is a definite relation between the pressure differential and the rate of flow, the tube may be made to serve as a metering device known as a ***venturi meter***. The venturi meter is used for measuring the rate of flow of both compressible and incompressible fluids.[9] In this section we shall consider the application of the venturi meter to incompressible fluids; its application to compressible fluids will be discussed in Sec. 11.10.

Writing the Bernoulli equation between sections (1) (inlet) and (2) (throat) of Fig. 11.19, we have, for the ideal case,

$$\frac{p_1}{\gamma} + z_1 + \frac{V_1^2}{2g} = \frac{p_2}{\gamma} + z_2 + \frac{V_2^2}{2g}$$

Substituting the continuity equation, $V_1 = (A_2/A_1)V_2$, we get for the ideal throat velocity

$$V_{2i} = \sqrt{\frac{1}{1 - (A_2/A_1)^2}}\sqrt{2g\left[\left(\frac{p_1}{\gamma} + z_1\right) - \left(\frac{p_2}{\gamma} + z_2\right)\right]}$$

As there is some friction loss between (1) and (2), the true velocity V_2 is slightly less than the ideal value given by this expression. Therefore we introduce a discharge coefficient C, so that the flow is given by CQ_i or

$$Q = A_2V_2 = CA_2V_{2i} = \frac{CA_2}{\sqrt{1 - (D_2/D_1)^4}}\sqrt{2g\left[\left(\frac{p_1}{\gamma} + z_1\right) - \left(\frac{p_2}{\gamma} + z_2\right)\right]} \tag{11.16}$$

In this situation the discharge coefficient C is identical to the velocity coefficient C_v since the coefficient of contraction $C_c = 1.0$. In the preceding equation it should be noted, by reference to Eq. (3.12) and Fig. 3.14, that if a differential manometer is used with piezometric connections at sections (1) and (2),

$$\left(\frac{p_1}{\gamma} + z_1\right) - \left(\frac{p_2}{\gamma} + z_2\right) = MR\left(\frac{s_M}{s_F} - 1\right) \tag{11.17}$$

where MR is the manometer reading, and s_M and s_F are the specific gravities of the manometer and flowing fluids, respectively.

[9] As mentioned earlier, if $\mathbf{M} < 0.1$, a compressible fluid can be treated as if it were incompressible without introducing much error.

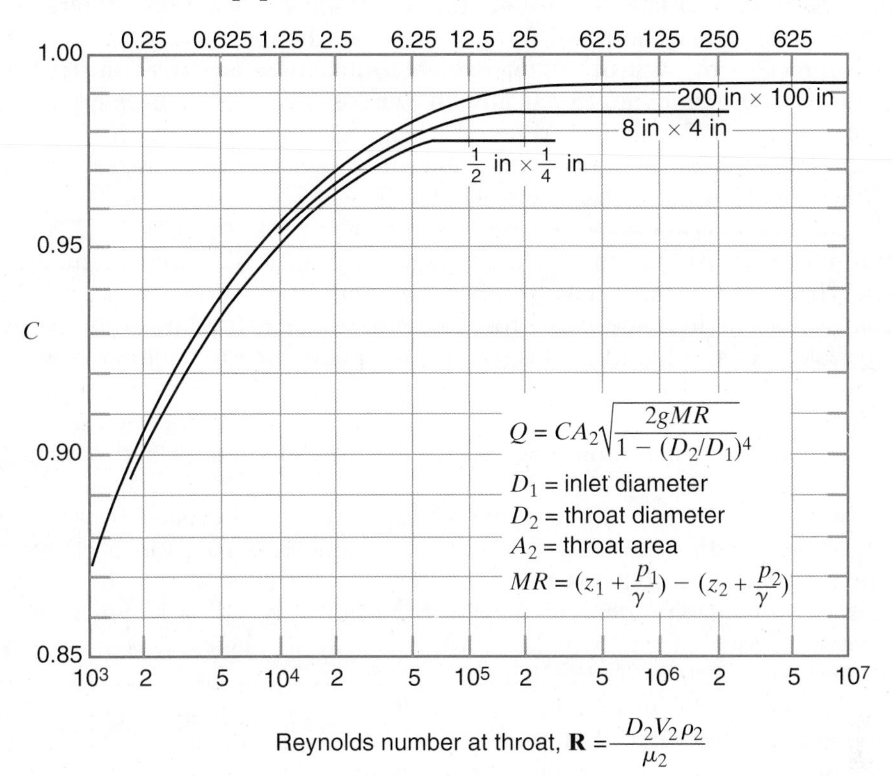

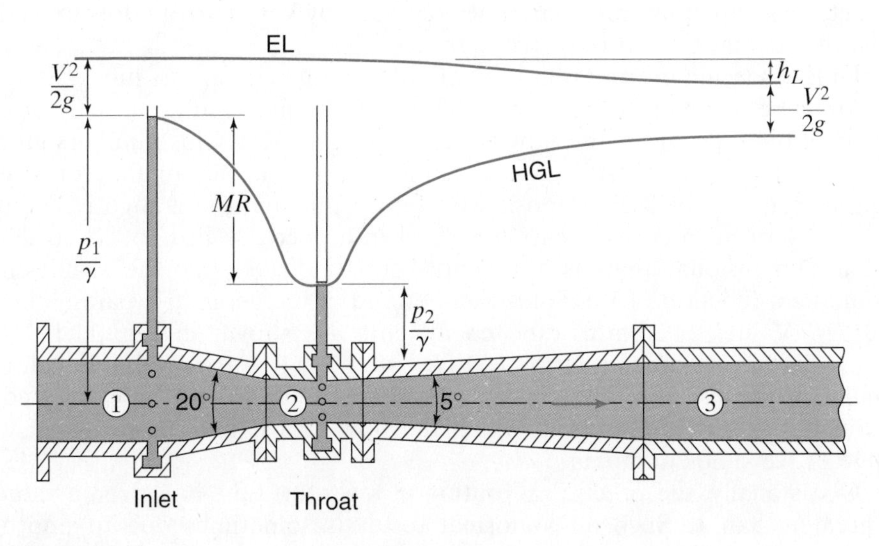

Figure 11.19
Venturi meter with conical entrance and flow coefficients for $D_2/D_1 = 0.5$.

The venturi tube provides an accurate means for measuring flow in pipelines. With a suitable recording device, the flow rate can be integrated so as to give the total quantity of flow. Aside from the installation cost, the only disadvantage of the venturi meter is that it introduces a permanent frictional resistance in the pipeline. Practically all this loss occurs in the diverging part between sections (2) and (3), as noted in Fig. 5.10, and is ordinarily from $0.1h$ to $0.2h$, where h is the static-head differential between the upstream section and the throat, as indicated in Fig. 11.19.

Values of D_2/D_1 may vary from $\frac{1}{4}$ to $\frac{3}{4}$, but a common ratio is $\frac{1}{2}$. A small ratio gives increased accuracy of the gage reading, but is accompanied by a higher friction loss and may produce an undesirably low pressure at the throat, sufficient in some cases to cause liberation of dissolved air or even vaporization of the liquid at this point. This phenomenon, called ***cavitation***, has been described in Sec. 5.11. The angles of convergence and divergence indicated in Fig. 11.19 are those considered optimum, though somewhat larger angles are sometimes used to reduce the length and cost of the tube.

For accuracy in use, the venturi meter should be preceded by a straight pipe whose length is at least 5 to 10 pipe diameters. The approach section becomes more important as the diameter ratio increases, and the required length of straight pipe depends on the conditions preceding it. For example, the vortex formed from two short-radius elbows in planes at right angles is not eliminated within 30 pipe diameters (see Sec. 8.23). Such a condition can be alleviated by the installation of straightening vanes preceding the meter. The pressure differential should be obtained from piezometer rings (Fig. 11.4) surrounding the pipe, with a number of suitable openings in the two sections. In fact, these openings are sometimes replaced by very narrow slots extending most of the way around the circumference.

Unless specific information is available for a given venturi tube, the value of C may be assumed to be about 0.99 for large tubes and about 0.97 or 0.98 for small ones, provided the flow is such as to give Reynolds numbers greater than about 10^5 (Fig. 11.19). A roughening of the surface of the converging section from age or scale deposit will reduce the coefficient slightly. Venturi tubes in service for many years have shown a decrease in C of the order of 1–2%. Dimensional analysis of a venturi tube indicates that the coefficient C should be a function of Reynolds number and of the geometric parameters D_1 and D_2. Values of venturi-tube coefficients are shown in Fig. 11.19. This diagram is for a diameter ratio of $D_2/D_1 = 0.5$, but is is reasonably valid for smaller ratios also. For best results a venturi meter should be calibrated by conducting a series of tests in which the flow rate is measured over a wide range of Reynolds numbers.

Occasionally, the precise calibration of a venturi tube has given a value of C greater than 1. Such an abnormal result is sometimes due to improper piezometer openings. But another possible explanation is that the α's at sections (1) and (2) are such that this is really so.

Sample Problem 11.4 Find the discharge rate of 20°C water through the venturi tube shown if $D_1 = 80$ cm, $D_2 = 40$ cm, $\Delta z = 200$ cm, and $y = 15$ cmHg. Assume Fig. 11.19 is applicable.

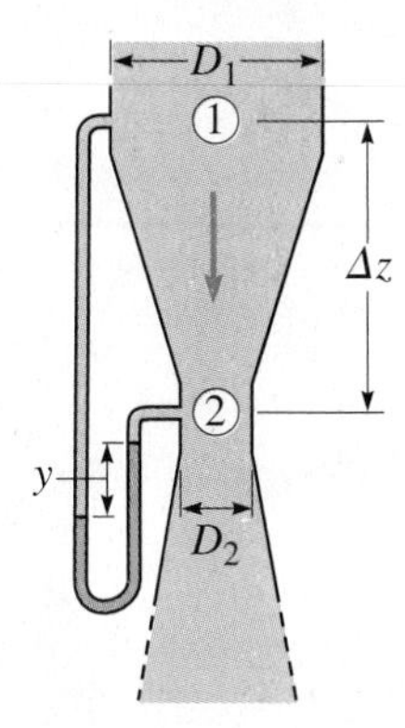

Figure S11.4

Solution

Table A.1 for water at 20°C: $\nu = 1.003 \times 10^{-6}$ m²/s

Venturi size is 80 cm × 40 cm = 31.5 in × 15.75 in.

This is about midway between the 8 in × 4 in and 200 in × 100 in curves on Fig. 11.19. So, from Fig. 11.19: Maximum $C \approx 0.988$; assume this value.

Eqs. (11.16) and (11.17):

$$Q = \frac{0.988\pi(0.40/2)^2}{\sqrt{1-(40/80)^4}}\sqrt{2(9.81)0.15\left(\frac{13.55}{1}-1\right)} = 0.779 \text{ m}^3/\text{s}$$

$$V_2 = \frac{Q}{A_2} = \frac{0.779}{\pi(0.40/2)^2} = 6.20 \text{ m/s}$$

Eq. (7.6): $$\mathbf{R} = \frac{D_2 V_2}{\nu} = \frac{0.40(6.20)}{1.003 \times 10^{-6}} = 2.47 \times 10^6$$

Check Fig. 11.19 for this **R**: $C = 0.988$. Hence $Q = 0.779$ m³/s ***ANS***

EXERCISES

11.7.1 The venturi meter of Fig. S11.4 has $D_1 = 8$ in, $D_2 = 4$ in, and $\Delta z = 1.5$ ft. Assume the discharge coefficients of Fig. 11.19 are applicable. Find the flow rate of 72°F water when the manometer containing carbon tetrachloride ($s = 1.59$) reads $y = 4$ in.

11.7.2 The venturi meter of Fig. S11.4 has $D_1 = 20$ cm, $D_2 = 10$ cm, and $\Delta z = 0.45$ m. Assume the discharge coefficients of Fig. 11.19 are applicable. Find the flow rate of 20°C water when the manometer containing carbon tetrachloride ($s = 1.59$) reads $y = 10$ cm.

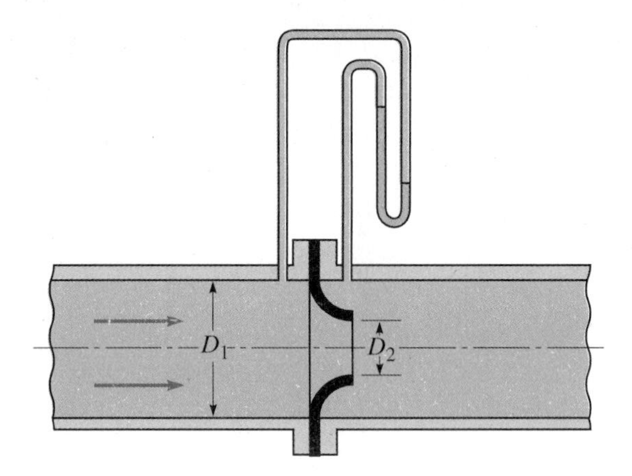

Figure 11.20
Flow nozzle.

11.7.3 The venturi meter of Fig. S11.4 has $D_1 = 90$ cm, $D_2 = 45$ cm, and $\Delta z = 2.25$ m. Assume the discharge coefficients of Fig. 11.19 are applicable. Find the flow rate of 20°C water when the mercury ($s = 13.55$) manometer reads $y = 16$ cm.

11.7.4 The venturi meter of Fig. S11.4 has $D_1 = 4$ in, $D_2 = 2$ in, and $\Delta z = 12$ in. Assume the discharge coefficients of Fig. 11.19 are applicable. Find the reading y of the mercury ($s = 13.55$) manometer when oil ($s = 0.92$) with a kinematic viscosity of 0.0005 ft²/s is flowing at a rate of 0.40 cfs.

11.8 FLOW NOZZLE

If the diverging discharge cone of a venturi tube is omitted, the result is a ***flow nozzle*** of the type shown in Fig. 11.20. This is simpler than the venturi tube and can be installed between the flanges of a pipeline. It will serve the same purpose, though at the expense of an increased frictional loss in the pipe. Although the venturi-meter equation [Eq. (11.16)] can be employed for the flow nozzle, it is more convenient and customary to include the correction for velocity of approach with the coefficient of discharge, so that

$$Q = KA_2\sqrt{2g\left[\left(\frac{p_1}{\gamma} + z_1\right) - \left(\frac{p_2}{\gamma} + z_2\right)\right]} \tag{11.18}$$

where K is called the ***flow coefficient*** and A_2 is the area of the nozzle throat. Comparison with Eq. (11.16) establishes the relation

$$K = \frac{C}{\sqrt{1 - (D_2/D_1)^4}} \tag{11.19}$$

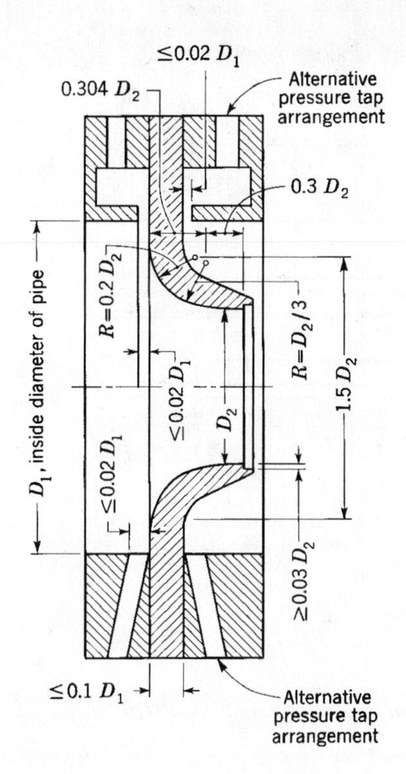

Figure 11.21
ISA flow nozzle.

Although there are many designs of flow nozzles, the ISA (International Standards Association) nozzle (Fig. 11.21) has become an accepted standard form in many countries. The quoted "nozzle diameter" is the throat diameter D_2. Values of K for various diameter ratios of the ISA nozzle are given in Fig. 11.22 as a function of Reynolds number. Note that in this case the Reynolds number is computed for the approach pipe rather than for the nozzle throat, which is a convenience since **R** in the pipe is frequently needed for other computations also.

As shown in Fig. 11.22, many of the values of K are greater than unity, which results from including the correction for approach velocity with the conventional coefficient of discharge. There have been many attempts to design a nozzle for which the velocity-of-approach correction would just compensate for the discharge coefficient, leaving a value of the flow coefficient equal to unity, principally using so-called ***long-radius nozzles***. Usually such a coefficient of unity is approached over only a limited range.

As in the case of the venturi meter, the flow nozzle should be preceded by at least 10 diameters of straight pipe for accurate measurement. Two alternative arrangements for the pressure taps are shown in Fig. 11.21.

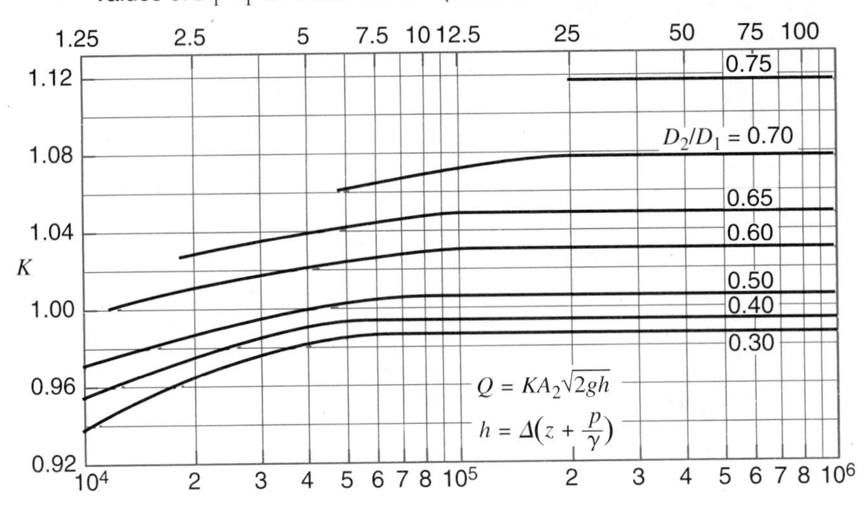

Figure 11.22
Flow coefficients for ISA nozzle. (*Adapted from ASME Flow Measurement,* 1959.)

SAMPLE PROBLEM **11.5** A 2-in ISA flow nozzle is installed in a 3-in pipe carrying water at 72°F. If a water-air manometer shows a differential of 2 in, find the flow.

Solution

This requires a trial-and-error type of solution. First assume a reasonable value of K. From Fig. 11.22, for $D_2/D_1 = 0.67$, and for the level part of the curve, $K = 1.06$.

$$A_1 = \frac{\pi}{4}\left(\frac{3}{12}\right)^2 = 0.0491 \text{ ft}^2; \quad A_2 = \frac{\pi}{4}\left(\frac{2}{12}\right)^2 = 0.0218 \text{ ft}^2$$

This air–water manometer is like Fig. 3.14*b* with $z_{AB} = 0$. For air and water, $s_M/s_F \approx 0.001$, so it can be neglected in Eq. (3.13), and

$$\Delta\left(\frac{p}{\gamma} + z\right) = MR = \frac{2}{12} = 0.1667 \text{ ft}$$

Eq. (11.18): $\quad Q = 1.06 \times 0.0218\sqrt{2(32.2) \times 0.1667} = 0.0746 \text{ cfs}$

With this first determination of Q,

$$V_1 = \frac{Q}{A} = \frac{0.0746}{0.0491} = 1.519 \text{ fps}$$

Then $$D_1'' V_1 = 3 \times 1.519 = 4.56$$

From Fig. 11.22, $K = 1.04$ and

$$Q = \frac{1.04}{1.06} \times 0.0746 = 0.0732 \text{ cfs}$$

No further correction is necessary.

EXERCISE

11.8.1 A 10-cm ISA flow nozzle (Figs. 11.21 and 11.22) is used to measure the flow of 40°C water through a 20-cm pipe. What would be the reading on a mercury manometer for the following flow rates; (*a*) 1.5 L/s; (*b*) 15 L/s; (*c*) 150 L/s.

11.9 ORIFICE METER

An orifice in a pipeline, as in Fig. 11.23, may be used as a meter in the same manner as the venturi tube or the flow nozzle. It may also be placed on the end of the pipe so as to discharge a free jet. The flow rate through an orifice meter is commonly expressed as

$$Q = KA_o \sqrt{2g\left[\left(\frac{p_1}{\gamma} + z_1\right) - \left(\frac{p_2}{\gamma} + z_2\right)\right]} \tag{11.20}$$

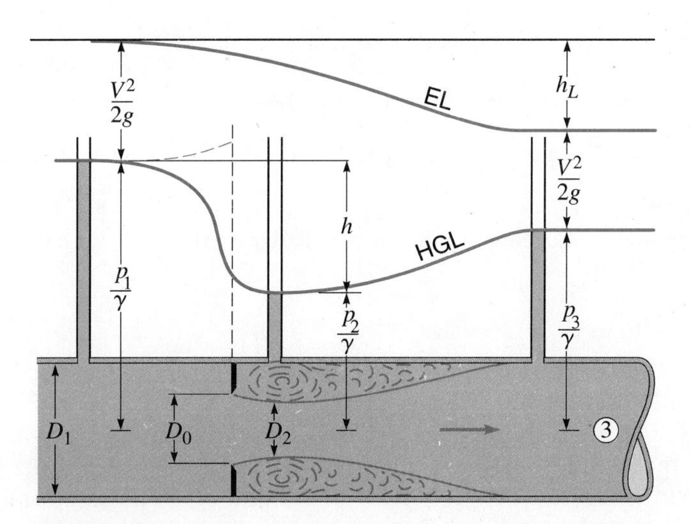

Figure 11.23
Thin-plate orifice in a pipe. (Scale distorted: the region of eddying turbulence will usually extend $4D_1$ to $8D_1$ downstream, depending upon the Reynolds number.)

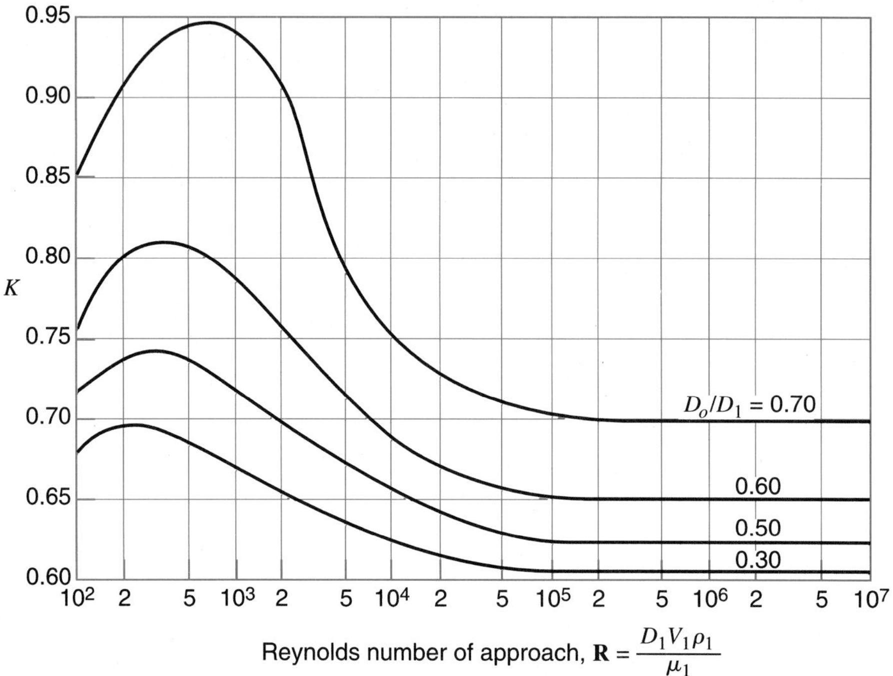

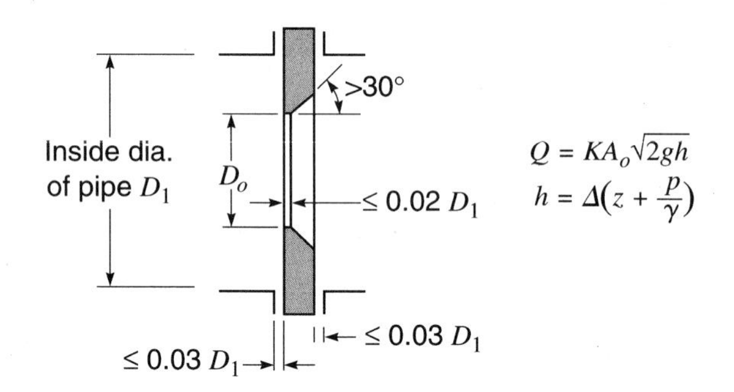

Figure 11.24
VDI orifice meter and flow coefficients for flange taps. (*Adapted from NACA Tech. Mem.* 952.)

This is the same form as Eq. (11.18), except that A_2 is replaced by A_o, the cross-sectional area of the orifice opening. Typical values of K for a standard orifice meter are given in Fig. 11.24. The variation of K with Reynolds number is quite different than the trend of the flow coefficients for venturi tubes and flow nozzles. At high Reynolds numbers K is essentially constant, but as the Reynolds number is lowered, an increase in the value of K for the orifice is noted, with maximum value of K occurring at Reynolds numbers between 200 and 600, depending on the D_o/D_1 ratio of the orifice. The lowering of the Reynolds number increases viscous action, which causes a decrease in C_v and an increase in C_c. The latter apparently predominates over the former until C_c reaches a maximum value of about 1.0. With a further decrease in Reynolds number, K then becomes smaller because C_v continues to decrease.

The difference between an orifice meter and a venturi tube or flow nozzle is that for both of the latter there is no contraction, so that A_2 is also the area of the throat and is fixed, while for the orifice A_2 is the area of the jet and is a variable and is less than A_o, the area of the orifice. For the venturi tube or flow nozzle the discharge coefficient is practically a velocity coefficient, while for the orifice it is much more affected by variations in C_c than it is by variations in C_v.

The pressure differential may be measured between a point about one pipe diameter upstream of the orifice and the vena contracta, approximately one-half the pipe diameter downstream. The distance to the vena contracta is not constant, but decreases as D_o/D_1 increases. The differential can also be measured between the two corners on each side of the orifice plate. These ***flange taps*** have the advantage that the orifice meter is self-contained; the plate may be slipped into a pipeline without the necessity of making piezometer connections in the pipe.

The particular advantage of an orifice as a measuring device is that it may be installed in a pipeline with a minimum of trouble and expense. Its principal disadvantage is the greater frictional loss it causes as compared with the venturi meter or the flow nozzle.

EXERCISE

11.9.1 Repeat Exer. 11.8.1 for a VDI orifice.

11.10 FLOW MEASUREMENT OF COMPRESSIBLE FLUIDS

Strictly speaking, most of the equations that have been presented in the preceding part of this chapter apply only to incompressible fluids, but, practically, they may be used for all liquids and even for gases and vapors where the pressure differential is small relative to the total pressure. As this is the condition usually encountered in the metering of all fluids, even compressible ones, the preceding treatment has extensive application. However, there are conditions in metering fluids where compressibility must be considered.

As in the case of incompressible fluids, equations may be derived for ideal frictionless flow and then a coefficient introduced to obtain a correct result.

The ideal condition that will be imposed on the compressible fluid is that the flow be isentropic, i.e., a frictionless adiabatic process (no transfer of heat, see Sec. 2.7). The latter is practically true for metering devices, as the time for the fluid to pass through is so short that very little heat transfer can take place.

Pitot Tubes

An expression applicable to pitot tubes for subsonic flow of compressible fluids can be derived by introducing the conditions at the upstream tip of the tube (that is, $V_2 = 0$ and $p_2 = p_0$) into Eq. (13.35), substituting the first expression for R from Eq. (13.4,) and making use of the perfect gas relationship, Eq. (2.5). Doing so gives for pitot tubes

$$\frac{V_1^2}{2} = c_p T_1\left[\left(\frac{p_2}{p_1}\right)^{(k-1)/k} - 1\right] = c_p T_2\left[1 - \left(\frac{p_1}{p_2}\right)^{(k-1)/k}\right] \tag{11.21}$$

The static pressure p_1 (see Sec. 11.2) may be obtained from the side openings of the pitot tube or from a regular piezometer, and the stagnation pressure p_0 $(= p_2$, Sec. 11.3) is indicated by the pitot tube itself. A coefficient must be applied if the side openings do not measure the true static pressure. Equation (11.21) does not apply to supersonic conditions, because a shock wave would form upstream of the stagnation point. In such a case a special analysis considering the effect of the shock wave is required.

Venturi Meters

To develop an expression applicable to compressible flow through venturi tubes, we take Eq. (13.35) and combine it with continuity ($g\dot{m} = G = \gamma_1 A_1 V_1 = \gamma_2 A_2 V_2$) to get

$$g\dot{m}_{\text{ideal}} = G_{\text{ideal}} = A_2\sqrt{2g\,\frac{k}{k-1}\,p_1\gamma_1\left(\frac{p_2}{p_1}\right)^{2/k}\frac{1-(p_2/p_1)^{(k-1)/k}}{1-(A_2/A_1)^2(p_2/p_1)^{2/k}}} \tag{11.22}$$

This equation can be transformed into an equation for the actual weight rate of flow through venturi tubes by introducing the discharge coefficient C (Fig. 11.19) and an expansion factor Y. The resulting equation is

$$g\dot{m} = G = CYA_2\sqrt{2g\gamma_1\,\frac{p_1 - p_2}{1-(D_2/D_1)^4}} \tag{11.23}$$

where C has the same value as for an incompressible fluid at the same Reynolds number and γ_1 may be replaced by p_1/RT_1 if desired. Values of Y for $k = 1.4$ are plotted in Fig. 11.25. For a venturi or nozzle throat where $C_c = 1$,

$$Y = \sqrt{\frac{[k/(k-1)](p_2/p_1)^{2/k}[1-(p_2/p_1)^{(k-1)/k}]}{1-(p_2/p_1)}}\sqrt{\frac{1-(D_2/D_1)^4}{1-(D_2/D_1)^4(p_2/p_1)^{2/k}}}$$

In Fig. 11.25 we see that for the venturi meter no values for Y are shown for p_2/p_1 ratios less than 0.528. This is so because, for air and other gases having adiabatic constant $k = 1.4$, the p_2/p_1 ratio will always be greater than 0.528 if the flow is subsonic, as is explained in Sec. 13.8.

Equation (11.23) is directly applicable to the flow of compressible fluids through venturi tubes where $C_c = 1.0$, provided the flow is subsonic.

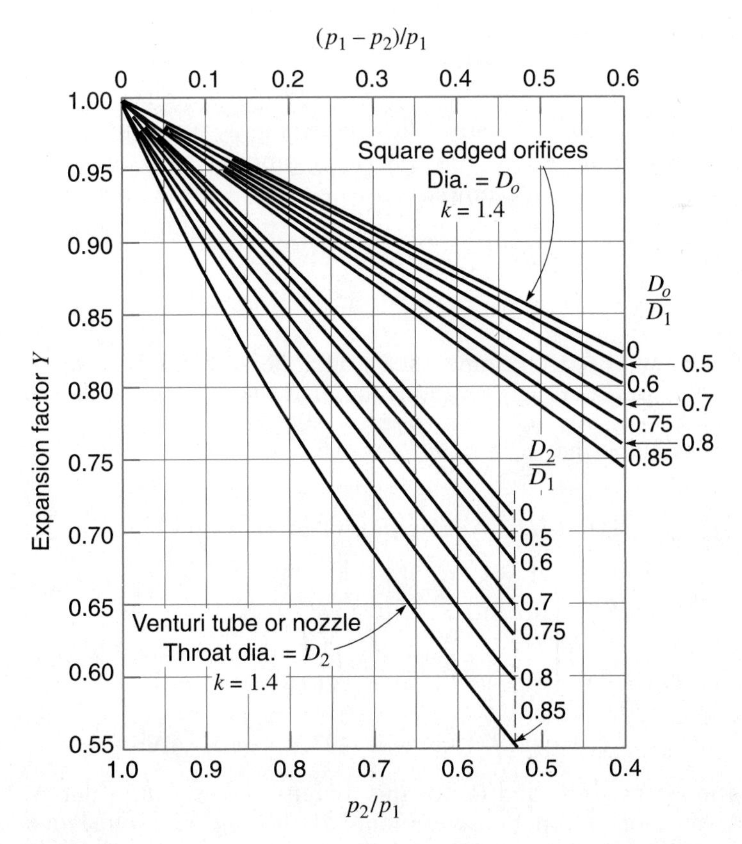

Figure 11.25
Expansion factors.

Flow Nozzles and Orifice Meters

Equation (11.23) can also be used for flow nozzles and orifice meters, though for flow nozzles C should be replaced by $K\sqrt{1-(D_2/D_1)^4}$ [from Eq. (11.19)], so that Fig. 11.22 can be used directly. For orifice meters the C of Eq. (11.23) should be replaced by $K\sqrt{1-(D_o/D_1)^4}$, D_2 should be replaced by D_o and A_2 should be replaced by A_o where D_o is the diameter of the orifice opening and A_o is its area.

For compressible fluids the C_c of an orifice meter depends on the ratio p_2/p_1; hence Y varies in a different manner than in the case of a venturi. Values of Y for orifice meters are shown in Fig. 11.25. In the case of an orifice meter the maximum jet velocity is the sonic velocity c, but this does not impose a limit on the rate of discharge because the jet area continues to increase with decreasing values of p_2/p_1. For this reason the values of Y for the orifice are extended in Fig. 11.25 to lower values of p_2/p_1.

Supersonic Conditions

The general case of flow measurement under supersonic conditions will not be discussed in this text. If supersonic flow occurs in a converging or converging–diverging nozzle attached to the end of a pipe or to a tank, Eqs. (13.43), (13.44), and (13.45) may be employed to compute ideal flow rates where the velocity of approach is negligible. This can be transformed into actual flow rates by introducing a proper flow coefficient.

SAMPLE PROBLEM 11.6 Determine the weight flow rate when air at 20°C and 700 kN/m^2 abs flows through a venturi meter if the pressure at the throat of the meter is 400 kN/m^2 abs. The diameters at inlet and throat are 25 and 12.5 cm respectively. Assume that $C = 0.985$.

Solution

$p_2/p_1 = \frac{400}{700} = 0.571$; $D_2/D_1 = 0.50$. Fig. 11.25: $Y \approx 0.72$

$$\text{Eq. (2.5):}\quad \gamma_1 = \frac{gp}{RT} = \frac{(9.81\ \text{m/s}^2)(700\ \text{kN/m}^2)}{[287\ \text{m}^2/(\text{s}^2\cdot\text{K})](273+20)\ \text{K}} = 0.0817\ \text{kN/m}^3$$

$$\text{Eq. (11.23):}\quad G = 0.985(0.72)\frac{\pi(0.125)^2}{4}\sqrt{2(9.81)0.0817\frac{700-400}{1-(0.5)^4}}$$

$$G = 0.1971\ \text{kN/s} = 197.1\ \text{N/s} \qquad \textit{\textbf{ANS}}$$

If the relation between C and $\mathbf{R}_2$ for this meter is known, the value of $\mathbf{R}_2$ for the computed value of G can be determined. If the assumed value of C does not correspond with this value of $\mathbf{R}_2$, a slight adjustment in the value of C can be made to give a more accurate answer.

EXERCISES

11.10.1 Natural gas, for which $R = 3100$ ft·lb/(slug·°R) and $k = 1.3$, flows through a venturi tube. The pipe diameter is 18 in and the throat diameter is 9 in. The initial pressure of the gas is 160 psia at 64°F. What is the weight flow rate of the gas if the throat pressure is 104 psia and the meter coefficient is 0.98?

11.10.2 Helium [$R = 12{,}420$ ft·lb/(slug·°R), $k = 1.66$] is stored in a tank at 60 psia and 76°F. The gas flows out through a $\frac{3}{4}$-in-diameter orifice for which $C_v = 0.98$ and $C_c = 0.62$ for liquids. Find the weight flow rate if $Y = 0.95$ and the pressure outside the tank is 44 psia.

11.10.3 A tank contains air at 200 psia and 100°F. It flows out through an orifice with an area of 1.5 in^2, and enters a space where the pressure is 80 psia. Assuming $C_d = 0.60$, find the weight flow rate.

11.10.4 A tank contains air at 1460 kN/m^2 abs and 44°C. It flows out through an orifice with an area of 12.5 cm^2, and enters a space where the pressure is 713 kN/m^2 abs. Assuming $C_d = 0.62$, find the weight flow rate.

11.10.5 Repeat Exer. 11.10.3, but discharging into a space where the pressure is 15 psia.

11.10.6 Repeat Exer. 11.10.4, but discharging into a space where the pressure is 105 kN/m^2 abs.

11.10.7 Find (*a*) the critical pressure and the corresponding throat velocity in a suitable nozzle, assuming air at 70°F and $p_1 = 100$ psia. Neglect the velocity of approach. Refer to Secs. 13.3 and 13.6. (*b*) What will the values be if $D_2/D_1 = 0.80$?

11.10.8 Find (*a*) the critical pressure and the corresponding throat velocity in a suitable nozzle, assuming air at 20°C and $p_1 = 700$ kN/m^2 abs. Neglect the velocity of approach. Refer to Secs. 13.3 and 13.6. (*b*) What will the values be if $D_2/D_1 = 0.80$?

11.11 THIN-PLATE WEIRS

Weirs have long been standard devices for the measurement of water flow. One category, known as ***thin-plate weirs***, includes those constructed from a thin plate, usually of metal and erected perpendicular to the flow. The upstream face of the weir plate should be smooth, and the plate must be strictly vertical. All thin-plate weirs are sharp-crested; the upstream edge is formed by a horizontal top surface usually less than $\frac{1}{16}$ in (1.6 mm) long in the flow direction, followed by a bevel on the downstream edge. Such a design

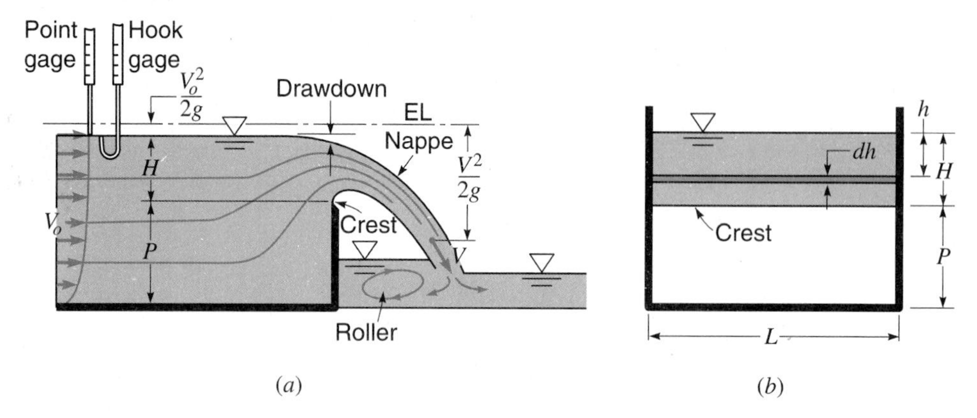

Figure 11.26
Flow over sharp-crested weir. (*a*) Side view. (*b*) Looking upstream.

causes the ***nappe*** (sheet of overflowing water) to spring clear as in Fig. 11.26*a*, with only a line contact at the crest, for all but the very lowest heads. If the nappe does not spring clear, the flow cannot be considered as true weir flow and the experimentally determined coefficients do not apply. The approach channel should be long enough so that a normal velocity distribution exists, and the water surface should be as free from waves as possible.

The weir crest may be straight and horizontal, or it may have a variety of shapes; a number of these are discussed below.

Suppressed Rectangular Weir

In its simplest form, water flows over the top of a plate with a straight and horizontal crest as shown in Fig. 11.26. The weir is as wide as the channel, so the width of the nappe is the same as the length of the crest. As there are no contractions of the stream at the sides, it is said that end contractions are ***suppressed***. It is essential that the sides of the channel upstream be smooth and regular. It is common to extend the sides of the channel downstream beyond the crest so that the nappe is confined laterally. The flowing water tends to entrain air from this enclosed space under the nappe, and unless this space is adequately ventilated, there will be a partial vacuum and perhaps all the air may eventually be swept out. The water will then cling to the downstream face of the plate, and the discharge will be greater for a given head than when the space is vented. Therefore venting of a suppressed weir is necessary if the standard formulas are to be applied.

The rate of flow is determined by measuring the water height H (head), above the crest, at a distance upstream from the crest equal to at least four times the maximum head to be employed. The amount of surface drawdown at the crest is typically about $0.15H$. The velocity at any point in the nappe is related to the energy line as shown in Fig. 11.26*a*.

To derive the flow equation for a rectangular weir having a crest of length L, consider an elementary area $dA = L\,dh$ in the plane of the crest, as shown in Fig. 11.26*b*. This elementary area is in effect a horizontal slot of length L and height dh. Neglecting velocity of approach, the ideal velocity of flow through this area will be equal to $\sqrt{2gh}$. The apparent flow through this area is

$$dQ_i = L\,dh\sqrt{2gh} = L\sqrt{2g}h^{1/2}\,dh$$

and this is to be integrated over the whole area, i.e., from $h = 0$ to $h = H$.

Performing this integration, we obtain an ideal Q_i, which is

$$Q_i = \sqrt{2g}\,L\int_0^H h^{1/2}\,dh = \tfrac{2}{3}\sqrt{2g}LH^{3/2}$$

The actual flow over the weir will be less than the ideal flow, because the effective flow area is considerably smaller than LH due to drawdown from the top and contraction of the nappe from the crest below (see Fig. 11.26*a*). Introducing a coefficient of discharge C_d to account for this,

$$Q = C_d\tfrac{2}{3}\sqrt{2g}LH^{3/2} \tag{11.24}$$

Dimensional analysis of weir flow leads to some interesting conclusions that provide a basis for an understanding of the factors that influence the coefficient of discharge. The physical variables that influence the flow Q over the weir of Fig. 11.26 include L, H, P, g, μ, σ, and ρ. Using the Buckingham Π theorem (Sec. 7.7), and without going into the details, we obtain

$$Q = \phi\left(\mathbf{W}, \mathbf{R}, \frac{P}{H}\right)L\sqrt{g}H^{3/2}$$

Thus, comparing this expression with Eq. (11.24), we conclude that C_d depends on **W**, **R**, and P/H. It has been found that P/H is the most important of these (Probs. 11.41 and 11.44). The Weber number **W**, which accounts for surface-tension effects, is important only at low heads. In the flow of water over weirs the Reynolds number is generally quite high, so viscous effects are generally insignificant. If one were to calibrate a weir for the flow of oil, however, **R** would undoubtedly affect C_d substantially. Typical values of C_d for sharp-crested weirs with water flowing range from about 0.62 for $H/P = 0.10$ to about 0.75 for $H/P = 2.0$.

Small-scale but precise experiments covering a wide range of conditions led T. Rehbock of the Karlsruhe Hydraulic Laboratory in Germany to the following expressions for C_d in Eq. (11.24):

BG units, H and P in ft:
$$C_d = 0.605 + \frac{1}{305H} + 0.08\frac{H}{P} \tag{11.25a}$$

SI units, H and P in m:
$$C_d = 0.605 + \frac{1}{1000H} + 0.08\frac{H}{P} \tag{11.25b}$$

These equations were obtained by fitting a curve to the plotted values of C_d for a great many experiments and are purely empirical. Capillarity is accounted for by the second term, while velocity of approach (assumed to be uniform) is responsible for the last term. Rehbock's formula has been found to be accurate within 0.5% for values of P from 0.33 to 3.3 ft (0.10 to 1.0 m) and for values of H from 0.08 to 2.0 ft (0.025 to 0.60 m) with the ratio H/P not greater than 1.0. It is even valid for ratios greater than 1.0 if the bottom of the discharge channel is lower than that of the approach channel so that backwater does not affect the head.

It is convenient to express Eq. (11.24) as

$$Q = C_W L H^{3/2} \tag{11.26}$$

where C_W, the ***weir coefficient***,[10] replaces $C_d \frac{2}{3}\sqrt{2g}$.

Using a value of 0.62 for C_d in Eq. (11.24), we can write

$$Q \approx \begin{cases} 3.32 L H^{3/2} & \text{in BG units} \\ 1.83 L H^{3/2} & \text{in SI units} \end{cases} \tag{11.27}$$

These equations give good results if $H/P < 0.4$, which is well within the usual operating range. If the velocity of approach V_0 is appreciable, a correction must be applied to the preceding equations either by changing the form of the equation or, more commonly, by changing the value of the coefficient.

Rectangular Weir with End Contractions

When the length L of the crest of a rectangular weir is less than the width of the channel, the nappe will have ***end contractions*** so that its width is less than L. Experiments have indicated that under the conditions depicted in Fig. 11.27 the effect of each side contraction is to reduce the effective width of the

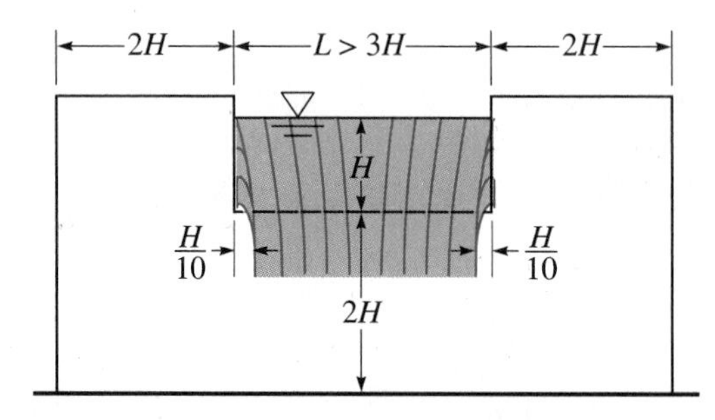

Figure 11.27
Limiting proportions of standard contracted weirs.

[10] Since C_W is not dimensionless, its value in BG units is different from that in SI units, as indicated in Eq. (11.27).

nappe by $0.1H$. Hence for such a situation the flow rate may be computed by employing any of the three preceding equations and substituting $(L - 0.1nH)$ for L, where n is the number of end contractions, normally 2 but sometimes 1. The results are called Francis formulas.

Cipolletti Weir

In order to avoid correcting for end contractions a Cipolletti weir is often used. It has a trapezoidal shape with side slopes of four vertical on one horizontal. The additional area adds approximately enough to the effective width of the stream to offset the lateral contractions.

V-notch, or Triangular, Weir

For relatively small flows the rectangular weir must be very narrow and thus of limited maximum capacity, or else the value of H will be so small that the nappe will not spring clear but will cling to the plate. For such a case the V-notch or triangular weir has the advantage that it can function for a very small flow and also measure reasonably large flows as well. The vertex angle is usually between 10° and 90° but rarely larger.

In Fig. 11.28 is a V-notch weir with a vertex angle θ. The ideal rate of discharge through an elementary area dA is $dQ = \sqrt{2gh}\, dA$. Now $dA = 2x\, dh$, and $x/(H - h) = \tan(\theta/2)$. Substituting in the foregoing, and introducing a coefficient of discharge C_d, the following result is obtained for the entire notch:

$$Q = C_d 2\sqrt{2g} \tan\frac{\theta}{2} \int_0^H (H - h)h^{1/2}\, dh$$

Integrating between limits and reducing, the fundamental equation for all V-notch weirs is obtained:

$$Q = C_d \frac{8}{15} \sqrt{2g} \tan\frac{\theta}{2} H^{5/2} \tag{11.28}$$

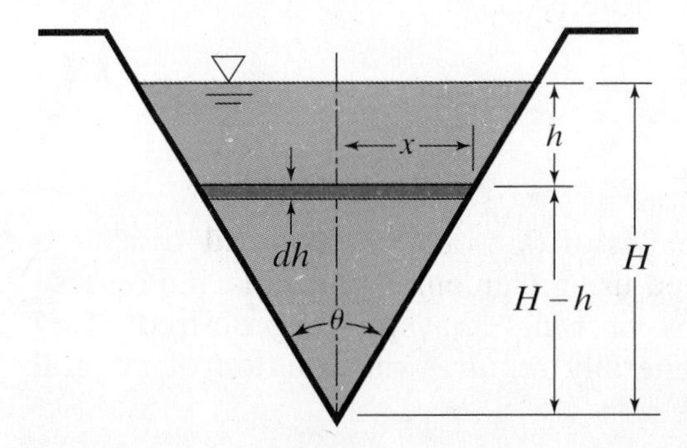

Figure 11.28
V-notch weir.

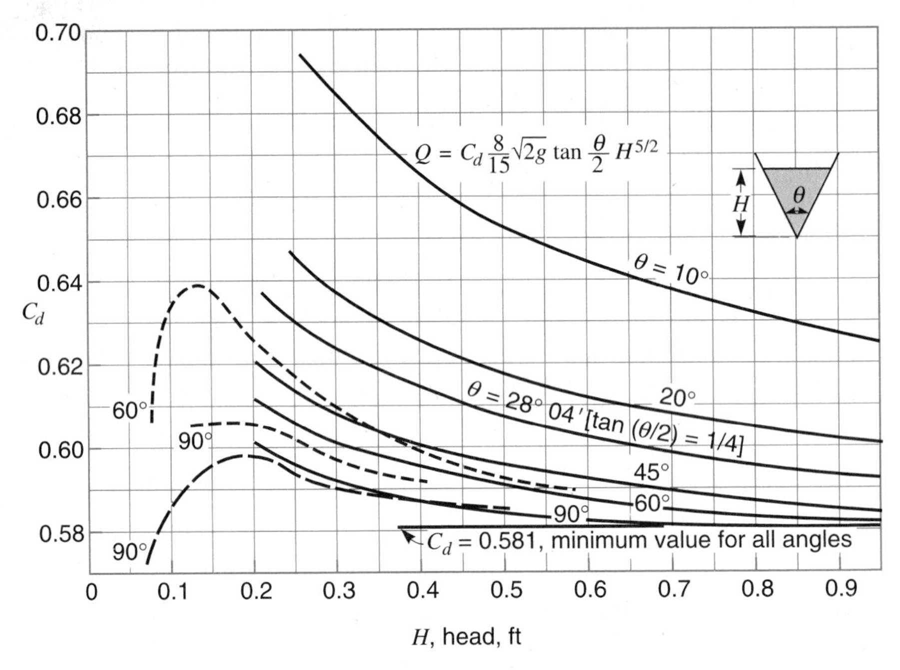

Figure 11.29
Coefficients for V-notch weirs obtained by three different experimenters.

For a given angle θ and assuming C_d is constant, this may be reduced to

$$Q = KH^{5/2} \tag{11.29}$$

The value of the constant K in English units will be different from that in SI units.

In Fig. 11.29 are presented experimental values for C_d for water flowing over V-notch weirs with central angles varying from 10° to 90°. The rise in C_d at heads less than 0.5 ft is due to incomplete contraction. At lower heads the frictional effects reduce the coefficient. At very low heads, when the nappe clings to the weir plate, the phenomenon can no longer be classed as weir flow and Eqs. (11.28) and (11.29) are inapplicable.

Proportional Weirs

While, for given simple geometric shapes of thin-plate weirs, head–discharge (H–Q) relations can easily be obtained using elementary calculus, the reverse problem of working out the shapes of weirs that produce desired H–Q relations is more challenging and generally requires the solution of integral equations. Recent research[11] has led to the development of new types of nonlinear weirs like the logarithmic weir ($Q \propto \log H$) and the quadratic weir ($Q \propto H^{1/2}$). These new types of weirs are called ***proportional weirs***. Many of these proportional weirs measure discharge more accurately then traditional weirs, and they can be used as efficient velocity controllers in sediment settling chambers and in industrial dosing facilities. An example of a linear proportional weir known as a Sutro Weir is given in Prob. 11.45.

Sample Problem 11.7 Water is flowing in a rectangular channel at a velocity of 3 fps and depth of 1.0 ft. Neglecting the effect of velocity of approach and employing Eq. (11.27), determine the height of a sharp-crested suppressed weir that must be installed to raise the water depth upstream of the weir to 4 ft.

Solution

$$L = \text{length of weir crest} = \text{width of channel}$$

Eq. (11.27): $$Q = AV = LyV = L(1)(3) = 3.33LH^{3/2}$$

$$H^{3/2} = \frac{3.0}{3.33} = 0.901, \quad H = 0.933 \text{ ft}$$

Fig. 11.26: P = height of weir required = $4.00 - 0.933 = 3.07$ ft ***ANS***

EXERCISES

11.11.1 A 2.5-ft-high rectangular sharp-crested weir extends across a 6-ft-wide rectangular channel. When the head is 1.12 ft, determine the flow rate by neglecting the velocity of approach.

11.11.2 A 0.75-m-high rectangular sharp-crested weir extends across a 2.2-m-wide rectangular channel. When the head is 32 cm, determine the flow rate by neglecting the velocity of approach.

11.11.3 Suppose the rectangular weir of Exer. 11.11.1 is contracted at both ends. (*a*) Find the flow rate for a head of 1.12 ft by the Francis formula. (*b*) What would be the maximum value of H for which the Francis formula could be used?

---

[11] K. Keshava Murthy and D. P. Giridhar, Improved Inverted V-Notch or Chimney Weir, *J. Irrigation and Drainage Engineering, ASCE,* vol. 116, no. 3, 1990; K. Keshava Murthy and M. N. S. Prakash, Practical Constant Accuracy Linear Weir, *J. Irrigation and Drainage Engineering, ASCE,* vol. 120, no. 3, 1994.

11.11.4 Suppose the rectangular weir of Exer. 11.11.2 is contracted at both ends. (*a*) Find the flow rate for a head of 32 cm by the Francis formula. (*b*) What would be the maximum value of H for which the Francis formula could be used?

11.11.5 (*a*) What is the rate of discharge of water over a 45° triangular weir when the head is 0.6 ft? (*b*) With the same head, what would be the increase in discharge obtained by doubling the notch angle, i.e., for a 90° weir? (*c*) What would be the head for a discharge of 1.5 cfs of water over a 60° triangular weir?

11.12 STREAMLINED WEIRS AND FREE OUTFALL

In contrast to thin-plate weirs, streamlined weirs have large dimensions in the flow direction and they must have rounded crests and/or approach edges (they are not sharp-crested). Shapes of streamlined weir cross sections include rectangular with a rounded upstream edge, triangular with a rounded crest, and hydrofoil shapes (similar to airfoils).

Streamlined weirs are usually built of concrete, giving them the advantage of being rugged and able to stand up well under field conditions.

Broad-Crested Rectangular Weir

The broad-crested weir (Fig. 11.30), as noted in Sample Prob. 10.8, is a critical-depth meter; that is, if the weir is high enough (Sec. 10.13) critical depth occurs on the crest of the weir. In Eq. (10.22) it was shown that for a rectangular channel $V_c = \sqrt{gy_c}$, while Eq. (10.26) stated that when the flow is critical $y_c = \frac{2}{3}E$. Employing these relations, we can write for the flow over a

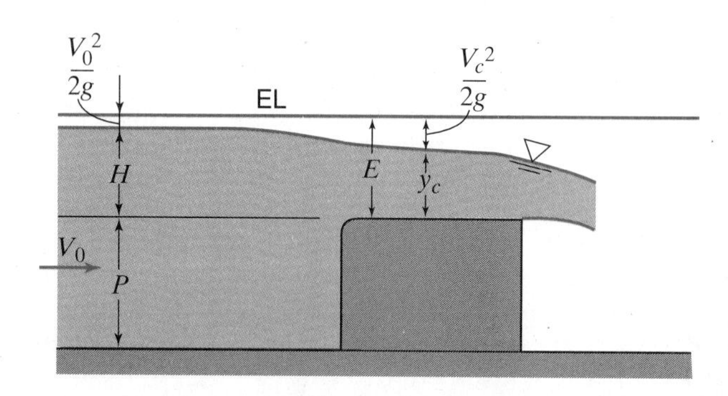

Figure 11.30
Broad-crested rectangular weir.

broad-crested weir,

$$Q = AV = (Ly_c)\sqrt{gy_c} = L\sqrt{g}y_c^{3/2} = L\sqrt{g}(\tfrac{2}{3})^{3/2}E^{3/2} \qquad (11.30)$$

where L is the width of the weir (length of crest), and E is the total head over the weir, equal to $H + V_0^2/2g$, in which H is the measured head and V_0 is the approach velocity from upstream. Let us now substitute Eq. (11.30) into Eq. (11.24), which is applicable to broad-crested weirs as well as sharp-crested suppressed rectangular weirs, since both have rectangular flow cross sections. This yields

$$C_d = \frac{1}{\sqrt{3}}\left(\frac{E}{H}\right)^{3/2} \qquad (11.31)$$

For very high weirs (that is, P/H large) the velocity of approach becomes small, so that $E \rightarrow H$ (Fig. 11.30) and thus $E/H \rightarrow 1$ and $C_d \rightarrow 1/\sqrt{3} = 0.577$. With a lower weir and the same flow rate the velocity of approach becomes larger and hence E/H increases in magnitude, resulting in larger values of C_d for lower broad-crested weirs. The actual value of C_d for a broad-crested weirs depends on the length of the weir and whether or not the upstream corner (edge) of the weir is rounded. The foregoing discussion, of course, assumes critical flow (Sec. 10.9) on the weir. If the approaching flow is supercritical, the presence of surface waves causes broad-crested rectangular weirs to be rather impractical for use as metering devices.

Other Streamlined Weirs

Many new streamlined weirs have been tested[12] and found not only to have high discharge coefficients but also to function efficiently under conditions of high submergence. Weir ***submergence*** σ is defined as the downstream water depth divided by the upstream water depth, i.e., $\sigma = h_d/h_u$. For sufficiently low values of σ the discharge over the weir is unaffected by the downstream depth; for increasing σ, when the discharge is first reduced by a prescribed small amount like 2 or 5% the submergence is defined to be the ***critical submergence*** σ_c.

The coefficient of discharge C_d of a streamlined weir, according to the International Standards Organization (ISO), is defined by the equation

$$Q = C_d(\tfrac{2}{3})^{3/2}\sqrt{g}LE^{3/2} \qquad (11.32)$$

where the variables are as for Eq. (11.30) (see also Figs. 11.30 and 10.13).

[12] N. S. Lakshmana Rao, Theory of Weirs, *Advances in Hydroscience,* V. T. Chow, ed., Academic Press, N.Y., vol. 10, 1975.

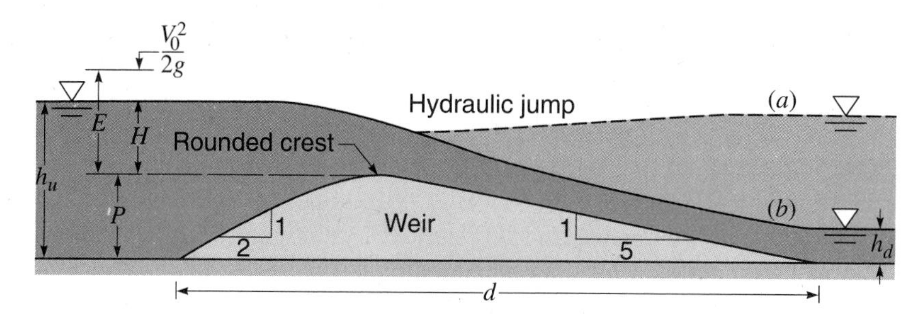

Figure 11.31
Streamlined triangular weir, $d/P = 7.5$, $C_d = 1.11$. Free surface at (*a*) critical submergence, $\sigma_c = 0.82$, and (*b*) $\sigma = 0.2$.

Note that this equation is the same as Eq. (11.30) with a discharge coefficient added.

Over 25 streamlined weirs have been classified into nine categories, depending on (*a*) whether σ_c is low (<0.65), medium (0.65–0.80), or high (>0.80), and (*b*) whether the weir discharge coefficient C_d is low (<1.0), medium (1.0–1.3), or high (>1.3). For example, several triangular-shaped streamlined weirs, and what are called hydrofoil weirs, fall into the category of medium C_d and high σ_c types. The streamlined triangular weir of Fig. 11.31 has a medium C_d and a high σ_c of 0.82 when we specify a discharge reduction of 6%.

Free Outfall

Another method for determining the flow rate in a rectangular channel is to measure the depth of flow y_b at the brink of a free outfall (Fig. 10.17). Substituting $y_c = y_b/0.72$ in Eq. (10.23) permits an approximate determination of the rate of flow per foot of channel width. This method can be used only if the flow is subcritical.

EXERCISES

11.12.1 What is the flow rate per unit width over a broad-crested weir that rises 1.0 ft above the bed of a horizontal channel when the head is 3.0 ft above the crest?

11.12.2 What is the flow rate per unit width over a broad-crested weir that rises 0.3 m above the bed of a horizontal channel when the head is 0.9 m above the crest?

11.12.3 Find the water depth just upstream of a 1.5-ft-high broad-crested weir in a channel 6 ft wide when the flow is 18 cfs.

11.12.4 Find the water depth just upstream of a 0.4-m-high broad-crested weir in a channel 1.8 m wide when the flow is 0.54 m^3/s.

11.12.5 Subcritical flow in a rectangular open channel 6 ft wide ends in a free outfall. If the water depth at the brink measures 2.70 ft, estimate the flow rate.

11.12.6 Subcritical flow in a rectangular open channel 2 m wide ends in a free outfall. If the water depth at the brink measures 0.75 m, estimate the flow rate.

11.13 OVERFLOW SPILLWAY

An overflow spillway is a section of dam designed to permit water to pass over its crest. Overflow spillways are widely used on gravity, arch, and buttress dams. The ideal spillway should take the form of the underside of the nappe of a sharp-crested weir (Fig. 11.32*a*) when the flow rate corresponds to

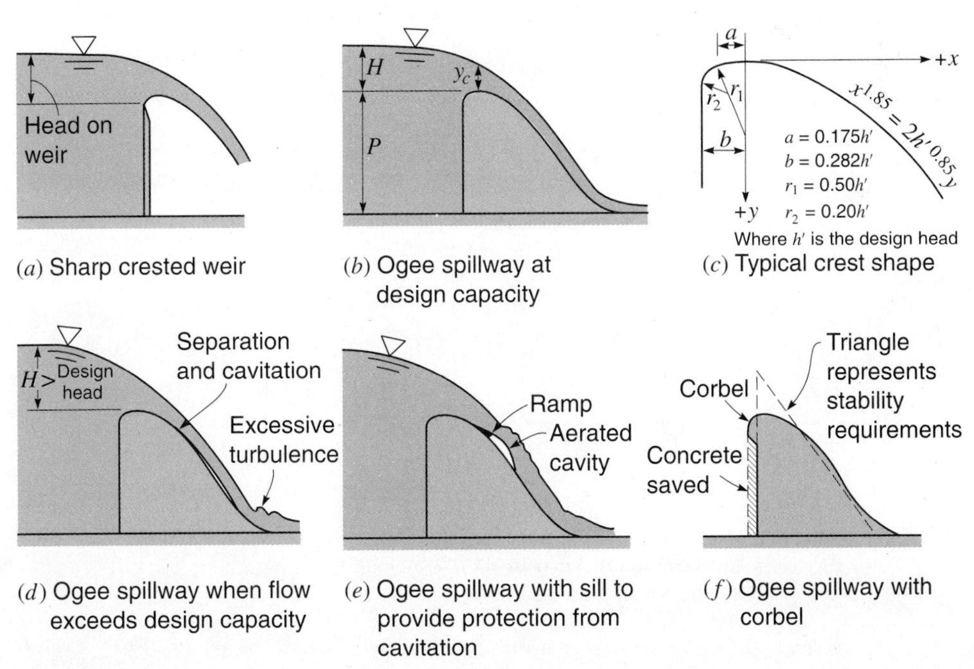

Figure 11.32
Characteristics of an ogee spillway.

the maximum design capacity of the spillway (Fig. 11.32*b*). Figure 11.32*c* shows an ***ogee spillway*** which closely approximates the ideal.[13] The reverse curve on the downstream face of the spillway should be smooth and gradual. A radius of about one-fourth of the spillway height has proved satisfactory.

The discharge of an overflow spillway is given by the weir equation, (11.26):

$$Q = C_w L H^{3/2} \tag{11.33}$$

where Q = discharge, cfs or m^3/s

C_w = weir coefficient

L = length of the crest, ft or m

H = head on the spillway (vertical distance from the crest of the spillway to the reservoir level), ft or m

The coefficient C_w varies with the design and head. For the standard overflow crest of Fig. 11.32*c* the variation of C_w is given in Fig. 11.33. Experimental

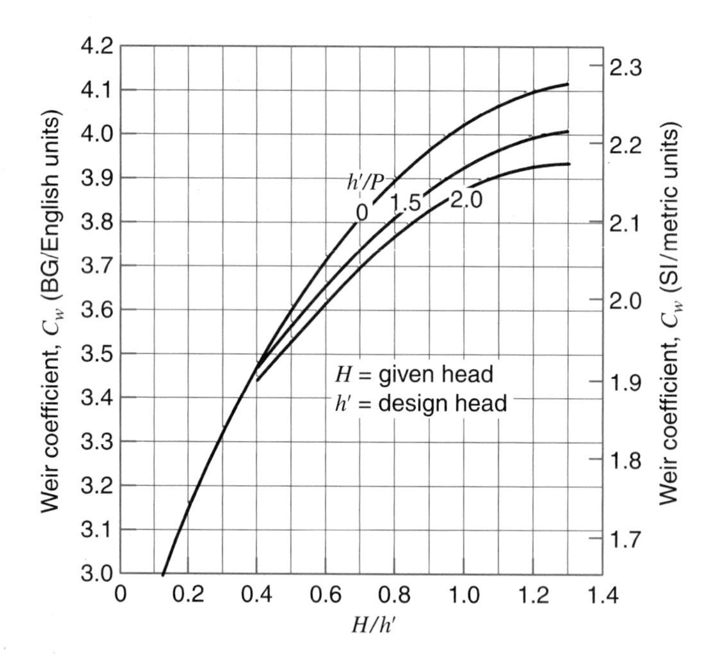

Figure 11.33
Variation of discharge coefficient with head for an ogee spillway crest such as shown in Fig. 11.32.

[13] Hydraulic Models as an Aid to the Development of Design Criteria, U.S. Waterways Expt. Sta., Bull. 37, Vicksburg, Miss., 1951.

models are often used to determine spillway coefficients. End contractions on a spillway reduce the effective length below the actual length L. Square-cornered piers disturb the flow considerably and reduce the effective length by the width of the piers plus about $0.2H$ for each pier. Streamlining the piers or flaring the spillway entrance minimizes the flow disturbance. If the cross-sectional area of the reservoir just upstream from the spillway is less than five times the area of flow over the spillway, the approach velocity will increase the discharge a noticeable amount. The effect of approach velocity can be accounted for by the equation

$$Q = C_w L\left(H + \frac{V_0^2}{2g}\right)^{3/2} \tag{11.34}$$

where V_0 is the approach velocity.

On high spillways, if the overflowing water breaks contact with the spillway surface, a vacuum will form at the point of separation (Fig. 11.32*d*) and cavitation may occur. Cavitation and vibration from the alternate making and breaking of contact between the water and the face of the spillway may result in serious structural damage. A ramp of proper shape and size when properly located (Fig. 11.32*e*) will direct the water away from the spillway surface to form a cavity.[14] To be effective, air must be freely admitted to the cavity. The result is that air is entrained in the water, the water bulks up, and when it returns to the spillway surface there is no problem with cavitation. On very high spillways these ramps may be used in tandem.

There are a number of other types of spillways including chute, side-channel, ski-jump, and shaft spillways.[15]

11.14 SLUICE GATE

The sluice gate shown in Fig. 11.34 is a device used to control the passage of water in an open channel. When properly calibrated, it may also serve as a means of flow measurement. As the lower edge of the gate opening is flush with the floor of the channel, contraction of the bottom surface of the issuing

[14] K. Zagustin and N. Castillejo, Model-Prototype Correlation for Flow Aeration in Guru Dam Spillway, *Proceedings International Association for Hydraulic Research,* vol. 3, 1983.

[15] For a brief discussion of the different types of spillways see: R. K. Linsley, J. B. Franzini, D. L. Freyberg, and G. Tchobanoglous, *Water Resource Engineering,* 4th ed., McGraw-Hill Book Co., New York, 1992.

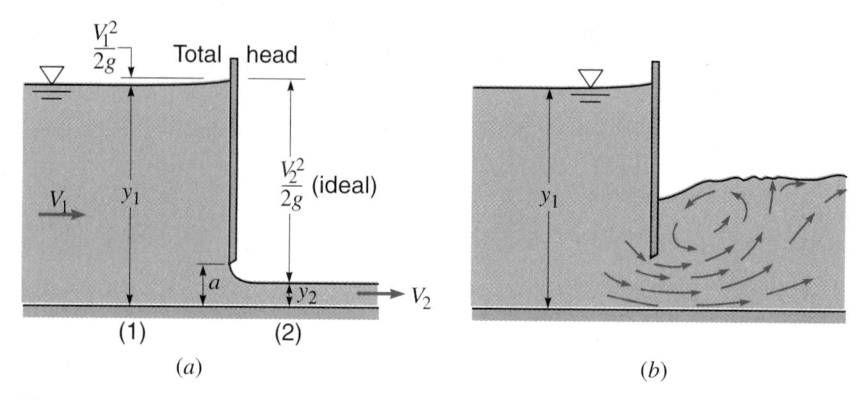

Figure 11.34
Flow through sluice gate. (*a*) Free flow ($y_2 = C_c a$). (*b*) Submerged flow.

stream is entirely suppressed. Side contractions will of course depend on the extent to which the opening spans the width of the channel. The complete contraction on the top side, however, because of the larger velocity components parallel to the face of the gate, will offset the suppressed bottom contraction, resulting in a coefficient of contraction nearly the same as for a slot with contractions at top and bottom.

Flow through a sluice gate differs fundamentally from flow through a slot in that the jet is not free but guided by a horizontal floor. Consequently, the final jet pressure is not atmospheric, but distributed hydrostatically in the vertical section. Writing the energy equation with respect to the stream bed as datum from point 1 to point 2 in the free-flow case (Fig. 11.34*a*) and neglecting head loss,

$$\frac{V_1^2}{2g} + y_1 = \frac{V_2^2}{2g} + y_2$$

from which, introducing continuity, the ideal velocity downstream is

$$V_{2_i} = \frac{1}{\sqrt{1 - (A_2/A_1)^2}} \sqrt{2g(y_1 - y_2)} \tag{11.35}$$

The actual flow rate $Q = C_d Q_i = C_c C_v (A V_{2_i})$, where $A = aB$ is the area of the gate opening and B is the width of the opening.

Absorbing the effects of flow contraction, friction, velocity of approach, and the downstream depth y_2 into an experimental flow coefficient, we obtain a simple discharge equation for flow under a sluice gate:

$$Q = K_s A \sqrt{2gy_1} \tag{11.36}$$

where K_s is defined as the ***sluice coefficient.***[16]

Values of K_s depend on a and y_1 (Fig. 11.34*a*) and are usually between 0.55 and 0.60 for free flow, but are significantly reduced when the flow conditions downstream are such as to produce submerged flow, as shown in Fig. 11.34*b*.

EXERCISES

11.14.1 A rectangular channel contains a sluice gate that extends across the full width of the channel with an opening of 0.24 ft. Given $C_d = 0.60$ and free flow, find the flow rate per unit width and the sluice coefficient when the upstream and downstream depths are 2.05 ft and 0.15 ft respectively.

11.14.2 A rectangular channel contains a sluice gate that extends across the full width of the channel with an opening of 0.10 m. Given $C_d = 0.60$ and free flow, find the flow rate per unit width and the sluice coefficient when the upstream and downstream depths are 0.85 m and 0.06 m respectively.

11.15 MEASUREMENT OF LIQUID SURFACE ELEVATION

The elevation of the surface of a liquid at rest may be determined through use of a piezometer column, manometer, or pressure gage (Sec. 3.6). These will also give accurate results when applied to a stationary liquid contained in a tank that is moving, provided the tank is not undergoing an acceleration (Sec. 3.11). Staff gages, such as those used at reservoirs, provide approximate liquid surface elevation data.

Various methods are used to measure the surface elevation of moving liquids. To determine the head H on a weir, the elevation difference between the crest of the weir and the liquid surface must be measured. In the field the elevation of the liquid surface is often determined through use of a ***stilling well*** connected by a pipe to the main liquid body. A ***float*** in the well is used

---

[16] Values of discharge coefficients for sluice and other types of gates may be found in Hunter Rouse (ed.), *Engineering Hydraulics,* pp. 536–543, John Wiley & Sons, Inc., New York, 1950.

to actuate a clock-driven liquid-level recorder so that a continuous record of the liquid-surface elevation is obtained. In the laboratory a ***hook gage*** or ***point gage*** (Fig. 11.26) is commonly employed for liquid-surface level determinations. The point gage is particularly suitable for fast-moving liquids where a hook gage would create a local disturbance in the liquid surface. In all liquid-surface level determinations care should be taken to make the measurements in regions where there is no curvature of streamlines; otherwise centrifugal effects will give a false reading of the piezometric head.

Other methods for determining the elevation of a liquid surface include the use of the sonic devices, electric gages, and bubblers. A ***sonic device*** is mounted at some convenient location above the liquid surface and the time required for a sound pulse to travel vertically downward to the liquid surface and return is indicative of the relationship between the elevation of the liquid surface and the device. ***Electrical gages*** include those with capacitive sensors and those with resistive sensors. In the resistive type two parallel bare-wire conductors are partially immersed in the liquid as a component of an electrical system. The electrical resistance between them is a function of the liquid depth. With proper circuitry and calibration, the device can be used to provide data on liquid-surface elevation. The sonic devices and the electrical gages are equally applicable to liquids at rest or in motion. The ***bubbler*** is used primarily for liquids at rest. In the bubbler system, the minimum pressure required to drive a gas into a liquid (i.e., to form bubbles) at a depth is a measure of the depth of the liquid. Knowing the elevation at which the bubbles are emitted permits determination of the elevation of the liquid surface.

11.16 OTHER METHODS OF MEASURING DISCHARGE

In addition to the foregoing "standard" devices for measuring the flow of fluids, there are a number of supplementary devices less amenable to exact theoretical analysis but worthy of brief mention. One of the simplest for measuring flow in a pipeline is the ***elbow meter***, which consists of nothing more than piezometer taps at the inner and outer walls of a 90° elbow in the line (see Sec. 8.23). The pressure difference, due to the centrifugal effects at the bend, will vary approximately as the velocity head in the pipe. Like other meters, the elbow should have sections of a straight pipe upstream and downstream and should be calibrated in place.

The ***rotameter*** (Fig. 11.35) consists of a vertical glass tube that is slightly tapered, in which the metering ***float*** is suspended by the upward motion of the fluid around it. Directional notches cut in the float keep it rotating and thus free of wall friction. The rate of flow determines the equilibrium height of the float, and the tube is graduated to read the flow directly. The rotameter

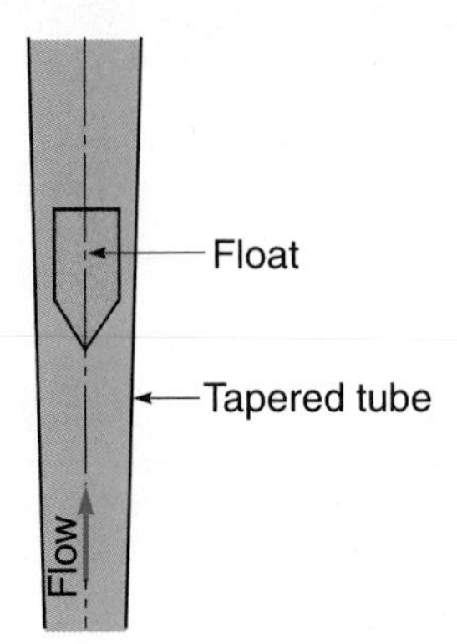

Figure 11.35
Rotameter type of flow meter.

is also used for gas flow, but the weight of the float and the graduation must be changed accordingly.

The ***inferential meter*** or ***turbine meter*** consists of a propeller or wheel with curved blades, shaped so that the flow of gas or liquid passing through it causes it to rotate. The rotational speed indicates the flow rate, after calibration. Such meters are often provided with guide vanes that may be adjusted to change the calibration.

Other techniques for measuring flow rate include the ***salt-velocity*** method. In this method a charge of concentrated salt is injected into the flow at an upstream station. Its arrival at a downward station is detected from conductivity measurements. In flowing between the two stations, the salt disperses and its arrival at the downstream station is spread out over a considerable period of time. The time of travel between the two stations is taken as the time from the instant of injection at the upstream station to the time at which the centroid of the conductivity-time curve passes the downstream station. Knowing the travel time and the distance, the velocity may be calculated, and then by multiplying by the cross-sectional area, we get the flow rate. In special circumstances other substances such as dye or tritium may be used instead of salt.

PROBLEMS

11.1 A hydrometer (Figs. 3.21 and 3.22*a*) is made in the form of an 8-in-long cylinder with a diameter of $\frac{3}{8}$ in. Attached to the end of the cylinder is a 1.2-in-diameter sphere. The entire device weighs 0.9 oz. What range of specific gravities can be measured with this device? When floating in a liquid with a density of 1.80 slugs/ft^3, how much of the cylinder will project above the surface?

11.2 A hydrometer (Figs. 3.21 and 3.22*a*) is made in the form of a 20-cm-long cylinder with a diameter of 6 mm. Attached to the end of the cylinder is a 22-mm-diameter sphere. The mass of the entire device is 8 g. What range of specific gravities can be measured with this device? When floating in a liquid having a density of 900 kg/m^3, how much of the cylinder will project above the surface?

11.3 Fifty cubic centimeters of 70°F water flows through a certain tube-type viscometer in 53.7 sec, whereas the same volume of 50°F oil ($s = 0.90$) flows through it in 850 sec. Find the absolute viscosity of the oil.

11.4 A liquid ($\rho = 889$ kg/m^3) under a head of 50 cm flows at a steady rate of 30 cm^3/min through a glass tube of diameter 2.0 mm and length 4.5 m. What is the absolute viscosity and the kinematic viscosity of the liquid? Express the answers in stokes and poises. Confirm by calculation that the flow is laminar.

11.5 A glass bead ($s = 2.60$) with a diameter of 16 mm falls through a liquid ($s = 1.59$) in a 10-cm-diameter tube. If the bead moves with a constant velocity of 14.5 cm/min, what is the absolute viscosity and the kinematic viscosity of the liquid?

11.6 A 3.6-in-diameter tube contains oil ($s = 0.93$) with a viscosity of 0.0065 $lb{\cdot}sec/ft^2$. What is the maximum size of steel sphere ($s = 7.84$) that will satisfy Stokes' law when released into this oil. What will be the fall velocity of this sphere?

11.7 A 10-cm-diameter tube contains oil ($s = 0.9$) with a viscosity of 0.25 $N{\cdot}s/m^2$. What is the maximum size of steel sphere ($s = 7.8$) that will satisfy Stokes' law when released into this oil. What will be the fall velocity of this sphere?

11.8 An 0.10-in-diameter sphere has a velocity of 0.005 fps when falling through liquid in a 1.0-in-diameter tube. What will be its fall velocities through the same liquid in tubes of diameter 0.50, 2.0, 4.0, and 10.0 in. Plot fall velocity vs tube diameter.

11.9 In Fig. S11.1, pressure gages *A* and *B* read 11.0 psi and 12.0 psi, respectively. If 60°F air is flowing, what is the velocity *u*? Atmospheric pressure is 27.0 inHg. Assume $C_I = 1.0$ and neglect compressibility effects.

11.10 Suppose the two pressure gages of Prob. 11.9 are replaced by a differential manometer containing water. What is the reading on the manometer?

11.11 The pitometers in Fig. P11.11 are connected to a mercury manometer, which reads 5.0 in. The velocity of flowing carbon tetrachloride ($s = 1.59$) is known to be 9.37 fps. What is C_I for this instrument?

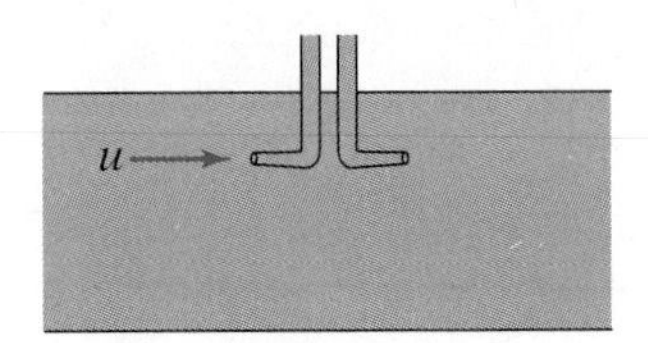

Figure P11.11

11.12 The pitometers in Fig. P11.11 are connected to a mercury manometer, which reads 12.0 cm. The velocity of flowing carbon tetrachloride ($s = 1.59$) is known to be 2.76 m/s. What is C_I for this instrument?

11.13 Given air at 60°C and standard atmospheric pressure is flowing in Fig. P11.11. The pitometers are attached to a manometer containing a liquid with $s = 0.90$. Plot the velocity u versus the manometer reading assuming $C_I = 0.93$.

11.14 Water at 90°F flows through a smooth, 12-in-diameter pipe. A Prandtl tube is placed on the center line of the pipe. The reading on a differential manometer attached to this Prandtl tube is 8 in of carbon tetrachloride ($s = 1.59$). What is the flow rate?

11.15 Assume the pipe diameter in Fig. 11.10 is 16 in, and the flow is laminar, i.e., $u = u_{max} - kr^2$, when $u_{max} = 0.1$ fps. Divide the circle into concentric rings with radii of 2, 4, 6, and 8 in, and compute the flow rate by the method of Fig. 11.10 by taking the velocities at radii of 1, 3, 5, and 7 in as representative of the rings. Compare the result with that obtained by integration.

11.16 Assume the pipe diameter in Fig. 11.10 is 48 cm, and the flow is laminar, i.e, $u = u_{max} - kr^2$, when $u_{max} = 30$ mm/s. Divide the circle into concentric rings with radii of 6, 12, 18, and 24 cm, and compute the flow rate by the method of Fig. 11.10 by taking the velocities at radii of 3, 9, 15, and 21 cm as representative of the rings. Compare the result with that obtained by integration.

11.17 Water flows at 12 fps through a horizontal 3-in-diameter pipe. A nozzle at the end of the pipe has a velocity coefficient of 0.98. If the pressure in the pipe is 7.5 psi, find (*a*) the velocity in the jet; (*b*) the rate of discharge; (*c*) the diameter of the jet; and (*d*) the head loss through the nozzle.

11.18 Water flows at 3.5 m/s through a horizontal 12-cm-diameter pipe. A nozzle at the end of the pipe has a velocity coefficient of 0.98. If the pressure in the pipe is 45 kN/m^2, find (*a*) the initial velocity of the jet; (*b*) the rate of discharge; (*c*) the diameter of the jet; and (*d*) the head loss through the nozzle.

11.19 In Fig. P11.19 the pressure at the base of the nozzle, point 1, is 21.0 psi, and the power available in the jet at point 2 is 3.42 hp. Find (*a*) the theoretical height to which the jet will rise; (*b*) the coefficient of velocity for this nozzle; (*c*) the head loss between points 1 and 2; and (*d*) the theoretical diameter of the jet at a point 20 ft above point 2.

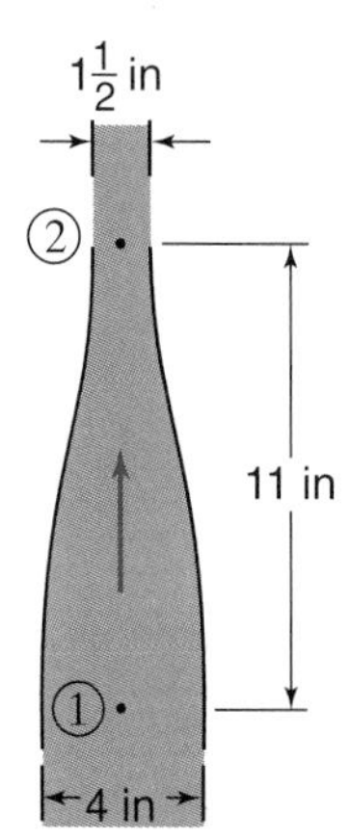

Figure P11.19

11.20 In Fig. P11.19 the pressure at the base of the nozzle, point 1, is 145 kPa, and the power available in the jet at point 2 is 2.55 kW. Find (*a*) the theoretical height to which the jet will rise; (*b*) the coefficient of velocity for this nozzle; (*c*) the head loss between points 1 and 2; and (*d*) the theoretical diameter of the jet at a point 6 m above point 2.

11.21 The loss of head due to friction in a nozzle, tube, or orifice may be expressed as $h_L = kV^2/2g$, where V is the actual velocity of the jet. (*a*) What is k for the three tubes in Fig. 11.16? (*b*) If these tubes discharge water under a head of 6 ft, what is the loss of head for each?

11.22 The loss of head due to friction in a nozzle, tube, or orifice may be expressed as $h_L = kV^2/2g$, where V is the actual velocity of the jet. (*a*) What is k for the three tubes in Fig. 11.16? (*b*) If these tubes discharge water under a head of 1.8 m, what is the loss of head for each?

11.23 Find the maximum theoretical head h at which the Borda tube of Fig. 11.17 will flow full if the liquid is 90°F water and the barometer reads 29.2 inHg. Assume $C_d = 0.72$ when the tube is flowing full.

11.24 Find the maximum theoretical head h at which the Borda tube of Fig. 11.17 will flow full if the liquid is 30°C water and the barometer reads 735 mmHg. Assume $C_d = 0.72$ when the tube is flowing full.

11.25 In Fig. P11.25 the orifice at the bottom of the large open tank has a diameter of 1.26 in. If the flow rate is 0.19 cfs, and the pitot tube registers a pressure of 16.8 psi, find the C_c and the C_v of the orifice. Neglect air resistance.

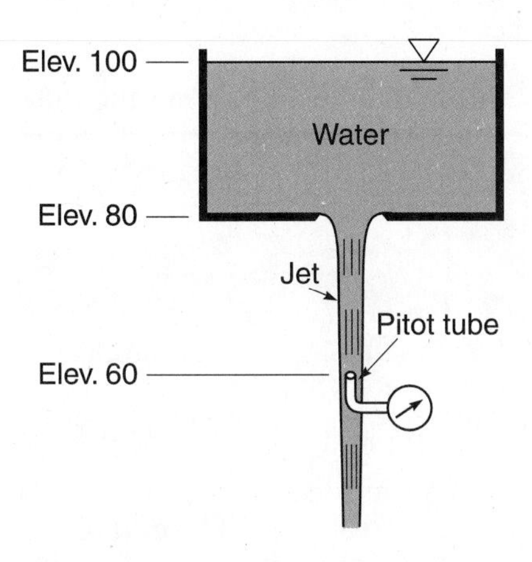

Figure P11.25

11.26 The venturi meter of Fig. S11.4 has $D_1 = 10$ in, $D_2 = 5$ in, and $\Delta z = 2$ ft. Assume the discharge coefficients of Fig. 11.19 are applicable and the diverging cone angle is 12°. (*a*) Find the flow rate of 72°F water when the mercury ($s = 13.55$) manometer reads $y = 5$ in. Find also the head loss (*b*) from inlet to throat, and (*c*) across the entire meter.

11.27 The venturi meter of Fig. S11.4 has $D_1 = 25$ cm, $D_2 = 12.5$ cm, and $\Delta z = 0.60$ m. Assume the discharge coefficients of Fig. 11.19 are applicable and the diverging cone angle is 12°. (*a*) Find the flow rate of 20°C water when the mercury ($s = 13.55$) manometer reads $y = 12.8$ mm. Find also the head loss (*b*) from inlet to throat, and (*c*) across the entire meter.

11.28 Repeat Prob. 11.26 when the venturi meter is set horizontally (i.e., $\Delta z = 0$), with all other data the same.

11.29 Repeat Prob. 11.27 when the venturi meter is set horizontally (i.e., $\Delta z = 0$), with all other data the same.

11.30 An 8 cm ISA flow nozzle (Figs. 11.21 and 11.22) is used to measure the flow of 40°C water through a 16-cm pipe. What would be the reading on a mercury manometer for the following flow rates: (*a*) 3.8 L/s; (*b*) 38 L/s; (*c*) 380 L/s. (*d*) For which of the preceding three flows is the mercury manometer practical?

11.31 A 7-in ISA flow nozzle is used to measure the flow of 20°F crude oil (s = 0.855) through a 10-in-diameter pipe. If a mercury manometer shows a reading of 6.9 in, find the flow rate.

11.32 Repeat Prob. 11.30 for a VDI orifice.

11.33 Repeat Prob. 11.31 for a VDI orifice.

11.34 An 8-in-diameter pipe carries 2.7 cfs of water at 72°F. Find the differential head and head loss across the following types of meters: (*a*) an 8-in by 4-in VDI orifice; (*b*) an 8-in by 4-in ISA flow nozzle; (*c*) an 8-in by 4-in venturi meter.

11.35 Air at 70°F and 104 psia flows through a venturi tube with a coefficient of 0.98. The pressure at the throat is 67.6 psia, the inlet area is 0.60 ft^2, and the throat area is 0.15 ft^2. Given k for the air is 1.40. (*a*) Using Eq. (11.22), find the ideal weight flow rate. (*b*) Evaluate Y from Fig. 11.25, and use it to find the actual weight flow rate. (*c*) What is the throat velocity?

11.36 Air at 20°C and 700 kN/m^2 abs flows through a venturi tube with a coefficient of 0.98. The pressure at the throat is 420 kN/m^2 abs, the inlet area is 0.060 m^2, and the throat area is 0.015 m^2. Given k for the air is 1.40. (*a*) Using Eq. (11.22), find the ideal weight flow rate. (*b*) Evaluate Y from Fig. 11.25, and use it to find the actual weight flow rate. (*c*) What is the throat velocity?

11.37 Using the same data as Prob. 11.35, what would be the value of Y and the weight flow rate for a square-edged orifice with C = 0.59?

11.38 Using the same data as Prob. 11.36, what would be the value of Y and the weight flow rate for a square-edged orifice with C = 0.62?

11.39 Air flows through a 20 cm by 10 cm venturi meter. At inlet the air temperature is 15°C and the gage pressure is 150 kN/m^2. Determine the flow rate if a mercury manometer reads 18 cm. Assume an atmospheric pressure of 101.3 kN/m^2 abs.

11.40 Air flows through a 5.0-cm-diameter orifice from a tank at 1500 kN/m^2 abs and 40°C into a space where the pressure is 500 kN/m^2 abs. (*a*) Compute the weight flow rate assuming C_d = 0.60. Refer to Sec. 13.8. Repeat for external pressures of (*b*) 750, (*c*) 1000, and (*d*) 1250 kN/m^2 abs.

11.41 Using the Rehbock formula, plot a family of curves of C_d versus P/H with H as a parameter. These curves give a complete picture of the variation of C_d for sharp-crested rectangular weirs. Include P/H values of 0.5, 1.0, 2.0, 5.0, 10.0 and H values of 0.2, 1.0, and 5.0 ft.

11.42 For the Cipolletti weir of Sec. 11.11, confirm that the side slopes of the trapezoid are $4V{:}1H$ by setting the reduction in discharge due to contraction equal to the increase in discharge due to the triangular area added.

11.43 Develop in general terms an expression for the percent of error in Q over a triangular weir if there is a small error in the measurement of the vertex angle. Assume there is no error in the weir coefficient. Compute the percent error in Q if there is a 1° error in the measurement of the total vertex angle of a triangular weir having a total vertex angle of 60°.

11.44 Plot C_d versus P/H for broad-crested weirs using Eq. (11.31). Include P/H values of 0.5, 1.0, 2.0, 5.0, 10.0, and H values of 0.2, 1.0, and 5.0 ft.

11.45 All the weir crests discussed in this chapter produce flow rates which vary as the head to some power greater than 1. In certain cases, such as in the outlet of a constant-velocity sedimentation chamber, it is desirable to employ a weir form in which Q varies directly with H. The proportional-flow weir is set flush with the bottom of the channel, as shown in the figure, while the sides taper inward, following the hyperbola $x\sqrt{y} = k$, a constant. Commencing with the head $h = H - y$, on the element of area $dA = 2x\,dy = 2(k/\sqrt{y})\,dy$, prove that the discharge equation for such a weir may be written as $Q = C_d \pi k \sqrt{2gH}$, and evaluate k in terms of the width B and the velocity V in the rectangular approach channel.

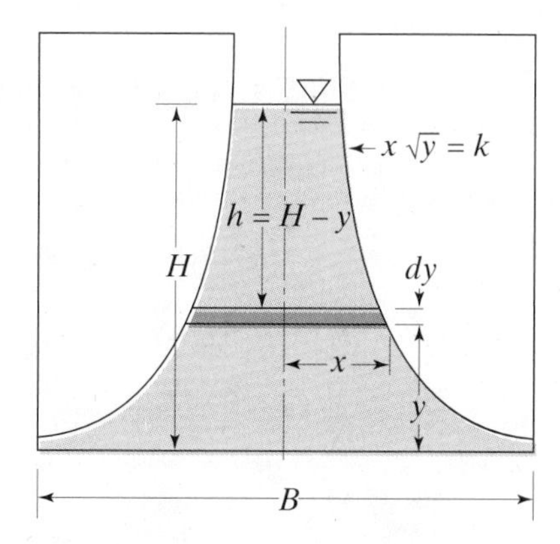

Figure P11.45

11.46 A 60° V-notch weir and a rectangular weir with end contractions (crest length = 2 ft) are both used to measure a flow rate of approximately 0.25 cfs. Assuming C_d is known to be minimum for each weir, compute the percentage of error in Q that would result from an error of 0.01 ft in the respective head measurements.

11.47 A 60° V-notch weir and a rectangular weir with end contractions (crest length = 0.6 m) are both used to measure a flow rate of approximately 7 L/S. Assuming C_d is known to be minimum for each weir, compute the percentage of error in Q that would result from an error of 5 mm in the respective head measurements.

11.48 An 8-ft-wide rectangular channel contains a sluice gate which extends across the full width of the channel with an opening of 0.6 ft. Assuming $C_d = 0.60$, $C_c = 0.62$, and free flow, find the flow rate and the sluice coefficient when the upstream depth is 5 ft.

11.49 A 2.0-m-wide rectangular channel contains a sluice gate which extends across the full width of the channel with an opening of 0.15 m. Assuming $C_d = 0.60$, $C_c = 0.62$, and free flow, find the flow rate and the sluice coefficient when the upstream depth is 1.25 m.

11.50 Refer to Sample Problem 5.11. (*a*) If $C_v = 0.98$, what is the flow rate? (*b*) If $C_c = 0.62$, what is the height of the opening? (*c*) What is K_s?

CHAPTER 12

Unsteady-Flow Problems

12.1 INTRODUCTION

This text deals mostly with steady flow, since the majority of cases of engineering interest are of this nature.[1] However, there are a number of cases of unsteady flow that are very important, some of which are discussed in this chapter. Earlier we learned that turbulent flow is unsteady in the strictest sense of the word, but if the mean temporal values are constant over a period of time, it is called mean steady flow. Here we shall consider cases where the mean temporal values continuously vary.

There are two main types of unsteady flow that we shall investigate here. The first is where the water level in a reservoir or pressure tank is steadily rising or falling, so that the flow rate varies continuously, but where change takes place slowly. The second is where the velocity in a pipeline is changed rapidly by the fast closing or opening of a valve.

In the first case, of slow change, the flow is subject to the same forces as have previously been considered. Fast changes, of the second type, require the consideration of elastic forces.

Unsteady flow also includes such topics as oscillations in connected reservoirs and in U tubes and such phenomena as tidal motion and flood waves in open channels. Likewise, the field of machinery regulation by servomechanisms is intimately connected with unsteady motion. However, we shall not consider any of these topics here.

[1] Where the unsteadiness is not too rapid, unsteady flow can usually be approximated by assuming the flow is steady at different rates over successive time periods of short duration.

12.2 DISCHARGE WITH VARYING HEAD

When flow occurs under varying head, the rate of discharge will continuously vary. Let us consider the situation depicted in Fig. 12.1 in which $\forall$ represents the volume of liquid contained in the tank at a particular instant of time. There is inflow at the rate Q_i and outflow at rate Q_o. The change in volume during a small time interval dt can be expressed as

$$d\forall = Q_i\,dt - Q_o\,dt$$

If A_s = area of the surface of the volume while dz is the change in level of the surface then $d\forall = A_s\,dz$. Equating these two expressions for $d\forall$,

$$A_s\,dz = Q_i\,dt - Q_o\,dt \tag{12.1}$$

Either Q_i or Q_o or both may be variable. The outflow Q_o is usually a function of z. For example, if liquid is discharged through an orifice or a pipe of area A under a differential head z, $Q_o = C_d A\sqrt{2gz}$, where C_d is a numerical discharge coefficient and z is a variable. If the liquid flows out over a weir or a spillway of length L, $Q_o = CLh^{3/2}$, where C is the appropriate coefficient and h is the head on the weir or spillway (Sec. 11.11). In either case z or h is the variable height of the liquid surface above the appropriate datum. The inflow Q_i commonly varies with time; however, such problems will not be considered here. We shall consider only the cases where $Q_i = 0$ or where Q_i = constant.

Rewriting Eq. (12.1) and integrating gives an expression for t, the time for the water level to change from z_1 to z_2. Thus

$$t = \int_{z_1}^{z_2} \frac{A_s\,dz}{Q_i - Q_o} \tag{12.2}$$

The right-hand side of this expression can be integrated if Q_i is zero or

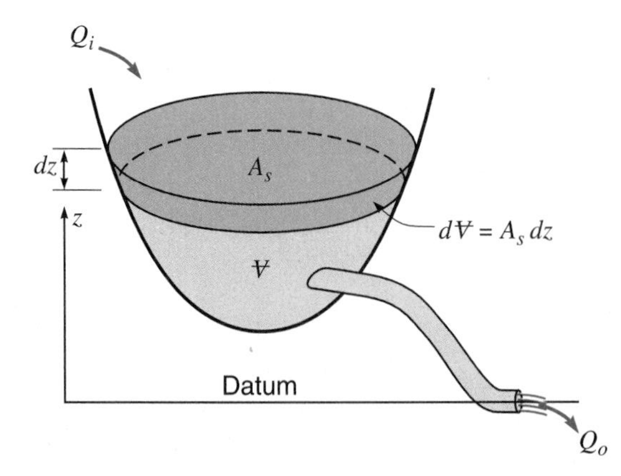

Figure 12.1

constant and if A_s and Q_o can be expressed as functions of z. In the case of natural reservoirs, the surface area cannot be expressed as a simple mathematical function of z, but values of it may be obtained from a topographic map. In such a case, Eq. (12.2) may be solved graphically by plotting values of $A_s/(Q_i - Q_o)$ against simultaneous values of z. The area under such a curve to some scale is the numerical value of the integral.

We must note here that instantaneous values for Q_o have been expressed in the same manner as for steady flow. This is not strictly correct, since for unsteady flow the energy equation should also include an acceleration head [see Eq. (12.6)]. The introduction of such a term renders the solution much more difficult. In cases where the value of z does not vary rapidly, no appreciable error will result if this acceleration term is disregarded. Therefore the equations can be written as for steady flow.

Sample Problem 12.1 The open wedge-shaped tank in Fig. S12.1 has a length of 15 ft perpendicular to the sketch. It is drained through a 3-in-diameter pipe 10 ft long whose discharge end is at elevation zero. The coefficient of loss at pipe entrance is 0.50, the total of the bend loss coefficients is 0.20, and f for the pipe is 0.018. Find the time required to lower the water surface in the tank from elevation 8 to 5 ft. Neglect the possible change of f with **R**, and assume that the acceleration effects in the pipe are negligible.

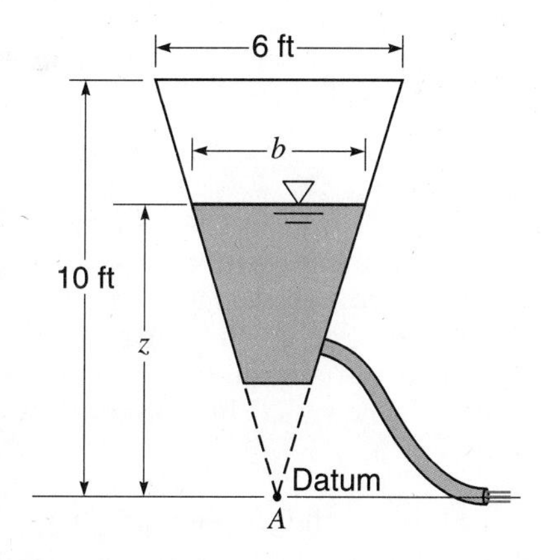

Figure S12.1

Solution

Energy Eq. (5.14) from water surface to jet at discharge:

$$z - \left[0.5 + 0.2 + 0.018\left(\frac{10}{0.25}\right)\right]\frac{V^2}{2g} = \frac{V^2}{2g}$$

$$z - 1.42\frac{V^2}{2g} = \frac{V^2}{2g}; \qquad V = 5.16z^{1/2}$$

$$Q_o = AV = \frac{\pi}{4}(0.25)^2 5.16z^{1/2} = 0.253z^{1/2}$$

Similar triangles: $\frac{b}{z} = \frac{6}{10}$; so $b = 0.6z$

So the area of the water surface is

$$A_s = 15b = 15(0.6)z = 9z$$

Eq. (12.2): $$t = \int_8^5 \frac{9z\,dz}{0 - 0.253z^{1/2}} = -\frac{9}{0.253}\int_8^5 z^{1/2}\,dz$$

$$= -35.5[\tfrac{2}{3}z^{3/2}]_8^5 = 271 \text{ sec} \qquad \textbf{\textit{ANS}}$$

Note: If the pipe had discharged at an elevation other than zero, the integral would have been different, because the head on the pipe would then have been $z + h$, where h is the vertical distance of the discharge end of the pipe below (h positive) or above (h negative) point A of the figure.

EXERCISES

12.2.1 (*a*) A ship lock has vertical sides. Water enters or leaves it through a conduit area of cross-sectional area A. The flow through this conduit is given by $Q = C_d A(2gz)^{1/2}$, where z is the variable difference in level between the water surface in the lock and that outside. Prove that the time required for the water level in the lock to drop from z_1 to z_2 is

$$t = \frac{2A_s}{C_d A\sqrt{2g}}(z_1^{1/2} - z_2^{1/2})$$

(*Note:* If the lock is being filled, the signs must be reversed.) (*b*) Suppose the lock is 400 ft long by 100 ft wide, and the discharge coefficient for the conduit is 0.50. If the water surface in the lock is initially 40 ft below the level of the surface of the water upstream, how large must the conduit be if the lock is to be filled in 10 min?

12.2.2 (*a*) Repeat Exer. 12.2.1(*a*). (*b*) Suppose the lock is 100 m long by 30 m wide, and the discharge coefficient for the conduit is 0.50. If the water surface in the lock is initially 12 m below the level of the surface of the water upstream, how large must the conduit be if the lock is to be filled in 10 min?

12.2.3 The tank in Fig. X12.2.3 has the shape of the frustum of a cone with a 2-ft^2 orifice ($C_d = 0.62$) in the bottom. Given $D_1 = 44$ ft, $D_2 = 20$ ft, $z_0 = 36$ ft, and that the water level outside the tank is constant at section 2, how long will it take the water level in the tank to drop from section 1 to section 2? (*Note:* The tank diameter $= ky$, and $y = z + h_2$, where z is the variable distance between surface levels.)

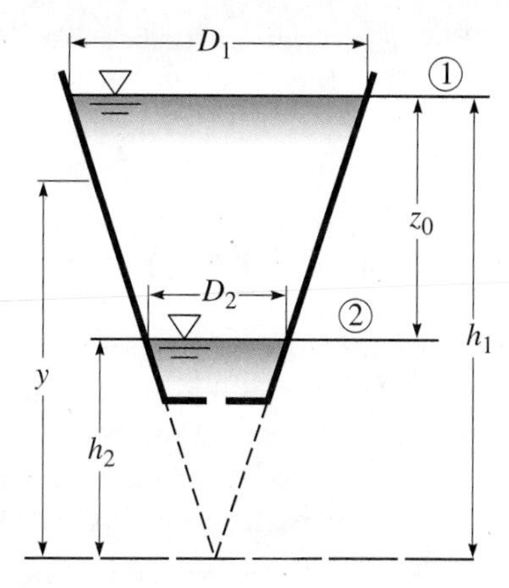

Figure X12.2.3

12.2.4 The cone-shaped tank in Fig. X12.2.3 has a 0.2-m² orifice ($C_d = 0.60$) in the bottom. Given $D_1 = 12$ m, $D_2 = 5.6$ m, $z_0 = 10.4$ m, and that the water level outside the tank is constant at section 2, how long will it take the water level in the tank to drop from section 1 to section 2? (*Note:* The tank diameter $= ky$, and $y = z + h_2$, where z is the variable distance between surface levels.)

12.2.5 Given the same tank and C_d as in Exer. 12.2.3, but now with the water surface outside constant at section 1 instead and the tank initially empty, how long will it take for the water level in the tank to rise from section 2 to section 1?

12.2.6 Given the same tank and C_d as in Exer. 12.2.4, but now with the water surface outside constant at section 1 instead and the tank initially empty, how long will it take for the water level in the tank to rise from section 2 to section 1?

12.2.7 The tank in Fig. X12.2.7 has vertical sides, $b_1 = 12$ ft, $b_2 = 24$ ft, $z_0 = 14$ ft, and its dimension normal to the plane of the paper is $a = 6$ ft. The vertical dividing plate has a submerged orifice 0.7 ft² in area ($C_d = 0.65$). How long will it take for the two water surfaces to equalize?

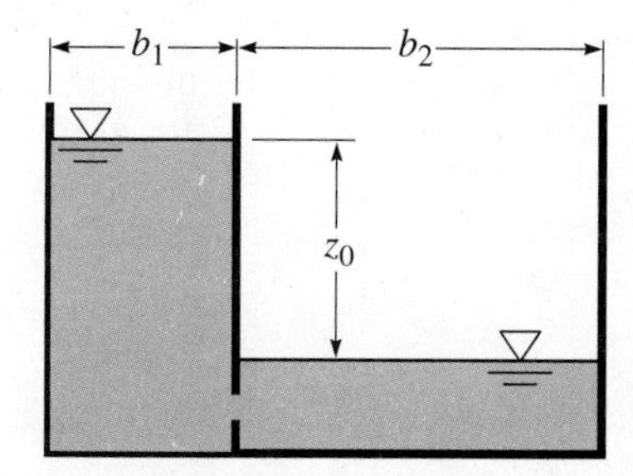

Figure X12.2.7

12.2.8 The tank in Fig. X12.2.7 has vertical sides, $b_1 = 3.6$ m, $b_2 = 7.2$ m, $z_0 = 4.2$ m, and its dimension normal to the plane of the paper is

$a = 1.8$ m. The vertical dividing plate has a submerged orifice 0.063 m^2 in area ($C_d = 0.65$). How long will it take for the two water surfaces to equalize?

12.3 UNSTEADY FLOW OF INCOMPRESSIBLE FLUIDS IN PIPES

When the flow in a pipe is unsteady, we shall show that the energy equation has a term, the ***accelerative head*** $h_a = (L/g)(dV/dt)$, which accounts for the effect of the acceleration of the fluid. Let us consider an elemental length of the flow in a pipe, as in Fig. 12.2. We shall follow the same procedure we used in Sec. 8.4 by writing $\Sigma F = ma$; however, in this situation, with unsteady flow, at a particular point in the flow at a particular instant, we express the acceleration as $V(dV/ds) + dV/dt$. This comes from the general expression for acceleration in unsteady flow [Eq. (4.29)]. Applying $\Sigma F = ma$ to the cylindrical fluid element of Fig. 12.2, we get for unsteady flow

$$F_1 - F_2 - dW \cos\theta - \tau_0 A_s = ma$$

i.e., $$pA - (p + dp)A - \rho g A\, ds\left(\frac{dz}{ds}\right) - \tau_0(2\pi r)ds = \rho A\, ds\left(V\frac{dV}{ds} + \frac{dV}{dt}\right)$$

Dividing by $\gamma = \rho g$, dividing by $A = \pi r^2$, and simplifying, we get

$$-\frac{dp}{\gamma} - dz - \frac{2\tau_0\, ds}{\gamma r} = \frac{V\,dV}{g} + \frac{ds}{g}\frac{dV}{dt} \tag{12.3}$$

Noting that $V\,dV = \frac{1}{2}d(V^2)$ this can be written

$$-\frac{dp}{\gamma} - dz - d\frac{V^2}{2g} = \frac{2\tau_0\, ds}{\gamma r} + \frac{ds}{g}\frac{dV}{dt} \tag{12.4}$$

This equation applies to unsteady flow of both compressible and incompressible real fluids. However, for compressible fluids γ is a variable, and a gas law equation relating γ to p and T must be introduced before integration.

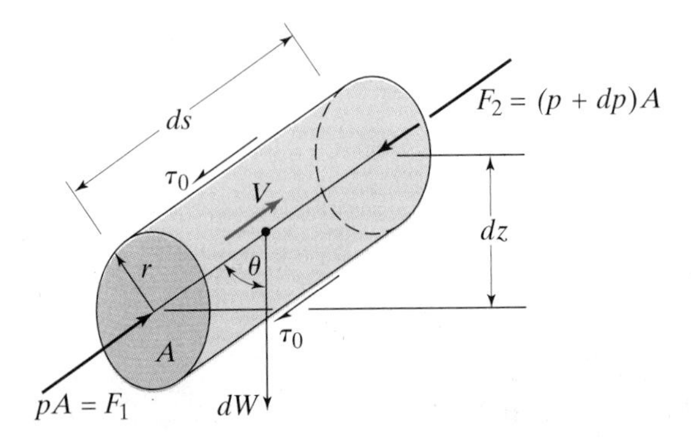

Figure 12.2
Element in pipe flow.

For incompressible fluids that we are considering here γ is a constant, so we can integrate directly. Integrating from some section 1 to another section 2, where the distance between them is L, we get

$$\frac{p_1 - p_2}{\gamma} + z_1 - z_2 + \frac{V_1^2 - V_2^2}{2g} = \frac{2\tau_0 L}{\gamma r} + \frac{L}{g}\frac{dV}{dt}$$

or, noting that $R_h = D/4 = r/2$,

$$\left(\frac{p}{\gamma} + z + \frac{V^2}{2g}\right)_1 - \left(\frac{p}{\gamma} + z + \frac{V^2}{2g}\right)_2 = \frac{\tau_0 L}{\gamma R_h} + \frac{L}{g}\frac{dV}{dt} \tag{12.5}$$

But when there is no acceleration ($a = 0$), i.e., the flow is steady and V is constant, from the above we see that the dV/dt term and the V^2 terms drop out, leaving

$$\left(\frac{p}{\gamma} + z\right)_1 - \left(\frac{p}{\gamma} + z\right)_2 = \frac{\tau_0 L}{\gamma R_h}$$

which, by comparison with Eqs. (8.5) and (8.6), indicates that the term $\tau_0 L/\gamma R_h = h_{L_{1-2}}$, the head loss over length L. Thus Eq. (12.5) becomes

$$\left(\frac{p}{\gamma} + z + \frac{V^2}{2g}\right)_1 - \left(\frac{p}{\gamma} + z + \frac{V^2}{2g}\right)_2 = h_L + \frac{L}{g}\frac{dV}{dt} \tag{12.6a}$$

or

$$H_1 - H_2 = h_L + h_a \tag{12.6b}$$

This is the same as the steady flow Eq. (5.14), with the addition of the last term, the accelerative head $h_a = (L/g)(dV/dt)$. Equation (12.6) states that the energy difference between sections 1 and 2 is equal to (a) that required to overcome the friction plus (b) that required to produce the acceleration. It is presumed that the head loss at any instant is equal to the steady-flow head loss for the flow rate at that instant. Experimental evidence indicates that this presumption is reasonably valid.

If the pipe consists of two or more pipes in series, an $(L/g)(dV/dt)$ term for each pipe should appear in the equation just as there would be a separate term for the head loss in each pipe. To clarify the discussion further, the simple case of unsteady flow of an incompressible fluid in a horizontal pipe is shown in Fig. 12.3. The left-hand sketch shows the steady-flow case, while unsteady flow is depicted in the two right-hand sketches. The analysis below the sketches indicates that, with the same instantaneous flow rates, the pressure is depressed at section 2 if the acceleration is positive or increased if it is negative.

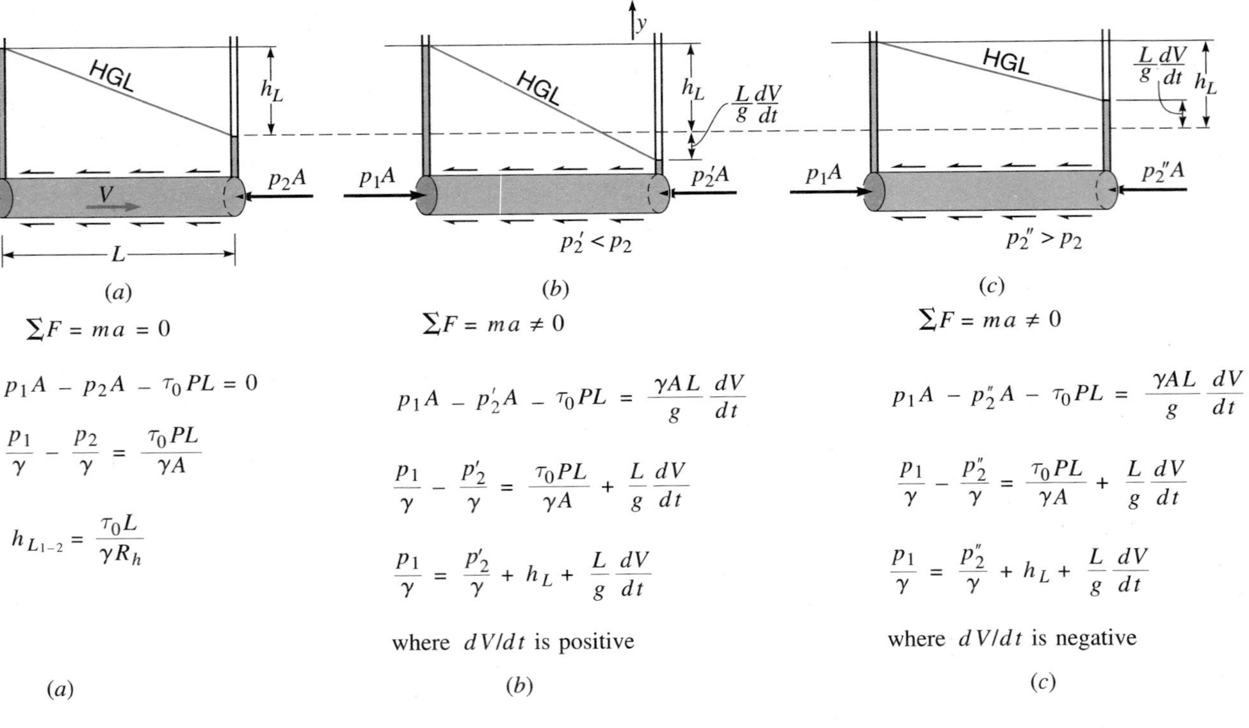

P = wetted perimeter; R_h = hydraulic radius.

Figure 12.3
Steady and unsteady flow of incompressible fluid in a horizontal pipe. (Flow is instantaneously equal in all three pipes.) (*a*) Steady flow ($dV/dt = 0$). (*b*) Unsteady flow (dV/dt is positive). (*c*) Unsteady flow (dV/dt is negative).

Sample Problem 12.2 Although the unrealistic assumptions of instantaneous change in pump speed and head are made in this example, it will serve to illustrate application of Eq. (12.6). When the centrifugal pump in the figure is rotating at 1650 rpm, the steady flow rate is 1600 gpm. Let us suppose that the pump speed can be increased instantaneously to 2000 rpm. Determine the flow rate as a function of time. Assume that the head developed by the pump is proportional to the square of the rotative speed (see Sec. 15.4).

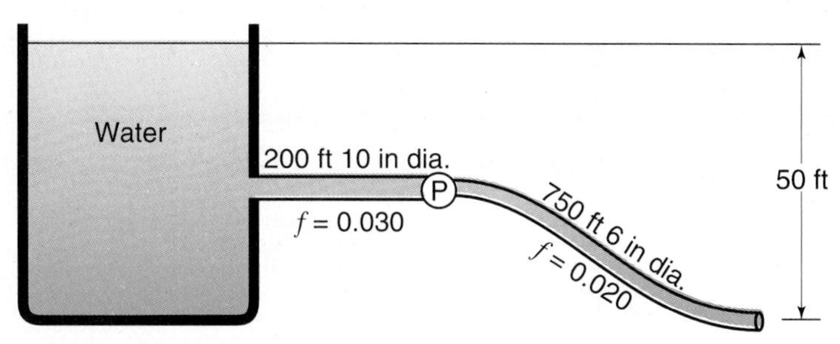

Figure S12.2

Solution

Writing the unsteady-flow energy equation,

$$50 - 0.5\frac{V_1^2}{2g} - f_1\frac{L_1}{D_1}\frac{V_1^2}{2g} + h_p - f_2\frac{L_2}{D_2}\frac{V_2^2}{2g} = \frac{V_2^2}{2g} + \frac{L_1}{g}\frac{dV_1}{dt} + \frac{L_2}{g}\frac{dV_2}{dt}$$

where the subscripts 1 and 2 refer to the 10- and 6-in-diameter pipes, respectively. Note that the accelerative head for each pipe depends on the respective L and dV/dt values.

From continuity: $V_1 = \dfrac{A_2V_2}{A_1} = \left(\dfrac{6}{10}\right)^2 V_2 = 0.36V_2$

Hence $$\frac{dV_1}{dt} = \frac{A_2}{A_1}\frac{dV_2}{dt} = 0.36\frac{dV_2}{dt}$$

Thus $$50 - 0.5\frac{(0.36V_2)^2}{2g} - 0.030\left(\frac{200}{10/12}\right)\frac{(0.36V_2)^2}{2g} + h_p - 0.020\left(\frac{750}{6/12}\right)\frac{V_2^2}{2g}$$

$$= \frac{V_2^2}{2g} + \frac{200}{g}(0.36)\frac{dV_2}{dt} + \frac{750}{g}\frac{dV_2}{dt}$$

Evaluating and combining terms: $$50 + h_p = 32.0\frac{V_2^2}{2g} + \frac{822}{g}\frac{dV_2}{dt} \qquad (a)$$

With the original steady-flow conditions ($dV/dt = 0$):

$$V_2 = \frac{Q}{A_2} = \frac{1600/449}{(\pi/4)(0.5)^2} = 18.16 \text{ fps}$$

and $$h_p = 32\frac{V_2^2}{2g} - 50 = 113.8 \text{ ft}$$

Given that $h_p \propto (\text{rpm})^2$, so after the speed is increased to 2000 rpm

$$h_p = 113.8\left(\frac{2000}{1650}\right)^2 = 167.2 \text{ ft}$$

Substituting into (a):
$$50 + 167 = 32\frac{V_2^2}{2g} + \frac{822}{g}\frac{dV_2}{dt}$$

Expressing the foregoing in terms of Q:
$$217 = 12.89Q^2 + 130.0\frac{dQ}{dt} \qquad (b)$$

Solving for dt and integrating, noting that at $t = 0$, $Q = 3.57$ cfs (1600 gpm):

$$\int_0^t dt = 130\int_{3.57}^{Q} \frac{dQ}{217 - 12.89Q^2}$$

$$t = 1.229 \ln \frac{4.10 + Q}{4.10 - Q} - 3.27$$

$$e^{0.814t+2.66} = \frac{4.10 + Q}{4.10 - Q}$$

Finally
$$Q = 4.10\,\frac{e^{0.814t+2.66} - 1}{e^{0.814t+2.66} + 1} \qquad \textbf{\textit{ANS}}$$

Notes:

1. As t gets larger, Q approaches 4.10 cfs (1840 gpm), the steady-state flow rate for the condition where $h_p = 169$ ft.
2. The speed of a pump cannot be changed instantaneously from one value to another, as was assumed in this example. To solve this problem correctly the operating characteristics of the pump and motor and the moment of inertia of the rotating system would have to be known.

EXERCISES

12.3.1 Verify that the neglect of the accelerative head was justified in Sample Prob. 12.1. Find its magnitude at $z = 5$ ft from values of V and t at $z = 4.5$ and 5.5 ft.

12.3.2 If the speed of the pump in Sample Prob. 12.2 is reduced instantaneously from 1650 to 1150 rpm, what is the deceleration of the flow rate immediately after the change in pump speed?

12.4 ESTABLISHMENT OF STEADY FLOW

Determining the time for the flow to become steady in a pipeline when a valve is suddenly opened at the end of the pipe can be accomplished through

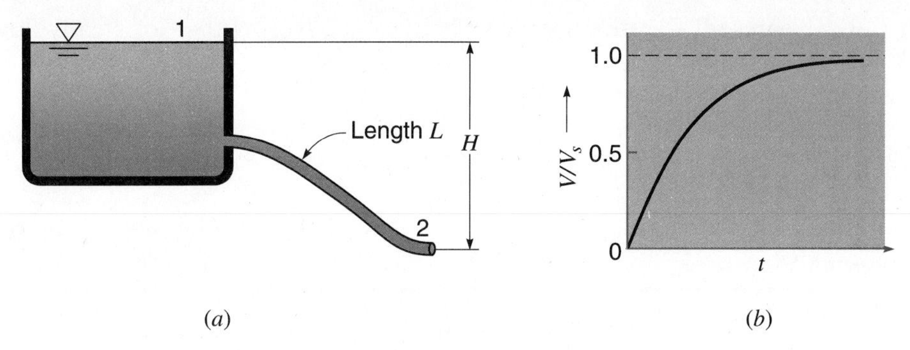

Figure 12.4
Establishment of steady flow (V_s = velocity at steady flow).

application of Eq. (12.6). Immediately after the valve is opened (Fig. 12.4), the head H is available to accelerate the flow. Thus flow commences, but as the velocity increases the accelerating head is reduced by fluid friction and minor losses. Let us assume the total head loss h_L can be expressed as $k_L V^2/2g$, where k_L is constant, although it may vary somewhat with velocity unless the pipe is very rough. Then

$$k_L = f\frac{L}{D} + \sum k$$

where $\sum k$ is the sum of the coefficients of all minor losses in the pipeline (Secs. 8.17–8.23). Writing Eq. (12.6) between sections 1 and 2 in Fig. 12.4, while noting that $p_1 = p_2 = V_1 = 0$ and $z_1 - z_2 = H$, gives

$$H - \frac{V_2^2}{2g} = h_L + \frac{L}{g}\frac{dV}{dt}$$

or

$$H = \left(\frac{fL}{D} + \sum k\right)\frac{V^2}{2g} + \frac{V^2}{2g} + \frac{L}{g}\frac{dV}{dt}$$

i.e.,

$$H = K\frac{V^2}{2g} + \frac{L}{g}\frac{dV}{dt}, \quad \text{where} \quad K = \frac{fL}{D} + 1 + \sum k$$

For the free discharge at point 2 in Fig. 12.4, note that in the expression for $K = k_L + 1$ the 1 represents the residual velocity head and $\sum k$ does *not* include a submerged discharge loss coefficient $k_d = 1$ (Sec. 8.19). If, however, there is submerged discharge into another quiescent reservoir, there is no residual velocity head so that $K = k_L$, but $\sum k$ now *does* include $k_d = 1$. As a result, the value of K is the same for both situations if minor losses are included. It is not quite the same if minor losses are neglected (Sec. 8.24), because the residual velocity head is not considered a minor loss.

Let us represent the steady-flow velocity by V_s. Noting that for steady flow $(dV/dt) = 0$, we get

$$V_s = \sqrt{\frac{2gH}{K}} \tag{12.7}$$

Substituting the value of H from this expression into the above energy equation gives

$$dt = \frac{2L}{K} \frac{dV}{V_s^2 - V^2}$$

Integrating and noting that the constant of integration = zero, since $V = 0$ at $t = 0$ and $\ln(V_s/V_s) = 0$, we get

$$t = \frac{L}{KV_s} \ln \frac{V_s + V}{V_s - V} \tag{12.8}$$

This equation indicates that V approaches V_s asymptotically and that equilibrium will be attained only after an infinite time (Fig. 12.4*b*), but it must be remembered that this is an idealized case. In reality there will be elastic waves and damping, so that true equilibrium will be reached in a finite time. Also, as noted earlier, Eqs. (12.7) and (12.8) can be applied to a submerged discharge as well as to a free discharge.

SAMPLE PROBLEM 12.3 Two large water reservoirs are connected to one another with a 10-cm-diameter pipe ($f = 0.02$) of length 15 m. The water-surface elevation difference between the reservoirs is 2.0 m. A valve in the pipe, initially closed, is suddenly opened. (*a*) Determine the times required for the flow to reach $\frac{1}{4}$, $\frac{1}{2}$, and $\frac{3}{4}$ of the steady-state flow rate. Assume the water-surface elevations remain constant. (*b*) Repeat for pipe lengths of 150 m and 1500 m with all other data remaining the same. In the first case $L/D = 15/0.10 = 150$, hence minor losses are significant (Sec. 8.24). Assume a square-edged entrance.

Solution

(*a*) For square-edged entrance and submerged discharge: $\sum k = 0.5 + 1.0 = 1.5$

$$K = k_L = \frac{fL}{D} + \sum k = 0.02 \frac{15}{0.10} + 1.5 = 4.5$$

For steady flow:

Eq. (12.7): $$V_s = \sqrt{\frac{2gH}{K}} = \sqrt{\frac{2(9.81)2}{4.5}} = 2.95 \text{ m/s}$$

For unsteady flow use Eq. (12.8):

$$t = \frac{L}{KV_s} \ln \frac{V_s + V}{V_s - V} = \frac{15}{4.5(2.95)} \ln \frac{2.95 + V}{2.95 - V} = 1.129 \ln \frac{2.95 + V}{2.95 - V}$$

For $Q = \frac{1}{4}Q_s$ substitute $V = \frac{1}{4}V_s$, etc.:

Q	V, m/s	$\frac{2.95 + V}{2.95 - V}$	ln	t, s	
$0.25Q_s$	0.74	1.667	0.511	0.577	
$0.50Q_s$	1.48	3.00	1.099	1.240	
$0.75Q_s$	2.21	7.00	1.946	2.197	***ANS***

(*b*) For the other two lengths the minor losses are insignificant (Sec. 8.24), and the results are as follows:

Q	$L = 150$ m	$L = 1500$ m	
$0.25Q_s$	2.18 s	7.04 s	
$0.50Q_s$	4.69 s	15.15 s	
$0.75Q_s$	8.30 s	26.84 s	***ANS***

EXERCISES

12.4.1 A 4-in-diameter drain pipe (flush entrance, $f = 0.020$) 3000 ft long will freely discharge reservoir water at an elevation 100 ft below the reservoir water surface. Initially there is no flow, since there is a plug in the pipe outlet. What will be the steady flow velocity in this pipe, and how long after the plug is removed will the flow velocity be half the steady velocity? See Sec. 8.24 regarding the neglect of minor losses.

12.4.2 A 150-mm-diameter drain pipe (flush entrance, $f = 0.030$) 500 m long will freely discharge reservoir water at an elevation 60 m below the reservoir water surface. Initially there is no flow, because a valve at the outlet end is closed. What will be the steady flow velocity in this pipe, and how long after the valve is suddenly opened will the flow velocity be half the steady velocity? See Sec. 8.24 regarding minor losses.

12.4.3 A 250-mm-diameter pipe (flush entrance, $f = 0.026$) 1000 m long can deliver water from a higher reservoir to a lower one (submerged discharge). The difference between the two water surface elevations is 75 m. After a valve near the downstream end is suddenly opened, what will be the steady flow velocity in this pipe, and how long after the valve is opened will the flow velocity be one-third and two-thirds of the steady velocity?

12.4.4 A 4-in-diameter drain pipe (flush entrance, $f = 0.022$) 300 ft long will freely discharge reservoir water at an elevation 40 ft below the reservoir water surface. Initially there is no flow, since there is a plug in the pipe outlet. What will be the steady flow velocity in this pipe, and how long after the plug is removed will the flow velocity be 50% and 75% of the steady velocity?

12.4.5 A 6-in-diameter pipe (reentrant entrance, $f = 0.028$) 475 ft long can deliver water from a higher reservoir to a lower one (submerged discharge). The difference between the two water surface elevations is 25 ft. After a valve near the downstream end is suddenly opened, what will be the steady flow velocity in this pipe, and how long after the valve is opened will the flow velocity be 90% of the steady velocity?

12.4.6 A 150-mm-diameter pipe (reentrant entrance, $f = 0.023$) 140 m long can deliver water from a higher reservoir to a lower one (submerged discharge). The difference between the two water surface elevations is 12 m. After a valve near the downstream end is suddenly opened, what will be the steady flow velocity in this pipe, and how long after the valve is opened will the flow velocity be 90% of the steady velocity?

12.5 VELOCITY OF PRESSURE WAVE IN PIPES

Unsteady phenomena, with rapid changes taking place, frequently involve the transmission of pressure in waves or surges. As shown in Sec. 13.3, the velocity or ***celerity*** of a pressure (sonic) wave is

$$c = \sqrt{\frac{E_v}{\rho}} = \sqrt{\frac{g}{\gamma}E_v} \tag{12.9}$$

where E_v is the volume (or bulk) modulus of elasticity of the medium. For water, a typical value of E_v is 300,000 psi (2.07×10^6 kN/m^2), and thus the velocity (or celerity) c of a pressure wave in water is about 4720 fps (1440 m/s), depending on the temperature.

For water in an elastic pipe, this velocity is reduced due to the stretching of the pipe walls. As explained in Sec. 13.3, E_v is replaced by the combined modulus E_c, such that

$$E_c = \frac{E_v}{1+\dfrac{D}{t}\dfrac{E_v}{E}} = \frac{1}{\dfrac{1}{E_v}+\dfrac{D}{tE}}$$

where D and t are the diameter and wall thickness of the pipe, respectively, and E is the modulus of elasticity of the pipe material. As the ratios D/t and E_v/E are dimensionless, any consistent units may be used in each.

The velocity of a pressure wave in an elastic pipe is then

$$c_P = \sqrt{\frac{E_c}{\rho}} = \sqrt{\frac{g}{\gamma}E_c} = \frac{c}{\sqrt{1+\dfrac{D}{t}\dfrac{E_v}{E}}} = \sqrt{\frac{g}{\gamma\left(\dfrac{1}{E_v}+\dfrac{D}{tE}\right)}} \tag{12.10}$$

Values[2] of the modulus of elasticity E for steel, cast iron, and concrete are about 30,000,000, 15,000,000, and 3,000,000 psi, respectively. Values of the volume (bulk) modulus E_v for various liquids are given in Appendix A, Table A.4.

For normal pipe dimensions the velocity of a pressure wave in a water pipe usually ranges between 2000 and 4000 fps (600 and 1200 m/s), but it will always be less than $c = 4720$ fps (1440 m/s), the velocity of a pressure wave in free water.

EXERCISES

12.5.1 Find the celerity of a pressure wave in open fresh water at 70°F. What will the celerity be if the water is contained in a cast iron pipe with 12.00-in inside diameter and $\frac{3}{8}$-in wall thickness? Express the celerity in the pipe as a percentage of the celerity in the open water.

12.5.2 Find the celerity of a pressure wave in open fresh water at 20°C. What will the celerity be if the water is contained in a cast iron pipe with 489-mm inside diameter and 14.3-mm wall thickness? Express the celerity in the pipe as a percentage of the celerity in the open water.

12.5.3 Find the celerity of a pressure wave in benzene (Appendix A, Table A.4) contained in an 8-in-diameter steel pipe having a wall thickness of 0.32 in.

12.5.4 Find the celerity of a pressure wave in benzene (Appendix A, Table A.4) contained in a 200-mm-diameter steel pipe having a wall thickness of 8.0 mm.

12.6 WATER HAMMER

In the preceding unsteady-flow cases in this chapter, the changes of velocity were presumed to take place slowly. But if the velocity of a liquid in a pipeline is abruptly decreased by a valve movement, the phenomenon encountered is called ***water hammer***. This is a very important problem in the case of hydroelectric plants, where the flow of water must be rapidly varied in proportion to the load changes on the turbine. Water hammer has burst large penstocks, causing great damage to hydraulic and power generating facilities, in addition to loss of life and power generation. Water hammer occurs in liquid-flow pressure systems whenever a valve is closed or opened, fully or partially. The terminology "water hammer" is perhaps misleading, since this phenomenon can occur with any liquid.

---

[2] Corresponding values of E for steel, cast iron, and concrete in SI units are 207×10^6, 103×10^6, and 20.7×10^6 kN/m^2, respectively.

Instantaneous Closure

Although it is physically impossible to close a valve instantaneously, such a concept is useful as an introduction to the study of real cases. For convenience let us start by considering steady flow in a horizontal pipe (Fig. 12.5*a*) with a partly open valve. Then let us assume that the valve at N

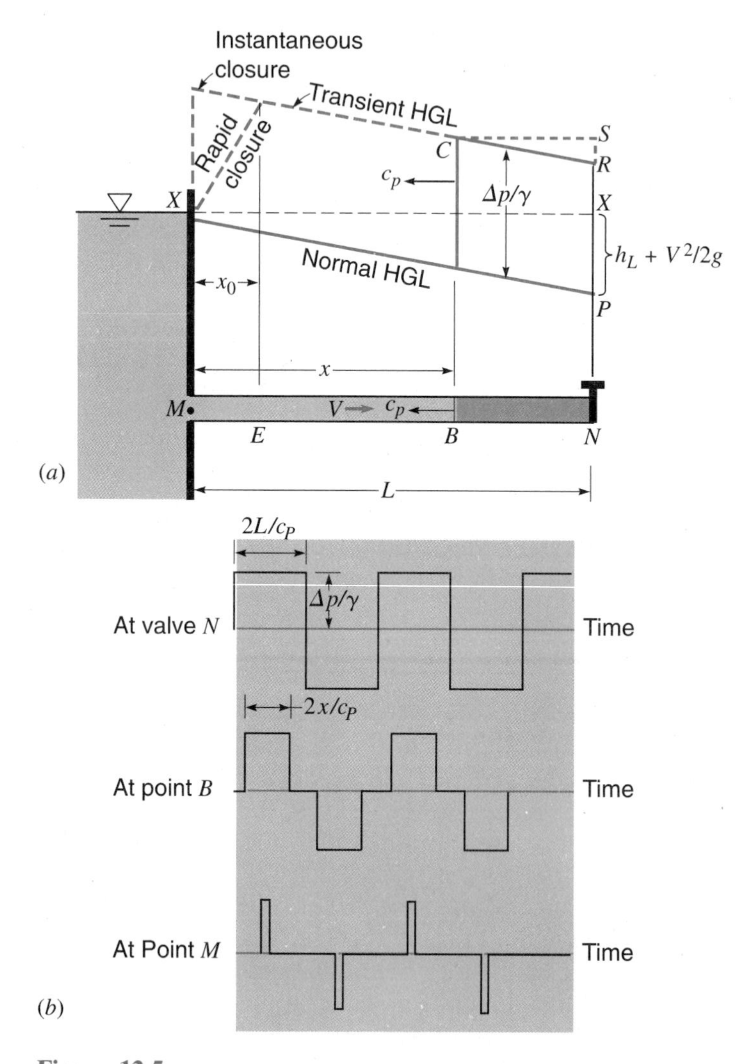

Figure 12.5
Water hammer with pipe friction and damping neglected.
(*a*) Valve N at the end of the pipe line is abruptly closed and the pressure wave has traveled part way up the pipe.
(*b*) Idealized water hammer pressure heads at N, B, and M as a function of time for instantaneous valve closure.

is closed completely and instantaneously. The lamina of liquid next to the valve will be compressed by the rest of the column of liquid flowing against it. At the same time the walls of the pipe surrounding this lamina will be stretched by the excess pressure produced. The next upstream lamina will then be brought to rest, and so on. The liquid in the pipe does not behave as a rigid incompressible body but the phenomenon is affected by the elasticity of both the liquid and the pipe. The cessation of flow and the resulting pressure increase move upstream along the pipe as a wave with the velocity c_p as given by Eq. (12.10).

After a short interval of time the liquid column *BN* will have been brought to rest at an increased pressure, while the liquid in the length *MB* will still be flowing with its initial velocity and initial pressure. When the pressure wave finally reaches the inlet at *M,* the entire mass in the length *L* will be at rest but will be under an excess pressure throughout. During travel of the pressure wave from *N* to *M,* there will be a transient hydraulic grade line parallel to the original steady flow grade line *XP* but at a height $\Delta p/\gamma$ above it, where Δp represents the water hammer pressure.

It is impossible for a pressure to exist at *M* that is greater than that due to depth *MX*, and so when the pressure wave arrives at *M,* the pressure at *M* drops instantly to the value it would have for zero flow. But the entire pipe is now under an excess pressure; the liquid in it is compressed, and the pipe walls are stretched. So some liquid starts to flow back into the reservoir and a wave of pressure unloading travels along the pipe from *M* to *N.* Assuming there is no damping, at the instant this unloading wave reaches *N,* the entire mass of liquid will be under the normal pressure indicated by the line *XP,* but the liquid is still flowing back into the reservoir. This reverse velocity will provide a suction or pressure drop at *N* that ideally will be as far below the normal, steady-flow pressure as the pressure an instant before was above it. Then a wave of low pressure or suction travels back up the pipe from *N* to *M.* Ideally, there would be a series of pressure waves traveling back and forth over the length of the pipe and alternating equally between high and low pressures. Actually, because of damping due to fluid friction and imperfect elasticity of liquid and pipe, the pressure extremes at any point in the pipe will gradually tend toward the pressure for the no-flow condition indicated by *XX* in Fig. 12.5*a.*

The time for a round trip of the pressure wave from *N* to *M* and back again is

$$T_r = 2\frac{L}{c_p} \qquad (12.11)$$

where *L* is the actual length of pipe, and not the length adjusted for minor losses as described in Sec. 8.17. So for an instantaneous valve closure the excess pressure at the valve remains constant for this length of time T_r, before it is affected by the return of the unloading pressure wave; and in like manner the pressure defect during the period of low pressure remains constant for

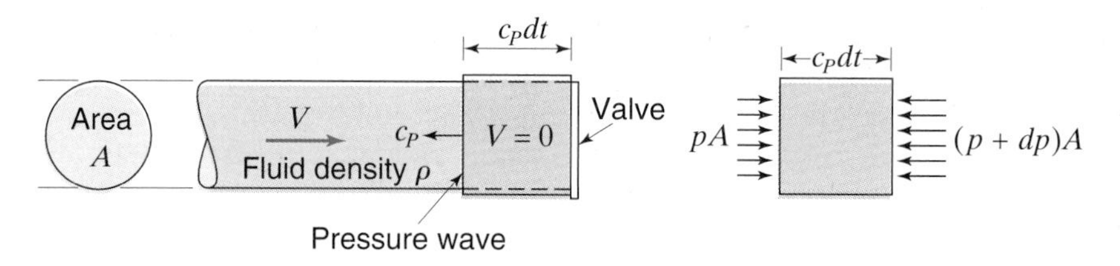

Figure 12.6
Definition sketch for analysis of water hammer in pipes.

the same length of time. At a distance x from the inlet, such as at B, the time for a round trip of a pressure wave is only $2x/c_p$, and so at that point the time duration of the excess or deficient pressure will be $2x/c_p$, as shown in Fig. 12.5*b*. At the inlet M, where $x = 0$, the excess pressure occurs for only an instant. Thus we find at points closer to the inlet M that the duration of the excess or deficient pressure is less, but its magnitude is the same.

We can calculate the pressure change caused by water hammer as follows. In Fig. 12.6 we see a close-up in the vicinity of the valve. If the valve is closed abruptly, a pressure wave travels up the pipe with a celerity c_P. In a short interval of time Δt an element of liquid of length $c_P\ \Delta t$ is decelerated. Therefore the mass m decelerated in the same time is $m = \rho \forall = \rho A(c_p\ \Delta t)$. Applying Newton's second law, $F\ \Delta t = m\ \Delta V$, recalling from Sec. 6.1 that $\Delta V = V_{\text{out}} - V_{\text{in}}$, and neglecting friction, we get

$$[pA - (p + \Delta p)A]\Delta t = (\rho A c_P\, \Delta t)\Delta V$$

from which

Instantaneous, partial or complete closure:
$$\Delta p = -\rho c_P\, \Delta V \tag{12.12}$$

which indicates the change in pressure Δp that results from an instantaneous change in flow velocity ΔV. Partially closing a valve will only partially change V, by $\Delta V = V_2 - V$, and Eq. (12.12) indicates that the pressure change is proportional to the velocity change.

We may also eliminate c_P between Eqs. (12.10) and (12.12) to obtain

Instantaneous, partial or complete closure:
$$\Delta p = -\Delta V\sqrt{\rho E_c} = -\Delta V\sqrt{\frac{\gamma}{g\left(\frac{1}{E_v} + \frac{D}{tE}\right)}} \tag{12.13}$$

We see that the pressure increase is independent of the length of the pipe and depends solely upon the celerity of the pressure wave in the pipe and the change in the velocity of the water. The total pressure at the valve immediately after closure is $\Delta p + p$, where p is the pressure in the pipe just upstream of the valve prior to closure.

In the case of instantaneous and complete closure of a valve the velocity is reduced from V to zero, that is, $\Delta V = -V$. Substituting this into Eqs. (12.12) and (12.13), we get

Instantaneous, *complete* closure:
$$\Delta p = \rho c_P V = V\sqrt{\frac{\gamma}{g\left(\frac{1}{E_v}+\frac{D}{tE}\right)}} \tag{12.14}$$

Consider now conditions at the valve as affected by both pipe friction and damping. In Fig. 12.5*a*, when the pressure wave from N has reached B, the water in BN will be at rest and for zero flow the hydraulic grade line CR should be a horizontal line. There is thus a tendency for the grade line to flatten out for the portion BN. Hence, instead of the transient gradient having the slope imposed by friction, as shown in the figure, it will approach the horizontal line CS as the pressure at C cannot change. Thus the pressure head at N will be raised to a slightly higher value than NR shortly after the valve closure.

This slight increase in pressure head at the valve over the theoretical value $c_P\,\Delta V/g$ has been borne out by tests. As the wave front (B in Fig. 12.5*a*) moves towards M, C moves with it and so RS gets larger. As a result, on the diagram of the pressure history at the valve in Fig. 12.7, the line ab is shown as sloping upward, and for the same reason ef may slope slightly downward, as all conditions are now reversed. Also, because of damping, the

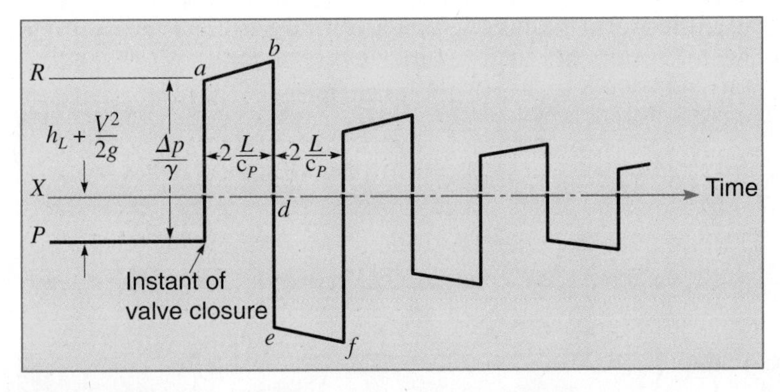

Figure 12.7
Pressure history at the valve N of Fig. 12.5 caused by instantaneous complete closure, considering pipe friction and damping.

waves will be of decreasing amplitude until the final equilibrium pressure is reached.

All of the preceding analysis assumes that the wave of low pressure (suction) will not cause the minimum pressure at any point to drop down to or below the vapor pressure. If it should do so, the water would separate and produce a discontinuity. When the pressure recovers and the water rushes back into the vapor cavity, its momentum and the resulting implosion on the valve could be very damaging.

SAMPLE PROBLEM 12.4 Water flows at 10 fps in a 400-ft-long steel pipe of 8-in diameter with 0.25-in-thick walls. Calculate the pulse interval and maximum pressure rise theoretically caused by instantaneously closing the end valve (a) completely and (b), partially, reducing the velocity to 6 fps.

Solution

Eq. (2.10): $$c_p = \frac{c}{\sqrt{1+\dfrac{D}{t}\dfrac{E_v}{E}}} = \frac{4720}{\sqrt{1+\dfrac{8}{0.25}\left(\dfrac{3\times10^5}{3\times10^7}\right)}} = 4110 \text{ fps}$$

Eq. (12.11): $$T_r = \frac{2(400)}{4110} = 0.195 \text{ sec} \qquad \textit{ANS}$$

(a) Eq. (12.14): $$\Delta p = \rho c_P V = \frac{62.4}{32.2}(4110)10 = 79{,}600 \text{ psf} \qquad \textit{ANS}$$

$$= 553 \text{ psi}$$

for which $$\Delta p/\gamma = 79{,}600/62.4 = 1276 \text{ ft of water}$$

For this pipe, the water hammer impact initially equivalent to the pressure of 1276 ft of water occurs about five times every second!

(b) With partial closure, c_P and T_r are unchanged.

Eq. (12.12): $$\Delta p = -\frac{62.4}{32.2}(4110)(6-10) = 31{,}800 \text{ psf} \qquad \textit{ANS}$$

Rapid Closure ($t_c < T_r$)

It is physically impossible for a valve to be closed instantaneously; so we shall now consider the real case where the valve is closed, completely or partially, in a finite time t_c which is more than zero but less than $T_r = 2L/c_P$. In

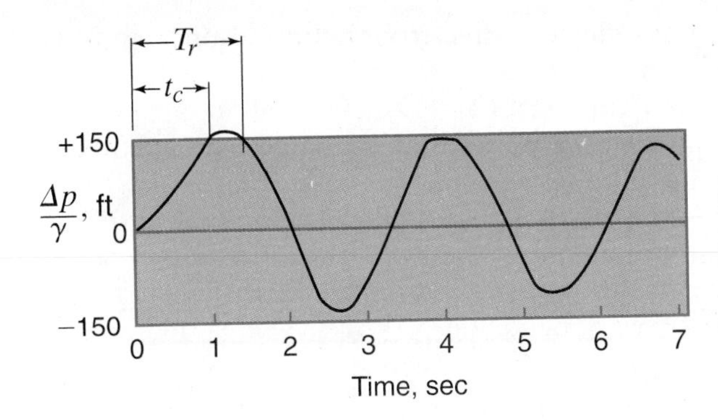

Figure 12.8
Rapid complete valve closure in time t_c (1 sec) less than T_r (1.40 sec). Actual measurements of pressure changes at the valve. (See footnote 3.)

Fig. 12.8 are shown actual pressure recordings, made at the valve, for such a case.[3] The slope of the curve during the time t_c depends entirely upon the operation of the valve and its varying effect upon the velocity in the pipe. But the maximum pressure rise is still the same as for instantaneous closure. The only differences are that it endures for a shorter period of time and the vertical lines of Fig. 12.7 are changed to the sloping lines of Fig. 12.8. The duration of the maximum pressure is $T_r - t_c$ (Fig. 12.8). If the time of valve closure were exactly T_r, the maximum pressure rise at the valve would still be the same, but the curves in Fig. 12.8 would all end in sharp points for both maximum and minimum values, since the time duration of maximum pressure would be reduced to zero.

Earlier we noted that at distance x from the inlet, such as B in Fig. 12.5, time for a round trip of a pressure wave is $2x/c_P$. Therefore, no matter how rapid the valve closure, since t_c cannot be zero there must be some distance x_0 such that $t_c = 2x_0/c_P$. Clearly

$$x_0 = \frac{c_P t_c}{2} \tag{12.15}$$

Thus, at the point E in Fig. 12.5, a distance x_0 from the inlet, the time for a round trip of a pressure wave is exactly t_c. At points closer to the inlet ($x < x_0$) this travel time is shorter. But, as we see in Fig. 12.8, it takes time t_c for the pressure at a point to rise from zero to its maximum value. Therefore, at points nearer to the inlet than E, the reflected relief (unloading) pressure wave front will arrive before the maximum pressure is achieved. This effect will reduce the maximum water hammer pressure at points where $x < x_0$.

[3] Figures 12.8 and 12.9 are from water-hammer studies made by the Southern California Edison Co. on an experimental pipe with the following data: $L = 3060$ ft, internal diameter = 2.06 in, $c_P = 4371$ fps, $V = 1.11$ fps, $\Delta p/\gamma = 150.7$ ft, $T_r =$ 1.40 sec, static head = 306.7 ft, head before valve closure = 301.6 ft, $h_L = 5.1$ ft. In Fig. 12.8 the time of closure = 1 sec and it will be noted that the actual rise in pressure head is slightly more than 151 ft. In Fig. 12.9 the time of closure = 3 sec.

Nearer the inlet, at smaller x, the reflected wave front arrives sooner and the reduction is greater. As a result, in any real case the maximum water hammer pressure experienced will be Δp [Eqs. (12.12)–(12.14)] at all points where $x > x_0$, and it will decrease from Δp at x_0 to zero at the inlet. The maximum pressure rise Δp cannot extend all the way to the reservoir intake. This modification of the transient hydraulic grade line is labeled "rapid closure" on Fig. 12.5. A linear variation of maximum transient pressure within x_0 is commonly assumed, and although this appears reasonable from Fig. 12.8 (for $t < t_c$), the actual shape of this curve depends on how the valve is operated.

Slow Closure ($t_c > T_r$)

The preceding discussion has assumed a closure so rapid (or a pipe so long) that there is an insufficient time for a pressure wave to make the round trip before the valve is closed. Slow closure will be defined as one in which the time of valve movement is greater than $T_r = 2L/c_P$. In this case the maximum pressure rise will be less than in the preceding, because the wave of pressure unloading will reach the valve before the valve is completely closed. This will prevent any further increase in pressure.

Thus in Fig. 12.9 the pressure at the valve would continue to rise if it were not for the fact that at time T_r a return unloading pressure wave reaches the valve and stops the pressure rise at a value of about 53 ft as contrasted with nearly three times that value in Fig. 12.8.[4] Viewed another way, this is equivalent to saying that the valve is to the left of E in Fig. 12.5, so the entire pipeline only experiences maximum transient pressures which vary approximately linearly for $x < x_0$, and the maximum value given by Eqs. (12.12)–(12.14) is never achieved.

Tests have shown that for slow valve closure, i.e., in a time greater than T_r, the excess pressure produced decreases uniformly from the value at the valve to zero at the intake. The maximum water-hammer pressure $\Delta p'$

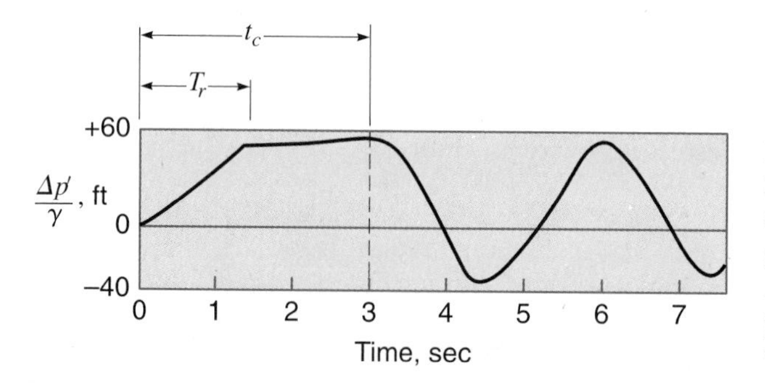

Figure 12.9
Slow complete valve closure in time t_c (3 sec) greater than T_r (1.40 sec). Actual measurements of pressure changes at the valve. (See footnote 3.)

[4] See footnote 3.

developed by gradual closure of a valve when $t_c > T_r$ is given approximately by

$$\frac{\Delta p'}{\Delta p} \approx \frac{L}{x_0}$$

from which

Slow, partial or complete closure:
$$\Delta p' \approx \frac{L}{x_0}\Delta p = \frac{2L}{c_p t_c}\Delta p = -\frac{2L\rho\,\Delta V}{t_c} \tag{12.16}$$

where t_c is the time of closure.

After an unloading pressure wave reaches the valve and the maximum pressure $\Delta p'$ is reached, elastic waves travel back and forth and pressure changes are very complex. They require a detailed step-by-step analysis that is beyond the scope of this text.[5] In brief, the method consists in assuming the valve movement to take place in a series of steps each of which produces a pressure Δp proportional to each ΔV.

SAMPLE PROBLEM 12.5 Assume that the elasticity and dimensions of the pipe in Fig. 12.5*a* are such that the celerity of the pressure wave is 3200 fps. Suppose that the pipe has a length of 2000 ft and a diameter of 4 ft. Find the maximum water hammer pressure at the valve, and the approximate maximum water hammer pressure at a point 300 ft from the reservoir, (*a*) if the valve is partially closed in 4.0 sec, thereby reducing the flow of water from 30 to 10 cfs; and (*b*) if water is initially flowing at 15 cfs and the valve is completely closed in 1.0 sec.

Solution

$$A = \pi(2)^2 = 12.57 \text{ ft}^2$$

Eq. (12.11): $$T_r = \frac{2(2000)}{3200} = 1.25 \text{ sec}$$

(*a*) $t_c = 4.0$ sec $> T_r$, so closure is "slow."

Eq. (12.15): $$x_0 = 0.5(3200)4.0 = 6400 \text{ ft}$$

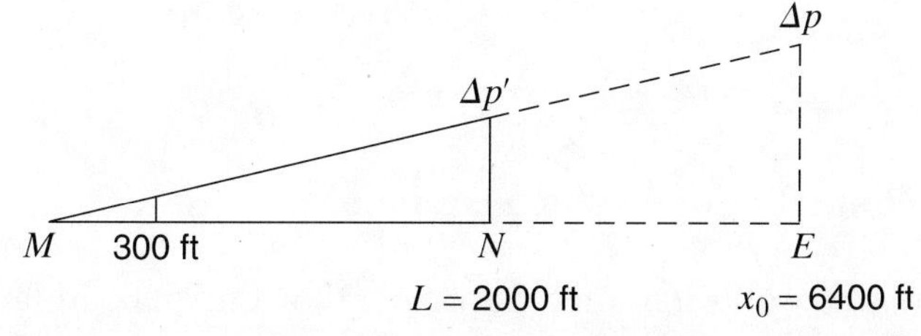

$$V = Q/A = 30/12.57 = 2.39 \text{ fps}; \quad V_2 = 10/12.57 = 0.796 \text{ fps}$$

[5] See, for example, E. B. Wylie and V. L. Streeter, *Fluid Transients in Systems,* Prentice-Hall, Inc., Englewood Cliffs, New Jersey, 1993.

Eq. (12.12): $$\Delta p = -\frac{62.4}{32.2}(3200)(0.796 - 2.39) = 9870 \text{ psf} = 68.5 \text{ psi}$$

At the valve,

Eq. (12.16): $$\Delta p' = \frac{2000}{6400}(68.5) = 21.4 \text{ psi} \qquad \textit{ANS}$$

At the point 300 ft from the reservoir, assuming a linear variation,

$$\Delta p = \frac{300}{2000}(21.4) = 3.21 \text{ psi} \qquad \textit{ANS}$$

(*b*) $$t_c = 1.0 \text{ sec} < T_r, \text{ so closure is ``rapid.''}$$

Eq. (12.15): $$x_0 = 0.5(3200)1.0 = 1600 \text{ ft}$$

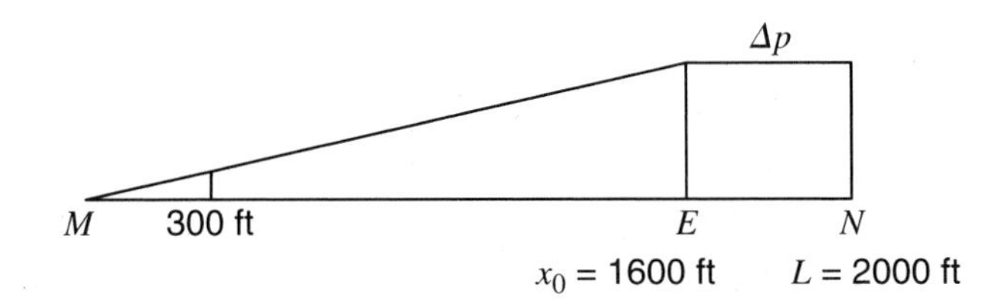

$$V = Q/A = 15/12.57 = 1.194 \text{ fps}$$

At the valve,

Eq. (12.14): $$\Delta p = \frac{62.4}{32.2}(3200)1.194 = 7400 \text{ psf} = 51.4 \text{ psi} \qquad \textit{ANS}$$

At the point 300 ft from the inlet, assuming a linear variation within x_0,

$$\Delta p = \frac{300}{1600}(51.4) = 9.64 \text{ psi} \qquad \textit{ANS}$$

Computer Techniques for Water Hammer

A stricter analysis of water hammer, based on Newton's second law of motion and continuity, results in two nonlinear partial differential equations each in two unknowns, V (or Q) and p (or p/γ), which are functions of x and t. These equations can take into account pipe friction and the slope of the pipe. Although there is no general solution to these equations, they can be solved by the ***method of characteristics*** using a finite difference solution on a computer. The two partial differential equations are combined in two particular ways to convert them into two ordinary differential equations, which may be written in finite difference form and solved simultaneously for the two unknowns. To do this, the pipe must be divided into a number of incremental lengths, initial values must be assigned to all the nodes, and boundary conditions must be prescribed at each end of the pipe.

A major advantage of such a computerized solution method is that it can incorporate complex boundary conditions. These may be specified variations in V or p, or relations between them, or connection to another pipeline or lines, or to a pump.

The details of these techniques are rather extensive, and may be found in more advanced texts.[6]

Protection from Water Hammer

A pipeline may be protected from damage by water hammer by connecting to it any of a variety of devices that keep the transient pressures within desired limits.

Slow-closing valves may be used to control the flow, and by-pass valves can divert sudden changes in flow. Automatic relief valves may be connected to the line to allow water to escape when the pressure exceeds a certain value, and similarly air valves may be fitted to admit air into the line if the pressure falls too low. Air chambers include so-called bladder accumulators and hydropneumatic surge arrestors. They contain compressed air or gas, which acts as a cushion to absorb pressure changes. Pump flywheels are used to absorb energy by increasing the inertia of the rotating element, and they thus increase the time required for incremental changes in the flow rate.

Surge tanks are standpipes connected to the pipeline that store liquid when pressures are high and return it to the line when pressures fall. The simple surge tank, which allows diversion of the flow when the downstream valve is closed, is explained in the following section. Other surge tanks admit the liquid through an orifice, which in addition helps to dissipate energy. Yet others, known as differential surge tanks, have multiple chambers with interconnections; the different water levels that occur in the various chambers help to dampen out the oscillations.

EXERCISES

12.6.1 Given that the pipe in Fig. 12.5*a* is 1500 ft long, that water is initially flowing through it at 2 fps, and that the pressure waves travel along it at 3750 fps, find the maximum water hammer pressure at the valve, and at points 500 ft and 1000 ft from the valve, (*a*) if the valve is completely closed instantaneously, and (*b*) if the valve is partially closed instantaneously, reducing the flow velocity to 0.6 fps.

[6] See footnote 5.

12.6.2 Given that the pipe in Fig. 12.5*a* is 300 m long, that water is initially flowing through it at 0.8 m/s, and that pressure waves travel along it at 1000 m/s, find the maximum water hammer pressure at the valve, and at points 100 m and 200 m from the valve, (*a*) if the valve is completely closed instantaneously, and (*b*) if the valve is partially closed instantaneously, reducing the flow velocity to 0.2 m/s.

12.6.3 Repeat Exer. 12.6.1 for valve closure times of 0.36 sec.

12.6.4 Repeat Exer. 12.6.2 for valve closure times of 0.33 s.

12.6.5 Repeat Exer. 12.6.1 for valve closure times of 1.00 sec.

12.6.6 Repeat Exer. 12.6.2 for valve closure times of 0.84 s.

12.7 SURGE TANKS

In a hydroelectric plant the flow of water to a turbine must be decreased very rapidly whenever there is a sudden drop in load. This rapid decrease in flow will result in high water-hammer pressures and may result in the need for a very strong and hence expensive pipe. There are several ways to handle a situation of this sort; one is by use of a ***surge tank***, or ***surge chamber***. A simple surge tank is a vertical standpipe connected to the pipeline as shown in Fig. 12.10. With steady flow V_0 in the pipe, the initial water level z_0 in the surge tank is below the static (no flow) level ($z = 0$), as shown in the figure. When the valve is suddenly closed, water rises in the surge tank. The water surface in the tank will then fluctuate up and down until damped out by fluid friction. The section of pipe upstream of the surge tank is in effect afforded protection from the high water-hammer pressures that would exist on valve closure if there were no tank.

An approximate analysis for this simple surge tank may be performed as follows. Initially, for steady conditions, indicated by the subscript 0, the amount by which the water level in the surge tank is below the static level is the sum of the head loss due to friction, the velocity head, and any minor losses in the line such as the entrance loss h'_e shown (see Chaps. 5 and 8). Therefore, as z_0 is negative, we have

$$-z_0 = \left(\frac{fL}{D} + 1 + \sum k\right)\frac{V_0^2}{2g} = K\frac{V_0^2}{2g} \tag{12.17}$$

where $\sum k$ is the sum of the coefficients for all minor losses in the line, and K is defined to be the total quantity in parentheses. For the condition soon after the valve is instantaneously closed, when the water is at some level z as it rises up the surge tank, we can write the energy equation (12.6) for unsteady

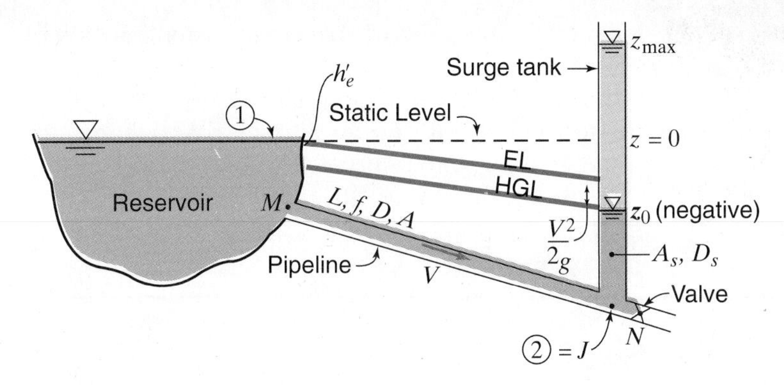

Figure 12.10
Definition sketch for surge tank analysis.

flow between the surface of the reservoir (point 1 on Fig. 12.10) and the end of the pipeline at the surge tank (point 2 = J, on the centerlines). Neglecting fluid friction in the surge tank and inertial effects in the surge tank, and letting h be the vertical distance from point 2 to elevation $z = 0$, at point 2 we have $p_2/\gamma = h + z$ and $z_2 = -h$, so that $(p/\gamma + z)_2 = h + z - h = z$. Substituting this and conditions at point 1 into Eq. (12.6), we therefore obtain

$$0 - \left(z + \frac{V^2}{2g}\right) = \left(\sum k + \frac{fL}{D}\right)\frac{V^2}{2g} + \frac{L}{g}\frac{dV}{dt}$$

In this equation z represents the level of the water surface in the surge tank measured positively upward from the static water level where $z = 0$ and $(L/g)(dV/dt)$ represents the accelerative head in the pipe between the reservoir and the surge tank (Sec. 12.3).

Using our definition of K, this equation can be written as

$$-z = K\frac{V^2}{2g} + \frac{L}{g}\frac{dV}{dz}\frac{dz}{dt} \tag{12.18}$$

With the valve completely closed, the continuity equation is

$$AV = A_s\frac{dz}{dt} \tag{12.19}$$

where A and A_s are the cross-sectional areas of the pipe and surge tank, respectively. Combining Eqs. (12.18) and (12.19) to eliminate dt, and then assuming f and K are constant while integrating and solving for V, yields

$$V^2 = \frac{2g}{K}\left(\frac{LA}{KA_s} - z\right) - Ce^{-(KA_s/LA)z} \tag{12.20}$$

This is the general equation relating the velocity in the pipe to the water-surface level in the surge tank over the time interval from valve closure to the top of the first surge. After that, the water changes its flow direction and so this equation is no longer valid. However, subsequent oscillations will be smaller due to damping by friction losses.

Before we use this equation to find the maximum surge height z_{max}, we must first take care of the unknown constant of integration C. We can do this as follows. The initial conditions for the time interval considered are $V = V_0$ and $z = z_0$ [Eq. (12.17)]. Substituting these into Eq. (12.20) gives

$$V_0^2 = \frac{2g}{K}\left(\frac{LA}{KA_s} + K\frac{V_0^2}{2g}\right) - Ce^{\left(\frac{KA_s}{LA}\right)K\frac{V_0^2}{2g}}$$

which simplifies to

$$\frac{2gLA}{K^2A_s} = Ce^{(K^2A_s/2gLA)V_0^2} \tag{12.21}$$

The conditions at the end of the time interval, at the top of the first surge, are $V = 0$ and $z = z_{max}$. Substituting these into Eq. (12.20) and rearranging gives

$$\frac{2g}{K}\left(\frac{LA}{KA_s} - z_{max}\right) = Ce^{-(KA_s/LA)z_{max}} \tag{12.22}$$

Finally, dividing Eq. (12.22) by Eq. (12.21) to eliminate C results in

$$1 - \frac{KA_s}{LA}z_{max} = e^{-\frac{KA_s}{LA}\left(K\frac{V_0^2}{2g}+z_{max}\right)} \tag{12.23}$$

Reviewing our result, we see that this equation is explicit for V_0, but is implicit for z_{max}, A_s, A, L, and K, and so therefore f. Also, we note that the right-hand side of this equation must always be positive because it is an exponential, so it follows that the left-hand side must be positive also, leading to

$$\frac{KA_s}{LA}z_{max} < 1 \tag{12.24}$$

Note that in all cases we can conveniently replace A_s/A by $(D_s/D)^2$, where D_s is the diameter of the surge tank.

If we wish to find V_0, we can rearrange Eq. (12.23) into

$$V_0^2 = -\frac{2g}{K}\left[z_{max} + \frac{LA}{KA_s}\ln\left(1 - \frac{KA_s}{LA}z_{max}\right)\right] \tag{12.25}$$

where the right-hand side will be positive because the ln term will be negative. If we wish to solve for z_{max}, A_s, or K (in order to find f), we must either use a programmable calculator with an equation solver, or manually use trial and error. For the manual method, because the right-hand side of Eq. (12.23) is usually quite small, it turns out that the limiting value given by Eq. (12.24) is a good indicator of the true answer. For example, when solving for z_{max}, Eq. (12.24) gives $z_{max} < LA/KA_s$, which serves as a good first estimate for trial and error.

Since this derivation neglected fluid friction and inertial effects in the surge tank, minor losses at the surge tank junction, and assumed constant f and instantaneous valve closure, the value of z_{max} as computed by Eq. (12.23) will be larger than the true value, and so the results provide a conservative estimate for preliminary design of simple surge tanks. For long pipelines, minor losses and the velocity head in the pipeline will be small compared with the friction head loss. If these are also neglected, according to Eq. (12.17) K must be replaced by fL/D. Then the results obtained will provide an even more conservative estimate.

More accurate analysis of surge tank phenomena is commonly conducted using computer programs. These use numerical methods to solve the differential and algebraic equations that represent the unsteady liquid motion, taking into account all the factors just mentioned, and including the actual rates of valve closure.

Surge tanks are usually open at the top and of sufficient height that they will not overflow. In some instances they are permitted to overflow if no damage will result. There are many types of surge tanks; other types were discussed at the end of Sec. 12.6.

The surge tank, in addition to providing protection against water-hammer pressures, fulfills another desirable function. That is, in the event of a sudden demand for increased flow, it can quickly provide some excess water, while the entire mass of water in a long pipeline is being accelerated. The acceleration of masses of liquids in pipelines was discussed in Sec. 12.4.

SAMPLE PROBLEM 12.6 You need to design a surge tank for a 1.5-m-diameter steel pipeline ($f = 0.018$) 800 m long that supplies water to a small power plant. The surge tank will be connected to the pipeline at a point where the center line is 35 m below the water surface in the reservoir. For a discharge of 12 m³/s, what is the smallest surge tank diameter that will prevent surges from exceeding 18.5 m above the reservoir surface? Neglect inertial effects and fluid friction in the surge tank only. The pipeline inlet is square-edged.

Solution

Sec. 8.18 for square-edged entrance: $k_e = 0.5$

Eq. (12.17): $$K = \frac{0.018(800)}{1.5} + 1 + 0.5 = 11.10$$

$$V_0 = Q_0/A = 12/\pi(1.5/2)^2 = 6.79 \text{ m/s}$$

Eq. (12.17): $z_0 = -11.1(6.79)^2/2(9.81) = -26.09$ m, possible cf 35 m.

From Eq. (12.24): $$A_s < \frac{LA}{Kz_{max}} = \frac{800\pi(1.5/2)^2}{11.10(18.5)} = 6.88 \text{ m}^2$$

: Programmed computing aids (Appx. B) could help solve problems marked with this icon.

Solving Eq. (12.23) by trial and error:

	Try A_s	*Left side*	*Right side*	
	6.88	0.000 64	0.089 94	
	6.80	0.012 26	0.092 49	
	6.00	0.128 47	0.122 39	
Last trial:	6.06	0.119 75	0.119 85	close enough

$$A_s = 6.06 \text{ m}^2 = (\pi/4)\, D_s^2; \quad D_s = 2.78 \text{ m} \qquad \textbf{\textit{ANS}}$$

EXERCISES

12.7.1 A 48-in-diameter steel pipe (f = 0.020, flush inlet) 4000 ft long delivers 275 cfs of water to a power plant. It is protected from water hammer by a 7-ft-diameter surge tank located at the downstream end just before the control valve. Without using trial and error, obtain a safe, preliminary estimate of the largest height (above the reservoir surface elevation) to which a surge would rise. Neglect all velocity heads and minor losses; in the surge tank (only) neglect fluid friction and inertial effects.

12.7.2 A 900-mm-diameter steel pipe (f = 0.018, flush inlet) 850 m long delivers 4 m^3/s of water to a power plant. It is protected from water hammer by a surge tank located at the downstream end just before the control valve. Without using trial and error, obtain a safe, preliminary estimate of the smallest surge tank diameter that would be required to prevent surges from exceeding 12 m above the reservoir water surface elevation. Neglect all velocity heads and minor losses; in the surge tank (only) neglect fluid friction and inertial effects.

PROBLEMS

12.1 (*a*) Outflow from a reservoir with vertical sides is over a spillway [$Q = C_W L H^{3/2}$, Eq. (11.26)]. Initially there is a steady inflow so that the height of the water surface above the spillway level is z_1. If the inflow is suddenly cut off, prove that the time required for the water level to fall from z_1 to z_2 is $t = (2As/C_W L)(z_2^{-0.5} - z_1^{-0.5})$. (*Note:* $z = H$.) (*b*) How long will it take theoretically for the outflow to cease entirely? What factors make this theoretical answer unrealistic? (*c*) Given the crest of the overflow spillway is 100 ft long, the value of C_W in Eq. (11.26) is 3.45, and the constant water surface area is 700,000 ft^2. With no inflow, find the time required for the water surface to fall from 3 ft to 1 ft above the spillway level.

12.2 Repeat parts (*a*) and (*b*) of Prob. 12.1. (*c*) Given the crest of the overflow spillway is 30 m long, the value of C_W in Eq. (11.26) is 1.91, and the constant water surface area is 65 000 m^2, with no inflow find the time required for the water surface to fall from 900 mm to 300 mm above the spillway level.

12.3 Figure P12.3 shows a tank with vertical sides containing liquid with a surface area of A_s. The liquid discharges through an orifice under a head z which varies from the initial height h to zero as the tank empties down to the orifice level. (*a*) Neglecting friction losses, what is the cumulative kinetic energy of the jet during the time required for the liquid surface to drop from h to zero? (*b*) How does this kinetic-energy summation compare with the total energy of the mass of fluid initially in the tank above the orifice level?

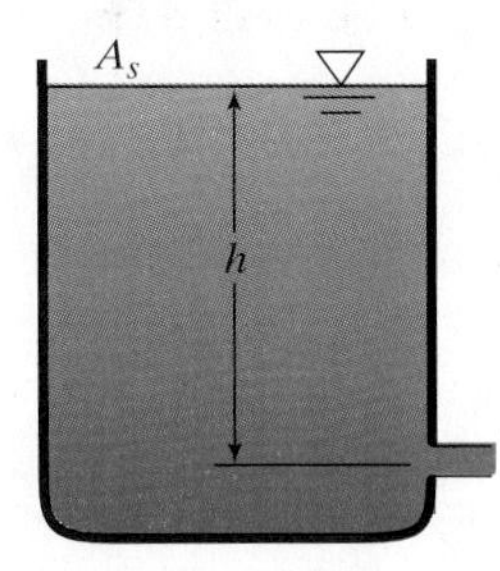

Firgure P12.3

12.4 How long will it take for the tank in Prob. 12.3 to empty down to the orifice level if A_s is 8 ft^2, h is 16 ft, and the jet diameter is 5 in? Plot a graph of h versus time, using increments of 4 ft.

12.5 How long will it take for the tank in Prob. 12.3 to empty down to the orifice level if A_s is 0.72 m^2, h is 5 m, and the jet diameter is 125 mm? Plot a graph of h versus time, using increments of 1 m.

12.6 Water enters a reservoir at such a rate that the height z of water above the level of the spillway crest is 3 ft. The spillway ($Q = C_W L H^{3/2}$, $H = z$) is 100 ft long, and the value of C_W is 3.50. The area of the water surface for various water levels is as follows:

z, ft:	3.00	2.50	2.00	1.50	1.25	1.00
A_s, ft^2:	850,000	800,000	700,000	550,000	510,000	470,000

If the inflow is suddenly reduced to 150 cfs, what will be the water level for equilibrium? How long will it take, theoretically, for equilibrium to be attained? How long will it take for the level to drop from 3 to 1 ft above the spillway level?

12.7 Work Prob. 12.6 using the same numbers but changing ft to m, ft^2 to m^2, and cfs to m^3/s.

12.8 A reservoir has an overflow spillway with a 40-ft-long crest, and a value of $C_W = 3.50$ [Eq. (11.26)]. For the range of water levels considered here, the area of the water surface is essentially constant at 600,000 ft^2. Initially, the water surface is 3 ft below the level of the spillway crest. If a flow of 500 cfs is suddenly discharged into this reservoir, what will be the height z of water in the reservoir (above spillway crest) for equilibrium? How much time will be required for this equilibrium height to be reached? How much time will be required for the water surface to reach 2 ft above the spillway level? (*Note:* This last part can be solved by integration after substituting x^3 for $z^{3/2}$ and consulting integral tables. However, it will be easier to solve it graphically, either by plotting and actually measuring the area under the curve, or by computing the latter by some method such as Simpson's rule.)

12.9 Attached to the tank in Fig. P12.9 is a flexible 1.5-in-diameter hose ($f = 0.020$) 250 ft long. The tank is hoisted in such a manner that $h = 25 + 2t$, where h is the head in feet and t is the time in seconds. (*a*) Find as accurately as you can the flow rate at $t = 10$ sec. (*b*) Suppose h were decreasing at the same rate. What would be the flow rate when $h = 45$ ft?

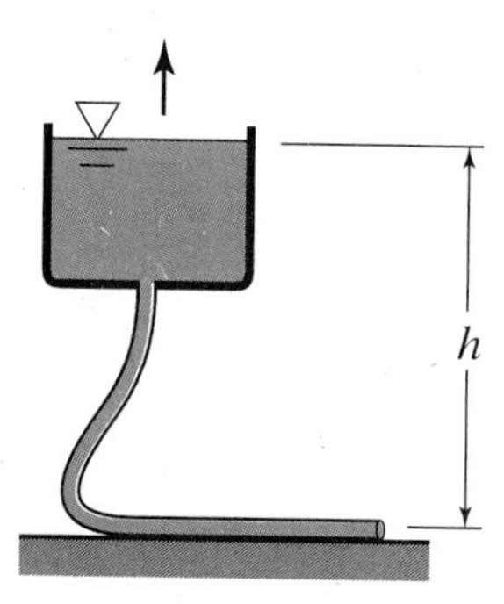

Figure P12.9

12.10 Attached to the tank in Fig. P12.9 is a flexible 40-mm-diameter hose ($f = 0.024$) 65 m long. The tank is hoisted in such a manner that $h = 10 + 0.5t$, where h is the head in meters and t is the time in seconds. (*a*) Find as accurately as you can the flow rate at $t = 10$ s. (*b*) Suppose h were decreasing at the same rate. What would be the flow rate when $h = 15$ m?

12.11 Repeat Prob. 12.9 for the case of a 6-in-diameter hose with all other data remaining the same.

12.12 Repeat Prob. 12.10 for the case of a 150-mm-diameter hose with all other data remaining the same.

12.13 Work Sample Prob. 12.2 for the case where the pipe lengths are 400 and 1500 ft rather than 200 and 750 ft. All other data are the same.

12.14 A 1-in-diameter smooth brass pipe 1000 ft long drains an open 4-ft-diameter cylindrical tank that contains oil having $\rho = 1.8$ slugs/ft^3 and $\mu = 0.0006$ lb·s/ft^2. The pipe discharges at elevation 100 ft. Find the time required for the oil level to drop from elevation 120 to elevation 108 ft.

12.15 A 25-mm-diameter smooth brass pipe 300 m long drains an open 1.2-m-diameter cylindrical tank which contains oil having $\rho = 950$ kg/m^3 and $\mu = 0.03$ N·s/m^2. The pipe discharges at elevation 30 m. Find the time required for the oil level to drop from elevation 36 to elevation 32.5 m.

12.16 Figure P12.16 depicts a large reservoir being drained with a pipe system ($f = 0.025$, $L = 2000$ ft, $D = 9$ in, $z_0 = 45$ ft). The flow rate is 3.5 cfs when the pump is initially rotating at 180 rpm. If the pump speed is increased instantaneously to 220 rpm, determine the flow rate as a function of time. Assume that the head h_p developed by the pump is proportional to the square of the rotative speed; that is $h_p \propto n^2$.

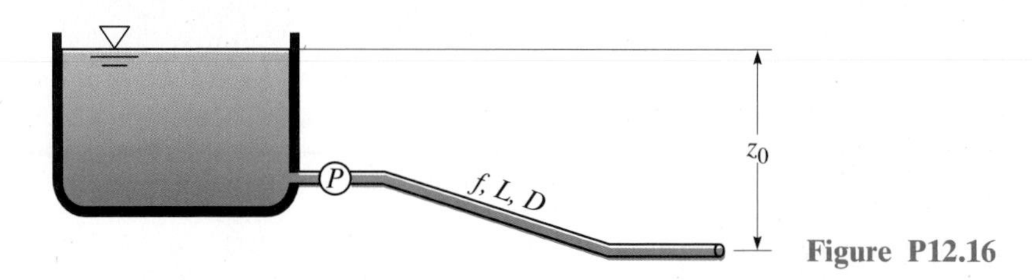

Figure P12.16

12.17 Figure P12.16 depicts a large reservoir being drained with a pipe system ($f = 0.022$, $L = 825$ m, $D = 250$ mm, $z_0 = 14$ m). The flow rate is 120 L/s when the pump is initially rotating at 200 rpm. If the pump speed is increased instantaneously to 250 rpm, determine the flow rate as a function of time. Assume that the head h_p developed by the pump is proportional to the square of the rotative speed; that is $h_p \propto n^2$.

12.18 (*a*) Repeat Prob. 12.16 with all the data the same except use an 18-in-diameter pipe rather than a 9-in pipe. (*b*) Repeat also for the case of an 8-in-diameter pipe.

12.19 (*a*) Repeat Prob. 12.17 with all the data the same except use a 500-mm-diameter pipe rather than a 250-mm pipe. (*b*) Repeat also for the case of a 200-mm-diameter pipe.

12.20 A reservoir is to be drained by a 6-in-diameter pipe (flush entrance, $f = 0.020$) 3500 ft long. The pipe outlet is at an elevation 110 ft below the reservoir water surface. Initially there is no flow since there is a plug at the pipe outlet. Plot Q versus t for time after the plug is removed.

12.21 Repeat Prob. 12.20 for the case where the pipe length is 400 ft rather than 3500 ft.

12.22 A reservoir is to be drained by a 200-mm-diameter pipe ($f = 0.028$) 700 m long. The elevation of the pipe outlet is 50 m below the reservoir water surface. Initially there is no flow because a valve at the pipe outlet is closed. Plot Q (L/s) versus t for time after the valve is suddenly opened.

12.23 A reservoir with its water surface at elevation 850 ft may be drained with a long pipe that discharges at elevation 50 ft. The pipe consists of 4000 ft of 4-in-diameter pipe ($f = 0.020$) followed by 500 ft of 2-in-diameter pipe ($f = 0.020$). Initially a valve at the discharge end is closed and the water in the pipe is at rest. The valve is then suddenly opened and as time ensues the flow rate gradually increases. Determine the time rate of change of the flow rate (cfs/sec) at the instant when the flow rate is 0.40 cfs.

12.24 The pipeline for draining a large reservoir consists of 200 ft of 6-in-diameter pipe ($f = 0.030$) followed by 500 ft of 10-in-diameter pipe

(f = 0.020). The elevation of the outlet is 100 ft below the reservoir water surface. A valve at the outlet, initially closed, is quickly opened. Derive an equation similar to Eq. (12.8) that is applicable to this situation, and plot flow rate versus time. Neglect minor losses.

12.25 The pipeline for draining a large reservoir consists of 60 m of 150-mm-diameter pipe (f = 0.030) followed by 150 m of 250-mm-diameter pipe (f = 0.020). The elevation of the outlet is 30 m below the reservoir water surface. A valve at the outlet, initially closed, is quickly opened. Derive an equation similar to Eq. (12.8) that is applicable to this situation, and plot flow rate (L/s) versus time. Neglect minor losses.

12.26 Water in a large reservoir may be released through a 12-in-diameter pipe (f = 0.020) 350 ft long. Losses at the pipe entrance are negligible; a nozzle (C_v = 0.95) at the outlet end produces a 6-in-diameter jet. The jet discharges at an elevation 50 ft below the reservoir water surface. Initially, there is a tight-fitting plug in the nozzle, which is then removed. For this situation derive an equation similar to Eq. (12.8) and plot flow rate versus time.

12.27 A large open tank containing oil (s = 0.85, μ = 0.0005 lb·s/ft^2) is connected to a 3-in-diameter smooth pipe 4000 ft long. The elevation of the discharge point is 10 ft below the liquid surface in the tank. A closed valve at the discharge end of the pipe is opened suddenly. Plot the ensuing flow rate versus time.

12.28 A large open tank containing oil (s = 0.82, ν = 6.5×10^{-5} m^2/s) is connected to a 150-mm-diameter pipe 450 m long. The elevation of the discharge point is 2.2 m below the liquid surface in the tank. A closed valve at the discharge end of the pipe is opened suddenly. Plot the ensuing flow rate (L/s) as a function of time.

12.29 A vertical pipe full of oil is allowed to drain by removing a plug from its lower end. (*a*) Obtain an equation for the varying flow velocity as a function of time t, the pipe diameter D, and the specific weight γ and viscosity μ of the oil. Assume that the head loss is given by the equation of established laminar flow and that surface-tension effects are negligible. (*b*) Obtain an equation for the distance x of the oil surface from the top of the pipe, in terms of the same variables. (*c*) Given that the pipe is 10 ft long, its diameter is 1 in, and the oil has specific gravity 0.88 and viscosity 0.004 lb·s/ft^2, find the time required to drain the pipe after the plug is removed. Check whether the laminar flow assumption is reasonable.

12.30 Repeat parts (*a*) and (*b*) of Prob. 12.29. (*c*) Given that the pipe is 2.8 m long, its diameter is 20 mm, and the oil has specific gravity 0.88 and viscosity 0.25 N·s/m^2, find the time required to drain the pipe after the plug is removed. Check whether the laminar flow assumption is reasonable.

12.31 (*a*) What is the celerity of a pressure wave in a 6-ft-diameter water pipe with 0.65-in steel walls? (*b*) If the pipe is 4200 ft long, what is the time required for a pressure wave to make the round trip from the valve? (*c*) If the initial water velocity is 7 fps, what will be the rise in pressure at the valve if the time of closure is less than the time of a round trip? (*d*) If the valve is closed at such a rate that the velocity in the pipe decreases uniformly with respect to time and closure is completed in a time $t_c = 5L/c_P$, approximately what will be the increase in pressure head at the valve when the first pressure unloading wave reaches the valve?

12.32 (*a*) What is the celerity of a pressure wave in a 1.8-m-diameter water pipe with 125-mm concrete walls? (*b*) If the pipe is 1250 m long, what is the time required for a pressure wave to make the round trip from the valve? (*c*) If the initial water velocity is 2.75 m/s, what will be the rise in pressure at the valve if the time of closure is less than the time of a round trip? (*d*) If the valve is closed at such a rate that the velocity in the pipe decreases uniformly with respect to time and closure is completed in a time $t_c = 5L/c_P$, approximately what will be the increase in pressure head at the valve when the first pressure unloading wave reaches the valve?

12.33 Using Eqs. (12.12) and (12.16) and the data for Figs. 12.8 and 12.9 as given in footnote 3, compute the water-hammer pressure for each case and compare the answers with the actual measurements. Also, for the given data, compute *f*.

12.34 Water at 15°C is flowing through a 300-mm-diameter welded-steel pipe 2400 m long that drains a reservoir under a head of 50 m. The pipe wall has a thickness of 8.5 mm. (*a*) If a valve at the end of the pipe is closed in 10 s, approximately what water-hammer pressure will be developed? (*b*) If the steady-state flow is instantaneously reduced to one-half its original value, what water-hammer pressure would you expect?

12.35 For the situation described in Sample Prob. 12.5, find the water-hammer pressure at the valve if a flow of 90 cfs is reduced to 30 cfs in 3.2 sec. Under these conditions, what would be the maximum water-hammer pressures at points 500 and 1500 ft from the reservoir?

12.36 In Fig. P12.36 the total length of pipe is 10,000 ft, its diameter is 42 in, and its wall thickness is 0.89 in. Assume $E = 32{,}000{,}000$ psi and $E_v = 300{,}000$ psi. If the initial velocity for steady flow is 8 fps and the valve at *G* is partially closed so as to reduce the flow to half of the initial velocity in 3.6 sec, find (*a*) the maximum pressure rise due to the water hammer; (*b*) the location of the point of maximum total pressure.

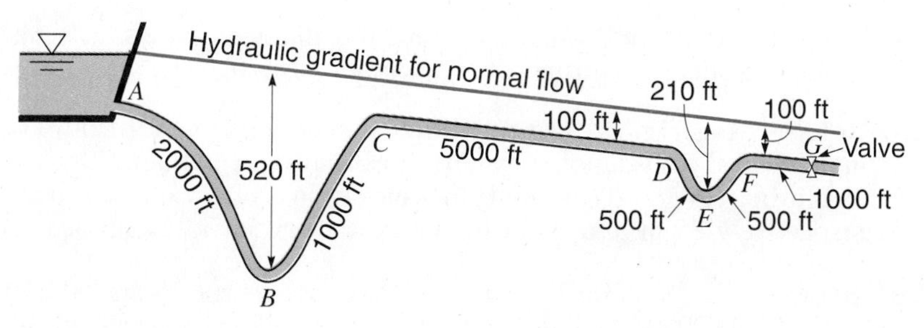

Figure P12.36

12.37 Refer to Fig. P12.36, but take all the dimensions given in feet to be in meters instead. This 10-km-long pipe has a diameter of 1.35 m and a wall thickness of 21 mm. Assume $E = 205$ GPa and $E_v = 2$ GPa. The initial steady flow velocity is 5 m/s. The valve at G is then partially closed so as to reduce the velocity to 1 m/s in 14 s. Find (*a*) the maximum pressure rise due to water hammer and (*b*) the location of the point of maximum total pressure.

12.38 Derive Eq. (12.20).

12.39 A 1.5-m-diameter steel pipeline 1000 m long (flush inlet, $f = 0.020$) carries water at Q_0 (m^3/s) from a reservoir to a power plant. When the valve at the outlet end is closed instantaneously, water rises in the 2.5-m-diameter simple surge tank immediately adjacent to the valve. Determine the maximum allowable initial discharge Q_0 so that the resulting surge will not rise more than 12 m above the reservoir surface. In the surge tank only, neglect the velocity head, minor losses, fluid friction, and inertial effects.

12.40 Repeat Prob. 12.39 while also neglecting the velocity head and minor losses in the pipeline.

12.41 Refer to Fig. 12.10. A 42-in-diameter steel pipe MN 3600 ft long (flush inlet, $f = 0.017$) supplies water to a small power plant. The discharge is 200 cfs, $JN = 100$ ft, and the elevations of J and the valve N are respectively 130 ft and 145 ft below the reservoir water surface. To protect against instantaneous closure of the valve, what height would be required for the simple 6.5-ft-diameter surge tank if it is not to overflow? In the surge tank only, neglect the velocity head, minor losses, fluid friction, and inertial effects.

12.42 Repeat Prob. 12.41 while also neglecting the velocity head and minor losses in the pipeline.

12.43 Repeat Prob. 12.41 for a surge tank diameter of 10 ft.

12.44 Using the data of Prob. 12.41, find the diameter of surge tank that will produce a surge requiring a tank height of 165 ft.

12.45 Using the data of Prob. 12.42, find the diameter of surge tank that will produce a surge requiring a tank height of 165 ft.

12.46 Refer to Fig. 12.10. A 1-m-diameter steel pipe MN 1070 m long (flush inlet, $f = 0.016$) supplies water to a small power plant. The discharge is 2.45 m^3/s, $JN = 20$ m, and the elevations of J and the valve N are respectively 25 m and 30 m below the reservoir water surface. To protect

---

: Programmed computing aids (Appx. B) could help solve problems marked with this icon.

against instantaneous closure of the valve, what height would be required for the simple 3.5-m-diameter surge tank if it is not to overflow? In the surge tank only, neglect the velocity head, minor losses, fluid friction, and inertial effects.

12.47 Repeat Prob. 12.46 while also neglecting the velocity head and minor losses in the pipeline.

: Programmed computing aids (Appx. B) could help solve problems marked with this icon.

CHAPTER 13

Steady Flow of Compressible Fluids[1]

Fluids that show appreciable variation in density are called ***compressible fluids***. The variation in density is mainly caused by variations in pressure and temperature. The study of such fluids in motion is sometimes called ***gas dynamics***.

When dealing with a flowing compressible fluid, if the density change is gradual and not more than a few percent, the flow may be treated as incompressible by using an average density for best results. However, if $\Delta\rho/\rho > 0.05$, the effects of compressibility must be considered. The purpose of this chapter is to investigate compressible-fluid problems that require such considerations. The discussion will be limited to steady one-dimensional flow of compressible fluids. Before proceeding further, the reader is advised to review Secs. 2.7, 2.8 and possibly 2.9.

13.1 THERMODYNAMIC CONSIDERATIONS

As a first step in understanding the flow of compressible fluids, it is advantageous in our discussion to review briefly some thermodynamic principles. The thermodynamic properties of a gas (Appendix A, Table A.5) include the gas constant R, specific heat c_p at constant pressure, specific heat c_v at constant volume, and the specific heat ratio $k = c_p/c_v$. The density ρ (or specific volume v) of a gas depends on the absolute pressure p and absolute

[1] The authors gratefully acknowledge the suggestions for revision to this chapter provided by Professor Michel A. Saad.

temperature T of the gas; for real gases at low pressures and moderate or high temperatures the relationship between these properties is closely defined by the perfect (ideal) gas law $p/\rho = pv = RT$, which is an equation of state (or property relation) that was discussed in Sec. 2.7. In this book we shall assume that all gases are perfect. Another fundamental equation introduced in Sec. 2.7 is pv^n = constant, which describes the changes of perfect gas properties from one state to another for a particular process.

The first law of thermodynamics was discussed in Sec. 5.4. The second law of thermodynamics deals with reversibility of processes. In general, a ***reversible process*** is one in which both the system *and its surroundings* can be returned precisely to their initial states without work being done. Processes involving friction, heat transfer, and mixing of gases are not reversible. Clearly reversible processes imply ideal fluids (Sec. 2.10). All real processes are irreversible, but some can be approximated well by reversible processes when the irreversibility is minor. For example, the flow through a converging nozzle where there is little friction and little or no heat transfer can be approximated as a reversible process. But flow in a pipeline is an irreversible process because of the pipe friction.

Related to irreversibility is ***entropy*** s, a property that measures the disorder, or the amount of energy unavailable for useful work, during a natural flow process. In real processes it always increases so that available energy decreases. Processes with constant entropy occur only in theory, but may be closely approximated in fact.

Process equations for perfect gases include the following:

For isothermal conditions ($n = 1$):	pv = constant
For isentropic conditions ($n = k$):	pv^k = constant

An ***isothermal*** process is one in which there is no change of temperature, while an ***adiabatic*** process is one in which no heat is added or removed. An adiabatic process that is reversible (entropy is constant) and for which k is a constant is known as an ***isentropic*** process. For adiabatic processes with friction, $n < k$ for expansion and $n > k$ for compression. Compressible-flow problems are solved in a manner similar to that used for incompressible flows, except that the equation describing the compressible process must be included.

The ***enthalpy*** h per unit mass of a gas is defined by $h = i + p/\rho$ and, for a perfect gas, $h = i + RT$, where i, the internal energy per unit mass due to the kinetic energy of molecular motion and the forces between molecules, is a function of temperature. Hence enthalpy is a composite energy property of a gas, and is a function of only temperature for a perfect gas.

The ***specific heat at constant pressure,*** c_p, is defined as the increase in enthalpy per unit of mass when the temperature of a gas is increased one degree with its pressure held constant. Thus,

$$c_p = \left(\frac{\partial h}{\partial T}\right)_{p\,=\,\text{constant}} \tag{13.1}$$

where h is the enthalpy per unit of mass.

The ***specific heat at constant volume***, c_v, is defined as the increase in internal energy per unit of mass when the temperature is increased one degree with its volume held constant. Thus,

$$c_v = \left(\frac{\partial i}{\partial T}\right)_{v\,=\,\text{constant}} \tag{13.2}$$

where i is the internal energy per unit mass.

For perfect gases these equations can be written as $dh = c_p\,dT$ and $di = c_v\,dT$. Now since $h = i + p/\rho = i + RT$, $dh = di + R\,dT$. Combining these relationships leads to

$$c_p - c_v = R \tag{13.3}$$

Combining the specific heat ratio $k = c_p/c_v$ with Eq. (13.3) gives

$$c_p = \frac{k}{k-1}R, \qquad c_v = \frac{R}{k-1} \tag{13.4}$$

SAMPLE PROBLEM 13.1 Compute the change in internal energy and the change in enthalpy of 15 kg of air if its temperature is raised from 20 to 30°C. The initial pressure is 95 kPa abs.

Solution

Properties of gases are given in Appendix A, Table A.5.

$$\Delta i = c_v(T_2 - T_1) = 716\,\frac{\text{N·m}}{\text{kg·K}}\,(30-20)\text{ K} = 7160\text{ J/kg (or N·m/kg)}$$

Change in internal energy $= \Delta i \times (15\text{ kg}) = 107\,400\text{ J}$

$$\Delta h = c_p(T_2 - T_1) = 1003(10) = 10\,030\text{ J/kg}$$

Change in enthalpy $= \Delta h \times (15\text{ kg}) = 150\,000\text{ J}$ ***ANS***

SAMPLE PROBLEM 13.2 Suppose that 15 kg of air ($T_1 = 20°\text{C}$) of Sample Prob. 13.1 were compressed isentropically to 40% of its original volume. Find the final temperature and pressure, the work required, and the changes in internal energy and enthalpy.

Solution

The following relations apply: $pv = RT$ and $pv^k = \text{constant}$, where $k = 1.40$ for air.

$$pv^k = pv\frac{v^k}{v} = \frac{RT}{v}v^k = RTv^{k-1} = \text{constant}$$

Since $R = \text{constant}$, $Tv^{k-1} = \text{constant}$

$$T_2 = T_1\left(\frac{v_1}{v_2}\right)^{k-1} = (273+20)\left(\frac{1.0}{0.4}\right)^{0.40} = 423 \text{ K} = 159°\text{C} \qquad \textbf{\textit{ANS}}$$

$$\frac{pv}{T} = R = \text{constant}, \qquad p_1 = 95 \text{ kPa abs (given in Sample Prob. 13.1)}$$

$$\frac{p_1 v_1}{T_1} = \frac{p_2 v_2}{T_2} \qquad \text{and} \qquad v_2 = 0.4v_1$$

$$\frac{95v_1}{293} = \frac{p_2(0.4v_1)}{423} \qquad p_2 = 343 \text{ kPa abs (or kN/m}^2\text{ abs)} \qquad \textbf{\textit{ANS}}$$

Since this is an adiabatic process, the work required is equal to the change in internal energy. This can be confirmed by computing the values of the pressure and corresponding volumes occupied by the gas during the isentropic process, plotting a pressure-vs-volume curve, and finding the area under the curve and thereby determining the work done on the fluid. Thus the work required is

$$\int_{s_1}^{s_2} F\,ds = \int_{s_1}^{s_2} (F/A)A\,ds = \int_{\text{vol 1}}^{\text{vol 2}} p\,d(\text{vol})$$

or

$$\Delta i = c_v(T_2 - T_1) = 716(423 - 293) = 92\ 900 \text{ J/kg}$$

$$\text{Work required} = \text{change in internal energy} = \Delta i \times 15 \text{ kg} = 1\ 393\ 000 \text{ J} \qquad \textbf{\textit{ANS}}$$

$$\Delta h = c_p(T_2 - T_1) = 1003(130) = 130\ 100 \text{ J/kg}$$

$$\text{Change in enthalpy} = \Delta h \times 15 \text{ kg} = 1\ 952\ 000 \text{ J} \qquad \textbf{\textit{ANS}}$$

EXERCISES

13.1.1 Compute the change in enthalpy of 15 slugs of oxygen if its temperature is increased from 120°F to 155°F.

13.1.2 Compute the change in enthalpy of 250 kg of oxygen if its temperature is increased from 50°C to 70°C.

13.1.3 Suppose the 15 slugs of oxygen of Exer. 13.1.1 were compressed isentropically to 80% of its original volume. Find the final temperature and pressure, the work required, and the change in enthalpy. Assume $T_1 = 120°\text{F}$ and $p_1 = 200$ psia.

13.1.4 Suppose the 250 kg of oxygen of Exer. 13.1.2 were compressed isentropically to 80% of its original volume. Find the final temperature and pressure, the work required, and the change in enthalpy. Assume $T_1 = 50°\text{C}$ and $p_1 = 1400 \text{ kN/m}^2$ abs.

13.1.5 Using the data of Sample Prob. 13.2 compute $\Delta(p/\rho)$ and thus show that $\Delta h = \Delta i + \Delta(p/\rho)$.

13.2 FUNDAMENTAL EQUATIONS APPLICABLE TO THE FLOW OF COMPRESSIBLE FLUIDS

The fundamental equations for the flow of compressible fluids have already been stated in Chaps. 4, 5, 6, and 7. For convenience, we restate them here.

Continuity

The expression for continuity for steady one-dimensional flow of a compressible fluid is

$$\dot{m} = \rho A V = \rho Q = \text{constant} \tag{13.5}$$

where $\dot{m}$ is the mass flow rate and $Q = AV$.

Energy Equation

For one-dimensional steady flow of a compressible fluid if there is no work interaction (no machine) between sections 1 and 2 the energy equation[2] is expressible as:

$$h_1 + \frac{V_1^2}{2} + q_H = h_2 + \frac{V_2^2}{2} \tag{13.6}$$

where q_H is the heat interaction per unit mass. It is convenient to combine the static enthalpy per unit mass h and the kinetic energy of the fluid into a single

[2] The reader should review Secs. 5.4 and 5.6 for a discussion of the energy equation as it applies to compressible fluids.

term called ***stagnation enthalpy*** or ***total enthalpy*** h_O, defined as

$$h_O = h + \frac{V^2}{2} \tag{13.7}$$

Impulse-Momentum Equation

The impulse-momentum equation (Sec. 6.1) for steady one-dimensional flow of a compressible fluid is

$$\sum \mathbf{F} = \rho_2 Q_2 \mathbf{V}_2 - \rho_1 Q_1 \mathbf{V}_1 \tag{13.8}$$

Euler Equation

For one-dimensional flow of an ideal fluid in a frictionless constant-area duct the Euler equation (Sec. 5.7) may be expressed as

$$\frac{dp}{\rho} + V\,dV = 0 \tag{13.9}$$

In Eqs. (13.6) and (13.9) the z terms were dropped, for in the flow of compressible fluids the z terms are almost always negligible compared with the other terms in the energy equation.

Mach Number

In Chap. 7 an important dimensionless parameter, the ***Mach number*** **M**, was mentioned

$$\mathbf{M} = \frac{V}{c} \tag{13.10}$$

where V is the velocity of flow and c is the ***speed of sound*** (***sonic*** or ***acoustic velocity***), i.e., the ***celerity*** at which a pressure wave will travel through a compressible fluid. If $\mathbf{M} < 1$, the flow is ***subsonic***; if $\mathbf{M} = 1$, the flow is ***sonic***; if $\mathbf{M} > 1$, the flow is ***supersonic***.

13.3 SPEED OF SOUND

Consider an elastic fluid at a pressure p and a density ρ at rest in a rigid pipe of cross-sectional area A as shown in Fig. 13.1*a*. Suppose a piston at one end is suddenly moved to the right with a small velocity dV. This will produce an

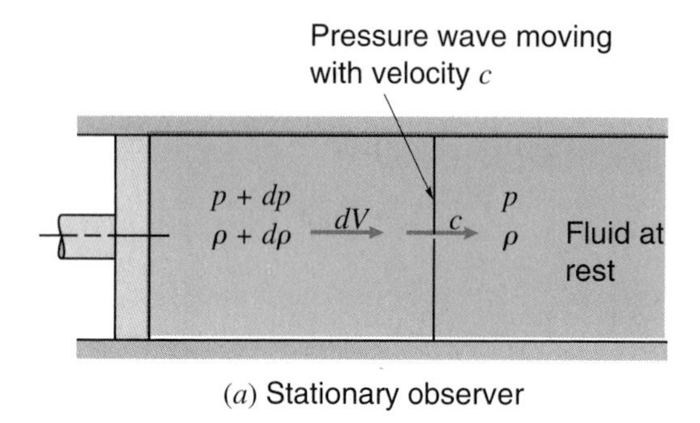

(*a*) Stationary observer

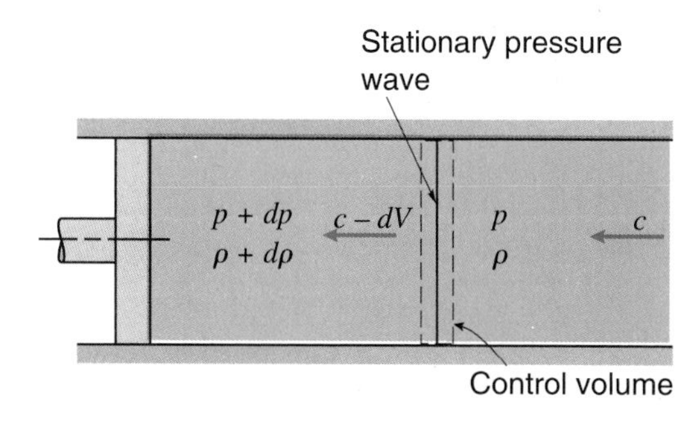

(*b*) Observer moving with pressure wave

Figure 13.1
Propagation of a weak pressure pulse.

infinitesimal pressure wave, which will travel through the fluid with a velocity (celerity) c, and the pressure and density of the fluid will change to $p + dp$ and $\rho + d\rho$. The fluid behind the wave is set in motion with a velocity dV. In Fig. 13.1*b* this process is shown relative to an observer moving with the wave. The continuity equation, applied to the control volume shown, gives

$$\rho A c = (\rho + d\rho)A(c - dV)$$

Since the amplitude of the sound wave is assumed infinitesimal, the term $d\rho\, dV$ is of second order and can be neglected. The above equation then becomes

$$\rho\, dV = c\, d\rho \tag{13.11}$$

The momentum equation, applied to a control volume, is

$$pA - (p + dp)A = \rho A c[(c - dV) - c]$$

which reduces to

$$dp = c\rho \, dV \tag{13.12}$$

Combining Eqs. (13.11) and (13.12) to eliminate dV gives

$$c^2 = \frac{dp}{d\rho} \tag{13.13}$$

A sonic, or pressure, wave travels through a fluid with such a high velocity that there is no time for any appreciable heat transfer from any heat of compression; moreover, the fluid friction is negligible and thus the process is isentropic. The speed of sound c can thus be written as

$$c = \sqrt{\left(\frac{\partial p}{\partial \rho}\right)_s} \tag{13.14}$$

where the subscript s indicates isentropic (constant entropy) conditions. For a perfect gas undergoing an isentropic process, pressure is related to density by

$$p = C\rho^k$$

Differentiating with respect to ρ gives

$$\left(\frac{\partial p}{\partial \rho}\right)_s = Ck\rho^{k-l} = kC\left(\frac{\rho^k}{\rho}\right) = k\left(\frac{p}{\rho}\right) = kRT$$

Substituting this expression into Eq. (13.14) gives the speed of sound in a perfect gas:

$$c = \sqrt{kRT} = \sqrt{\frac{kp}{\rho}} \tag{13.15}$$

This shows that for perfect gas the sonic speed is a function of its absolute temperature.

The ***isentropic compressibility*** of a fluid is defined as

$$K_s \equiv \frac{1}{\rho}\left(\frac{\partial \rho}{\partial p}\right)_s$$

But in liquids and solids, changes in pressure generally produce small changes in temperature, so that $(\partial p/\partial \rho)_s \approx (\partial p/\partial \rho)_T$. Also, the bulk modulus of elasticity

$$E_v = -v\left(\frac{\partial p}{\partial v}\right)_T = \rho\left(\frac{\partial p}{\partial \rho}\right)_T \approx \rho\left(\frac{\partial p}{\partial \rho}\right)_s$$

so that

$$\left(\frac{\partial p}{\partial \rho}\right)_s \approx \frac{E_v}{\rho} \tag{13.16}$$

Combining Eqs. (13.14) and (13.16) gives the speed of sound in terms of the modulus of elasticity as

$$c = \sqrt{\frac{E_v}{\rho}} \tag{13.17}$$

The foregoing analysis has considered the pipe to be rigid. In reality the pipe is elastic, and the stretching of the pipe walls due to the pressure wave makes the modulus of the combination less than that of the fluid alone.

This new combined modulus will be expressed by E_c, and we shall let $dv = dv' + dv''$, where dv' is due to compression of the fluid and dv'' is due to stretching of the pipe wall. Thus $E_c = -v\,dp/(dv' + dv'')$, from which

$$\frac{1}{E_c} = -\frac{dv'}{v\,dp} - \frac{dv''}{v\,dp} \tag{13.18}$$

The first term on the right is seen to be $1/E_v$ (Sec. 2.5). From the concept of hoop tension, the increment of stress in the wall of the pipe is $r\,dp/t$, where r is the radius of the pipe and t is its thickness. If the circumference is stretched an amount dl, the unit deformation is $dl/2\pi r$. Since $dl = 2\pi\,dr$, the increment of unit deformation becomes dr/r. From these relations, recalling that the modulus of elasticity E of a solid = (increment of stress) ÷ (increment of unit deformation), we obtain $dp = Et\,dr/r^2$. For unit length of pipe and unit mass of fluid, $v = \pi r^2$, and the increase in volume per unit length of pipe for a unit mass of fluid is equal to the increase in area, so that $dv'' = 2\pi r\,dr$. Substituting these quantities for the three items in the last term of Eq. (13.18) gives $dv''/(v\,dp) = 2r/Et = D/Et$, where D is the pipe diameter. Therefore

$$\frac{1}{E_c} = \frac{1}{E_v} + \frac{D}{Et} \tag{13.19}$$

Solving for E_c gives

$$E_c = \frac{E_v}{1 + (D/t)(E_v/E)} \tag{13.20}$$

The celerity c_1 of a pressure wave in an elastic fluid in an elastic pipe is then

$$c_1 = \sqrt{\frac{E_c}{\rho}} = \sqrt{\frac{E_c}{\rho}\frac{E_v}{E_v}} = c\sqrt{\frac{E_c}{E_v}} = c\sqrt{\frac{1}{1 + (D/t)(E_v/E)}} \tag{13.21}$$

From Eqs. (13.20) and (13.21) we see that for a rigid pipe (i.e., $E = \infty$), E_c reduces to E_v and c_1 reduces to c.

EXERCISES

13.3.1 Use Sec. 13.3 to calculate the sonic velocity in air at sea level and at elevations 5000, 10,000, 20,000, and 30,000 ft. Assume standard atmosphere (Appendix A, Table A.3).

13.3.2 Repeat Exer. 13.3.1 for sea level, 2000 and 10 000 m, expressing the answers in SI units.

13.4 ADIABATIC FLOW (WITH OR WITHOUT FRICTION)

If heat transfer q_H is zero, the flow is adiabatic. Hence the energy Eq. (13.6) may be written as

$$h_1 + \frac{V_1^2}{2} = h_2 + \frac{V_2^2}{2} \tag{13.22}$$

Since $\Delta h = c_p \, \Delta T$, we get for adiabatic flow

$$V_2^2 - V_1^2 = 2(h_1 - h_2) = 2c_p(T_1 - T_2) \tag{13.23}$$

From Eq. (13.4), $c_p = kR/(k-1)$ and for a perfect gas $pv = RT$. Substituting these into Eq. (13.23) gives for adiabatic flow

$$V_2^2 - V_1^2 = \frac{2k}{k-1}(p_1 v_1 - p_2 v_2) = \frac{2k}{k-1} RT_1\left(1 - \frac{T_2}{T_1}\right) \tag{13.24}$$

The preceding equations are valid for flow either with or without friction. It should be noted that in these equations the temperatures and pressures are absolute.

Equation (13.22) can be written as

$$c_p T_1 + \frac{V_1^2}{2} = c_p T_2 + \frac{V_2^2}{2} = c_p T_O \tag{13.25}$$

where T_O is the ***stagnation temperature*** (where V is zero, see Sec. 5.8). Thus, in adiabatic flow, the stagnation temperature is constant along a streamline regardless of whether or not the flow is frictionless.

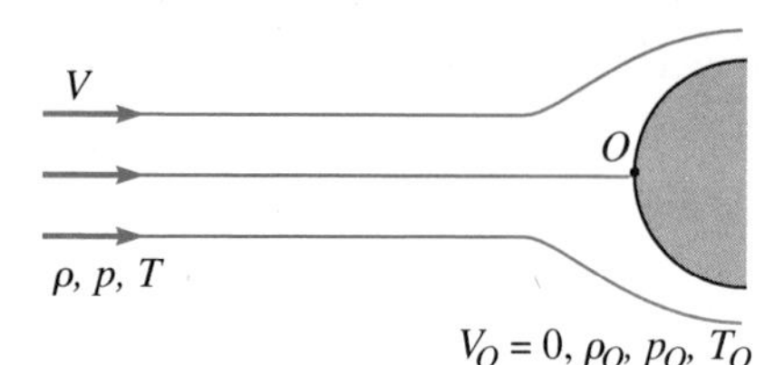

Figure 13.2
Stagnation point

13.5 STAGNATION PROPERTIES

In this section we develop relations between static and stagnation properties (Sec. 5.8). Figure 13.2 shows a stagnation point in compressible flow. At the stagnation point O the velocity V_O is zero and the increased pressure causes a rise in density and temperature.

Equation (13.25) can be used to define the ***stagnation temperature*** T_O:

$$\frac{T_O}{T} = 1 + \frac{V^2}{2c_P T}$$

But from Eqs. (13.10) and (13.15), $V^2/(kRT) = \mathbf{M}^2$, and from Eq. (13.4), $c_p = kR/(k-1)$; therefore the ratio of the stagnation temperature to the static temperature becomes

$$\frac{T_O}{T} = 1 + \frac{k-1}{2}\mathbf{M}^2 \tag{13.26}$$

This relationship is valid for adiabatic flow and for isentropic flow. Note also that T_O is the same for all points in the flow, provided that the flow is adiabatic.

When a perfect gas flows isentropically, its pressure and density are related to temperature in the following ways:

$$\frac{p_O}{p} = \left(\frac{T_O}{T}\right)^{k/(k-1)}, \qquad \frac{\rho_O}{\rho} = \left(\frac{T_O}{T}\right)^{1/(k-1)}$$

where p_O and ρ_O are the ***stagnation pressure*** and ***stagnation density***, respectively. By combining these with Eq. (13.26), pressure and density can be expressed in terms of Mach number:

$$\frac{p_O}{p} = \left(1 + \frac{k-1}{2}\mathbf{M}^2\right)^{k/(k-1)} \tag{13.27}$$

and

$$\frac{\rho_O}{\rho} = \left(1 + \frac{k-1}{2}\mathbf{M}^2\right)^{1/(k-1)} \tag{13.28}$$

Table 13.1
One-dimensional isentropic flow functions with $k = 1.4$

M	T/T_o	p/p_o	ρ/ρ_o
0	1.000 00	1.000 00	1.000 00
0.10	0.998 00	0.993 03	0.995 02
0.20	0.992 06	0.972 50	0.980 28
0.30	0.982 32	0.939 47	0.956 38
0.40	0.968 99	0.895 61	0.924 27
0.50	0.952 38	0.843 02	0.885 17
0.60	0.932 84	0.784 00	0.840 45
0.70	0.910 75	0.720 93	0.791 58
0.80	0.886 53	0.656 02	0.739 99
0.90	0.860 59	0.591 26	0.687 04
1.00	0.833 33	0.528 28	0.633 94
1.10	0.805 15	0.468 35	0.581 70
1.20	0.776 40	0.412 38	0.531 14
1.30	0.747 38	0.360 91	0.482 90
1.40	0.718 39	0.314 24	0.437 42
1.50	0.689 66	0.272 40	0.394 98
1.60	0.661 38	0.235 27	0.355 73
1.70	0.633 71	0.202 59	0.319 69
1.80	0.606 80	0.174 04	0.286 82
1.90	0.580 72	0.149 24	0.256 99
2.00	0.555 56	0.127 81	0.230 05

p_o and ρ_o are the same at all points in the flow, provided that the flow is isentropic.

Stagnation property ratios are given numerically in Table 13.1 and presented qualitatively in Fig. 13.3, as functions of **M**.

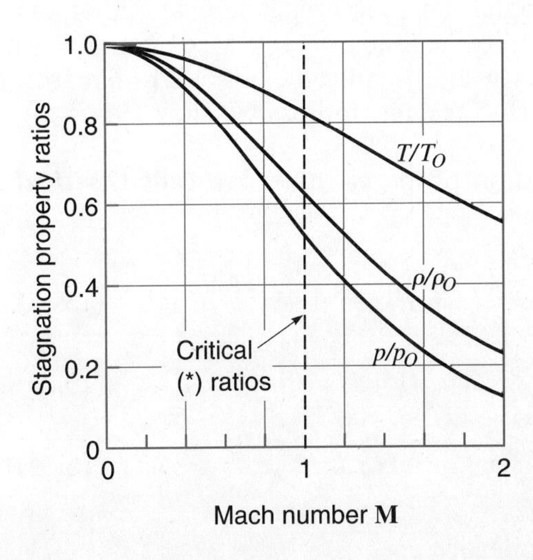

Figure 13.3
Stagnation property ratios for one-dimensional flow of an ideal gas with $k = 1.4$ (see also Table 13.1).

The speed of sound also is a function of temperature:

$$\frac{c_O}{c} = \sqrt{\frac{T_O}{T}}$$

Therefore sonic speed can be expressed as a function of the Mach number:

$$\frac{c_O}{c} = \left(1 + \frac{k-1}{2}\mathbf{M}^2\right)^{1/2} \tag{13.29}$$

where c is the speed of sound in the gas, while c_O is the speed of sound when the gas is at the stagnation temperature. When a flow is decelerated adiabatically to zero velocity, the temperature of the fluid becomes T_O. However, the fluid will not necessarily exist at the isentropic stagnation state. Unless deceleration is accomplished isentropically, the actual stagnation pressure and stagnation density fall below the isentropic stagnation values.

If we expand Eq. (13.27) by the binomial theorem, we obtain

$$p_O = p\left[1 + \frac{k}{2}\mathbf{M}^2 + \frac{k}{8}\mathbf{M}^4 + \frac{k(2-k)}{48}\mathbf{M}^6 + \cdots\right]$$

But since

$$\frac{1}{2}kp\mathbf{M}^2 = \frac{1}{2}kp\frac{V^2}{kRT} = \frac{1}{2}\rho V^2$$

this expression becomes

$$p_O = p + \frac{1}{2}\rho V^2\left[1 + \frac{\mathbf{M}^2}{4} + \frac{(2-k)\mathbf{M}^4}{24} + \cdots\right] \tag{13.30}$$

or

$$\frac{p_O - p}{\frac{1}{2}\rho V^2} = 1 + \frac{\mathbf{M}^2}{4} + \frac{(2-k)\mathbf{M}^4}{24} + \cdots \tag{13.31}$$

where $\frac{1}{2}\rho V^2$ is the ***dynamic pressure.***

Equation (13.30) is identical with Eq. (5.27); it is applicable only if $\frac{1}{2}(k-1)\mathbf{M}^2 < 1$. In determining the stagnation pressure, it indicates that the error caused by neglecting compressibility depends on the Mach number of the flow. At low Mach numbers the error is insignificant, but as $\mathbf{M}$ increases, the error becomes sizable. Note also that for incompressible flow ($\mathbf{M} = 0$), $p_O = p + \frac{1}{2}\rho V^2$.

When $\mathbf{M} = 1$, the static-to-stagnation property ratios are called ***critical ratios***. They are

$$\frac{T^*}{T_O} = \frac{2}{k+1} = 0.833 \quad (\text{for } k = 1.4) \tag{13.32}$$

$$\frac{p^*}{p_O} = \left(\frac{2}{k+1}\right)^{k/(k-1)} = 0.528 \quad (\text{for } k = 1.4) \tag{13.33}$$

$$\frac{\rho^*}{\rho_O} = \left(\frac{2}{k+1}\right)^{1/(k-1)} = 0.634 \quad (\text{for } k = 1.4) \tag{13.34}$$

where the asterisks identify critical properties.

Sample Problem 13.3 Find the Mach number, stagnation pressure, and stagnation temperature of nitrogen flowing at 600 ft/sec if the pressure and temperature in the undisturbed flow field are 100 psia and 200°F, respectively. (See Appx. A, Table A.5, for properties of gases.)

Solution

Eq. (13.15): $c = \sqrt{kRT} = \sqrt{(1.4)(1773)(660)} = 1280 \text{ ft/sec}$

Eq. (13.10): $\mathbf{M} = \dfrac{V}{c} = \dfrac{600}{1280} = 0.469$ ***ANS***

From Eq. (13.15): $\rho = \dfrac{kp}{c^2} = \dfrac{1.4(100 \times 144)}{(1280)^2} = 0.0123 \text{ slug/ft}^3$

From Eq. (13.30):

$$p_O = 100(144) + \tfrac{1}{2}(0.0123)(600)^2\left[1 + \tfrac{1}{4}(0.469)^2 + \frac{(2-1.4)}{24}(0.469)^4 + \cdots\right]$$

$$= 14{,}400 + 2214(1 + 0.0550 + 0.001\,21 + \cdots)$$

$$= 16{,}740 \text{ lb/ft}^2 = 116.2 \text{ psia}$$ ***ANS***

Applying Eq. (13.25): $$c_pT_1 + \frac{V_1^2}{2} = c_pT_O$$

$$6210(660) + \frac{(600)^2}{2} = 6210T_O$$

$$T_O = 689°\text{R} = 229°\text{F}$$ ***ANS***

Alternatively, from Eq. (13.27),

$$p_O = p\left(1 + \frac{k-1}{2}\mathbf{M}^2\right)^{k/(k-1)} = 100[1 + 0.2(0.469)^2]^{1.4/0.4} = 116.3 \text{ psia}$$ ***ANS***

and from Eq. (13.26)

$$T_O = T\left(1 + \frac{k-1}{2}\mathbf{M}^2\right) = 660[1 + 0.2(0.469)^2] = 689°\text{R}.$$ ***ANS***

A second alternative is to use Table 13.1. At $\mathbf{M} = 0.469$, we find by interpolation that $p/p_O = 0.859$ and $T/T_O = 0.958$. Therefore $p_O = 100/0.859 = 116.4$ psia and $T_O = 660/0.958 = 689°$R.

EXERCISES

13.5.1 Air flows past an object at 600 ft/sec. Determine the stagnation pressures and temperatures in the standard atmosphere at elevations of sea level, 5000 and 30,000 ft.

13.5.2 Air flows past an object at 200 m/s. Determine the stagnation pressures and temperatures in the standard atmosphere at elevations of sea level, 2000 and 10 000 m.

13.5.3 Air at 250 psia is moving at 500 ft/sec in a high-pressure wind tunnel at a temperature of 100°F. Find the stagnation pressure and temperature. Note the magnitude of the sonic velocity for the 250-psia 100°F air.

13.5.4 Air at 1750 kPA abs is moving at 150 m/s in a high-pressure wind tunnel at a temperature of 40°C. Find the stagnation pressure and temperature. Note the magnitude of the sonic velocity for the 1750-kPa 40°C air.

13.6 ISENTROPIC FLOW

Frictionless adiabatic flow is referred to as isentropic (constant entropy) flow. Such flow does not occur in nature. However, flow through a nozzle or flow in a free stream of fluid over a reasonably short distance may be considered isentropic because there is very little heat transfer and fluid-friction effects are small. Equations for isentropic flow can be derived by substituting $pv^k =$ constant in Eq. (13.24). The resulting equations for isentropic flow are

$$\frac{V_2^2 - V_1^2}{2} = \frac{p_1}{\rho_1}\frac{k}{k-1}\left[1 - \left(\frac{p_2}{p_1}\right)^{(k-1)/k}\right] = \frac{p_2}{\rho_2}\frac{k}{k-1}\left[\left(\frac{p_1}{p_2}\right)^{(k-1)/k} - 1\right] \tag{13.35}$$

Often, using Eq. (2.4), p/ρ is replaced by RT. Equation (13.35) may also be derived by integrating the Euler equation $dp/\rho + V\,dV = 0$ along a stream tube, noting that $pv^k =$ constant.

The relation between the temperatures at two points along a streamline in isentropic flow can be derived by equating the expressions for $V_2^2 - V_1^2$ from Eqs. (13.24) and (13.35). The result is

$$T_2 = T_1 - \frac{p_1}{R\rho_1}\left[1 - \left(\frac{p_2}{p_1}\right)^{(k-1)/k}\right] = T_1 - T_1\left[1 - \left(\frac{p_2}{p_1}\right)^{(k-1)/k}\right]$$

from which

$$\frac{T_2}{T_1} = \left(\frac{p_2}{p_1}\right)^{(k-1)/k} \tag{13.36}$$

This is the isentropic relation between temperature and pressure; once again, temperatures and pressures are absolute. This equation was introduced in Sec. 2.7.

EXERCISES

13.6.1 Air at a pressure of 150 psia and a temperature of 100°F expands in a suitable nozzle to 15 psia. (*a*) If the flow is frictionless and adiabatic and the initial velocity is negligible, find the final velocity by Eq. (13.35). (*b*) Find the final temperature at the end of the expansion through use of Eq. (13.23).

13.6.2 Air at a pressure of 1000 kPa abs and a temperature of 40°C expands in a suitable nozzle to 100 kPa abs. (*a*) If the flow is frictionless and adiabatic and the initial velocity is negligible, find the final velocity by Eq. (13.35). (*b*) Find the final temperature at the end of the expansion through use of Eq. (13.23).

13.6.3 Carbon dioxide flows isentropically. At a point in the flow the velocity is 50 ft/sec and the temperature is 125°F. At a second point on the same streamline the temperature is 80°F. What is the velocity at the second point? Check if your answer is valid.

13.6.4 Carbon dioxide flows isentropically. At a point in the flow the velocity is 15 m/s and the temperature is 50°C. At a second point on the same streamline the temperature is 25°C. What is the velocity at the second point? Check if your answer is valid.

13.7 EFFECT OF AREA VARIATION ON ONE-DIMENSIONAL COMPRESSIBLE FLOW

In steady flow the velocity of an *incompressible* fluid varies inversely with the cross-sectional area of the flow stream. This is not the case with a compressible fluid because variations in density will also influence the velocity. Moreover, the behavior of a compressible fluid, when there is a change in cross-sectional area, depends on whether the flow is ***subsonic*** ($\mathbf{M} < 1$) or ***supersonic*** ($\mathbf{M} > 1$). We shall now examine this phenomenon. In this discussion we shall confine our remarks to ideal flow.

Taking the logarithm of the continuity equation (13.5) and differentiating gives

$$\frac{dA}{A} + \frac{d\rho}{\rho} + \frac{dV}{V} = 0 \tag{13.37}$$

Noting that $c^2 = dp/d\rho$ for constant entropy [Eq. (13.13)], the Euler equation (13.9) for an isentropic flow of an *ideal* fluid may be expressed as

$$\frac{dp}{d\rho}\frac{d\rho}{\rho} + V\,dV = c^2\frac{d\rho}{\rho} + V\,dV = 0 \tag{13.38}$$

Combining this with Eq. (13.37) to eliminate $d\rho/\rho$, using Eq. (13.10) to replace V/c with $\mathbf{M}$, and rearranging, we get

$$\frac{dV}{V} = \frac{1}{\mathbf{M}^2 - 1}\frac{dA}{A} \tag{13.39}$$

From this equation, we can arrive at some significant conclusion as follows:

1. For subsonic flow ($\mathbf{M} < 1$):

   If $dA/A < 0$, $dV/V > 0$ (decreasing area causes velocity to increase)

   If $dA/A > 0$, $dV/V < 0$ (increasing area causes velocity to decrease)

2. For supersonic flow ($\mathbf{M} > 1$):

   If $dA/A < 0$, $dV/V < 0$ (decreasing area causes velocity to decrease)

   If $dA/A > 0$, $dV/V > 0$ (increasing area causes velocity to increase)

3. For sonic flow ($\mathbf{M} = 1$): $\dfrac{dA}{A} = 0$

Thus it is seen that subsonic and supersonic flows behave *oppositely* if there is an area variation. To accelerate a flow at subsonic velocity, a converging passage is required, just as in the case of an incompressible flow. To accelerate a flow at supersonic velocity, however, a diverging passage is required.

For sonic velocity it is noted that $dA/A = 0$. This condition occurs at the throat of a converging or a converging-diverging passage. However, the flow will be sonic at the throat only if the pressure differential between the upstream region and the throat is large enough to accelerate the flow sufficiently. At modest pressure differentials the velocity at the throat will be subsonic ($\mathbf{M} < 1$). As the pressure differential between the upstream region and the throat is increased, the velocity at the throat will increase until sonic velocity ($\mathbf{M} = 1$) occurs there. With further increase in pressure differential, the flow rate will increase (due to density increase) but the velocity at the throat will remain sonic. This is examined further in Secs. 13.8 and 13.9. Supersonic flow ($\mathbf{M} > 1$) will occur in the diverging section of a converging-diverging passage only if the flow at the throat is sonic. If the flow at the throat is subsonic, the flow in the diverging section will also be subsonic and the velocity will decrease in the diverging section. In Fig. 13.4 we see the

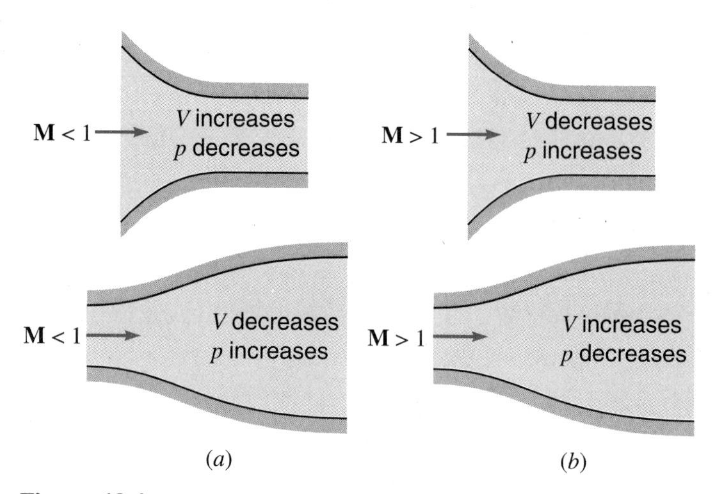

Figure 13.4
Effect of area variation on compressible flow. (*a*) Subsonic flow. (*b*) Supersonic flow.

behavior of subsonic and supersonic flow through converging and diverging passages.

EXERCISE

13.7.1 Show in detail the development of Eq. (13.39) from Eqs. (13.38) and (13.37).

13.8 COMPRESSIBLE FLOW THROUGH A CONVERGING NOZZLE

Let us now consider one-dimensional flow of a compressible fluid through the converging nozzle of Fig. 13.5. Since there is little opportunity for heat transfer between sections 1 and 2 and since frictional effects are small, we shall assume isentropic conditions. If the velocity of approach is negligible, p_1 becomes the stagnation pressure p_O in Eq. (13.35), which can then be expressed as

$$\frac{V_2^2}{2} = \frac{p_2}{\rho_2}\frac{k}{k-1}\left[\left(\frac{p_O}{p_2}\right)^{(k-1)/k} - 1\right] \tag{13.40}$$

Noting from Sec. 13.3 that $c_2 = \sqrt{kp_2/\rho_2}$, Eq. (13.40) can be rearranged to give

$$\left(\frac{V_2}{c_2}\right)^2 = \mathbf{M}_2^2 = \frac{2}{k-1}\left[\left(\frac{p_O}{p_2}\right)^{(k-1)/k} - 1\right] \tag{13.41}$$

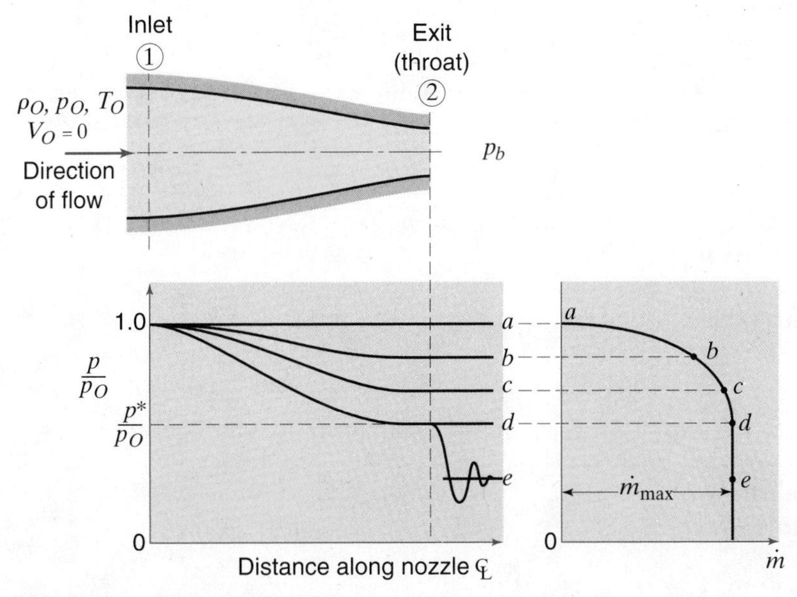

Figure 13.5
Characteristics of compressible flow through a converging nozzle.

Alternatively, this can be obtained by rearranging Eq. (13.27). From Eq. (13.41), we note that the velocity of flow at the throat (section 2 of the figure) depends on the p_O/p_2 ratio. If there is a large enough pressure differential between p_O and the ***back pressure*** p_b, sonic velocity will occur at section 2. From the discussion of Sec. 13.7, it is recognized that supersonic flow is impossible in this situation. If the flow through the throat is subsonic, the pressure at the throat is identical with that downstream of the nozzle ($p_2 = p_b$, the back pressure). If the flow through the throat is sonic, the pressure at the throat may be equal to, but is generally greater than, that downstream ($p_2 \geqslant p_b$).

Let us now assume the condition of sonic flow at the throat (that is, $\mathbf{M}_2 = 1.0$). Substituting $\mathbf{M}_2 = 1.0$ into Eq. (13.41) and rearranging, we get the ***critical pressure ratio***[3]

$$\frac{p_2^*}{p_O} = \left(\frac{2}{k+1}\right)^{k/(k-1)} \tag{13.42}$$

which is the same as (13.33). This critical pressure ratio exists whenever the pressure differential between p_1 and p_2 is great enough to create sonic velocity at the throat. If the flow through the throat is subsonic, then p_2/p_O will be larger than the ratio given by Eq. (13.42). We note that p_2/p_O can never be smaller than p_2^*/p_O.

The rate of flow through the nozzle of Fig. 13.5 may be found by substituting V_2 from Eq. (13.40) into $\dot{m} = \rho_2 A_2 V_2$. Making use of the isentropic relation between the p's and ρ's and rearranging, we get

Subsonic flow at throat:
$$\dot{m} = A_2\sqrt{\frac{2k}{k-1}p_O\rho_O\left[\left(\frac{p_2}{p_O}\right)^{2/k} - \left(\frac{p_2}{p_O}\right)^{(k+1)/k}\right]} \tag{13.43}$$

This expression is applicable so long as $p_2/p_O > p_2^*/p_O$ as given by Eq. (13.42). Thus it applies to the condition where there is subsonic flow at the throat.

The maximum rate of flow through the nozzle occurs when there is sonic flow at the throat. Substituting $p_2/p_O = p_2^*/p_O = [2/(k+1)]^{k/(k-1)}$ into Eq. (13.43) gives

Sonic flow at throat:
$$\dot{m}_{\max} = A_2\sqrt{p_O\rho_O k\left(\frac{2}{k+1}\right)^{(k+1)/(k-1)}} \tag{13.44}$$

Introducing Eq. (2.4), this equation may also be expressed as

Sonic flow at throat:
$$\dot{m}_{\max} = \frac{A_2 p_O}{\sqrt{T_O}}\sqrt{\frac{k}{R}\left(\frac{2}{k+1}\right)^{(k+1)/(k-1)}} \tag{13.45}$$

[3] In all equations here the pressure p is absolute pressure.

In this equation the expression under the second radical depends only on the properties of the gas. Thus a simple device for metering compressible flows is a converging nozzle at whose throat the sonic velocity is produced. It should be noted that with subsonic flow, as p_O is increased, the value of p_2/p_O becomes smaller and approaches p_2^*/p_O. At the point where p_2/p_O first reaches the value of p_2^*/p_O the flow at the throat changes from subsonic to sonic. At this point, a threshold point, Eq. (13.43) for subsonic flow and Eqs. (13.44) and (13.45) for sonic flow give the same result. As p_O is increased beyond the threshold point, p_2/p_O maintains the value of p_2^*/p_O and the velocity at the throat remains sonic. However, the flow rate increases directly with p_O, as can be seen from Eq. (13.45).

For air ($k = 1.4$), according to Eqs. (13.32) and (13.33), $T_2^*/T_O = 0.833$ and $p_2^*/p_O = 0.528$. Isentropic conditions have been assumed in the preceding equations; hence the flows represent those for an ideal fluid. The flows for real fluids through converging nozzles are only slightly less than those given by these equations.

Sample Problem 13.4 Air at 80°F flows from a large tank through a converging nozzle of 2.0-in exit diameter. The discharge is to an atmospheric pressure of 13.5 psia. Determine the mass flow rate through the nozzle for pressures within the tank of 5, 10, 15, and 20 psig. Assume isentropic conditions. Plot $\dot{m}$ as a function of p_O. Assume that the temperature within the tank is 80°F in all cases.

Solution

From Eq. (13.42), the critical pressure ratio for air is $p_2^*/p_O = 0.528$. If the flow at the throat is subsonic, $p_2/p_O > p_2^*/p_O$. Thus for subsonic flow at the throat, $p_2/p_O > 0.528$ and $p_b = p_2$. So $p_b/p_O > 0.528$ and $p_O < p_b/0.528$.

Since $p_b = 13.5$ psia, the flow at the throat will be subsonic if $p_O < 25.6$ psia (12.1 psig) and sonic if $p_O > 25.6$ psia (12.1 psig).

To find the flow rate for conditions where $p_O < 12.1$ psig (subsonic flow at throat), we use Eq. (13.43). Substituting the appropriate value of p_O into the equation and noting that for this condition $p_2 = p_b = 13.5$ psia, we get

For $p_O = 5$ psig (18.5 psia): $\dot{m} = 0.0374$ slug/sec ***ANS***

For $p_O = 10$ psig (23.5 psia): $\dot{m} = 0.0525$ slug/sec ***ANS***

To find the flow rate for conditions where $p_O > 12.1$ psig (sonic flow at the throat) we use Eq. (13.45). We get

For $p_O = 15$ psig (28.5 psia): $\dot{m}_{max} = 0.0634$ slug/sec ***ANS***

For $p_O = 20$ psig (33.5 psia): $\dot{m}_{max} = 0.0745$ slug/sec ***ANS***

Substituting $p_O = 25.6$ psia in Eq. (13.43) for subsonic flow gives $\dot{m} =$ 0.0571 slug/sec as does Eq. (13.44) or Eq. (13.45). This is the threshold point at which the flow in the throat changes from subsonic to sonic. When $p_O =$ 25.6 psia the flow rate as found from Eq. (13.45) is

$$(\dot{m}_{\max})_{p_O=25.6\text{ psia}} = \frac{(0.0218)144(25.6)}{\sqrt{540}}\sqrt{\frac{1.4}{1715}\left(\frac{2}{24}\right)^{2.4/0.4}} = 0.0571\text{ slug/sec}$$

As p_O is increased beyond 12.1 psig, sonic flow prevails at the throat and the flow rate increases linearly with $(p_O)_{\text{abs}}$ as indicated by Eq. (13.45). The variation of the flow rate with p_O is shown in Fig. S13.4. Other information concerning various

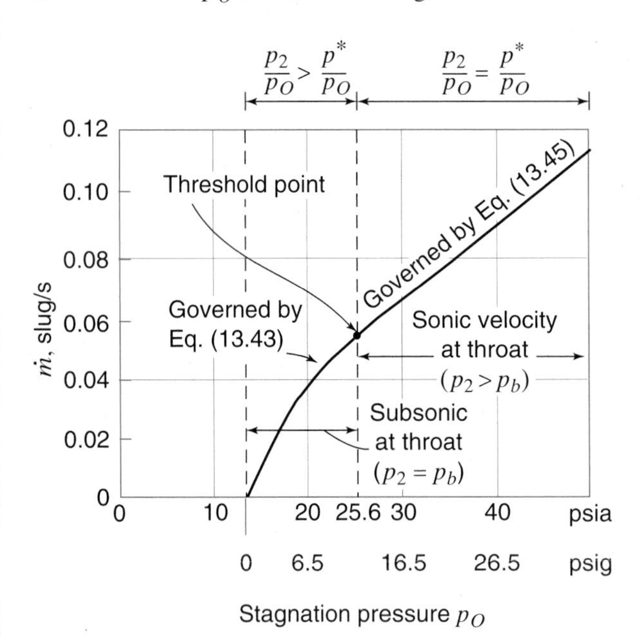

Figure S13.4

aspects of this problem can be found by applying the equations that have been presented, namely, the gas law ($pv = RT$ or $\rho = p/RT$), the process equation (pv^k = constant or p/ρ^k = constant), continuity ($\dot{m} = \rho AV$), and the energy equation, Eq. (13.40). Applying these, for example, for the case where $p_O =$ 5 psig (18.5 psia) yields $\rho_O = 0.002\,88$ slug/ft³, $\rho_2 = 0.002\,30$ slug/ft³, $V_2 =$ 747 ft/sec, $T_2 = 494°$R, and $p_2 = 13.50$ psia. Note that in this case $p_2 = p_b$.

EXERCISES

13.8.1 Start with Eq. (13.40) and derive Eq. (13.43).

13.8.2 Differentiate Eq. (13.43) with respect to p_2/p_O and set to zero to find the value of p_2/p_O for which $\dot{m}$ is a maximum. The answer should correspond to Eq. (13.42).

13.8.3 Carbon dioxide within a tank at 40 psia and 80°F discharges through a convergent nozzle into a 14.2-psia atmosphere. Find the velocity, pressure, and temperature at the nozzle outlet. Assume isentropic conditions.

13.8.4 Carbon dioxide within a tank at 280 kPa abs and 25°C discharges through a convergent nozzle into a 98 kPA abs atmosphere. Find the velocity, pressure, and temperature at the nozzle outlet. Assume isentropic conditions.

13.8.5 In Exer. 13.8.3 if the pressure and temperature within the tank had been 20 psia and 100°F, what would have been the velocity, pressure, and temperature at the nozzle outlet? Assume isentropic conditions.

13.8.6 In Exer. 13.8.4 if the pressure and temperature within the tank had been 140 kPa abs and 40°C, what would have been the velocity, pressure, and temperature at the nozzle outlet? Assume isentropic conditions.

13.9 ISENTROPIC FLOW THROUGH A CONVERGING–DIVERGING NOZZLE

If a converging nozzle has attached to it a diverging section, it is possible to attain supersonic velocities in the diverging section. This will happen if sonic flow exists in the throat. In such a case the gas or vapor will continue to expand in the diverging section to lower pressures and the velocity will continue to increase. The flow through such a converging–diverging nozzle is shown in Fig. 13.6. If there is not enough pressure differential to attain sonic

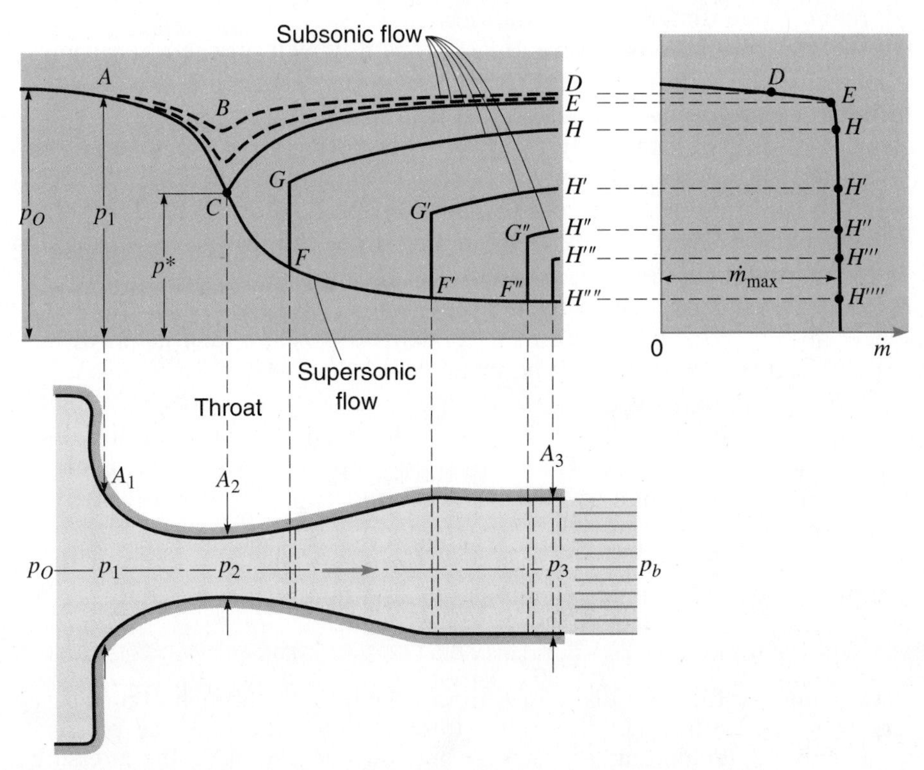

Figure 13.6
Flow of compressible fluid through a converging-diverging nozzle.

velocity at the throat, the gas or vapor will behave in much the same manner as a liquid, with acceleration in the region up to the throat and deceleration in the diverging section beyond the throat. A plot of the pressure in the flow for such a condition is shown by the dashed lines *ABD* of Fig. 13.6.

Suppose in Fig. 13.6 that the back pressure p_b is gradually reduced while p_O remains constant. In such a case $p_3 = p_b$, and the pressure at the throat (2) decreases while the velocity at the throat increases until the limiting sonic velocity is reached at the throat when the pressure plot is *ACE*. If the back pressure is further reduced to *H*, the pressure plot is *ACFGH*, the jump from *F* to *G* being a pressure shock, or ***normal shock wave*** (normal to the approaching flow, i.e., one-dimensional), which is analogous to the hydraulic jump, or standing wave, often seen in open channels conveying water. Through the shock wave the velocity is reduced abruptly from supersonic to subsonic, while at the same time the pressure jumps as shown by the lines *FG*, *F'G'*, and *F"G"*. The flow through a shock wave is not isentropic, since part of the kinetic energy is converted into heat.

Further reduction of the back pressure p_b causes the shock wave to move farther downstream until at some value given by *H'''* the shock wave is located at the downstream end of the nozzle. If p_b is lowered below the level of *H'''*, the shock wave occurs in the flow field downstream of the nozzle exit. Such flow fields are either two- or three-dimensional and cannot be described by the foregoing one-dimensional equations.

If the back pressure is lowered to *H''''*, the flow will proceed isentropically to supersonic throughout the entire region downstream from the throat, the velocity will increase continuously from 0 to its maximum value at 3 and the pressure will drop continuously from 0 to 3. As long as p_b is above *H''''* then $p_3 = p_b$; but if p_b drops below *H''''* then $p_3 > p_b$ and supersonic flow occurs through the entire length of the divergent portion of the nozzle.

If the back pressure is above *E* in Fig. 13.6, the flow rate through the converging–diverging nozzle is given by Eq. (13.43). In this instance the p_2 of Eq. (13.43) must be replaced by the p_b of Fig. 13.6. If the back pressure is below *E* in Fig. 13.6, critical pressure, as defined by Eq. (13.42), will occur at the throat and the flow rate will be given by Eq. (13.44) or Eq. (13.45).

If the stagnation pressure p_O is increased, the sonic velocity may be shown to remain unaltered, but since the density of the gas is increased, the rate of discharge will be greater. The converging nozzle and the converging–diverging nozzle are alike insofar as discharge capacity is concerned. The only difference is that with the converging–diverging nozzle, a supersonic velocity may be attained at discharge from the device, while with the converging nozzle, the sonic velocity is the maximum value possible.

SAMPLE PROBLEM 13.5 Air discharges from a large tank through a converging–diverging nozzle with a 2.0-in-diameter throat. Within the tank the pressure is 50 psia and the temperature is 80°F, while outside the tank the pressure is 13.5 psia. The nozzle is to operate with supersonic flow throughout its diverging section with a 13.5 psia pressure at its outlet. Find the required diameter of the nozzle outlet. Determine the mass flow rate and the velocities and temperatures at sections 2 and 3. Assume isentropic flow.

Solution

The pressure at the throat must be such that sonic velocity will occur there. Hence from Eq. (13.33), $p_2 = p_2^* = 0.528p_O = 0.528(50) = 26.4$ psia.

The velocity V_3 at the outlet may be found from Eq. (13.35) with $V_O = 0$:

$$\frac{V_3^2}{2} = RT_O\frac{k}{k-1}\left[1-\left(\frac{p_3}{p_O}\right)^{(k-1)/k}\right] = 1715(540)\frac{1.4}{0.4}\left[1-\left(\frac{13.5}{50}\right)^{0.4/1.4}\right]$$

$$= 1{,}012{,}000 \text{ ft·lb/slug (or ft}^2/\text{sec}^2)$$

from which $V_3 = 1420$ ft/sec. ***ANS***

The mass flow rate is computed from Eq. (13.45):

$$\dot{m}_{\max} = \frac{0.0218(50\times 144)}{\sqrt{540}}\sqrt{\frac{(1.4)}{1715}\left(\frac{2}{2.4}\right)^{2.4/0.4}} = 0.1117 \text{ slug/sec} \qquad \textbf{\textit{ANS}}$$

The temperature at 3 may be determined by using Eq. (13.23) with $V_O = 0$:

$$V_3^2 - V_O^2 = 2c_p(T_O - T_3)$$

$$(1420)^2 = 2(6000)(540 - T_3)$$

$$T_3 = 371°\text{R} = -89°\text{F} \qquad \textbf{\textit{ANS}}$$

From the perfect-gas law $p_3/\rho_3 = RT_3$:

$$\rho_3 = \frac{13.5\times 144}{1715(371)} = 0.003\,05 \text{ slug/ft}^3$$

Isentropic flow between 2 and 3 may be assumed, since the shock wave does not occur within that region. Thus $p_2/\rho_2^{1.4} = p_3/\rho_3^{1.4}$:

$$\frac{26.4}{\rho_2^{1.4}} = \frac{13.5}{(0.003\,05)^{1.4}}$$

$$\rho_2 = 0.004\,93 \text{ slug/ft}^3$$

The velocity at 2 may now be computed:

$$V_2 = \frac{\dot{m}}{\rho_2 A_2} = \frac{0.1117}{0.004\,93(0.0218)} = 1040 \text{ ft/sec} \qquad \textbf{\textit{ANS}}$$

The temperature at 2 can be found from Eq. (13.23):

$$(1420)^2 - (1040)^2 = 2(6000)(T_2 - 371)$$

$$T_2 = 450°\text{R} = -10°\text{F} \qquad \textbf{\textit{ANS}}$$

The area at 3 is computed from

$$A_3 = \frac{\dot{m}}{\rho_3 V_3} = \frac{0.1117}{0.003\ 05(1420)} = 0.0257\ \text{ft}^2$$

Finally, $D_3 = 2.17$ in, the required outlet diameter. ***ANS***

Check for sonic velocity at throat:

$$c_2 = \sqrt{kRT_2} = \sqrt{1.4(1715 \times 450)} = 1040\ \text{ft/sec} = V_2$$

With sonic velocity at the throat, if $D_3 < 2.17$ in, there will be supersonic flow throughout the divergent portion of the nozzle and a shock wave will occur in the flow field downstream of the nozzle exit. If $D_3 > 2.17$ in, with sonic flow in the throat, in order to satisfy pressure conditions, a shock wave will occur in the tube somewhere between the throat and the nozzle exit.

EXERCISES

13.9.1 Air enters a converging–diverging nozzle at a pressure of 120 psia and a temperature of 90°F. Neglecting the entrance velocity and assuming a frictionless process, find the Mach number at the cross section where the pressure is 35 psia.

13.9.2 Air enters a converging–diverging nozzle at a pressure of 830 kPa abs and a temperature of 32°C. Neglecting the entrance velocity and assuming a frictionless process, find the Mach number at the cross section where the pressure is 240 kPa abs.

13.9.3 Work Sample Prob. 13.5 with all data the same except for the pressure within the tank, which is 100 rather than 50 psia.

13.9.4 Air is to flow through a converging–diverging nozzle at 0.6 slug/sec. At the throat the pressure, temperature, and velocity are to be 20 psia, 100°F, and 500 ft/sec respectively. At outlet the velocity is to be 200 ft/sec. Determine the throat diameter. Assume isentropic flow.

13.9.5 Air in a tank at a pressure of 140 psia and 70°F flows out into the atmosphere through a 1.00-in-diameter converging nozzle. (*a*) Find the mass flow rate. (*b*) If a diverging section with an outlet diameter of 1.50 in were attached to the converging nozzle, what then would be the flow rate? Neglect friction.

13.9.6 Air in a tank at a pressure of 950 kPa abs and 20°C flows out into the atmosphere through a 2.5-cm-diameter converging nozzle. (*a*) Find the mass flow rate. (*b*) If a diverging section with an outlet diameter of 4 cm were attached to the converging nozzle, what then would be the flow rate? Neglect friction.

13.10 ONE-DIMENSIONAL SHOCK WAVE

In Fig. 13.7 is shown a one-dimensional shock wave where the approaching supersonic flow (subscript 1) changes to subsonic flow (subscript 2). This

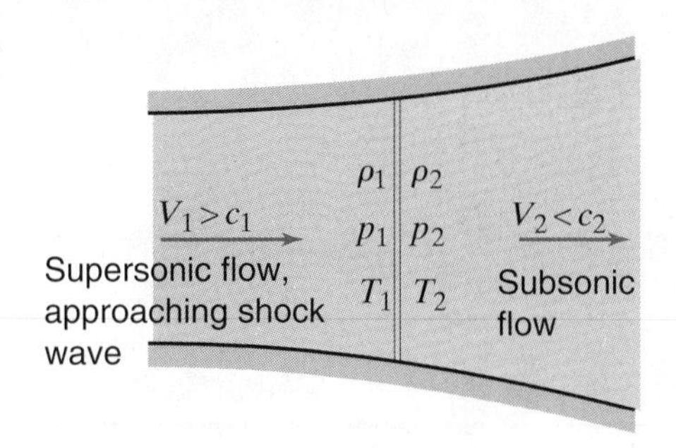

Figure 13.7
A one-dimensional normal shock wave.

phenomenon is accompanied by a sudden rise in pressure, density, and temperature. Applying the impulse-momentum principle to the fluid in the shock wave, we get

$$\sum F_x = p_1 A_1 - p_2 A_2 = \dot{m}(V_2 - V_1) \tag{13.46}$$

Substituting the continuity conditions ($\dot{m} = \rho_1 A_1 V_1 = \rho_2 A_2 V_2$) and noting that $A_1 = A_2$, we get

$$p_2 - p_1 = \rho_1 V_1^2 - \rho_2 V_2^2 \tag{13.47}$$

which is the pressure jump across the wave.

The flow across the shock wave may be considered adiabatic, and can be expressed as

$$V_2^2 - V_1^2 = \frac{2k}{k-1}(p_1 v_1 - p_2 v_2) \tag{13.48}$$

This is identical with Eq. (13.24). It is suggested that the reader review the development of this equation, in Sec. 13.4.

Equations (13.47) and (13.48) may be solved simultaneously and rearranged algebraically to give some significant relationships. Several such relations are as follows:

$$\frac{p_2}{p_1} = \frac{2k\mathbf{M}_1^2 - (k-1)}{k+1} \tag{13.49}$$

$$\frac{T_2}{T_1} = \frac{[2k\mathbf{M}_1^2 - (k-1)][2 + (k-1)\mathbf{M}_1^2]}{(k+1)^2\mathbf{M}_1^2} \tag{13.50}$$

$$\frac{V_2}{V_1} = \frac{\rho_1}{\rho_2} = \frac{(k-1)\mathbf{M}_1^2 + 2}{(k+1)\mathbf{M}_1^2} \tag{13.51}$$

and

$$\mathbf{M}_2^2 = \frac{2 + (k-1)\mathbf{M}_1^2}{2k\mathbf{M}_1^2 - (k-1)} \tag{13.52}$$

These equations permit one to find the physical properties of the flow on the two sides of the one-dimensional shock wave. These equations, of

Table 13.2
One-dimensional normal shock functions for an ideal gas with $k = 1.4$

$\mathbf{M}_1$	$\mathbf{M}_2$	T_2/T_1	p_2/p_1	ρ_2/ρ_1
1.00	1.000 00	1.000	1.000	1.000
1.10	0.911 77	1.065	1.245	1.169
1.20	0.842 17	1.128	1.513	1.342
1.30	0.785 96	1.191	1.805	1.516
1.40	0.739 71	1.255	2.120	1.690
1.50	0.701 09	1.320	2.458	1.862
1.60	0.668 44	1.388	2.820	2.032
1.70	0.640 55	1.458	3.205	2.198
1.80	0.616 50	1.532	3.613	2.359
1.90	0.595 62	1.608	4.045	2.516
2.00	0.577 35	1.688	4.500	2.667
2.10	0.561 28	1.770	4.978	2.812
2.20	0.547 06	1.857	5.480	2.951
2.30	0.534 41	1.947	6.005	3.085
2.40	0.523 12	2.040	6.553	3.212
2.50	0.512 99	2.138	7.125	3.333
2.60	0.503 87	2.238	7.720	3.449
2.70	0.495 63	2.343	8.338	3.559
2.80	0.488 17	2.451	8.980	3.664
2.90	0.481 38	2.563	9.645	3.763
3.00	0.475 19	2.679	10.333	3.857
4.00	0.434 96	4.047	18.500	4.571
5.00	0.415 23	5.800	29.000	5.000

course, are applicable only if $\mathbf{M}_1 > 1$; that is, the oncoming flow must be supersonic. It will be seen that the shock wave is analogous to the hydraulic jump in open-channel flow (Sec. 10.18). Properties across a shock are given numerically in Table 13.2 and presented qualitatively in Fig. 13.8, in terms of $\mathbf{M}_1$.

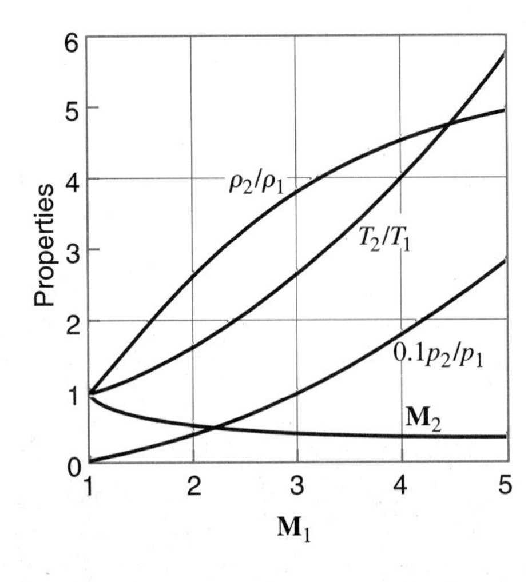

Figure 13.8
Properties across a shock wave for one-dimensional flow of an ideal gas with $k = 1.4$ (see also Table 13.2).

Sample Problem 13.6 A normal shock wave occurs in the flow of air where $p_1 = 10$ psia (70 Pa abs), $T_1 = 40$°F (5°C), and $V_1 = 1400$ ft/sec (425 m/s). Find p_2, V_2, and T_2.

Solution

In BG units:

$$\rho_1 = \frac{p_1}{RT_1} = \frac{10(144)\ \text{lb/ft}^2\ \text{abs}}{1715\ \text{ft·lb/(slug·°R)}(460+40)\text{°R}} = 0.001\ 68\ \text{slug/ft}^3$$

$$c_1 = \sqrt{kRT_1} = \sqrt{1.4 \times 1715 \times (460+40)} = 1096\ \text{ft/sec}$$

$$\mathbf{M}_1 = \frac{V_1}{c_1} = \frac{1400}{1096} = 1.28$$

From (Eq. (13.49): $\quad \dfrac{p_2}{p_1} = 1.75, \quad p_2 = 17.5$ psia

From Eq. (13.51): $\quad \dfrac{V_2}{V_1} = 0.675, \quad V_2 = 945$ ft/sec ***ANS***

$$\rho_1 V_1 = \rho_2 V_2, \quad \rho_2 = \frac{0.001\ 68}{0.675} = 0.002\ 49\ \text{slug/ft}^3 \qquad \textbf{\textit{ANS}}$$

$$pv = \frac{p}{\rho} = RT, \quad T_1 = \frac{p_2}{\rho_2 R} = 590\text{°R} = 130\text{°F} \qquad \textbf{\textit{ANS}}$$

In SI units:

$$\rho_1 = \frac{p_1}{RT_1} = \frac{70\ \text{N/m}^2\ \text{abs}}{287\ \text{N·m/(kg·K)}(273+5)\ \text{K}} = 8.8 \times 10^{-4}\ \text{kg/m}^3$$

$$c = \sqrt{kRT_1} = \sqrt{1.4 \times 287 \times (273+5)} = 334\ \text{m/s}$$

$$\mathbf{M}_1 = \frac{V_1}{c_1} = \frac{425}{334} = 1.27$$

From Eq. (13.49): $\quad \dfrac{p_2}{p_1} = 1.75, \quad p_2 = 122.5$ Pa (or N/m²) abs

From Eq. (13.51): $\quad \dfrac{V_2}{V_1} = 0.675, \quad V_2 = 286$ m/s ***ANS***

$$\rho_1 V_1 = \rho_2 V_2, \quad \rho_2 = \frac{8.8 \times 10^{-4}}{0.675} = 1.3 \times 10^{-3}\ \text{kg/m}^3 \qquad \textbf{\textit{ANS}}$$

$$T_2 = \frac{p_2}{\rho_2 R} = \frac{122.5}{(1.3 \times 10^{-3})287} = 328\ \text{K} = 55\text{°C} \qquad \textbf{\textit{ANS}}$$

EXERCISES

13.10.1 Just downstream of a normal shock wave the pressure, velocity, and temperature are 52 psia, 400 ft/sec and 120°F. Compute the Mach number upstream of the shock wave. Consider (*a*) air and (*b*) carbon dioxide as the working fluids.

13.10.2 Just downstream of a normal shock wave the pressure, velocity, and temperature are 360 kPa abs, 110 m/s and 50°C. Compute the Mach number upstream of the shock wave. Consider (*a*) air and (*b*) carbon dioxide as the working fluids.

13.11 THE OBLIQUE SHOCK WAVE

When the velocity of a body through any fluid, whether a liquid or a gas, exceeds that of a sound wave in the same fluid, the flow conditions are entirely different from those for subsonic flow. Thus, instead of streamlines such as are shown in Fig. 4.12, the conditions are as shown in Fig. 13.9, which is a schlieren photograph[4] of supersonic flow past a sharp-nosed model in a wind tunnel. It could also represent a projectile in flight through still air. A conical compression or shock wave extends backward from the tip, as may be seen by the strong density gradient revealed as a bright shadow in the photograph. A streamline in the undisturbed fluid is unaffected by the solid boundary or by a moving projectile until it intersects a shock-wave front, when it is abruptly changed in direction, proceeding roughly parallel to the nose form. Where the conical nose is joined to the cylindrical body of the

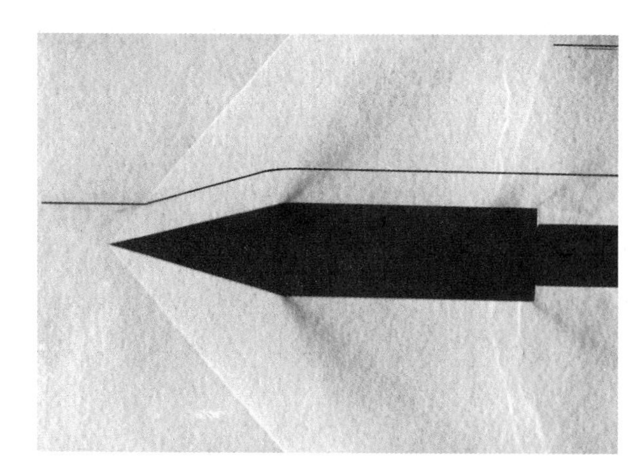

Figure 13.9
Schlieren photograph of head wave on 30° (total angle) cone at $\mathbf{M} = 1.88$. (*Photo by Guggenheim Aeronautical Laboratory, California Institute of Technology.*)

[4] The schlieren method employ the fact that the refraction of light is disturbed by a pressure (density) gradient. Details of the optics of flow visualization are discussed in Sec. 11.6 of Michel A. Saad, *Compressible Fluid Flow,* 2d ed., Prentice-Hall, Englewood Cliffs, New Jersey, 1993.

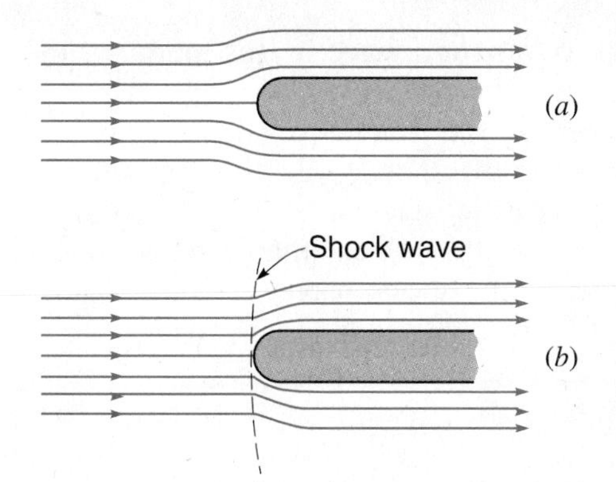

Figure 13.10
Flow pattern upstream of a body.
(*a*) Subsonic flow. (*b*) Supersonic flow.

model, dark shadows may be seen, representing rarefaction waves. The streamlines are again changed in direction through this region, becoming parallel to the main flow again. A typical streamline has been superimposed on the photograph. In Fig. 13.10 the difference between a subsonic and supersonic flow pattern upstream of a body is shown.

The reason why streamlines are unaffected in front of a projectile is that the body travels faster than the disturbance can be transmitted ahead. This is illustrated in Fig. 13.11. Consider a point source of an infinitesimal disturbance moving at a supersonic velocity V through a fluid. At the instant when this source passes through the point A_0, the disturbance commences to

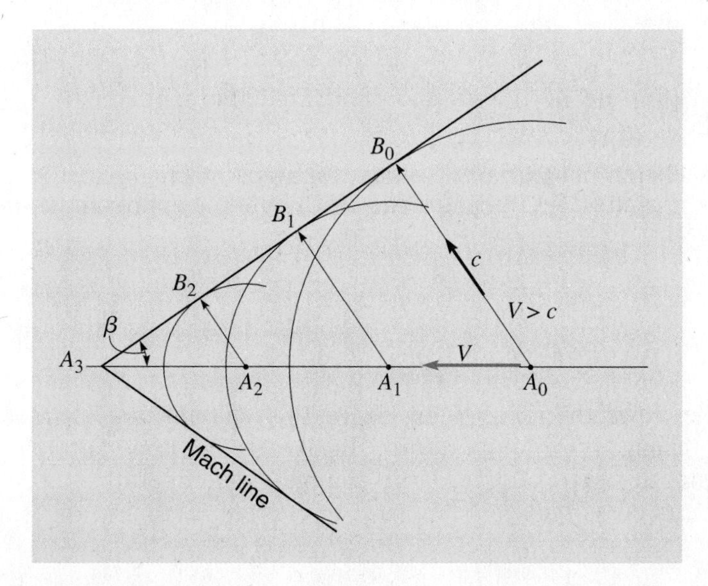

Figure 13.11
Schematic diagram of a disturbance moving with supersonic velocity.

radiate in all directions with the velocity c of a sound wave in this medium. In successive instants the source passes through the points A_1, A_2, and A_3, the last of which may represent the position of the source at the instant of observation. While the source has covered the distance A_0A_3 with velocity V, the sound wave, traveling at the slower acoustic velocity c, has progressed only as far as radius A_0B_0. Similar termini of disturbances emanating from A_1 and A_2 form a straight-line envelope, which is the shock wave. The angle β is called the ***Mach angle***, and it is seen that

$$\sin \beta = \frac{A_0B_0}{A_0A_3} = \frac{c}{V} = \frac{1}{\mathbf{M}} \tag{13.53}$$

where $\mathbf{M}$, the dimensionless velocity ratio V/c, is the Mach number.

In the case of the finite projectile of Fig. 13.9, because of the conical nose that follows the tip of the projectile, the shock wave leaves the tip at an angle with the main flow which exceeds the Mach angle. Appropriate corrections[5] can be applied, however, and the shock-wave angle from such a sharp-nosed object can serve as an accurate means of measuring supersonic velocities. For a blunt-nosed object (Fig. 13.10) the shock wave is a combination of partly normal and partly oblique shock waves.

EXERCISES

13.11.1 Assuming the tip of the model in Fig. 13.9 to be a point source of infinitesimal disturbance, find the air velocity if the temperature is $-60°$F and $k = 1.4$. If the actual Mach number is 1.38, what is the percentage error involved in the preceding assumption?

13.11.2 A schlieren photograph of a bullet shows a Mach angle of 40°. The air is at a pressure of 14 psia and 50°F. Find the approximate speed of the bullet.

13.11.3 A schlieren photograph of a bullet shows a Mach angle of 25°. The air is at a pressure of 9.5 kPa abs and 20°C. Find the approximate speed of the bullet.

13.12 ISOTHERMAL FLOW

Most of the flows discussed thus far in this chapter are characterized by substantial changes in temperature. In isentropic flow (pv^k = constant) through a nozzle, for example, large changes in temperature occur because the variation in cross-sectional flow area causes substantial changes in V, p, ρ, and T.

---

[5] Donald A. Gilbrech, *Fluid Mechanics*, p. 405, Wadsworth Publishing Co., Belmont, California, 1965.

Let us now examine some aspects of isothermal (T = constant) flow. An example application is gas flow through buried pipelines, where the ground serves as a heat sink to maintain pipeline temperatures constant. In this case $pv = RT$ = constant so that p/ρ = constant and $I_1 = I_2$. As a result $h_1 = h_2$ and Eq. (13.6) reduces to

$$\frac{V_1^2}{2} + q_H = \frac{V_2^2}{2} \tag{13.54}$$

where q_H is the heat interaction per unit mass.

This equation is applicable to both real and ideal compressible fluids. For the case of isothermal compressible flow of a real fluid in a pipe of uniform diameter (Sec. 13.13) there is a drop in pressure along the pipe because of fluid friction, and thus a decrease in density. To maintain continuity, the velocity must increase along the pipe and thus the kinetic energy also increases. Reference to Eq. (13.54) shows that, with $A_1 = A_2$, heat energy must be added to the fluid to maintain isothermal flow in a pipe of uniform diameter.

13.13 ISOTHERMAL FLOW IN A CONSTANT-AREA DUCT

To obtain some understanding of the characteristics of compressible flow with friction, let us consider isothermal flow in a pipe of uniform diameter. The flow of a gas through a long, uninsulated pipe in an isothermal environment may be assumed to approximate isothermal flow, since the temperature rise created by fluid friction is very small. If the energy Eq. (5.21) is rederived considering also the friction force due to the shear stress acting on an elemental length dx of pipe wall, with area $2\pi r\,dx$, the result is

$$\frac{dp}{\rho} + g\,dz + V\,dV + \frac{2\tau}{\rho r}\,dx = 0$$

Integrating between points 1 and 2 and comparing the result with Eq. (5.14) reveals that the new last term represents the head loss caused by fluid friction, and

$$\frac{2\tau}{\gamma r} = \frac{h_L}{L}$$

But from the Darcy–Weisbach equation (8.10) we also have

$$\frac{h_L}{L} = \frac{f}{D}\frac{V^2}{2g}$$

and so

$$\frac{dp}{\rho} + g\,dz + V\,dV + \frac{f\,dx}{D}\frac{V^2}{2} = 0$$

As we are here dealing with a gas, we may disregard variations in z, so that the energy equation becomes

$$\frac{dp}{\rho} + V\,dV + \frac{f\,dx}{D}\frac{V^2}{2} = 0 \tag{13.55}$$

Recalling from Eqs. (13.10) and (13.15) that the Mach number $\mathbf{M} = V/c$, where $c = \sqrt{kp/\rho} = \sqrt{kRT}$ = constant for isothermal flow, we can restate Eq. (13.55) as follows:

$$\frac{dp}{p} + k\mathbf{M}^2\left(\frac{d\mathbf{M}}{\mathbf{M}} + f\frac{dx}{2D}\right) = 0 \tag{13.56}$$

It has been observed that the f values given in Fig. 8.11 are applicable to compressible flow if $\mathbf{M} < 1$. For $\mathbf{M} > 1$, the values for f are about one-half of those of Fig. 8.11.

From continuity ρAV = constant, and hence ρV = constant, since we are dealing with a duct of uniform cross section. But from the perfect-gas law ($\rho = p/RT$), we get $pV = RT$ = constant for isothermal flow. Substituting this expression (pV = constant) into Eq. (13.56) and rearranging, we get for isothermal flow in a duct of constant cross section.

$$\frac{dp}{p} = -\frac{dV}{V} = -\frac{d\mathbf{M}}{\mathbf{M}} = -\frac{k\mathbf{M}^2}{1 - k\mathbf{M}^2} f\frac{dx}{2D} \tag{13.57}$$

This equation shows that, when $\mathbf{M} < 1/\sqrt{k}$, the pressure will decrease in the direction of flow. On the other hand, if $\mathbf{M} > 1/\sqrt{k}$, the pressure increases in the direction of flow.

The mass flow rate $\dot{m} = \rho AV$; hence the Reynolds number may be expressed as

$$\mathbf{R} = \frac{DV\rho}{\mu} = \frac{\dot{m}D}{\mu A} \tag{13.58}$$

At usual pressures the viscosity μ of a gas depends only on temperature and is constant for a given temperature; therefore for steady isothermal flow ($\dot{m}$ = constant and T = constant), the Reynolds number (and therefore also f) is constant along the entire length of a uniform-diameter pipe for any given flow. From the perfect-gas law, $\rho = p/RT$, and from continuity, $V = \dot{m}/\rho A$. Hence $V = \dot{m}RT/pA$. Substituting these values in Eq. (13.55) and rearranging, we obtain

$$-\left(\frac{2A^2}{\dot{m}^2RT}\right)p\,dp = \frac{f}{D}\,dx + \frac{2dV}{V}$$

For isothermal flow in a pipe of uniform diameter this equation is readily integrated for a length $L = x_2 - x_1$ to give

$$p_1^2 - p_2^2 = \frac{\dot{m}^2RT}{A^2}\left(f\frac{L}{D} + 2\ln\frac{p_1}{p_2}\right) \tag{13.59}$$

where p_1/p_2 has replaced V_2/V_1. That this substitution is valid can be seen by combining ρV = constant from continuity with $p/\rho = RT$ = constant for isothermal flow, or simply by integrating Eq. (13.57). Equation (13.59) may be used to find $\dot{m}$ if all other values are given, but if $\dot{m}$ and other values at (1) are given, p_2 may be found by successive approximation. In most cases the last term is small compared with fL/D and may be neglected as a first approximation. If it proves to be significant, a second solution, using an approximate value of p_1/p_2, may be made if greater accuracy is desired. It may be noted that p and A involve the *same area units,* and so, for numerical work in BG units it is usually more convenient to use p in pounds per square inch and A in square inches.

There is a restriction to Eq. (13.59) similar to that discussed in Sec. 13.8 and inferred in Eq. (13.57). To illustrate, disregarding the logarithmic term, Eq. (13.59) can be expressed as $p_2^2 = p_1^2 - NL$, where N is a constant for the given conditions and L is the distance along the pipe from section 1 to section 2. According to the equation as L gets larger, p_2 will get smaller until it eventually drops to zero which, of course, is physically impossible. Actually what happens is that as L gets larger, p gets smaller as does ρ; and since ρV = constant, V gets larger. However, there is a limit to how large V can get. This occurs when $\mathbf{M} = 1/\sqrt{k}$ as can be seen by examining Eq. (13.57). Thus Eq. (13.59) is applicable as long as $\mathbf{M} < 1/\sqrt{k}$. Another way of stating this is that for isothermal flow in a pipe of constant diameter there is a maximum length of pipe for which the given isothermal flow will proceed continuously. If the pipe exceeds this limiting length, there is a ***choking*** of the flow that limits the mass flow rate.

Another way of expressing Eq. (13.59) is to divide it by p_1^2, substitute $\rho_1 A_1 V_1$ for $\dot{m}$, and note that $\mathbf{M}_1/\mathbf{M}_2 = p_2/p_1$ for this situation. Upon rearrangement, we get

$$\frac{\mathbf{M}_1^2}{\mathbf{M}_2^2} = 1 - k\mathbf{M}_1^2\left(2 \ln \frac{\mathbf{M}_2}{\mathbf{M}_1} + f\frac{L}{D}\right) \tag{13.60}$$

Equation (13.60) is a particularly useful form of the isothermal-flow equation, since it can be handily employed to determine the maximum length of pipe such that the maximum possible velocity (limiting condition) will occur at the downstream end of the pipe. This is shown in the following example.

SAMPLE PROBLEM 13.7 Air flow isothermally at 65°F through a horizontal 10-in by 14-in rectangular duct at 3.0 slug/sec. If the pressure at a section is 80 psia, find the pressure at a second section 500 ft downstream from the first. Assume the duct surface is very smooth; hence the lowest curve of Fig. 8.11 may be used to determine f.

▪: Programmed computing aids (Appendix B) could help solve problems marked with this icon.

Solution

First determine the applicability of Eq. (13.59):

$$R_h = \frac{A}{P} = \frac{140}{48} = 2.92 \text{ in} = 0.243 \text{ ft}$$

Interpolating in Table A.2: Viscosity μ of air at 65°F is 3.78×10^{-7} lb·sec/ft^2.

$$\mathbf{R}_1 = \frac{DV\rho}{\mu} = \frac{\dot{m}D}{\mu A} = \frac{\dot{m}(4R_h)}{\mu A} = \frac{3.0(4 \times 0.243)}{3.78 \times 10^{-7}(10 \times 14)/144} = 7.93 \times 10^6$$

$$\mathbf{R}_2 = \mathbf{R}_1, \quad \text{since} \quad \rho_1 V_1 = \rho_2 V_2 \quad \text{and} \quad \mu_1 = \mu_2$$

From Fig. 8.11, $f = 0.0083$.

$$\rho_1 = \frac{p}{RT} = \frac{80 \times 144}{1715(460 + 65)} = 0.012\ 79 \text{ slug/ft}^3$$

$$V_1 = \frac{\dot{m}}{\rho_1 A} = \frac{3.0}{(0.012\ 79)\frac{140}{144}} = 242 \text{ ft/sec}$$

$$c = \sqrt{kRT} = \sqrt{1.4 \times 1715 \times 525} = 1123 \text{ ft/sec}$$

$$\mathbf{M}_1 = \frac{V_1}{c} = \frac{242}{1123} = 0.216$$

The limiting value of $\mathbf{M}_1$ is $1/\sqrt{1.4} = 0.845$. Substituting into Eq. (13.60),

$$\frac{(0.216)^2}{(0.845)^2} = 1 - 14(0.216)^2\left[2 \ln \frac{0.845}{0.216} + 0.0083 \frac{L}{4(0.243)}\right]$$

$$L = 1359 \text{ ft}$$

Thus Eq. (13.59) applies for all values of $L < 1359$ ft. Substituting $L = 500$ ft in Eq. (13.59) and neglecting the usually small logarithmic term,

$$(80 \times 144)^2 - p_2^2 = \frac{(3.0)^2(1715)(525)}{(140/144)^2}\left[0.0083 \times \frac{500}{4(0.243)}\right]$$

from which $p_2 = 68.1$ psia. Substituting this value of p_2 into Eq. (13.59) and considering the logarithmic term yields $p_2 = 67.1$ psia. ***ANS***

Repeating the process again will give a more accurate answer. Iteration can be avoided by using mathematics software or a programmable calculator with an equation (root) solving capability.

SAMPLE PROBLEM 13.8 For the case of Sample Prob. 13.7 with a duct length of 500 ft, compute the thermal energy (heat) that must be added to the fluid to maintain isothermal conditions.

Solution

Since the flow is isothermal, $p_1/\rho_1 = p_2/\rho_2 = RT = $ constant; $p_1 = 80$ psia and from Sample Prob. 9.7, $p_2 = 67.1$ psia. Thus $\rho_1/\rho_2 = 80/67.1 = 1.192$ and $V_2/V_1 = 1.192$, since $\rho V = $ constant from continuity.

So $V_2 = 1.192(242) = 289$ ft/sec; applying Eq. (13.54), $q_H = (289)^2/2 - (242)^2/2 = 12{,}380$ ft·lb/slug or 15.91 Btu/slug of air. Since $\dot{m} = 3.0$ slug/sec, the rate at which heat must be added to the fluid is $3.0(15.91) = 47.7$ Btu/s. ***ANS***

Note that if $q_H > 15.91$ Btu/slug of air, $T_2 > T_1$, and if $q_H < 15.91$ Btu/slug of air, $T_2 < T_1$.

EXERCISES

13.13.1 Air flows isothermally through a long horizontal pipe of uniform diameter. At a section where the pressure is 100 psia, the velocity is 120 ft/sec. Because of fluid friction the pressure at a distant point is 40 psia. (*a*) What is the increase in kinetic energy per slug of air? (*b*) What is the amount of thermal energy in Btu per slug of air that must be transferred in order to maintain the temperature constant? (*c*) Is this heat transferred to the air in the pipe or removed from it? (*d*) If the temperature of the air is 100°F and the diameter of the pipe is 3 in find the total heat transferred in Btu per hour.

13.13.2 Air flows isothermally through a long horizontal pipe of uniform diameter. At a section where the pressure is 700 kPa abs, the velocity is 35 m/s. Because of fluid friction the pressure at a distant point is 280 kPa abs. (*a*) What is the increase in kinetic energy per kg of air? (*b*) What is the amount of thermal energy in J/kg of air that must be transferred in order to maintain the temperature constant? (*c*) Is this heat transferred to the air in the pipe or removed from it? (*d*) If the temperature of the air is 40°C and the diameter of the pipe is 7.5 cm find the total heat transferred in joules per hour.

13.14 ADIABATIC FLOW IN A CONSTANT-AREA DUCT

The flow of fluids with friction in well-insulated pipes is a case that approaches adiabatic flow, i.e. $q_H = 0$. The usual applications of insulation are to pipes conveying steam or refrigerating fluids, such as ammonia vapor. In some situations such pipes are short, the pressure drops are relatively small, and the problem can be solved as if the fluid were incompressible. However, there are situations where the effects of compressibility must be considered.

Let us consider the case of steady adiabatic flow with friction in a duct of uniform cross section. Substituting $V_1 = \dot{m}/\rho_1 A$ and $V_2 = \dot{m}/\rho_2 A$ into Eq. (13.24), we get

$$\frac{\dot{m}^2}{\rho_2{}^2 A^2} - \frac{\dot{m}^2}{\rho_1{}^2 A^2} = \frac{2k}{k-1}(p_1 v_1 - p_2 v_2) \tag{13.61}$$

where $\dot{m}$ is the mass rate of flow and A is the cross-sectional area of the duct. This equation can be rewritten as

$$\frac{\dot{m}^2}{\rho_1{}^2A^2} + \frac{2k}{k-1}p_1v_1 = \frac{\dot{m}^2}{\rho_2{}^2A^2} + \frac{2k}{k-1}p_2v_2 = C \tag{13.62}$$

Where C is a constant evaluated from known conditions at section 1. Thus, in general,

$$pv = \frac{p}{\rho} = \frac{k-1}{2k}\left(C - \frac{\dot{m}^2}{\rho^2A^2}\right)$$

or

$$p = \frac{k-1}{2k}\left(C\rho - \frac{\dot{m}^2}{A^2\rho}\right) \tag{13.63}$$

Hence

$$dp = \frac{k-1}{2k}\left(C + \frac{\dot{m}^2}{A^2\rho^2}\right)d\rho$$

Multiplying both sides by ρ and integrating,

$$\int_1^2 \rho\, dp = \frac{k-1}{2k}\left[C\left(\frac{\rho_2{}^2 - \rho_1{}^2}{2}\right) + \frac{\dot{m}^2}{A^2}\ln\frac{\rho_2}{\rho_1}\right] \tag{13.64}$$

The value of $\int_1^2 \rho\, dp$ may be obtained from Eq. (13.64) inasmuch as ρ may be found for any flow $\dot{m}$ and any value of p by Eq. (13.63). Practically, it will be better to assume values of ρ and find the corresponding values of p by Eq. (13.62).

Next let us rewrite Eq. (13.55). Multiplying each term by ρ^2 so that the first term becomes $\rho\, dp$, and substituting $\rho V = \dot{m}/A$ and $\rho\, dV = -V\, d\rho$ (from the differential of ρAV = constant), we obtain

$$\rho\, dp - \left(\frac{\dot{m}}{A}\right)^2\frac{d\rho}{\rho} + \frac{f}{2D}\left(\frac{\dot{m}}{A}\right)^2 dx = 0$$

Integrating this (assuming f to be constant)[6] and rearranging with $x_2 - x_1 = L$,

$$f\frac{L}{D}\frac{1}{2}\left(\frac{\dot{m}}{A}\right)^2 = -\int_1^2 \rho\, dp - \left(\frac{\dot{m}}{A}\right)^2 \ln\frac{\rho_1}{\rho_2} \tag{13.65}$$

Since $\int_1^2 \rho\, dp$ may be evaluated by Eq. (13.64) it is possible to solve Eq. (13.65) to obtain a value of f if p_1 and p_2 have been measured for a known distance L, or the value of L may be found for any assumed values of p_1 and p_2 (or preferably ρ_1 and ρ_2) if f is given or assumed.

[6] In adiabatic flow in a duct of constant area the Reynolds number is not constant along the pipe because of changes in viscosity caused by variations in temperature. Thus, since **R** varies, f must vary. For most situations, however, f may be assumed to have a constant value without introducing much error.

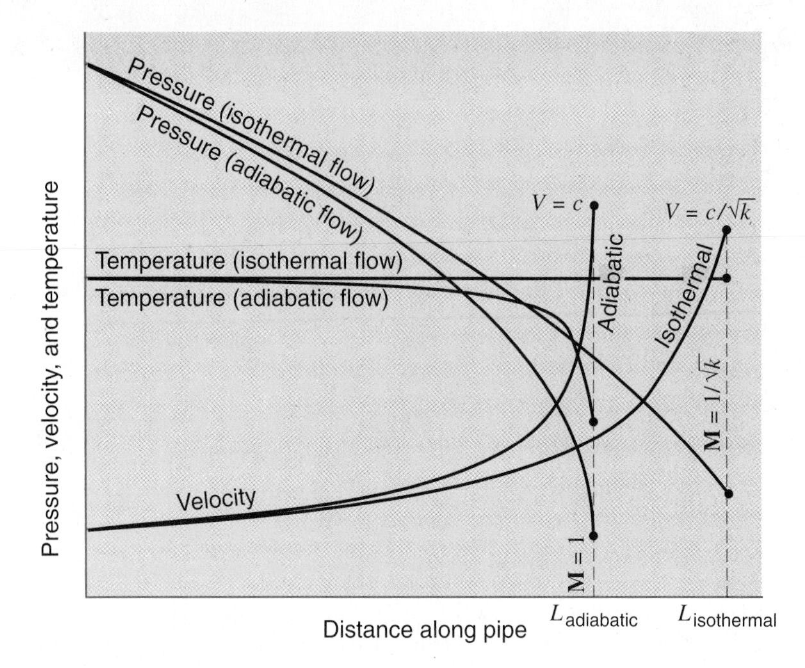

Figure 13.12
Subsonic flow of a compressible fluid in a constant-area duct.

Thus it is possible through successive calculations to plot a curve such as that in Fig. 13.12 for any assumed flow and initial conditions where p_2 represents any pressure along the pipe at any distance x_2. However, as in the case of isothermal flow, there is a minimum value of p_2 where the velocity has attained its maximum value. It can be proved that in adiabatic flow, critical conditions occur when $\mathbf{M} = 1.0$.

The equations in this section [Eqs. (13.61)–(13.65)] apply equally well to supersonic flow. However, the key characteristics of the flow are then reversed, as will be explained.

Adiabatic flow with friction is called ***Fanno flow***, and can be represented on an enthalpy–entropy (h-s) diagram as in Fig. 13.13. Enthalpy h and entropy s were discussed in Sec. 13.1.

In the subsonic region, the ***Fanno line*** approaches asymptotically the stagnation enthalpy line as the Mach number decreases. Thus the extreme left end of the Fanno line is nearly horizontal.

The point on the Fanno line that corresponds to maximum entropy represents conditions of sonic velocity and Mach 1. If a gas entering a duct is flowing at subsonic velocity, friction will have the effect of accelerating the flow so that sonic velocity is approached; likewise, if the flow at the entrance is supersonic, the gas will be decelerated, also approaching Mach 1. In each case, when Mach 1 is reached choking of the flow occurs.

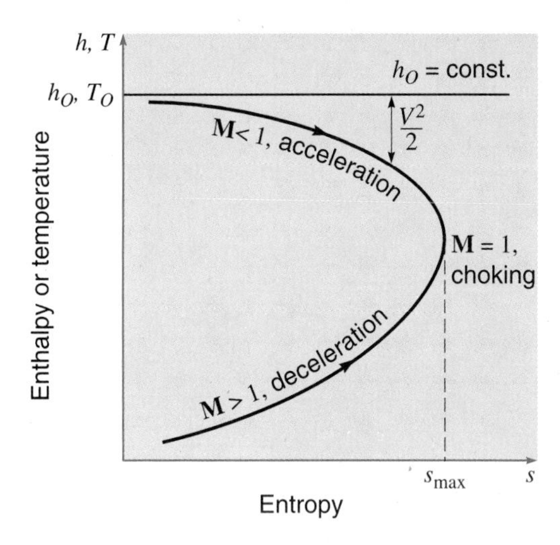

Figure 13.13
Fanno line on an h-s diagram.

All points on a Fanno line relate to the same stagnation temperature and to the same flow rate per unit area. If the cross-sectional area of a duct changes, or if the mass rate of flow changes in a duct, a different Fanno line then applies.

Flow of a fluid in the path described by a Fanno line always represents an entropy increase. In subsonic flow along a Fanno line there is a reduction in enthalpy, temperature, and density, and so an increase in velocity. Also, according to momentum principles, there must be a decrease in pressure. In supersonic flow along a Fanno line there is an increase in enthalpy, so that the velocity decreases, and therefore there is an increase in density and an increase in pressure. In both subsonic flow and supersonic flow, friction reduces the stagnation pressure. Table 13.3 summarizes the direction of changes in these properties. A fluid flowing at subsonic velocity tends to expand as a result of friction; on the other hand, a gas flowing supersonically tends to contract.

Table 13.3
Changes of flow properties in the downstream direction for adiabatic flow in a constant-area duct

Flow	$d\mathbf{M}$	dh	dT	$d\rho$	dV	dp	dp_O	ds
Subsonic ($\mathbf{M}<1$)	+	−	−	−	+	−	−	+
Supersonic ($\mathbf{M}>1$)	−	+	+	+	−	+	−	+

\+ indicates increase
− indicates decrease

Sample Problem 13.9 Air flows adiabatically through a 10-in by 14-in rectangular duct at 3.0 slug/sec. At a certain section the pressure is 80 psia and the temperature is 65°F (same data as Sample Prob. 13.7). Find the distance along the duct to the section where $\rho_2 = 0.80\rho_1$. Assume the duct surface is very smooth; hence the lowest curve of Fig. 8.11 may be used to determine f.

Solution

From Sample Prob. 13.7:

$$\rho_1 = 0.012\ 79\ \text{slug/ft}^3, \quad V_1 = 242\ \text{ft/sec}, \quad \mathbf{R}_1 = 7.93\times10^6, \quad f = 0.0083$$

From Eq. (13.62):

$$C = \frac{3.0^2}{(0.012\ 79\times140/144)^2} + \frac{2(1.4)}{0.4}\left(\frac{80\times144}{0.012\ 79}\right) = 6{,}360{,}000\ \text{ft}^2/\text{sec}^2$$

Equation (13.63) yields

$$p_2 = \frac{0.4}{2.8}\left(6{,}360{,}000\times0.8\times0.012\ 79 - \frac{3.0^2}{(140/144)^2(0.8\times0.012\ 79)}\right)$$
$$= 9170\ \text{lb/ft}^2\ \text{abs} = 63.7\ \text{psia}$$

From Eq. (13.64), $\int_1^2 \rho\, dp = -27.1\ \text{lb}^2\text{·sec}^2/\text{ft}^6$. Putting this value in Eq. (13.65) and substituting for f and (from Sec. 8.3) for $D = 4A/P = 0.972$ ft yields

$$L = 614\ \text{ft} \qquad \textit{ANS}$$

Thus at a section 614 ft downstream from the first section, $\rho_2 = 0.8\rho_1 = 0.010\ 24\ \text{slug/ft}^3$ and $p_2 = 63.7$ psia. Also,

$$V_2 = \frac{\dot{m}}{\rho_2 A} = \frac{3.0}{0.01024\times140/144} = 301\ \text{ft/sec}$$

$$T_2 = \frac{p_2}{\rho_2 R} = \frac{9170}{0.010\ 24\times1715} = 522°\text{R} = 62°\text{F}$$

By assuming other values of ρ_2, one can get a complete picture of the flow at various sections along the pipe.

13.15 COMPARISON OF FLOW TYPES

The two types of compressible flow in pipelines that have just been discussed, isothermal and adiabatic flow, are special cases, which are often approximated in practice and which are amenable to mathematical treatment. For fluids that do not follow the perfect-gas laws, such as wet steam, the preceding equations are only rough approximations. A more complicated thermodynamic treatment is necessary than is within the scope of this text.

In Fig. 13.12 curves are plotted to scale for the flow of air through a constant-area duct for both isothermal and adiabatic conditions, assuming the same initial values for each. Inspection of this diagram shows that for small pressure drops from the initial, such as up to $\Delta p/p_1 = 0.10$ (or $p_2/p_1 = 0.90$), there is very little difference between the two curves. Thus, for such a situation where $p_2/p_1 > 0.90$, adiabatic flow in a pipe can be analyzed as isothermal flow without introducing much error. The flow of gas in a pipe is rarely either isothermal or adiabatic. Isothermal flow requires that heat be transferred into the flowing fluid from the surrounding atmosphere at just the right rate, and this rate must increase along the length of the pipe. If the rate of heat transfer is less than this required amount, the performance curves will lie between the isothermal and the adiabatic curves in Fig. 13.12. Heat transfer is proportional to the temperature difference between the fluid and the surrounding atmosphere. If these temperatures are denoted as T_f and T_a, respectively, the heat transfer is some function of $T_a - T_f$. Isothermal flow is possible only if T_a is greater than T_f.

If the fluid in the pipe is very much colder than the surroundings, its absorption of heat might be such as to cause the pressure in the pipe to be higher than for isothermal flow. On the other hand, for example, air at a high temperature might be discharged directly from a compressor into a pipe so that T_f might be greater than T_a, which would cause heat to flow from the fluid in the pipe to the surrounding atmosphere. In this case the pressure along the pipe would decrease even faster than for adiabatic flow.

Because the energy equation for compressible fluids does not contain a term for friction and the momentum equation which contains a term for friction[7] does not include any term for heat transfer, it is seen that there is no simple analytical solution of such cases. An approximate approach to such problems is to divide the entire length of pipe into short reaches and employ the equations of incompressible flow using average values of density and velocity within each reach. This step-by-step method will give results that are approximately correct if the lengths of the reaches are made small enough. Small reach lengths are particularly important in the regions where the curves of Fig. 13.12 are sharply curved.

13.16 CONCLUDING REMARKS

In this chapter we have seen that in order to solve problems of compressible flow, equations describing the flow process of the gas must be combined with

[7] Friction is accounted for as one of the forces acting on the fluid element in the equation $\sum F = \rho Q \, \Delta V$.

the energy equation and the continuity principle. Hence thermodynamics is commonly involved in compressible-flow problems and the expressions relating the various physical parameters are generally quite complicated. In the discussion we have restricted ourselves to one-dimensional flow, and have not considered multidimensional flow, which is of course an extremely important topic, especially when dealing with aircraft and missiles. The intent here was merely to provide an introduction to the flow of compressible fluids. Among the references cited at the end of the book will be found some excellent treatments of compressible flow.

PROBLEMS

13.1 Using the data of Sample Prob. 13.2, determine the work done in compressing the air by finding the area under a pressure-vs-volume curve. Compute and tabulate volumes and pressures using volume increments that are 10% of the original volume.

13.2 Find the stagnation pressure and temperature in air flowing at 88 ft/sec if the static pressure and static temperature are 14.7 psia and 50°F respectively.

13.3 Derive Eq. (13.35) for isentropic flow by integrating the Euler equation.

13.4 Carbon dioxide flows isentropically. At a point in the flow the velocity is 50 ft/sec, the temperature is 125°F, and the pressure is 20 psia. Determine the pressure and temperature on the nose of a streamlined object placed in the flow at that point.

13.5 Air flows at 150°F from a large tank through a 1.5-in-diameter converging nozzle. Within the tank the pressure is 85 psia. Calculate the flow rate for back pressures of 10, 30, 50, and 70 psia. Assume isentropic conditions. Plot $\dot{m}$ as a function of p_b. Assume that the temperature within the tank is 150°F in all cases. Compute also the temperature at the nozzle outlet for each condition.

13.6 Air flows at 65°C from a large tank through a 4-cm-diameter converging nozzle. Within the tank the pressure is 600 kPa abs. Calculate the flow rate for back pressures of 50, 200, 350, and 500 kPa abs. Assume isentropic conditions. Plot $\dot{m}$ as a function of p_b. Assume that the temperature within the tank is 65°C in all cases. Compute also the temperature at the nozzle outlet for each condition.

13.7 Air flows at 25°C from a large tank through a 10-cm-diameter converging nozzle. Within the tank the pressure is 50 kPa abs. Calculate the flow rate for pressures of 30, 20, and 10 kPa abs. Assume isentropic conditions. Plot $\dot{m}$ as a function of p_b. Assume that the temperature within the tank is 25°C in all cases. Compute also the temperature at the nozzle outlet for each condition.

13.8 Air within a tank at 120°F flows isentropically through a 2-in-diameter convergent nozzle into a 14.2-psia atmosphere. Find the flow rate for air pressures within the tank of 5, 10, 20, 40, and 50 psia.

13.9 Refer to Sample Prob. 13.4. If the pressure in the tank is 5 psig, confirm by calculator that $\dot{m} = 0.0374$ slug/sec, $p_2 = 13.50$ psia, and $T_2 = 494°R$.

13.10 Air discharges from a large tank through a converging–diverging nozzle. The throat diameter is 3.0 in, and the exit diameter is 4.0 in. Within the tank the air pressure and temperature are 40 psia and 150°F, respectively. Calculate the flow rate for back pressures of 39, 38, 36, and 30 psia. Assume no friction.

13.11 Air discharges from a large tank through a converging–diverging nozzle. The throat diameter is 7.5 cm, and the exit diameter is 10 cm. Within the tank the air pressure and temperature are 290 kPa abs and 65°C, respectively. Calculate the flow rate for back pressures of 280, 270, 250, and 200 kPa abs. Assume no friction.

13.12 Repeat Exer. 13.9.5 for the case where the air within the tank is at 20 psia. Assume all other data to be the same.

13.13 Repeat Exer. 13.9.6 for the case where the air within the tank is at 140 kPa abs. Assume all other data to be the same.

13.14 Air discharges from a large tank through a converging–diverging nozzle with a 2.5-cm-diameter throat into the atmosphere. The gage pressure and temperature in the tank are 700 kPa and 40°C, respectively, the barometric pressure is 995 millibars. (*a*) Find the nozzle-tip diameter required for p_3 to be equal to the atmospheric pressure. For this case, what are the flow velocity, sonic velocity, and Mach number at the nozzle exit? (*b*) Determine the value of p_b that will cause the shock wave to be located at the nozzle exit.

13.15 The pressure, velocity, and temperature just upstream of a normal shock wave in air are 10 psia, 2200 fps and 23°F. Determine the pressure, velocity, and temperature just downstream of the wave.

13.16 Air flows isothermally in a long pipe. At one section the pressure is 90 psia, the temperature is 80°F, and the velocity is 100 fps. At a second section some distance from the first the pressure is 15 psia. Find the energy head loss due to friction, and determine the thermal energy that must have been added to or taken from the fluid between the two sections. The diameter of the pipe is constant.

13.17 Air flows isothermally in a long pipe. At one section the pressure is 600 kPa abs, the temperature is 25°C, and the velocity is 30 m/s. At a second section some distance from the first the pressure is 100 kPa abs. Find the energy head loss due to friction, and determine the thermal energy that must have been added to or taken from the fluid between the two sections. The diameter of the pipe is constant.

: Programmed computing aids (Appendix B) could help solve problems marked with this icon.

13.18 Refer to Sample Prob. 13.7. Neglecting the logarithm term in Eq. (13.59), find the pressures and velocities at sections 100, 300, and 800 ft downstream of the section where the pressure is 80 psia. Plot the pressure and velocity as a function of distance along the pipe.

13.19 Carbon dioxide flows isothermally at 100°F through a horizontal 6-in-diameter pipe. At this temperature $\mu = 4.0 \times 10^{-7}$ lb·sec/ft^2. The pressure changes from 150.0 to 140.0 psig in a 100-ft length of pipe. Determine the mass flow rate if the atmospheric pressure is 14.5 psia and e for the pipe is 0.002 ft.

13.20 Carbon dioxide flows isothermally at 40°C through a horizontal 15-cm-diameter pipe. At this temperature $\mu = 1.95 \times 10^{-5}$ N·s/m^2. The pressure changes from 1000 to 930 kPa gage in a 30-m length of pipe. Determine the mass flow rate if the atmospheric pressure is 100 kPa and e for the pipe is 0.60 mm.

13.21 Methane gas is to be pumped through a 24-in-diameter welded-steel pipe connecting two compressor stations 25 mi apart. At the upstream station the pressure is not to exceed 60 psia, and at the downstream station it is to be at least 20 psia. Determine the maximum possible rate of flow (in cubic feet per day at 60°F and 1 atm). Assume isothermal flow at 60°F.

13.22 Refer to Sample Prob. 13.9. Find the distance along the pipe to (*a*) where $\rho_2 = 0.9\rho_1$; (*b*) where $\rho_2 = 0.7\rho_1$; (*c*) where subsonic adiabatic flow ends. Compute the corresponding values of p, V, T, and $\mathbf{M}$ and plot the first three as a function of distance along the pipe.

13.23 Air flows adiabatically at 3.0 slug/sec in a 12-in-diameter horizontal pipe. At a certain section the pressure is 150 psia and the temperature is 140°F. Determine the distance along the pipe to the section where $\rho_2 = 0.80\rho_1$. Assume $e/D = 0.0004$.

13.24 Air flows adiabatically at 50 kg/s in a 30-cm-diameter horizontal pipe. At a certain section the pressure is 1000 kPa abs and the temperature 60°C. Determine the distance along the pipe to the section where $\rho_2 = 0.80\rho_1$. Assume $e/D = 0.0004$.

■: Programmed computing aids (Appendix B) could help solve problems marked with this icon.

CHAPTER 14

Ideal Flow Mathematics

In this chapter we discuss various mathematical methods for describing the flow of imaginary ideal (frictionless) fluids. This subject is often referred to as ***hydrodynamics***. It is a vast subject, so that the presentation here provides only an introduction, but it does give a good idea of the possibilities of a rigorous mathematical approach to flow problems.

Even though such an approach does not consider all the real properties of fluids, the results often closely approximate the behavior of real fluids. This is because there are numerous situations in which friction plays only a minor role. For example, in Chaps. 8 and 9, we noted, for fluids of low viscosity, that the viscosity affects only a thin region at the fluid boundaries. Also, in Chaps. 8–10 we often saw that turbulence and separation of the boundary layer occur far more readily with decelerating flows, and that accelerating flows generally have thin boundary layers. For such flows, mathematical analysis of ideal fluids yields results, often elegant, that can and do provide many useful and important insights into real fluid behavior.

To concentrate on fundamentals, after Sec. 14.1 we shall limit our discussions to incompressible fluids and to two-dimensional, steady flow fields. In Sec. 14.8 we shall see, rather interestingly, how the same methods can be applied to the flow of a *real* fluid through porous media.

14.1 DIFFERENTIAL EQUATION OF CONTINUITY

In Chap. 4 a very practical, but special, form of the equation of continuity was presented. For some purposes a more general three-dimensional form is desired. Also, in that chapter the concept of the flow net was explained

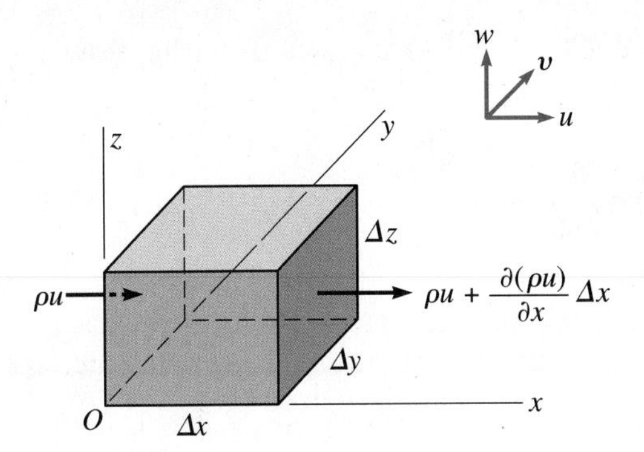

Figure 14.1

largely on an intuitive basis. To reach a more fundamental understanding of the mechanics of the flow net, it is necessary to consider the differential equations of continuity and irrotationality (Sec. 14.2) that give rise to the orthogonal network of streamlines and equipotential lines.

Aside from application to the flow net, the differential form of the continuity equation has an important advantage over the one-dimensional form that was derived in Sec. 4.7 in that it is perfectly general for two- or three-dimensional fluid space and for either steady or unsteady flow. Some of the equations in this section only will also be applicable to compressible flow.

Figure 14.1 shows three coordinate axes x, y, z mutually perpendicular and fixed in space. Let the velocity components in these three directions be u, v, w, respectively. Consider now a small parallelepiped, having sides Δx, Δy, Δz. In the x direction the rate of mass flow into this box through the left-hand face is approximately $\rho u\ \Delta y\ \Delta z$, this expression becoming exact in the limit as the box is shrunk to a point. The corresponding rate of mass flow out of the box through the right-hand face is $\{\rho u + [\partial(\rho u)/\partial x]\ \Delta x\}\ \Delta y\ \Delta z$. Thus the net rate of mass flow into the box in the x direction is $-[\partial(\rho u)/\partial x]\ \Delta x\ \Delta y\ \Delta z$. Similar expressions may be obtained for the y and z directions. The sum of the rates of mass inflow in the three directions must equal the time rate of change of the mass in the box, or $(\partial\rho/\partial t)\ \Delta x\ \Delta y\ \Delta z$. Summing up, applying the limiting process, and dividing both sides of the equation by the volume of the parallelepiped, which is common to all terms, we get

Unsteady compressible flow:
$$-\frac{\partial(\rho u)}{\partial x}-\frac{\partial(\rho v)}{\partial y}-\frac{\partial(\rho w)}{\partial z}=\frac{\partial\rho}{\partial t} \tag{14.1}$$

which is the equation of continuity in its most general form. This equation as well as the other equations in this section are, of course, valid regardless of whether the fluid is a real one or an ideal one. If the flow is steady, ρ does not vary with time, but it may vary in space. Since $\partial(\rho u)/\partial x = \rho(\partial u/\partial x) + u(\partial\rho/\partial x)$, it follows that for steady flow the equation may be written as

Steady compressible flow:
$$u\frac{\partial\rho}{\partial x}+v\frac{\partial\rho}{\partial y}+w\frac{\partial\rho}{\partial z}+\rho\left(\frac{\partial u}{\partial x}+\frac{\partial v}{\partial y}+\frac{\partial w}{\partial z}\right)=0 \tag{14.2}$$

In the case of an incompressible fluid (ρ = constant), whether the flow is steady or not, the equation of continuity becomes

Steady incompressible flow:
$$\frac{\partial u}{\partial x}+\frac{\partial v}{\partial y}+\frac{\partial w}{\partial z}=0 \qquad (14.3)$$

For two-dimensional flow, application of the same procedure to an elemental volume in polar coordinates yields for steady flow the following equations:

Steady compressible flow:
$$\frac{1}{r}(\rho v_r)+\frac{\partial}{\partial r}(\rho v_r)+\frac{\partial}{r\,\partial\theta}(\rho v_t)=0 \qquad (14.4)$$

Steady, incompressible flow:
$$\frac{v_r}{r}+\frac{\partial v_r}{\partial r}+\frac{\partial v_t}{r\,\partial\theta}=0 \qquad (14.5)$$

where v_r and v_t represent the velocities in the radial and tangential[1] directions, respectively.

SAMPLE PROBLEM 14.1 Assuming ρ to be constant, do the following flows satisfy continuity? (*a*) $u = -2y$, $v = 3x$; (*b*) $u = 0$, $v = 3xy$; (*c*) $u = 2x$, $v = -2y$.

Solution

From Eq. (14.3): Continuity for incompressible fluids is satisfied if $\partial u/\partial x + \partial x/\partial y = 0$.

(*a*) $\frac{\partial(-2y)}{\partial x}+\frac{\partial(3x)}{\partial y}=0+0=0$ Continuity is satisfied ***ANS***

(*b*) $\frac{\partial(0)}{\partial x}+\frac{\partial(3xy)}{\partial y}=0+3x\neq 0$ Continuity is not satisfied ***ANS***

(*c*) $\frac{\partial(2x)}{\partial x}+\frac{\partial(-2y)}{\partial y}=2-2=0$ Continuity is satisfied ***ANS***

Note: If (*b*) did indeed describe a flow field, the fluid must be compressible.

EXERCISES

14.1.1 Which of the following incompressible flows satisfy continuity?
(*a*) $u = 2$
(*b*) $u = 2$, $v = 3$

[1] In this chapter v_t is used to represent the tangential component of velocity.

(*c*) $u = 2 + 3x, \quad v = 4$
(*d*) $u = 2y, \quad v = 3x$
(*e*) $u = 2y, \quad v = -3x$
(*f*) $u = 3xy, \quad v = 1.5x^2$

14.1.2 Which of the following incompressible flows satisfy continuity?
(*a*) $u = 3y, \quad v = 0$
(*b*) $u = 3x, \quad v = 3y$
(*c*) $u = 3x, \quad v = -3y$
(*d*) $u = 4 + 2x, \quad v = -6 - 2y$
(*e*) $u = 4xy + x^2, \quad v = -2xy - 2y^2$
(*f*) $u = -2xy + 2x^2, \quad v = 4xy - y^2$
(*g*) $u = -2xy - 2x^2 + 2y^2, \quad v = 4xy - x^2 + y^2$

14.2 ROTATIONAL AND IRROTATIONAL FLOW

The discussion in the remainder of this chapter is restricted to incompressible fluids. ***Irrotational flow*** may be briefly described as flow in which each element of the moving fluid suffers no *net* rotation from one instant to the next, with respect to a given frame of reference. The classic example of irrotational motion (although not a fluid) is that of the carriages on a Ferris wheel used for amusement rides. Each carriage describes a circular path as the wheel revolves, but does not rotate with respect to the earth. In irrotational flow, however, a fluid element may deform as shown in Fig. 14.2*a*, where the axes of the element rotate equally toward or away from each other. As long as the algebraic average rotation is zero, the motion is irrotational.

In Fig. 14.2*b* is depicted an example of rotational flow. In this case there is a net rotation of the fluid element. Actually, the deformation of the element in Fig. 14.2*b* is less than that of Fig. 14.2*a*.

Let us now express the condition of irrotationality in mathematical terms. It will help to restrict the discussion at first to two-dimensional motion in the

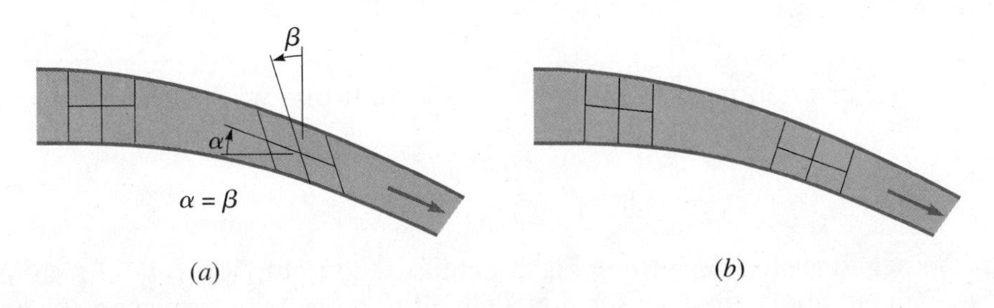

Figure 14.2
Two-dimensional flow along a curved path.
(*a*) Irrotational flow. (*b*) Rotational flow.

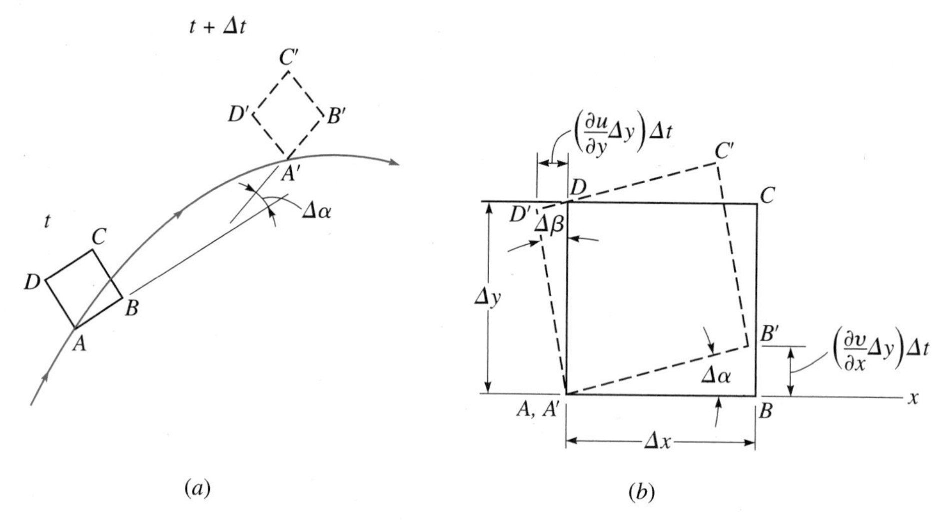

Figure 14.3

xy plane. Consider a small fluid element moving as depicted in Fig. 14.3*a*. During a short time interval Δt, the element moves from one position to another and in the process it deforms as indicated. Superimposing A' on A, defining an x axis along AB, and enlarging the diagram, we get Fig. 14.3*b*. The angle $\Delta\alpha$ between AB and $A'B'$ can be expressed from Fig. 14.3*b* as

$$\Delta\alpha = \frac{BB'}{\Delta x} = \frac{[(\partial v/\partial x)\,\Delta x]\,\Delta t}{\Delta x} = \frac{\partial v}{\partial x}\,\Delta t$$

Hence the rate of rotation of the edge of the element that was originally aligned with AB is

$$\omega_\alpha = \frac{\Delta\alpha}{\Delta t} = \frac{\partial v}{\partial x}$$

Likewise
$$\Delta\beta = \frac{DD'}{\Delta y} = \frac{[-(\partial u/\partial y)\,\Delta y]\,\Delta t}{\Delta y} = -\frac{\partial u}{\partial y}\,\Delta t$$

and the rate of rotation of the edge of the element that was originally aligned with AD is

$$\omega_\beta = \frac{\Delta\beta}{\Delta t} = -\frac{\partial u}{\partial y}$$

with the negative sign because $+u$ is directed to the right. The rate of rotation of the element about the z axis is now defined to be ω_z, the average of ω_α and ω_β; thus

$$\omega_z = \frac{1}{2}\left(\frac{\partial v}{\partial x} - \frac{\partial u}{\partial y}\right) \tag{14.6}$$

But the criterion we originally stipulated for irrotational flow was that the rate of rotation be zero. Therefore we have

Irrotational flow in xy plane:

$$\frac{\partial v}{\partial x} - \frac{\partial u}{\partial y} = 0 \tag{14.7}$$

In three-dimensional flow there are corresponding expressions for the components of angular-deformation rates about the x and y axes. Finally, for the general case, irrotational flow is defined to be that for which

$$\omega_x = \omega_y = \omega_z = 0 \tag{14.8}$$

In Sec. 14.6 we shall see that the primary significance of irrotational flow is that it is defined by a velocity potential.

Sample Problem 14.2 Determine whether the following flows are rotational or irrotational: (*a*) $u = -2y$, $v = 3x$; (*b*) $u = 0$, $v = 3xy$; (*c*) $u = 2x$, $v = -2y$.

Solution

Using Eq. (14.7):

(*a*) $\frac{\partial(3x)}{\partial x} - \frac{\partial(-2y)}{\partial y} = 3 + 2 \neq 0$ Flow is rotational ***ANS***

(*b*) $\frac{\partial(3xy)}{\partial x} - \frac{\partial(0)}{\partial y} = 3y - 0 \neq 0$ Flow is rotational ***ANS***

(*c*) $\frac{\partial(-2y)}{\partial x} - \frac{\partial(2x)}{\partial y} = 0 - 0 = 0$ Flow is irrotational ***ANS***

EXERCISES

14.2.1 Which of the flows of Exer. 14.1.1 are irrotational?

14.2.2 Which of the flows of Exer. 14.1.2 are irrotational?

14.3 CIRCULATION AND VORTICITY

To get a better understanding of the character of a flow field, we should acquaint ourselves with the concept of ***circulation***.[2] Let the streamlines of

[2] In Secs. 9.10 and 9.11 the concept of circulation is utilized to develop an expression for lift force on an air foil.

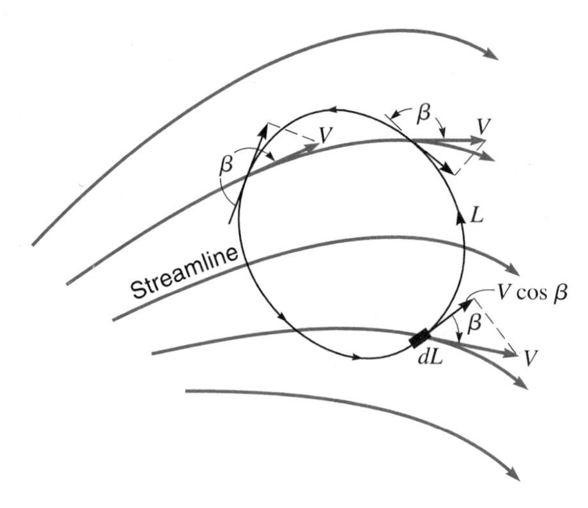

Figure 14.4
Circulation around a closed path in a two-dimensional field.

Fig. 14.4 represent a two-dimensional ***flow field***, while L represents any closed path in this field. The circulation Γ (gamma) is defined mathematically as a ***line integral*** of the velocity about a closed path. Thus

$$\Gamma = \oint_L \mathbf{V} \cdot d\mathbf{L} = \oint_L V \cos \beta \, dL \tag{14.9}$$

where $\mathbf{V}$ is the velocity in the flow field at the element $d\mathbf{L}$ of the path, and β is the angle between $\mathbf{V}$ and the tangent to the path (in the positive direction along the path) at that point. Equation (14.9) is analogous to the common equation in mechanics for work done as a body moves along a curved path while the force makes some angle with the path. The only difference here is the substitution of a velocity for a force.

Evaluation of Eq. (14.9) about any closed curve generally involves a tedious step-by-step integration. Some valuable information is acquired, however, by evaluating the circulation of the two-dimensional flow field of Fig. 14.5 by taking the line integral around the boundary of the indicated element. Since the element is of differential size, the resulting circulation is also differential. Thus, starting at A and proceeding counterclockwise,

$$d\Gamma = \frac{u_A + u_B}{2} dx + \frac{v_B + v_C}{2} dy - \frac{u_C + u_D}{2} dx - \frac{v_D + v_A}{2} dy \tag{14.10}$$

where the values of u_A, u_B, u_C, u_D and v_A, v_B, v_C, v_D are as indicated in Fig. 14.5. Substituting these values into Eq. (14.10), expanding, combining terms, and disregarding those of higher order yields

$$d\Gamma = \left(\frac{\partial v}{\partial x} - \frac{\partial u}{\partial y}\right) dx \, dy \tag{14.11}$$

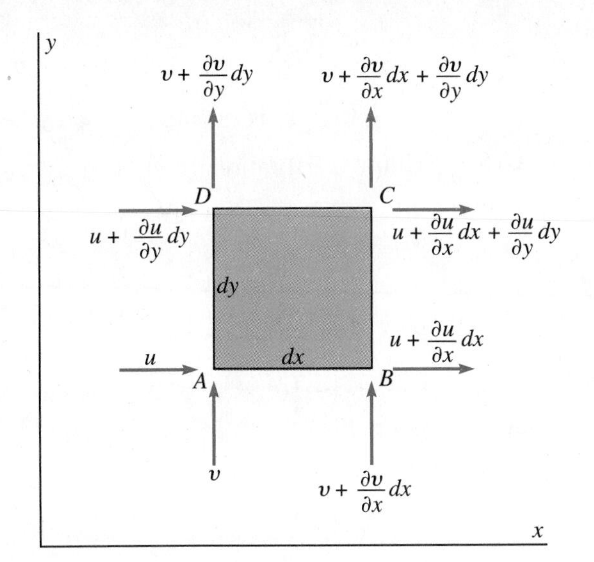

Figure 14.5

The ***vorticity*** ξ (xi) is defined as the circulation per unit of enclosed area. Thus

$$\xi = \frac{d\Gamma}{dx\,dy} = \frac{\partial v}{\partial x} - \frac{\partial u}{\partial y} \tag{14.12}$$

Comparing Eq. (14.12) with Eq. (14.7), we see that an irrotational flow is one for which the vorticity $\xi = 0$. Similarly, if the flow is rotational, $\xi \neq 0$.

Using a similar procedure for polar coordinates, we find

$$\xi = \frac{\partial v_t}{\partial r} + \frac{v_t}{r} - \frac{1}{r}\frac{\partial v_r}{\partial \theta} \tag{14.13}$$

Sample Problem 14.3 Check the following incompressible flows for continuity and determine the vorticity of each: (*a*) $v_t = 6r$, $v_r = 0$; (*b*) $v_t = 0$, $v_r = -5/r$.

Solution

Applying Eqs. (14.5) and (14.13):

(*a*) $\frac{0}{r} + \frac{\partial(0)}{\partial r} + \frac{1}{r}\frac{\partial(6r)}{\partial \theta} = 0$ Continuity is satisfied ***ANS***

$\xi = \frac{\partial(6r)}{\partial r} + \frac{6r}{r} - \frac{1}{r}\frac{\partial(0)}{\partial \theta} = 6 + 6 - 0 = 12$ ***ANS*** (Flow is rotational) (*b*) $-\frac{5/r}{r}+\frac{\partial(-5r^{-1})}{\partial r}+\frac{1}{r}\frac{\partial(0)}{\partial\theta} = -\frac{5}{r^2}+\frac{5}{r^2}+0 = 0$

Continuity is satisfied **ANS**

$$\xi = \frac{\partial(0)}{\partial r}+\frac{0}{r}-\frac{1}{r}\frac{\partial(-5/r)}{\partial\theta} = 0$$ **ANS** (Flow is irrotational)

EXERCISES

14.3.1 Find the vorticity of each of the flows in Exer. 14.1.1.

14.3.2 Find the vorticity of each of the flows in Exer. 14.1.2

14.4 THE STREAM FUNCTION

The ***stream function*** ψ (psi), based on the continuity principle, is a mathematical expression that describes a flow field. In Fig. 14.6 are shown two adjacent streamlines of a two-dimensional flow field. Let $\psi(x, y)$ represent the streamline nearest the origin. Then $\psi + d\psi$ is representative of the second streamline. Since there is no flow across a streamline, we can let ψ be indicative of the flow carried through the area from the origin O to the first streamline. And thus $d\psi$ represents the flow carried between the two streamlines of Fig. 14.6. From continuity, referring to the triangular fluid element of Fig. 14.6, we see that for an incompressible fluid

$$d\psi = -v\,dx + u\,dy \tag{14.14}$$

The total derivative $d\psi$ may also be expressed as

$$d\psi = \frac{\partial\psi}{\partial x}dx + \frac{\partial\psi}{\partial y}dy \tag{14.15}$$

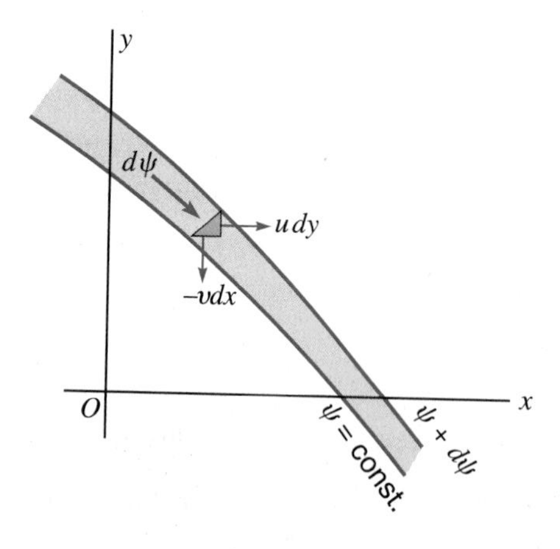

Figure 14.6
Stream function.

Comparing these last two equations, we note that

$$u = \frac{\partial \psi}{\partial y} \quad \text{and} \quad v = -\frac{\partial \psi}{\partial x} \tag{14.16}$$

Thus, if ψ can be expressed as a function of x and y, we can find the velocity components (u and v) at any point of a two-dimensional flow field by application of Eqs. (14.16). Conversely, if u and v are expressed as functions of x and y, we can find ψ by integrating Eq. (14.14). However, it should be noted that since the derivation of ψ is based on the principle of continuity, *it is necessary that continuity be satisfied for the stream function to exist.* Also, since vorticity was not considered in the derivation of ψ, *the flow need not be irrotational for the stream function to exist.*

The equation of continuity

$$\frac{\partial u}{\partial x} + \frac{\partial v}{\partial y} = 0$$

may be expressed in terms of ψ by substituting the expressions for u and v from Eqs. (14.16); doing so, we get

$$\frac{\partial}{\partial x}\left(\frac{\partial \psi}{\partial y}\right) - \frac{\partial}{\partial y}\left(\frac{\partial \psi}{\partial x}\right) = 0, \quad \text{or} \quad \frac{\partial^2 \psi}{\partial x\, \partial y} = \frac{\partial^2 \psi}{\partial y\, \partial x}$$

which shows that, if $\psi = \psi(x, y)$, the derivatives taken in either order give the same result and that a flow described by a stream function automatically satisfies the continuity equation.

EXERCISES

14.4.1 The flow of an incompressible fluid is defined by $u = 2$, $v = 8x$. Does a stream function exist for this flow? If so, determine the expression for the stream function.

14.4.2 For each of the flows of Exer. 14.1.1, write an expression for the stream function if one exists.

14.4.3 For each of the flows of Exer. 14.1.2, write an expression for the stream function if one exists.

14.5 BASIC FLOW FIELDS

In this section we shall discuss several basic flow fields that are commonly encountered. Though these flow fields imply an ideal fluid, they closely depict the flow of a real fluid outside the zone of viscous influence provided there is

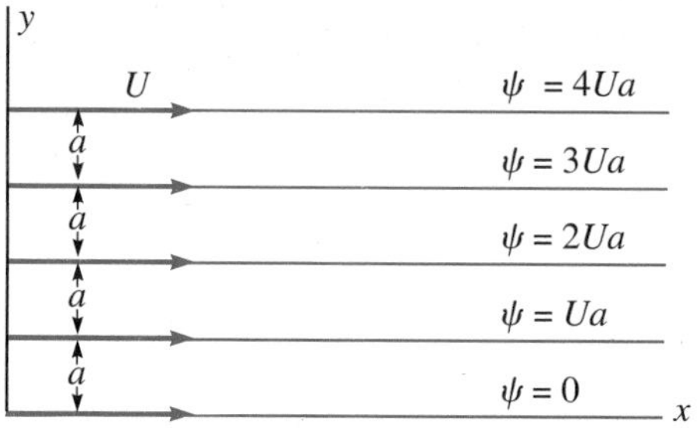

Figure 14.7
Rectilinear flow field.

no separation of the flow from the boundaries (see Sec. 4.10). The simplest of all flows is that in which the streamlines are straight, parallel, and evenly spaced as indicated in Fig. 14.7. In this case $v = 0$ and $u =$ constant. Thus, from Eq. (14.14), $d\psi = u\,dy$, and hence $\psi = Uy$, where U is the velocity of flow. If the distance between streamlines is a, the values of ψ for the streamlines are as indicated in Fig. 14.7.

Another flow field of general interest is that of a ***source*** or a ***sink***. In the case of a source, the flow field consists of radial streamlines symmetrically spaced as shown in Fig. 14.8. If q is the ***source strength***, or rate of flow from the source, it is at once apparent that $\psi = q\theta/2\pi$. Customarily, for this case, the $\psi = 0$ streamline is defined as that coincident with the direction of the x axis. From inspection of the flow field it is obvious that $v_t = 0$ and $v_r = q/2\pi r$. Thus $v_r \to 0$ as $r \to \infty$. For a sink (inward flow), the stream function is expressible as $\psi = -q\theta/2\pi$.

Flow fields may be combined by superposition to give other fields of importance. For example, let us combine a source and sink of equal strength with a rectilinear flow. Let $2a$ be the distance between the source and sink. Referring to Fig. 14.9 and defining θ_1 and θ_2 as shown, we can write for the combined field

$$\psi = Uy + \frac{q\theta_1}{2\pi} - \frac{q\theta_2}{2\pi} \tag{14.17}$$

Transforming the last two terms of this equation to cartesian coordinates by

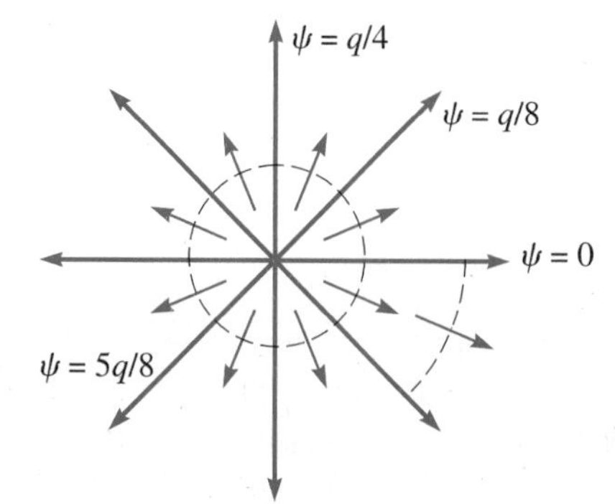

Figure 14.8
Source flow field.

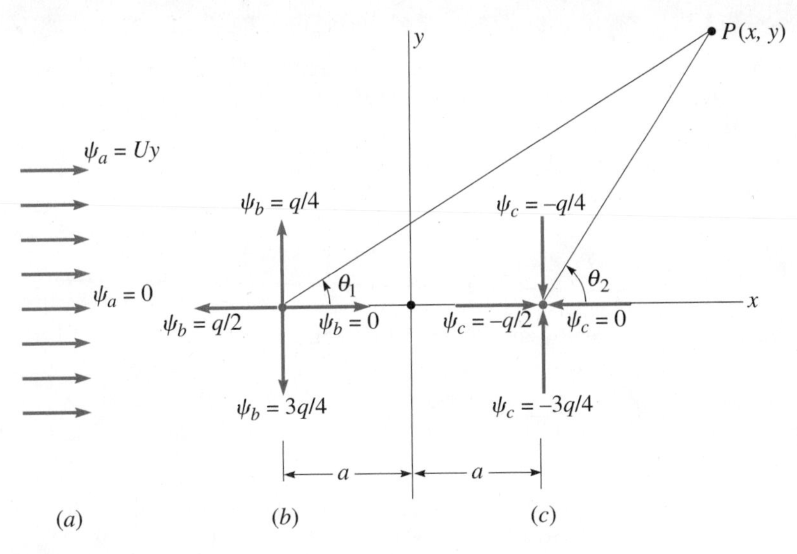

Figure 14.9
Superposition of flow fields. (*a*) Uniform rectilinear flow. (*b*) Source. (*c*) Source and sink of equal strengths and at a distance $2a$ apart along the x axis.

replacing the θ's with appropriate trigonometric functions, we get

$$\psi = Uy + \frac{q}{2\pi}\left(\arctan\frac{y}{x+a} - \arctan\frac{y}{x-a}\right) \tag{14.18}$$

This equation will permit us to plot streamlines by determining values of ψ at various points in the flow field having coordinates (x, y). Lines of constant ψ are streamlines. The resulting flow field for this case is shown in Sample Prob. 14.4. The $\psi = 0$ line produces a closed curve (oval), and thus the flow field represents ideal flow past a body of that shape. By using different values of a and different relationships between U and q, it is possible to describe a whole array of two-dimensional flow fields about ovals of various shapes. As $2a$, the distance between the source and sink gets smaller, the oval approaches a circle. However, when $a = 0$, the flow field reduces mathematically to uniform rectilinear flow since the source and sink will cancel each other out. The location of stagnation points S may be found by differentiating Eq. (14.18) to obtain an expression for $u = \partial\psi/\partial y$ and then to determine the values of x for which $u = 0$.

The flow field of Sample Prob. 14.4 is for an ideal fluid and, of course, does not represent the flow picture for a real fluid, where there may be separation[3] with the formation of a wake on the downstream side of the body (Fig. 9.12). However, on the upstream side of the body where the boundary layer is thin, the flow of a real fluid is well represented by this example.

---

[3] Refer to Sec. 9.6 for a discussion of the conditions under which separation will take place on the back side of a solid body.

SAMPLE PROBLEM 14.4 A flow field for a source and sink of equal strength is combined with a uniform rectilinear flow. Given $U = 0.80$, $q = 2\pi$, $a = 2$, plot the flow field.

Solution

Eq. (14.18):
$$\psi = 0.80y + \arctan\frac{y}{x+2} - \arctan\frac{y}{x-2}$$

Letting
$$A = \frac{y}{x+2} \quad \text{and} \quad B = \frac{y}{x-2}$$

we compute ψ for one of the symmetric quadrants:

				Degrees			Radians		
x	y	$\frac{y}{x+2} = A$	$\frac{y}{x-2} = B$	arctan A	arctan B	$0.8y$	arctan A	arctan B	ψ
0	2	$\frac{2}{2}$	$-\frac{2}{2}$	45°00′	135°00′	1.60	0.78	2.36	0.00
0	3	$\frac{3}{2}$	$-\frac{3}{2}$	56°19′	123°41′	2.40	0.98	2.16	1.22
0	4	$\frac{4}{2}$	$-\frac{4}{2}$	63°26′	116°34′	3.20	1.11	2.04	2.27
2	2	$\frac{2}{4}$	∞	26°34′	90°00′	1.60	0.46	1.57	0.49
2	3	$\frac{3}{4}$	∞	36°54′	90°00′	2.40	0.64	1.57	1.47
5	1	$\frac{1}{7}$	$\frac{1}{3}$	8°08′	18°26′	0.80	0.14	0.32	0.62
5	2	$\frac{2}{7}$	$\frac{2}{3}$	15°55′	33°42′	1.60	0.28	0.59	1.29
8	1	$\frac{1}{10}$	$\frac{1}{6}$	5°43′	9°28′	0.80	0.10	0.17	0.73
8	2	$\frac{2}{10}$	$\frac{2}{6}$	11°19′	18°26′	1.60	0.19	0.32	1.47

Plotting curves of constant ψ, we obtain Fig. S14.4 ***ANS***

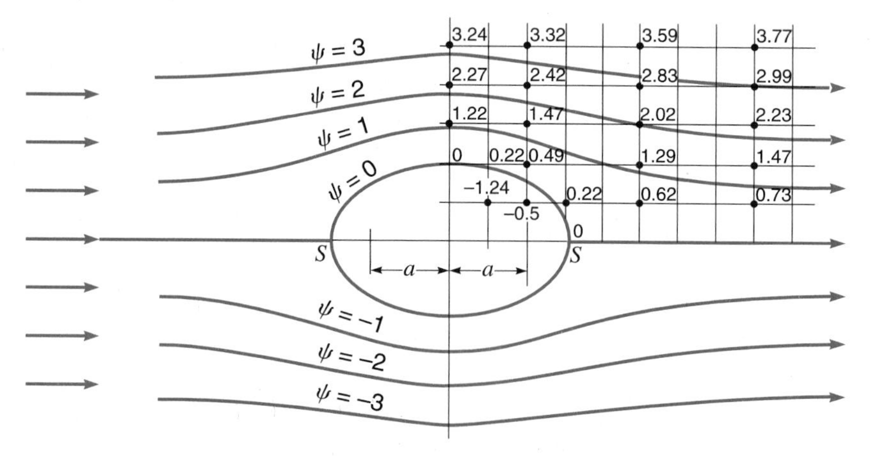

Figure S14.4
Flow field for source and sink of equal strength in a uniform rectilinear flow field. Note that point S represents a stagnation point.

Suppose we define a ***doublet*** as a source–sink combination for which $2qa = m$, a constant. Permitting a to approach zero, we see from the geometry of Fig. 14.9 that $r(\theta_2 - \theta_1) = 2a \sin \theta$. So $q(\theta_1 - \theta_2) = -(2aq \sin \theta)/r = -(m \sin \theta)/r$, which when substituted into Eq. (14.17) gives the stream function of the doublet imposed on the uniform field as

$$\psi = Uy - \frac{m \sin \theta}{2\pi r} \tag{14.19}$$

Taking $\psi = 0$ to determine the form of the closed-body contour and noting that $y = r \sin \theta$, we get

$$0 = Ur \sin \theta - \frac{m \sin \theta}{2\pi r}$$

or

$$r_{\psi = 0} = R = \sqrt{\frac{m}{U2\pi}} = \text{constant}$$

Therefore the closed-body contour for this case is a circle with radius R, and Eq. (14.19) is the stream function for two-dimensional flow about a circular cylinder. This flow was also discussed in Sec. 9.10, with lift and circulation. Other basic flow fields of interest include the forced vortex ($\psi = \omega r^2/r$), the free vortex ($\psi = \Gamma \ln r/2\pi$), and various combinations such as the source in rectilinear flow (see Prob. 14.14), the source and sink of equal strength, and the doublet in rectilinear flow with circulation (see Fig. 9.18).

EXERCISES

14.5.1 A source discharging 25 m^3/s per m is located at the origin and a uniform flow with a velocity of 5 m/s from left to right is superimposed on the source flow. Determine the stream function of the flow (a) in polar and (b) in rectangular coordinates.

14.5.2 For the flow of Exer. 14.5.1, (a) find the location of the stagnation points and (b) find the velocity at $x = 3$ m, $y = 4$ m.

14.5.3 Refer to Exer. 14.5.1 Use Bernoulli's theorem to find the difference in pressure head between point A (−14 m, 0) and point B(0, 2 m).

14.5.4 A source discharging 13 cfs/ft is at (−1, 0) and a sink taking in 13 cfs/ft is at (+1, 0). If a uniform flow with velocity 8 fps from left to right is superimposed on the source–sink combination, what is the length of the resulting closed body contour?

14.6 VELOCITY POTENTIAL

Let us define the potential

$$-d\phi = u\,dx + v\,dy \tag{14.20}$$

Mathematically, this is termed an "exact" differential, and therefore the function $\phi(x, y)$ exists, if

$$\frac{\partial u}{\partial y} = \frac{\partial v}{\partial x} \tag{14.21}$$

But the total derivative is defined to be

$$d\phi = \frac{\partial \phi}{\partial x}\,dx + \frac{\partial \phi}{\partial y}\,dy \tag{14.22}$$

By comparing (14.20) with (14.22) we see that in cartesian coordinates

$$u = -\frac{\partial \phi}{\partial x} \quad \text{and} \quad v = -\frac{\partial \phi}{\partial y} \tag{14.23}$$

For two-dimensional flow, ϕ (phi) with conditions (14.23) is termed the ***velocity potential*** function. In polar coordinates, the corresponding expressions are

$$v_r = -\frac{\partial \phi}{\partial r} \quad \text{and} \quad v_t = -\frac{1}{r}\frac{\partial \phi}{\partial \theta} \tag{14.24}$$

The use of a minus sign in Eq. (14.20) led to the minus signs in the expressions (14.23), which indicate that the velocity potential decreases in the direction of flow, i.e., flow moves from areas of high potential to low potential. Some authors prefer the opposite, and so change these signs.

Differentiating Eqs. (14.23), we get

$$\frac{\partial u}{\partial y} = -\frac{\partial^2 \phi}{\partial y\,\partial x} \quad \text{and} \quad \frac{\partial v}{\partial x} = -\frac{\partial^2 \phi}{\partial x\,\partial y}$$

Since the right-hand sides of these two last quantities are equal, this satisfies the requirement (14.21), which, from Eq. (14.12), proves that $\xi = 0$. Thus it follows that *if a flow is irrotational* ($\xi = 0$) *then a velocity potential exists, and vice versa.* Because of the existence of a velocity potential, such flow is often referred to as ***potential flow***.

The rotation of fluid particles requires the application of torque, which in turn depends on shearing forces. Such forces are possible only in a viscous

fluid. In inviscid (or ideal) fluids there can be no shears and hence no torques.
If we substitute Eqs. (14.23) into the continuity Eq. (14.3), we get

$$\frac{\partial^2 \phi}{\partial x^2} + \frac{\partial^2 \phi}{\partial y^2} = 0 \tag{14.25}$$

This is the Laplace equation, named after the French mathematician and astronomer, Marquis Pierre Simon de Laplace (1749–1827). It is possibly the best known of all partial differential equations, important also in solid mechanics and thermodynamics. For fluids, if a function ϕ satisfies Laplace's equation, the resulting flow must be irrotational.

EXERCISES

14.6.1 If $\phi = y + 2x^2$ is the velocity potential function for a two-dimensional flow, is it irrotational? Does it satisfy the Laplace equation? If not, suggest why.

14.6.2 Given the stream function $\psi = 10x - 7y$, is this a potential flow? If it is, determine the velocity potential function. Does it satisfy the Laplace equation?

14.6.3 For the following stream functions, determine if a potential function exists, find the potential function if it does, and determine if the Laplace equation in ϕ is satisfied: (*a*) $\psi = 3xy + 2x$; (*b*) $\psi = 3xy + 2x^2$.

14.6.4 For the following stream functions, determine if a potential function exists, find the potential function if it does, and determine if the Laplace equation in ϕ is satisfied; (*a*) $\psi = 6xy$; (*b*) $\psi = x \sin y$.

14.7 ORTHOGONALITY OF STREAMLINES AND EQUIPOTENTIAL LINES

From Eqs. (14.14) and (14.20), we have

$$d\psi = -v\,dx + u\,dy$$

and

$$d\phi = -u\,dx - v\,dy$$

Along a streamline, $\psi =$ constant, so $d\psi = 0$, and from the first equation we get $dy/dx = v/u$. Along an equipotential line, $\phi =$ constant, so $d\phi = 0$, and from the second equation we get $dy/dx = -u/v$. Geometrically, this tells us that the streamlines and equipotential lines are ***orthogonal***, or

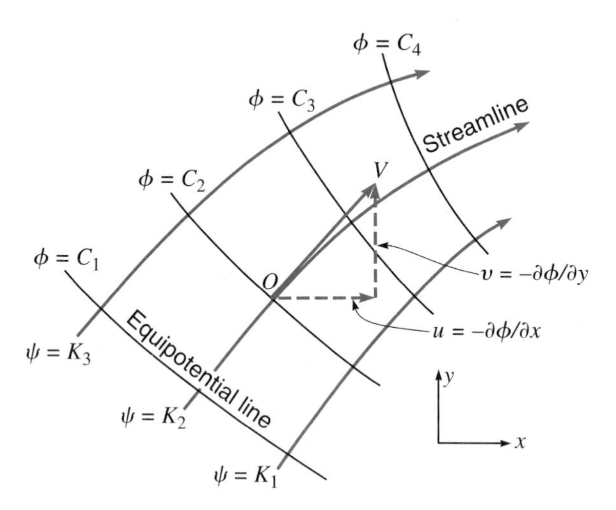

Figure 14.10
Flow net.

everywhere perpendicular to each other. As a result, the stream function and the velocity potential are known as ***conjugate functions***.

The equipotential lines $\phi = C_i$ and the streamlines $\psi = K_i$, where the C_i and the K_i have equal increments between adjacent lines, form a network of intersecting perpendicular lines that is called a ***flow net*** (Fig. 14.10; see also Secs. 4.9 and 4.10). The small quadrilaterals must evidently become squares as their size approaches zero, if the x and y scales are the same, since from Eqs. (14.16) and (14.23) $|u| = |\partial\phi/\partial x| = |\partial\psi/\partial y|$, or for finite increments, $|\Delta\phi/\Delta x| = |\Delta\psi/\Delta y|$. The difference in value of the stream function between adjacent streamlines is called the ***strength*** of the stream tube bounded by the two streamlines, and it represents the two-dimensional flow through the tube.[4]

Referring to Fig. 14.10, the maximum velocity at any point O is seen to be tangential to the streamline. This velocity is given by $V = -\partial\phi/\partial s$, where s is measured along the streamline. Because ϕ does not vary along an equipotential line, perpendicular to s, the partial derivative $\partial\phi/\partial s$ is equal to the total derivative $d\phi/ds$. The expression $\partial\phi/\partial s$ is also known as the ***gradient*** of the velocity potential. Thus the velocity is often written in convenient vector notation as $\mathbf{V} = -\text{grad}\ \phi$, which holds for either two- or three-dimensional flow.[5] The absolute velocity may always be written as the

[4] Consider flow in the direction of the x axis in a stream tube (of unit thickness perpendicular to the plane of the paper) bounded by $\psi = K_1$ and $\psi = K_2$. Let y_1 and y_2 represent the graphical locations of the two streamlines. The flow through the tube is then given by

$$Q = \int_{y_1}^{y_2} u\, dy = -\int_{y_1}^{y_2} \frac{\partial\psi}{\partial y}\, dy = -\int_{K_1}^{K_2} d\psi = K_1 - K_2$$

[5] grad ϕ is in the direction of increasing ϕ.

vector sum of its components; thus, in three dimensions,

$$V = |\mathbf{V}| = \sqrt{u^2 + v^2 + w^2} \tag{14.26}$$

Stream functions can exist in the absence of irrotationality, and potential functions are possible even though continuity is not satisfied. But, since lines of ϕ and ψ are required to form an orthogonal network, a flow net can only exist if irrotationality (the condition for the existence of ϕ) and continuity (the condition for the existence of ψ) are satisfied. The Laplace equation [Eq. (14.25)] was derived assuming the existence of velocity potentials and the satisfaction of continuity. Thus, if a given flow satisfies the Laplace equation, a flow net can be constructed for that flow. Because of the irrotationality requirement such potential flows are usually those of ideal fluids.

Sample Problem 14.5 An incompressible flow is defined by $u = 2x$ and $v = -2y$. Find the stream function and potential function for this flow and plot the flow net.

Solution

Check continuity:

Eq. (14.3): $$\frac{\partial u}{\partial x} + \frac{\partial v}{\partial y} = 2 - 2 = 0$$

Hence continuity is satisfied, and it is possible for a stream function to exist:

Eq. (14.14): $$d\psi = -v\,dx + u\,dy = 2y\,dx + 2x\,dy$$

Integrating: $$\psi = 2xy + C_1 \qquad \textit{ANS}$$

Check to see if the flow is irrotational:

Eq. (14.7): $$\frac{\partial v}{\partial x} - \frac{\partial u}{\partial y} = 0 - 0 = 0$$

Hence $\xi = 0$, the flow is irrotational, and a potential function exists:

Eq. (14.20): $$d\phi = -u\,dx - v\,dy = -2x\,dx + 2y\,dy$$

Integrating: $$\phi = -(x^2 - y^2) + C_2 \qquad \textit{ANS}$$

Letting $\psi = 0$ and $\phi = 0$ pass through the origin, we get $C_1 = C_2 = 0$.

The location of lines of equal ψ can be found by substituting values of ψ into the expression $\psi = 2xy$. Thus for $\psi = 60$, $x = 30/y$. This line is plotted (in the upper right-hand quadrant) on the adjoining figure. In a similar fashion lines of equal potential can be plotted. For example, for $\phi = -60$ we have $-(x^2 - y^2) = -60$ and $x = \pm\sqrt{y^2 + 60}$. This line is also plotted on the figure. The flow net depicts ***flow in a corner***. Mathematically the net will plot symmetrically in all four quadrants.

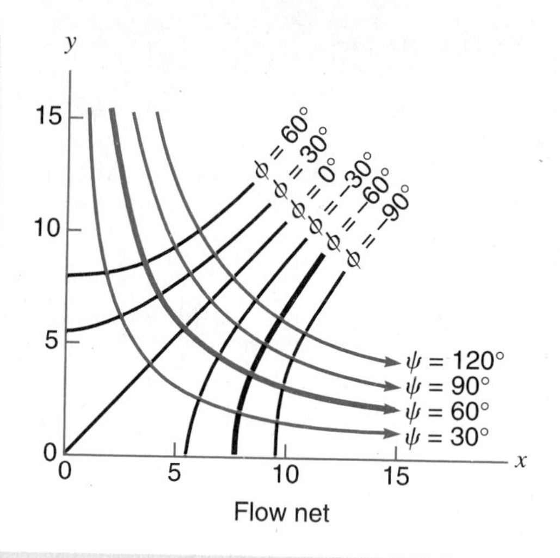

Figure S14.5

EXERCISES

14.7.1 Which of the flows in Exer. 14.1.1 can be described by a flow net? Write expressions for the stream functions and the potential functions.

14.7.2 Which of the flows in Exer. 14.1.2 can be described by a flow net? Write expressions for the stream functions and the potential functions.

14.8 FLOW THROUGH POROUS MEDIA

An exception where a real fluid satisfies the conditions for potential flow is that of ***laminar flow through porous media***. In such a case the velocity head is negligible and the energy Eq. (5.14) may be written as

$$\left(\frac{p_1}{\gamma} + z_1\right) - \left(\frac{p_2}{\gamma} + z_2\right) = h_L$$

where the head loss h_L is directly proportional to velocity for laminar flow (see Sec. 8.1). Taking the differential, we get

$$-d\left(\frac{p}{\gamma} + z\right) = dh_L = \frac{V}{K}\,ds$$

where ds is the distance along the streamline, and $1/K$ is a constant of proportionality. Hence

$$V = -K\frac{d(p/\gamma + z)}{ds} = -\frac{d\phi}{ds} \tag{14.27}$$

Thus we see for porous media flow that $\phi = K(p/\gamma + z)$. The constant K is known as the ***hydraulic conductivity***; it has the units of velocity, and has high values for gravels and sands and low values for clays. The quantity $p/\gamma + z$ is called the ***hydraulic head***, usually represented by the variable h, and dh/ds is the ***hydraulic gradient***, which is negative because h decreases in the direction of flow.

Equation (14.27) is known as ***Darcy's law***, in honor of Henry Darcy (1803–1858), a French engineer. He was the first to propose such a relationship, in the form $V = K\, dh_L/ds$, published in 1856, based on his experiments with vertical flow through sand used for water filtration. As a result, he is recognized as the father of groundwater hydraulics. Darcy's law today remains an empirical law, based only on experimental evidence. Subsequently it has been observed that the proportionality does not hold if the flow is not laminar,[6] which seldom occurs with groundwater flow.

A distinction must be made between the ***specific discharge*** (or apparent velocity or Darcy velocity) $V = Q/A$, where A is the total cross-sectional area of the media, and V_p, the ***average pore velocity***. This is defined by

$$V_p = \frac{V}{n} \tag{14.28}$$

where n is the porosity of the medium, equal to the volume of the pores divided by the total volume.

The top of the saturated groundwater zone, where not confined, is known as the ***water table***. There the pressure is atmospheric, or zero gage pressure, and so $p/\gamma + z = h = z$. From elevations of surface water bodies, and of water levels in unpumped wells, contours of water table elevations can be prepared. Since the elevations of the water table contours correspond to the hydraulic head h, it follows from the orthogonality principle that where the groundwater flow is nearly horizontal the flow will be perpendicular to the water table contours. This, then, is another useful application for flow nets, helping determine groundwater flow directions. Furthermore, from Eq. (14.27) and an estimate of K, groundwater flow velocities can be calculated.

Flow nets are used in many other situations to estimate flow rates through soils, such as through earth dams and under impervious dams.[7]

Seepage and groundwater hydraulics are important for economic, environmental, and other reasons. For example, contaminants in groundwater are transported by the groundwater flow, and because flow velocities are

[6] For flow through a granular medium, the flow is laminar if the Reynolds number has a value less than 1.0, where the Reynolds number is defined as $d_{10}V\rho/\mu$, where d_{10} is the effective particle size (the 10% finer than value), V is the Darcy velocity defined by Eq. (14.27), and ρ and μ are the properties of the fluid that is passing through the medium. At higher Reynolds numbers (larger particles, larger gradient, and lower fluid viscosity) the head loss tends to become proportional to V^2.

[7] H. R. Vallentine, *Applied Hydrodynamics,* 2d ed., Chap. 3, Butterworth & Co. Ltd., London, 1967.

usually very slow, more or less in the range of a foot per day to a foot per year, it can take a long time to identify contamination, and to remedy it. Overpumping of groundwater, say for agricultural use, lowers the water table and increases the cost of pumping. In coastal areas it can lead to contamination by saltwater intrusion. Water table lowering is used beneficially to dewater construction sites.

Darcy's law is used in a number of other important areas besides the analysis of groundwater flow. These include the approximate analysis of unsaturated soil moisture movement, the flow of oil in petroleum reservoirs, and for the design of chemical and ceramic filters and related processes.

SAMPLE PROBLEM 14.6 Groundwater is flowing through an aquifer, 100 ft thick and 5000 ft wide, that has a porosity of 0.2 and a hydraulic conductivity of 0.1 ft/day. If the hydraulic gradient is −0.035, what is the flow rate through the aquifer in ft^3/day? To assess the speed of migration of contaminants, find how long on average it takes a particle of water to travel 100 ft?

Solution

Eq. (14.27): $V = -K(dh/ds) = -0.1(-0.035) = 0.0035 \text{ ft/day}$

$$Q = AV = 100(5000)0.0035 = 1750 \text{ ft}^3\text{/day} \qquad \textbf{\textit{ANS}}$$

Eq. (14.28): $V_p = V/n = 0.0035/0.2 = 0.0175 \text{ ft/day}$

$$t = s/V_p = 100/0.0175 = 5714 \text{ days} = 15.6 \text{ yr} \qquad \textbf{\textit{ANS}}$$

EXERCISES

14.8.1 The estimated average depth of saturated flow in a sloping water table aquifer (porosity 0.25, $K = 5$ ft/day) is 11.8 ft. The distance between 1-ft water table contours is 34 ft. What is the flow rate through a 1-ft width of aquifer, and what is the average pore velocity?

14.8.2 The estimated average depth of saturated flow in a sloping water table aquifer (porosity 0.23, $K = 1.6$ m/d) is 3.8 m. The distance between 1-m water table contours is 37 m. What is the flow rate through a 1-m width of aquifer, and what is the average pore velocity?

14.8.3 From dye tests between two wells, groundwater is estimated to travel at 0.034 ft/day. If the corresponding water table slope is 0.0018, and the average porosity of the aquifer is 0.15, estimate its hydraulic conductivity.

14.8.4 From dye tests between two wells, ground water is estimated to travel at 0.0094 m/d. If the corresponding water table slope is 0.0016, and the average porosity of the aquifer is 0.18, estimate its hydraulic conductivity.

PROBLEMS

14.1 Given a flow defined by $u = 3 + 2x$. If this flow satisfies continuity, what can be said about the density of the fluid?

14.2 Why are Eqs. (14.2) and (14.3) applicable to real fluids as well as ideal fluids?

14.3 Derive Eqs. (14.4) and (14.5).

14.4 Sketch streamlines ($\psi = 0, 1, 2, 3$) for the following flow fields, note the values of u and v, and verify that continuity is satisfied in all cases: (*a*) $\psi = 10y$; (*b*) $\psi = -20x$; (*c*) $\psi = 10y - 20x$.

14.5 A flow field is described by the equation $\psi = 1.2xy$. Sketch the streamlines in one quadrant for $\psi = 0, 1, 2, 3, 4$.

14.6 Plot the streamlines in the upper right-hand quadrant for the flow defined by $\psi = 1.5x^2 + y^2$ and determine the value of the velocity at $x = 4$, $y = 2$.

14.7 The components of the velocities of a certain flow system are

$$u = -\frac{Q}{2\pi}\frac{x}{x^2 + y^2} + By + C$$

$$v = -A\frac{y}{x^2 + y^2} + Dx + E$$

(*a*) Calculate a value of A consistent with continuous flow. (*b*) Sketch the streamlines for this flow system, assuming $B = C = D = E = 0$.

14.8 A flow field is described by $\psi = x^2 - y$. Sketch the streamlines for $\psi = 0$, 1, and 2. Derive an expression for the velocity at any point in the flow field and determine the vorticity of the flow.

14.9 A source of strength 8π is located at (2, 0). Another source of strength 16π is located at (−3, 0). For the combined flow field produced by these two sources: (*a*) find the location of the stagnation point; (*b*) plot the $\psi = 0$, $\psi = 4\pi$, $\psi = 8\pi$ lines; (*c*) find the values of ψ at (0, 2) and at (3, −1); (*d*) find the velocity at (−2, 5).

14.10 Using the method described in Sample Prob. 14.4, plot the boundary of the body and a set of streamlines for a steady two-dimensional flow past a body such as that of Fig. 4.12, for $b = 15$ m using a scale of 1 cm = 2 m.

14.11 Combine the uniform flow defined by $U = 16$ ft/s with the doublet $2qa = m$, where $q = 10$ cfs/ft and $a = 2$ in. Sketch the streamlines for $\psi = -3, -2, -1, -\frac{1}{2}, 0, \frac{1}{2}, 1, 2,$ and 3 cfs/ft. Use a scale of 1 in = 1 in.

14.12 A flow is defined by the stream function $\psi = 15r \sin\theta - 30 \ln r - (20/r)\sin\theta$. Sketch this flow field. Calculate the velocities at $r = 3$ for $\theta = 0°, 45°, 90°, 150°, 210°,$ and $315°$.

14.13 Given is the two-dimensional flow described by $u = x^2 + 2x - 4y$, $v = -2xy - 2y$. (*a*) Does this satisfy continuity? (*b*) Compute the vorticity. (*c*) Plot the velocity vectors for $0 < x < 5$ and $0 < y < 4$ and sketch the general flow pattern. (*d*) Find the location of all stagnation points in the entire flow field. (*e*) Find the expression for the stream function.

14.14 The flow around the body of Fig. 4.12 may be considered as that due to the sum of two velocity potentials, $\phi_1 = -Ux$, representing an undisturbed flow of velocity U in the x direction, and $\phi_2 = -S \ln r$, representing the radial flow from a source located inside the body behind the stagnation point. To relate U and S, it is observed that the total flow $2\pi S$ from the source (which is hydrodynamically equivalent to the body itself) must be equal to the flow of the main stream that is not passing through the body of width b, or $2\pi S = Ub$. This gives

$$\phi_2 = -\frac{Ub}{2\pi} \ln r$$

(*a*) The distance from the stagnation point to the source is determined by setting the radial velocity from the source, $v_r = -\partial\phi/\partial r$, equal and opposite to the undisturbed velocity U. Prove that this establishes the source at a distance $b/2\pi$ behind the stagnation point. The absolute velocity at any point of the field may be determined by the vector sum of the components U and v_r.

There follows an ingenious method of plotting the boundary of such a streamlined body, as shown in Fig. P14.14. Suppose that the streamlines in

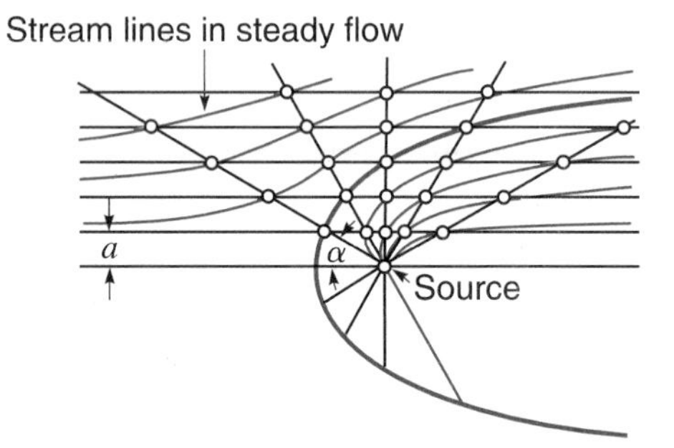

Figure P14.14

the undisturbed flow are spaced a distance a apart, where $b/2a = n$, an integer. Next divide the upper half of the source into n radial sectors, each of angle α, that is, $n\alpha = \pi$. Then the undisturbed flow between the x axis and the first streamline is associated with the source flow in the first sector from the stagnation point. Thus the intersection of the first streamline with the first line must be a point on the boundary of the body, through which there can be no flow. Similarly, the intersection of the horizontal line at $2a$ with the radial line at 2α forms another point, and so on. Further streamlines can be plotted by connecting successive intersections of the original horizontal lines with the radial lines, recognizing that the same flow must exist between any adjacent pair of streamlines. Thus the intersection

of a horizontal line ea above the axis with a radial line at $f\alpha$ from the stagnation point must lie on a streamline which is $(e-f)a$ distant from the axis in the undisturbed region, where e and f are integers.
(*b*) Assume a value of $U = 20$ ft/sec and a two-dimensional flow past a streamlined body for which $b = 36$ ft. Compute the distance from the source to the stagnation point and to the surface of the body at a radius 90° to the axis. What is the value of the source velocity at the latter point?
(*c*) What is the magnitude of the velocity of the fluid along the surface at the 90° point? (Compare with the results of Sample Prob. 4.2). What is its direction relative to the axis?

14.15 Solve Prob. 14.14 in SI units; for parts (*b*) and (*c*) use $U = 5$ m/s and $b = 12$ m.

14.16 Find the distance to the surface of the body and the two velocities called for in Prob. 14.14 for an angle of 30°. (Compare with Sample Prob. 4.2).

14.17 Find the distance to the surface of the body and the two velocities called for in Prob. 14.15 for an angle of 30°. (See Sample Prob. 4.2).

14.18 An ideal fluid flows in a two-dimensional 90° bend. The inner and outer radii of the bend are 0.6 and 1.6 ft. (*a*) Sketch the flow net and estimate the velocity at the inner and outer walls of the bend if the velocity in the 1.0-ft-wide straight section is 8 ft/sec. (*b*) Develop an analytic expression for the stream function, in this case noting that $v_t = -\partial\psi/\partial r$ and $v_r = r^{-1}\,\partial\psi/\partial\theta$. Determine the inner and outer velocities accurately.

14.19 An ideal fluid flows in a two-dimensional 90° bend. The inner and outer radii of the bend are 0.2 and 0.6 m. (*a*) Sketch the flow net and estimate the velocity at the inner and outer walls of the bend if the velocity in the 0.4-m-wide straight section is 2.5 m/s. (*b*) Develop an analytic expression for the stream function, in this case noting that $v_t = -\partial\psi/\partial r$ and $v_r = r^{-1}\,\partial\psi/\partial\theta$. Determine the inner and outer velocities accurately.

14.20 For the two-dimensional flow of a frictionless incompressible fluid against a flat plate normal to the initial velocity, the stream function is given by $\psi = -2axy$, while its conjugate function, the velocity potential, is

$$\phi = a(x^2 - y^2)$$

where a is a constant and the flow is symmetrical about the yz plane (Fig. 14.1). By direct differentiation, demonstrate that these functions satisfy Eq. (14.25). Using a scale of 1 in = 1 unit of distance, plot the streamlines given by $\psi = \pm 2a, \pm 4a, \pm 6a, \pm 8a$, and the equipotential lines given by $\phi = 0, \pm 2a, \pm 4a, \pm 6a, \pm 8a$. Observe that this flow net also gives the ideal flow around an inside square corner. Compare your results with Sample Prob. 14.5 and note the effect of changing the sign of ψ and ϕ.

14.21 In Prob. 14.20 determine the velocity components u and v, and demonstrate that they satisfy the differential equations for continuity and irrotational flow. In which direction is the flow? Prove that the absolute velocity is given by $V = 2ar$, where r is the radius to the point from the origin. Now assume that the linear scale is 1 in = 1 ft. Determine the constant a

such that the flow net of Prob. 14.20 will represent a flow of 3 ft^2/s between any two adjacent streamlines. What are the dimensions of a? Draw curves of equal velocity for values of 3, 6, 9, 12 ft/s. How does the velocity vary along the surface of the plate?

14.22 A cylindrical drum with a 2-ft radius is securely held in position in an open channel of rectangular section. The channel is 10 ft wide, and the flow rate is 240 cfs. Water flows beneath the drum as shown in Fig P14.22. Sketch the flow net, and determine from flow net measurements the pressure at the points indicated along the wetted drum surface. Neglect fluid friction. Sketch the pressure distribution, and by numerical integration determine an approximate value of the horizontal thrust on the cylinder. (Compare with Exer. 6.4.1).

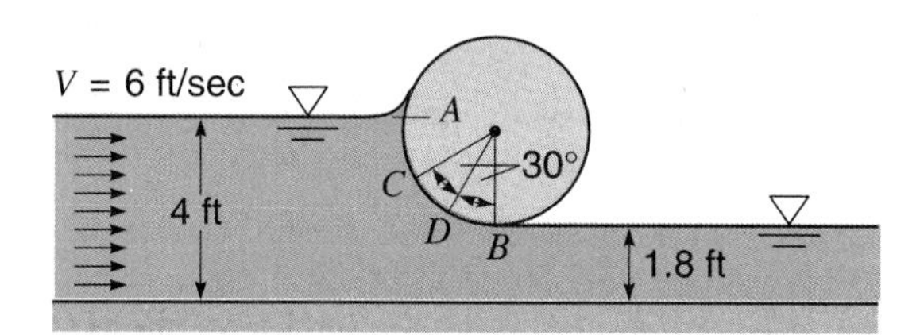

Figure P14.22

14.23 Refer to Sample Prob. 6.1. Sketch a flow net. Using the given dimensions in BG units through application of Bernoulli's principle, determine the approximate pressure distribution along the channel bottom and around the curved structure. By numerical integration estimate the magnitude of the horizontal and vertical components of the force of the water on the structure.

14.24 Work Prob. 14.23 using the dimension as given in SI units in Sample Prob. 6.1.

14.25 The three-dimensional counterpart of the flow in Probs. 14.20 and 14.21 is that of flow along the y axis approaching the plate in the xz plane. As the flow must be symmetrical about the y axis, the traces of stream and equipotential surfaces in the xy plane will be representative of those in all planes containing the y axis. The velocity potential is now given by $\phi = -a(0.5x^2 - y^2)$, and the stream function by $\psi = -ax^2y$. Notice that these functions no longer satisfy Eq. (14.25). Why not? Again plot streamlines and equipotential lines for the values given in Prob. 14.20. The velocities u and v may still be determined by Eq. (14.23). Prove that the absolute velocity for this case is given by $V = a\sqrt{x^2 + 4y^2}$. With the value of $a = 1.5\ s^{-1}$ found in Prob. 14.21, draw curves of equal velocity for values of 3, 6, 9, 12 ft/s. How does the velocity vary along the surface? What is the total flow between any two adjacent stream surfaces?

14.26 For the two-dimensional flow around any angle α, the velocity potential and stream function are given in polar coordinates as $\phi = -ar^{\pi/\alpha} \cos(\pi\theta/\alpha)$ and $\psi = -ar^{\pi/\alpha} \sin(\pi\theta/\alpha)$, respectively. Prove that the functions given in Prob. 14.20 are a specialization of these expressions for

$\alpha = \pi/2$. Take the case of $\alpha = 3\pi/2$, and plot streamlines and equipotential lines for the values given in Prob. 14.20. Compare the velocity at the corner with that at the corner in Prob. 14.20.

14.27 Superimpose a point source ($Q = 100$ cfs) on a rectilinear flow field ($U = 20$ ft/s). Plot the body contour at $\theta = 30°, 60°, 90°, 120°, 150°, 180°$ using a scale of 1 in = 1 ft. Compute the velocities along the body contour at these points. Determine the pressures at these points assuming $\rho = 1.94$ slug/ft^3 with zero pressure in the undisturbed rectilinear flow field. What is the velocity and pressure in the combined flow field at the following points? *Hint*: Refer to Prob. 14.14.

(*a*) $\theta = 45°$, $r = 4.0$ ft
(*b*) $\theta = 90°$, $r = 2.0$ ft
(*c*) $\theta = 90°$, $r = 4.0$ ft
(*d*) $\theta = 135°$, $r = 2.0$ ft

CHAPTER 15

Hydraulic Machinery—Pumps[1]

There are various types of fluid machinery. Among them are those that transfer fluid energy (***torque converters***), those that convert mechanical energy to fluid energy (***pumps***), and those that convert fluid energy to mechanical energy (***turbines***). The conversion of mechanical energy to fluid energy is accomplished by pumps for the case of incompressible fluids, and by ***blowers***, ***fans***, and ***compressors*** for the case of compressible fluids. In this chapter we shall confine our discussion to pumps. In Secs. 6.11–6.14 there is a discussion of flow through rotating machinery and rotating conduits; we suggest the reader review that material before proceeding further. Since pumps are much more prevalent than turbines, we discuss pumps first. We shall deal with ***centrifugal and axial-flow pumps***, but shall not discuss positive displacement pumps such as reciprocating piston and rotary types.

In our discussion of hydraulic machinery, emphasis will be placed on the selection of pumps (and turbines) for particular situations. To accomplish this, we shall first describe the various types of pumps (and turbines) and discuss their performance characteristics. Finally we shall show how pumps (and turbines) fit into the hydraulics of the systems in which they operate. We shall find that cavitation plays an important role in the selection of pumps and turbines.

---

[1] In this chapter and the next, because of the difficulty of carrying everything in both sets of units, for the most part we shall deal with pumps and turbines using conventional English (BG) units in conformance with practice in the United States. However, a substantial number of sample problems, exercises and problems will be expressed in SI metric units.

15.1 DESCRIPTION OF CENTRIFUGAL AND AXIAL-FLOW PUMPS

The rotating element of a centrifugal pump is called the ***impeller*** (Fig. 15.1). The impeller may be shaped to force water outward in a plane at right angles to its axis (***radial flow***), to give the water an axial as well as radial velocity (***mixed flow***), or to induce a spiral flow on coaxial cylinders in an axial direction (***axial flow***). Radial-flow and mixed-flow machines are commonly referred to as centrifugal pumps, while axial-flow machines are called axial-flow pumps or ***propeller pumps***. Radial and mixed-flow impellers may be either open or closed. The open impeller consists of a hub to which vanes are attached, while the closed impeller has plates (or ***shrouds***) on each side of the vanes. The open impeller does not have as high an efficiency as the closed impeller, but it is less likely to become clogged and hence is suited to handling liquids containing solids.

Radial-flow pumps are provided with a ***spiral casing***, often referred to as

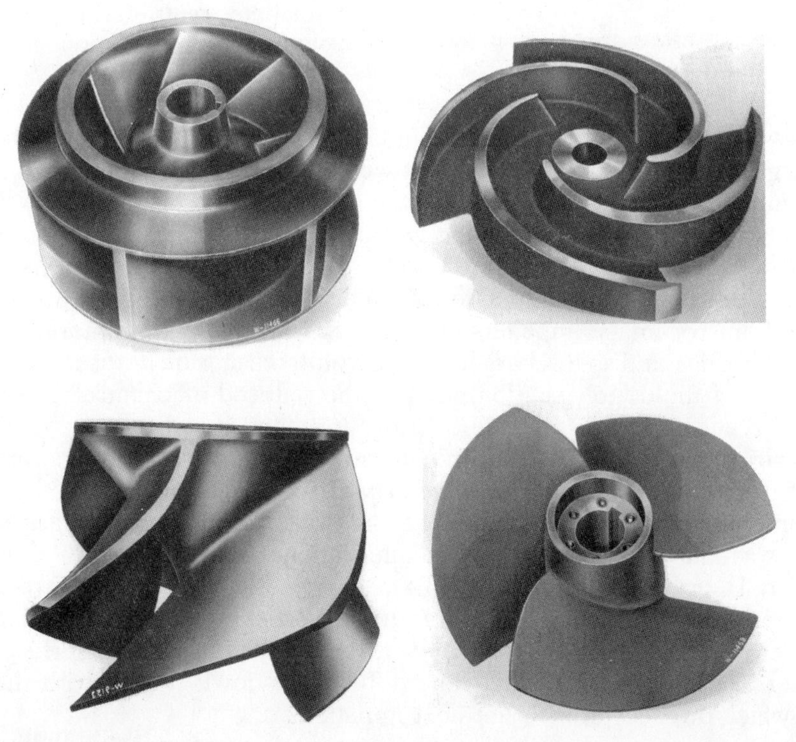

Figure 15.1
Types of pump impellers. Top left: closed or shrouded radial. Top right: open or unshrouded radial. Bottom left: mixed flow. Bottom right: propeller. (*Worthington Pump Co.*)

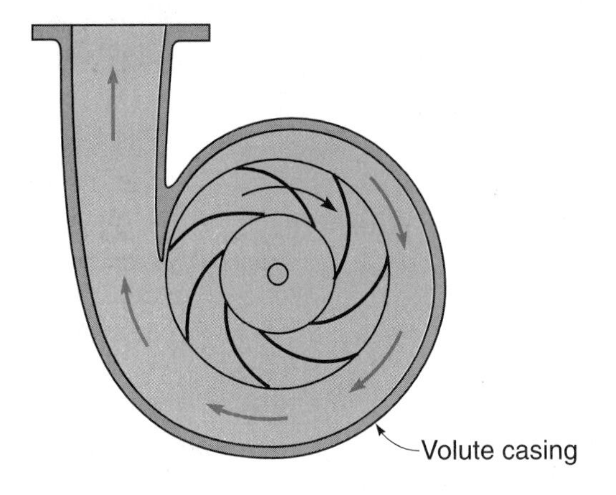

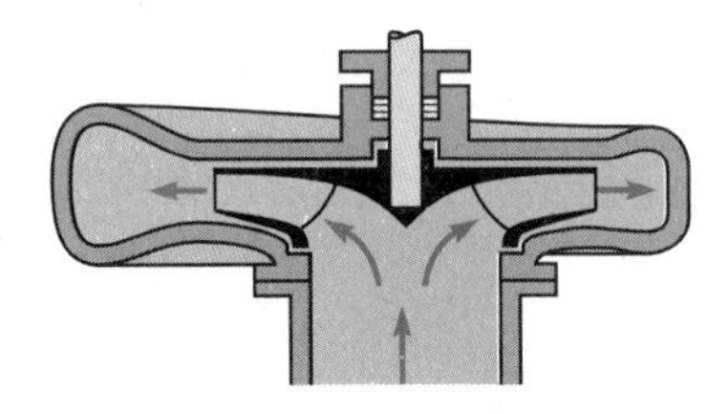

Figure 15.2
Radial-flow centrifugal pump with volute casing.

a ***volute casing*** (Fig. 15.2), which guides the flow from the impeller to the discharge pipe. The ever increasing flow cross-section around the casing tends to maintain a constant velocity within the casing. This helps to provide relatively smooth flow conditions at exit from the impeller. Some pumps have diffuser vanes in lieu of a volute casing. Such pumps are known as ***turbine pumps***. Some radial pumps are of the double-suction type. They have identical, mirror-image impellers placed back to back. Water enters the pump from both sides and is discharged into a volute casing or diffuser vanes. The advantage of the double-suction pump is the reduced mechanical friction that results because the thrust on the bearings is balanced.

Typical centrifugal-flow and axial-flow pump installations are shown in Fig. 15.3. Pumps may be ***single-stage*** or ***multi-stage***. A single-stage pump has only one impeller, while a multi-stage has two or more impellers arranged in such a way that the discharge from one impeller enters the eye of the next impeller. Deep-well pumps (Fig. 15.4), a type of turbine pump, are usually multi-stage, having several impellers on a vertical shaft suspended from a prime mover, usually an electric motor, located at the ground surface. Each impeller discharges into a fixed-vane diffuser, or bowl, coaxial with the drive shaft, which directs water to the next impeller.

Proper arrangement of the suction and discharge piping is necessary if a centrifugal pump is to operate at best efficiency. For economy, the diameter of the pump casing at suction and discharge is often smaller than that of the pipe to which it is attached. If there is a horizontal reducer between the

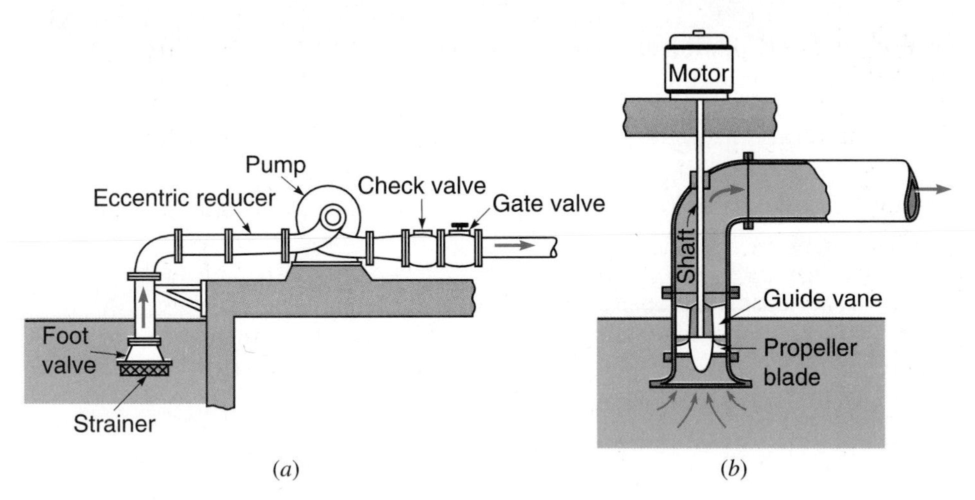

Figure 15.3
(*a*) Typical centrifugal pump installation. (*b*) Typical axial-flow pump installation.

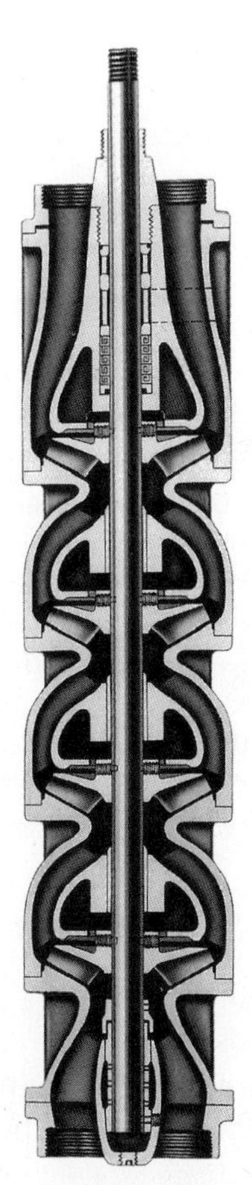

Figure 15.4
Deep-well multistage mixed-flow turbine pump.
(*Courtesy of Byron Jackson Company*).

suction and the pump, an ***eccentric reducer*** (Fig. 15.3*a*) should be used to prevent air accumulation. A ***foot valve*** (check valve) may be installed in the suction pipe to prevent water from leaving the pump when it is stopped. The discharge pipe is usually provided with a check valve and a gate valve. The ***check valve*** prevents backflow through the pump if there is a power failure. Suction pipes taking water from a sump or reservoir are usually provided with a screen to prevent entrance of debris that might clog the pump.

Axial-flow pumps (Fig. 15.3*b*) usually have only two to four blades and, hence, large unobstructed passages that permit handling of water containing debris without clogging. The blades of some large axial-flow pumps are adjustable to permit setting the pitch for the best efficiency under existing conditions.

15.2 HEAD DEVELOPED BY A PUMP

The head h developed by a pump is determined by measuring the pressures on both the suction and discharge sides of the pump, computing the velocities by dividing the measured discharge by the respective cross-sectional areas, and noting the difference in elevation between the suction and discharge sides. The net head h delivered by the pump to the fluid is

$$h = H_d - H_s = \left(\frac{p_d}{\gamma} + \frac{V_d^2}{2g} + z_d\right) - \left(\frac{p_s}{\gamma} + \frac{V_s^2}{2g} + z_s\right) \qquad (15.1)$$

where the subscripts d and s refer to the discharge and suction sides of the pump, as shown in Fig. 15.5. If the discharge and suction pipes are the same size, the velocity heads cancel out, but frequently the intake pipe is larger than the discharge pipe. It should be noted that h, the head put into the fluid by the pump, was previously referred to as h_M in Sec. 5.4 and h_p in Sec. 8.25.

The official test code provides that the head developed by a pump be the difference between the total energy heads at the intake and discharge flanges. However, flow conditions at the discharge flange are usually too irregular for accurate pressure measurement, and it is more reliable to measure the pressure at 10 or more pipe diameters away from the pump and to add an estimated pipe friction head for that length of pipe. On the intake side, ***prerotation*** sometimes exists in the pipe near the pump, and this will cause the pressure reading on a gage to be different from the true average pressure at that section.

15.3 PUMP EFFICIENCY

As liquid flows through a pump, only part of the energy imparted to the shaft of the impeller is transferred to the flowing liquid. There is friction in the

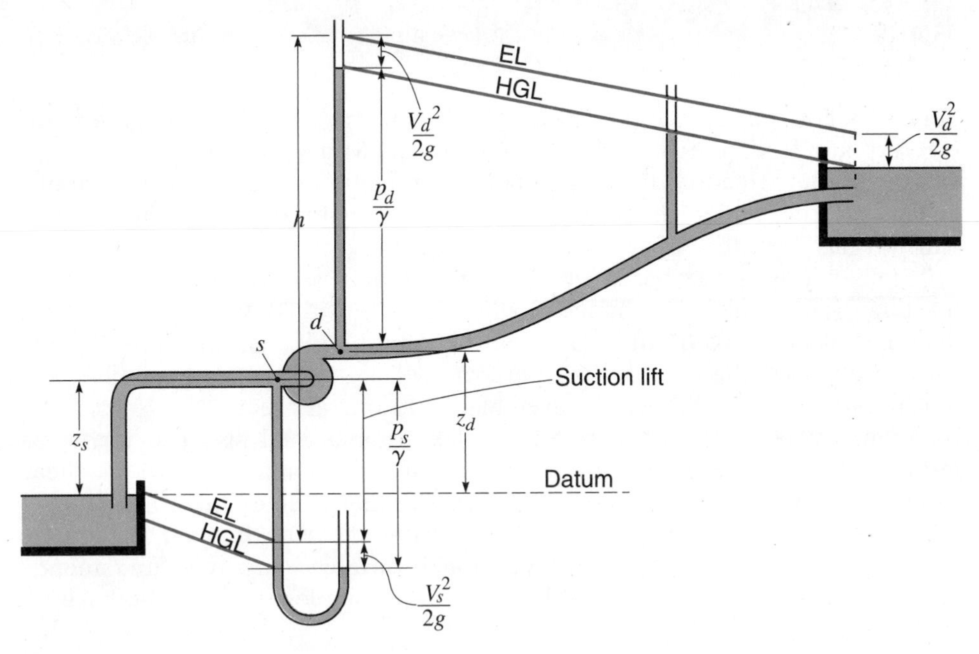

Figure 15.5
Head developed by a pump. In this case p_s/γ is negative.

bearings and packings, not all liquid passing through the pump is effectively acted upon by the impeller, and there is substantial loss of energy due to fluid friction which has a number of components including shock loss at entry to the impeller,[2] fluid friction as the fluid passes through the space between the vanes or blades, and head loss as the fluid leaves the impeller. The efficiency of a pump is quite sensitive to the conditions under which it is operated, as will be discussed in Sec. 15.5.

The efficiency η of a pump is given by

$$\eta = \frac{\text{power delivered to the fluid}}{\text{power put into the shaft (brake power)}} = \frac{\gamma Q h}{T\omega} \tag{15.2}$$

where γ, Q, and h are defined in the usual fashion; T is the torque exerted on the shaft of the pump by the motor that drives the shaft, and ω is the rate of rotation of the shaft in radians per second.

---

[2] Shock loss occurs when the flow does not enter the impeller smoothly. This results in separation of the flow from the impeller blade.

15.4 SIMILARITY LAWS FOR PUMPS

Similarity laws permit the prediction of the performance of a prototype pump (or turbine) from the test of a scaled model. Moreover, and of particular value in pump selection, these laws permit prediction of the performance of a given machine under different conditions of operation from those under which it has been tested.

Similarity laws are based on the concept that two geometrically similar machines (i.e., same scale change in all three dimensions) with similar velocity diagrams at entrance to and exit from the rotating element are ***homologous***. This means that their streamline patterns will be geometrically similar, i.e., that their behaviors will bear a resemblance to one another.

Similarly laws can be derived by dimensional analysis. The most significant variables[3] affecting the operation of a turbomachine are the head h, the discharge Q, the rotative speed n, the diameter of the rotor D, and the acceleration due to gravity g. Thus, from the Buckingham Π-theorem (Sec. 7.7), since there are five dimensional variables and two fundamental dimensions (L and T), there will be three dimensionless groups. We have

$$f(h, Q, n, D, g) = 0$$

Upon grouping these variables into dimensionless quantities, we get

$$f'\left(\frac{Q}{nD^3}, \frac{g}{n^2D}, \frac{h}{D}\right) = 0$$

Laboratory tests on turbomachines have demonstrated that the second dimensionless quantity is inversely proportional to the third. These can be combined to give

$$\frac{g}{n^2D} = K\frac{D}{h} \quad \text{and} \quad K = \frac{gh}{n^2D^2}$$

Thus

$$f''\left(\frac{Q}{nD^3}, \frac{gh}{n^2D^2}\right) = 0 \tag{15.3}$$

From Eq. (15.3), we find

$$Q \propto nD^3 \quad \text{or} \quad Q = K_Q nD^3 \tag{15.4}$$

and, assuming g is constant,[4]

$$h \propto n^2D^2 \quad \text{or} \quad h = K_h n^2D^2 \tag{15.5}$$

[3] If it is desired to relate the operation of one pump to another with different fluids in each then kinematic viscosity is a significant variable.

[4] If one were on the moon, or in some gravitational field different than that of the earth, g must be included in Eqs. (15.5) and (15.6).

Since power $P \propto Qh$, we get

$$P \propto n^3D^5 \quad \text{or} \quad P = K_P n^3 D^5 \tag{15.6}$$

For any one design of a pump, K_Q, K_h, and K_P can be evaluated, preferably from test data, and then used to predict the performance of homologous pumps. Similarity laws are of great practical value, but care must be exercised when applying them. Thus, in comparing two machines of different sizes, the two must be homologous and the variation in the values of h, D, and n should not be too large. For example, a machine that operates satisfactorily at low speeds may cavitate at high speeds. The values of K in each of Eqs. (15.4)–(15.6) change somewhat as h, D, and n are varied, because the efficiencies of homologous machines are not identical. Large machines are usually more efficient than smaller ones, because their flow passages are larger. Also, efficiency usually increases with speed of rotation, because power output varies with the cube of the speed while mechanical losses increase only as the square of the speed. As a result, Eqs. (15.4)–(15.6) are approximate.

An empirical equation suggested by Moody,[5] originally intended for application to homologous reaction turbines, that gives fairly reasonable results for estimating the efficiency of a prototype pump from the test of a geometrically similar (model) pump is

$$\frac{1-\eta_p}{1-\eta_m} \approx \left(\frac{D_m}{D_p}\right)^{1/5} \tag{15.7}$$

Sample Problem 15.1 Develop a similarity equation for torque.

Solution

Power $= \gamma Qh = T\omega$

For water, with $g = 32.2$ ft/s^2, we can neglect γ.

Thus, since $Q = K_Q nD^3$ and $h = K_h n^2 D^2$, we get

$$(K_Q nD^3)(K_h n^2 D^2) = T\omega, \quad \text{but} \quad \omega = 2\pi n/60$$

Hence

$$T = K_Q K_h \left(\frac{60}{2\pi n}\right) n^3 D^5 = K_T n^2 D^5 \qquad \textbf{\textit{ANS}}$$

Sample Problem 15.2 A small pump serving as a model, when tested in the laboratory at 3600 rpm, delivered 3.0 cfs at a head of 125 ft. (*a*) If the efficiency of this model pump is 84%, what is the horsepower input to this pump?

[5] See J. H. T. Sun, "Hydraulic Machinery", page 21.26, in V. J. Zapporo and H. Hasen (Eds.), *Davis' Handbook of Applied Hydraulics,* 4th ed., McGraw-Hill, New York, 1993.

(*b*) Predict the speed, capacity and horsepower input to the prototype pump if it is to develop the same head as the model pump and the model pump has a scale ratio of 1:10. Assume the efficiency of the prototype pump is 90%.

Solution

(*a*) From Eq. (15.2): $P_m = T\omega = \gamma Q h / 550\eta$ hp

$$P_m = 62.4(3)125/[550(0.84)] = 50.6 \text{ hp} \qquad \textbf{\textit{ANS}}$$

(*b*) From Eq. (15.5):

$$n_p = n_m(h_p/h_m)^{1/2}(D_m/D_p) = 3600(1)^{1/2}(1/10) = 360 \text{ rpm} \qquad \textbf{\textit{ANS}}$$

Eq. (15.4): $Q_p = Q_m(D_p/D_m)^3 n_p/n_m = 3 \times 10^3 \times 1/10 = 300$ cfs ***ANS***

Eq. (15.6): $P_p = 62.4(300)125/[550(0.90)] = 4730$ hp ***ANS***

Note: To satisfy homologous conditions (streamlines geometrically similar) the larger prototype pump must operate at a smaller rotative speed than the model pump.

EXERCISES

15.4.1 A model centrifugal pump has a scale ratio of 1:15. The model when tested at 3600 rpm, delivered 0.10 m^3/s of water at a head of 40 m with an efficiency of 80%. Assuming the prototype has an efficiency of 88%, what will be its speed, capacity, and power requirement at a head of 50 m?

15.4.2 A $19\frac{1}{2}$-in-diameter centrifugal-pump impeller discharges 20 cfs at a head of 100 ft when running at 1200 rpm. (*a*) If its efficiency is 85%, what is the brake horsepower, i.e. what is the horsepower input to the shaft of the pump? (*b*) If the same pump were run at 1500 rpm, what would be *h*, *Q*, and the brake horsepower for homologous conditions?

15.4.3 An axial-flow pump delivers 300 L/s at a head of 6.0 m when rotating at 1800 rpm. (a) If its efficiency is 80%, how many kilowatts of power must the shaft deliver to the pump? (*b*) If this same pump were operated at 1500 rpm, what would be *h*, *Q*, and the power delivered by the shaft for homologous conditions?

15.4.4 In Sample Prob. 15.2 the model with a scale ratio of 1:10 has a maximum efficiency of 84%. Using Eq. (15.7), calculate the maximum efficiency of the prototype pump and compare the result with the data given in the sample problem. Note that Eq. (15.7) is empirical, and will provide only approximate values.

15.5 PERFORMANCE CHARACTERISTICS OF PUMPS AT CONSTANT SPEED

The efficiency of a pump varies considerably, depending upon the conditions under which it must operate. Because of this, when selecting a pump for a

Table 15.1
Operating speeds of constant-speed electric motors

Pairs of poles	60-cycle		50-cycle	
	Synchronous	**Induction**[a]	**Synchronous**	**Induction**[a]
1	3600 rpm	3500 rpm	3000 rpm	2900 rpm
2	1800	1750	1500	1450
3	1200	1160	1000	960
4	900	870	750	720
5	720	695	600	575
6	600	580	500	480
8	450	435	375	360
10	360	350	300	290
12	300	290	250	240

[a] These values for the operating speed of induction motors are approximate. The speed of induction motors is usually 2–3% lower than that of synchronous motors.

given situation, it is important for the pump selector to have information regarding the performance of various pumps among which the selection is to be made. The pump manufacturer usually has information of this type, as determined by laboratory tests, for what are called shelf items, i.e., standard pumps. Large-capacity pumps, however, are sometimes custom-made. Often a model of such a pump is made and tested before final design of the prototype pump.

Though some centrifugal pumps are driven by variable-speed motors, the usual mode of operation of a pump is at constant speed. Typical speeds of constant-speed electric motors are given in Table 15.1. The ***pump characteristic curve*** (head versus capacity) and other performance curves for a typical mixed-flow centrifugal pump are shown in Fig. 15.6. This particular pump has a ***normal capacity*** of 10,500 gpm when developing a normal head of 60 ft at an operating speed of 1450 rpm. What is referred to as the "normal" capacity corresponds to the ***point of optimum efficiency*** or ***BEP*** (best efficiency point). Similar curves for a typical axial-flow pump are shown in Fig. 15.7. Curves such as those shown in Figs. 15.6 and 15.7 are usually determined by pump manufacturers through laboratory testing. Inspection of these two figures shows the remarkable difference in the characteristics of these two pumps. It can be seen that the efficiency of both pumps drops rather rapidly when the flow rate at which they are pumping exceeds the optimum. This is particularly true in the case of the axial-flow pump.

The shape of the impellers and vanes and their relationship to the pump casing cause variations in the intensity of shock loss, fluid friction, and turbulence. These vary with head and flow rate, and are responsible for the wide variation in pump characteristics. The ***shutoff head*** is that which is developed when there is no flow. In the case of the mixed-flow centrifugal

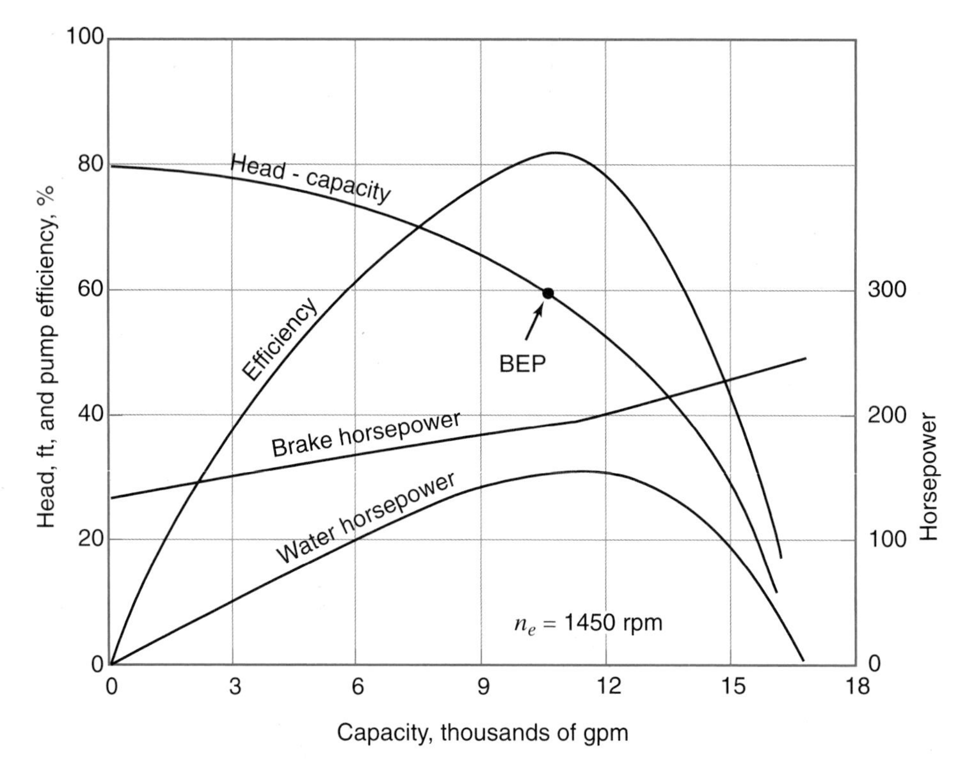

Figure 15.6
Characteristic curves for a typical mixed-flow centrifugal pump.

pump (Fig. 15.6), the shutoff head is about 10% greater than the normal head, that which occurs at the point of optimum efficiency, while in the case of the axial-flow pump (Fig. 15.7), the shutoff head may be as much as three times the normal head.

15.6 PERFORMANCE CHARACTERISTICS AT DIFFERENT SPEEDS AND SIZES

The choice of a pump for a given situation will depend on the rotative speed of the motor used to drive the pump. If the characteristic curve for a pump at a given rotative speed is known, the relation between head and capacity at different rotative speeds can be derived approximately through use of Eqs. (15.4) and (15.5). For example, in Fig. 15.8, assume the characteristic curve (Curve 1) of a pump when operating at n_1 rpm is given. The characteristic curves at different speeds of operation such as n_2 and n_3 can be

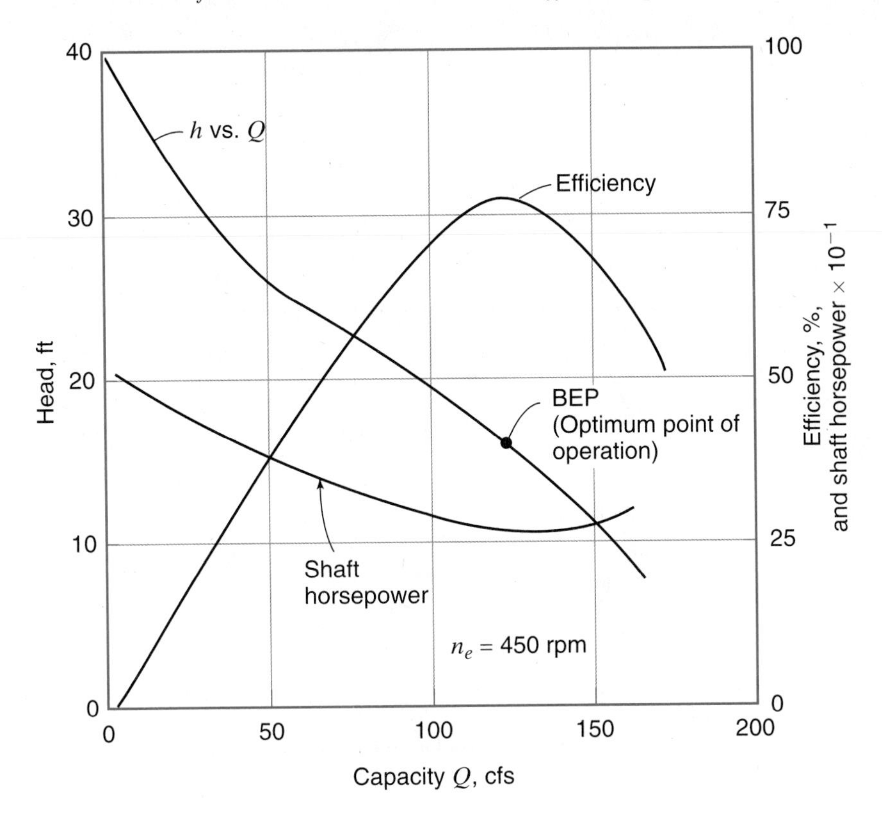

Figure 15.7
Characteristic curves for a typical axial-flow pump.

approximately derived by transferring points on Curve 1 to corresponding points on Curves 2 and 3 respectively. Thus, from Eqs. (15.4) and (15.5), we have for Curve 2, $Q_2 = Q_1(n_2/n_1)$ and $h_2 = h_1(n_2/n_1)^2$. Similar expressions can be used to develop Curve 3. Superimposed on Fig. 15.8 are lines of equal efficiency as determined by test. From this, it can be seen that efficiency drops off rather rapidly as one moves away from the BEP. Characteristic curves for pumps of different size, all operating at the same speed, can also be approximately derived making use of Eqs. (15.4) and (15.5). Such curves are shown in Fig. 15.9. Corresponding points were transferred through use of $Q_2 = Q_1(D_2/D_1)^3$ and $h_2 = h_1(D_2/D_1)^2$. Curves of equal efficiency have also been superimposed on Fig. 15.9.

Characteristic curves for pumps having similar geometric shape, of different size, operating at various constant speeds, can be approximately[6] developed through use of Eqs. (15.4) and (15.5), in which case $Q_2 = (n_2/n_1)(D_2/D_1)^3$ and $h_1 = (n_2/n_1)^2(D_2/D_1)^2$.

[6] The accuracy drops off with large variations in D and n.

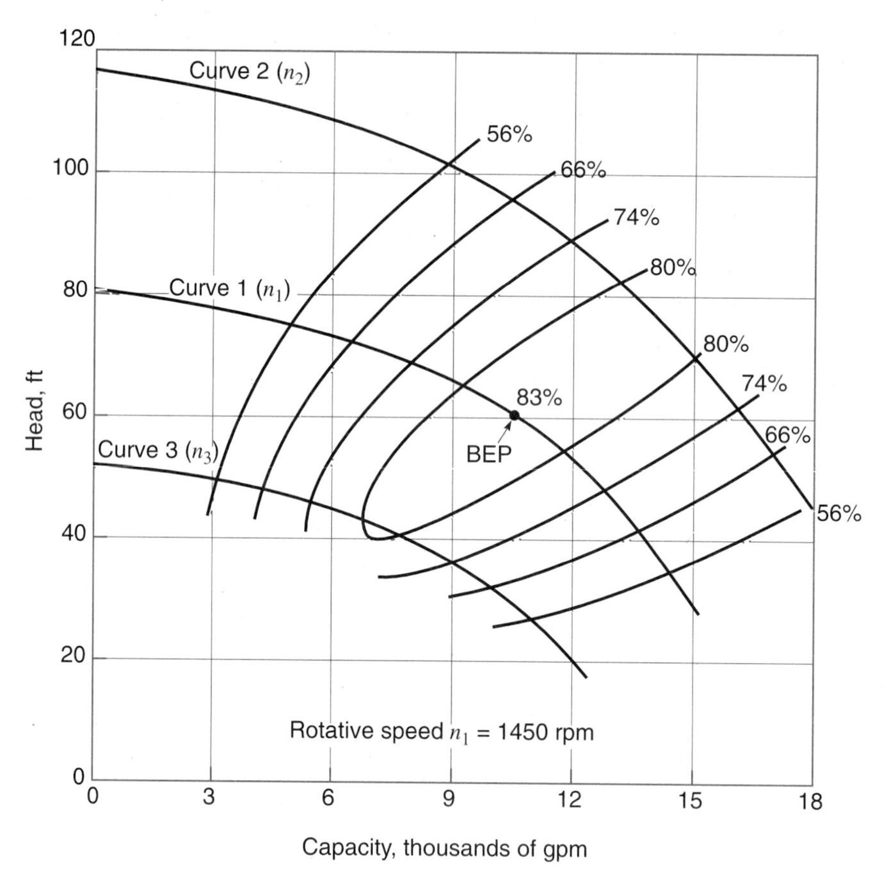

Figure 15.8
Characteristic performance curves of a typical mixed-flow centrifugal pump (Fig. 15.6) at various speeds of rotation with contours of equal efficiency.

SAMPLE PROBLEM 15.3 In Fig. 15.8, $n_1 = 1450$ rpm. Estimate the rotative speeds n_2 and n_3.

Solution

Using values of h for $Q = 0$ gpm, we have $h_1 = 80$ ft. $h_2 \approx 117$ ft and $h_3 \approx 52$ ft.

Eq. (15.5): $$h \propto n^2$$

Thus:

$$\frac{80}{117} = \left(\frac{1450}{n_2}\right)^2, \quad \text{from which } n_2 = 1450\left(\frac{117}{80}\right)^{1/2} = 1753 \approx 1750 \text{ rpm} \qquad \textbf{\textit{ANS}}$$

and

$$\frac{52}{80} = \left(\frac{n_3}{1450}\right)^2, \quad \text{from which } n_3 = 1450\left(\frac{52}{80}\right)^{1/2} = 1169 \approx 1170 \text{ rpm} \qquad \textbf{\textit{ANS}}$$

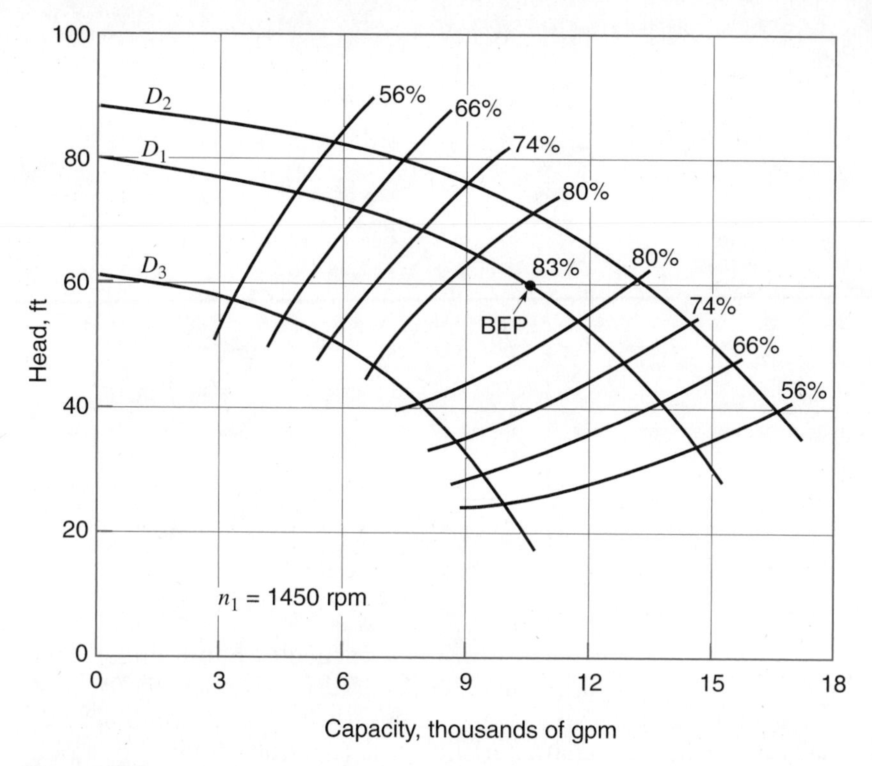

Figure 15.9
Characteristic performance curves with contours of equal efficiency for typical homologous mixed-flow centrifugal pumps having impellers of different size. D_1 is the same pump as in Fig. 15.6.

EXERCISES

15.6.1 Refer to Fig. 15.8. Transfer the point (h = 70 ft and Q = 7500 gpm) to a point on Curve 3 if n_1 = 1450 rpm and n_3 = 1160 rpm. Check your result with Fig. 15.8.

15.6.2 Refer to Fig. 15.9. Transfer the point (h = 70 ft and Q = 7500 gpm) to a point on the D_2 curve if D_1 = 16.7 in and D_2 = 17.5 in. Check your result with Fig. 15.9.

15.6.3 Plot h versus Q for the pump of Fig. 15.7 for the case where n = 600 rpm.

15.7 OPERATING POINT OF A PUMP

The manner in which a pump operates depends not only on the pump performance characteristics, but also on the characteristics of the system in

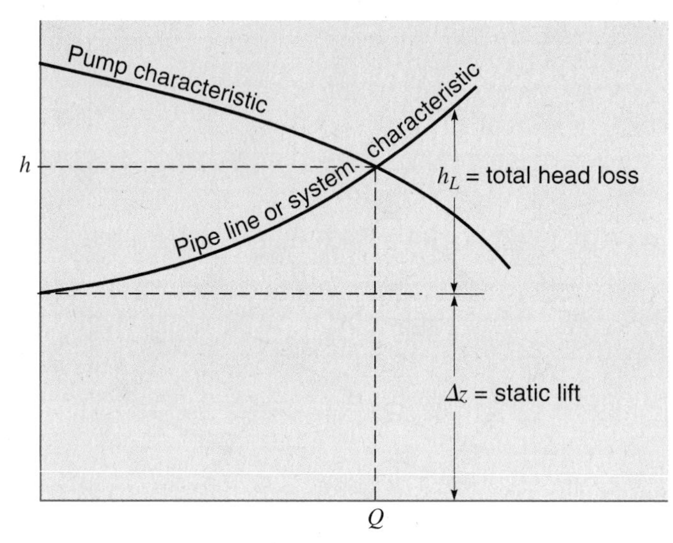

Figure 15.10 Graphical method for finding the operating point of a pump and pipeline.

which it is to operate. In Fig. 15.10, for a particular pump under consideration, we show the pump operating characteristics (h versus Q) for a selected speed of operation, usually close to the speed that gives optimum efficiency. We also show the system characteristic curve (i.e., the required pumping head versus Q). In this case, the pump is delivering liquid through a piping system with a static lift of Δz. The head that the pump must develop is equal to the static lift plus the total head loss in the piping system (approximately proportional to Q^2). The actual ***pump-operating head*** and flow rate are determined by the intersection of the two curves.

The particular values of h and Q determined by this intersection may or may not be those for the maximum efficiency. If they are not, this means the pump is not exactly suited to the specific conditions. In Figs. 15.8 and 15.9 we see that efficiency drops off as one moves away from the BEP. Hence it is important to select a pump such that the intersection of the pump performance and system characteristic curves intersect near the BEP. Changing the speed of the pump may help. Further discussion of the behavior of pumps and their relationship to the systems in which they operate is presented in Secs. 15.12 and 15.13.

SAMPLE PROBLEM 15.4 The pump whose characteristic curve is shown in Fig. 15.6 when operating at 1450 rpm is used to pump water from reservoir A to reservoir B through an 18-in-diameter pipe ($f = 0.032$) 500 ft long. Neglecting minor losses, find the flow rate for the following conditions: (a) reservoir water surface elevations are identical; (b) water surface elevation of reservoir B is 20 ft higher than that in reservoir A; (c) water surface elevation of reservoir B is 65 ft higher than that in reservoir A. Assume f does not change with flow rate. Also find the efficiency by referring to Fig. 15.8.

Solution

The pump characteristic curve is given by Fig. 15.6. Note that this curve is the same as Curve 1 of Fig. 15.8.

Energy equation: $$0 + h - h_L = \Delta z$$

from which we find the equation of the system characteristic:

$$h = \Delta z + h_L = \Delta z + f\frac{L}{D}\frac{V^2}{2g}$$

$$= \Delta z + (0.032)\frac{500}{1.5}\frac{Q^2}{[(3.14)(1.5)^2/4]^2 2(32.2)}$$

$$= \Delta z + 0.0531Q^2 \qquad \text{(Q expressed in cfs)}$$

Coordinates of system curves:

		(a)	(b)	(c)
Q, cfs	Q, gpm	h_L, ft	$20 + h_L$	$65 + h_L$
0	0	0.0	20.0	65.0
10	4890	5.3	25.3	70.3
20	9780	21.2	41.2	86.2
30	14,670	47.8	67.8	(112.8)[7]

By plotting the system curves on Fig. 15.8 we find h and Q at the points of intersection. To find pump efficiency we note the location of the points of intersection and interpolate between contours of equal efficiency.

$$\left.\begin{array}{l} (a)\ h = 41 \text{ ft and } Q = 13{,}600 \text{ gpm}, \eta = 66\% \\ (b)\ h = 52 \text{ ft and } Q = 12{,}000 \text{ gpm}, \eta = 79\% \\ (c)\ h = 72 \text{ ft and } Q = 6000 \text{ gpm}, \eta = 62\% \end{array}\right\} \qquad \textit{ANS}$$

EXERCISES

15.7.1 Repeat Sample Prob. 15.4 for the case where the length of the 18-in-diameter pipe is 1000 ft.

15.7.2 Repeat Sample Prob. 15.4 for the case where the pipe diameter is 12 in rather than 18 in, all other data remaining the same.

[7] For plotting purposes only. The pump cannot develop that much head when operating at 1450 rpm.

15.8 SPECIFIC SPEED OF PUMPS

Specific speed is a number that defines the type of pump (radial-flow, mixed-flow, or axial-flow). Traditionally in the United States, specific speed N_s of a pump has been expressed as

$$N_s = \frac{n_e\sqrt{Q}}{h^{3/4}} = \frac{n_e\sqrt{\text{gpm}}}{h^{3/4}} \tag{15.8a}$$

where the values of n_e (rpm), Q (gpm), and h (ft) are those that occur at the *point of optimum operating efficiency,* commonly referred to as the BEP. Computed values of specific speed for a given pump throughout its entire operating range from zero discharge (shutoff head) to maximum discharge would give values from zero to some very large number. The only value that has any real significance is that corresponding to values of head, discharge, and speed at the point of maximum efficiency.

Equation (15.8*a*) is derived by eliminating D from Eqs. (15.4) and (15.5) in such a way that n becomes a term in the numerator of the resulting expression. It is left to the reader to work out this derivation (see Exer. 15.8.1). For large-capacity pumps, specific speed has sometimes been calculated using Q expressed in cfs rather than gpm, in which case $(N_s)_{\text{cfs}} = 0.0472(N_s)_{\text{gpm}}$.

In the SI, specific speed of pumps is defined by

$$(N_s)_{\text{SI}} = \frac{\omega_e\sqrt{Q}}{(gh)^{3/4}} \tag{15.8b}$$

where the values of ω_e (rad/s), Q (m^3/s), and h (m) are those that occur at the BEP. The acceleration due to gravity g is 9.81 m/s^2. In this form, specific speed is dimensionless. However, rotative speed is expressed in radians per second, which is rather cumbersome. The relation between SI specific speed for pumps and the traditional mode of expressing specific speed is $(N_s)_{\text{SI}} = 0.000\,368(N_s)_{\text{gpm}}$.

Figure 15.11 shows several typical impellers in section and their corresponding specific speeds. Radial-flow impellers generally have specific speeds [Eq. (15.8*a*)] between 500 and 5000, mixed-flow between 4000 and 10,000 and axial-flow from 9000 to 15,000. In the SI the corresponding values of specific speed are approximately 0.2–2.0 for radial-flow, 1.5–3.7 for mixed-flow, and 3.3–5.5 for axial-flow pumps. Two impellers having the same geometric shape have the same specific speed though their sizes may differ. The curves of Fig. 15.11 show the variation of peak efficiency with specific speed. It should be noted that pumps with specific speed below 800 tend to be inefficient.

Equations (15.8) indicate that pumping against high heads requires a

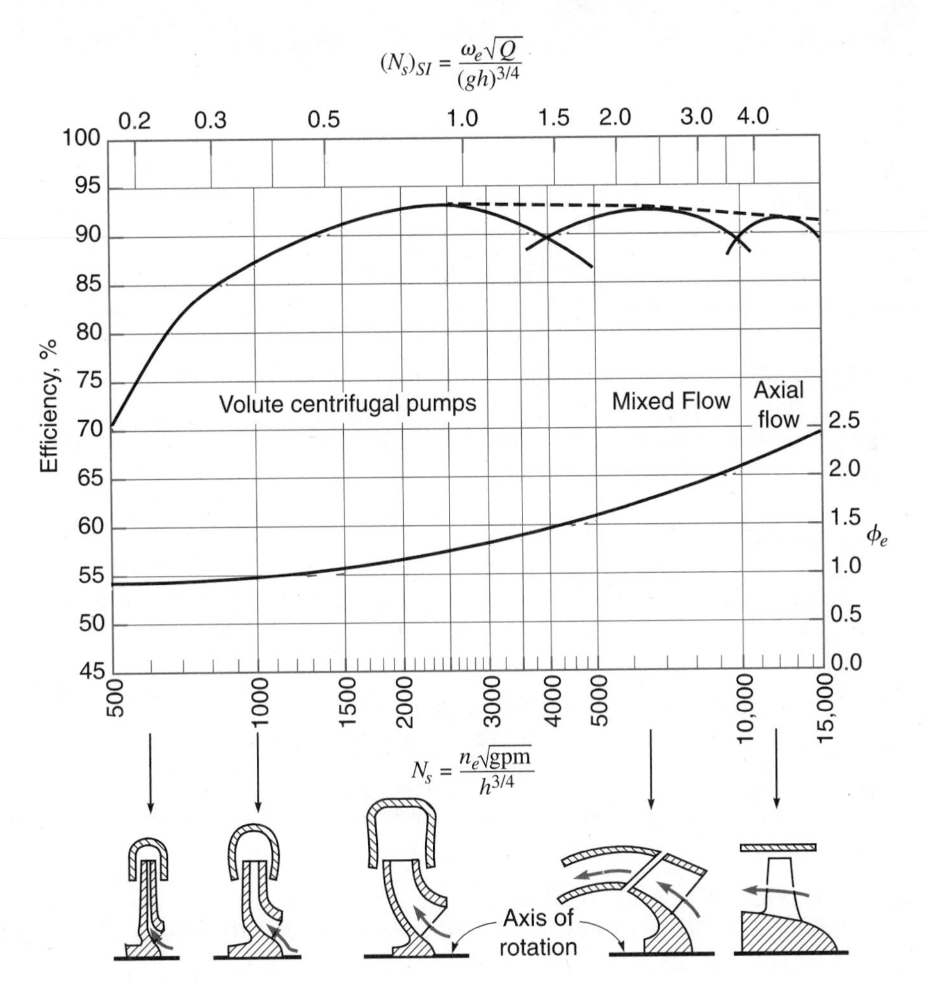

Figure 15.11
Optimum efficiency and typical values of ϕ_e for water pumps as a function of specific speed.

low-specific speed pump. This will be discussed further in Sec. 15.10. For very high heads and low discharges, the required specific speed may fall below the values for normal design and result in a pump with a low efficiency. To overcome this problem, the head may be distributed among a number of pumps in series, or a multi-stage unit may be used. The head per stage is generally limited to about 400 ft, although some pumps in use develop more than 600 ft of head per stage. A multi-stage pump is usually less expensive than a series of individual pumps, but this may be offset by the very high pressures developed in the multi-stage pump. When several pumps are spaced more or less uniformily along a pipeline, excessive pressures in the system can be avoided.

SAMPLE PROBLEM 15.5 What is the specific speed of the pump whose performance characteristics are given in Fig. 15.6? What type of pump is this?

Solution

At its BEP this pump has a capacity of 10,500 gpm while developing a head of 60 ft at a rotative speed of 1450 rpm.

Hence: $$N_s = \frac{n_e\sqrt{\text{gpm}}}{h^{3/4}} = \frac{1450\sqrt{10{,}500}}{60^{3/4}} = 6890 \qquad \textit{ANS}$$

From Fig. 15.11: This is a mixed-flow centrifugal pump. ***ANS***

EXERCISES

15.8.1 Derive the expression for the specific speed of a pump by eliminating D from Eqs. (15.4) and (15.5) in such a way that n becomes a term in the numerator.

15.8.2 What is the specific speed of the pump whose performance characteristics are given in Fig. 15.7? What type of pump is this?

15.8.3 What are the specific speeds of the pumps on the Colorado River Aqueduct at the Hayfield, Gene, and Iron Mountain Plants? (See Sec. 15.14.)

15.8.4 Find the specific speed of a 10-stage pump that develops a total head of 600 ft at a capacity of 1600 gpm when operating at maximum efficiency at a rotative speed of 900 rpm.

15.8.5 A pump is to discharge 0.8 m^3/s at a head of 40 m when running at 300 rpm. What type of pump will be required?

15.9 PERIPHERAL-VELOCITY FACTOR

For a pump impeller (Fig. 15.1) or a turbine runner (Fig. 16.7), the ratio of the peripheral velocity to $\sqrt{2gh}$ is referred to as the ***peripheral-velocity factor***, denoted by ϕ. Thus, for a pump,

$$u_2 = \phi\sqrt{2gh} \tag{15.9}$$

where u_2 is the peripheral speed of the impeller. For an axial-flow pump it is the vane-tip speed that is used in Eq. (15.9).

For any machine its peripheral velocity might be any value from zero up to some maximum under a given head, depending on the operating speed, and ϕ would consequently vary through a wide range. But the speed that is of most practical significance is that at which the efficiency is a maximum. The value of ϕ at the speed of maximum efficiency is denoted as ϕ_e.

At maximum efficiency,

$$u_2 = \frac{2\pi r n_e}{60} = \frac{\pi D n_e}{60} = \phi_e \sqrt{2gh}$$

Thus,

$$D = \frac{60\sqrt{2g}\ \phi_e \sqrt{h}}{n_e}$$

which in BG units reduces to

$$D = \frac{153.3\phi_e \sqrt{h}}{n_e} \tag{15.10}$$

Typical values of ϕ_e for various types of pumps as a function of specific speed are given in Fig. 15.11. This curve, together with Eq. (15.10), can be used to estimate the diameter of the impeller of a pump if the specific speed of the pump is known.

SAMPLE PROBLEM 15.6 Find the value of ϕ_e for the Eagle Mountain pumps mentioned in Sec. 15.14. How does this computed value of ϕ_e compare with the value given in Fig. 15.11?

Solution

Eq. (15.10): $D = 153.3\phi_e \sqrt{h}/n_e$

$$(81.6/12) = 153.3\phi_e \sqrt{440}/450$$

From which: $\phi_e = 0.95$ ***ANS***

$$\text{Specific speed } N_s = \frac{450\sqrt{200(448.8)}}{440^{3/4}} = 1404$$

From Fig. 15.11, $\phi_e = 1.0$, which is a close check. ***ANS***

SAMPLE PROBLEM 15.7 Estimate the diameter of the impeller of the pump whose operating characteristic is shown in Fig. 15.6. See Sample Prob. 15.5.

From Sample Prob. 15.5: Specific speed $N_s = 6890$

From Fig. 15.11 for $N_s = 6890$: $\phi_e = 1.7$

Eq. (15.10): $D = 155.3\phi_e\sqrt{h}/n$

$D = 153.3(1.7)\sqrt{60}/1450$

$D = 1.40\text{ ft} = 16.70\text{ in}$ ***ANS***

EXERCISES

15.9.1 Estimate the diameter of the impeller of the pump whose operating characteristics are shown in Fig. 15.7.

15.9.2 Estimate the diameter of the impeller of a pump whose BEP is defined by $Q = 5000$ gpm at a head of 82 ft when operating at 3000 rpm.

15.9.3 At maximum efficiency, a four-stage pump delivers 400 L/s against a head of 300 m at a rotative speed ω of 125.6 rad/s. Estimate the diameter of the impellers if all four impellers are identical. Calculate $(N_s)_{SI}$, convert to N_s, and then obtain an estimated value of D.

15.10 CAVITATION IN PUMPS

An important factor in the satisfactory operation of a pump is the avoidance of cavitation (Sec. 5.11), both for the sake of good efficiency and for the prevention of impeller damage. As liquid passes through the impeller of a pump, there is a change in pressure. If the absolute pressure of the liquid drops to the vapor pressure, cavitation will occur. The region of vaporization hinders the flow and places a limit on the capacity of the pump. As the fluid moves further into a region of higher pressure, the bubbles collapse and the implosion of the bubbles may cause pitting of the impeller.[8] Cavitation is most likely to occur near the point of discharge (periphery) of radial-flow and mixed-flow impellers, where the velocities are highest. It may also occur on the suction side of the impeller, where the pressures are the lowest. In the case of an axial-flow pump, the blade-tip is the most vulnerable to cavitation.

For pumps, a ***cavitation parameter*** has been defined as

$$\sigma = \frac{(p_s)_{\text{abs}}/\gamma + V_s^2/2g - p_v/\gamma}{h} = \frac{\text{NPSH}}{h} \tag{15.11}$$

where subscript s refers to values at the pump intake (i.e., suction side of the pump), h is the head developed by the pump, and p_v is the vapor pressure.

[8] The pitting of a turbine runner is shown in Fig. 16.15. Similar phenomena occur in pump impellers if cavitation takes place.

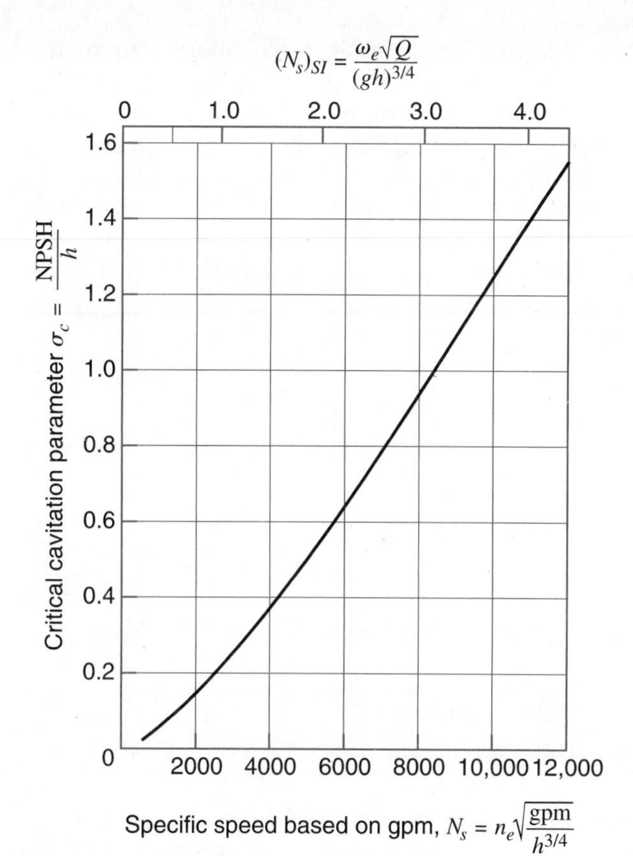

Figure 15.12
Approximate values of critical cavitation parameter σ_c as a function of specific speed.

As the latter is normally given in absolute units, it follows that p_s must also be absolute pressure. NPSH, the numerator of Eq. (15.11), is referred to as the ***net positive suction head.***

There is a critical value of σ, denoted as σ_c, below which there is a drop in pump efficiency indicative of the onset of cavitation. The value of σ_c depends on the type of pump and the conditions of operation. For a given pump, the value of σ_c varies with flow rate. Approximate values of σ_c for centrifugal pumps operating under normal conditions near optimum efficiency are shown in Fig. 15.12. In important installations the value of σ_c should be determined experimentally in a model study. For modeling purposes, a useful parameter, the suction specific speed S, has been defined as

$$S = \frac{n\sqrt{\text{gpm}}}{\text{NPSH}^{3/4}} \tag{15.12}$$

If model and prototype are operated at identical values of N_s and S, similarity of flow and cavitation is achieved provided the model and prototype are geometrically similar to one another.

To prevent the occurence of cavitation, a pump must operate in such a manner that σ is larger than σ_c. This can be accomplished by selecting the proper type and size of pump and speed of operation and by setting the pump at the proper point and elevation in the system. Through inspection of Eq. (15.11), it can be seen that small values of σ result from large values of h, the head developed by the pump. Hence, for any given situation, there is a limiting value of h above which σ will be less than σ_c, resulting in cavitation.

By writing an energy equation for the situation depicted in Fig. 15.5 in terms of absolute pressure from the surface of the source reservoir to the suction side of the pump, we obtain

$$\frac{(p_0)_{\text{abs}}}{\gamma} - h_L = z_s + \frac{(p_s)_{\text{abs}}}{\gamma} + \frac{V_s^2}{2g}$$

or

$$\frac{(p_s)_{\text{abs}}}{\gamma} + \frac{V_s^2}{2g} = \frac{(p_0)_{\text{abs}}}{\gamma} - h_L - z_s$$

Substituting the right-hand side of the above expression into Eq. (15.11), we get

$$\sigma = \frac{\dfrac{(p_0)_{\text{abs}}}{\gamma} - h_L - z_s - \dfrac{p_v}{\gamma}}{h} \tag{15.13}$$

In the above equations $(p_0)_{\text{abs}}/\gamma$ is the absolute pressure on the surface of the source reservoir. If the liquid is drawn from a closed reservoir, $(p_0)_{\text{abs}}/\gamma$ could be either greater or less than the atmospheric pressure. In the usual case the reservoir is open to the atmosphere, and $(p_0)_{\text{abs}}/\gamma = p_{\text{atm}}/\gamma$.

The expression for σ [Eq. (15.13)] indicates that σ will tend to be small (hence there will be a tendency toward cavitation) in the following situations: (*a*) high head; (*b*) low atmospheric pressure, i.e. high elevation; (*c*) large head loss between source reservoir and the pump; (*d*) large value of z_s, i.e., with the pump at a relatively high elevation compared with the elevation of the reservoir water surface; and (*e*) large value of vapor pressure, i.e. high temperature and/or a very volatile liquid such as gasoline being pumped. With (*a*), we can limit the required head by using multistage pumps. For a given liquid to be pumped at a certain elevation and temperature, we have no control over items (*b*) and (*e*) above. However, we can do something about (*c*) and (*d*). When we design the layout of a pumping system, we can reduce the tendency toward cavitation by minimizing h_L [item (*c*)] by placing the pump close to the source reservoir, and by setting the pump at a low elevation [item (*e*)] relative to the reservoir water surface.

Introducing the critical value of σ into Eq. (15.13) and rearranging, we get an expression for $(z_s)_{\text{max}}$, the highest elevation at which a pump can be safely

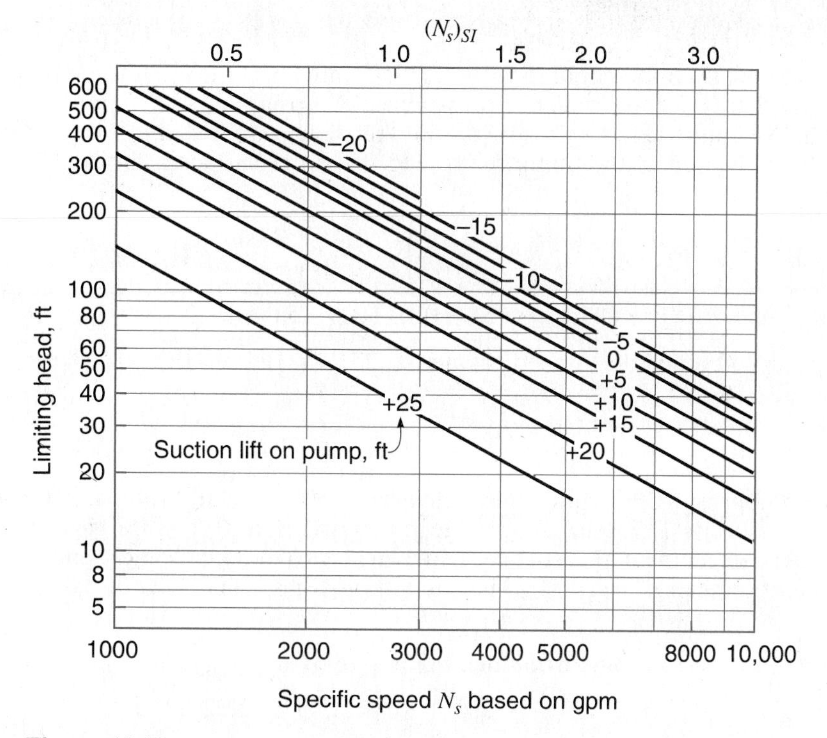

Figure 15.13
Recommended limiting heads for single-stage, single-suction pumps as a function of specific speed and suction lift, at sea level with water temperature of 80°F.

set to guard against cavitation:

$$(z_s)_{\max} = \frac{(p_0)_{\text{abs}}}{\gamma} - \frac{p_v}{\gamma} - \sigma_c h - h_L \tag{15.14}$$

As long as z_s is less than $(z_s)_{\max}$, there should be no problem with cavitation, assuming, of course, that an accurate value of σ_c is known for the given conditions of operation. Pump manufacturers often run tests on pumps to determine values of σ_c for different conditions of operation.

Based on experience with pumps, in Fig. 15.13 ***limiting heads*** are shown for the prevention of cavitation in single-stage single-suction pumps as a function of specific speed and ***suction lift***, the elevation difference between the energy line at suction and the center of the impeller as indicated in Fig. 15.5. A positive suction lift indicates that the impeller is above the energy line. The curves of Fig. 15.13 are applicable to water at 80°F under sea-level atmospheric pressure. These curves can readily be shifted for other conditions: different temperatures, different elevations, and different liquids.

SAMPLE PROBLEM 15.8 A pump with a critical value of σ of 0.10 is to pump against a head of 500 ft. The barometric pressure is 14.3 psia, and the vapor pressure is 0.5 psia. Assume friction losses in the intake piping are 5 ft. Find the maximum allowable elevation of the pump relative to the water surface at intake.

Solution

$$p_{atm}/\gamma = (14.3)144/62.4 = 33.0 \text{ ft}$$

$$p_v/\gamma = (0.5)144/62.4 = 1.154 \text{ ft}$$

Eq. (15.14): $(z_s)_{max} = 33.0 - 1.154 - 0.10(500) - 5.0 = -23.2 \text{ ft}$

The pump should be placed at least 23.2 ft below the reservoir water surface. ***ANS***

SAMPLE PROBLEM 15.9 A pump is delivering 7500 gpm at 140°F and the barometric pressure is 13.8 psia. Determine the reading on a pressure gage in inches of mercury vacuum at the suction flange when cavitation is incipient if σ_c = 0.085 and the head h = 240 ft. The suction pipe has a diameter of 2.0 ft.

Solution

Table A.1 at 140°F: $\gamma = 61.38$ pcf, $p_v/\gamma = 6.67$ ft

$$V_s = Q/A_s = (7500/449)/(\pi 2^2/4) = 5.32 \text{ ft/sec}$$

Let p = gage pressure at suction flange: $(p_s)_{abs} = p_{at} + p = (13.8 + p)$ psia

Eq. (15.11): $0.085 = [(13.8 + p)(144/61.38) + 5.32^2/2(32.2) - 6.67]/240$

$$p = -2.45 \text{ psi}$$

$$-2.45(29.9 \text{ inHg}/14.7 \text{ psia}) = 4.98 \text{ inHg vacuum}$$ ***ANS***

EXERCISES

15.10.1 A mixed-flow pump, located at sea level, with a specific speed of 6000 is to be used to pump 80°F water from a reservoir. The head to be developed is 40 ft. What is the greatest elevation above the reservoir water surface that the pump can be placed such that cavitation will not occur in the pump? Assume the head loss from the reservoir to the pump is 1.5 ft.

15.10.2 Solve Exer. 15.10.1 for the case where the pump is located at elevation 10,000 ft with a water temperature of 50°F. Assume all other data remain the same.

15.10.3 A pump with a critical value of σ of 0.20 is to pump against a head of 20 m. The barometric pressure is 98.5 kPa abs, and the vapor pressure is 5.2 kPa abs. Friction losses from the reservoir to the pump are 0.5 m. Find the maximum allowable height of the pump relative to the water surface in the reservoir.

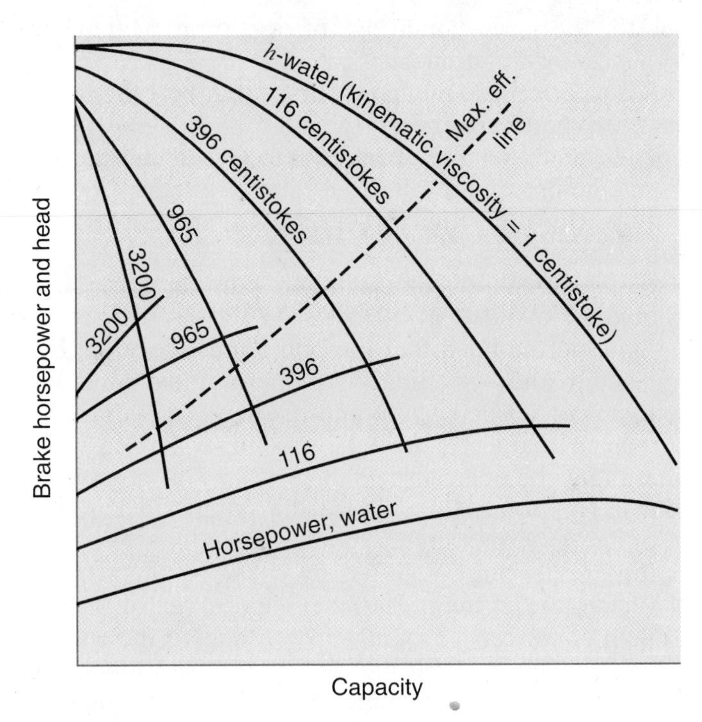

Figure 15.14
Centrifugal pump with viscous oils.

15.11 VISCOSITY EFFECT

Centrifugal pumps are also used to pump liquids with viscosities different from that of water. Figure 15.14 shows the actual test curves of performance for a very extreme range of viscosities, from water to an oil with a kinematic viscosity 3200 times that of water. It is seen that, as the viscosity is increased, the head–capacity curve becomes steeper and the power required increases. The dashed line indicates the points of maximum efficiency for each curve. It is seen that both the head and capacity at the point of maximum efficiency decrease with increasing viscosity. As these are accompanied by an increase in brake horsepower, there is a marked decrease in efficiency. For example, for the situation depicted in Fig. 15.14, if the optimum efficiency of the pump when pumping water is 0.85, its optimum efficiency is only 0.47 when pumping a liquid whose viscosity is 116 times that of water and only 0.18 when pumping a liquid whose viscosity is 396 times that of water.

15.12 SELECTION OF PUMPS

In selecting a pump for a given situation, we have a variety of pumps to choose among. We strive to select a pump that will operate at a relatively high efficiency for the given conditions of operation. Manufacturers provide pump performance information such as that in Figs. 15.6 and 15.7. The engineer's task is to select the pump or pumps that best fits in with the system characteristics. Alternatives to be investigated include specific speed N_s, size D of the impeller, and speed n of operation. Other alternatives include the use of multi-stage pumps, pumps in series, pumps in parallel, etc. Even, under certain conditions, throttling the flow in the system can result in savings in energy.

In all cases the possibility of cavitation must be carefully investigated. Cavitation should be avoided at all costs. Problems with cavitation can be eliminated by limiting the head that the pump must develop, by selecting the proper type of pump, and by setting the pump at a low enough elevation.

The objective is to select a pump and its speed of operation such that the pump's performance characteristics relate to the system in which it operates in such a fashion that the operating point (Sec. 15.7) is close to the BEP (best operating point). This tends to optimize the pump efficiency, resulting in a minimization of energy expenditure.

The operating point can be shifted by changing the pump characteristic curve, by changing the system characteristic curve, or by changing both curves. The pump curve can be changed by changing the speed of operation of a given pump or by selecting a different pump with different performance characteristics (Fig. 15.15*a*). In some instances it is helpful to trim the

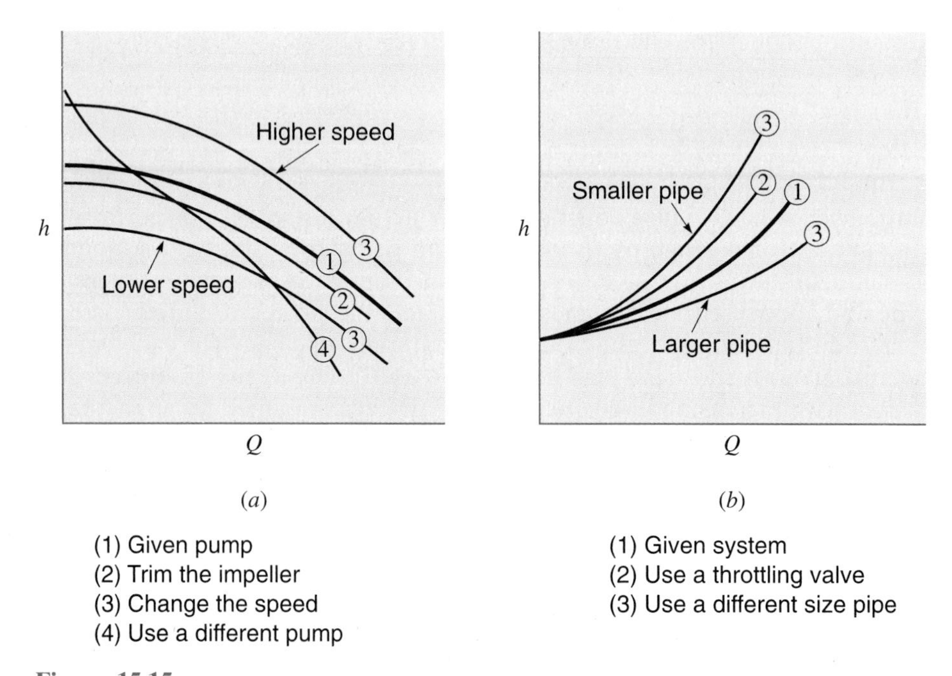

Figure 15.15
Changing the operating point by (*a*) changing the pump characteristic curve or (*b*) changing the system characteristic curve.

impeller, i.e., reduce its diameter somewhat, perhaps 5% or so, by grinding it down. The smaller impeller is installed in the original casing. The system characteristic curve can be changed by changing the pipe size or by throttling the flow (Fig. 15.15*b*).

A complication that often occurs is that the energy levels at the two ends of the system do not remain constant, such as that created by fluctuating reservoir levels. In such a case it is difficult to achieve a high efficiency for all modes of operation. In extreme cases a variable-speed motor is sometimes used.

Sample Problem 15.10 Water at 50°F is to be pumped between reservoirs *A* and *B* whose water surfaces are at elevations 5910 ft and 6060 ft, respectively. The water is to be delivered at 20 cfs through a 24-in-diameter pipe of length 2000 ft. Assume $f = 0.03$. (*a*) Find the specific speed of the most suitable pump if the speed of operation is to be 600 rpm. (*b*) If this pump is installed at a point 300 ft from reservoir *A* at the same elevation as the water surface in reservoir *A*, will it be safe against cavitation?

Solution

(*a*)
$$V = Q/A = 20/\pi(1)^2 = 6.37 \text{ ft/sec}$$

$$h = (6060 - 5910) + 0.03 \frac{2000}{2} \frac{(6.37)^2}{2(32.2)} = 168.9 \text{ ft}$$

Eq. (15.8*a*):
$$N_s = \frac{n_e \sqrt{\text{gpm}}}{h^{3/4}} = \frac{600\sqrt{20 \times 449}}{(168.9)^{3/4}} = 1214 \qquad \textit{ANS}$$

(*b*) Eq. (15.14):
$$(z_s)_{\max} = \frac{(p_0)_{\text{abs}}}{\gamma} - \frac{p_v}{\gamma} - \sigma_c h - h_L$$

Tables A.3 and A.1:
$$\frac{(p_0)_{\text{abs}}}{\gamma} = \frac{11.8(144)}{62.4} = 27.3 \text{ ft}$$

and
$$\frac{p_v}{\gamma} = 0.41 \text{ ft} \approx 0.4 \text{ ft}$$

Fig. 15.12:
$$\sigma_c \approx 0.08, \qquad \sigma_c h \approx 0.08(168.9) \approx 15.2 \text{ ft}$$

Eq. (8.10):
$$h_L = f \frac{L}{D} \frac{V^2}{2g} = 0.03 \frac{300}{2} \frac{(6.37)^2}{2(32.2)} = 2.8 \text{ ft}$$

$$(z_s)_{\max} = 27.3 - 0.4 - 13.5 - 2.8 = 10.5 \text{ ft}$$

Yes, the pump will be safe from cavitation. It can be placed as much as 10.5 ft above the water surface elevation in reservoir *A*. ***ANS***

Sample Problem 15.11 Repeat Sample Prob. 15.10 for a pump where the rotative speed is 1200 rpm rather than 600 rpm, with all other data remaining the same.

Solution

Eq. (15.8*a*): $N_s \propto n_e$; hence $N_s = (1200/600)1214 = 2428$

From Fig. 15.12, when $N_s = 2428$: $\sigma_c \approx 0.18$

Thus: $\sigma_c h \approx 0.18(168.9) \approx 30.4$ ft

And $(z_s)_{\max} = 27.3 - 0.4 - 30.4 - 2.8 \approx -6.4$ ft

To prevent cavitation, the pump must be placed 6.4 ft or more below the water surface elevation in reservoir *A*. ***ANS***

By comparing the results of Sample Probs. 15.10 and 15.11, we find that rotative speed can greatly influence the specific speed of the pump that should be selected at a given installation. We also observe that an increase in operating speed results in a pump with a higher specific speed that is more vulnerable to cavitation.

SAMPLE PROBLEM 15.12 Find the approximate diameters of the pumps of Sample Probs. 15.10 and 15.11.

Solution

Eq. (15.10):
$$D = \frac{153.3\phi_e\sqrt{h}}{n_e}$$

Fig. 15.11: $N_s = 1214$, $\phi_e \approx 0.9$, and $N_s = 2428$, $\phi_e \approx 1.2$

for $N_s = 1214$:
$$D = \frac{153.3(0.9)\sqrt{168.9}}{600} = 2.99 \text{ ft} = 35.9 \text{ in}$$

for $N_s = 2428$:
$$D = \frac{153.3(1.2)\sqrt{168.9}}{1200} = 1.99 \text{ ft} = 23.9 \text{ in}$$

Thus the faster speed permits us to employ a smaller pump, but it is more vulnerable to cavitation (i.e., it must be set at a lower elevation).

EXERCISES

15.12.1 Repeat Sample Prob. 15.10 for the case where the reservoir water surface elevations are 120 ft and 160 ft respectively, with all other data remaining the same.

15.12.2 Repeat Sample Prob. 15.10 for the case where the flow rate is 50 cfs rather than 20 cfs, with all other data remaining the same.

15.12.3 Water flows at 80°C through a 100-mm-diameter pipe ($f = 0.024$) of length 300 m from a large pressurized tank ($p = 200$ kPa, water surface elevation 60.0 m) to discharge into the atmosphere at elevation 68.0 m. If the flow rate is to be 20 L/s, find the head that a pump, rotating at 1200 rpm placed in the pipeline, must develop and determine its specific speed. Assume the pressure in the tank remains constant and neglect minor losses. This pump is placed 6.0 m above the water level in the tank. Will cavitation occur if the head loss to the pump is 0.8 m?

15.13 PUMPS OPERATING IN SERIES AND IN PARALLEL

In Fig. 15.16*a, b,* and *c* we summarize our discussion thus far by showing: (*a*) the general shape of the characteristic curves for the three different types of pumps, (*b*) the effect on performance of changing the rotative speed of a given pump, and (*c*) the effect on performance of different sizes of pump impellers. Figures 15.16*d* and 15.16*e* show the performance curves for two

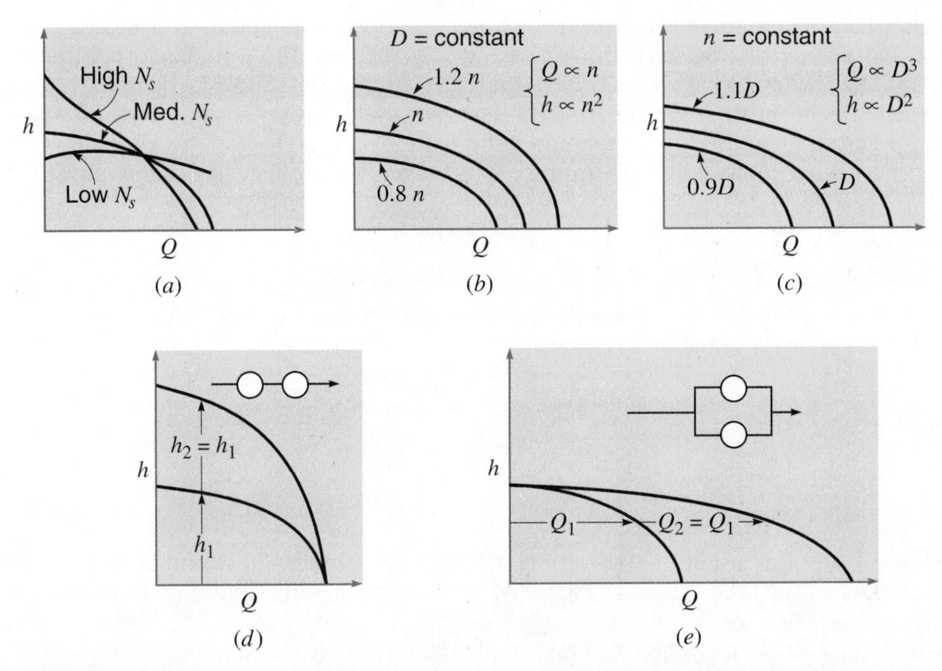

Figure 15.16
Pumping alternatives. (*a*) Different pumps with different characteristics. (*b*) A particular pump at different speeds. (*c*) Homologous pumps of different size. (*d*) Two identical pumps in series. (*e*) Two identical pumps in parallel. (*Note*: In series or parallel the pumps need not be identical, but their performance characteristics should be close to one another.)

identical pumps in series and for two identical pumps in parallel. By examining Figs. 15.16*d* and 15.16*e,* we see that pumps in series tend to increase head, while pumps in parallel tend to increase flow rate.

When pumps are installed in series or parallel, it is very important that they have reasonably similar, or better yet identical, head–capacity characteristics throughout their range of operation; otherwise, one pump will carry most of the load and, under certain conditions, all of the load, with the other pump acting as a hindrance rather than a help. In fact, in parallel, if the performance characteristics of the pumps are quite different, a condition of backflow can occur in one of the pumps. Finally, one must always be sure that the selected pump (or pumps) will not encounter cavitation problems over the full range of operating conditions.

The mode of operation for any pumping system is best determined by plotting the pumping characteristics and the pipe system characteristics on the same diagram (Fig. 15.10). The point at which the two curves intersect indicates what will take place. Several aspects of the relationship between the pump and system characteristics are demonstrated in the following sample problem.

SAMPLE PROBLEM 15.13 Two reservoirs A and B are connected with a long pipe that has characteristics such that the head loss through the pipe is expressible as $h_L = 20Q^2$, where h_L is in feet and Q is the flow rate in 100's of gpm. The water surface elevation in reservoir B is 35 ft above that in reservoir A. Two identical pumps are available for use to pump the water from A to B. The characteristic curve of each pump when operating at 1800 rpm is given in the following table.

Operation at 1800 rpm

Head, ft	Flow rate, gpm
100	0
90	110
80	180
60	250
40	300
20	340

At the optimum point of operation, the pump delivers 200 gpm at a head of 75 ft. Determine the specific speed N_s of the pump and find the rate of flow under the following conditions: (*a*) A single pump operating at 1800 rpm; (*b*) two pumps in series, each operating at 1800 rpm; (*c*) two pumps in parallel, each operating at 1800 rpm.

Solution

Eq. (15.7): $$N_s = \frac{n_e\sqrt{\text{gpm}}}{h^{3/4}} = \frac{1800(200)^{1/2}}{75^{3/4}} = 999 \qquad \textbf{\textit{ANS}}$$

The head–capacity curves for the pumping alternatives are plotted, and so is the h versus Q curve for the pipe system. In this case $h = \Delta z + h_L = 35 + 20Q^2$.

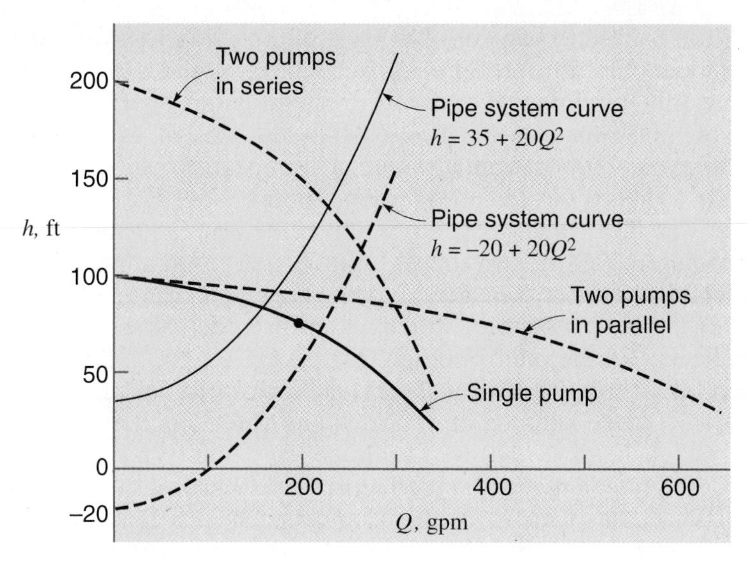

Figure S15.13

The answers are found at the points of intersection of the curves. They are as follows: (*a*) single pump, 156 gpm; (*b*) two pumps in series, 224 gpm; (*c*) two pumps in parallel, 170 gpm. ***ANS***

If Δz had been greater than 100 ft, neither the single pump nor the two pumps in parallel would have delivered any water. If Δz had been −20 ft (i.e., with the water surface elevation in reservoir *B* 20 ft below that in *A*, the flows would have been: (*a*) 212 gpm; (*b*) 258 gpm; (*c*) 232 gpm.

EXERCISES

15.13.1 Suppose the pumps of Sample Prob. 15.13 were operated at 1500 rpm. What then would have been the flow rates for (*a*) a single pump; (*b*) two pumps in series; (*c*) two pumps in parallel. All other data to remain the same.

15.13.2 Two pumps whose characteristics are given in Sample Prob. 15.13 are to be used in parallel. They must develop a head $h = 35 + 20Q^2$ as in the sample problem. (*a*) What is the flow rate with both pumps operating at 1800 rpm? (*b*) The speed of one of the pumps is reduced until it no longer delivers water. At approximately what speed will this happen?

15.14 PUMP INSTALLATIONS

A few examples of pump installations are presented here as illustrations of modern practice.

The Byron Jackson Company has built pumps with as many as 54 stages. Water has been lifted to heights of several thousand feet by multi-stage pumps. Ingersoll-Rand produced a 6-in 10-stage pump operating at 3750 rpm that delivers 1600 gpm at a head of 6000 ft, the shutoff head being 7000 ft. The Worthington Corporation installed a pump at Rocky River to deliver 279.5 cfs at a head of 238.84 ft when running at 327 rpm. The brake horsepower was 8259, giving an efficiency of 91.7%. The impeller diameter was approximately 7.54 ft, and the width at outlet approximately 0.72 ft, with an ***eye*** (inlet) diameter of 4.24 ft.

On the Colorado River Aqueduct, which delivers water from the Colorado River to the metropolitan Los Angeles area, the Worthington corporation built three pumps for the Hayfield plant to deliver 200 cfs each at a head of 444 ft when running at 450 rpm, and three similar pumps for Eagle Mountain with a head of 440 ft. The impeller diameters are approximately 81.6 in, and the eye diameters 34 in. The Byron Jackson Company built three pumps for the Gene plant to deliver 200 cfs each at a head of 310 ft when running at 400 rpm. The impeller diameters are 78 in. At the Intake plant, where the head is 294 ft, the impeller diameters are 76 in. The Allis-Chalmers Company built three pumps for the Iron Mountain plant to deliver 200 cfs each at a head of 146 ft when running at 300 rpm.

A typical moderate-sized large-capacity pumping plant is that at Cartersville, Georgia where the Johnston Pump Company installed a two-stage mixed-flow vertical-shaft pump that delivers 95,000 gpm against a head of 95 ft when operating at 394 rpm. At Marineland of the Pacific in Palos Verdes, California, three-stage Johnston vertical turbine pumps are used to pump salt water to tanks housing marine life. Because of the corrosive action of the sea water, the pump bowls are of iron with vitreous enamel coating. All moving parts that come in contact with the sea water are constructed of corrosion-resistant material.

A noteworthy pumping project is that at Grand Coulee on the Columbia River, where pumps with an impeller diameter of $167\frac{3}{8}$ in operating at their BEP deliver 1250 cfs against a head of 344 ft when operating at 200 rpm. The maximum efficiency is 90.8%. The shutoff head is 422 ft. These pumps when operating at a head of 270 ft deliver 1650 cfs at an efficiency of 84%.

The Allis-Chalmers Company has built a combination reversible pump–turbine for the Hiwasee plant of the Tennessee Valley Authority, which as a pump will deliver 3900 cfs at a head of 205 ft at the point of maximum efficiency while requiring approximately 100,000 bhp. The impeller diameter is 266 in, and it runs at 105.9 rpm. The maximum capacity is 5200 cfs at 135-ft head.

In Italy is a pump built in Switzerland that discharges 250,000 gpm, or 558 cfs, at a head of 787 ft at 450 rpm. It requires 62,000 bhp.

A hot-oil pump to deliver 875 gpm at a head of 8600 ft with 19 stages has been built by the Byron Jackson Company; this is one of the highest head pumps in existence.

The Harvey O. Banks Delta Pumping Station, named after the first Director of Water Resources for the State of California, lifts water 244 ft from the Sacramento–San Joaquin Delta to the California Aqueduct, which delivers water 350 miles to Southern California. There are five units that deliver 1067 cfs each and two units that deliver 350 cfs each for a total plant capacity of over 6000 cfs.

One of the world's largest pumping installations is the Edmonston Pumping Plant of the State of California water project. This plant accepts water from the California Aqueduct, and lifts water over the Tehachapi mountains. At this plant there are 14 four-stage vertical-shaft centrifugal pumps. Each is capable of delivering 315 cfs against a head of 1970 ft when rotating at 600 rpm. Their maximum efficiency is 92%. The maximum energy requirements for this plant are approximately 6×10^9 kWh per year.

PROBLEMS

15.1 The diameter of the discharge side of a pump is 100 mm, and that of the intake pipe is 120 mm. The pressure gage at discharge reads 240 kPa and the pressure gage at intake reads 60 kPA. If $Q = 45$ L/s of water and the pump efficiency is 0.82, find the power delivered to the pump by the drive shaft. The intake of the pump is 350 mm below the discharge side. Neglect the effects of pre-rotation in the entry pipe and assume smooth flow at discharge from the pump.

15.2 The diameter of the discharge side of a pump is 6 in, and that of the intake pipe is 8 in. The pressure gage at discharge reads 30 psi, and the vacuum gage at intake reads 10 inHg. If $Q = 3.0$ cfs of water and the brake horsepower is 35.0, find the efficiency of the pump. The intake and discharge are at the same elevation.

15.3 When operating at its BEP, the pump of Fig. 15.6 develops 60 ft of head at a capacity of 10,500 gpm. (*a*) Calculate the horsepower delivered to the water by the pump. (*b*) If the brake horsepower is 192 hp, what is the efficiency of the pump at its BEP? Does your answer to (*b*) agree with Fig. 15.6?

15.4 The pump of Fig. 15.6 delivers 26 ft of head at a capacity of 15,000 gpm. The pump efficiency under these conditions of operation is 44%. Calculate the water horsepower and the brake horsepower, and check to see if your answers agree with Fig. 15.6.

15.5 A centrifugal pump delivers 40 gpm of gasoline ($s = 0.72$) when developing a net head of 15 ft. If the shaft power is 0.126 hp, what is the efficiency of the pump?

15.6 A pump delivers 1000 L/s of water at a head of 8.5 m. If the efficiency of the pump is 68%, what is the shaft power?

15.7 A centrifugal pump with an 18-in-diameter impeller is rated at 25 cfs under a head of 100 ft when running at 1200 rpm. (*a*) If its efficiency is 85%, what is the brake horsepower? (*b*) If the same pump is run at 1800 rpm, what would be its rating in terms of h, Q, and the brake horsepower? Assume the efficiency does not change with the change in speed.

15.8 A centrifugal pump with a 300-mm-diameter impeller is rated at 40 L/s against a head of 25 m when rotating at 1750 rpm. What would be the rating of a pump of identical geometric shape with a 200-mm-diameter impeller? Assume pump efficiencies and pump speeds are identical.

15.9 At its optimum point of operation, a given centrifugal pump delivers 3.2 m^3/s of water against a head of 25 m when rotating at 1450 rpm. (*a*) If the efficiency is 82%, what is the brake power of the drive shaft? (*b*) If a homologous pump with an impeller diameter one half as large is rotating at 1200 rpm, what would be the discharge, head, and shaft power at its point of optimum efficiency? Assume both pumps operate at the same efficiency.

15.10 Using Eq. (15.7), find the approximate efficiency of the smaller pump of Prob. 15.9.

15.11 At its BEP, a given pump develops 30 ft of head when operating at 720 rpm. The flow rate is 4.2 cfs. Find the head and flow rate for a homologous pump operating at 600 rpm if its diameter is 90% that of the given pump.

15.12 Plot h versus Q for the pump of Fig. 15.6 for the case where n = 1200 rpm.

15.13 Plot h versus Q for the pump of Fig. 15.7 for the case where n = 600 rpm.

15.14 A centrifugal pump is installed to deliver water from a reservoir of water surface elevation zero to another of elevation 300 ft. The 12-in-diameter suction pipe (f = 0.020) is 100 ft long and the 10-in-diameter discharge pipe (f = 0.026) is 5000 ft long. The pump characteristic at 1200 rpm is defined by $h_p = 375 - 24Q^2$, where the pump head h_p is in ft and Q is in cfs. Calculate the rate at which this pump will deliver water, assuming the setting is low enough to avoid cavitation.

15.15 A pump whose characteristics are shown on Fig. 15.7 when operating at 450 rpm is used to pump water from reservoir A to reservoir B through a 24-in-diameter pipe (f = 0.028) of length 170 ft. Neglecting minor losses, find the flow rate and pump efficiency for the following conditions: (*a*) reservoir water surface elevations identical; (*b*) water surface in reservoir B is 10 ft higher than that in reservoir A; (*c*) reservoir water surface elevation in reservoir B is 10 ft lower than that in reservoir A.

15.16 Repeat Prob. 15.15 using the same data except consider minor losses: square entrance and submerged discharge. Neglect bend losses.

15.17 A pump whose characteristics are shown in Fig. 15.8, operating at 1450 rpm, is placed in a 24-in-diameter pipe (f = 0.032) of length 5000 ft. The discharge end of the pipe is 30.0 ft lower than the water surface elevation in the reservoir. The water discharges freely into the atmosphere. (*a*) Find the flow rate. (*b*) What would be the flow rate if the pump had not been installed in the pipeline?

15.18 Repeat Prob. 15.17 with all data the same except operate the pump at 1200 rpm rather than 1450 rpm. Note the pump efficiency under this mode of operation.

15.19 The characteristic curve for a pump operating at 1200 rpm is given by the following:

Head, m	Q, L/s
20	0
15	165
10	250

This pump is installed in a pipeline to augment the flow from reservoir A to reservoir B. When the water surface elevation in reservoir A exceeds that in reservoir B by 15 m, the flow rate is 150 L/s. What will be the flow rate when the elevation difference is 3 m? Assume the system curve is parabolic.

15.20 All dimensions of a model pump are one-third as large as the corresponding dimensions of a prototype pump. The model pump operating at 1800 rpm delivers 3.0 cfs at a head of 35 ft. Find the speed and capacity of the prototype if it is to develop a head of 35 ft. Find the specific speed of the model pump and the prototype. They should be the same. Assume the efficiency of the model and prototype are the same.

15.21 All dimensions of a model pump are one-third as large as the corresponding dimensions of a prototype pump. If the model pump operating at 1800 rpm delivers 3.0 cfs at a head of 35 ft, find the speed and head of the prototype when it delivers 3.0 cfs. Find also the specific speed of the model pump and the prototype. They should be the same. Assume the efficiency of the model and prototype are the same.

15.22 All dimensions of a model pump are one-third as large as the corresponding dimensions of the prototype pump. If the model operating at 1800 rpm delivers 3.0 cfs at a head of 35 ft, find the speed and capacity of the prototype if is is to develop a head of 140 ft. Assume the efficiency of the prototype and the model are the same. Find also the specific speed of the model and the prototype.

15.23 A centrifugal pump is required to deliver 1600 gpm against a head of 350 ft at a rotative speed of 1150 rpm. Would a two-stage pump be more efficient than a single-stage?

15.24 A pump is to deliver 2.0 m^3/s against a head of 160 m when operating at 300 rpm. What type of pump will be required?

15.25 At its optimum point of operation, a given centrifugal pump with an impeller diameter of 500 mm delivers 3.2 m^3/s of water against a head of 25 m when rotating at 1450 rpm. (*a*) If its efficiency is 82%, what is the brake power of the drive shaft? (*b*) If a homologous pump with a diameter of 800 mm is rotating at 1200 rpm, what would be the discharge, head and shaft power? Assume both pumps operate at the same efficiency. (*c*) Compute the specific speed of both pumps.

15.26 A three-stage pump is rated at 7500 gpm against a head of 600 ft at a speed of 1200 rpm. The three impellers are identical. Find the N_s of this pump and the approximate diameter of the impellers.

15.27 A four-stage pump is designed to deliver 65 L/s against a head of 120 m when operating at 1450 rpm. The four impellers are identical. Find the specific speed of this pump. What type of pump is it? Estimate the diameter of the impellers.

15.28 Estimate the diameter of the impeller of the model pump depicted in Exer. 15.4.1.

15.29 Approximately what would be the diameter of the impeller of a pump that delivers at its BEP 150 L/s of water at a head of 45 m when operating at 1750 rpm.

15.30 Approximately how much head will be developed by a pump when operating at its point of optimum efficiency if it delivers 6500 gpm when running at 1200 rpm? The diameter of the impeller is 15.0 in.

15.31 Calculate the values of ϕ_e for the pumps of Sample Prob. 15.2 and compare these values with those given in Fig. 15.11. Perform these calculations for both the model and the prototype, given the model diameter is 7.4 in.

15.32 At maximum efficiency, the pump at Rocky River delivers 280 cfs at a head of 238 ft when operating at 327 rpm. Calculate the specific speed of this pump and provide an estimate of the diameter of the impeller.

15.33 At the Grand Coulee project on the Columbia River there are several identical pumps, each with an impeller diameter of $167\frac{3}{8}$ in. The rotative speed of these pumps is 200 rpm and the maximum efficiency is 90.8% when discharging 1250 cfs at a head of 344 ft. Find the specific speed and calculate ϕ_e. How does the calculated value of ϕ_e compare with the value shown in Fig. 5.11?

15.34 Approximately what head would you expect a pump (impeller diameter = 10.0 in) to develop at 1200 rpm if its specific speed is 3500?

15.35 A given axial-flow pump delivers 300 L/s at a head of 6 m when operating at 1800 rpm. If this pump were operating on the moon at the same speed, what head would it develop when delivering water at 300 L/s? Assume the gravitational constant of the moon is one-sixth that of the earth. Assume cavitation does not occur.

15.36 Is the pump in the preceding problem likely to cavitate? Why?

15.37 A pump with a critical value of σ_c of 0.20 is to pump against a head of 200 ft. The barometric pressure is 14.3 psia, and the vapor pressure of the water is 0.8 psia. Assume the friction losses in the intake piping are 4 ft. Find the maximum allowable height of the pump relative to the water surface at intake to assure that cavitation will not occur.

15.38 A boiler feed pump delivers water at 200°F, which it draws from an open hot well with a frictional loss of 2 ft in the intake pipe. The barometric pressure is 29 inHg, and the value of σ_c for the pump is 0.16. What must be the elevation of the water surface in the hot well relative to that of the pump intake? The total pumping head is 140 ft.

15.39 Repeat the previous problem with all data identical except for the total pumping head. Consider three cases as follows: total pumping heads of 40 ft, 100 ft, and 180 ft.

15.40 Suppose a pump was pumping water at a head of 30 ft with a water temperature of 100°F and a barometric pressure of 14.2 psia. At intake to the pump the pressure is 17 inHg vacuum and the pipeline velocity is 16 ft/sec. What are the values of NPSH and σ?

15.41 A centrifugal pump with $(N_s)_{SI} = 1.48$ is to pump 30 L/s of gasoline ($s = 0.82$, vapor pressure = 32 kPa) from an open tank. The pump impeller rotates at 157 rad/s. Assuming the free surface in the tank is at elevation 100.0 m and the barometric pressure is 101 kPa abs, what is the highest elevation at which the pump centerline may be placed without cavitation problems? The suction pipe ($f = 0.022$) is 40 m long and has a diameter of 120 mm. Assume square entrance conditions where the pipe takes off from the tank. Express the answer in terms of h.

15.42 At rated capacity a four-stage centrifugal pump rotates at 1200 rpm and delivers 360 L/s of water against a head of 180 m. What diameter of impellers should be used? The suction and discharge flanges have inside diameters of 300 mm and 320 mm, respectively, and both are located 0.7 m below the pump centerline. At what gage pressure at the suction flange may cavitation be expected? Water temperature is 34°C and barometric pressure is 98 kPa abs.

15.43 Rework Prob. 15.42 for the case of a three-stage pump. Assume speed, flow rate, and total head are unchanged.

15.44 A pump with a critical value of σ_c of 0.18 is to pump against a head of 80 m. The barometric pressure is 98.5 kPa abs, and the vapor pressure is 5.4 kPa abs. Assume the friction losses in the intake are 1.2 m. Find the maximum allowable elevation of the pump relative to the water surface at intake.

15.45 Determine the specific speed of a centrifugal pump that is rated at 3500 gpm under a head of 70 ft at 1750 rpm. What would be the head and capacity of this pump if operated at 1160 rpm? For each rotative speed note the maximum tolerable suction lift as recommended in Fig. 15.13.

15.46 Water at 60°F flows by gravity between two reservoirs. The water surface in reservoir A is at elevation 5010 ft and that in reservoir B is at elevation 4980 ft. The water flows through a 24-in-diameter pipe ($f = 0.026$) of length 8000 ft. (*a*) What is the flow rate? Neglect minor losses. (*b*) If one wishes to double the flow rate using a pump operating at 1800 rpm, what specific speed pump would you recommend? (*c*) Approximately what would be the diameter of the impeller of the pump? (*d*) If the pump were set at the upper end of the pipeline, at what elevation must the pump be set to safeguard against cavitation if the head loss from reservoir A to the pump is 1.5 ft?

15.47 Repeat Prob. 15.46 for the case where the pump is to operate at 600 rpm rather than 1800 rpm, with all other data remaining the same.

15.48 Water at 170°F is pumped through a 15-in-diameter pipe ($f = 0.028$) of length 2000 ft from a large pressurized tank (p = 30 psi) whose water surface is at elevation 200 ft to a point of discharge in the atmosphere at elevation 242 ft. (a) If the flow rate is 5800 gpm, find the head that a pump, placed in the pipeline and rotating at 1200 rpm, must develop and determine the specific speed of the pump. Assume the pressure in the tank remains constant and neglect minor losses. (b) Approximately what will be the diameter of the impeller? (c) At what elevation must the pump be set to safeguard against cavitation if the head loss from the lower reservoir to the pump is 1.6 ft?

15.49 A pump manufacturer is asked to provide a pump that will deliver 84,500 gpm against a head of 225 ft. Several speeds of operation are considered, namely 225 rpm, 600 rpm, 1200 rpm, and 1800 rpm. Determine the specific speeds of pumps that will operate at these speeds and estimate the diameter of each of their impellers. What factors might one consider in making a choice among them?

15.50 You are asked to select a single-stage or multi-stage pump to deliver 10 cfs against a head of 300 ft when operating at 450 rpm. Specify the number of stages, the specific speed, and the approximate diameter of the impeller.

15.51 You are asked to select a single-stage or multi-stage pump to deliver 300 cfs against a head of 10 ft when operating at 450 rpm. Specify the number of stages, the specific speed and the approximate diameter of the impeller.

15.52 A submersible centrifugal pump in a water well is to be set 120 ft below the land surface. The steel discharge pipe from the pump to the surface has a diameter of 6.0 in. The water surface in the well at maximum drawdown is 106 ft below the land surface. How many stages (N_s = 2650) would you recommend for the pump if the discharge from the well at maximum drawdown is 800 gpm and the pump operates at 1450 rpm?

15.53 A pump is required to deliver 830 L/s against a head of 230 m. If the minimum desirable specific speed from an efficiency standpoint is $(N_s)_{\mathrm{SI}}$ = 0.35 and the motor speed is 1500 rpm, how many stages would you recommend? Also, what is the maximum permissible suction lift?

15.54 In planning a pumping station an engineer decided to use two identical pumps operating in parallel. These pumps, when both are operating, must deliver a total of 200 cfs against a head of 40 ft. Calculate the specific speed of each of these pumps and estimate the diameter of the impellers.

15.55 In planning a pumping station, an engineer decided to use two identical pumps operating in series. These pumps, when both are operating, must deliver 200 cfs against a total head of 40 ft. Calculate the specific speed of each of these pumps and estimate the diameter of the impellers.

15.56 A very long pipe connects two reservoirs whose water levels are the same. Identical pumps A and B are connected to the pipe in parallel. The operating characteristics of each of the pumps are shown in Fig. 15.6. The pumps are operated at 1450 rpm. When only one pump is operating, the flow rate was found to be 13,200 gpm. (a) How much flow will there be when both pumps operate? (b) Find also the rate of delivery (with one and two pumps) if the pump speeds are reduced to 1200 rpm?

15.57 Select the specific speed of the pump or pumps required to convey 1.0 cfs of water against a head of 115 ft. The pump rotative speed is 1750 rpm. Consider the following cases: a single pump; two pumps in parallel; three pumps in parallel; two pumps in series; three pumps in series.

15.58 The pump of Fig. 15.6 is placed in an 8-in-diameter pipe ($f = 0.030$) 150 ft long that is used to lift water from one pond to another. The difference in water surface elevations between the two ponds varies from 20 ft to 50 ft. Plot a curve showing delivery rate versus water surface elevation difference.

15.59 Under normal operating conditions, a centrifugal pump with an impeller diameter of 20.0 in delivers water at a head of 170 ft with an efficiency of 70% at 1800 rpm. Compute the peripheral velocity of the impeller, the specific speed, and the flow rate.

15.60 A centrifugal pump driven by an electric motor lifts water a total height of 200 ft. The pump efficiency is 78% and the motor efficiency is 88%. The lift is through 1220 ft of 6-in-diameter pipe and the pumping rate is 350 gpm. If $f = 0.022$ and power costs 70 mils/kWh, what is the cost of pumping a million gallons of water? (A mil is one-thousandth of a dollar = 0.10 cent)

15.61 What is the specific speed of the four-stage pumps mentioned in the last paragraph of Sec. 15.14? Approximately what is the diameter of the impellers?

CHAPTER 16

Hydraulic Machinery—Turbines

In this chapter we shall deal with hydraulic (i.e. water driven) turbines. We shall not deal with steam or gas turbines, since their theory is outside the scope of this book. Water-driven turbines are used primarily for the development of hydroelectric energy. Turbines extract energy from flowing water and convert it to mechanical energy to drive electric generators. ***Hydroelectric power***, as it is called, is an important source of energy. In the United States about 15% of the electric energy is derived from hydropower, though in some countries, such as Norway and Brazil, over 70% of the electric energy is developed at hydroelectric plants.

16.1 HYDRAULIC TURBINES

There are two basic types of hydraulic turbines. In the ***impulse turbine*** a free jet of water impinges on the revolving element of the machine, which is exposed to atmospheric pressure. In a ***reaction turbine***, flow takes place under pressure in a closed chamber. Although the energy delivered to an impulse turbine is all kinetic, while the reaction turbine utilizes pressure energy as well as kinetic energy, the action of both turbines depends on a change in the momentum of the water so that a dynamic force is exerted on the rotating element, or ***runner***. The runner of a reaction turbine is similar in design, but not identical to a pump impeller. By rotating either rotor (i.e. centrifugal pump or reaction turbine) in the reverse direction, the machine assumes the function of its counterpart. The efficiency, however, will be small in the reverse mode because the geometric shape of the rotating element and its surrounding casing will not be optimum.

Turbines are operated at constant speed. In the United States, 60-cycle (cycles/sec or Hz) electric current is most common, and under such conditions the rotative speed n of a turbine is revolutions per minute is given by $N = 7200/n$, where N is the number of poles in the generator and must be an even integer. Most 60-Hz generators have from 12 to 96 poles. In many parts of the world, 50-cycle current is used, in which case $N = 6000/n$. The power demand of an electric distribution system varies throughout the day; it is usually higher during daylight hours and lower at night. Consequently there is a variation in the "load" on the system. In a large system the variation in load can be accomodated by varying the number of generators in operation. Even so, the load on a single generator may vary with time. Hence, if the generator is driven by a hydraulic turbine, in order to maintain constant speed, there must be some way in which the flow passing through the turbine can be adjusted to regulate the power output of the turbine. Speed regulation will be discussed briefly in later sections of this chapter.

16.2 IMPULSE TURBINES

The impulse turbine (Fig. 16.1) is sometimes called an impulse wheel or ***Pelton wheel***, so called in honor of Lester A. Pelton (1829–1908), who contributed much to its development in the early gold-mining days in California. The wheel (or runner) has a series of split buckets located around its periphery (Fig. 16.2). The buckets may be individually cast and bolted to the central spider, or, more commonly, the entire runner is cast as a single unit. When the jet strikes the dividing ridge of the bucket, it is split into two parts which discharge at both sides of the bucket. Each split bucket has a notch that enables the bucket to attain a position nearly tangent to the direction of the jet before the bucket lip intercepts the jet. Only one jet is used on small tubines, but two or more jets impinging at different points

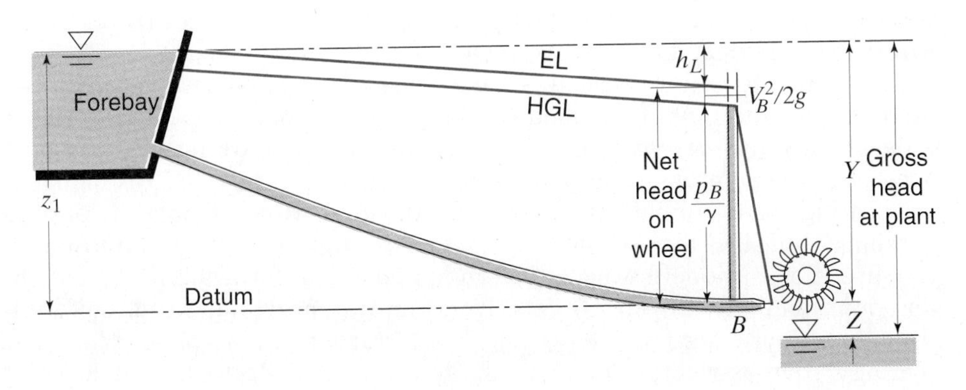

Figure 16.1
Definition sketch for the impulse turbine.

Figure 16.2
Impulse wheel (runner) at Big Creek-2A plant of Southern California Edison Co. Static head = 2418 ft, net head = 2200 ft, n = 250 rpm, pitch diameter = 162 in. (This original runner for 50-cycle generation has been replaced by a 300-rpm runner for 60-cycle generation.)

around the wheel are often used on large units. The jets are usually produced by a needle nozzle (Sec. 16.5).

Since water hammer (Sec. 12.6) might occur in the supply pipe if the nozzle were suddenly closed, some nozzles are provided with a bypass valve that opens whenever the needle valve is closed quickly. The same effect can be obtained with a jet deflector, whose position can be adjusted to deflect the jet away from the wheel when the load drops. A governor is required to actuate the nozzle and bypass or deflector units. The governing mechanism is provided with a compensating arrangement that prevents over-travel of the governor to provide very sensitive regulation.

The generator rotor is usually mounted on a horizontal shaft between two bearings with the runner installed on the projecting end of the shaft. This is known as a ***single-overhung*** installation. Often runners are installed on both sides of the generator (***double-overhung*** construction; Fig. 16.3) to equalize the bearing loads. Impulse turbines are provided with housings to prevent splashing, but the air within the housing is substantially at atmospheric pressure. Some modern wheels are mounted on a vertical axis below the generator and are driven by jets from several nozzles spaced uniformly around the periphery of the wheel. This arrangement simplifies shaft and bearing design as well as details of the jets and jet deflectors, though piping for the multiple nozzles becomes complicated.

For good efficiency the width of the bucket should be 3–4 times the jet

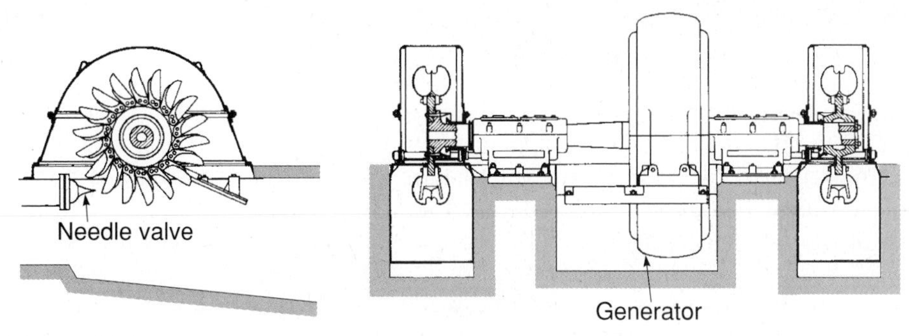

Figure 16.3
Double-overhung impulse wheel installation.

diameter, and the wheel diameter is usually 15–20 times the jet diameter.[1] The ***wheel diameter***, also referred to as the ***pitch diameter***, is the diameter of the ***pitch circle***, the circle to which the centerline of the jet is tangent. The diameters of impulse turbines range up to about 15 ft (5 m). Theoretically maximum efficiency would result if a bucket completely reversed the relative velocity of the jet. This is not possible because the water must be deflected to one side to avoid interfering with the following bucket, and the bucket angle β_2 is usually about 165° (Fig. 16.4*b*).

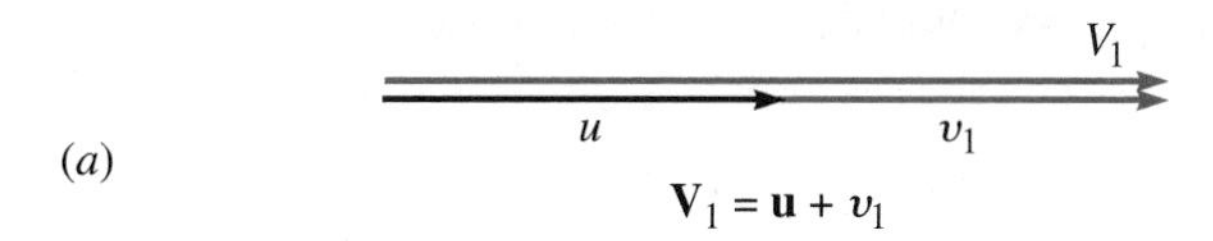

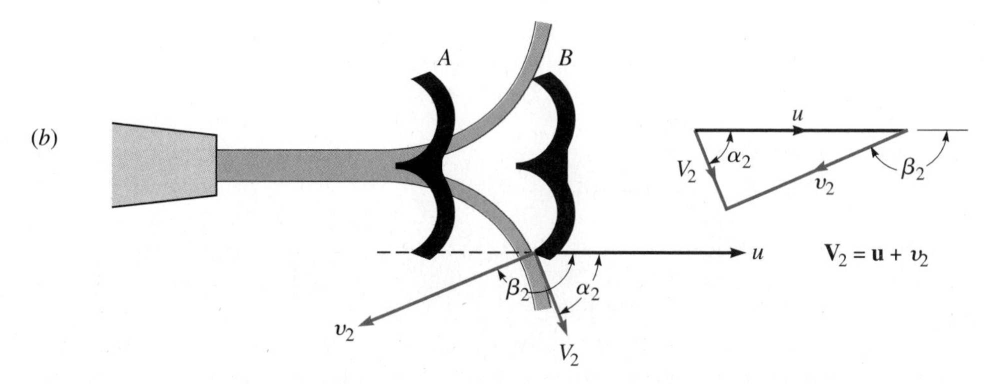

Figure 16.4
Velocity vector diagrams at (*a*) entrance to the bucket and (*b*) at exit from the bucket.

[1] The ratio of wheel diameter to jet diameter varies over a wide range, from 10 to 80 or more, depending on the spacing of the buckets.

16.3 ACTION OF THE IMPULSE TURBINE

Figure 16.4*a* shows a portion of a rotating impulse turbine with a velocity u at the center line of the buckets. A plan view of one of the buckets is shown in Fig. 16.4*b*. Position A represents the instant of entry of water into the bucket, while position B indicates the moment the water leaves the bucket. The true path of the water is shown, and its velocity changes from V_1 at entry to V_2 at exit. A vector diagram of the velocities at entry is shown in Fig. 16.4*a*. The vector $\boldsymbol{v}$ represents the velocity of the water relative to the bucket. Assuming friction to be negligible, the magnitude v of the water velocity relative to the bucket remains constant, but its direction at discharge must be tangential to the bucket (Fig. 16.4*b*). Applying the impulse-momentum principle, the force exerted by the water on the bucket in the direction of motion is

$$F = \rho Q(V_1 - V_2 \cos \alpha_2) \tag{16.1}$$

where Q is the discharge of the nozzle. In terms of relative velocities,

$$F = \rho Q(v - v \cos \beta_2) = \rho Q(V_1 - u)(1 - \cos \beta_2) \tag{16.2}$$

since the relative velocity $v = V_1 - u$.

The power transmitted to the buckets from the water is the product of the force and the velocity of the body on which the force is acting. Hence,

$$P = Fu = \rho Q(V_1 - u)(1 - \cos \beta_2)u \tag{16.3}$$

From Eq. (16.3), we observe that there is no power developed when $u = 0$ or when $u = V_1$. For a given turbine and jet the maximum power occurs at an intermediate u, which can be found by differentiating Eq. (16.3) and equating to zero. Thus

$$\frac{dP}{du} = \rho Q(1 - \cos \beta_2)(V_1 - u - u) = 0 \tag{16.4}$$

from which $u = V_1/2$. Thus the greatest hydraulic efficiency (neglecting fluid friction) occurs when the peripheral speed of the wheel is half of the jet velocity. Tests of impulse turbines show that, because of energy losses, the best operating conditions occur when u/V_1 is between 0.43 and 0.48.

In the preceding analysis we have assumed that the buckets are shaped at the split in such a manner that the bucket angle at entrance is 0°; i.e., the water enters the bucket tangentially to the bucket. In actuality, there is usually a small bucket angle of less than 10°, in which case the V_1 in Eq. (16.1) should be replaced by $V_1 \cos \alpha_1$. A small error, usually less than two percent is introduced by assuming $\alpha_1 = 0°$; hence we commonly make that assumption. Of somewhat greater importance is the fact that in the preceding analysis, which is idealized, we neglected fluid friction as the flow passes through the buckets. Because of fluid friction, the velocity v of the water relative to the buckets gets smaller as the water flows through the buckets.

The velocity v_2 of the water relative to the buckets at exit is usually between $0.8v_1$ and $0.9v_1$. In the development of Eqs. (16.1)–(16.4) we assumed $v = v_1 = v_2$. This assumption leads to substantial inaccuracy. To achieve accurate results, the drop in v as it passes through the buckets should be taken into consideration. Sample Probs. 16.1 and 16.2 illustrate the effect of neglecting the decrease in velocity as the water flows through the buckets.

Sample Problem 16.1 Consider an impulse turbine with a pitch diameter of 10.0 ft and a bucket angle β_2 of 160°. If the jet velocity V_1 is 200 ft/sec, the jet diameter 2.0 in and the rotating speed 240 rpm, find (*a*) the force on the buckets, (*b*) the torque on the runner, and (*c*) the horsepower transferred to the runner. Assume that the velocity relative to the bucket does not change (i.e., assume $v_2 = v_1$).

Solution

$$u = \omega r = 240\left(\frac{2\pi}{60}\right)5 = 125.6 \text{ ft/sec}$$

$$V_1 = 200 \text{ ft/sec} = v_1 + u = v_1 + 125.6 \text{ ft/sec}; \quad v_1 = 74.4 \text{ ft/sec}$$

V_1 = 200 ft/sec

u v_1

u 20° 160° V_2 v_2

$$v_2 = v_1 = 74.4 \text{ ft/sec}, \quad v_2 \cos 20° = 69.9 \text{ ft/sec}$$

Hence: $\Delta v_x = v_{2x} - v_{1x} = -69.9 - 74.4 = -144.3 \text{ ft/sec}$

Note also, $V_{2x} = u - v_2 \cos 20° = 125.6 - 69.9 = 55.7 \text{ ft/sec}$

Thus $\Delta V_x = V_{2x} - V_{1x} = 55.7 - 200.0 = -144.3 \text{ ft/sec}$

Hence $\Delta V_x = \Delta v_x$

(*a*) $F = \rho Q(\Delta V_x) = \rho Q(\Delta v_x)$,

where $Q = A_j V_j = \pi\left(\frac{1}{12}\right)^2 \times 200 = 4.36 \text{ ft}^3\text{/sec}$

Thus $F = 1.94(4.36)144.3 = 1220 \text{ lb}$ ***ANS***

(*b*) $T = Fr = (1220)5 = 6100 \text{ ft} \cdot \text{lb}$ ***ANS***

(*c*) $\text{HP} = \frac{T\omega}{550} = \frac{(6100)240(2\pi/60)}{550} = 278.6 \text{ hp}$ ***ANS***

or $\text{HP} = \frac{Fu}{550} = \frac{(1220)125.6}{550} = 278.6 \text{ hp}$

SAMPLE PROBLEM 16.2 Repeat Sample Prob. 16.1 for the case where $v_2 = 0.9v_1$. Calculate the percent error made by assuming $v_2 = v_1$ in Sample Prob. 16.1.

Solution

From the solution to Sample Prob. 16.1:

$$u = 125.6 \text{ ft/s}, \quad v_1 = 74.4 \text{ ft/sec} \quad \text{and} \quad Q = 4.36 \text{ cfs}$$

$$v_2 = 0.9v_1 = 67.0 \text{ ft/sec}$$

$$\Delta v_x = v_{2x} - v_{1x} = -67.0 \cos 20° - 74.4 = -137.4 \text{ ft/sec}$$

(*a*) $F = 1.94(4.36)137.4 = 1162 \text{ lb}$ ***ANS***

(*b*) $T = Fr = (1162)5 = 5810 \text{ ft·lb}$ ***ANS***

(*c*) $\text{HP} = Fu/550 = (1162)125.6/550 = 265.4 \text{ hp}$ ***ANS***

(*d*) $\text{Error} = \dfrac{278.6 - 265.4}{265.4} = 0.050, \text{ or } 5.0\%$ ***ANS***

EXERCISES

16.3.1 In Sample Prob. 16.1 subtract the horsepower of the absolute velocity of the water at bucket exit from the horsepower of the jet at bucket entrance. What does the result represent? Why?

16.3.2 Repeat Sample Prob. 16.1 for the case where $v_2 = 0.8v_1$ and calculate the percent error made by assuming $v_2 = v_1$ in Sample Prob. 16.1.

16.3.3 Consider an impulse turbine with a pitch diameter of 2.0 m and a bucket angle β_2 of 165°. If the jet velocity is 80 m/s, the jet diameter 112 mm, and the rotative speed 375 rpm, find (*a*) the force on the buckets, (*b*) the torque on the runner, and (*c*) the power transferred to the runner. Assume that the velocity relative to the bucket does not change (i.e., assume $v_2 = v_1$).

16.4 HEAD ON AN IMPULSE TURBINE AND EFFICIENCY

The gross head for a power plant is the difference in elevation between headwater and tailwater, or $Y + Z$ in Fig. 16.1. The pressure within the case of an impulse wheel is atmospheric, and consequently this is the pressure at which the jet is discharged. Thus the portion Z is unavailable; so the gross head on the wheel itself is Y only. This is also called the ***static head***. It is impractical to set the wheel too near the surface of the tailwater because it might then be submerged with any rise in level of the latter.

The nozzle is considered an integral part of the impulse turbine. Hence the ***net head,*** or ***effective head*** h, i.e., the head at the base B of the nozzle in Fig. 16.1, is the static head minus the pipe friction losses. Thus, for net head, we have

$$h = \frac{p_B}{\gamma} + \frac{V_B^2}{2g} = z_1 - h_L \tag{16.5}$$

The energy, or head, supplied at the nozzle is expended in four ways. Some energy is lost in fluid friction in the nozzle [Eq. (11.14)], a portion is expended in fluid friction over the buckets, kinetic energy is carried away in the water discharged from the buckets, and the rest is available to the buckets. Thus

$$h = \left(\frac{1}{C_v^2} - 1\right)\left[1 - \left(\frac{A_j}{A_B}\right)^2\right]\frac{V_j^2}{2g} + k\frac{v_2^2}{2g} + \frac{V_2^2}{2g} + h'' \tag{16.6}$$

where C_v = coefficient of velocity of the nozzle (Sec. 11.6)
A_j = cross-sectional flow area of the jet
A_B = cross-sectional flow area of the pipe upstream of the nozzle
V_j = jet velocity (this is the same as V_1 in Fig. 16.4)
v_2 = velocity of water relative to bucket at exit from bucket
V_2 = absolute velocity of water leaving the bucket
h'' = energy head directly available to the buckets (i.e., transferred to the runner)

Typical values of k, the ***bucket friction loss coefficient,*** vary from 0.2 to about 0.6. The greater part of the energy delivered to the buckets is used in driving the generator, but some of it is used in overcoming mechanical friction in the bearings and in windage loss (air friction).

Since the nozzle is considered an integral part of the impulse turbine, the ***hydraulic efficiency*** η' of the impulse turbine is the ratio of the power transferred directly to the turbine buckets to the power in the flow at the base of the nozzle. Thus, for impulse turbines:

$$\eta' = \frac{\gamma Q h''}{\gamma Q h} = \frac{h''}{h} \tag{16.7}$$

The ***overall efficiency*** (or simply ***efficiency***) η of an impulse turbine is considerably less than the hydraulic efficiency η' because of friction in the bearings and windage (air resistance).

The efficiency of an impulse turbine is given by

$$\eta = \frac{\text{Shaft power}}{\text{Input power}} = \frac{T\omega}{\gamma Q h} \tag{16.8}$$

where h is the head as defined by Eqs. (16.5) and (16.6).

SAMPLE PROBLEM 16.3 Demonstrate that when taking account of fluid friction in the buckets of an impulse wheel by using $kv_2^2/2g$ with values of k ranging from 0.2 to 0.6, this is nearly equivalent to $v_2 = 0.8\text{–}0.9$ times v_1.

Solution

$$\frac{v_1^2}{2g} - h_L = \frac{v_2^2}{2g}, \qquad \text{where} \quad h_L = k\frac{v_2^2}{2g}$$

Thus:

$$\frac{v_1^2}{2g} - k\frac{v_2^2}{2g} = \frac{v_2^2}{2g} \quad \text{and} \quad v_1^2 - kv_2^2 = v_2^2$$

from which:

$$v_2 = v_1/\sqrt{1+k}$$

$$k = 0.2, \qquad v_2 = 0.91v_1$$

$$k = 0.6, \qquad v_2 = 0.79v_1$$

SAMPLE PROBLEM 16.4 Water is delivered from a reservoir through a 4-ft-diameter pipe ($f = 0.025$) 10,000 ft long to a nozzle that emits an 8-in-diameter jet that impinges on an impulse wheel. The surface of the reservoir is at an elevation 2420 ft higher than the nozzle. The impulse wheel is connected to a 20-pole generator in a 60-cycle system. The wheel has a diameter of 106 in and a bucket angle $\beta_2 = 165°$. Assuming the head loss through the nozzle can be expressed as $0.05V_j^2/2g$, where V_j is the jet velocity, find the following: velocity of flow in the pipe, jet velocity, flow rate, speed of the bucket, value of u/V_1, net head, force F on the bucket, torque on the wheel, shaft horsepower, horsepower at the base of the nozzle, and overall turbine efficiency. Assume $v_2 = v_1$, neglect bearing friction and windage.

Solution

Energy equation from surface of reservoir to jet neglecting minor loss at pipe entrance:

$$2420 - 0.025\frac{10{,}000}{4}\frac{V^2}{2g} - 0.05\frac{V_j^2}{2g} = \frac{V_j^2}{2g}$$

Continuity:

$$V_j = (48/8)^2V, \qquad V_j = 36V$$

$$2420 - 62.5\frac{V^2}{2g} - 0.05\frac{(36V)^2}{2g} = \frac{(36V)^2}{2g}$$

$$2420 = \frac{V^2}{2g}(62.5 + 64.8 + 1296) = \frac{V^2}{2g}(1423.3)$$

$V = \sqrt{2g(2420/1423)} = 10.47$ ft/sec, velocity in the pipe. ***ANS***

$V_j = 36(10.47) = 377$ ft/sec ***ANS***

$Q = AV = \pi 2^2(10.47) = 131.5$ cfs ***ANS***

$$N = 7200/n \quad \text{gives} \quad n = 7200/N = 7200/20 = 360 \text{ rpm}$$

$$\omega = 2\pi n/60 = 6.28(360)/60 = 37.7 \text{ rad/sec}$$

$u = \omega r = 37.7\left(\frac{106/2}{12}\right) = 166.5$ ft/sec ***ANS***

Hence: $u/V_1 = 166.5/377 = 0.44$ ***ANS***

Net head

$$h = \frac{p_B}{\gamma} + \frac{V_B^2}{2g} = 2420 - f\frac{L}{D}\frac{V^2}{2g} = 2420 - 62.5\frac{10.47^2}{2g} = 2314 \text{ ft} \qquad \textbf{\textit{ANS}}$$

Force on bucket

$$F = \rho Q(V_1 - u)(1 - \cos\beta); \quad \beta = 165°, \quad \cos\beta = -0.966$$
$$= 1.94(131.5)(377 - 166.5)(1 + 0.966) = 105{,}600 \text{ lb} \qquad \textbf{\textit{ANS}}$$

Torque on wheel

$$T = Fr = 105{,}600(53/12) = 466{,}300 \text{ ft·lb} \qquad \textbf{\textit{ANS}}$$

Shaft power $\quad P = \dfrac{T\omega}{550} = \dfrac{466{,}300(37.7)}{550} = 31{,}960 \text{ hp} \qquad$ ***ANS***

Power at base of nozzle (i.e., power input to entire turbine, including nozzle)

$$P = \frac{\gamma Q h}{550} = \frac{62.4(131.5)2314}{550} = 34{,}520 \text{ hp} \qquad \textbf{\textit{ANS}}$$

Overall efficiency

$$\eta = \frac{\text{Output (or shaft) power}}{\text{Input power}} = \frac{T\omega}{\gamma Q h} = \frac{31{,}960}{34{,}520} = 0.926 = 92.6\% \qquad \textbf{\textit{ANS}}$$

EXERCISES

16.4.1 Refer to Sample Prob. 16.4. (*a*) Find the head h'' delivered to the buckets of the turbine if the absolute velocity at discharge from the buckets is 60 ft/sec. Assume a bucket loss coefficient of 0.30. (*b*) Calculate the hydraulic efficiency of the flow through the turbine buckets. (c) What is the overall efficiency of the turbine?

16.4.2 Repeat Sample Prob. 16.4 for the case of a 4-in-diameter jet, with all other data remaining the same.

16.4.3 Refer to Sample Prob. 16.1. Calculate the hydraulic efficiency of the flow through the turbine buckets. Note, this is different than the hydraulic efficiency η' of the turbine because the nozzle is considered an integral part of the turbine.

16.5 NOZZLES FOR IMPULSE TURBINES

In any turbine, in order to maintain a constant speed of rotation, it is necessary that the flow rate be varied in proportion to the load on the machine and for the impulse wheel this is done by varying the size of the jet. This is accomplished by varying the position of the needle in the needle nozzle. The shape of both nozzle tip and needle should be such as to cause a minimum friction loss for all positions of the needle and also such as to avoid cavitation damage to the needle at any position.

An important feature in attaining high efficiency in an impulse wheel is that the jet be uniform, the ideal being to have all particles of water moving in parallel lines with equal velocities and with no spreading out of the jet. Air friction retards the water on the outside of the jet, and the needle causes the velocity in the center to be slightly reduced. Careful design of both nozzle tip and needle will minimize these effects, and a gain in turbine efficiency of several percent has been made by improved nozzle design producing better jets. Values of C_v for needle nozzles vary from about 0.95 when partly closed to permit one-half of maximum flow to 0.99 at the fully opened position.

For a given pipeline there is a unique jet diameter that will deliver maximum power to a jet. Refer to Fig. 16.1 and note that the power of the jet issuing from the nozzle may be expressed as

$$P_{\text{jet}} = \gamma Q \frac{V_j^2}{2g} \tag{16.9}$$

where V_j is the jet velocity [equal to V_1 of Eq. (16.1)]. As the size of the nozzle opening is increased, the flow rate Q gets larger while the jet velocity V_j gets smaller. Hence, from Eq. (16.9) and from the preceding statement, we must conclude that there is some intermediate size of nozzle opening (and hence of jet diameter) that will provide maximum power to the jet. This is best illustrated by a sample problem.

SAMPLE PROBLEM 16.5 A 6-in-diameter pipe ($f = 0.020$) 1000 ft long delivers water from a reservoir with a water-surface elevation of 500 ft to a nozzle at elevation 300 ft. The jet from the nozzle is used to drive a small impulse turbine. If the head loss through the nozzle can be expressed as $0.04V_j^2/2g$ [Eq. (11.14)], find the jet diameter that will result in maximum power in the jet. Neglect the head loss at the entrance to the pipe from the reservoir. Evaluate the power in the jet.

Solution

Energy equation:

$$500 - 0.02\frac{1000}{0.5}\frac{V_p^2}{2g} - 0.04\frac{V_j^2}{2g} = 300 + \frac{V_j^2}{2g}$$

If we define the pipe diameter and velocity as D_p and V_p and the jet diameter and velocity as D_j and V_j, from continuity we get

$$A_pV_p = A_jV_j, \quad D_p^2V_p = D_j^2V_j$$

Since the pipe diameter $D_p = 0.50$ ft,

$$0.25V_p = D_j^2V_j, \quad \text{and} \quad V_p = 4D_j^2V_j$$

Substituting this expression for V_p in the energy equation gives

$$200 = \frac{V_j^2}{2g}(1.04 + 640D_j^4)$$

Assuming different values for D_j, we can compute corresponding values of V_j and Q, and then the jet power can be computed using Eq. (16.9). The results are as follows:

D_j, in	D_j, ft	V_j, fps	A_j, ft^2	$Q = A_jV_j$, cfs	P_{jet}, hp
1.0	0.083	111	0.0054	0.60	12.8
1.5	0.125	105	0.0122	1.28	24.2
2.0	0.167	91	0.0218	2.00	29.8
2.5	0.208	76	0.0338	2.57	26.2
3.0	0.250	60	0.0491	2.94	18.8
4.0	0.333	38	0.0873	3.29	8.4
6.0	0.500	18	0.1964	3.49	1.9

Thus a 2-in-diameter jet is the optimum; it will have about 30 hp.

An alternate procedure for solving this problem is to set up an algebraic expression for the power of the jet, P_{jet}, as a function of the jet diameter, D_j, and differentiate P_{jet} with respect to D_j and equate to zero to find the value of D_j for which P_{jet} is a maximum.

EXERCISES

16.5.1 Repeat Sample Prob. 16.5 for the case where the pipe length is 10,000 ft. All other data are to remain the same.

16.5.2 Repeat Sample Prob. 16.5 for the case where the pipe diameter is 12 in. All other data are to remain the same.

16.6 REACTION TURBINES

As mentioned in Sec. 16.1, a reaction turbine is one in which flow takes place in a closed chamber under pressure. The flow through a reaction turbine may be radially inward, axial, or mixed (partially radial and partially axial). There are two types of reaction turbines in general use, the ***Francis turbine*** and the ***axial-flow*** (or ***propeller***) ***turbine***. The Francis turbine is named after James B. Francis (1815–1892), an eminent American hydraulic engineer who designed, built and tested the first efficient inward-flow turbine in 1849. All inward flow turbines are known as Francis turbines, both in the United States

Figure 16.5
Scroll case.

and Europe, though they have developed into very different forms since the original.

In the usual Francis turbine water enters the ***scroll case*** (Fig. 16.5) and moves into the runner through a series of ***guide vanes*** (Fig. 16.6) with contracting passages that convert pressure head to velocity head. These vanes, known as ***wicket gates***, are adjustable so that the quantity and direction of flow can be controlled. They are operated by moving a ***shifting ring*** to which each gate is attached. Constant rotative speed of the runner under varying load is achieved by a governor that actuates a mechanism that adjusts the gate openings. A relief valve or a surge tank (Sec. 12.7) is generally necessary to prevent serious water hammer pressures.

Flow through the usual Francis runner (Fig. 16.7) is at first inward in the radial direction, gradually changing to axial. Francis turbines are usually mounted on a vertical axis. The scroll case of a Francis turbine is designed to decrease the cross-sectional area in proportion to the decreasing flow rate passing a given section of the casing and thus to maintain constant velocity in the casing so that flow enters the guide vanes uniformily around the periphery of the vanes. Large turbines are sometimes provided with an outside guide-vane assembly known as the ***stay-ring***. This consists of vanes fixed in position, known as ***stay vanes***. The vanes of the stay ring serve as columns to aid in supporting the weight of the generator above, and also they direct the flow smoothly to the inner guide-vane assembly. For heads under about 40 ft (12 m), Francis turbines are often used with an open flume setting (Fig. 16.8), in which case there is no need for a scroll case.

The propeller turbine (Fig. 16.9), an axial-flow machine with its runner

Figure 16.6
Guide-vane assembly showing wicket gates and control mechanism.

Figure 16.7
One of four vertical-shaft, reaction turbine runners for Itauba hydroelectric power station, Brazil. Each unit is designed to generate 128 000 kW at 150 rpm under a net head of 87.6 m. (*Voith*)

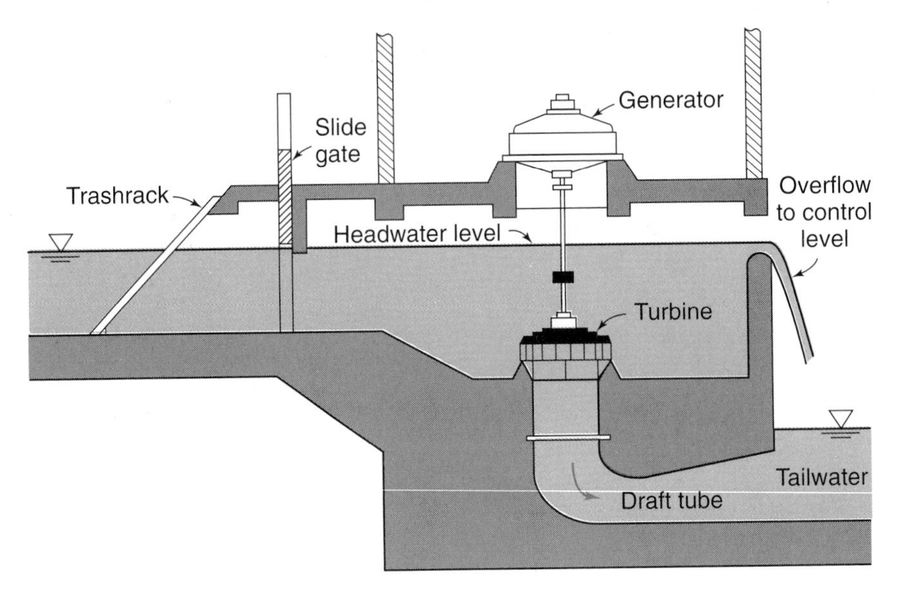

Figure 16.8
Typical open flume setting for a reaction turbine at low heads.

Figure 16.9
Kaplan turbine (a propeller turbine with adjustable blades) at Watts Bar Dam. It develops 42,000 hp at 94.7 rpm under a net head of 52 ft. Flow is downward.

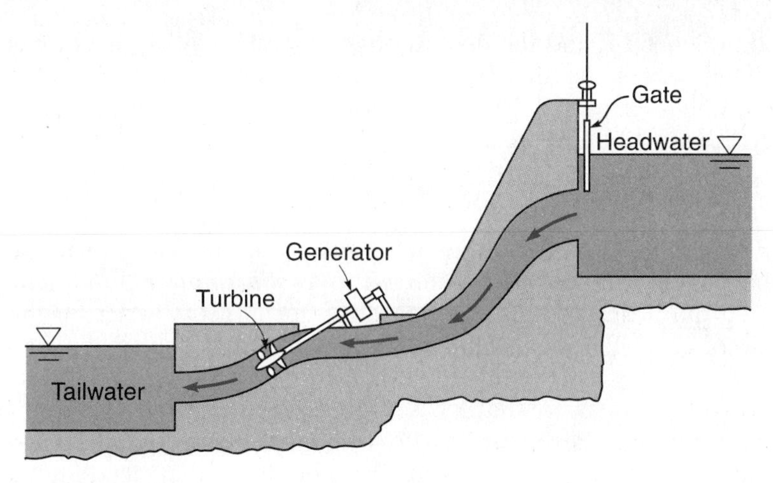

Figure 16.10
Tube turbine (a type of propeller turbine).

confined in a closed conduit, is commonly set on a vertical axis, though it may be set on a horizontal or slightly inclined axis (Fig. 16.10). The usual runner has four to eight blades mounted on a hub, with very little clearance between the blades and the conduit wall. The blades have free outer ends like a marine propeller. Adjustable gates upstream of the runner are used to regulate the flow. These gates are actuated by a governor to maintain constant speed. A ***Kaplan turbine*** is a propeller turbine with movable blades whose pitch can be adjusted to suit existing operating conditions. The adjustment is accomplished by a mechanism in the runner hub, which is actuated hydraulically by the governor in synchronization with guide-vane adjustments. With an axial-flow turbine, the generator may be set outside the water passageway as in Fig. 16.10, or it may be placed in a streamlined watertight steel housing mounted in the center of the passageway. Two other types of propeller turbine include the ***Deriaz***, an adjustable-blade, diagonal-flow turbine where the flow is directed inwards as it passes through the blades, and the ***tube turbine*** (Fig. 16.10), an inclined-axis type that is particularly well adapted to low-head installations, since the water passages can be formed directly in the concrete structure of a low dam.

16.7 ACTION OF THE REACTION TURBINE[2]

A general understanding of the action of a Francis turbine can be obtained by considering the flow through a radial-flow runner (Fig. 6.9). This is

[2] Before reading further, we recommend that you review Sec. 6.11, including Sample Prob. 6.6 in particular.

discussed in Sec. 6.11, and the discussion leads to Eq. (6.28), an expression for torque. This represents the torque exerted on the runner by the flowing water. Multiplying Eq. (6.28) by ω, the rotative speed of the runner in radians per second, leads to an expression for power:

$$P = T\omega = \rho Q\omega(r_1 V_1 \cos \alpha_1 - r_2 V_2 \cos \alpha_2) \tag{16.10}$$

where α_1 and α_2 are defined in Fig. 6.9. This represents the power transferred to the shaft of the runner and is referred to as ***shaft power***. Fluid friction, the fact that all the water passing through does not effectively act on the blades, and friction in the bearings slightly reduce the actual power delivered to the generator.

An analysis similar to that of Sec. 6.11 may be used for a mixed-flow runner in which the direction of flow changes from radial to axial. The analysis must, however, be modified to account for the fact that the inflow and outflow vectors do not lie in a plane perpendicular to the axis of the runner. Analysis of a propeller turbine is rather complex, and will not be presented here.

SAMPLE PROBLEM 16.6 A Francis turbine operating at 300 rpm develops a shaft horsepower of 1125 hp when the flow is 120 cfs. Find the torque transmitted from the flowing water to the shaft.

Solution

Shaft power $= T\omega$

$$\frac{T\left(\dfrac{2\pi n}{60}\right)}{550} = 1125 \text{ hp}, \qquad \text{where} \quad n = 300 \text{ rpm}$$

$$T = \frac{550 \times 1125}{\left(\dfrac{2\pi}{60}\right)300} = 19{,}700 \text{ ft}\cdot\text{lb} \qquad \textbf{\textit{ANS}}$$

EXERCISES

16.7.1 A Kaplan turbine operating at 327.3 rpm develops a shaft power of 300 kW when the flow is 5.5 m^3/s. Find the torque transmitted from the flowing water to the shaft.

16.7.2 A torque of 7500 ft·lb is transferred from the flowing water to a reaction turbine whose shaft power is 285.5 hp. How many poles must the generator have to develop 50-cycle electricity?

16.8 DRAFT TUBES AND EFFECTIVE HEAD ON REACTION TURBINES

To operate properly, reaction turbines must have a submerged discharge. The water, after passing through the runner, enters the ***draft tube*** (Figs. 16.8, 16.11, and 16.12), which directs the water to the point of discharge (point 2 in Fig. 16.11). The draft tube is an integral part of a reaction turbine, and its design criteria are usually specified by the turbine manufacturer. The draft tube has two functions. One is to enable the turbine to be set above tailwater level without losing any head thereby. A reduced pressure is produced at the upper end of the draft tube (point 1 in Fig. 16.11), which places a limit on the height z_1 above tailwater at which the turbine runner can be set.

The second function of the draft tube is to reduce the head loss at submerged discharge to thereby increase the net head available to the turbine runner. This is accomplished by using a gradually diverging tube whose cross-sectional area at discharge is considerably larger than the cross-sectional area at entrance to the tube. Applying the energy equation to the draft tube of Fig. 16.11 and letting z_1 signify the elevation of the entrance (or top) of the draft tube above the surface of the water in the tailrace, the absolute pressure

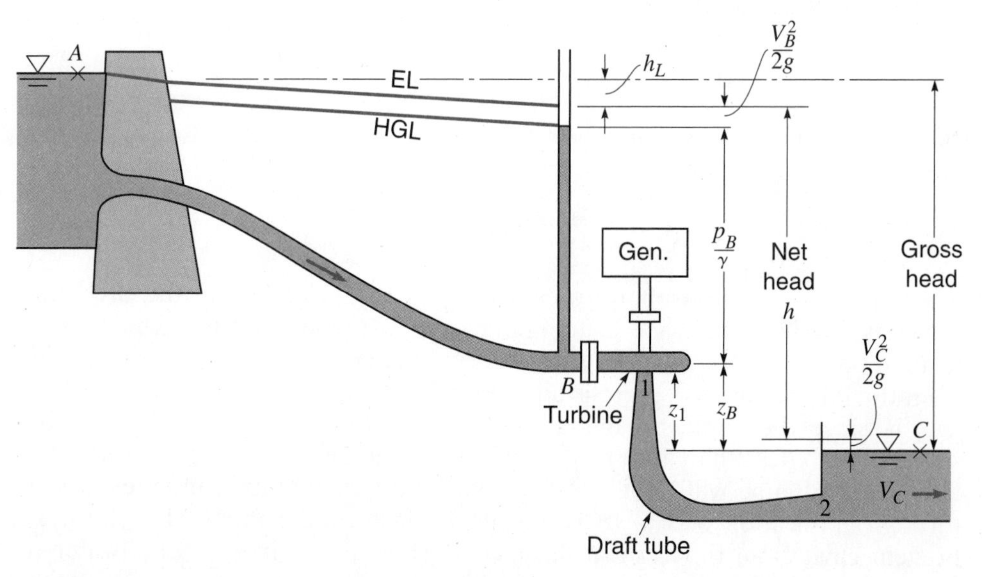

Figure 16.11
Net head h on a reaction turbine.

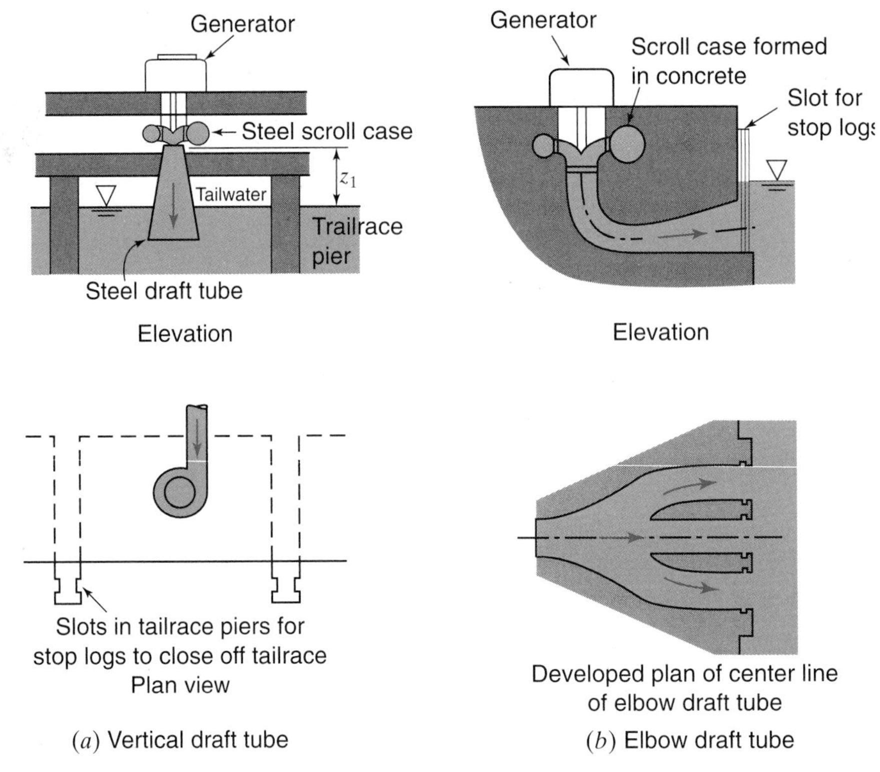

Figure 16.12
Two types of draft tubes for reaction turbines.

head at that section is given by

$$\frac{(p_1)_{\text{abs}}}{\gamma} = \frac{p_{\text{atm}}}{\gamma} - z_1 - \frac{V_1^2}{2g} + h_L + \frac{V_2^2}{2g} \tag{16.11}$$

where p_{atm} is the atmospheric pressure, h_L is the head loss in the diverging tube, and $V_2^2/2g$ is the kinetic-energy head at exit from the tube. The velocity head at point 2 can be reduced by increasing the cross-sectional area A_2 at exit from the tube. However, there is a practical upper limit to this because of tube length, since the angle of divergence of the tube must be kept reasonably small (usually less than 10°) to prevent or at least minimize separation of the flow from the tube wall. The head loss in the diverging tube can be estimated through application of Eq. (8.59) using the loss coefficients of Fig. 8.19. To prevent cavitation, the vertical distance z_1 (Fig. 16.11) from the tailwater to the draft tube inlet should be limited (Sec. 16.12) so that at no point within the draft tube or turbine will the absolute pressure drop to the vapor pressure of the water.

For a reaction turbine the ***net head*** h is the difference between the energy level just upstream of the turbine and that of the tailrace. Thus in Fig. 16.11 the net head on the turbine is $h = H_B - H_C$, or

$$h = \left(z_B + \frac{p_B}{\gamma} + \frac{V_B^2}{2g}\right) - \frac{V_C^2}{2g} \tag{16.12}$$

where z_B is defined as the ***draft head*** and V_C is the velocity in the tailrace. In most instances $V_C^2/2g$ is very small and may be neglected. By comparing Fig. 16.11 with Fig. 16.1, it is apparent that, for the same setting, the net head on a reaction turbine will be greater than that on a Pelton wheel. The difference is of small importance in a high-head plant, but it is important for a low-head plant.

The draft tube is considered an integral part of a reaction turbine; thus the ***effective head*** h' available to act on the runner of a reaction turbine is

$$h' = h - k'\frac{(V_1 - V_2)^2}{2g} - \left[\frac{V_2^2}{2g} - \frac{V_C^2}{2g}\right] \tag{16.13}$$

where h is the net head as defined in the foregoing, and the other two terms refer to the head loss in the draft tube and the loss at submerged discharge from the tube. The head h'' that is actually extracted from the water by the runner is smaller than h' by an amount equal to the ***shock loss*** at entry to the runner and the hydraulic friction losses in the scroll case, guide vanes, and runner. Shock loss can be quite large if either the magnitude or direction of the velocity changes abruptly when the water leaves the guide vanes and enters the runner. The shock loss and the hydraulic losses in the turbine depend on the conditions under which the turbine is operating.

Sample Problem 16.7 Refer to Fig. 16.11. A draft tube leading from the discharge side of the turbine to the submerged discharge into the tailrace consists of a 40-ft-long pipe ($f = 0.03$) of constant diameter 3 ft. The flow rate is 130 cfs. If this draft tube were replaced by a diverging tube of length 40 ft whose diameter changed uniformly from 3 ft to 8 ft over the 40-ft length, how much additional effective head would be developed by the replacement draft tube? Neglect head loss due to the curvature in the bend and assume the velocity V_C in the tailrace is negligible.

Solution

Original draft tube:

Head loss due to draft tube

$$h_L = f\frac{L}{D}\frac{V^2}{2g}, \qquad \text{where} \quad V = \frac{Q}{A} = \frac{130}{\pi(1.5)^2} = 18.4\ \text{ft/sec}$$

Thus:
$$h_L = 0.03 \frac{40}{3} \frac{(18.4)^2}{2(32.2)} = 2.10 \text{ ft}$$

Head loss at discharge

$$h_{L_x} = \frac{V_2^2}{2g} - \frac{V_C^2}{2g} = \frac{18.4^2}{2(32.2)} - 0 = 5.26 \text{ ft}$$

Thus total head loss = 2.10 ft + 5.26 ft = 7.36 ft

Replacement draft tube:
Head loss due to draft tube

Eq. (8.59):
$$h_L = k' \frac{(V_1 - V_2)^2}{2g}, \quad \text{where} \quad V_1 = 18.4 \text{ ft/sec}$$

and
$$V_2 = \left(\frac{D_1}{D_2}\right)^2 V_1 = \left(\frac{3}{8}\right)^2 18.4 = 2.59 \text{ ft/sec}$$

Obtain k' from Fig. 8.19*a* (we shall assume the solid line applies):

$$\tan \alpha = \frac{(8 - 3)}{40} = \frac{5}{40} = 0.125, \quad \alpha = 7.1°, \quad k' = 0.13$$

Thus:
$$h_L = 0.13 \frac{(18.4 - 2.6)^2}{2(32.2)} = 0.50 \text{ ft}$$

Head loss at discharge

$$h_{L_x} = \frac{V_2^2}{2g} - \frac{V_C^2}{2g} = \frac{2.6^2}{2(32.2)} - 0 = 0.10 \text{ ft}$$

Thus: Total head loss = 0.50 ft + 0.10 ft = 0.60 ft

Additional effective head available with the replacement draft tube

$$= 7.36 - 0.60 = 6.76 \text{ ft} \qquad \textit{ANS}$$

Note: 5.26 − 0.10 = 5.16 ft of the additional effective head is attributable to the decrease in the head loss at discharge. An extra 6.76 ft of effective head is quite significant if we are dealing with a low head installation.

EXERCISES

16.8.1 Repeat Sample Prob. 16.7 for the case where the flow rate is 90 cfs, with all other data remaining the same.

16.8.2 A Francis turbine operating at 300 rpm develops a shaft power of 1125 hp when the flow is 120 cfs. What is the net head on the runner if the turbine efficiency is 85%?

16.8.3 A Kaplan turbine operating at 327.3 rpm develops a shaft power of 300 kW when the flow is 5.5 m^3/s. What is the net head on the runner if the turbine efficiency is 92%?

16.9 EFFICIENCY OF TURBINES

Equation (16.8) for the efficiency of impulse turbines is equally applicable to reaction turbines. Hence, in general, the efficiency of turbines is defined by:

$$\eta = \frac{\text{power delivered to the shaft (brake power)}}{\text{power taken from the water}} = \frac{T\omega}{\gamma Q h} \tag{16.14}$$

where T is the torque delivered to the shaft by the turbine [Eq. (6.28)], ω is the rotative speed in radians per second, Q is the flow rate, and h is the net head on the turbine [Eq. (16.5) for impulse wheels and Eq. (16.12) for reaction turbines].

The efficiencies of various types of turbines change with load as shown in Fig. 16.13. The impulse turbine maintains high efficiency over a wide range of loads, with significant decrease in efficiency occurring when the load drops below about 30% of normal load.[3] The efficiency of propeller turbines is very sensitive to load, and a significant drop in efficiency is experienced as the

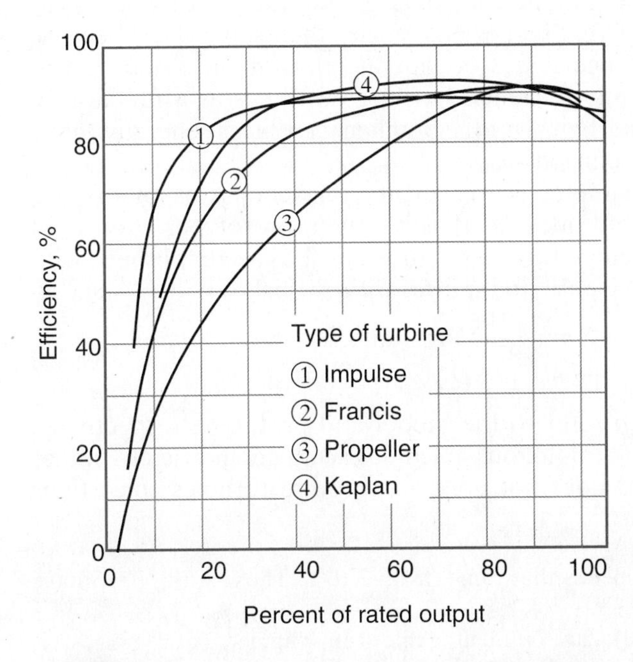

Figure 16.13
Efficiency versus load for typical turbines.

[3] Normal load refers to the conditions of operation for which efficiency is a maximum.

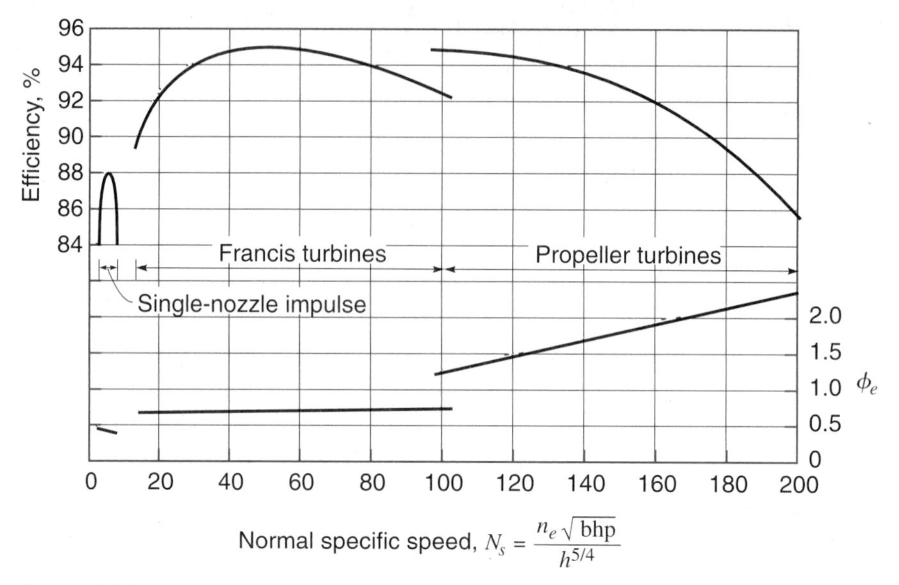

Figure 16.14
Maximum turbine efficiency and typical values of ϕ_e as functions of specific speed.

load falls below normal load. In contrast, the Kaplan turbine, with its adjustable blades, maintains high efficiency over a wide range of load.

Large turbines have somewhat higher efficiencies than small ones, because mechanical friction and windage losses do not increase at the same rate as hydraulic losses. The maximum efficiencies of large turbines as a function of specific speed are shown in Fig. 16.14.[4] Smaller turbines generally have somewhat smaller maximum efficiencies.

The effect of size (i.e., diameter D) on turbine efficiency is of importance in transferring test results on small models to their prototypes. For both Francis turbines and propeller turbines, this can be accomplished with reasonable accuracy by the ***Moody***[5] ***step-up formula***, which is

$$\frac{1-\eta_m}{1-\eta_p} \approx \left(\frac{D_p}{D_m}\right)^{1/5} \tag{16.15}$$

where the subscripts m and p refer to the model and prototype respectively. This equation applies only to homologous machines (with geometrically similar flow patterns, Sec. 15.4), and does not apply to impulse turbines since their

[4] Impulse wheels have specific speeds that range from 2 to 8. Those with $N_s \approx 5$ have maximum efficiency.

[5] Lewis F. Moody (1880–1953), an eminent American engineer and professor, contributed greatly to our understanding of similitude and cavitation as applied to hydraulic machinery.

efficiency is nearly independent of size. For the case of the Francis turbine the diameter D is the diameter of the runner at flow entrance, while the blade-tip diameter represents D for the axial-flow turbine.

Sample Problem 16.8 A small Francis runner with a diameter of 2.0 ft is tested and found to have an efficiency of 0.893 when operating under most efficient conditions. Approximately what would be the efficiency of a homologous runner having a diameter of 6.0 ft?

Solution

Eq. (16.15): $$\frac{1-\eta_m}{1-\eta_p} = \left(\frac{D_p}{D_m}\right)^{1/5} = \left(\frac{6}{2}\right)^{1/5} = 3^{0.2} = 1.246$$

$$1-\eta_p = (1-0.893)/1.246$$

From which: $$\eta_p = 0.914 = 91.4\% \qquad \textit{ANS}$$

16.10 SIMILARITY LAWS FOR REACTION TURBINES

In Sec. 15.4 we developed similarity laws for homologous pumps and pointed out that these laws can be used to predict performance of a prototype pump from the test of a scaled model. These laws also hold true for reaction turbines; however, they are expressed in different ways than for pumps. For a turbine, we are usually interested in its operation under a certain head, which is fixed by the natural characteristics of the site at which it is to be located. Hence, when dealing with turbines, it is convenient to have h in the similarity expressions.[6] We can accomplish this by rearranging Eq. (15.5) to get

$$n \propto h^{1/2}/D \quad \text{or} \quad n = K_n h^{1/2}/D \tag{16.16}$$

Substituting Eq. (16.16) into Eq. (15.4), we get

$$Q \propto nD^3 \propto (h^{1/2}/D)D^3 \propto h^{1/2}D^2$$

Hence, for homologous reaction turbines,

$$Q = K_Q h^{1/2} D^2 \tag{16.17}$$

[6] In applying these similarity laws it would be more appropriate to use the actual head across the runner, h'', rather than the net head h. However, it is more convenient to use h, and in so doing, the results are quite accurate except for the case of very-low-head turbines with heads less than about 10 ft.

This expression can also be developed using another approach. Note in Fig. 6.9 that $V_{r_1} \propto V \propto \sqrt{2gh} \propto h^{1/2}$ for a given value of g, and $A_{r_1} = \pi D_1 Z \propto D_1^2$ for geometrically similar runners where V_{r_1} is the radial component of the velocity V at $r = r_1$, A_{r_1} is the effective area through which V_{r_1} is passing, and Z is the width of the runner at entrance (i.e., at $r = r_1$). From the above, substituting D for D_1 since they are the same, we get $Q = A_{r_1}V_{r_1} \propto D^2h^{1/2}$, or $Q = K_Q h^{1/2}D^2$.

Since power $P = \gamma Qh$, and from Eq. (16.17) $Q \propto h^{1/2}D^2$, for a given value of γ we get $P \propto h^{1/2}D^2(h) \propto h^{3/2}D^2$. Hence, for homologous reaction turbines with a given value of γ,

$$P = K_P h^{3/2}D^2 \tag{16.18}$$

Equations (16.16)–(16.18) can be used to determine the relationship between homologous reaction turbines.

SAMPLE PROBLEM 16.9 A given turbine runs at a maximum efficiency of 87% when discharging 200 cfs at 150 rpm under a net head of 81 ft. This turbine drives an electric generator that is developing 60-cycle current. (*a*) How many poles are required for this generator? (*b*) What would be the speed, flow rate, and brake horsepower of this same turbine under a net head of 40.5 ft for homologous conditions? (*c*) How many poles would be required for the generator to satisfy the conditions given in (*b*)?

Solution

(*a*) $N = \dfrac{7200}{n} = \dfrac{7200}{150} = 48$ poles,

$$\text{bhp} = \frac{\eta\gamma Qh}{550} = \frac{0.87(62.4)(200)81}{550} = 1600 \text{ hp} \qquad \textbf{\textit{ANS}}$$

(*b*) To solve for the speed, we make use of Eq. (16.16):

Eq. (16.16): $n \propto h^{1/2}/D$

So:
$$\frac{n_2}{n_1} = \left(\frac{h_2}{h_1}\right)^{1/2}\left(\frac{D_1}{D_2}\right) = \left(\frac{40.5}{81}\right)^{1/2}\left(\frac{D}{D}\right)$$

$$n_2 = n_1(\tfrac{1}{2})^{1/2} = 150(0.707) = 106 \text{ rpm} \qquad \textbf{\textit{ANS}}$$

Eq. (16.17): $Q \propto h^{1/2}D^2$

So:
$$\frac{Q_2}{Q_1} = \left(\frac{h_2}{h_1}\right)^{1/2}\left(\frac{D_2}{D_1}\right)^2 = \left(\frac{40.5}{81}\right)^{1/2}\left(\frac{D}{D}\right)^2$$

$$Q_2 = Q_1(\tfrac{1}{2})^{1/2} = 200(0.707) = 114 \text{ cfs} \qquad \textbf{\textit{ANS}}$$

Eq. (16.18): $P \propto h^{3/2}D^2$

So: $$\frac{P_2}{P_1} = \left(\frac{h_2}{h_1}\right)^{3/2}\left(\frac{D_2}{D_1}\right)^2 = \left(\frac{40.5}{81}\right)^{3/2} \times 1$$

$$P_2 = 1600(\tfrac{1}{2})^{3/2} = 566 \text{ hp} \qquad \textit{ANS}$$

(*c*) $N = 7200/n_2 = 7200/106 = 68$ poles *ANS*

Note: If the number of poles does not turn out to be an even number, the rotative speed must be changed somewhat. For example, suppose the head was 66 ft, with all other data unchanged. Then $n_2 = 150(66/81)^{1/2} = 135.4$ rpm. For this case the number of poles $N = 7200/135.4 = 53.2$, which is impossible. So, we shall use 54 poles, in which case $n = 7200/N = 7200/54 = 133.3$ rpm.

Sample Problem 16.10 A turbine homogolous to the turbine of Sample Prob. 16.9 has all dimensions one-half as large. For this turbine, find the speed of rotation, flow rate, and brake horsepower when operating under the same net head of 81 ft.

Solution

Eq. (16.16): If h does not change, $n \propto D^{-1}$

Hence: $n_2 = (2/1)n_1 = (2/1)150 = 300$ rpm *ANS*

Eq. (16.17): If h does not change, $Q \propto D^2$

Hence: $Q_2 = (1/2)^2 Q_1 = (1/4)200 = 50$ cfs *ANS*

Eq. (16.18): If h does not change, $P \propto D^2$

Hence: $P_2 = (1/2)^2 P_1 = (1/4)1600 = 400$ hp *ANS*

In these two sample problems we have looked at two different situations. In Sample Prob. 16.9 we took a given turbine and moved it to a different site where the net head was smaller. In so doing, we found a decrease in speed, flow rate and shaft horsepower. In Sample Prob. 16.10 we compared two turbines that are geometrically similar, but of different size operating under the same net head. We found that in this situation, to maintain homologous conditions (similar flow patterns), the smaller turbine must rotate at a greater speed. However, the flow rate and shaft horsepower are substantially smaller.

EXERCISES

16.10.1 At maximum efficiency a turbine runs at 150 rpm, discharges 10.0 m^3/s, and develops a shaft power of 2600 kW under a net head of 30 m. (*a*) What is its efficiency? (*b*) What would be the revolutions per minute, the flow rate and brake power of this same turbine under a net head of 48 m for homologous conditions? (c) How many poles would be required for 60-cycle electricity if the net head is 48 m?

16.10.2 At maximum efficiency of 93% a turbine runs at 300 rpm and discharges 310 cfs under a net head of 64 ft. This turbine drives a generator that is developing 50-cycle electricity. (a) How many poles are required for this generator? (b) What is the brake horsepower?

16.10.3 If the turbine of Exer. 16.10.2 is operated under a head of 48 ft, find the speed, flow rate and brake power for homologous conditions. Be sure to use an even number of poles. *Hint:* It may not be possible to have perfectly homologous conditions.

16.11 PERIPHERAL VELOCITY FACTOR AND SPECIFIC SPEED OF TURBINES

In Sec. 15.9 we developed Eq. (15.9), an expression for the ***peripheral velocity factor*** ϕ applicable to pumps. For turbines, the expression is the same, the only difference being that in the development of the equation we use u_1 rather than u_2. Thus, for turbines,

$$u_1 = \phi\sqrt{2gh} \tag{16.19}$$

where u_1 is the tangential velocity of a point on the periphery of the rotating element. As with pumps, for any given turbine, ϕ varies over a wide range during different conditions of operation. The value of ϕ that is of most importance is that for conditions of maximum efficiency. Hence the ϕ most commonly used in Eq. (16.19) is that which occurs at maximum efficiency, and this is referred to as ϕ_e (Fig. 16.14).

For convenience, we repeat Eq. (15.10) here:

$$D = \frac{153.3\phi_e\sqrt{h}}{n} \tag{16.20}$$

This equation is equally applicable to reaction turbines, and can be used to estimate the diameter of the runners for Francis and propeller-type turbines.

Traditionally, in the British Gravitational system the ***specific speed*** of a turbine[7] has been expressed as

$$N_s = \frac{n_e\sqrt{\text{bhp}}}{h^{5/4}} \tag{16.21}$$

[7] In the metric system the specific speed of turbines has been expressed in various forms, but will not be dealt with here. We will define specific speed of turbines using BG units only (Eq. 16.21).

where the values of n_e (rpm), bhp (shaft or brake horsepower), and h (net head in ft) are those that occur at maximum efficiency. Two turbines of identical geometric shape, but of different size, will have the same specific speed. Hence, as with pumps, N_s serves to classify turbines as to type. In fact, N_s might better be called the *type characteristic,* or some similar name, because it indicates the type of turbine (Fig. 16.14). Impulse wheels have low specific speeds, Francis turbines have medium values of N_s, and propeller turbines have high values. Typical values of maximum efficiency and values of ϕ_e for the different types of turbines are shown in Fig. 16.14. Values of ϕ_e vary approximately as follows:

Impulse wheels[8]	0.43–0.48
Francis turbines	0.7–0.8
Propeller turbines	1.4–2.0

Sample Problem 16.11 Assume the Francis turbine of Sample Prob. 16.6 is operating at maximum efficiency. Find the specific speed of the turbine and estimate the diameter of its runner.

Solution

Eq. (16.21): $$N_s = \frac{n_e\sqrt{\text{bhp}}}{h^{5/4}} = \frac{300\sqrt{1125}}{97.2^{5/4}} = 33.0 \qquad \textbf{\textit{ANS}}$$

From Fig. 16.14: $\phi_e = 0.72$

Eq. (16.20): $$D = \frac{153.3\phi_e\sqrt{97.2}}{300} = 3.63\text{ ft} \qquad \textbf{\textit{ANS}}$$

EXERCISES

16.11.1 Find the specific speed of an impulse wheel operating at a maximum efficiency of 92.6% under a net head of 2314 ft at 360 rpm if the flow rate from the nozzle is 131.5 cfs. Estimate the diameter of the runner. *Note:* This is the turbine of Sample Prob. 16.4.

16.11.2 Find the specific speed of a turbine that operates at a maximum efficiency of 90% at 150 rpm under a net head of 81 ft with a flow rate of 200 cfs. Estimate the diameter of the runner. What type of turbine is this?

[8] Note that the ϕ_e for impulse wheels is approximately equal to the ratio of u/V_1 of Sec. 16.3.

16.11.3 Find the specific speed of a turbine that operates at a maximum efficiency of 83% at 450 rpm under a net head of 20 ft with a flow rate of 100 cfs. Estimate the diameter of the runner. What type of turbine is this?

16.12 CAVITATION IN TURBINES

Cavitation (Sec. 5.11) is undesirable because it results in pitting (Fig. 16.15), mechanical vibration, and loss of efficiency. The high heads on impulse turbines create extremely high velocities of flow. If the nozzle and buckets are not properly shaped for the particular flow conditions under which the turbine is operating, separation of the flow from the boundaries may cause regions of low pressure and result in cavitation. Low-specific-speed impulse wheels ($N_s \approx 2$) can operate at heads as high as 2000 ft (600 m) without cavitation, while high-specific-speed impulse wheels ($N_s \approx 8$) will experience cavitation at heads in the vicinity of 400 ft (120 m). Figure 16.16 shows recommended

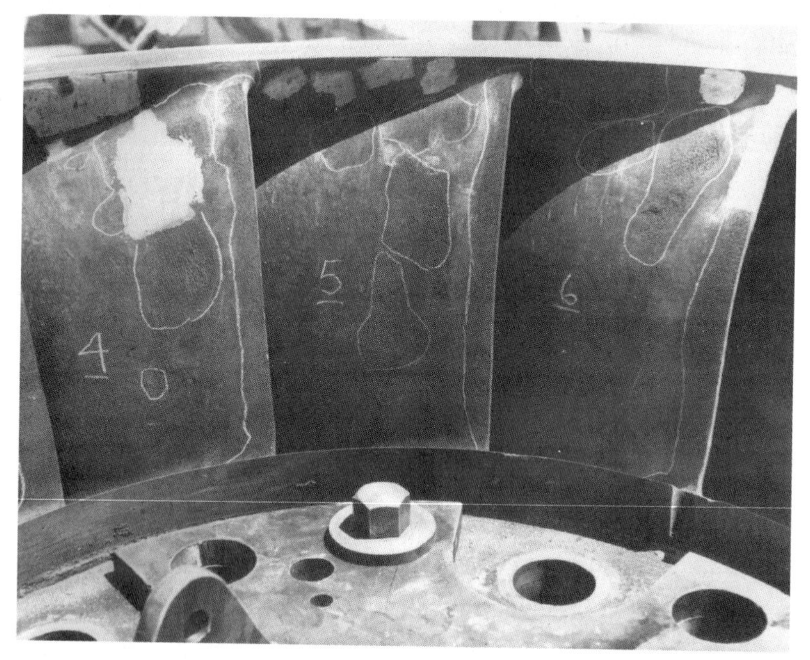

Figure 16.15
Cavitation pitting of Francis wheel and scroll case at Mammoth Pool Powerhouse after $2\frac{1}{2}$ years of operation. Conditions of service were relatively severe. The turbine develops 88,000 hp at an effective head of 950 ft at 360 rpm. The bright shiny areas are stainless-steel welds that have withstood cavitation pitting for over a year. (*Courtesy of Southern California Edison Co.*)

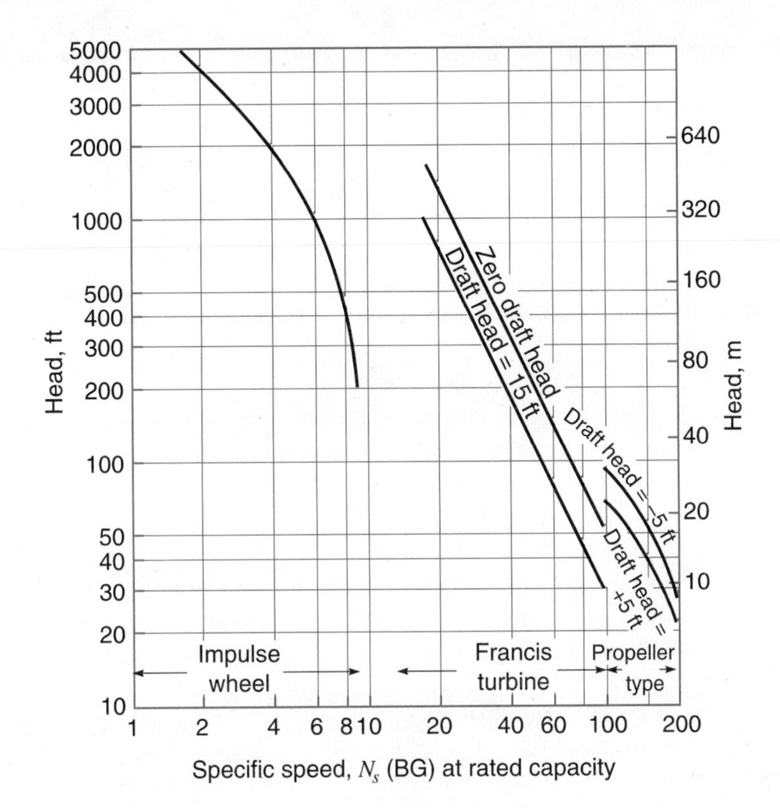

Figure 16.16
Recommended limits of specific speed for turbines under various effective heads at sea level with water temperature at 80°F. (*After Moody*)

limits of safe specific speed of turbines for various heads and settings based on experience at existing power plants.

In reaction turbines the most likely place for the occurrence of cavitation is on the back sides of the runner blades near their trailing edges. Cavitation may be avoided by designing, installing, and operating a turbine in such a manner that at no point will the local absolute pressure drop to the vapor pressure of the water.

In comparing the cavitation characteristic of reaction turbines, it is convenient to define a ***cavitation parameter*** σ as

$$\sigma = \frac{p_{\text{atm}}/\gamma - p_v/\gamma - z_B}{h} \tag{16.22}$$

where z_B and h are as defined in Fig. 16.11. The term $p_{\text{atm}}/\gamma - p_v/\gamma$ represents the height to which water will rise in a water barometer. At sea level with 80°F water, $p_{\text{atm}}/\gamma - p_v/\gamma = 32.8$ ft (10.0 m). At higher elevations and at higher water temperatures it is smaller than 32.8 ft (10.0 m).

The critical value σ_c of the cavitation parameter is the value of σ below which cavitation will occur. By inspection of Eq. (16.22), we see that σ tends to get smaller at low values of p_{atm} (high elevations), high values of p_v (high water temperatures), large values of z_B (large distances of the turbine setting above tailwater), and high values of head h. The critical value σ_c can be determined experimentally for a given turbine by noting the operating conditions under which cavitation first occurs as evidenced by the presence of noise, vibration, or loss of efficiency.

By substituting the critical condition for cavitation ($\sigma = \sigma_c$) into Eq. (16.22), we obtain

$$h = \frac{p_{\text{atm}}/\gamma - p_v/\gamma - z_B}{\sigma_c} \tag{16.23}$$

or

$$z_B = p_{\text{atm}}/\gamma - p_v/\gamma - \sigma_c h \tag{16.24}$$

The first of these equations gives the maximum head under which a reaction turbine should operate in order to be safe from cavitation, while the second equation indicates the maximum value of z_B (Fig. 16.11) to assure freedom from cavitation. To be on the safe side, z_B is usually limited to 15 ft (4.5 m). Typical values of σ_c as a function of specific speed are given in Fig. 16.17.

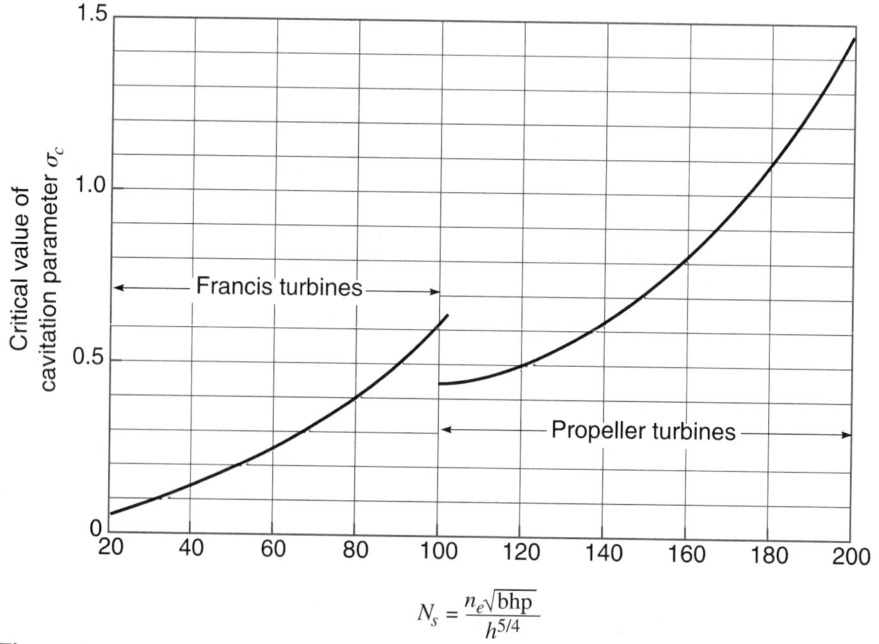

Figure 16.17
Typical values of the critical cavitation parameter σ_c for reaction-type turbines.

SAMPLE PROBLEM 16.12 Find the maximum permissible head under which a Francis turbine ($N_s = 70$) can operate if it is set 10 ft above tailwater at an elevation of 5000 ft with water temperature at 60°F.

Solution

At 5000 ft elevation: $\dfrac{p_{\text{atm}}}{\gamma} = \dfrac{12.23(144)}{62.4} = 28.22$ ft (from Table A.3)

Table A.1 60°F: $\dfrac{p_v}{\gamma} = 0.59$ ft

Fig. 16.17 for Francis turbine with $N_s = 70$: $\sigma_c = 0.32$

Eq. (16.24): $10 = 28.22 - 0.59 - 0.32(h)$

$h = 55.1$ ft (maximum permissible head to assure against cavitation) ***ANS***

SAMPLE PROBLEM 16.13 Repeat Sample Prob. 16.12, assuming the Francis turbine is at sea level with a water temperature of 80°F. Check the result against Fig. 16.16.

Solution

At sea level: $\dfrac{p_{\text{atm}}}{\gamma} = \dfrac{14.70(144)}{62.4} = 33.91$ ft (from Table A.3)

Table A.1 at 80°F: $\dfrac{P_v}{\gamma} = 1.17$ ft

Fig. 16.17 for Francis turbine with $N_s = 70$: $\sigma_c = 0.32$

Eq. (16.24): $10 = 33.91 - 1.17 - 0.32(h)$

$h = 71.1$ ft (maximum permissible to assure against cavitation) ***ANS***

This checks closely with the information given on Fig. 16.16.

EXERCISES

16.12.1 Using Eq. (16.24) and Fig. 16.17 for σ_c calculate the maximum permissible head under which a Francis turbine ($N_s = 40$) can operate if it is set 12 ft above tailwater. The installation is at sea level and the water temperature is 80°F. How does your answer compare to Fig. 16.16?

16.12.2 (*a*) Find the maximum permissible head under which an axial-flow turbine ($N_s = 160$) can operate if it is set 5 ft below tailwater. The installation is at elevation 3150 ft and the water temperature is 65°F. (*b*) What would be the maximum permissible head if the turbine had been set 5 ft above tailwater?

16.12.3 Repeat Exer. 16.12.1 for the case where the turbine is at elevation 10,000 ft rather than at sea-level. In this case one cannot compare the result with the curves presented in Fig. 16.16, because of the change in p_{atm}.

16.13 SELECTION OF TURBINES

Inspection of Eq. (16.21) indicates that at high heads for a given speed and power output, a low-specific-speed machine such as an impulse wheel is required. On the other hand, an axial-flow turbine with a high N_s is indicated for low heads. An impulse turbine may, however, be suitable for a low-head installation if the flow rate (or power requirement) is small, but often under such conditions the efficiency of the impulse wheel is quite low. Impulse wheels have been used for heads as low as 50 ft if the capacity is small, but they are more commonly employed for heads greater than 500 or 1000 ft. The limiting head for Francis turbines is about 1500 ft because of possible cavitation and the difficulty of building casings to withstand such high pressures. By choosing a high speed of operation, and hence a high-specific-speed turbine, the runner size and thus first cost are reduced. However, there is some loss of efficiency at high specific speeds.

In the selection of a turbine or turbines at a given installation, options are available with respect to the number and type (N_s) of turbines. Generally it is considered good practice to have at least two turbines at an installation so that the plant can continue operation while one of the turbines is shut down for repairs or inspection. The head h is determined primarily by topography, and the flow Q by the hydrology of the watershed and characteristics of the reservoir. In selecting a turbine for a given installation, freedom from cavitation must be verified. Some of the factors influencing the choice of turbines are apparent in the following sample problem.

SAMPLE PROBLEM 16.14 Two or more identical turbines are to be selected for an installation where the net head is 350 ft and the total flow is to be 600 cfs. Select turbines for this installation assuming 90% efficiency for all turbines.

Solution

The total available power is

$$\frac{\gamma Q h \eta}{550} = \frac{62.4(600)350(0.90)}{550} = 21{,}400 \text{ hp}$$

Assume two turbines at an operating speed of 75 rpm (96-pole generators for 60-cycle electricity).

Eq. (16.21):
$$N_s = \frac{75\sqrt{21400/2}}{350^{5/4}} = 5.12$$

Thus, if the operating speed is 75 rpm, use two impulse turbines with $N_s = 5.12$, in which case from Fig. 16.14 $\phi_e \approx 0.45$. The required wheel diameter of these turbines is found from Eq. (16.20):

$$D = \frac{153.3\sqrt{350} \times 0.45}{75} = 17.2 \text{ ft}$$

A wheel diameter of 17.2 ft is quite large; a smaller size is possible by increasing the rotative speed. If $n = 100$ rpm, $N_s = 5.12(100/75) = 6.8$ and $D = 17.2 \times (75/100) = 12.9$ ft. Other combinations of N_s and D could be used with other speeds; however, in accordance with Fig. 16.14, N_s should be less than about 7.0 if impulse wheels are selected. Another possible solution is four identical turbines with $N_s = 5.8$ and $D = 10.7$ ft operating at 120 rpm. ***ANS***

Finally, let us explore the possibility of using Francis turbines. Assume two Francis turbines operating at 600 rpm (12-pole generators for 60-cycle electricity).

Eq. (16.21):
$$N_s = \frac{600\sqrt{21{,}400/2}}{350^{5/4}} = 41$$

According to Fig. 16.16, turbines with $N_s = 41$ and $h = 350$ ft will be safe against cavitation only if they are set one or more feet below tailwater. This may be impractical in some topographic situations. ***ANS***

To provide greater safeguard against cavitation, we might select a lower-specific-speed machine, but then its efficiency may not be so good as indicated by Fig. 16.14. A good choice would be two Francis turbines with $N_s = 30.8$ operating at 450 rpm. The required diameter of these turbines would be about 4.8 ft, assuming $\phi = 0.75$. ***ANS***

There are actually an infinite number of alternatives. The things to watch out for are: (*a*) freedom from cavitation (Fig. 16.16); (*b*) reasonably high efficiency (Fig. 16.14); and (*c*) size not too large [Eq. (16.19)]. Flexibility of choice is achieved through variation in the number of units (and hence brake horsepower per unit) and in the operating speed. Variation in the draft-head setting also provides some flexibility.

EXERCISES

16.13.1 (*a*) If $Q = 40$ cfs and $h = 70$ ft, would it be possible to use a single impulse wheel with a specific speed less than 7.0 at a site where 60-Hz current is to be generated? Assume a turbine efficiency of 90% and that the generator is to have no more than 96 poles. (*b*) If a single impulse wheel can be used under these conditions, calculate the approximate diameter of the turbine runner for the range of possible speeds of operation.

16.13.2 (*a*) If $Q = 40$ cfs and $h = 70$ ft, would it be possible to use a single Francis turbine with a specific speed of at least 20 at a site where 60-Hz current is to be generated? Assume a turbine efficiency of 90% and assume the generator is to have no less than 12 poles. (*b*) If a single Francis turbine can satisfy these conditions, calculate the approximate diameter of the turbine runner for the range of possible speeds of operation.

16.13.3 Specify the specific speed and the approximate runner diameter at a site where $Q = 2350$ cfs and $h = 480$ ft. Use a 40-pole generator for 60-cycle current. Assume a turbine efficiency of 90%. These are the specifications for some of the turbines at Hoover Dam.

16.13.4 (*a*) What is the maximum head for which a propeller-type turbine can be used if the flowrate is 40 cfs? Assume a turbine efficiency of 90% with specific speeds for propeller-type turbines varying from 100 to 200. Also, assume a 12-pole generator with 50-Hz current. (*b*) Find the maximum head for which a 20-pole generator could have been used with the same turbine efficiency and 50-Hz current. Neglect the possibility of cavitation.

16.14 PUMP TURBINE

In recent years the ***pump–turbine hydraulic machine*** has been developed. It is very similar in design and construction to the Francis turbine. When water enters the rotor at the periphery and flows inward the machine acts as a turbine. With water entering at the center (or ***eye***) and flowing outward, the machine acts as a pump. The direction of rotation is, of course, opposite in the two cases. The pump turbine is connected to a motor generator, which acts as either a motor or generator depending on the direction of rotation.

The pump turbine is used at pumped-storage hydroelectric plants, which pump water from a lower reservoir to an upper reservoir during off-peak load periods so that water is available to drive the machine as a turbine during the time that peak power generation is needed.

An example of a pump turbine is that at the Kisenyama Pumped Storage Project of the Kansai Electric Company in Japan. There are two identical pump turbines at that installation. Under the normal range of operating conditions each machine has the following characteristics:

As a turbine ($n = 225$ rpm):

Develops 322,000 hp at maximum net head of 722 ft.
Develops 238,000 hp at minimum net head of 607 ft.

As a pump ($n = 225$ rpm):

Delivers 3880 cfs at minimum net head of 649 ft.
Delivers 3040 cfs at maximum net head of 755 ft.

When operating as a turbine, these machines have a specific speed $N_s = 35$

in the BG system of units. As a pump, in the BG system, these machines have $N_s = 2080$.

At Bad Creek in South Carolina the Duke Power Company has four vertical-shaft pump turbines. When pumping against a head of 378 m at 300 rpm, the flow rate is 100 m^3/s. As a turbine each unit develops 360 MW under a head of 367 m at 300 rpm. The corresponding values of specific speed in BG units are 1808 when acting as a pump and 27 when acting as a turbine.

16.15 TURBINE INSTALLATIONS

A few examples of turbine installations are presented here as illustrations of modern practice.

Impulse Turbines

The impulse wheel is especially adapted for use under high heads. For heads above 1500 ft it is the only type that can be used. Some interesting installations include the following.

At Dixence, Switzerland, a double-overhung unit operates under a gross head of 5735 ft and a net head of 5330 ft, giving a jet velocity of approximately 580 fps. Each single-nozzle wheel delivers 25,000 hp, or 50,000 hp for the unit, and runs at 500 rpm. The jet diameters are 3.71 in, and the pitch diameter of the wheels is 10.89 ft. Also in Switzerland, the Fully plant in Valais operates under a gross head of 5410 ft and a net head of 4830 ft. There are four wheels of 3000 hp each running at 500 rpm. The jet diameters are 1.5 in, and the wheel diameters are 11.67 ft.

A multiple-nozzle wheel at Pragnières, France, delivers 100,000 hp at 428 rpm under a gross head of 3920 ft.

In North America there are a number of plants operating under heads of between 2000 and 2500 ft. At the Big Creek-2A plant of the Southern California Edison Co. is a double-overhung single-jet impulse turbine of 137-in diameter with an 8.4-in jet on each wheel and a similar unit of 135.25-in diameter with one 8.5-in jet on each wheel. Each unit runs at 300 rpm, and the maximum static head with a full reservoir is 2418 ft. With both units carrying full load, the net head is 2200 ft, and the maximum output is 65,100 bhp for the first unit and 67,300 bhp for the second. At the Balch plant of the Pacific Gas and Electric Co., is a double-overhung unit with 115-in-diameter wheels and one 7.5-in jet on each wheel. Its runs at 360 rpm and develops 49,000 hp under a net head of 2243 ft. The Colgate plant in California has two vertical-shaft impulse wheels, each with six nozzles operating under a head of 413 m at 180 rpm and delivering 175 000 kW.

The Kitimat plant in British Columbia of the Aluminum Company of Canada has three four-nozzle vertical-shaft units that operate at 327 rpm under a net head of 2500 ft. Each unit is rated at 140,000 hp.

One of the highest-head plants in South America is at Los Molles in Chile, where an impulse wheel develops 11,500 hp at 1000 rpm under a head of 3460 ft. At Cementos El Cairo in Colombia are two single-overhung impulse turbines operating under a head of 1900 ft. Each unit develops 3000 hp at 900 rpm.

In Peru at the Paucartambo plant of Cerro de Pasco Corp., there is a single-overhung double-jet impulse wheel rated at 28,000 hp under a net head of 1580 ft operating at 450 rpm.

Francis Turbines

The Francis turbine is especially adapted for heads ranging from 50 to 1000 ft, and has been used on heads as high as 1500 ft.

Among some interesting installations of Francis turbines are the following. At Fionnay, Switzerland, a Francis turbine operates under a head of 1490 ft and delivers 63,200 hp at 750 rpm. In Austria a head of 1430 ft is used for a 77,700-hp turbine at 500 rpm. In Norway a head of 1360 ft is used to develop 69,000 hp at 500 rpm. In Italy a turbine running at 1000 rpm develops 20,140 hp under a head of 1320 ft. At Oak Grove, Oregon, a Francis turbine under a head of 850 ft delivers 35,000 hp at 514 rpm. These are all low-specific-speed reaction turbines.

Examples of higher specific speeds are a Francis turbine in France, which delivers 154,000 hp at 187.5 rpm under a head of 336 ft and another one, also in France, which runs under a head of 233 ft and develops 135,000 hp at 150 rpm. At Conowingo on the Susquehanna River 54,000-hp units run at 81.8 rpm under a head of 89 ft. The Conowingo runners are 18 ft in diameter. Other examples of large reaction turbines are the four vertical-shaft Francis turbines at the Kootenay Canal Power Station in British Columbia, Canada. These are each designed to develop 196,000 hp when operating at 128.6 rpm under a net head of 268 ft. At Grand Coulee Dam on the Columbia River in the State of Washington there are six large Francis turbines ($N_s = 55$). Each unit develops over 800,000 hp under a head of 285 ft at 72 rpm. Each runner has a diameter of 32 ft and weighs 550 tons. Water is delivered to each unit at a rate as high as 30,000 cfs on a 40-ft diameter pipeline. The turbines operate at heads ranging from 220 to 355 ft.

A unique installation is the Itaipu hydroelectric plant on the Parana River in South America. This is a joint project between Brazil and Paraguay. There are18 turbine units operating under a head of about 122 m. Nine of the units develop 50-cycle electricity using 66-pole generators (90.9 rpm), while the other nine units develop 60-cycle electricity using 78-pole generators. Another interesting installation is at the Trinity River Power Plant in California. The turbine is designed to permit the use of either of two interchangeable runners. The high-head runner ($N_s = 30$) is rated at 85,000 hp when operating at 200 rpm under a head of 426 ft. The low-head runner is rated at 70,000 hp when operating at 200 rpm under a head of 334 ft.

Propeller turbines

Propeller turbines, both fixed-blade and adjustable-blade (Kaplan turbine), are best suited for low-head installations, usually for heads less than 100 ft. A large fixed-blade turbine at the St. Lawrence Power Dam operates under a head of 81 ft at 94.7 rpm and delivers 79,000 hp. At Wheeler Dam in Alabama a fixed-blade propeller unit delivers 45,000 hp at 85.7 rpm under a head of 48 ft. Kaplan turbines are represented by two units at Camargos in Brazil. Each unit develops 35,700 hp under a head of 88.6 ft at 150 rpm. At Bonneville, on the Columbia River, are units that have 280-in-diameter runners. They deliver 66,000 hp at 75 rpm under 50 ft of head. A typical tube turbine would be the one at Ludwigswehr, Germany, where a four-blade runner with movable blades on a horizontal axis develops 0.78 MW under a head of 4.5 m at 125.5 rpm. Another example of a tube turbine is at Hausen, Germany, where a four-blade runner on an inclined axis develops 0.94 MW at 200 rpm under a head of 5.0 m.

PROBLEMS

16.1 Repeat Sample Prob. 16.1 for rotative speeds of 80, 160, 300, and 360 rpm, with all other data remaining the same. Also find the hydraulic efficiency of the flow through the buckets for each speed. Plot horsepower as ordinate versus rotative speed as abcissa to find the approximate speed at which maximum power will be developed. At maximum power, approximately what is the value of u/V_1?

16.2 Same as Prob. 16.1, except also consider hydraulic friction in the buckets by letting $v_2 = 0.85v_1$.

16.3 Consider an impulse wheel with a pitch diameter of 2.65 m and a bucket angle of 168°. If the jet velocity is 55.0 m/s, the jet diameter is 100 mm, and the rotative speed is 300 rpm, find (*a*) the force on the buckets, (*b*) the torque on the runner, and (*c*) the power transferred to the runner. Assume $v_2 = 0.85v_1$.

16.4 An impulse turbine with a pitch diameter of 4.0 ft and a bucket angle of 160° is rotating at 514.7 rpm. The nozzle produces a jet with a velocity of 240 ft/sec and diameter of 4 in. Find (*a*) the force on the buckets, (*b*) the torque on the runner, (*c*) the horsepower transferred to the runner, and (*d*) the efficiency η of the turbine under these conditions of operation. Assume the head loss through the nozzle can be expressed as $0.05V_j^2/2g$ and assume the velocity relative to the bucket does not change (i.e., assume $v_2 = v_1$). Neglect bearing friction and windage.

16.5 Work Prob. 16.4 where the speed of rotation is 300 rpm, all other data remaining the same.

16.6 (*a*) Solve Sample Prob. 16.1 for blade angles of 160°, 165°, and 170° for the case of an impulse turbine whose pitch diameter is 8 ft rather than 10 ft. All other data are to remain the same. (*b*) Repeat the above for the case where $v_2 = 0.80v_1$.

16.7 Repeat Sample Prob. 16.4 for the case of a 6-in-diameter jet, with all other data remaining the same.

16.8 A torque of 157,600 ft·lb is transmitted to a turbine that is developing a brake horsepower of 6000. How many poles does the generator have if it is producing 50-cycle electricity?

16.9 A 1.5-in-diameter water jet with a velocity of 120 ft/sec impinges on a bucket of a stationary impulse wheel that has a pitch diameter of 8.0 ft. If the bucket blade angle is 165°, what torque must be applied to the turbine shaft to prevent it from rotating? Consider two cases: (*a*) $v_2 = v_1$; (*b*) $v_2 = 0.9v_1$.

16.10 Water is delivered from a reservoir through a 1200-mm-diameter pipe ($f = 0.022$) 3200 m long to a nozzle that emits a 200-mm-diameter jet. The jet impinges on the buckets of an impulse turbine, pitch diameter 2.5 m, which drives a 16-pole generator that produces 50-Hz electricity. The surface of the reservoir providing the water is at an elevation 800 m higher than the nozzle. Assume that the head loss through the nozzle can be expressed as $0.05V_j^2/2g$ where V_j is the jet velocity. Find the following: velocity of flow in the pipe, jet velocity, flow rate, speed of the buckets, value of u/V_1, net head, force F on the bucket, torque on the wheel, shaft power, power at the head of the nozzle, and overall turbine efficiency. Assume $\beta_2 = 165°$ and $v_2 = v_1$; neglect bearing friction and windage.

16.11 The diameter of a Pelton wheel is 6 ft. The velocity of the 4-in-diameter jet is 400 ft/sec and the bucket angle is 165°. Assuming a k value of 0.21, for bucket speeds of 100, 200, 300, and 400 ft/sec find the torque in ft·lb, the horsepower, and the hydraulic efficiency of the buckets. Neglect bearing friction and windage.

16.12 An impulse turbine is supplied by a 1200-mm-diameter steel pipe (e = 0.10 mm) 500 m long. The jet diameter of the single nozzle is 400 mm and its loss coefficient is 0.05. When the water level in the reservoir is 150 m above the nozzle of the turbine, the brake power is measured to be 7900 kW. What is the turbine efficiency?

16.13 Using the data of Prob. 16.11, find values of (*a*) head loss in bucket friction, (*b*) velocity head at discharge, and (*c*) total head loss for bucket speeds of 180, 190, 200, and 210 fps. At what bucket speed is the discharge loss a minimum? At what bucket speed is the total head loss a minimum?

16.14 A 36-in pipeline ($f = 0.020$) 10,000 ft long connects a reservoir whose water surface elevation is 1800 ft to a nozzle at elevation 1000 ft. The jet from the nozzle is used to drive an impulse wheel. (*a*) If the head loss through the nozzle is expressible as $0.04V_j^2/2g$, determine the horsepower that will be developed by jets having diameters of 8 in, 10 in, and 12 in. (*b*) On the basis of the answers to (*a*), approximately what would be the horsepower and diameter of the jet that gives the maximum horsepower?

16.15 Using the data of Prob. 16.14, express the jet power analytically in terms of D_j and V_j, differentiate, and equate to zero to find the value of D_j for which the jet power is a maximum. Evaluate the value of the maximum jet power. *Hint:* Express power as $P = f(x)$, where $x = D_j^2$.

16.16 In the test of a Francis turbine, a pressure of 140 kPa was measured at the flange at the entrance to a spiral turbine-case where the diameter is 600 mm. Neglecting the small velocity head in the tailrace, find the net head on the turbine if the flow rate was 2.5 m^3/s and the flange was 3.0 m above the tailrace.

16.17 A reaction turbine is supplied with water through a 1500-mm-diameter pipe ($e = 0.10$ mm) that is 50 m long. The water surface in the reservoir is 27 m above the draft-tube inlet that is 4.1 m above the water level in the tailrace. If the turbine efficiency is 92% and the discharge is 12 m^3/s, what is the power output of the turbine in kilowatts? By how much must the discharge be increased to increase power production by 500 kW? Assume the velocity in the tailrace can be neglected.

16.18 A draft tube leading from the discharge side of a turbine to the submerged discharge in the tailrace as in Fig. 16.11 consists of a pipe ($f = 0.025$) of constant diameter 0.8 m and length 10.0 m. The flow rate is 9.2 m^3/s. If this draft tube were to be replaced by a diverging tube 12.0 m long whose diameter changed uniformly from 0.8 m to 2.5 m over the 12.0 m length, how much additional head would be developed by the replacement draft tube? Neglect head loss due to curvature in the bend.

16.19 At its upper end a draft tube has a diameter of 24.5 in where it joins the discharge side of the turbine at a point 11.0 ft above the surface of the water in the tailrace. The discharge end of the draft tube has a diameter of 42 in and the velocity in the tailrace is negligible. The total head loss in the draft tube is $0.15V_2^2/2g$ plus the submerged discharge loss of $V_3^2/2g$, where subscripts 2 and 3 refer to the upper and lower ends of the draft tube, respectively. (*a*) When the flow is 38.8 cfs, what is the pressure at the upper end of the draft tube? (*b*) Suppose the draft tube were of uniform diameter; what then would be the pressure at the upper end of the tube? (*c*) How much head is saved by the diverging tube? Assume the draft tube has a length of 18 ft and $f = 0.020$.

16.20 A model turbine (one-twentieth of prototype size) has a maximum hydraulic efficiency of 86.2%. Estimate the efficiency of the prototype utilizing the Moody step-up formula.

16.21 A turbine runs at 150 rpm, discharges 200 cfs, and develops 1600 bhp under a net head of 81 ft. (*a*) What is its efficiency? (*b*) What would be the revolutions per minute, Q, and brake horsepower of the same turbine under a net head of 162 ft for homologous conditions?

16.22 If a turbine homologous to that of Prob. 16.21 has a runner of twice the diameter, what would be the revolutions per minute, Q, and brake horsepower under the same head of 81 ft?

16.23 A 2.6-m-diameter reaction turbine is to be operated at 200 rpm under a net head of 30 m. A 1:10 model of this turbine is built and tested in the laboratory. If the model is operated at 600 rpm, under what net head should it be tested to simulate normal operating conditions in the prototype?

16.24 A small Francis turbine ($N_s = 30$, $D = 2$ ft) is tested and found to have an efficiency of 0.893 when operating under optimum conditions. Approximately what would be the maximum efficiency of a homologous runner ($N_s = 30$) with a diameter of 6 ft.

16.25 A 12-ft-diameter reaction turbine is to be operated at 100 rpm under a net head of 96 ft. A 1 : 8 model of this turbine is built and tested in the laboratory. If the model is operated at 450 rpm, under what net head should it be tested to simulate normal operating conditions?

16.26 A 1:8 model of a 12-ft-diameter turbine is operated at 600 rpm under a net head of 54.0 ft. Under this mode of operation the bhp and Q of the model were observed to be 332 hp and 62 cfs, respectively. From the above data compute (*a*) the specific speed of the model and the value of ϕ_e, (*b*) the efficiency and shaft torque of the model, (*c*) the efficiency of the prototype, and (*d*) the flow rate and horsepower of the prototype if it is operated at 450 rpm under a net head of 200 ft.

16.27 Calculate the specific speed of the turbine whose runner is shown in Fig. 16.2 and estimate the runner diameter.

16.28 Find the specific speed of the turbine whose runner is shown in Fig. 16.7 and estimate the runner diameter.

16.29 Calculate the specific speed of the turbine shown in Fig. 16.9 and estimate the runner diameter.

16.30 Find the specific speed of the propeller turbine mentioned in Exer. 16.7.1. Express the answer in BG units.

16.31 Find the specific speed of a turbine that runs at a maximum efficiency of 90% at 300 rpm under a net head of 81 ft with a flow rate of 50 cfs. Estimate the runner diameter.

16.32 In Sec. 16.15 a double-overhung impulse turbine at Dixence, Switzerland is mentioned. Each wheel of the two-wheel unit develops 25,000 hp at 500 rpm under a head of 5330 ft. (*a*) What is the specific speed of these wheels? (*b*) Estimate their diameter and compare your answer with their actual diameter of 10.89 ft.

16.33 In Sec. 16.15 a Francis turbine on the Susquehanna River is mentioned. This turbine develops 54,000 hp at 81.8 rpm under a head of 89 ft. (*a*) What is the specific speed of this turbine? (*b*) Estimate the runner diameter and compare your answer with the actual diameter of 18 ft.

16.34 A Kaplan turbine at Rock River, Illinois develops 800 hp at 80 rpm under a head of 7 ft. (*a*) What is the specific speed of this turbine? (*b*) Estimate the runner diameter and compare your answer with the actual diameter of 136 in.

16.35 Consider the case of a Francis turbine having a specific speed of 40 that is set 10 ft above tailwater elevation. Perform calculations using Fig. 16.17 and Eq. (16.22) to find the maximum permissible head under which this turbine should operate in order to be safe from cavitation. Check your calculated result against the information shown on Fig. 16.16. Do they agree?

16.36 Consider the case of a propeller turbine having a specific speed of 150 that is set 5 ft below tailwater. Perform calculations using Fig. 16.17 and Eq. (16.22) to find the maximum permissible head under which this turbine should operate in order to be safe from cavitation. Check your calculated result against the information shown in Fig. 16.16. Do they agree?

16.37 At its maximum efficiency of 93% a turbine delivers 3000 hp to the shaft under a head of 72 ft when operating at 300 rpm. Find the following: (*a*) the flow rate through the turbine; (*b*) the specific speed of the turbine; (*c*) the approximate diameter of the turbine runner; (*d*) the elevation at which the turbine should be set to be safe against cavitation if it is installed at sea level.

16.38 A propeller turbine operates at a maximum efficiency of 92% under a head of 30 ft at 450 rpm and develops a shaft power of 620 hp. Find (*a*) the flow rate through the turbine, (*b*) the specific speed of the turbine, (*c*) the approximate diameter of the turbine runner, and (*d*) how far above tailwater elevation the turbine can be set and still be safe from cavitation. Assume the turbine is at an elevation of 100 ft.

16.39 Repeat part (*d*) of Prob. 16.38 for the case where the turbine is installed at elevation 5000 ft.

16.40 A turbine whose specific speed is 80 is to operate under a head of 50 ft at an elevation where the atmospheric pressure is 12.8 psia. The water temperature is 50°F. (*a*) If this turbine is set 6 ft above tailwater, will it be safe from cavitation? (*b*) What is the highest permissible elevation of this turbine with respect to tailwater when operating under a head of 60 ft?

16.41 (*a*) A turbine is to be installed at a point where the available head is 175 ft and the available flow will average 1000 cfs. What type of turbine would you recommend? Specify the operating speed and the number of generator poles for 60-cycle electricity if a turbine with the highest tolerable specific speed to safeguard against cavitation is selected. Assume a draft head of 10 ft and 90% turbine efficiency. Approximately what size of turbine runner is required? (*b*) For the same conditions, select a set of two identical turbines to be operated in parallel. Specify the speed and size of the units.

16.42 What is the least number of identical turbines that can be used at a powerhouse where the available head is 1200 ft and $Q = 1650$ cfs? Assume turbine efficiency is 90% and speed of operation is 138.5 rpm. Specify the size and specific speed of the units.

16.43 A six-jet impulse turbine operating at 300 rpm develops 60,000 hp under a net head of 1060 ft. The runner has a diameter of 6.0 ft. How large a homologous runner would be needed for a single-jet machine operating under the same head and developing the same horsepower?

16.44 An impulse turbine ($N_s \approx 4$) develops 100,000 hp under a head of 2000 ft. (*a*) For 60-cycle electricity calculate the turbine speed (rpm), wheel diameter (ft), and number of poles in the generator. (*b*) Solve the problem for a six-nozzle unit using the same N_s, bhp, and head. In both instances assume $\phi_e = 0.45$. Neglect cavitation.

16.45 A multinozzle impulse turbine is to be designed to develop 60,000 hp at 300 rpm under a head of 1200 ft. How many nozzles should this turbine have? Specify the approximate wheel diameter for this design. How large would a single-jet machine have to be to satisfy these requirements?

16.46 It is desired to develop 15,000 bhp under a head of 1000 ft. Make any necessary assumptions and estimate the diameter of the wheel required and the rotative speed.

16.47 A single hydraulic turbine is to be selected for a power site with a net head of 100 ft. The turbine is to produce 25,000 hp at maximum efficiency. What speed (rpm) and diameter should this turbine have if (*a*) a Francis turbine is selected; (*b*) a propeller turbine is selected? What are the highest "settings" (above or below tailwater) that should be recommended for each of these machines for them to run cavitation-free at their points of maximum efficiency?

16.48 For 50-cycle electricity how many poles would you recommend for a generator that is connected to a turbine operating under a design head of 3000 ft with a flow of 80 cfs? Assume turbine efficiencies as given in Fig. 16.14 and be sure the turbine is free of cavitation.

16.49 Specify the type, speed, and size of a single turbine to be installed at a site with an effective head of 48 m, a maximum draft head of 2 m, and a flow rate of 5 m^3/s. How would your recommendation change if the available flow was 50 m^3/s?

16.50 It is desired to develop 300,000 hp under a head of 49 ft and to operate at 60 rpm. (*a*) If turbines with a specific speed of approximately 150 are to be used, how many units would be required? (*b*) If Francis turbines with a specific speed of 80 were to be used, how many units would be required?

16.51 Select two, four, and six identical turbines for an installation where $h = 400$ ft and total $Q = 300$ cfs. Develop 60-cycle electricity using either 36- or 72-pole generators. Be sure your selection is free of cavitation. Assume the turbine efficiency is 90%.

16.52 A turbine is to be installed at a point where the net available head is 35 m and the available flow will average 23 m^3/s. What type of turbine would you recommend? Specify the operating speed and number of generator poles for 60-cycle electricity if a turbine with the highest tolerable specific speed to safeguard against cavitation is selected. Assume a draft head of 3 m and 90% turbine efficiency. Approximately what size of turbine runner is required?

16.53 A turbine is to be installed where the net available head is 185 ft, and the available flow will average 900 cfs. What type of turbine would you recommend? Specify the operating speed and number of generator poles for 60-cycle electricity if a turbine with the highest tolerable specific speed that will safeguard against cavitation is selected. Assume the turbine is set 5 ft above tailwater. Assume turbine efficiency is 90%. Approximately what size of runner is required?

APPENDIX A

Fluid and Geometric Properties

Table	Contents
A.1	Physical properties of water at standard sea-level atmospheric pressure
A.2	Physical properties of air at standard sea-level atmospheric pressure
A.3	The ICAO standard atmosphere
A.4	Physical properties of common liquids at standard sea-level atmospheric pressure
A.5	Physical properties of common gases at standard sea-level atmospheric pressure
A.6	Areas of circles
A.7	Properties of areas
A.8	Properties of solid bodies

Table A.1
Physical properties of water at standard sea-level atmospheric pressure[a]

Temperature, T	Specific weight, γ	Density, ρ	Viscosity,[b] μ	Kinematic viscosity,[b] ν	Surface tension, σ	Saturation vapor pressure, p_v	Satur'n vapor pressure head, p_v/γ	Bulk modulus of elasticity, E_v
°F	lb/ft^3	slugs/ft^3	10^{-6} lb·s/ft^2	10^{-6} ft^2/s	lb/ft	psia	ft abs	psi
32°F	62.42	1.940	37.46	19.31	0.005 18	0.09	0.20	293,000
40°F	62.43	1.940	32.29	16.64	0.005 14	0.12	0.28	294,000
50°F	62.41	1.940	27.35	14.10	0.005 09	0.18	0.41	305,000
60°F	62.37	1.938	23.59	12.17	0.005 04	0.26	0.59	311,000
70°F	62.30	1.936	20.50	10.59	0.004 98	0.36	0.84	320,000
80°F	62.22	1.934	17.99	9.30	0.004 92	0.51	1.17	322,000
90°F	62.11	1.931	15.95	8.26	0.004 86	0.70	1.61	323,000
100°F	62.00	1.927	14.24	7.39	0.004 80	0.95	2.19	327,000
110°F	61.86	1.923	12.84	6.67	0.004 73	1.27	2.95	331,000
120°F	61.71	1.918	11.68	6.09	0.004 67	1.69	3.91	333,000
130°F	61.55	1.913	10.69	5.58	0.004 60	2.22	5.13	334,000
140°F	61.38	1.908	9.81	5.14	0.004 54	2.89	6.67	330,000
150°F	61.20	1.902	9.05	4.76	0.004 47	3.72	8.58	328,000
160°F	61.00	1.896	8.38	4.42	0.004 41	4.74	10.95	326,000
170°F	60.80	1.890	7.80	4.13	0.004 34	5.99	13.83	322,000
180°F	60.58	1.883	7.26	3.85	0.004 27	7.51	17.33	318,000
190°F	60.36	1.876	6.78	3.62	0.004 20	9.34	21.55	313,000
200°F	60.12	1.868	6.37	3.41	0.004 13	11.52	26.59	308,000
212°F	59.83	1.860	5.93	3.19	0.004 04	14.70	33.90	300,000
°C	kN/m^3	kg/m^3	N·s/m^2	10^{-6} m^2/s	N/m	kN/m^2 abs	m abs	10^6 kN/m^2
0°C	9.805	999.8	0.001 781	1.785	0.0756	0.61	0.06	2.02
5°C	9.807	1000.0	0.001 518	1.519	0.0749	0.87	0.09	2.06
10°C	9.804	999.7	0.001 307	1.306	0.0742	1.23	0.12	2.10
15°C	9.798	999.1	0.001 139	1.139	0.0735	1.70	0.17	2.14
20°C	9.789	998.2	0.001 002	1.003	0.0728	2.34	0.25	2.18
25°C	9.777	997.0	0.000 890	0.893	0.0720	3.17	0.33	2.22
30°C	9.764	995.7	0.000 798	0.800	0.0712	4.24	0.44	2.25
40°C	9.730	992.2	0.000 653	0.658	0.0696	7.38	0.76	2.28
50°C	9.689	988.0	0.000 547	0.553	0.0679	12.33	1.26	2.29
60°C	9.642	983.2	0.000 466	0.474	0.0662	19.92	2.03	2.28
70°C	9.589	977.8	0.000 404	0.413	0.0644	31.16	3.20	2.25
80°C	9.530	971.8	0.000 354	0.364	0.0626	47.34	4.96	2.20
90°C	9.466	965.3	0.000 315	0.326	0.0608	70.10	7.18	2.14
100°C	9.399	958.4	0.000 282	0.294	0.0589	101.33	10.33	2.07

[a] In these tables, if (for example, at 32°F) μ is given as 37.46 and the units are 10^{-6} lb·s/ft^2 then $\mu = 37.46 \times 10^{-6}$ lb·s/ft^2.

[b] For viscosity, see also Figs. 2.3 and 2.4.

Table A.2
Physical properties of air at standard sea-level atmospheric pressure[a]

Temperature T	Density, ρ	Specific weight, γ	Viscosity,[b] μ	Kinematic viscosity,[b] ν
°F	**slug/ft^3**	**lb/ft^3**	**10^{-6} lb·s/ft^2**	**10^{-3} ft^2/s**
−40°F	0.002 940	0.094 60	0.312	0.106
−20°F	0.002 807	0.090 30	0.325	0.116
0°F	0.002 684	0.086 37	0.338	0.126
10°F	0.002 627	0.084 53	0.345	0.131
20°F	0.002 572	0.082 77	0.350	0.136
30°F	0.002 520	0.081 08	0.358	0.142
40°F	0.002 470	0.079 45	0.362	0.146
50°F	0.002 421	0.077 90	0.368	0.152
60°F	0.002 374	0.076 40	0.374	0.158
70°F	0.002 330	0.074 95	0.382	0.164
80°F	0.002 286	0.073 57	0.385	0.169
90°F	0.002 245	0.072 23	0.390	0.174
100°F	0.002 205	0.070 94	0.396	0.180
120°F	0.002 129	0.068 49	0.407	0.189
140°F	0.002 058	0.066 20	0.414	0.201
160°F	0.001 991	0.064 07	0.422	0.212
180°F	0.001 929	0.062 06	0.434	0.225
200°F	0.001 871	0.060 18	0.449	0.240
250°F	0.001 739	0.055 94	0.487	0.280
°C	**kg/m^3**	**N/m^3**	**10^{-6} N·s/m^2**	**10^{-6} m^2/s**
−40°C	1.515	14.86	14.9	9.8
−20°C	1.395	13.68	16.1	11.5
0°C	1.293	12.68	17.1	13.2
10°C	1.248	12.24	17.6	14.1
20°C	1.205	11.82	18.1	15.0
30°C	1.165	11.43	18.6	16.0
40°C	1.128	11.06	19.0	16.8
60°C	1.060	10.40	20.0	18.7
80°C	1.000	9.81	20.9	20.9
100°C	0.946	9.28	21.8	23.1
200°C	0.747	7.33	25.8	34.5

[a] In these tables, if (for example, at −40°F) μ is given as 0.312 and the units are 10^{-6} lb·s/ft^2 then $\mu = 0.312 \times 10^{-6}$ lb·s/ft^2.
[b] For viscosity, see also Figs. 2.3 and 2.4.

Table A.3
The ICAO[a] standard atmosphere[b]

Elevation above sea level	Temperature, T	Absolute pressure, p	Specific weight, γ	Density, ρ	Viscosity, μ	Kinematic viscosity, ν	Speed of sound, c	Gravitational acceleration, g
ft	°F	psia	lb/ft³	slug/ft³	10^{-6} lb·s/ft²	10^{-3} ft²/s	ft/s	ft/s²
0	59.000	14.695 9	0.076 472	0.002 376 8	0.373 72	0.157 24	1116.45	32.1740
5,000	41.173	12.228 3	0.065 864	0.002 048 1	0.363 66	0.177 56	1097.08	32.158
10,000	23.355	10.108 3	0.056 424	0.001 755 5	0.353 43	0.201 33	1077.40	32.142
15,000	5.545	8.297 0	0.048 068	0.001 496 1	0.343 02	0.229 28	1057.35	32.129
20,000	−12.255	6.758 8	0.040 694	0.001 267 2	0.332 44	0.262 34	1036.94	32.113
25,000	−30.048	5.460 7	0.034 224	0.001 066 3	0.321 66	0.301 67	1016.11	32.097
30,000	−47.832	4.372 6	0.028 573	0.000 890 65	0.310 69	0.348 84	994.85	32.081
35,000	−65.607	3.467 6	0.023 672	0.000 738 19	0.299 52	0.405 75	973.13	32.068
40,000	−69.700	2.730 0	0.018 823	0.000 587 26	0.296 91	0.505 59	968.08	32.052
45,000	−69.700	2.148 9	0.014 809	0.000 462 27	0.296 91	0.642 30	968.08	32.036
50,000	−69.700	1.691 7	0.011 652	0.000 363 91	0.296 91	0.815 89	968.08	32.020
60,000	−69.700	1.048 8	0.007 217 5	0.000 225 61	0.296 91	1.316 0	968.08	31.991
70,000	−67.425	0.650 87	0.004 448 5	0.000 139 20	0.298 36	2.143 4	970.90	31.958
80,000	−61.976	0.406 32	0.002 736 6	0.000 085 707	0.301 82	3.521 5	977.62	31.930
90,000	−56.535	0.255 40	0.001 695 2	0.000 053 145	0.305 25	5.743 6	984.28	31.897
100,000	−51.099	0.161 60	0.001 057 5	0.000 033 182	0.308 65	9.301 8	990.91	31.868
km	°C	kPa abs	N/m³	kg/m³	10^{-6} N·s/m²	10^{-6} m²/s	m/s	m/s²
0	15.000	101.325	12.013 1	1.225 0	17.894	14.607	340.294	9.806 65
1	8.501	89.876	10.898 7	1.111 7	17.579	15.813	336.43	9.803 6
2	2.004	79.501	9.865 2	1.006 6	17.260	17.147	332.53	9.800 5
3	−4.500	70.121	8.908 3	0.909 25	16.938	18.628	328.58	9.797 4
4	−10.984	61.660	8.025 0	0.819 35	16.612	20.275	324.59	9.794 3
5	−17.474	54.048	7.210 5	0.736 43	16.282	22.110	320.55	9.791 2
6	−23.963	47.217	6.461 3	0.660 11	15.949	24.161	316.45	9.788 2
8	−36.935	35.651	5.143 3	0.525 79	15.271	29.044	308.11	9.782 0
10	−49.898	26.499	4.042 4	0.413 51	14.577	35.251	299.53	9.775 9
12	−56.500	19.399	3.047 6	0.311 94	14.216	45.574	295.07	9.769 7
14	−56.500	14.170	2.224 7	0.227 86	14.216	62.391	295.07	9.763 6
16	−56.500	10.352	1.624 3	0.166 47	14.216	85.397	295.07	9.757 5
18	−56.500	7.565	1.186 2	0.121 65	14.216	116.86	295.07	9.751 3
20	−56.500	5.529	0.866 4	0.088 91	14.216	159.89	295.07	9.745 2
25	−51.598	2.549	0.390 0	0.040 08	14.484	361.35	298.39	9.730 0
30	−46.641	1.197	0.178 8	0.018 41	14.753	801.34	301.71	9.714 7

[a] International Civil Aviation Organization; see Sec. 2.9

[b] In these tables, if (for example, at 0 ft) μ is given as 0.373 72 and the units are 10^{-6} lb·s/ft² then $\mu = 0.373\,72 \times 10^{-6}$ lb·s/ft².

Table A.4
Physical properties of common liquids at standard sea-level atmospheric pressure[a]

Liquid	Temperature, T	Density, ρ	Specific gravity,[b] s	Viscosity,[c] μ	Surface tension, σ	Vapor pressure, p_v	Bulk modulus of elasticity, E_v	Specific heat, c
	°F	slug/ft³	—	10^{-6} lb·s/ft²	lb/ft	psia	psi	ft·lb/(slug·°R) = ft²/(s²·°R)
Benzene	68°F	1.70	0.88	14.37	0.002 0	1.45	150,000	10,290
Carbon tetrachloride	68°F	3.08	1.594	20.35	0.001 8	1.90	160,000	5,035
Crude oil	68°F	1.66	0.86	150	0.002	—	—	—
Gasoline	68°F	1.32	0.68	6.1	—	8.0	—	12,500
Glycerin	68°F	2.44	1.26	31,200	0.004 3	0.000 002	630,000	14,270
Hydrogen	−430°F	0.143	0.074	0.435	0.000 2	3.1	—	—
Kerosene	68°F	1.57	0.81	40	0.001 7	0.46	—	12,000
Mercury	68°F	26.3	13.56	33	0.032	0.000 025	3,800,000	834
Oxygen	−320°F	2.34	1.21	5.8	0.001	3.1	—	~5,760
SAE 10 oil	68°F	1.78	0.92	1,700	0.002 5	—	—	—
SAE 30 oil	68°F	1.78	0.92	9,200	0.002 4	—	—	—
Fresh water	68°F	1.936	0.999	21.0	0.005 0	0.34	318,000	25,000
Seawater	68°F	1.985	1.024	22.5	0.005 0	0.34	336,000	23,500
	°C	kg/m³	—	10^{-3} N·s/m²	N/m	kN/m² abs	10^6 N/m²	N·m/(kg·K) = m²/(s²·K)
Benzene	20°C	876	0.88	0.65	0.029	10.0	1 030	1720
Carbon tetrachloride	20°C	1 588	1.594	0.97	0.026	12.1	1 100	842
Crude oil	20°C	856	0.86	7.2	0.03	—	—	—
Gasoline	20°C	680	0.68	0.29	—	55.2	—	2100
Glycerin	20°C	1 258	1.26	1494	0.063	0.000 014	4 344	2386
Hydrogen	−257°C	73.7	0.074	0.021	0.002 9	21.4	—	—
Kerosene	20°C	808	0.81	1.92	0.025	3.20	—	2000
Mercury	20°C	13 550	13.56	1.56	0.51	0.000 17	26 200	139.4
Oxygen	−195°C	1 206	1.21	0.278	0.015	21.4	—	~964
SAE 10 oil	20°C	918	0.92	82	0.037	—	—	—
SAE 30 oil	20°C	918	0.92	440	0.036	—	—	—
Fresh water	20°C	998	0.999	1.00	0.073	2.34	2 171	4187
Seawater	20°C	1 023	1.024	1.07	0.073	2.34	2 300	3933

[a] In these tables, if (for example, for benzene at 68°F) μ is given as 1.437 and the units are 10^{-6} lb·s/ft² then $\mu = 1.437 \times 10^{-6}$ lb·s/ft².
[b] Relative to pure water at 60°F.
[c] For viscosity, see also Figs. 2.3 and 2.4.

Table A.5

Physical properties of common gases at standard sea-level atmospheric pressure[a]

Gas	Chemical formula	Molecular weight, M	Density ρ	Viscosity,[b] μ	Gas constant, R	Specific heat, c_p	Specific heat, c_v	Specific heat ratio, $k = c_p/c_v$
at 68°F	—	—	slug/ft³	10^{-6} lb·s/ft²	ft·lb/(slug·°R) = ft²/(s²·°R)	ft·lb/(slug·°R) = ft²/(s²·°R)		—
Air		29.0	0.002 31	0.376	1,715	6,000	4,285	1.40
Carbon dioxide	CO_2	44.0	0.003 54	0.310	1,123	5,132	4,009	1.28
Carbon monoxide	CO	28.0	0.002 26	0.380	1,778	6,218	4,440	1.40
Helium	He	4.00	0.000 323	0.411	12,420	31,230	18,810	1.66
Hydrogen	H_2	2.02	0.000 162	0.189	24,680	86,390	61,710	1.40
Methane	CH_4	16.0	0.001 29	0.280	3,100	13,400	10,300	1.30
Nitrogen	N_2	28.0	0.002 26	0.368	1,773	6,210	4,437	1.40
Oxygen	O_2	32.0	0.002 58	0.418	1,554	5,437	3,883	1.40
Water vapor	H_2O	18.0	0.001 45	0.212	2,760	11,110	8,350	1.33
at 20°C	—	—	kg/m³	10^{-6} N·s/m²	N·m/(kg·K) = m²/(s²·K)	N·m/(kg·K) = m²/(s²·K)		—
Air		29.0	1.205	18.0	287	1 003	716	1.40
Carbon dioxide	CO_2	44.0	1.84	14.8	188	858	670	1.28
Carbon monoxide	CO	28.0	1.16	18.2	297	1 040	743	1.40
Helium	He	4.00	0.166	19.7	2077	5 220	3 143	1.66
Hydrogen	H_2	2.02	0.0839	9.0	4120	14 450	10 330	1.40
Methane	CH_4	16.0	0.668	13.4	520	2 250	1 730	1.30
Nitrogen	N_2	28.0	1.16	17.6	297	1 040	743	1.40
Oxygen	O_2	32.0	1.33	20.0	260	909	649	1.40
Water vapor	H_2O	18.0	0.747	10.1	462	1 862	1 400	1.33

[a] In these tables, if (for example, for air at 68°F) μ is given as 0.376 and the units are 10^{-6} lb·s/ft² then $\mu = 0.376 \times 10^{-6}$ lb·s/ft².

[b] For viscosity, see also Figs. 2.3 and 2.4.

Table A.6
Areas of circles

Diameter	Area	
in	**in²**	**ft²**
0.25	0.049 087	0.000 340 88
0.5	0.196 35	0.001 363 5
1.0	0.785 40	0.005 454 2
2.0	3.141 6	0.021 817
3.0	7.068 6	0.049 087
4.0	12.566	0.087 266
6.0	28.274	0.196 35
8.0	50.265	0.349 07
9.0	63.617	0.441 79
10.0	78.540	0.545 42
12.0	113.097	0.785 40

m	**m²**
0.05	0.001 963 5
0.10	0.007 854 0
0.15	0.017 671
0.20	0.031 416
0.25	0.049 087
0.30	0.070 686
0.50	0.196 35
1.00	0.785 40
1.50	1.767 15
2.00	3.141 59

Table A.7
Properties of areas

Shape	Sketch	Area	Location of centroid	I_c or I^a
Rectangle	b, h, I_c, y_c	bh	$y_c = \frac{h}{2}$	$I_c = \frac{bh^3}{12}$
Triangle	h, b, I_c, y_c	$\frac{bh}{2}$	$y_c = \frac{h}{3}$	$I_c = \frac{bh^3}{36}$
Circle	D, I_c, y_c	$\frac{\pi D^2}{4}$	$y_c = \frac{D}{2}$	$I_c = \frac{\pi D^4}{64}$
Semicircle	r, D, I, y_c	$\frac{\pi D^2}{8}$	$y_c = \frac{4r}{3\pi}$	$I = \frac{\pi D^4}{128}$
Ellipse	b, h, I_c, y_c	$\frac{\pi bh}{4}$	$y_c = \frac{h}{2}$	$I_c = \frac{\pi bh^3}{64}$
Semiellipse	h, b, I, y_c	$\frac{\pi bh}{4}$	$y_c = \frac{4h}{3\pi}$	$I = \frac{\pi bh^3}{16}$
Parabola	b, h, I, x_c, y_c	$\frac{2bh}{3}$	$x_c = \frac{3b}{8}$ $y_c = \frac{3h}{5}$	$I = \frac{2bh^3}{7}$

[a] Parallel axis theorem: $I = I_c + Ay_c^2$.

Table A.8
Properties of solid bodies

Body	Sketch	Volume	Surface area	Location of center of mass
Cylinder	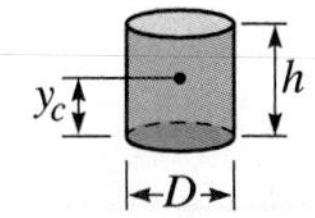	$\frac{\pi D^2 h}{4}$	$\pi D h + \frac{\pi D^2}{2}$	$y_c = \frac{h}{2}$
Cone	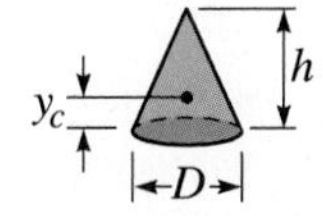	$\frac{1}{3}\left(\frac{\pi D^2 h}{4}\right)$	$\frac{\pi D h}{2}\left(1 + \frac{D^2}{4h^2}\right)^{1/2} + \frac{\pi D^2}{4}$	$y_c = \frac{h}{4}$
Sphere	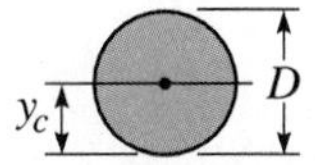	$\frac{\pi D^3}{6}$	πD^2	$y_c = \frac{D}{2}$
Hemisphere	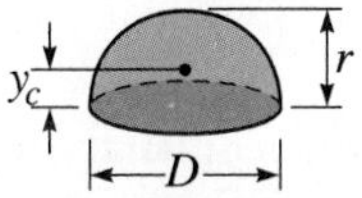	$\frac{\pi D^3}{12}$	$\frac{\pi D^2}{2} + \frac{\pi D^2}{4} = \frac{3\pi D^2}{4}$	$y_c = \frac{3r}{8}$
Paraboloid	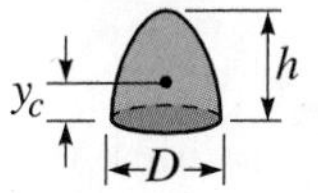	$\frac{1}{2}\left(\frac{\pi D^2 h}{4}\right)$		$y_c = \frac{h}{3}$
Half ellipsoid	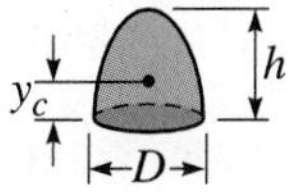	$\frac{2}{3}\left(\frac{\pi D^2 h}{4}\right)$		$y_c = \frac{3h}{8}$

APPENDIX B

Programming and Computer Applications

The solution of many problems in fluid mechanics and hydraulics requires repetitive calculations, so that using programmed procedures can save considerable time and tedious labor.

The purpose of this appendix is to introduce students to the various programming procedures available, and to give some idea of the characteristics and capabilities of each. Because they all make use of advanced technology, students should remain aware that such procedures continue to evolve quite rapidly.

The most common types of programmable aids used to help solve problems in fluid mechanics and hydraulics are:

- Programmable scientific calculators (and equation solvers)
- Spreadsheets
- Generic mathematics software
- Other generic software
- Applications software
- Programming languages

Each of these types is described further below.

Before deciding to adopt any of these aids, it is important to be aware of the start up time required to learn how to prepare and/or use such programs. This can vary from a few hours to a number of weeks or more, involving considerable effort in some cases. As a result, for most introductory problems (including most of those in Chaps. 1–7) it is not time-effective to use such aids if one is not already familiar with them. Furthermore, it is not wise to use preprogrammed procedures without properly understanding the systems simulated, how the programs work, and the nature of the assumptions involved. The use of most of these procedures is more appropriate with more advanced problems, as in Chaps. 8 and thereafter. The authors strongly recommend that students *first* concentrate on mastering the principles involved by solving a larger number of simpler problems using manual procedures, with pencil, paper, and a basic scientific calculator,[1] and by making a few manual repetitions where necessary.

B.1 PROGRAMMABLE SCIENTIFIC CALCULATORS

Attractive features of the programmable scientific calculator are that it is an extension of a very familiar tool, the basic (not programmable) scientific calculator, that its small size makes it easily portable, to classes, to examinations, and into the field, and that its response times are usually very rapid. Although obviously less powerful than a personal computer (PC), the capabilities of some programmable scientific calculators are very extensive, and continue to increase rapidly.

In fluid mechanics and hydraulics the most common need is for help in solving equations. The equations that arise are often nonlinear, they may involve transcendental functions (exponential, logarithmic, and/or trigonometric), and they are often implicit (the equation cannot be rearranged to have only one occurrence of the unknown variable). Many problems are governed by a simultaneous set of n such equations in n unknowns.

Earlier programmable calculators had small memories, and so could execute only short programs and store perhaps only one equation. Also, they could solve only explicit equations, and these had to be converted into unfamiliar codes before entry. Nevertheless, ways were devised to facilitate the solution of many previously tedious problems in fluid mechanics and hydraulics.[2]

More recent programmable scientific calculators, such as the HP 48G, 48GX, 48S, and 48SX, have built-in preprogrammed equation solvers. These will solve a single equation for one unknown, and the equation may be linear or nonlinear, transcendental or not, and explicit or implicit; the solver uses an iterative optimization procedure to accurately find the value of the unknown variable. If the equation has multiple roots, it will find the root closest to the guessed value that the user must provide.

Although the equation is stored in the form of codes, it may conveniently be entered with a so-called "equation writer," which presents the equation on the screen in the same form as we write it on paper. Equations may be

[1] A basic scientific calculator is here defined to be one that is not programmable and does not have automatic equation solving capabilities.

[2] T. E. Croley, *Hydrologic and Hydraulic Computations on Small Programmable Calculators,* Iowa Institute of Hydraulic Research, University of Iowa, Iowa City, 1977.

stored in a directory in the calculator's memory, and recalled at any time for editing or problem solving. Each equation may include many different variables, and when solving them the user assigns the known values and designates the unknown to be solved. This is of great convenience because it enables the same equation to be used to solve a variety of problems.

An alternative method of solution is to plot the function and then identify the root or roots.

When problems are governed by a simultaneous set of n nonlinear equations in n unknowns, these cannot be solved using present (1997) equation solvers unless the equations can be combined into one. This is often possible by substitution, although the result may be rather cumbersome. In such cases the equations also can be "linked," so that once the combined equation is solved, the solver can then conveniently use the same variable values to solve the other equations in the set, one by one for the remaining (eliminated) unknowns.

Example applications to specific equations are described in Secs. 8.14, 8.24, and 10.3.

B.2 SPREADSHEETS

Nowadays most engineering students are familiar with spreadsheets like Excel, Lotus 1-2-3, or Quattro Pro. These continue to become more powerful and convenient for scientific applications, and are now able to be programmed and to handle some quite complex functions. Because of their tabular organization, they are particularly suitable for solving problems that are traditionally solved in a stepwise and/or iterative fashion that is best tabulated; many appropriate applications are noted throughout this book, a good example being to the solution of gradually varied flow profiles in open channels (Sample Prob. 10.9).

Spreadsheets can of course also be set up to solve trial-and-error problems. For example, the user inputs an estimate of the unknown variable, and in subsequent columns the spreadsheet evaluates the left- and right-hand sides of the governing equation. The user can then repetitively try various values for the unknown until the two sides become equal. This is particularly useful where part of the input must come from reading a chart that is not available in equation form (see Sample Probs. 11.4 and 11.5). Some spreadsheets include solvers, but these are generally less capable and less convenient to use than those described in the next section.

Data from spreadsheets can be exported to graphing software to generate various plots and visual representations of the tabulated data.

B.3 GENERIC MATHEMATICS SOFTWARE

Software for solving mathematics problems can be very usefully applied to solving engineering problems. Mathematica is probably the largest such program, with extensive capabilities, often very involved. Other such packages, less involved to varying degrees, include Derive, GAMS, Maple, Mathcad, MATLAB, Theorist, and TK Solver. Specialized packages are also available for symbolic manipulations, statistics, numerical methods, and optimization.

Most of these software packages are available in a variety of versions or forms, some of which include inexpensive but limited-capability student editions. Prospective users are advised to check carefully whether the capabilities they need (such as to solve simultaneous nonlinear equations as described in this book) are included in the version they are planning to use. Some brands have the needed capabilities in optional supplemental packages.

Compared with programmable calculators, generic mathematics software has all the potential advantages that come with a PC, such as greater computing power and the ability to produce fancy reports. Certain types of problems mentioned earlier, involving a set of simultaneous nonlinear equations that cannot be reduced to one equation by substitution (see, e.g., Sample Prob. 8.12), can only be solved by PC-based software or by manual iteration. Generic software has the disadvantages of a PC also, which include slower bootup times than a calculator and less convenient accessibility. Accessibility is improving, however, with increasingly portable and affordable laptop computers.

For this book, example applications of generic mathematics software have been described using Mathcad. This was chosen because it is a more basic package particularly geared to student use. (Note, however, that the version 6 *student* edition does not include the capability to solve simultaneous nonlinear equations.) In Mathcad equations are entered and presented in the same form that we normally write them; in many other packages they must be entered in the form of less familiar "codes." One simply lists the known and estimated variable values on a "worksheet," lists the equations in a designated "solve block," and the solutions for the unknown variables appear below; learning time is relatively short. Note, however, that for larger numbers of simultaneous equations the success of the procedure tends to become more sensitive to the estimated values; if at first it is not successful, different estimates should be tried. In addition to solving equations, Mathcad also has a convenient report writing capability that includes equations, it readily plots graphs, it has a spreadsheet equivalent, and it has optional modules for special applications.

B.4 OTHER GENERIC SOFTWARE

A variety of generic software packages have been developed for the solution of conditions that vary throughout a region, such as temperature, pressure, or stress. In particular, these packages can be used to solve flow fields, such as flow through passages of various shapes, flow around immersed bodies, and flow through porous media such as aquifers and earth dams. These packages use relatively advanced procedures, such as the finite difference method (FDM), the finite element method (FEM), or the boundary element method (BEM), to solve the differential equations governing the flow field by discretizing the flow region and/or its boundaries. Each procedure has its advantages and disadvantages such as: simplicity, flexibility, and convenience of formulating the problem to be solved. Computational fluid mechanics and computational fluid dynamics (CFD) are general terms used for any program developed to specifically solve flow fields.

Some of these packages can solve flow fields with a free surface, known as "free surface flows." A free surface may be part of a problem because the liquid surface forms one of the boundaries, such as for waves or the ground water table, or because an internal surface forms, as around cavities produced by cavitation (Sec. 5.11). The fact that the position of the free surface usually depends on the solution of the flow field, and vice versa, adds to the complexity of the problem.

B.5 APPLICATIONS SOFTWARE

Much software for particular applications in fluid mechanics, hydraulic engineering, and hydrology, has been developed by individual authors, private companies, research institutions, and government agencies.

A textbook with accompanying software devoted to fluid mechanics problems has been prepared by Olfe (see Ref. 48 in Appendix C). A so-called electronic book on fluid mechanics and hydraulics (Ref. 22), entirely on diskette, is based on Mathcad so that the user can interactively change numerical values and immediately see the changed results.

Applications software packages prepared commercially or by government agencies include the following.

Pipeline models

These may be for steady or unsteady flow (including surges and waterhammer) in single pipes or branching pipes, which may incorporate pumps, turbines, and valves. They include: ALGEBRA, GENEQUA, IMPREM, LIQT, PipeCalc, SINGLE, SUR, SURGE5, WH, WHAMO.

Water distribution network models

These are often capable of handling pressure zones, valves, pumps, storage, and graphical presentations. They include: AFT-Fathom, Autowater, Aquanet, CYBERNET, Click & Go, EPAnet, Faast, FlowMaster, Hydroflo II,

Hydronet, KYPIPE, Micro Hardy Cross, PIPENET, Stoner, Tdhnet, Wadiso, Water, WaterCAD, Watermax, Waterworks, Watnet, Watsys, Wgraph.

Models of storm drains and natural drainage channels

These often can simulate both free surface and pressure flows, adverse flows, and junction losses, and are usually compatible with AutoCAD or similar computer-aided drafting (CAD) packages. They include: Boss, CDS, CIA, CulvertMaster, DAMBRK (N.O.A.A.), Dodson, DWOPER (N.O.A.A.), Eagle Point, Engenious, FEQ, FLDWAV (N.O.A.A.), FLDWY (U.S. S.C.S.), FlowMaster, FLOWPROF, HEC-2 and HEC-RAS (U.S. Army Corps of Engineers), HY-8, Hydra, Hydraflow, Hydrain-HYCLV, Intelisolve, MOUSE, Networx, PIPECAP, Pizer, Quickpipe, SECTION, SoftDesk, StormCAD, Streamline, SWMM (U.S. E.P.A.), THYSYS, WSPRO (U.S.G.S.), WSP2 (U.S. S.C.S.), XP Software.

Sanitary sewer system models

These models have capabilities similar to those for storm drains. They include: CAPS, Eagle Point, Hydra, MODS, Pizer, and SewerCAD.

Other related areas for which applications software packages have been developed include: potential flow; laminar and turbulent flow (CFD packages); water hammer and surges; jet and spillway flows; air entrainment; shallow-water hydrodynamics; reservoir hydrodynamics; coastal hydraulics; natural and urban hydrology; seepage; groundwater hydraulics and transport; water quality of rivers, impoundments, and estuaries; and stream sedimentation.

B.6 PROGRAMMING LANGUAGES

Instead of using commercial software of the various types described in the preceding sections, an engineer can learn a programming language and then prepare "custom-made" programs to solve specific problems.

Assuming one knows a programming language, which takes quite a while to learn, it still requires a considerable amount of time to write programs. Because there are so many problems that should be solved with computing aids, it would be very inefficient to require students to write programs to solve them. Therefore we have not included such assignments or sample codes in this book, and have instead encouraged the mastery of programmable calculators, spreadsheets, and generic mathematics software (Secs. B.1–B.3). While the ability to write computer programs is important, we believe that in most instances it is best done during graduate studies and for research or professional projects. Students interested to pursue programming for fluid mechanics may see example programs in other texts (Refs. 16, 54, 55, 65 and 66).

The two major programming languages in current use by engineers in this field are FORTRAN and C. Each has its strengths and weaknesses.

FORTRAN stands for FORmula TRANslation. It was released in 1957, to provide a high-level language for scientific computing. It uses a notation that simplifies the programming of mathematical formulas, as its name suggests. Long the preferred language of civil and mechanical engineers, it continues to be the most convenient language to use for programming numerical methods. In the area of large scientific programs, FORTRAN is also the preferred language for parallel processing.

The programming language C, released around 1970, uses a low level of programming abstraction (is more detailed), which gives it a more flexible and therefore more powerful programming capability than FORTRAN. As a result, it has broader abilities, being suitable for graphics and for data bases and transaction processing in addition to scientific computation. However, its extra power and breadth comes at the cost of requiring more convoluted programming; hence C is harder to learn than FORTRAN and it is harder for a casual programmer to follow the logic of C.

As a result, there is a wealth of scientific and engineering software that has been written in FORTRAN, and there are many graphics packages written in C that are attractive to scientists and engineers. Although the latest revision, FORTRAN 95, can reportedly do the same graphics as C, scientists would naturally prefer to use what is readily available, and ways have been developed to do this. Now one can execute a software package that includes both languages. In a modular mode, a main FORTRAN program, for example, can call a subroutine written in C.

It seems to be regularly remarked that FORTRAN is a dying language, even though this has now been said for many decades. It will not die, because it is the most convenient language for advanced numerical operations, because there is such a vast inventory of FORTRAN-based software, and because all high-powered, high-level computational software is written in FORTRAN. Many of these software packages like ANSYS, SPICE, and FEM and CFD programs contain *millions* of lines of FORTRAN code that will not translate into other languages. I-DEAS, for example, has over ten million lines.

The language BASIC, which stands for Beginner's All-purpose Symbolic Instruction Code, is somewhat like FORTRAN. BASIC was developed in the mid-1960s, after FORTRAN, to provide an alternative that was easier to learn and to use (but less efficient to execute). Much BASIC code looks like a simplified subset of FORTRAN, but it resembles the English language more closely. To facilitate "debugging" (correcting errors), statements in early versions of BASIC are converted into machine language and executed on a line-by-line basis, rather than compiling the entire program before execution.

Another important tool for program development is the recently developed Visual Basic. Not related to the early BASIC just discussed, Visual Basic is a visual language based on an integrated graphical user interface (GUI) and a compiled language more like a simplified C. It works in a Windows environment, where it is a powerful tool for developing user-friendly applications.

B.7 SUMMARY

Of the above programmable aids, probably the most important for students of fluid mechanics to learn to use, particularly in a second course, are: spreadsheets, equation solvers on programmable calculators, and a simple and flexible mathematics software package like Mathcad.

Note that the mathematics software is the only one of these three aids that can perform the functions of all of them. Although advanced scientific calculators can solve a system of linear equations, and can solve a single nonlinear (and implicit) equation, they cannot (in 1997) automatically solve a system of nonlinear equations such as we sometimes meet in applied fluid mechanics. While computer software can do this, computers usually are not so accessible as a calculator, and by using the proper procedures calculators can solve the majority of problems. It is clear that students need to be familiar with the abilities and advantages of each.

In addition, students should take every opportunity to see demonstrations of, and learn more about, the capabilities of the various types of generic and applications software described above.

Students interested in graduate studies or research are also advised to learn a programming language or two, probably FORTRAN and/or C.

APPENDIX C

References

There is a great volume of literature available on the various aspects of fluid mechanics and hydraulics. The results of original research may be found in papers published in technical journals. A list of books covering various topics of fluid mechanics and its engineering applications is presented here for the convenience of the student. This list by no means includes all the important books that have been written; the intent here is merely to provide a representative list. Students are encouraged to "probe deeper" and to widen their horizons by further reading.

1. Anderson, D. A., J. C. Tannehill, and R. H. Pletcher. *Computational Fluid Mechanics and Heat Transfer.* McGraw-Hill, New York, 1984.
2. Anderson, J. D. *Fundamentals of Aerodynamics,* 2d ed. McGraw-Hill, New York, 1991.
3. Barenblatt, G. I. *Dimensional Analysis,* translated from Russian by P. Makinen. Gordon and Breach, New York, 1987.
4. Batchelor, G. K. *An Introduction to Fluid Dynamics.* Cambridge University Press, Cambridge, England, 1967, reprinted 1992.
5. Bertin, J. J., and M. L. Smith. *Aerodynamics for Engineers,* 2d ed. McGraw-Hill, New York, 1989.
6. Böhme, G. *Non-Newtonian Fluid Mechanics.* Elsevier, New York, 1987.
7. Bos, M. G., J. A. Replogle, and A. J. Clemens. *Flow Measuring Flumes for Open Channel Systems.* Wiley, New York, 1984.
8. Brater, E. F., H. W. King, J. E. Lindell, and C. Y. Wei. *Handbook of Applied Hydraulics,* 7th ed. McGraw-Hill, New York, 1996.
9. Chaudhry, M. H. *Applied Hydraulic Transients,* 2d ed. Van Nostrand-Reinhold, New York, 1987.
10. Chaudhry, M. H. *Open Channel Flow.* Prentice-Hall, Englewood Cliffs, NJ, 1993.
11. Cheremisinoff, N. P. *Fundamentals of Wind Engineering.* Ann Arbor Science, Ann Arbor, MI, 1978.

12. Chorin, A. J., and J. E. Marsden. *A Mathematical Introduction to Fluid Mechanics,* 3d ed. Springer-Verlag, New York, 1993.
13. Chow, V. T. *Open-Channel Hydraulics.* McGraw-Hill, New York, 1959.
14. Churchill, S. W. *Viscous Flows—The Practical Use of Theory.* Butterworth, Boston, 1988.
15. Eggleston, D. M., and F. S. Stoddard. *Wind Turbine Design.* Van Nostrand-Reinhold, New York, 1987.
16. Evett, J. B., and C. Liu. *Fundamentals of Fluid Mechanics.* McGraw-Hill, New York, 1987.
17. Fischer, H. B., E. J. List, R. C. Y. Koh, J. Imberger, and N. H. Brooks. *Mixing in Inland and Coastal Waters.* Academic Press, New York, 1979.
18. Fox, R. W., and A. T. McDonald. *Introduction to Fluid Mechanics,* 4th ed. Wiley, New York, 1992.
19. Freeze, R. A., and J. A. Cherry. *Groundwater.* Prentice-Hall, Englewood Cliffs, NJ, 1979.
20. French, R. H. *Open-Channel Hydraulics.* McGraw-Hill, New York, 1985.
21. Garay, P. N. *Pump Application Desk Book,* 2d ed. Fairmont Press, Lilburn, GA, 1993.
22. Giles, R. V., J. B. Evett, and C. Liu. *Schaum's Interactive Outline: Fluid Mechanics and Hydraulics.* MathSoft, Cambridge, MA, and McGraw-Hill, New York, 1995.
23. Graf, W. H. *Hydraulics of Sediment Transport.* McGraw-Hill, New York, 1971, reprinted by Water Resource Publications, Littleton, CO, 1984.
24. Granger, R. A. *Experiments in Fluid Mechanics.* Holt, Rinehart and Winston, New York, 1988.
25. Hamrock, B. J. *Fundamentals of Fluid Film Lubrication.* McGraw-Hill, New York, 1994.
26. Henderson, F. M. *Open Channel Flow.* Macmillan, New York, 1966.
27. Herring, J. R., and J. C. McWilliams (eds.). *Lecture Notes on Turbulence.* World Scientific, Singapore, 1989.
28. Hinze, J. O. *Turbulence,* 2d ed. McGraw-Hill, New York, 1975.
29. Horvath, I. *Hydraulics in Water and Wastewater Treatment Technology.* Wiley, New York, 1994.
30. Hydraulic Institute. *Pipe Friction Manual,* 3d ed. New York, 1961.
31. Hydraulic Institute. *Standards of the Hydraulic Institute,* 14th ed. New York, 1983.
32. Idel'chik, I. E. *Handbook of Hydraulic Resistance,* 2d ed. Hemisphere, New York, 1986.
33. Kirchoff, R. H. *Potential Flows—Computer Graphic Solutions.* Marcel Dekker, New York, 1985.
34. Kline, S. J. *Similitude and Approximation Theory.* Springer-Verlag, New York, 1986.
35. Kuethe, A. M., and C.-Y. Chow. *Foundations of Aerodynamics,* 4th ed. Wiley, New York, 1986.
36. Lamb, H. *Hydrodynamics,* 6th ed. Cambridge University Press, Cambridge, England, 1932, reprinted 1993.
37. Langhaar, H. L. *Dimensional Analysis and the Theory of Models.* Wiley, New York, 1951, reprinted by Kreiger, 1980.
38. Liggett, J. A. *Intermediate Fluid Mechanics.* McGraw-Hill, New York, 1994.

39. Linsley, R. K., J. B. Franzini, D. L. Freyberg, and G. Tchobanoglous. *Water Resources Engineering,* 4th ed. McGraw-Hill, New York, 1992.
40. Liptak, B. G. (ed.). *Flow Measurement.* Chilton Book Co., Radnor, PA, 1993.
41. Lobanoff, V. S., and R. R. Ross. *Centrifugal Pumps—Design and Application,* 2d ed., Gulf Publishing Co., Houston, Texas, 1992.
42. Merzkirch, W. *Flow Visualization,* 2d ed. Academic Press, New York, 1987.
43. Moran, M. J., and H. N. Shapiro. *Fundamentals of Engineering Thermodynamics,* 2d ed. Wiley, New York, 1992.
44. Morris, H. M., and J. M. Wiggert. *Applied Hydraulics in Engineering,* 2d ed. Wiley, New York, 1972.
45. Munson, B. R., D. F. Young, and T. H. Okiishi. *Fundamentals of Fluid Mechanics,* 2d ed. Wiley, New York, 1994.
46. Novak, P., and J. Cábelka. *Models in Hydraulic Engineering—Physical Principles and Design Applications.* Pitman, Boston, 1981.
47. Nunn, R. H. *Intermediate Fluid Mechanics.* Hemisphere, New York, 1989.
48. Olfe, D. B. *Fluid Mechanics Programs for the IBM PC.* McGraw-Hill, New York, 1987.
49. Pai, S.-I., and S. Luo. *Theoretical and Computational Dynamics of a Compressible Flow.* Van Nostrand-Reinhold, New York, 1991.
50. Parker, S. P. *Fluid Mechanics Source Book.* McGraw-Hill, New York, 1988.
51. Peyret, R., and T. D. Taylor. *Computational Methods for Fluid Flow.* Springer-Verlag, New York, 1983.
52. Raudkivi, A. J. *Loose Boundary Hydraulics,* 3d ed. Pergamon, New York, 1990.
53. Reynolds, W. C., and H. C. Perkins. *Engineering Thermodynamics,* 2d ed. McGraw-Hill, New York, 1977.
54. Roberson, J. A., J. J. Cassidy, and M. H. Chaudhry. *Hydraulic Engineering.* Houghton Mifflin, Boston, 1988.
55. Roberson, J. A., and C. T. Crowe. *Engineering Fluid Mechanics,* 5th ed. Wiley, New York, 1995.
56. Rouse, H., and S. Ince. *History of Hydraulics.* Dover, New York, 1963.
57. Saad, M. *Compressible Fluid Flow,* 2d ed. Prentice-Hall, Englewood Cliffs, NJ, 1993.
58. Schetz, J. A. *Boundary Layer Analysis.* Prentice-Hall, Englewood Cliffs, NJ, 1993.
59. Schlichting, H. *Boundary Layer Theory,* 7th ed. McGraw-Hill, New York, 1987.
60. Sharpe, G. J. *Solving Problems in Fluid Dynamics.* Longman Scientific and Technical, Harlow, England, 1994.
61. Sherman, F. S. *Viscous Flow.* McGraw-Hill, New York, 1990.
62. Soo, S. L. *Particulates and Continuum in Multiphase Fluid Dynamics.* Hemisphere, New York, 1989.
63. Stepanoff, A. J. *Centrifugal and Axial Flow Pumps,* 2d ed. Wiley, New York, 1957, reprinted by Kreiger, 1993.
64. Stoker, J. J. *Water Waves—The Mathematicial Theory with Applications.* Wiley, New York, 1959, reprinted 1992.

65. Street, R. L., G. Z. Watters, and J. K. Vennard. *Elementary Fluid Mechanics,* 7th ed. Wiley, New York, 1996.
66. Streeter, V. L., and E. B. Wylie. *Fluid Mechanics,* 8th ed. McGraw-Hill, New York, 1985.
67. Sullivan, J. A. *Fluid Power—Theory and Applications,* 3d ed. Prentice-Hall, Englewood Cliffs, NJ, 1989.
68. Sutton, G. P. *Rocket Propulsion Elements,* 6th ed. Wiley, New York, 1992.
69. Todd, D. K. *Groundwater Hydrology,* 2d ed. Wiley, New York, 1980.
70. Tullis, J. P. *Hydraulics of Pipelines: Pumps, Valves, and Cavitation Transients.* Wiley, New York, 1989.
71. Turton, R. K. *Rotodynamic Pump Design.* Cambridge University Press, Cambridge, England, 1994.
72. Vallentine, H. R. *Applied Hydrodynamics,* 2d ed. Butterworth, London, 1967.
73. Walski, T. M. *Analysis of Water Distribution Systems.* Van Nostrand-Reinhold, New York, 1984.
74. Watters, G. Z. *Analysis and Control of Unsteady Flow in Pipelines,* 2d ed. Butterworth, Stoneham, MA, 1984.
75. White, F. M. *Fluid Mechanics,* 3d ed. McGraw-Hill, New York, 1994.
76. White, F. M. *Viscous Fluid Flow,* 2d ed. McGraw-Hill, New York, 1991.
77. Wylie, E. B. and V. L. Streeter. *Fluid Transients in Systems.* Prentice-Hall, Englewood Cliffs, NJ, 1993.
78. Yalin, M. S. *River Mechanics.* Pergamon, Oxford, 1992.
79. Zapporo, V. J., and H. Hasen (eds.). *Davis' Handbook of Applied Hydraulics,* 4th ed. McGraw-Hill, New York, 1993.
80. Zaruba, J. *Water Hammer in Pipe-line Systems.* Elsevier, New York, 1993.

APPENDIX D

Answers to Exercises

CHAPTER 1

1.5.1 All are F or MLT^{-2}.
1.5.2 All are L^3/T.
1.5.3 All are L^2/T.
1.5.4 All are L.
1.5.5 All are M/T.
1.5.6 (*a*) 8.33 lb; (*b*) 37.0 N; (*c*) 3.70×10^6 dyne.
1.5.7 (*a*) 2.20 lb; (*b*) 9.80 N; (*c*) 980×10^3 dyne.
1.5.8. (*a*) 23.2 cfs; (*b*) 0.657 m^3/s.
1.5.9. (*a*) 20.8 m/s; (*b*) 68.4 ft/sec.

CHAPTER 2

2.3.1 1.429 slugs/ft^3, 0.700 ft^3/slug, 0.737.
2.3.2 0.00373 slug/ft^3, 268 ft^3/slug, 1.600.
2.3.3 1.835 kg/m^3, 0.545 m^3/kg, 1.500.
2.3.4 2.44 slugs/ft^3, 1.261, 12.37 kN/m^3.
2.3.5 1.677 slugs/ft^3.
2.3.6 775 kg/m^3.
2.3.7 0.0920 lb/ft^3.
2.3.8 12.26 N/m^3.
2.3.9 1028.2 mm.
2.5.1 1.34%.
2.5.2 (*a*) −0.01765 ft^3/slug; (*b*) 0.485 ft^3/slug; (*c*) 66.3 lb/ft^3.
2.5.3 (*a*) 3.51% decrease; (*b*) 3.64% increase.
2.5.4 2500 MN/m^2.
2.5.5 9600 psi.

2.5.6 44.1 MPa.
2.5.7 19.91 in.
2.5.8 497 mm.
2.6.1 9.57 kN/m^3, 9.63 kN/m^3.
2.6.2 0.0962% increase, 0.1200 lb.
2.6.3 4.86 m^3.
2.7.1 6600 ft^3.
2.7.2 0.0452 lb/ft^3, 713 $ft^3/slug$, 2858 $ft{\cdot}lb/(slug{\cdot}{}^\circ R)$.
2.7.3 260 $m^2/(s^2{\cdot}K)$, oxygen.
2.7.4 0.233 pcf, 0.00724 $slug/ft^3$, 138.2 $ft^3/slug$.
2.7.5 36.0 N/m^3, 3.67 kg/m^3, 0.273 m^3/kg.
2.7.6 (*a*) 0.001527 lb/ft^3; (*b*) 14.00 psia, 0.0688 lb/ft^3; (*c*) 0.0703 lb/ft^3.
2.7.7 (*a*) 0.245 N/m^3; (*b*) 98.5 kPa abs, 11.11 N/m^3; (*c*) 11.36 N/m^3.
2.7.8 (*a*) 2.61 N/m^3, 20.3 kPa abs; (*b*) 9.14 N/m^3, 81.1 kPa abs; (*c*) 11.75 N/m^3.
2.8.1 Methane 21.0 psi, N_2 21.0 psi, equal compressibilities.
2.8.2 Methane 120.0 kPa, N_2 126.0 kPa, methane is more compressible.
2.8.3 362 psia.
2.8.4 2390 kPa abs.
2.8.5 (*a*) 40.0 kPa abs; (*b*) 25.8 kPa abs, –77.7°C.
2.11.1 16.76 $mN{\cdot}s/m^2$, 18.03×10^{-6} m^2/s.
2.11.2 (*a*) 3.33; (*b*) 290; (*c*) 1.782.
2.11.3 4.8×10^{-6} ft^2/sec, 4.5×10^{-7} m^2/s.
2.11.4 375°F.
2.11.5 1:53.7, 15.5:1.
2.11.6 390 N.
2.11.7 120 lb.
2.11.8 (*a*) 3.44 N; (*b*) 4.59 N.
2.11.9 0.006 55 $kN{\cdot}m/s$, 6.55 J/s, 22.4 Btu/hr, 4.83 $ft{\cdot}lb/sec$, 0.008 78 hp.
2.11.10 362 lb.
2.11.11 158.7 Btu/hr.
2.11.12 $\pi\mu\omega d^4/(32\ \Delta h)$.
2.11.13 At $y = 0$ in, $du/dy = 20\ sec^{-1}$, $\tau = 0.209\ lb/ft^2$
At $y = 3$ in, $du/dy = 15\ sec^{-1}$, $\tau = 0.1566\ lb/ft^2$
At $y = 6$ in, $du/dy = 10\ sec^{-1}$, $\tau = 0.1044\ lb/ft^2$
At $y = 9$ in, $du/dy = 5\ sec^{-1}$, $\tau = 0.0522\ lb/ft^2$
At $y = 12$ in, $du/dy = 0\ sec^{-1}$, $\tau = 0\ lb/ft^2$
2.12.1 About 17.94 mm.
2.12.2 About 5.93 in.
2.12.3 0.1535 in, cf. 0.127 in from Fig. 2.8
2.12.4 30.5 mm.
2.12.5 0.276 in.
2.13.1 198.6°F.
2.13.2 123.3 mbar abs.

CHAPTER 3

3.2.1 5330 psi.
3.2.2 40 200 kPa.
3.2.3 7.77 kN/m^3, 792 kg/m^3.
3.2.4 55.2 lb/ft^3, 1.714 $slugs/ft^3$, 0.885.
3.3.1 20 kPa (= kN/m^2), 78.9 kPa.
3.3.2 0.932 psi, 4.83 psi.
3.3.3 27,700 ft.

3.3.4 5.43 psi.
3.4.1 169.6 kPa, 1696 mbar, 24.6 psi.
3.4.2 36.8 kPa abs.
3.4.3 9.19 psia.
3.4.4 67.7 kPa abs.
3.4.5 58.8 psi.
3.4.6 404 kPa.
3.4.7 33.14 ft, 30.35 ft.
3.5.1 35.72 ft.
3.5.2 5.88 m.
3.6.1 2155 mbar abs, 1.620 m Hg abs.
3.6.2 40.6 in.
3.6.3 27.6 psi, 27.8 psi.
3.6.4 (*a*) 9.19 ft of water, 3.98 psi; (*b*) 85.2 in.
3.6.5 (*a*) 2.76 m, 27.0 kPa; (*b*) 2.13 m.
3.6.6 About 0.09 psi ($p_B > p_A$), 65.7 inHg, 58.9 ft glycerin.
3.6.7 34.8 psi.
3.6.8 280 kPa.
3.6.9 43.1 ft, 6.04 ft.
3.6.10 46.1 kPa Vac or 346 mmHg Vac, 481 mm.
3.8.1 $0.75d$.
3.8.2 $(6a^2 + 8ad + 3d^2)/[6(a + \frac{2}{3}d)]$.
3.8.3 $0.5d$.
3.8.4 $5d/8$.
3.8.5 11,000 lb, 3.06 ft.
3.8.6 (*a*) 2496 lb, 2.67 ft; (*b*) 3744 lb, 3.44 ft; (*c*) 127,296 lb, 102.0 ft.
3.8.7 (*a*) 392 kN, 2.67 m; (*b*) 589 kN, 3.44 m; (*c*) 20 000 kN, 102.0 m.
3.8.8 (*a*) 1248 lb, 2.67 ft; (*b*) 2500 lb, 4.33 ft.
3.8.9 1177 kN for any θ; 0.750 m at 90°, 0.650 m at 60°, 0.375 m at 30°, 0 m at 0°.
3.8.10 5510 lb, 4.85 ft.
3.8.11 30.3 kN, 1.830 m.
3.8.12 QED.
3.8.13 $3b/8$.
3.8.14 3.67 ft, $x_p = 1.111$ ft.
3.8.15 1.100 m, $x_p = 0.333$ m.
3.8.16 392 lb, 147.0 lb, 245 lb; (*a*) 21.0 psi; (*b*) 7.88 psi.
3.8.17 53.1 kN, 3.16 m.
3.8.18 4660 ft·lb.
3.8.19 (*a*) 1.80 m; (*b*) $B_x = 77.4$ kN/m, $B_y = 0$, $N_x = 17.59$ kN/m.
3.9.1 2390 psi.
3.9.2 (*a*) 706 kN at 3.44 m depth; (*b*) 609 kN at 1.061 m to right of *N*; (*c*) 933 kN through intersection of F_x and F_y, at 40.8° to horizontal.
3.9.3 (*a*) 43,800 lb at 7.54 ft depth; (*b*) 35,900 lb at 2.25 ft to right of *N*, (*c*) 56,700 lb through intersection of F_x and F_y, at 39.4° to horizontal.
3.9.4 3120 lb/ft at 6.67 ft depth, 1980 lb/ft at 3.40 ft to left of *AB*.
3.9.5 60.1 kN/m at 2.33 m depth, 37.9 kN/m at 1.123 m to left of *AB*.
3.9.6 14,400 lb/ft at 5 ft below top, 8640 lb/ft at 3.0 ft to left of *AB*.
3.9.7 263 kN/m at 1.75 m below top, 150 kN/m at 1.0 m to left of *AB*.
3.9.8 1710 kN/m, 77 700 kPa, 11,270 psi.
3.9.9 (*a*) 3920 lb, 1046 lb; (*b*) 4120 lb, 1354 lb; (*c*) 10,860 lb, 1.257 lb; (*b*) 3950 lb, 523 lb.
3.9.10 561 lb.
3.10.1 0.875, 10.26%.
3.10.2 161.3 mm, 99.4 mm, 58.2 mm; no, $dy/ds = -7000/(28.3s^2)$.
3.10.3 0.223 ft³, 87.1 lb/ft³.
3.10.4 340 lb.

3.10.5 6.06 kN.
3.10.6 2.53 in.
3.10.7 (*a*) 185.5 mm; (*b*) 19.63 L.
3.10.8 (*a*) 0.962 in deep; (*b*) 4.95 in deep.
3.10.9 (*a*) 6.66 kN, 15.41%; (*b*) 18.00 kN.
3.10.10 (*a*) 31.5 lb/ft^3; (*b*) yes; (*c*) 242 lb·ft↶.
3.10.11 (*a*) 5240 N/m^3; (*b*) yes; (*c*) 298 N·m↶.
3.11.1 0.392 psi.
3.11.2 3180 Pa.
3.11.3 2.70 psi.
3.11.4 33.5 kPa.
3.11.5 192 ft^3.
3.11.6 260 m^3.
3.11.7 240 psf; top left 240 psf, bottom right 250 psf, bottom left 490 psf.
3.11.8 43.2 kPa; bottom right 32 kPa, top left 43.2 kPa, bottom left 75.2 kPa.

CHAPTER 4

4.3.1 (*a*) Unsteady, nonuniform; (*b*) unsteady (but steady with respect to rotating frame), nonuniform; (*c*) steady, uniform; (*d*) almost steady, nonuniform; (*e*) unsteady, nonuniform; (*f*) unsteady, uniform.
4.5.2 6.67 fps, 3.92 fps.
4.5.3 0.0884 cfs, 39.7 gpm, 0.273 slug/sec, 8.77 lb/sec.
4.5.4 12.72 L/s, 0.012 72 m^3/s, 20.2 kg/s, 0.1984 kN/s.
4.5.5 0.0511, 0.012 77, and 0.003 19 ft/sec.
4.5.6 0.015 92, 0.003 98, and 0.009 95 m/s.
4.7.1 147.7 kg/m^3; no, need $\mathcal{V}$ between the two sections and need $\partial\rho/\partial t$.
4.7.2 −0.000 229 lb/ft^3 per sec.
4.7.3 450 cfs.
4.7.4 15.70 m^3/s.
4.10.1 0.497 m/s at *D*, 3.40 m/s at *B*.
4.10.2 0.23 m/s.
4.10.3 (*a*) 13.9 ft/sec; (*b*) 10.4 ft/sec.
4.10.4 (*a*) 4.8 m/s; (*b*) zero at stagnation point.
4.12.1 5.39 L/T.
4.12.2 1.204 L/T^2; converging.
4.12.3 $a_x = 2x$, $a_y = 2y$; 3.61 L/T, 6.32 L/T^2.
4.12.4 $a_x = 4x$, $a_y = y$; 6.32 L/T, 12.17 L/T^2.
4.12.5 $a_x = 2xy$, $a_y = 2y^2 + x^2y$; 8.49 L/T, 32.3 L/T^2.
4.12.6 $a_x = 6xy$, $a_y = 6y^2 + 4x^2y$, $a_z = 16z$; 8.25 L/T, 37.7 L/T^2.
4.12.7 1.389 and 0 ft/sec^2.
4.12.8 0.320 and 0 m/s^2.
4.12.9 0.239 ft/sec, 0.0570 ft/sec^2 at 2 ft; 0.1061 ft/sec, 0.007 51 ft/sec^2 at 3 ft.
4.12.10 0.1910 m/s, 0.1459 m/s^2 at 0.5 m; 0.0477 m/s, 0.004 56 m/s^2 at 1 m.
4.12.11 Approx. 57 ft/sec^2, 29 ft/sec^2, 64 ft/sec^2.
4.13.1 16.16 L/T, 5.39 L/T^2.
4.13.2 108.3 L/T, 368 L/T^2.
4.13.3 1.891 and 1.500 ft/sec^2.
4.13.4 0.564 and 0.500 m/s^2.

CHAPTER 5

5.1.1 1.031.
5.1.2 $\frac{10}{9}$.
5.1.3 $\frac{54}{35}$.

5.5.1 0.226 Btu/lb, 0.226°R.
5.5.2 52 J/N, 0.1218 K.
5.5.3 (*a*) 80.4 ft; (*b*) 88.0 ft.
5.5.4 (*a*) 26.5 m; (*b*) 28.8 m.
5.5.5 (*a*) 58.9 ft; (*b*) 74.1 ft.
5.5.6 (*a*) 19.50 ft, from *B* to *A*; (*b*) 13.33 ft, from *A* to *B*.
5.5.7 1.58 m, from *B* to *A*.
5.6.1 0, 0.0316 Btu/lb gain.
5.6.2 0, 7.34 J/N loss.
5.6.3 1.161 Btu/lb, 1990 ft·lb/slug loss.
5.6.4 269 J/N, 232 N·m/kg loss.
5.7.1 2.80 cfs, −20.0 ft.
5.7.2 0.252 m^3/s, −6.20 m.
5.8.1 26.6 psi.
5.8.2 284 kN/m^2.
5.8.3 6.37 psi.
5.10.1 25.4 hp.
5.10.2 6.83 kW.
5.10.3 234 hp, 5090 hp.
5.10.4 1383 kW, 10 350 kW.
5.10.5 9380 hp.
5.11.1 42.4 cfs.
5.11.2 0.480 m^3/s.
5.11.3 1.179 ft.
5.11.4 0.371 m.
5.11.5 1.749 m^3/s.
5.14.1 15,760 hp.
5.14.2 13 730 kW.
5.14.3 0.202 m^3/s, −5.10 m.
5.14.4 8.40 cfs, −16.00 ft.
5.14.5 8.60 m
5.14.6 28.8 ft.
5.14.7 −26.2 ft, 33.8 ft, 71.8 ft.
5.14.8 0.0926 m, 15.09 m, 29.8 m.
5.15.1 74.1 cfs per ft.
5.15.2 3.30 m^3/s per m.
5.15.3 9.51 ft.
5.15.4 2.41 m.
5.16.1 39.5° or 78.0°.
5.16.2 2.21 m/s, 3.13 m/s, 3.84 m/s, 4.43 m/s.
5.16.3 6.03 ft/sec.
5.17.1 10.87 psi.
5.17.2 111.7 kN/m^2.
5.18.1 245 ft (any fluid); (*a*) 0.1294 psi; (*b*) 106.1 psi; (*c*) 78.3 psi.
5.18.2 267 kPa.
5.18.3 171.4 rpm for both.
5.19.1 0.1372 m^3/s, 12.38 m/s at 16°, 85.4 kPa.
5.19.2 5.41 cfs, 35.7 ft/sec at 16°, 8.24 psi.

CHAPTER 6

6.3.1 $\frac{4}{3}$.
6.3.2 1.014.
6.4.1 567 lb to the right.
6.4.2 11.0 kN/m to the right.

6.4.3 742 lb/ft to the right.
6.4.4 273 lb/ft to the right.
6.4.5 42.9 kN to the right.
6.5.1 3860 lb to the right; 17.02 ft.
6.5.2 3.98 kN to the right; 28.5 m.
6.5.3 5.27 lb to the right.
6.5.4 6880 lb at 45° towards the SE.
6.5.5 31.2 kN at 45° towards the SE.
6.5.6 10.83°.
6.6.1 $(2\gamma A/g)V^2 \sin(\theta/2)$
6.6.2 (*a*) 135.9 lb; (*b*) 86.5 lb; (*c*) 161.1 lb at 32.5° below horizontal.
6.6.3 (*a*) 1117 N; (*b*) 712 N; (*c*) 1325 N at 32.5° below horizontal.
6.6.4 (*a*) 125.2 lb; (*b*) 63.8 lb; (*c*) 140.5 lb at 27.0° below horizontal.
6.6.5 (*a*) 1051 N; (*b*) 569 N; (*c*) 1195 N at 28.5° below horizontal.
6.6.6 95.5 lb.
6.6.7 785 N.
6.6.8 (*a*) 60.7 psi; (*b*) 5.53 psi.
6.6.9 (*a*) 313 kN/m^2; (*b*) 25.0 kN/m^2.
6.8.2 (*a*) 942 lb; (*b*) 5890 lb.
6.8.3 (*a*) 6.13 kN; (*b*) 28.1 kN.
6.8.4 F_x = 996 lb, F_y = 252 lb.
6.8.5 26.8°, 29.0°, 35.7°, 51.3°, 99.4 fps.
6.9.1 99.5 lb.
6.9.2 377 N.
6.10.1 1441 lb.
6.10.2 2.23 kN.
6.11.1 36.2 ft·lb; 65.8 hp.
6.11.2 47.7 N·m; 71.6 kW.
6.14.1 928 rpm.
6.14.2 0.0756 cfs; 15.70 ft·lb; zero hp.
6.14.3 0.002 23 m^3/s; 21.8 N·m; zero power.
6.14.4 1414 rpm.
6.14.5 1770 rpm.
6.15.1 (*a*) 1.498 lb; (*b*) 0.848 psf; (*c*) 0.0365 hp.
6.15.2 (*a*) 16.69 N; (*b*) 6.56 N/m^2; (*c*) 27.3 W.
6.15.3 (*a*) 29.3 ft/sec; (*b*) 48.1 W.
6.15.4 (*a*) 7.60 m/s; (*b*) 28.8 W.
6.15.5 (*a*) 150.0 cfs; (*b*) 45.4 lb.

CHAPTER 7

7.4.1 264,000.
7.4.2 1700.
7.4.3 25 atm.
7.4.4 7240 ft/sec > sonic velocity, so model is unsuitable.
7.4.5 4.15 ft/sec, 0.231.
7.4.6 2.19 m/s, 4050 N.
7.4.7 35.8 ft/sec.
7.4.8 2350 ft/sec.
7.7.1 (*a*) $\tau/(\rho V^2)$; (*b*) $\sigma/(\rho L V^2)$.
7.7.2 (*a*) $(\Delta p)g/(\gamma V^2)$; (*b*) $F/(\rho L^2 V^2)$.
7.7.3 $P = CT\omega$.
7.7.4 $\Pi_1 = \mathbf{R}$, $\Pi_2 = \tau/(\rho V^2)$, so $\tau = \rho V^2 \phi(\mathbf{R})$.
7.7.5 $\Pi_1 = F_D/(\rho D^2 V^2)$, $\Pi_2 = \mathbf{R}$, $\Pi_3 = \mathbf{F}$, so $F_D = \rho D^2 V^2 \phi(\mathbf{F}, \mathbf{R})$.

CHAPTER 8

8.2.1 1.080 ft/sec.
8.2.2 0.284 m/s.
8.2.3 Laminar.
8.3.1 2.88 in.
8.3.2 Zero.
8.5.1 0.005 94 ft/ft.
8.5.2 0.012 71 m/m.
8.5.3 0.1877 psf.
8.5.4 9.13 N/m^2.
8.6.1 0.746 cfs, 2660 ft.
8.6.2 $0.707r_0$.
8.6.3 $V = \frac{2}{3}V_c$.
8.7.1 21.8 ft.
8.7.2 0.0977 m.
8.8.1 (a) 0.000 138 7 psf, 0.398 psf; (*b*) 0.804 in, 0.1608.
8.9.1 0.0387 in.
8.9.2 (*a*) 0.000 957 in; (*b*) 0.000 558 in.
8.9.3 (*a*) 0.0376 mm; (*b*) 0.0227 mm.
8.9.4 0.013 40 in.
8.9.5 0.526 mm.
8.9.6 (*a*) 8.51 μm; (*b*) Yes.
8.10.1 0.973 m^3/s.
8.10.2 6.17 ft/ft; 36.5 psf; 72.6 ft/sec.
8.10.3 0.776.
8.11.1 (*a*) 0.0399, 0.0404, 0.0397; (*b*) 0.0259, 0.0257, 0.0266; (*c*) 0.0180, 0.0178, 0.0178; (*d*) For **R** = 20 000 (3.2%).
8.11.2 Left side = 8.74, right side = 8.68.
8.11.3 Left side = 8.74, right side = 8.72.
8.12.1 (*a*) 0.0241 mm; (*b*) 0.019 67 mm.
8.12.2 (*a*) 3790 lb; (*b*) 485 lb.
8.13.1 (*a*) 0.0357 ft, 8.15 ft; (*b*) 0.582 ft, 133 ft; (*c*) 2.01 ft, 459 ft.
8.13.2 (*a*) 7.55 psi; (*b*) 205 ft.
8.13.3 0.1394 cfs.
8.13.4 35.2 L/s.
8.14.1 (*a*) and (*b*) 0.414 m^3/s.
8.14.2 (*a*), (*b*), and (*c*) 0.1393 cfs.
8.14.3 (*a*), (*b*), and (*c*) 35.2 L/s.
8.15.1 103.2.
8.15.2 98.9.
8.16.1 0.0800 ft/ft, 0.013 96 hp/ft.
8.16.2 0.0813 m/m, 28.0 W/m.
8.19.1. (*a*) 18.75; (*b*) 0.938; (*c*) 0.0469.
8.19.2 (*a*) 15.00; (*b*) 0.600; (*c*) 0.0300.
8.19.3 5.23 ft.
8.19.4 1.474 m.
8.19.5 11.31 ft, 4.51 psi.
8.19.6 3.17 m, 29.2 kN/m^2.
8.21.1 (*a*) 0.946 m; (*b*) 1.613 m; (*c*) 1.048 m, 0.282 m.
8.21.2 (*a*) 1.186 cfs; (*b*) 1.294 cfs.
8.23.1 6.15 ft.
8.23.2 2.64 m
8.24.1 5.12 cfs, 75.3 psi.
8.24.2 94.8 L/s, 586 kN/m^2.

8.24.3 (*a*) 2.30 psi; (*b*) −1.882 psi.
8.24.4 (*a*) 20.6 kN/m^2; (*b*) −15.39 kN/m^2.
8.24.5 (*a*) 20.6 cfs; (*b*) 12.33 cfs.
8.24.6 Q = 19.81 cfs, f = 0.0304.
8.25.1 72,900 hp.
8.25.2 216 hp.
8.25.3 143.8 kW.
8.25.4 30.1 hp.
8.25.5 20.3 kW.
8.25.6 10.53 hp.
8.25.7 8.25 kW.
8.25.8 4.58 cfs.
8.25.9 (*a*) 112.4 ft; (*b*) 76.5 hp.
8.25.10 (*a*) 34.1 m; (*b*) 56.4 kW.

CHAPTER 9

9.3.1 0.0615 and 0.0435 lb/ft^2.
9.3.2 (*a*) 0.552 N/m^2, 3.03 mm; (*b*) 0.391 N/m^2, 4.28 mm; (*c*) 1.405 N.
9.3.3 3470; fair agreement with 2000.
9.4.1 (*a*) 0.0337 lb/ft^2; (*b*) 0.0236 lb/ft^2; (*c*) 0.0209 lb/ft^2.
9.4.2 (*a*) 1.999 N/m^2, 9.52 mm; (*b*) 1.740 N/m^2, 16.57 mm; (*c*) 3.92 N; 2.8–4.5 times greater.
9.4.3 65.9 hp.
9.4.4 (*a*) 17.53 ft/sec; (*b*) 19.35 ft/sec.
9.4.5 0.002 93, 0.003 00 (within 2.6%).
9.5.1 0.1124 lb.
9.5.2 (*a*) 1.728 lb; (*b*) 1.084 in.
9.5.3 (*a*) 0.214 mm; (*b*) 415 N.
9.7.1 0.374 lb·sec/ft^2.
9.7.2 8.41 N·s/m^2.
9.7.3 0, 1.250, 5.28, 12.50 N.
9.7.4 3.35 hp, 0.669 hp.
9.7.5 94.0 lb, about one-third of the friction drag.
9.7.6 17.68 ft.
9.7.7 5.15 m.
9.7.8 6630 N, 33.1 m/s.
9.8.1 8.47 ft·lb.
9.8.2 (*a*) 2455 Hz; (*b*) 2452 Hz.
9.9.1 86.6 lb/ft.
9.10.1 At −23.4°; 9690 lb.
9.10.2 At −21.3°; 83.9 kN.
9.12.1 (*a*) 0.279; (*b*) 3.70 m/s; (*c*) 3200 N.
9.13.1 (*a*) 116.7 ft/sec; (*b*) 13.37 hp; (*c*) 140.0 ft^2/sec; (*d*) 22.3 ft^2/sec.
9.13.2 2.58°.
9.13.3 2.58°.
9.14.1 (*a*) −451 ft/sec^2; (*b*) −471 ft/sec^2.
9.14.2 (*a*) −96.0 m/s^2; (*b*) −102.9 m/s^2.

CHAPTER 10

10.3.1 82.7, 308, 703, 1298 cfs.
10.3.2 970 cfs.
10.3.3 2.18 m^3/s.

10.3.4 1.251 m.
10.3.5 0.000 648, or 3.42 ft/mile.
10.3.6 0.005 53.
10.3.7 0.017 83.
10.3.8 0.014 81.
10.3.9 8730 cfs.
10.3.10 146.6 m^3/s.
10.4.1 13.55 ft/sec, 2.58 ft.
10.4.2 5.22, 5.76, 6.30, 6.62, 7.02 (max) m/s.
10.5.1 2.78%.
10.5.2 3.45%.
10.5.3 23.6%; No.
10.6.1 9.78 ft; 15.36 ft versus 18 ft.
10.6.2 (*a*) 98.5 cfs; (*b*) 84.9 cfs; (*c*) Form (*a*) has 16% more capacity yet requires only 80% of the lumber.
10.6.3 (*a*) 3.32 m^3/s; (*b*) 2.86 m^3/s; As Exer. 10.6.2.
10.7.1 Approx. 237 cfs, approx. 3.55 ft/sec.
10.7.2 (*a*) Approx. 33.0 cm; (*b*) approx. 35.4 cm. Larger n caused greater depth.
10.8.1 (*a*) 6.60 cfs; (*b*) 0.000 190 2.
10.8.2 (*a*) 0.225 m^3/s; (*b*) 0.001 019.
10.8.3 8.11×10^{-4} cfs/ft; 76.6.
10.8.4 0.282 L/s per m; 281.
10.9.1 834 cfs.
10.9.2 81.4 m^3/s.
10.9.3 0.168 m; 0.216 m^3/s per m.
10.10.1 1.204 ft, 6.23 ft/sec, 0.003 08.
10.11.1 1.550 ft, 0.003 05.
10.11.2 0.437 m, 0.003 13.
10.11.3 (*a*) 6.42 ft/mile; (*b*) subcritical.
10.11.4 2.44 m.
10.13.1 0.414 m.
10.13.2 0.769-ft rise.
10.13.3 0.145-m rise.
10.13.4 0.070-ft drop.
10.15.1 1377 ft downstream.
10.15.2 640 m downstream.
10.15.3 0.009 88.
10.15.4 0.011 41.
10.17.1 By S_2 or S_3 or C_3 profile.
10.17.2 M_2.
10.17.3 M_2.
10.17.4 A_2.
10.17.5 A_2.
10.17.6 Increase; S_3 after sluice gate.
10.17.7 Sketch $M_1 > y_0 = 0.279$ ft $> y_c = 0.1009$ ft.
10.17.8 Sketch $M_1 > y_0 = 0.0865$ m $> y_c = 0.0311$ m.
10.17.9 $M_1 > y_0 = 4.26$ ft $> y_c = 3.52$ ft; damming action, therefore y_c on dam.
10.18.1 (*a*) 0.492 ft; (*b*) 7.28 ft, 165.2 hp.
10.18.2 (*a*) 0.1467 m; (*b*) 2.81 m, 221 kW.
10.18.3 85.7 cfs, 1.906 ft.
10.20.1 13.10 ft/sec.
10.21.1 16.31 mm.
10.22.1 2.18 ft.
10.22.2 (*a*) 1.518 m; (*b*) expanding.
10.23.1 1.5 m diameter (theoretically 1.409 m).

CHAPTER 11

11.1.1 1.420 slugs/ft^3.
11.1.2 1.699 g/cm^3 or 1699 kg/m^3.
11.1.3 76.7 pcf; surface tension.
11.1.4 11 940 N/m^3; surface tension.
11.1.5 0.386 lb·sec/ft^2.
11.1.6 20.4 N·s/m^2.
11.1.7 0.000 0227 ft^2/sec.
11.1.8 72.9 sec.
11.1.9 49.1 sec.
11.1.10 0.0806 lb·sec/ft^2; 0.0647.
11.3.1 0.444 m/s.
11.3.2 14.84 ft/sec.
11.3.3 4.25 m/s.
11.3.4 4.03 ft/sec.
11.3.5 1.156 m/s.
11.3.6 1.531 ft/sec.
11.3.7 0.470 m/s.
11.6.1 0.606, 0.618, 0.981.
11.6.2 (*a*) 0.606; (*b*) 0.969; (*c*) 0.625.
11.6.3 (*a*) 0.609; (*b*) 0.977; (*c*) 0.623.
11.6.4 39.6 ft/sec, 3.46 cfs.
11.6.5 39.3 ft/sec, −18.00 ft.
11.6.6 10.90 ft; separation, $p_o = p_a$, smaller Q.
11.6.7 28.3 ft/sec, −6.68 ft; less V, more p with friction.
11.6.8 13.43 m/s, −6.19 m.
11.6.9 4.39 m.
11.6.10 10.43 L/s.
11.7.1 0.316 cfs.
11.7.2 8.58 L/s.
11.7.3 1.019 m^3/s.
11.7.4 4.84 in.
11.8.1 (*a*) 0.000 1548 m; (*b*) 0.014 65 m; (*c*) 1.465 m.
11.9.1 (*a*) 0.000 354 m; (*b*) 0.0383 m; (*c*) 3.83 m.
11.10.1 169.4 lb/sec.
11.10.2 0.1394 lb/sec.
11.10.3 5.34 lb/sec.
11.10.4 31.6 N/s.
11.10.5 5.87 lb/sec.
11.10.6 36.6 N/s.
11.10.7 (*a*) 52.8 psia, 1030 ft/sec; (*b*) 52.8 psia, about 1045 ft/sec.
11.10.8 (*a*) 370 kN/m^2 abs, 313 m/s; (*b*) 370 kN/m^2 abs, about 318 m/s.
11.11.1 23.6 cfs.
11.11.2 0.729 m^3/s.
11.11.3 (*a*) 22.7 cfs; (*b*) 2.00 ft.
11.11.4 (*a*) 0.708 m^3/s; (*b*) 0.733 m.
11.11.5 (*a*) 0.293 cfs; (*b*) 137.5% increase; 1.018 ft.
11.12.1 18.92 cfs/ft.
11.12.2 1.715 m^3/s per m.
11.12.3 2.46 ft.
11.12.4 0.705 m.
11.12.5 247 cfs.
11.12.6 6.66 m^3/s.
11.14.1 1.597 cfs/ft, 0.579.
11.14.2 0.237 m^3/s per m, 0.581.

CHAPTER 12

12.2.1 210 ft^2.
12.2.2 15.64 m^2.
12.2.3 791 sec.
12.2.4 605 s.
12.2.5 1276 sec.
12.2.6 962 s.
12.2.7 98.4 sec.
12.2.8 97.6 s.
12.3.1 −0.004 52 ft.
12.3.2 −0.450 cfs/sec.
12.4.1 5.96 ft/sec, 3.05 sec.
12.4.2 3.41 m/s, 1.593 s.
12.4.3 3.76 m/s, 1.772 s, 4.11 s.
12.4.4 11.00 ft/sec, 1.407 sec, 2.49 sec.
12.4.5 7.57 ft/sec, 6.58 sec.
12.4.6 3.20 m/s, 5.61 s.
12.5.1 4879 ft/sec, 3761 ft/sec, 77.1%.
12.5.2 1478 m/s, 1126 m/s, 76.2%.
12.5.3 3354 ft/sec.
12.5.4 1020 m/s.
12.6.1 (*a*) All 100.9 psi; (*b*) all 70.7 psi.
12.6.2 (*a*) All 800 kPa; (*b*) all 600 kPa.
12.6.3 (*a*) 100.9 psi, 100.9 psi, 74.8 psi; (*b*) 70.7 psi, 70.7 psi, 52.3 psi.
12.6.4 (*a*) 800 kPa, 800 kPa, 485 kPa; (*b*) 600 kPa, 600 kPa, 364 kPa.
12.6.5 (*a*) 80.7 psi, 53.8 psi, 26.9 psi; (*b*) 56.5 psi, 37.7 psi, 18.84 psi.
12.6.6 (*a*) 571 kPa, 381 kPa, 190.5 kPa; (*b*) 429 kPa, 286 kPa, 142.9 kPa.
12.7.1 65.3 ft.
12.7.2 1.837 m.

CHAPTER 13

13.1.1 2.85×10^6 ft·lb.
13.1.2 4.55 MJ.
13.1.3 174.2°F, 273 psia, 3.15×10^6 ft·lb, $+4.42 \times 10^6$ ft·lb.
13.1.4 80.2°C, 1913 kPa abs, 4.89 MJ, +6.85 MJ.
13.1.5 37 300 J/kg, 130 200 J/kg checks closely.
13.3.1 1116, 1097, 1077, 1037, and 995 ft/sec.
13.3.2 340, 332, and 299 m/s.
13.5.1 17.89 psi, 89.0°F; 14.98 psi, 71.2°F; 5.59 psi, −17.8°F.
13.5.2 127.9 kPa abs, 34.9°C; 101.5 kPa abs, 21.9°C; 35.7 kPa abs, −30.0°C.
13.5.3 284 psia, 121°F, 1160 ft/sec.
13.5.4 1979 kPa abs, 51.2°C, 355 m/s.
13.6.1 1800 ft/sec, −170°F.
13.6.2 551 m/s, −111.1°C.
13.6.3 682 ft/sec.
13.6.4 208 m/s.
13.8.3 825 ft/sec, 22.0 psia, 13.68°F.
13.8.4 251 m/s, 153.8 kPa abs, −11.60°C.
13.8.5 644 ft/sec, 14.2 psia, 59.6°F.
13.8.6 201 m/s, 98 kPa abs, 16.51°C.
13.9.1 1.453.
13.9.2 1.459.
13.9.3 2.56 in, 0.224 slug/sec, (2) 1039 ft/sec, −10.00°F, (3) 1681 ft/sec, −155.3°F.

13.9.4 8.57 in.
13.9.5 (*a*) and (*b*) 0.0790 slug/sec.
13.9.6 (*a*) and (*b*) 1.101 kg/s.
13.10.1 (*a*) impossible situation; (*b*) 3.12.
13.10.2 (*a*) impossible situation; (*b*) 4.15.
13.11.1 1299 ft/sec, 3.98%,
13.11.2 1722 ft/sec.
13.11.3 812 m/s.
13.13.1 (*a*) 37,800 ft·lb/slug; (*b*) 48.6 Btu/slug; (*c*) added; (*d*) 15,450 Btu/hr.
13.13.2 (*a*) 3220 N·m/kg; (*b*) 3220 J/kg; (*c*) added; (*d*) 13.95×10^6 J/h.

CHAPTER 14

14.1.1 (*a*), (*b*), (*d*), (*e*).
14.1.2 (*a*), (*c*), (*d*), (*e*), (*g*).
14.2.1 (*a*), (*b*), (*c*), (*f*).
14.2.2 (*b*), (*c*), (*d*), (*g*).
14.3.1 (*a*) 0; (*b*) 0; (*c*) 0; (*d*) 1; (*e*) −5; (*f*) 0.
14.3.2 (*a*) −3; (*b*) 0; (*c*) 0; (*d*) 0; (*e*) $-2y - 4x$; (*f*) $4y + 2x$; (*g*) 0.
14.4.1 Yes. $-4x^2 + 2y$.
14.4.2 (*a*) $2y$; (*b*) $2y - 3x$; (*d*) $-1.5x^2 + y^2$; (*e*) $1.5x^2 + y^2$.
14.4.3 (*a*) $1.5y^2$; (*c*) $3xy$; (*d*) $4y + 6x + 2xy$; (*e*) $x^2y + 2xy^2$; (*g*) $x^3/3 - 2x^2y - xy^2 + 2y^3/3$.
14.5.1 (*a*) $12.5\theta/\pi + 5r \sin \theta$; (*b*) $(12.5/\pi) \arctan (y/x) + 5y$.
14.5.2 (*a*) $x = -0.796$ m; $y = 0$ ($r = 0.796$, $\theta = 180°$); (*b*) 5.51 m/s.
14.5.3 0.342 m.
14.5.4 2.46 ft.
14.6.1 Yes; no, because continuity is not satisfied.
14.6.2 Yes; $7x + 10y$; yes.
14.6.3 (*a*) Yes, $2y + 1.5y^2 - 1.5x^2$, yes; (*b*) no, yes.
14.6.4 (*a*) Yes, $3(y^2 - x^2)$, yes; (*b*) no, yes.
14.7.1 (*a*) $\psi = 2y$, $\phi = -2x$; (*b*) $\psi = 2y - 3x$, $\phi = -3y - 2x$.
14.7.2 (*c*) $\psi = 3xy$, $\phi = -1.5x^2 + 1.5y^2$; (*d*) $\psi = 4y + 6x + 2xy$, $\phi = 6y - 4x - x^2 + y^2$; (*g*) $\psi = x^3/3 - 2x^2y - xy^2 + 2y^3/3$, $\phi = 2x^3/3 + x^2y - 2xy^2 - y^3/3$.
14.8.1 1.735 ft^3/day per ft, 0.588 ft/day.
14.8.2 0.1643 m^3/d per m, 0.1880 m/d.
14.8.3 2.83 ft/day.
14.8.4 1.058 m/d.

CHAPTER 15

15.4.1 268 rpm, 25.1 m^3/s, 14 000 kW.
15.4.2 267 hp; 156.3 ft, 25 cfs, 521 hp.
15.4.3 22.1 kW; 8.64 m, 360 L/s, 38.1 kW.
15.4.4 89.9%.
15.6.1 44.8 ft, 6000 gpm.
15.6.2 76.9 ft, 8640 gpm.
15.6.3 Shutoff 71.1 ft, 0 cfs; BEP 28.4 ft, 162.6 cfs; 75 cfs at 23 ft becomes 100 cfs at 40.9 ft, etc.
15.7.1 $h = \Delta z + 0.1061Q^2$; $h \approx 56$ ft with $Q \approx 11{,}300$ gpm, $h \approx 63$ ft with $Q \approx 9700$ gpm, $h \approx 76$ ft with $Q \approx 4800$ gpm. Approx. pump efficiencies: 81%, 81%, and 54%, respectively.
15.7.2 $h = \Delta z + 0.403Q^2$; $h \approx 72$ ft with $Q \approx 6500$ gpm, $h \approx 74$ ft with $Q \approx 5500$ gpm, $h \approx 77$ ft with $Q \approx 2500$ gpm. Approx. pump efficiencies: 66%, 60%, and 40%, respectively.
15.8.1 Raise Eq. (15.4) to the $\frac{1}{2}$ power and raise Eq. (15.5) to the $\frac{3}{4}$ power. Divide the first equation by the second to eliminate *D*, and find $N_s = nQ^{1/2}/h^{3/4}$.

15.8.2 $N_s = 13{,}165$: axial-flow pump.
15.8.3 1395, 1622, and 2140, respectively; all are radial-flow pumps.
15.8.4 $N_s = 1670$, radial-flow pump.
15.8.5 $(N_s)_{SI} = 0.318$; radial-flow pump.
15.9.1 Approx. 38 in.
15.9.2 Approx. 10 in.
15.9.3 Approx. 24.8 in = 630 mm.
15.10.1 Approx. 6.2 ft above the reservoir water surface. Approximate answer using Fig. 15.13: 10 ft above water surface.
15.10.2 Approx. 3.8 ft or more below the elevation of the reservoir water surface.
15.10.3 Place the pump no more than 5.01 m above the water surface.
15.12.1 $N_s = 2670$, no cavitation if pump is less than about 18 ft higher than the water surface.
15.12.2 $N_s = 1357$, no cavitation if pump is at least about 12 ft lower than the water surface.
15.12.3 10.9 m, $(N_s)_{SI} = 0.534$, no more than about 3.9 m above the water surface; hence cavitation will occur (6 m > 3.9 m).
15.13.1 Approx. 124 gpm, 172 gpm, 142 gpm.
15.13.2 Approx. 170 gpm; 1650 rpm.

CHAPTER 16

16.3.1 278 hp. Horsepower transmitted to the runner, because $v_2 = v_1$.
16.3.2 1102 lb, 5510 ft·lb, 251.7 hp, 10.69%.
16.3.3 63 100 N, 63 100 N·m, 2480 kW.
16.4.1 1942 ft, 88%, 84%.
16.4.2 2.67 ft/sec, 385 ft/sec, 33.5 cfs, 166.5 ft/sec, 0.432, 2413 ft, 27,900 lb, 123,300 ft·lb, 8450 hp, 9170 hp, 92.1%.
16.4.3 90.6%.
16.5.1 About 1.0-in-diameter jet with about 8.95 hp.
16.5.2 About 167 hp with jet diameter of slightly less than 5 in.
16.7.1 8750 N·m.
16.7.2 30 poles.
16.8.1 3.23 ft of additional head.
16.8.2 97.2 ft.
16.8.3 6.04 m.
16.10.1 88.3%, 189.7 rpm, 12.65 m^3/s, 5260 kW, 38 poles.
16.10.2 20 poles, 2093 hp.
16.10.3 22 poles: 273 rpm, 282 cfs, 1573 hp. 24 poles: 250 rpm, 258 cfs, 1209 hp.
16.11.1 4.01, approx. 8.8 ft.
16.11.2 25.1, approx. 6.4 ft, Francis turbine.
16.11.3 146.0, approx. 2.71 ft, propeller-type turbine.
16.12.1 Approx. 157 ft.
16.12.2 Approx. 42.5 ft, approx. 30.2 ft.
16.12.3 Approx. 77 ft.
16.13.1 (*a*) It is possible, $N_s = 6.25 < 7.0$. (*b*) Range of N_s: from 6.25 (96 poles) to 6.99 (86 poles). Range of D: from approx. 6.8 ft to approx 6.1 ft.
16.13.2 (*a*) It is possible, $N_s = 50.1 > 20$. (*b*) Range of N_s: from 20.1 (30 poles) to 50.1 (12 poles). Range of D: from approx. 3.7 ft to approx. 1.54 ft.
16.13.3 $N_s = 27.2$, D = approx 13.1 ft.
16.13.4 21.8 ft, 11.06 ft.

Index

Page numbers in the index that are followed by an *n* refer to footnotes on those pages.